Toward a Theory on Biological-Physical Interactions in the World Ocean

NATO ASI Series

Advanced Science Institutes Series

A Series presenting the results of activities sponsored by the NATO Science Committee, which aims at the dissemination of advanced scientific and technological knowledge, with a view to strengthening links between scientific communities.

The Series is published by an international board of publishers in conjunction with the NATO Scientific Affairs Division

A Life Sciences	Plenum Publishing Corporation
B Physics	London and New York
C Mathematical	Kluwer Academic Publishers
and Physical Sciences	Dordrecht, Boston and London
D Behavioural and Social Sciences	
E Applied Sciences	
F Computer and Systems Sciences	Springer-Verlag
G Ecological Sciences	Berlin, Heidelberg, New York, London,
H Cell Biology	Paris and Tokyo

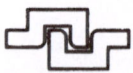

Toward a Theory on Biological-Physical Interactions in the World Ocean

edited by

B. J. Rothschild

University of Maryland,
Center for Environmental and Estuarine Studies,
Solomons, Maryland, U.S.A.

Kluwer Academic Publishers

Dordrecht / Boston / London

Published in cooperation with NATO Scientific Affairs Division

Proceedings of the NATO Advanced Research Workshop on
Toward a Theory on Biological-Physical Interactions in the World Ocean
Castéra–Verduzan, France
1–5 June 1987

Library of Congress Cataloging in Publication Data

Toward a theory on biological-physical interactions
  in the world ocean.

  (NATO ASI series. Series C, Mathematical and
physical sciences ; no. 239)
  Papers from the NATO Advanced Research Workshop
"Toward a Theory on the Biological-Physical
Interactions in the World Ocean", held in
Castéra-Verduzan, France.
  Includes index.
  1. Marine ecology--Congresses. 2. Marine productivity
--Congresses. 3. Oceanography--Congresses.
I. Rothschild, Brian J., 1934-   . II. NATO Advanced
Research Workshop "Toward a Theory on the Biological-
Physical Interactions in the World Ocean" (1987 :
Castéra-Verduzan, France) III. Series.
QH541.5.S3T68 1988   574.5'2636   88-8950

ISBN-13: 978-94-010-7859-7   e-ISBN-13: 978-94-009-3023-0
DOI: 10.1007/ 978-94-009-3023-0

Published by Kluwer Academic Publishers,
P.O. Box 17, 3300 AA Dordrecht, The Netherlands.

Kluwer Academic Publishers incorporates the publishing programmes of
D. Reidel, Martinus Nijhoff, Dr W. Junk, and MTP Press.

Sold and distributed in the U.S.A. and Canada
by Kluwer Academic Publishers,
101 Philip Drive, Norwell, MA 02061, U.S.A.

In all other countries, sold and distributed
by Kluwer Academic Publishers Group,
P.O. Box 322, 3300 AH Dordrecht, The Netherlands.

CONTENTS

Toward a Theory on Biological-Physical Interactions in the World Ocean: Introduction

Opportunities to increase our understanding of the complexities of biological productivity of the sea were explored at a NATO Advanced Research Workshop "Toward a Theory on the Biological Physical Interactions in the World Ocean" held at the Château de Bonas, Castéra-Verduzan, France.

The Workshop noted major recent advances including an increased understanding of the contribution to biological production of various major groups of organisms; physical and biological mesoscale dynamics; and the kinds of sampling technology necessary to harmonize observations of physical and biological phenomena. At the same time, the Workshop also noted that much needed to be done to move more rapidly toward enhancing theory on biological and physical interactions in the world ocean.

While not diminishing the importance of traditionally focused research, major components of biological productivity seem to be receiving relatively little research attention. These components include inter alia a) the relations among physical variability (particularly in submesoscale ranges), population dynamics (of both plants and animals), and production (as conventionally defined, e.g., "primary production" or "grazers"); b) the contribution of secondary production to total biological production (in terms of both the very small organisms and larger organisms); and c) the contribution of genetic variability to longer-term population change.

The Workshop noted that integrating research in well-studied areas with research in areas receiving relatively little attention can be greatly facilitated by enhancing existing theory on the biological dynamics of the sea, particularly regarding to the interaction between population dynamics and physical variability. At the same time, the Workshop noted that theoretical enhancement needs to be fueled by a rejuvenated data-collection strategy that would terminate "the decades of undersampling of biological-physical phenomena."

The Workshop agreed that the development of a complete and rigorous theory or a network of component biological-physical theories would take many years of concentrated effort, placing an immediate urgency on committing greatly enhanced resources now to couple theoretical development, numerical modeling, and data collection. Without such development, our understanding of an important part of the biosphere, and the quality of critical public-sector decisions on environmental management, impacts of climate, and resource harvesting will be held at stake.

The Workshop considered that the development of theory required a) creating new opportunities and incentives for physicists, biologists, chemists, and mathematicians to work together on the

B. J. Rothschild (ed.), Toward a Theory on Biological-Physical Interactions in the World Ocean, 1–6.
© 1988 by Kluwer Academic Publishers.

mechanisms that drive variability in biological production of the sea;
b) undertaking a intensive effort to terminate the undersampling of
biological and related physical phenomena, using satellite, acoustic,
optic, and other technology as well as modern statistical techniques
that are now available or potentially available; c) initiating a
"crash program" for the further development of sensors and measuring
devices particularly oriented to recording biological and physical
interactions; and d) developing intense pilot programs in the field
to learn about the transformations of nutrient pulses through
secondary and higher- order production, particularly with regard to
the physical and population-dynamics setting of such transformations.

The Workshop felt that the oceanographic community is at a point
in time, unique since IDOE days, where the physical and biological
observations of the sea can again be brought into harmony, a challenge
that is urgent, and a goal that is achievable.

The details of a theory linking the dynamics of biological
productivity of the sea with atmospheric and oceanic physics require
considerable elaboration and increased resolution of the biological-
physical system. To increase system resolution, two major problems
need to be put in perspective. The first is that the biophysical
system is represented by an infinitude of dimensions which represent
time-and-space scales that vary over many orders of magnitude. The
second is the nature of the interactions among these dimensions seems
to change more or less continually owing to the nonlinearities in the
biodynamic system and other complex factors.

As a consequence, it is difficult, in the absence of criteria, to
determine which time-and-space scales are important. Without such
criteria, conclusions are easily generated which can only be
applicable to special or correlative cases, while conclusions relative
to a general interpretation of system causality are difficult to
attain. Even if resources were available to increase the sampling
intensity, available technology is neither generally available, nor
sufficiently tested for full deployment, and as a consequence the
efficiencies of advanced technology cannot be fully realized. Future
emphasis on technological development of biological and physical
measurement devices is of the highest priority.

We are left with an image considerable complexity. Yet to what
extent is this complexity a function of the biodynamic system per se,
and to what extent is it a function of the quality of the match
between the problems that need to be resolved and the array of
research that has been accomplished?

There are no concrete answers to such questions. On the other
hand, we can observe that a component of the perceived complexity
stems from the fact that only portions of the system are relatively
well studied. In particular, the critical importance of the
interactions among non-linear population dynamics, the physical
environment, and population genetics has been suppressed by
theoretical considerations that have been generally restricted to
primary production or to aggregating of taxonomically similar
populations (and size groups to a more limited extent), paralleled by
the fact that at the same time, empirical enterprise has tended to

focus on primary production and on temporal-and-spatial physical scales often greater and to the exclusion of those known to be biologically important.

Any criticism of population aggregation technique seems to point in the direction of considering each and every population, a task which would be impossible, just in terms of practical considerations. Accordingly, we are left with the requirement for alternative integrating principles and criteria based upon individual population properties. Clearly these need to implicate the fundamental sources of population variation, the density-dependent population dynamics variables of growth, mortality, and reproduction with the physical variables of temperature, motion and light. Without these criteria it will be difficult to understand either the short- or long-term effects of climate and meteorological conditions on productivity, or how these are related to anthropogenic effects of pollution and fishing, for example.

To sort out the differences between attempting to understand productivity using techniques of aggregating populations according to taxonomic or metric affinities and attempting to understand productivity using techniques aimed at the most fundamental definition of the interactions between physical and population dynamics, the Workshop considered various aspects of the biology and physics of the sea from the perspective of individual participants.

As a point of departure, examples of how physical processes place plant cells, nutrients, and light in juxtaposition, setting in motion the processes generating primary productivity were considered. For example, Woods, shows one implementation of such a juxtaposition. He noticed "hot spots" in chlorophyll indices, deduced from satellite imagery, and inferred that these hot spots, accompanied by streaks and whorls of enhanced chlorophyll concentration, result from mesoscale jets generated by the isopycnal potential vorticity gradient. Joyce discusses another example of the juxtaposition of plant cells, light, and nutrient-rich water, showing the dynamics of how a doming nutricline in the center of a warm core ring increases the volume of nutrient- rich water in the photic zone and thereby contributes to the elaboration of primary production. As another example he shows how internal waves contribute to short-term periodicities in the volume of nutrient-rich water in the photic zone and hence to periodicities in primary production. Taking another consideration of the juxtaposition of phytoplankton cells, light and nutrients, Wolf and Woods consider a one-dimensional Lagrangian ensemble model of phytoplankton production taking account the fate of individual cells. They show, in the context of the model, the evolution and subsidence of the spring bloom. Another approach to considering the effects of the vertical distribution of plant cells' exposure to illumination is considered by Prieur and Legendre, taking specific account of the transmission of irradiance and the physics of photon capture. That there are parallels between theoretical studies and observation is shown by Strass and Woods who describe the spatial-temporal evolution of the near-surface chlorophyll maximum and its relation to vertical stability and isopycnal spacing based upon batfish SEA ROVER tracks.

They note in particular the relation between satellite and surface
measurements of chlorophyll indices.

In another form of "synoptic survey", Dandonneau reports on ship-
of-opportunity observations in portions of the oligotrophic Pacific
Ocean in which he finds temporal and spatial production "hot spots"
suggesting that pulses of nutrient injection into the upper layers of
the sea may have been "averaged out" and therefore not observed in
classical studies. The evolution of our understanding of physical-
biological interactions both from a theoretical and observational
point of view and the resolution of measurement difficulties, which
have only become recently understood, are exemplified by Herbland's
concern with the nature of pigment measurement as an index of primary
production. This general problem no doubt contributes to Lasker's
criticism that methods of studying the production of the sea are not
yet adequate to measure primary production or the transfer of primary
production to secondary and higher production levels, particularly
fish.

The physical-biological consideration of secondary production is
generally similar to consideration of primary production in the sense
that physical conditions bring predators and grazers into
juxtaposition with their food (rather than nutrients). Boicourt shows
that organisms can utilize the often rich complexities of their
kinetic environment to generate specific trajectories "over-the-
bottom" which are independent of the trajectories of any particular
lamina of water. Boucher discusses as well the consequences of
vertical and horizontal movements on spatial-temporal distributions of
plankton. But to maintain any specific trajectory, organisms need to
account for the partially correlated and random components of flow as
pointed out by Yamazaki and Osborn in their documentation of the
variety of hardly understood turbulent structures that influence
biological interactions among predator and prey organisms. The
consequential effects of juxtaposing plant cells and nutrients;
grazers and plant cells; and predators and prey in a setting of
varying physical conditions is dealt with by several participants.
For example, Cushing reports that the "Great Slug", the pool of
unusually cool and fresh water that drifted across the north Atlantic
for twenty years caused a decrease in zooplankton production and
affected the recruitment to many fish stocks. While the precise
nature of the linkages between the "Great Slug" and other anomalous
events are not completely understood, it is easy to see how it might
have modified various production processes. To show how delicately
tuned such mechanisms might be, Le Fèvre and Frontier discuss another
consequential effect on secondary production, in particular the
allocation of trophic exchange between small and large heterotrophs.
They describe a physical setting for alternative pathways of community
metabolism. In the vicinity of the shelf break, variability in
production is dominated by nutrient injection which stimulates the
classic phytoplankton-herbivore food chain. In contrast, in the
vicinity of ergoclines variability in production is dominated by the
functioning of the accumulation biotype where microheterotrophs are
fed upon by microphagous zooplankton, the importance of the microbial

pathways as alternative to the classical food chain only being recognized in the last two decades. Goldman considers in more detail the operation of microbial loop in the so-called oligotrophic ocean showing how the dynamics of more classical notions of trophic transfer interact with those at the microbial level.

As the size of the adult organisms increase, the problems become seemingly more tangible, but no less complex. Mullin considers the copepods, the most common of the larger heterotrophic plankton. He observes that the most important components of production are related to reproduction and food availability which is clearly influenced by physical dynamics suggesting the importance of the interaction between physics and population dynamics as affecting the biodynamics of copepods. His review further suggests that study of this interaction is impeded by inadequate sampling methodology. Copepods, as is the case with many other forms of plankton, have multiple-stage life histories. Nival, Carlotti and Sciandra develop a multiple-stage life history model to study the interactions of the feeding regime with life-history structure. They show how nutrition and predation interact in the transfer of survivors from one life history stage to the next. Among the plankton, ichthyoplankton representing many commercial species, are of considerable economic importance. Yet the ichthyoplankton can be hardly understood without understanding the dynamics of their predators and prey. Jones and Henderson develop a model of these dynamics for a haddock-like fish larva. They point out a remarkable requirement for survival of the ideal haddock-like larval; the larva must increase its body weight by a factor of 2500 during the first 80 days of life. Jones and Henderson link the growth of larvae to the growth of larval food and show the criticality of larvae being "at the right place at the right time" if they are to survive. Taking another view of the problem Cohen, Sissenwine, and Laurence consider the interrelationships of larval nutrition, predation on larvae, and the physical environment relative to larval mortality.

In many studies it is often implicitly assumed that populations are homogeneous genetic units. This is often not the case as pointed out by Battaglia and Bisol and Powers et al., reflecting that understanding variability in populations, particularly that which occurs over long periods of time needs to be supported by an understanding of genetic variability.

With respect to adult stages of fish and fishing and its relation to the affects of environmental perturbations, it is necessary to understand the relation between the number of recruits and the number of spawning fish. Of the many subtleties to this problem, Daan considers the relation between density-dependent and density-independent control. This question of recruitment-stock relationship and its integration with environmental change in the ecosystem is perhaps best reflected by the variation in population structure, examples of which are given by Sherman along with an account of major studies of these interactions. Troadec reflects upon the resource harvesting setting and increased priorities on research necessitated by the new "Ocean Regime."

As stated earlier, advancing a materially revised or enhanced theory on the interaction of physics and biology in the world ocean requires both theoretical and empirical advances. The necessary transitions are described by Reeve. He looks at the past, present, and future seeing the "past" as sampling with inefficient nets and descriptive laboratory studies which nevertheless led to the great synthesis; the "present" as recognizing particles and advanced net systems and acoustic- and image-recognition techniques as well as taking account of the importance of coelenterates and chaetognaths and large experimental containers; and the "future" in developing new theory that interrelates biology and physics with the instrumentation and tools that can verify the theory.

The theoretical considerations can take two pathways. On one hand, existing notions can be extended. On the other hand, different views of the problem such as Steele's attempt to resolve the many scales of interaction or Rothschild's attempt to define an "ecological quantum change" might bear fruit. Whether any theoretical approach will actually be fruitful will depend upon the extent to which the theory can be verified contemporaneously or at a future date. In retrospect many past studies of oceanographic phenomena had the "dreamlike quality" described Munk and Farmer owing to inadequate empirical material. Yet as pointed out by Dickey and Farmer and Huston techniques and ideas are "on the shelf" which can do much to alleviate the undersampling problem.

These examples reflect many of the concerns critical to understanding the interactions of biology and physics. Their scientific importance is paralleled by their practical importance to fishing, decision-making in the control of anthropogenic substances and activity, and forecasting long-term climatic effects.

The Workshop concluded that in the next few decades great progress can be made in moving toward a better theoretical understanding of the interaction of biology and physics in the world ocean, but that the rapidity of moving ahead will be directly related to the incentives for biologists and physicists to work together. Thus, the participants have highlighted their major concerns regarding the integration of the production system, taking account of the physical setting and reflecting what seem to be fruitful changes in emphasis and direction.

In addition to NATO and the NATO Science Committee, extensive assistance, collaboration, and financial support was received from IFREMER. Support was also received from the NSF, Ocean Sciences Division, and NOAA, NMFS. Thanks are due to Pamela Weddle for typing the manuscript and compiling the index.

BJR

SCALE UPWELLING AND PRIMARY PRODUCTION

John Woods
Robert Hooke Institute
Oxford University
England

ABSTRACT. Upwelling due to vortex contraction on the anticyclonic
flank of transient mesoscale jets is fast enough and sustained for
long enough to effect substantial local increase in primary
production. Dynamical constraints limit upwelling to patches with
horizontal dimensions of about ten kilometres, similar to those of
primary production "hot spots" observed in satellite images. *In situ*
surveys suggest that the distribution of these mesoscale patches of
high plankton concentration strongly influences the large scale
variation of primary production. The latter can be estimated from
the statistics of mesoscale upwelling events. Given the new
understanding of mesoscale dynamics, those statistics can be computed
using geostrophic turbulence theory, provided the large scale
distribution of isopycnic potential vorticity Q is known. (The
relevant properties of Q are summarized in an appendix.) The seasonal
climatology of Q in the euphotic zone is described, and it is shown
how inter-annual variations can be predicted by means of a model of
ocean circulation and mixed layer dynamics. A multi-year programme
of experiments in the North Atlantic has been undertaken to test the
theory. This has involved a series of high resolution sections
extending 2,000 km between the Azores and Greenland, and synoptic
mapping of mesoscale structure at the inter-gyre front. The phase
relationships between distributions of Q, temperature, velocity and
concentrations of particles and chlorophyll in the maps are
consistent with the theory. The distributions of Q, upwelling and
chlorophyll in the sections supports the hypothesis that large scale
variation of primary production is best viewed in terms of the
statistics of mesoscale events.

1. INTRODUCTION

The aim of the research reported in this paper is to improve our
understanding of physical processes that influence horizontal
variation in primary production on scales of kilometres to
megametres. Motivation comes from the need to incorporate primary
production into computer models of climate change associated with

7

B. J. Rothschild (ed.), Toward a Theory on Biological-Physical Interactions in the World Ocean, 7–38.

variation of atmospheric carbon dioxide (Brewer *et al.*, 1986). In such models, the large-scale variations are described explicitly by the equations, but the influence of variation on scales smaller than the model grid must be parametrized in terms of the larger structure. Computer limitations make it unlikely that the grid scale will be less than a few tens of kilometres, so it will be necessary to parametrize the effects (on atmospheric CO_2) of primary production patches with smaller scale. That is worrying because primary production is dominated by patchiness on scales of 1-100 km (Steele 1978). The principal oceanographic features in this *mesoscale* range are internal waves and fronts (Woods 1980). Fasham and Pugh (1976) have considered the influence of internal waves on primary production; Le Fèvre (1986) has reviewed biology at fronts.

The present paper focuses on transient mesoscale fronts because (unlike internal waves) they are accompanied by the persistent

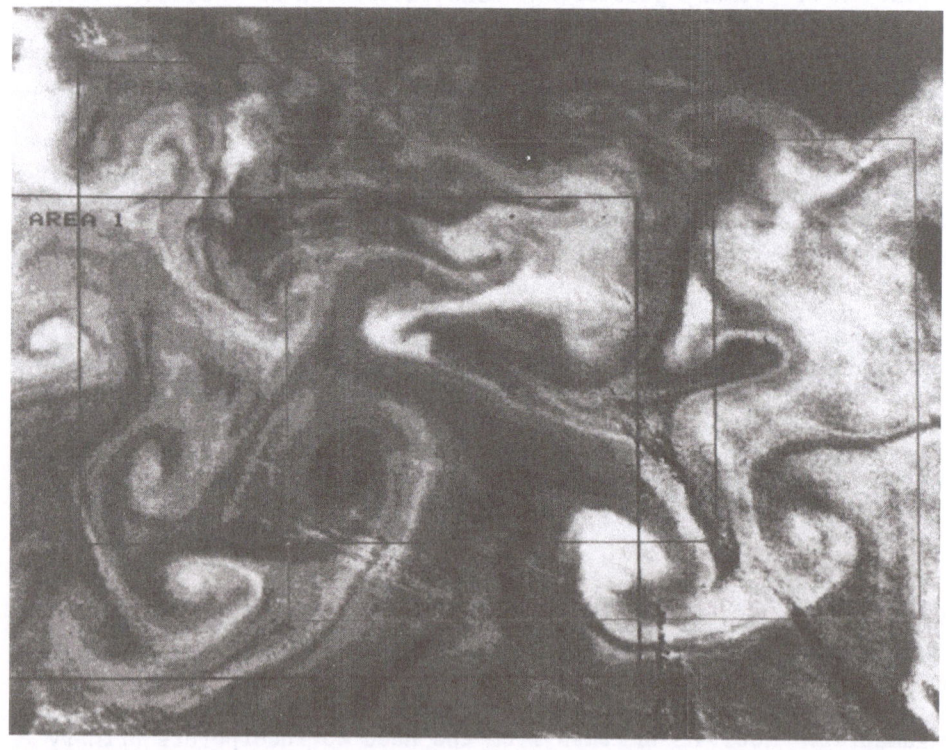

Fig. 1. High resolution satellite image of chlorophyll distribution during the spring plankton bloom (reproduced from Gower *et al.*, 1980 with permission).

Eulerian upwelling needed to achieve the observed spatial modulation of chlorophyll concentration. Mesoscale variability is the main feature in satellite colour images (Gower *et al.*, 1980); even the large scale variation seen in CZCS composites (Esaias *et al.*, 1986) comprises a modulation of the spatial concentration of chlorophyll with a characteristic pattern of filaments drawn out of mesoscale spot sources, as does the variability detected *in situ* (Fasham *et al.*, 1985; Strass and Woods 1988). So model predictions will be sensitive to the method of parametrizing the unresolved plankton patchiness. It follows that high priority must be given to improving our understanding of the 1-100 km patchiness in primary production, and to discovering how to estimate its grid-scale-mean effect on atmospheric CO_2 from the larger variations resolved in models.

2. HYPOTHESIS

The hypothesis of this paper is that the key to understanding plankton patchiness lies in the dynamics of mesoscale jets. Specifically it will be argued that upwelling associated with vortex stretching at mesoscale jets strongly modulates primary production. Furthermore, it is suggested that the large scale distribution of chlorophyll deduced from satellite colour images reflects the distribution of the kinetic energy in these jets. This approach provides dynamical underpinning for statistical theories of primary production, which have hitherto been based on dimensional arguments (*e.g.* Denman and Platt 1975).

Upwelling plays a key role in primary production, the classical example being coastal upwelling induced by offshore Ekman transport which raises nutrients to the surface (Denman and Powell 1984). It is the aim of this paper to show that mesoscale upwelling in mid-ocean can have the same effect and that it is central to the interpretation of satellite colour images and *in situ* surveys of chlorophyll distribution. We start by considering how upwelling affects primary production, refering to the companion paper by Strass and Woods (1988). Then it is shown that the sustained vertical motion required for a spectacular increase in production is achieved at mesoscale jets simulated by a dynamical model based on dynamical conditions typical of the euphotic zone. The model predicts the horizontal dimensions of upwelling patches, which appear to be consistent with observation. It also explains the streaky character of plankton patchiness revealed by high resolution satellite images of ocean colour.

In order to relate this understanding of plankton patchiness to the large scale distribution of primary production, it is necesary to consider the physical conditions controlling mesoscale jet formation. The key condition is the horizontal variation of density stratification, which is best described by the isopycnic gradient of potential vorticity (IPVG), the properties of which are summarized in

the appendix. The seasonal climatology of IPV in the euphotic zone
of the North Atlantic reveals features that hold implications for the
distribution of primary production, if the hypothesis of this paper
is correct, namely that the statistics of upwelling at mesoscale jets
strongly influence large scale distribution of production.

3. THE EFFECT OF UPWELLING ON PRIMARY PRODUCTION

It is beyond the power of today's computers to simulate the
distribution of plankton shown in Figure 1 in terms of a model of
mesoscale upwelling and circulation. But we can consider the
specification for such a model. It would include the response of the
mixed layer depth to the mesoscale motion. Plankton growth would
best be simulated in terms of the statistics of an ensemble of
particles each of which grow independently as it is advected along
the meandering jet, rising and following as it passes through regions
of upwelling and downwelling, and interacting with the diurnally
deepening mixed layer. Within ten years it should be possible to
make such simulations, assuming that the speed of the fastest
computers continues to increase by a factor of ten every six years.
Meanwhile, several of the key ingredients of the model have been
tested, and they provide a basis for our present understanding of the
way in which mesoscale upwelling influences primary production. The
following components have been studied:

(a) Ross (1987) has modelled the interaction of transient vertical
motion on mixed layer depth and the diurnal jet. He showed that this
interaction produces the horizontal dispersion of particles in the
diurnal boundary layer which has been documented by dye diffusion
experiments (Okubo 1971). Thus particles that are upwelled into the
diurnal boundary layer are dispersed horizontally.

(b) MacVean and Woods (1980) have modelled vortex stretching in the
mixed layer (without entrainment), showing how the latter thins on
the anti-cyclonic side of an intensifying mesoscale jet. So, as
upwelling raises particles and nutrients towards the surface, the
mixed layer draws away from them, delaying the onset of the surface
hot spot in an otherwise oligotrophic regime.

(c) Wolf and Woods (1988) have used the one-dimensional Lagrangian-
ensemble method pioneered by Woods and Onken (1982) to model the
response to upwelling of plankton growing in an oligotrophic regime.
They showed that if upwelling is slower than the fall speed of the
particles (about 1 m/d according to Walsby and Reynolds 1980) their
enhanced growth is sufficient to prevent the nutricline from rising
into the mixed layer, and the result is intensification of the deep
chlorophyll maximum (as observed by Fasham *et al.*, 1985 and Strass
and Woods 1988). Whether fast or slow, sustained or not, upwelling
increases total production in the oligotrophic regime by providing
more nutrients.

(d) The response of the spring bloom to mesoscale upwelling is also being studied with the Lagrangian ensemble model. Growth is light-limited until nutrients become depleted and the physical environment affects growth mainly through variation of mixed layer depth. So upwelling modulates the spring bloom by reducing mixed layer depth earlier than the seasonal trend, producing a vigorous crop before herbivores are ready to graze on it. Perhaps that is why ungrazed phytoplankton products occasionally reach the seabed in abundance? Modulation of the spring bloom affects the timing of the transition to oligotrophy, but not the total production.

4. MESOSCALE JET DYNAMICS

Mesoscale jets are formed in the deep ocean euphotic zone by a confluence in the geostrophic flow acting on an isopycnic potential vorticity gradient. (See the appendix for a summary of potential vorticity.) The confluence kinematically sharpens the IPVG and the baroclinicity increases, giving a geostrophic jet which becomes progressively faster and narrower until a limit is reached. There is a limit to how much of the baroclinic available potential energy in the catchment of water converging into the jet that can be converted into geostrophic jet kinetic energy. The limit is reached when the ageostrophic vertical circulation forced by vortex stretching has flattenned the isopycnals, thereby cancelling the baroclinicity created kinematically by the externally imposed confluence. Consequently mesoscale jets have a universal form controlled by the profile of IPVG in the catchment.

   The time taken to reach the energetic limit is controlled by the imposed confluence. The confluence between transient eddies (order 10^{-5} s^{-1}), produces limiting mesoscale jets in a few days even in regions of weak eddy kinetic energy (see maps by Dantzler 1977). This time is much shorter than the persistence of the confluence, which is related to the eddy time scale of about one month. So the mesoscale jets form quickly between eddies then persist for several weeks as an Eulerian feature of the eddy field. The confluence between gyres (order 10^{-6} s^{-1}) produce mesoscale jets in a few months, which is just fast enough for completion during the summer, after spring subduction has established an IPVG profile and therefore restocked the euphotic zone with baroclinic available potential energy. The persistence of mesoscale jets formed by the confluence between gyres is not limited by changes in the confluence (which is permanent), but may be limited by upstream variation of IPVG in the catchment. In both cases (inter-eddy and inter-gyre), the confluence rate determines how quickly the jet forms, but the IPVG profile determines the structure of the jet, including the vertical displacement of isopycnals, which once established become a steady Eulerian feature of the jet.

12

4.1 A Computer Model of Mesoscale Jet Upwelling

As in other aspects of oceanography it is difficult to collect an experimental data set with sufficient information to achieve a satisfactory dynamical account of mesoscale jets. So the description given above rests on the results of theoretical studies stimulated and constrained by observations. It is not possible to make much progress by mathematical analysis alone (*e.g.* DeFant 1929) because the dynamical equations that control the development of high Rossby number mesoscale jets are non-linear. However, they can be solved by numerical integration using a computer; the principal constraint then becomes the speed and capacity of the computer. The main difficulty has been to integrate the equations of motion with a bandwidth sufficient to resolve the changing structure of a front as its width decreased from hundreds of kilometres to hundreds of metres. Despite the growing power of computers it is still necessary to use mathematical "tricks" to produce a satisfactory simulation of a mesoscale front. The following example illustrates the state of the art.

Figure 2a shows a mesoscale jet simulated by the two-dimensional primitive equation model of Bleck, Onken and Woods (1988), in which a barotropic deformation characteristic of the geostrophic eddy field (10^{-5} s^{-1}) acts for three days on an initially 800 km broad baroclinic zone, with an IPVG maximum at the top of the seasonal thermocline. Figure 2b shows the distribution of (Eulerian) upwelling after 2.2 days of confluence.

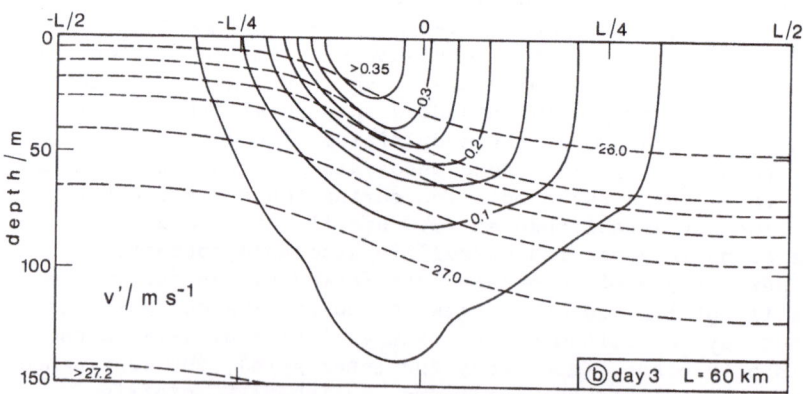

Fig. 2a. Cross-section through a simulated geostrophic jet. Jet speed (m/s) and depth variation of isopycnals (from Bleck *et al.* 1988).

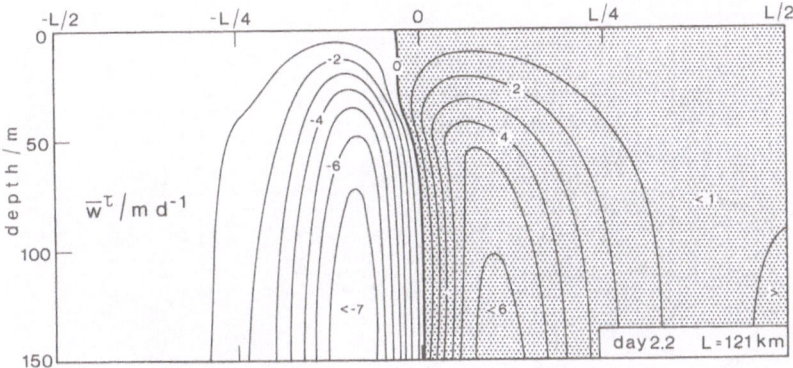

Fig. 2b. Cross-section through a simulated geostrophic jet.
Eulerian vertical speed (m/d) (upwelling shaded) (from Bleck *et al*.
1988).

Mesoscale jets are unstable. They soon develop meanders, with
downstream wavelengths in the range 10 to 100 km (Figure 3a).
Despite the rather simple velocity field, the particle displacement
during meander growth produces complex patterns of isotherms and Q
contours on isopycnic surfaces (Figure 3b,c). (It is important to
bear this point in mind when interpreting satellite images: the
existence of eddy-like windings in scalar distributions does not
provide evidence of closed eddy-like circulation.) The curvature in
these meanders contributes additional vorticity, with downstream
structure. As they grow, the vorticity increases giving patchy
upwelling with horizontal dimensions of a few kilometres both along
and across the direction of the mean stream (Figure 3d). We shall
see below how these mesoscale patches of upwelling generate primary
production "hot spots".

4.2 The Response of Primary Production to Mesoscale Upwelling

We can use these results to interpret the chlorophyll pattern in
Figure 1. The white patches (high chlorophyll concentration) are
isolated "hot spots" forced by mesoscale upwelling; the black patches
indicate sites where mesoscale downwelling has impeded the onset of
the spring bloom; and the long streaks are the result of advection by
meandering mesoscale jets. (The pattern cannot be interpreted in
terms of the redistribution of chlorophyll from an initially large
scale gradient.) Similar mesoscale patchiness during the spring bloom
has been observed *in situ* (see Figure 4 of the companion paper by
Strass and Woods 1988).

14

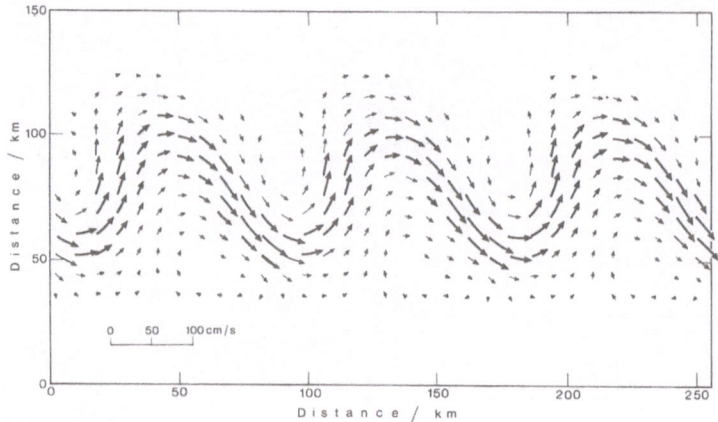

Fig. 3a. Velocity distribution on an isopycnic surface with mean depth 20 metres in a simulated meandering mesoscale jet. (Dr. R. Onken, IfM Kiel, unpublished).

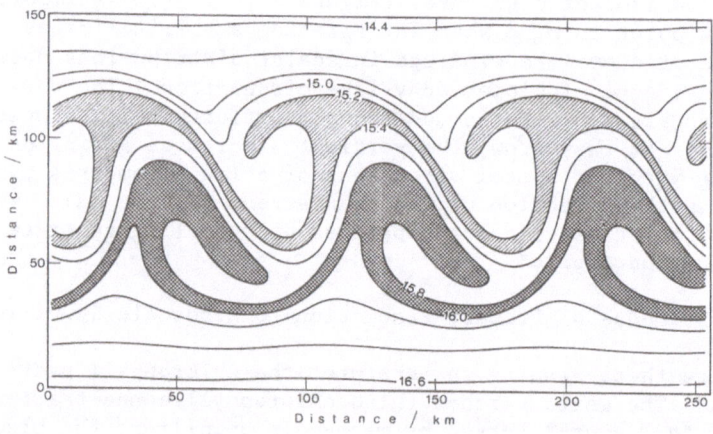

Fig. 3b. Temperature distribution on an isopycnic surface with mean depth 20' metres in a simulated meandering mesoscale jet. (Dr. R. Onken, IfM Kiel, unpublished).

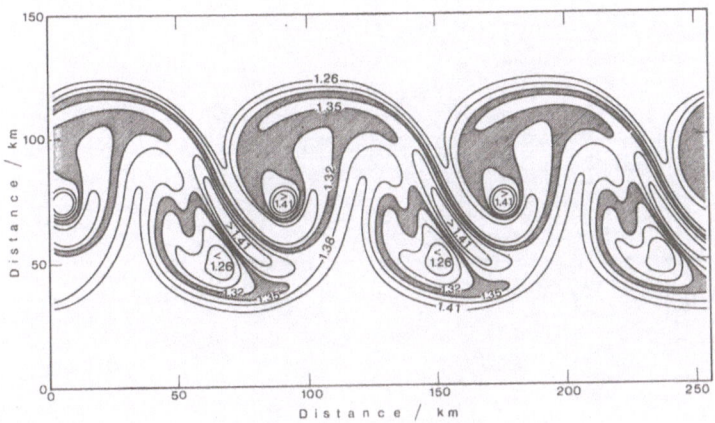

Fig. 3c. Distribution of isopycnic potential vorticity on an isopycnic surface with mean depth 20 metres in a simulated meandering mesoscale jet. (Dr. R. Onken, IfM Kiel, unpublished).

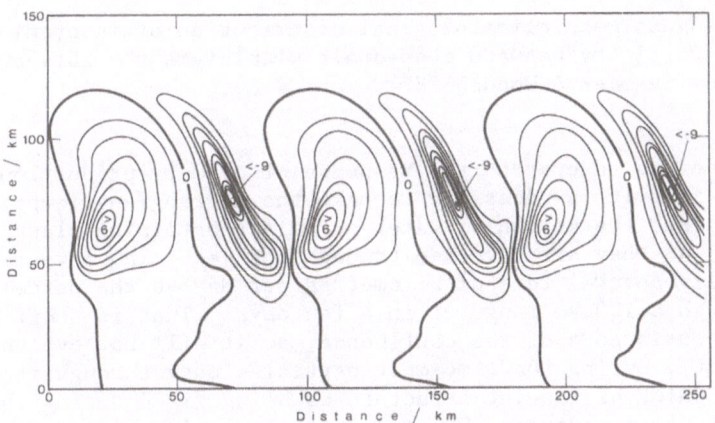

Fig. 3d. Distribution of vertical motion (m/d, upwelling positive) on an isopycnic surface with mean depth 20 metres in a simulated meandering mesoscale jet. (Dr. R. Onken, IfM Kiel, unpublished). The asymmetry is explained in the appendix (see "vortex stretching").

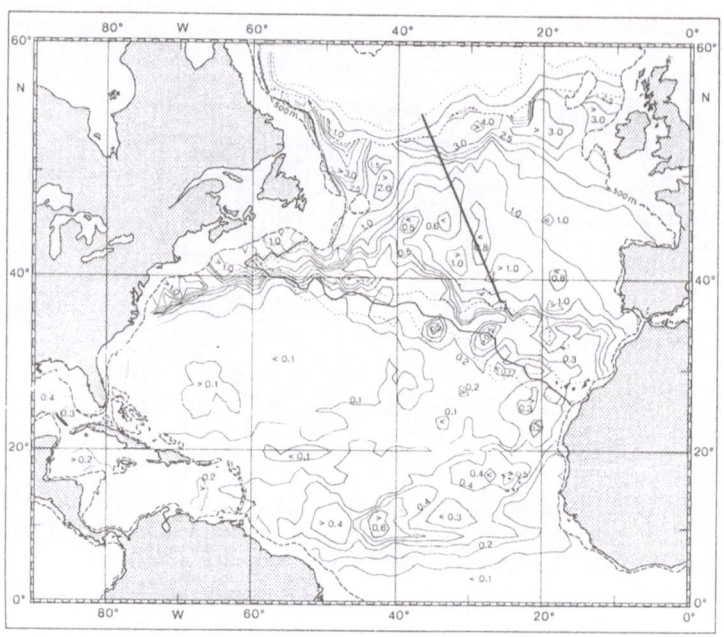

Fig. 4. August mean climatological distribution of isopycnic potential vorticity between isopycnals with sigma-t = 26.4 and 26.6 kg/m^2 (from Stammer & Woods 1987).

In order to understand the response of primary production to mesoscale jets it is necessary to take the Lagrangian viewpoint, that is to consider the motion of water parcels (containing plankton and nutrients) as they are advected through the jet. The advection speed is typically several tens of kilometres per day so the parcel will pass through a 100 km long jet in a few days. That is short compared with the persistence of the confluence, so it will not evolve significantly during the time most particles pass through it. We can therefore think of the jet structure as being fixed during the rapid transit of water parcels. We are particularly interested in the vertical component of motion of the water parcels. To a good approximation the particles pass through the jet without changing density, so they flow along isopycnic surfaces; they move upwards as they enter regions of anticyclonic vorticity where isopycnals dome, and they move downwards as they enter regions of cyclonic vorticity where isopycnals dip.

According to the Wolf and Woods (1988) model of oligotrophic upwelling, primary production increases exponentially with upwelling speed, provided it is less than the sinking speed of the plankters. Particles move upwards at sites where the isopycnic surface they flow along is domed. Let us stay with the Lagrangian frame of reference to follow a nutrient rich water parcel flowing towards such a dome. If the vertical speed is less than the fall speed of the plankters (about 1 m/d) the nutrients will be consumed by enhanced production in the seasonal thermocline (*i.e.* in the deep chlorophyll maximum) before the water parcel is entrained into the mixed layer. But if the vertical speed is much faster the increased primary production does not have time to consume all the nutrients in the water parcel before it becomes entrained into the mixed layer, and the remaining nutrients are rapidly diffused vertically through the mixed layer, as are the plankters that were in the water parcel. The brilliant sunlight in the mixed layer powers rapid growth which speedily consumes all the nutrients that were in the parcel.

Now consider the same process from the Eulerian perspective; that is, focussing on the changes that occur at a location fixed relative to the structure of the jet (which is assumed to be changing only slowly). Let us chose a site in the seasonal thermocline on the upstream side of an anti-cyclonic dome. A continuous stream of water parcels flow upslope into the site where they lose their inorganic nutrients to faster primary production induced by the brighter sunlight. The parcels flow away from the site with enhanced chlorophyll and depleted inorganic nutrient concentrations. The net effect is that nutrients are stripped out from particles passing through the site at a rate determined by their horizontal speed and the downstream slope of the isopycnic surface; the rate of nutrient consumption determines the source strength of chlorophyll at the site. That is our explanation of the patches of high chlorophyll concentration seen in Figure 6 of Strass and Woods (1988).

5. CLIMATOLOGY OF ISOPYCNIC POTENTIAL VORTICITY IN THE EUPHOTIC ZONE

The theoretical studies of mesoscale jets have shown that the precise location where they form and the time taken to reach maturity is determined by the eddy or gyre confluence, but their speed and structure, including the amplitude of isopycnal doming induced by vortex stretching, is determined by the initial distribution of isopycnic potential vorticity in the catchment from which water parcels flow through the jet. The fastest jets with largest vertical displacement of isopycnals will occur in regions where the large scale IPVG is greatest. A direct experimental test of that theoretical prediction based on synoptic maps will be described later. Here we consider the possibility that the correlation between IPVG and primary production might also be valid statistically on the large scale. If so, the potential enstrophy spectrum of geostrophic turbulence theory (see appendix) can be used to predict the concentration of mesoscale upwelling events.

In order to test that conjecture we must compare large scale climatological maps of IPVG with corresponding maps of primary production. Of course primary production occurs in the seasonal boundary layer of the ocean where the stratification (and therefore isopycnic potential vorticity) is destroyed by convection each winter and re-established the following spring. So it is necessary to use monthly (rather than annual) mean maps of potential vorticity. The first such maps have recently been published by Stammer and Woods (1987). Figure 4 shows their August mean map of Q_S between the isopycnals with sigma-theta = 26.4 and 26.6. Note the bunching of Q_S contours around 52°N, showing that each summer an IPVG front forms in the euphotic zone between the subtropical and subarctic gyres, and that a similar bunching occurs near the Azores. These fronts are well known sites of enhanced primary production. We have explored the inter-gyre front (see below) and Fasham et al., (1985) have explored the Azores front. The fact that such coherent patterns emerge after processing thousands of bathythermograph and hydrographic profiles collected randomly over thirty years indicates that the IPVG climate has a strong signature in the euphotic zone. The fact that regions of strong climatological IPVG coincide with known sites of enhanced chlorophyll concentration supports the hypothesis that the large scale distribution of primary production in the open ocean depends substantially on the distribution of mesoscale upwelling.

5.1 The Source of Potential Vorticity in the Euphotic Zone

Encouraged by that success we can go one stage further and ask whether the inter-annual variation in the large scale distribution of primary production might be related to different patterns of potential vorticity established each spring. If so, can we hope to predict the potential vorticity distribution each year? In order to address those questions we must discuss the annual source of potential vorticity in the euphotic zone.

5.1.1. *A Lagrangian view*. Isopycnic potential vorticity (Q) is a measure of static stability and is therefore only defined in the thermocline (see Appendix). The mixed layer is statically unstable because of the upward heat flux. A water parcel in the mixed layer has well-defined temperature, salinity and density, but not Q. A value of Q can be computed for the water parcel only after the two isopycnals defining its upper and lower bounds have both been subducted from the mixed layer into the thermocline. Subduction of isopycnals occurs when the mixed layer gets shallower during the heating season, abandoning water parcels in the newly formed thermocline. The Q of those water parcels depends on the Lagrangian correlation of mixed layer depth and density as they both decrease (Woods 1985). Lagrangian integration of a mixed layer model can predict the source Q as a function of the surface fluxes of wind stress, vorticity (wind stress curl), heat and water. Figure 5 shows the results of a 5-year Lagrangian integration in the Sargasso Sea (Woods and Barkmann 1986). Conservation of Q is best included in a

model based on the Lagrangian approach, in which a one-dimensional
model is integrated along the trajectory of water parcels following
the large scale circulation.

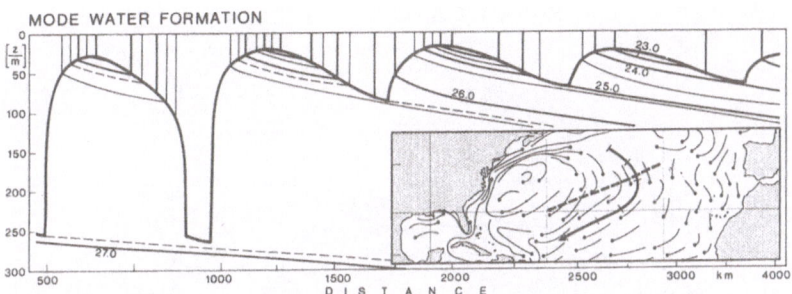

Fig. 5. Barotropic Lagrangian integration of a one-dimensional model
of the mixed layer in the Sargasso Sea showing the vernal subduction
of isopycnals, whose spacing determines isopycnic potential vorticity
in the euphotic zone. The heavy line marks the bottom of the mixed
layer depth. (Woods and Barkmann 1986).

5.1.2. *An Eulerian View.* The Eulerian view is needed to explain
how the spatial patterns of Q_S are established each year in the
seasonal thermocline. The distribution of Q_S between a given pair of
isopycnals depends only on the latitude and the Sverdrupian spacing,
which can be determined from the climatological mean spacing. So the
question is answered if it can be discovered how the regional pattern
of spacing between a pair of isopycnals is established each year in
the seasonal thermocline.

Consider first the annual behaviour of a single isopycnic
surface, which slopes gently through the thermocline, then rises
vertically through the mixed layer to outcrop in a line running
across the ocean surface. The outcrop line migrates annually. It
lies closest to the equator at the end of the cooling season and
closest to the pole at the end of the heating season (see Figure 1 of
the accompanying paper by Strass and Woods 1988). The migration is
distorted by the large scale gyre circulation (Figure 6). Poleward
warm and equatorward cold currents advance and delay the seasonal
change in the mixed layer depth respectively. The outcrop line
advancing poleward in the heating season is vertically above the
leading edge of the isopycnic surface in the thermocline. At the end
of the cooling season the surface lies exclusively in the permanent

thermocline; as heating proceeds it extends towards the pole in the
newly established seasonal thermocline. The maximum extent is
attained in August, when the outcrop line is two thousand kilometres
North of its March position. The extent of the migration measures
the extent of the seasonal thermocline on that isopycnal.

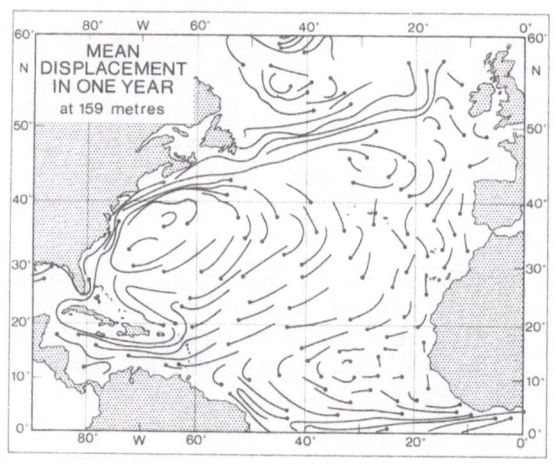

Figure 6. Mean circulation at a depth of 159 metres shown by annual
displacement of particles.(redrawn from Sarmiento and Bryan 1982)

The depth distribution on the isopycnic surface depends on (1)
the subduction depth of the isopycnal at each point along the outcrop
line as it advances poleward, and (2) subsequent advection and
baroclinic adjustment (Gill 1982). Monthly mean maps of depth on
isopycnic surfaces are given in the atlas of Bauer and Woods (1984).
    Now consider a second isopycnal extending poleward in the same
way during the heating season. The initial spacing between it and
the first surface depends on the different depths at which the two
surfaces were subducted. For the Eulerian perspective, the difference
is calculated in the direction of the advancing front. If the two
surfaces have only a small density difference (*e.g.* 0.2 kg/m^3, there
is only a short time interval between the passage of the two outcrop
lines past a point in the ocean, and at any moment the outcrop lines
are not far apart (see the broken lines in Figure 4). Thus the
regional distribution of Q in the euphotic zone is determined by the

spacing between two isopycnic surfaces established as they are subducted one after the other when their outcrop lines pass each location in the span of the seasonal migration. Where the relative vorticity happens to be negligible during subduction, the spacing has the Sverdupian value h_S; otherwise it starts stretched by the relative vorticity. The variance of this stretching deviation from the Sverdrupian value due to source enstrophy can, in principle, be estimated from geostrophic turbulence theory.

5.2 Correlation Between Temperature and Potential Vorticity

When a water parcel leaves the mixed layer to lie between a pair of isopycnals in the seasonal thermocline it has the temperature of the mixed layer during subduction. The Q of that water parcel depends on the difference between the mixed layer depths as the two isopycnals are subducted. Inspection of climatological maps of Q and T on isopycnals in the seasonal thermocline reveals a broad correlation, with high Q being found at higher latitudes, where the temperature is lower. That is because as subduction proceeds through the heating season (a) the location of subduction (directly below the outcrop of the isopycnal) migrates poleward towards cooler waters (and higher f), and (b) the depth of subduction (*i.e.* the mixed layer depth) decreases less rapidly, giving smaller h. The Q:T correlation would be stronger if the mixed layer did not warm up as the subduction location migrates poleward. But if we monitor the changes *at the migrating subduction location* the seasonal rise in temperature is normally less than the geographical decrease (defined as the temperature at the start of the heating season). The Q:T correlation is particularly strong in the seasonal thermocline at the front between the sub-Tropical and Sub-Arctic gyres (Figure 11c). Strong temperature differences between water masses are usually accompanied by strong differences between their values of Q in the euphotic zone. That correlation allows us to interpret satellite infrared images in terms of Q. For example, the dark regions in Figure 7 represent high temperature and high Q, and the light regions low temperature and Q. The cyclonic eddy centred on [37°W, 47°N] is entraining water and cold water in spiral bands from a broad catchment, creating a temperature and IPVG front that sharpens in the confluence generating unstable meanders.

6. TESTING THE THEORY

It is necessary to test the theory that plankton patches are formed by mesoscale upwelling at sites of enhanced IPVG. Computer models have been developed to illustrate the consequences of the theory as an aid to understanding. They also make predictions that can be compared with observations. Effective testing requires that both model and experimental techniques be developed until they yield equivalent products. Much of the work of my research group at the University of Kiel has been directed to that goal.

6.1 Computer Models

Our computer models were designed to meet the stringent
requirements posed by the need to make realistic predictions. Most
dynamical models do not conserve Q: ours used an isopycnic grid to
ensure that Q is accurately conserved (Bleck et al., 1988). Most
models of the upper ocean incorporate a simple treatment of solar
heating: ours incorporates a high resolution multispectral model
developed by Woods et al., (1985). Most mixed layer models do not
resolve the diurnal variation needed to control plankter motion: ours
does so (Woods and Barkmann 1986), and led to some controversy among
physical oceanographers until the predicted diurnal cycle was
observed in time series of microstructure profiles (Gregg et al.,
1986).

6.2 Observations in the North Atlantic

Our experimental apparatus was the Kiel "Sea Rover" system, based
on a combination of the Canadian batfish and the acoustic doppler
current profiler (Figure 8). The data were analysed to give profiles
of temperature, salinity, density, current speed, solar irradiance
and chlorophyll concentration. The horizontal spacing between the
profiles is about 0.5 km; average (and standard deviation) profiles
were computed at intervals of 10 km and each degree of latitude.
Profiles of Q_S were computed by averaging the raw profiles over one
degree of latitude. Data reports including these products are
available from the Institut fuer Meerskunde, Kiel (e.g. Bauer et al.,
1985).
    The experimental strategy was based two patterns: (1) long
sections, and (2) synoptic maps.

6.3 Long Sections

A section 2000 km long between the Azores and Greenland (indicated in
Figure 4) was first run in 1981 and repeated each year from 1983 to
1987 at different phases of the heating season. The aim was to
obtain statistics at different seasons to test the theory in the
context of the large scale climatological analysis of Stammer and
Woods (1987). Figure 9 shows the variation along the line of
temperature and (f/h) measured between 26.4 and 26.6 kg/m^3. Averages
of the (f/h) values in each degree of latitude give the Sverdrupian
potential vorticity (because the sample is large enough to have
negligible residual relative vorticity): these synoptic values are
close to the climatological values plotted in Figure 4. The variance
in that interval arises from a combination of enstrophy and potential
enstrophy. High values of Q advected from the west by the branches
of the North Atlantic current are seen at 47°N, 49°N and 50.5°N.
Estimates of Q become unreliable at the Northern end of the section
near the outcrops of the isopycnic surfaces.

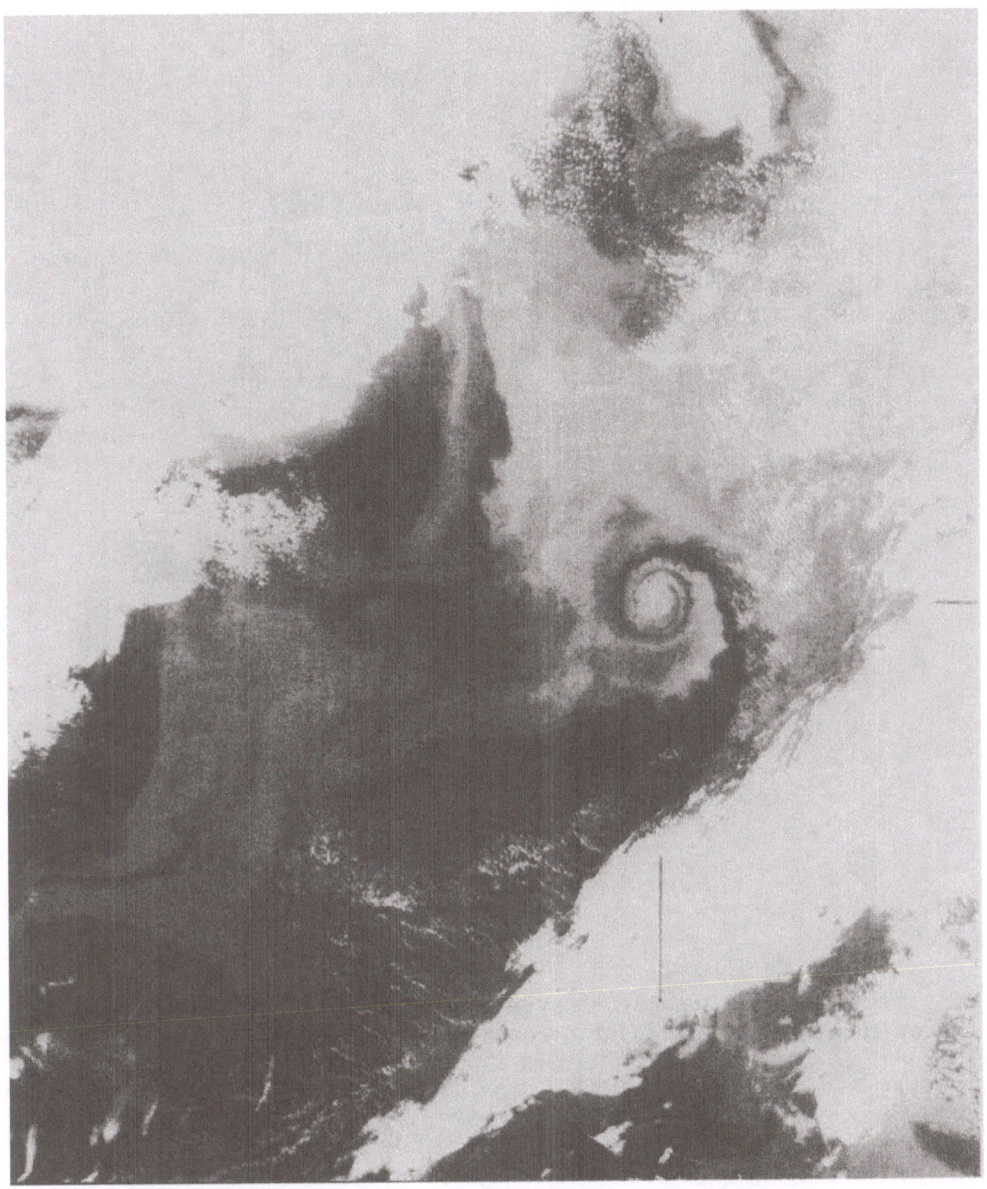

Fig. 7. Infrared satellite image of an eddy centred on 37°W, 47°N on 22 October 1980, showing spiral entrainment of warm and cold water masses (AVHRR image No. 272/06B reproduced by courtesy of Dundee University receiving station)

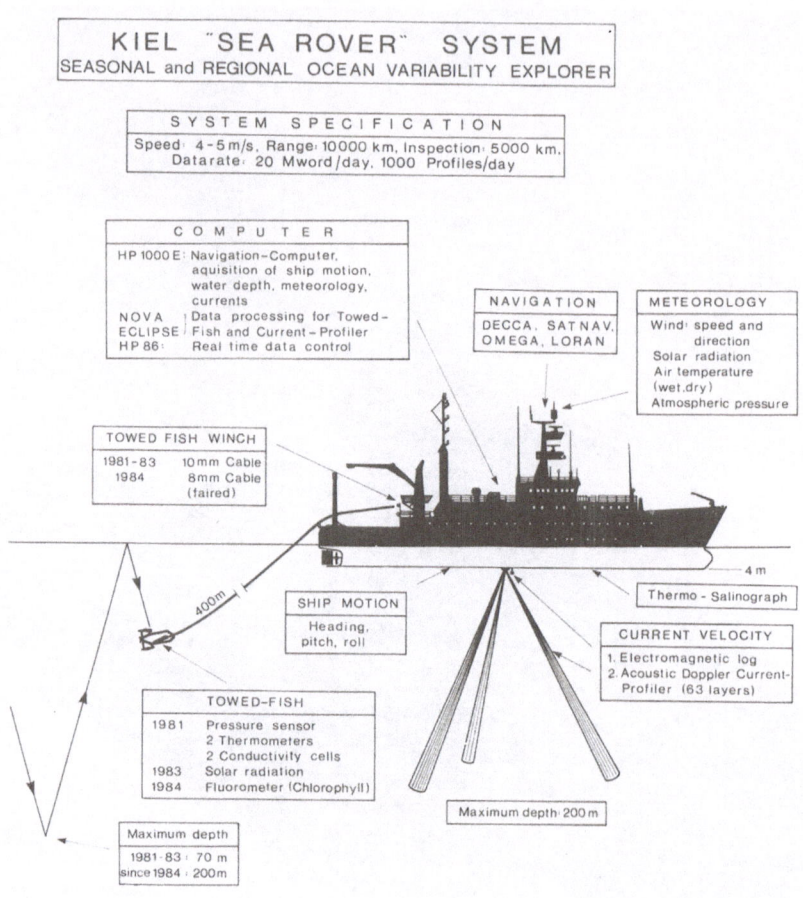

KIEL "SEA ROVER" SYSTEM
SEASONAL and REGIONAL OCEAN VARIABILITY EXPLORER

SYSTEM SPECIFICATION
Speed: 4-5 m/s, Range: 10000 km, Inspection: 5000 km,
Datarate: 20 Mword/day, 1000 Profiles/day

COMPUTER
HP 1000 E: Navigation-Computer,
aquisition of ship motion,
water depth, meteorology,
currents
NOVA : Data processing for Towed-
ECLIPSE: Fish and Current-Profiler
HP 86: Real time data control

NAVIGATION
DECCA, SATNAV,
OMEGA, LORAN

METEOROLOGY
Wind: speed and
direction
Solar radiation
Air temperature
(wet,dry)
Atmospheric pressure

TOWED FISH WINCH
1981-83   10mm Cable
1984   8mm Cable
           (faired)

400m

4 m

Thermo - Salinograph

SHIP MOTION
Heading,
pitch, roll

CURRENT VELOCITY
1. Electromagnetic log
2. Acoustic Doppler Current-
   Profiler (63 layers)

TOWED-FISH
1981   Pressure sensor
       2 Thermometers
       2 Conductivity cells
1983   Solar radiation
1984   Fluorometer (Chlorophyll)

Maximum depth: 200 m

Maximum depth
1981-83 : 70 m
since 1984 : 200m

Fig. 8. The Kiel "Sea Rover" system used to collect data to test the
theory that primary production correlates with IPVG.

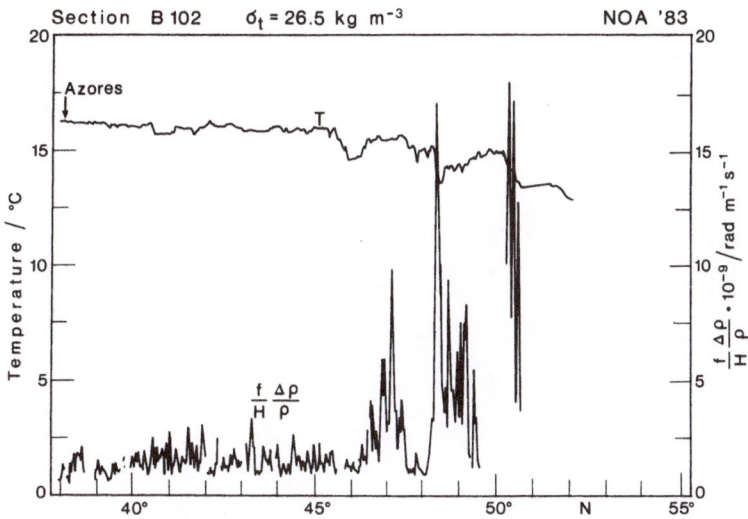

Fig. 9. Distribution of temperature (T) and potential vorticity
(f/h) measured along the line between the Azores and Greenland marked
in Figure 4 (from Stammer 1986).

6.4 Synoptic Maps

In 1981 and 1983 we mapped the distributions of physical and
biological variables in a box at the polar front between the subtropical and sub-arctic gyres, where there are climatological maxima
in IPVG and summer primary production. These were designed to
provide information about the detailed structure of Q and
phytoplankton concentration on the scales that the models predict
will exhibit strong correlation. Figure 10a gives an example of the
temperature and velocity fields measured during the 1983 survey. The
"synoptic" map shows about half of a meander in the inter-gyre front,
with a cut-off anticyclone centred on ($32^{o}20'$W, $52^{o}20'$N). The
corresponding map of attenuation (related to particle concentration)
in the top 30 metres is shown in Figure 10b. It is seen that the
highest concentrations are found in the anti-cyclone identified above
and in the anticyclonic bend centred on ($34^{o}45'$W, $52^{o}15'$N). Closer
examination by Horch (1987) has related the former to mesoscale

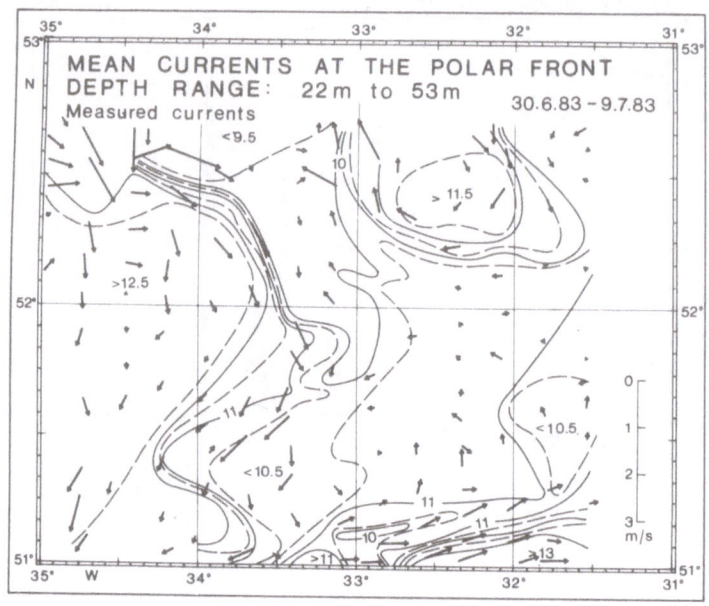

Fig. 10a. A synoptic survey of current vectors and isotherms on sigma-t = 26.8 at the North Atlantic inter-gyre front 30th June - 9th July 1983

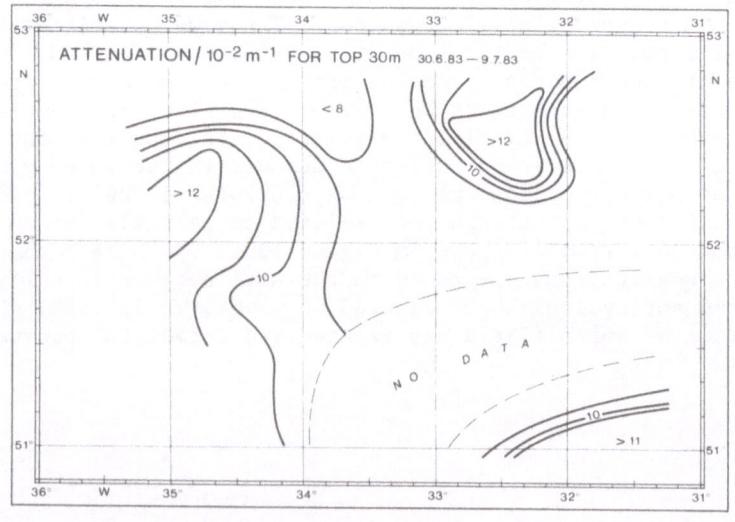

Fig. 10b. Distribution of turbidity measured in the top 30 metres during the 1983 synoptic survey (Horch 1987).

upwelling on secondary fronts associated with the IPVG between water masses of different T-S and Q winding around the anti-cyclone (as in Figure 7). The range of turbidities in Figure 10b extends over the full span of Jerlov's (1976) scale, from the clearest to the most turbid water. Mesoscale upwelling at the inter-gyre front produces patches of very high phytoplankton concentration close to the surface with an amplitude that is detectable by satellite ocean colour image.

Figure 11 illustrates the progress that has been made in measuring synoptic distributions and correlations of potential vorticity Q. The first ever synoptic map of Q (based on the full absolute vorticity) is shown in Figure 11a. The corresponding climatological distribution of QS at this site in August is shown in Figure 11b for comparison. The measured (synoptic) correlation of temperature and Q on the isopycnal with sigma-t = 26.9 kg/m^3 is shown in Figure 11c. These results (see Fischer, *et al.*, 1988) show that it is now possible to produce synoptic maps of potential vorticity (and of the relative vorticity and spacing between isopycnals which were used to produce Figure 11a) on the scales of plankton patchiness.

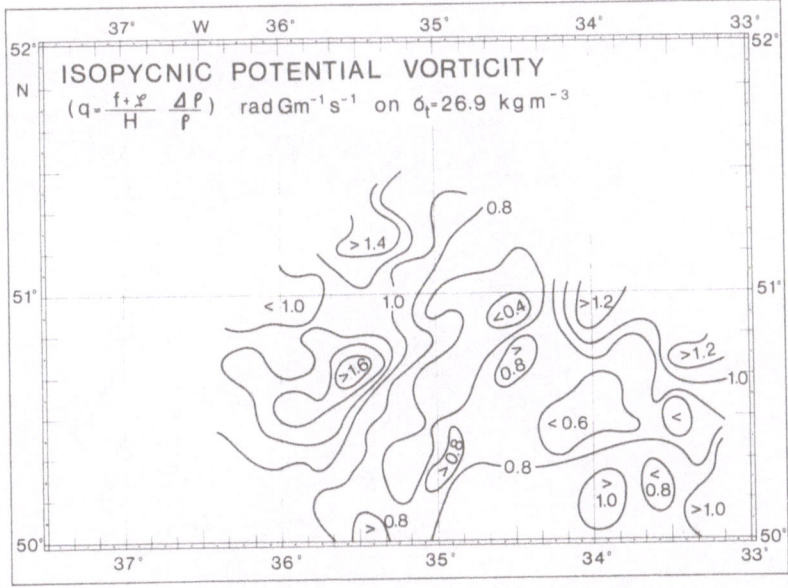

Fig. 11a. Synoptic map of potential vorticity at the North Atlantic inter-gyre front, showing a local maximum in IPVG at a mesoscale front (from Fischer *et al.*, 1988).

28

7. EVIDENCE IN OBSERVATIONS PUBLISHED BY OTHER AUTHORS

It is also possible to identify structures consistent with those
predicted by the present theory in the figures of earlier papers by
other authors. Attention is drawn especially to the pioneering
investigation of Fasham *et al.* (1985), who surveyed the distribution
of chlorophyll and density in the Azores front, which can now be
identified as a summer IPVG front which passes south of the Azores
and extends upstream for over 2000 km to the Gulf Stream extension
(see Figure 4). They show a mesoscale jet (their Figure 11) with
large amplitude meanders (their Figure 15). The IPVG at this jet is
indicated by a change of spacing between isopycnals (the change is so
extensive that it cannot be attributed to vortex stretching). The
chlorophyll maximum in the thermocline occurs at the IPVG maximum.
It is not possible from the published figures to identify the maximum
with anticyclonic vorticity: the horizontal spacing between survey
legs was too large to resolve the 10 km patches expected on such a
meandering jet. Nevertheless, the mesoscale spatial variation in
chlorophyll concentration is so strong that it masks larger scale

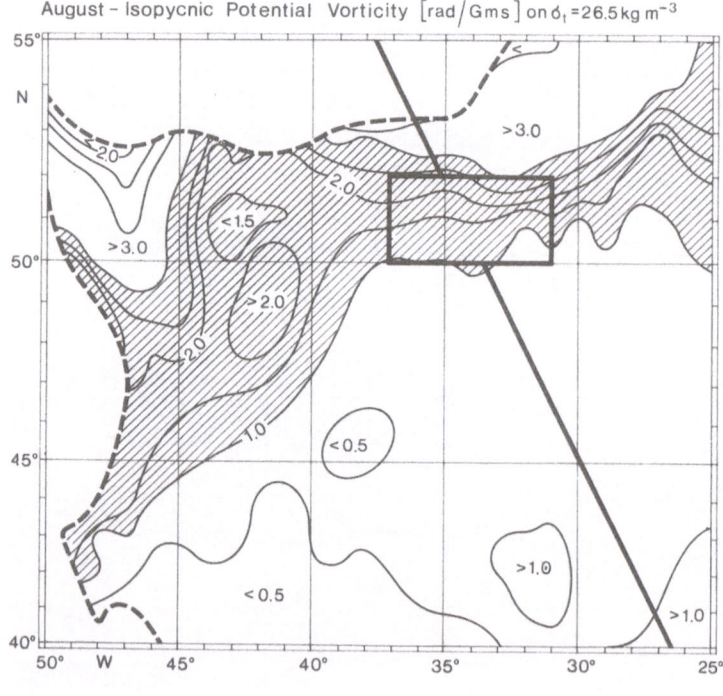

August - Isopycnic Potential Vorticity [rad/Gms] on $\sigma_t = 26.5$ kg m^{-3}

Fig. 11b. August mean climatology of isopycnic potential
vorticity at the experimental site (from Stammer and Woods 1987)

gradient. To summarize, the evidence in Fasham *et al.* (1985) is consistent with the dynamical theory presented in this paper, and with the hypothesis that large scale variation of primary production can only be described by the statistics of the intense mesoscale variability.

8. CONCLUSION

This paper presented a dynamical theory for phytoplankton patchiness, which not only explains many features of the observed finestructure of chlorophyll concentration, but also opens the way to a statistical approach to predicting the large scale distribution of primary production and its inter-annual variability. The theory extends classical one-dimensional treatment of the role of density stratification in controlling primary production to three dimensions by adopting the isopycnic gradient of potential vorticity as the

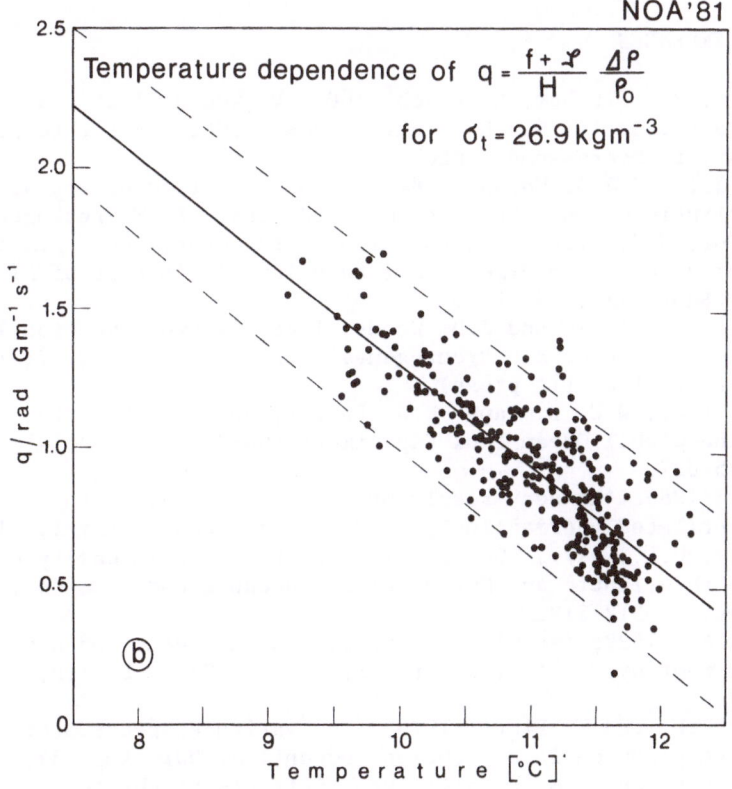

Fig. 11c. Correlation between potential vorticity and temperature on the isopycnal sigma-t = 26.9 (from Fischer *et al.*, 1988).

dynamical variable. The theory is supported by evidence in the published literature and in the new results presented here and in the companion paper by Strass and Woods (1988). Further analysis of the data collected during Kiel *Sea Rover* surveys in 1983 and 1985 will permit more refined tests, and the theory will be refined by the continuing modelling programme based on the *Lagrangian ensemble* method.

9. ACKNOWLEDGEMENTS

I acknowledge with thanks the contributions of my colleagues and students in the Abteilung Regionale Ozeanographie, Institut fuer Meereskunde an der Universitaet Kiel to the work reported in this paper (they are mentioned individually at the appropriate places in the text and figure captions). The research was supported by the German Research Society (DFG) through the following contracts: SFB 133/TP-B1 and Wo 254/10.

10. REFERENCES

Bauer, J., J. Fischer, H. Leach, and J.D. Woods: 1985, 'Sea Rover data report 1 - North Atlantic summer 1981', Berichte No. 143, Inst. f. Meereskunde, Kiel.

Bauer, J., and J.D. Woods: 1984, 'Isopycnic atlas of the North Atlantic Ocean', Berichte No. 132, Inst. f. Meereskunde, Kiel.

Batchelor, G.K.: 1969, 'Computation of the energy spectrum in homogeneous two-dimensional turbulence', *Physics of Fluids* 12(Suppl.II), 233-239.

Bleck, R., R. Onken and J.D. Woods: 1988, 'A two-dimensional model of mesoscale frontogenesis in the ocean', *Q. J. roy. meteor. Soc.* (in press).

Brewer, P.G., W.W. Bruland, R.W. Eppley, and J.J. McCarthy: 1986, 'The Global Ocean Flux Experiment (GOFS)', *EOS* 67(44), 827-832, 835-837.

Cox, M.: 1985, 'An eddy resolving numerical model of the ventilated thermocline', *J. Physical Oceanography* 15, 1312-24.

Dantzler, H.L.: 1977, 'Geographic variations in intensity of the North Atlantic and North Pacific oceanic eddy fields', *Deep-Sea Res.* 7, 512-519.

DeFant, A.: 1929, 'Stable stratification in oceans and associated current systems', Veroeff. Inst. Meer. Univ. Berlin, Nerfolge A, 19.

Denman, K.L. and T. Platt: 1975, 'The variance spectrum of phytoplankton in a turbulent ocean', *J. Mar. Res.* 34, 593-601.

Denman, K.L. and T.M. Powell: 1984, 'Effects of physical processes on plankonic ecosystems in the ocean', *Oceanogr. Mar. Biol. Ann. Rev.* 22, 125-168.

Esaias, W., G.C. Feldman, C.R. McClain, and J.A. Elrod: 1986,
    'Monthly satellite-derived phytoplankton pigment distribution
    for the North Atlantic ocean basin', *EOS* **67**, 835-7.
Fasham, M.J.R., T. Platt, B. Irwin, and K. Jones: 1985, 'Factors
    affecting the spatial pattern of the deep chlorophyll maximum in
    the region of the Azores front', *Prog. Oceanog.* **14**, 129-165.
Fasham, M.J.R. and P.R. Pugh: 1976, 'Observations of the horizontal
    coherence of chlorophyll a and temperature', *Deep-Sea Res.* **23**,
    527-538.
Fischer, J., H. Leach, and J.D. Woods: 1988, 'A synoptic map of
    isopycnic potential vorticity in the seasonal thermocline at the
    North Atlantic Polar Front', (in preparation).
Gill, A.E.: 1982, *Atmosphere-Ocean Dynamics*, Academic Press, London.
Gower, J.F.R., K.L. Denman, and R.J. Holyer: 1980, 'Phytoplankton
    patchiness indicates the fluctuations spectrum of mesoscale
    oceanic structure', *Nature* **288**, 157-159.
Gregg, M.C., H. Peters, J.C. Wesson, N.S. Oaker, and T.J. Shay:
    1986, 'Intensive measurements of turbulence and shear in the
    equatorial undercurrent', *Nature* **318**, 140-144.
Horch, A.: 1987, 'Doctorate Thesis', Kiel Univ.
Hoskins, McIntyre and Robertson: 1985, 'On the use and
    significance of isentropic potential vorticity maps', *Q. J. roy.
    Meteor. Soc.* **111**, 877- 946.
Jerlov, N.G.: 1976, *Marine Optics*, Elsevier, Amsterdam 231 pp.
Leach, H., P.J. Minnett, and J.D. Woods: 1985, 'The GATE
    Lagrangian Batfish experiment', *Deep-Sea Res.* **32**, 575-597.
Le Fèvre, J.: 1986, 'Aspects of the biology of frontal systems', *Adv.
    Mar. Biol.* **23**, 163-299.
MacVean, M.K. and J.D. Woods: 1980, 'Redistribution of scalars
    during upper ocean frontogenesis: a numerical model', *Q. J. roy.
    meteor. Soc.* **106**, 293-311.
Okubo, A.: 1971, 'Oceanic diffusion diagrams', *Deep-Sea Res.* **18**,
    789-802.
Pingree, R.D., P.R. Pugh, P.M. Holligan, and G.R. Forster: 1975,
    'Summer phytoplankton blooms and red tides along tidal fronts in
    the approaches to the English Channel', *Nature* **258**, 672-677.
Rhines, P.B.: 1979, 'Geostrophic turbulence', *Ann. Rev. Fluid
    Mechanics* **11**, 401-441.
Rhines, P.B., and W.R. Young: 1982, 'Homogenization of potential
    vorticity in planetary gyres', *J. Fluid Mech.* **122**, 347-367.
Robinson, A.R.: 1983, *Eddies in Marine Science*, Springer-Verlag,
    Berlin 609 pp.
Ross, H.: 1987, *Der einfluss des Triftsrtomes auf die horizontale
    Dispersion in der planetarischen Grenzschicht1*, Diplomarbeit,
    IfM, Kiel 189 pp.
Sarmiento, J. and K. Bryan: 1982, 'An ocean transport model for
    the North Atlantic', *J. Geophys. Res.* **87**, 394-408.
Stammer, D.: 1986, *Die jahreszeitliche Veraenderlichkeit der
    isopyknischen potentiellen Vorticity in der Warmwassersphere des
    Nordatlantiks*, Diplomarbeit Inst. f. Meereskunde Kiel, 140 pp.

Stammer, D. and J.D. Woods: 1987, *Isopycnic potential vorticity atlas of the North Atlantic ocean*, Kiel Institut fuer Meereskunde Report No. 165, 108 pp.

Steele, J.: 1978, *Spatial patterns in plankton communities*, Plenum, New York.

Stommel, H. and F. Schott: 1977, 'The beta spiral and the determination of the absolute velocity field from hydrographic station data', *Deep-Sea Research* **24**, 325-329.

Strass, V. and J.D. Woods: 1988, 'Horizontal and seasonal variation of density and chlorophyll profiles between the Azores and Greenland', in B.J. Rothschild (ed.), *Towards a Theory on Biological-Physical Interactions in the World Ocean*, Kluwer Academic Publishers, Dordrecht, pp. 113-136.

Walsby, A.F. and C.S. Reynolds: 1980, 'Sinking and floating', Ch. 10 in I. Morris (ed.), *The Physiological Ecology of Phytoplankton*, Blackwell, Oxford.

Welander, P.: 1971, 'Thermocline Problem', *Phil. Trans. Roy. Soc. London* **A270**, 69-73.

Wolf, U. and J.D. Woods: 1988, 'Lagrangian simulation of primary production in the physical environment - the deep chlorophyll maximum and nutricline', in B.J. Rothschild (ed.), *Towards a Theory on Biological- Physical Interactions in the World Ocean*, Kluwer Academic Publishers, Dordrecht, pp. 51-70.

Woods, J.D.: 1980, 'Do waves limit turbulent diffusion in the ocean?', *Nature* **288**, 219-224.

Woods, J.D.: 1985, 'The physics of thermocline ventilation', in J.C.J. Nihoul (ed.), *Coupled Ocean-Atmosphere Modelling*, Elsevier, Amsterdam, pp. 543-590.

Woods, J.D. and W. Barkmann, 1986, 'A lagrangian mixed layer model of Atlantic 18 degC water formation', *Nature* **319**, 574-576.

Woods, J.D., J. Fischer, and R. Onken: 1986, 'Thermohaline intrusions created isopycnically at oceanic fronts are inclined to isopycnals', *Nature* **322**, 446-449.

Woods, J.D., A. Horch, and W. Barkmann: 1985, 'Solar heating of the oceans', *Quart. J. roy. Meteor. Soc.* **110**, 633-656.

Woods, J.D. and P.J. Minnett: 1979, 'Analysis of mesoscale thermoclinicity with an example from the tropical thermocline during GATE', *Deep-Sea Res.* **26**, 85-96.

Woods, J.D. and R. Onken: 1982, 'Diurnal variation and primary production in the ocean - preliminary results of a Lagrangian ensemble model', *J. Plankton Res.* **4**, 735-736.

11. APPENDIX: ISOPYCNIC POTENTIAL VORTICITY

The density stratification in the thermocline is classically described by the Brunt-Väisälä frequency, N

$$N^2 := g/\rho \ (d\rho/dz) \tag{1}$$

The stratification is linearly related to the isopycnic potential vorticity, Q (Hoskins, McIntyre and Robertson 1985)

$$Q := (\zeta + f) \cdot N^2/g \tag{2}$$

The relative vorticity is given by:

$$\zeta := dV/dx - dU/dy, \tag{3}$$

with current velocity components (U,V) measured along isopycnals (simplified in practice to horizontal components).
The component of planetary vorticity orthogonal to isopycnals (in practice vertical):

$$f := 2\Omega \sin\phi \tag{4}$$

where Ω is the Earth's rotation rate and ϕ is the latitude.
It is convenient to express Q in finite difference form, *i.e.* the value for a water parcel (or vortex tube) bounded above and below by isopycnals with mean density ρ, density difference $\Delta\rho$ and spacing h:

$$Q := \frac{\zeta + f}{h} \cdot \frac{\Delta\rho}{\rho} \tag{5}$$

The Rossby number of the water parcel measures the ratio of its relative to planetary vorticity,

$$R := \zeta/f \tag{6}$$

11.1 A Conservation Law For Stratification

The isopycnic potential vorticity is a conservative property of any small water parcel circulating adiabatically around the ocean between a specified pair of isopycnals. That Lagrangian conservation law can be transformed to an Eulerian form relating the stratification at different locations in the thermocline to the circulation and the sources and sinks of Q.

11.2 Sverdrup Balance

According to Sverdrup's theory of large scale ocean circulation, meridional motion keeps the relative vorticity small compared with the Coriolis frequency, *i.e.* R < 1.

$$\beta V = f \cdot dW/dz, \tag{7}$$

The stratification law then reduces to conservation of Sverdrupian potential vorticity Q_S

$$Q_S := (f/g) \cdot (NS)^2, \tag{8}$$

or in finite difference form

$$Q_S := (f/gh_S) \cdot (\Delta\rho/\rho) \tag{9}$$

where the Brunt-Väisälä frequency of the water parcel and the spacing between its bounding isopycnals have the Sverdrupian values N_S and h_S respectively.

11.3 Relative Vorticity

Significant relative vorticity is generated only by transient motions with time and space scales so small that they do not adjust to the Sverdrup balance. The transient geostrophic eddies with horizontal scales of order 100 km have relative vorticity of order 0.01f (Robinson 1983). Larger relative vorticities are found at even smaller scales, where the motion has the form of jets rather than eddies, but is still nearly geostrophic.

When geostrophic jets are sufficiently swift and narrow, they have substantial relative vorticity because of the large horizontal shear. Consider, for example, a 10 km wide jet flowing East at 1 m/s at 50°N.

$$dU/dy = +/- \ 10^{-4} \ s^{-1}, \ dV/dx = 0 \tag{10}$$

$$\varsigma = +/- \ 10^{-4} \ s^{-1}, \ f = 10^{-4} \ s^{-1}$$

$$R = 1.$$

Following meteorological practice, we give the name *mesoscale* to motions with $R \sim 1$.

Planetary vorticity is, by definition, cyclonic. So cyclonic relative vorticity is positive and anticyclonic relative vorticity is negative. In the Northern Hemisphere, when one looks downstream along the axis of a mesoscale jet, cyclonic vorticity lies on the left of a jet, and anticyclonic vorticity on the right.

11.4 Vortex Stretching

Conservation of Q requires that the spacing h between the isopycnals defining the upper and lower boundaries of a water parcel changes when the absolute vorticity (ς + f) of a water parcel changes:

$$h = h_S \cdot (1 + \varsigma/f) = h_S \cdot (1 + R) \tag{11}$$

where h_S is the spacing that the parcel would have at that latitude if it were in Sverdrup balance, with negligible relative vorticity (R = 0). The spacing h varies in time as the large scale circulation carries the parcel to different latitudes, and as it flows through transient eddies and jets. This temporal modulation of h is called vortex stretching. It was first identified in the tropics where f is small so a modest relative vorticity produces a large change in h (Woods and Minnett 1979; Leach, Minnett and Woods 1985).

The spacing h must remain positive because by definition Q exists only for statically stable water parcels. So, while cyclonic vorticity may increase without limit, anticyclonic vorticity cannot become more negative than -f. This gives a lateral asymmetry to mesoscale jets, which can become narrow on the cyclonic side but must remain broad on the anticyclonic side. This dynamical constraint determines the horizontal scale of upwelling (see Figure 3d) and therefore of plankton patches (Figure 1).

The parcel conserves potential vorticity by becomes thicker or thinner when its absolute vorticity changes. Its rate of change of thickness depends on the rate of change of absolute vorticity,

$$dh/dt = h_S \cdot (1 + R) \cdot dR/dt \tag{12}$$

As the parcel becomes thicker, conservation of mass requires a convergence of the flow [U,V] along the isopycnals

$$dU/dx + dV/dy = -dW/dz = -(1/h) \cdot (dh/dt) \tag{13}$$

Classical theories of the large scale circulation of the thermocline (*e.g.* Welander 1971) are based on this mass balance with the assumpion that the relative vorticity is negligible (Sverdrup balance).

We are concerned with the opposite assumption, namely that the rate of change of planetary vorticity is negligible compared with the rate of change of relative vorticity. That is true for high Rossby number mesoscale jets which are so small that water parcels passing through them do not experience significant change of f, but do experience rapid change of relative vorticity. So, neglecting changes in f, we can can write:

$$dW/dz = -[dU/dx + dV/dy] = dR/dt \tag{14}$$

This ageostrophic motion due to vortex stretching play a vital role in controlling plankton patchiness. The vertical component W controls mesoscale upwelling, giving the "hot spots" of vigorous primary production seen in Figure 1. The horizontal components (U,V) shape the streaks of chlorophyll drawn out from those hot spots by the geostrophic jets whose relative vorticity caused the vortex stretching. They also create intrusive tongues inclined to isopycnals (Pingree *et al.*, 1975) characteristic of thermohaline finestructure at mesoscale fronts (Woods *et al.*, 1986).

11.5 Vertical Motion

In order to compute the vertical motion due to vortex stretching we divide the thermocline into a set of layers, separated by isopycnic surfaces. The profile of vertical motion is computed by integrating the last equation down from the top of the thermocline

$$W(z) = \int_{z_m}^{z} [dR/dt]\ dz \tag{15}$$

The vertical speed of the isopycnal at the top of the thermocline (depth z_m) can be similarly calculated from the mass convergence in the mixed layer, with water parcels bounded above by the sea surface and below by the top of the thermocline, remembering to include the injection of relative vorticity by the wind stress curl.

11.6 Synoptic Maps of Isopycnic Potential Vorticity

Consider a pair of isopycnals in the thermocline. If all the water parcels between those isopycnals have the same value of Q, then according to the invertibility principle (Hoskins *et al.*, 1985) they are all dynamically equivalent. Under those conditions, a synoptic map of the spacing h between the pair of isopycnals would depend only on the latitude and the regional variation of relative vorticity. In reality, that ideal situation is never achieved; the water parcels between a pair of isopycnals do not all have the same Q. They have differences which originated in their respective source regions. Synoptic maps of Q-contours show patterns (Figures 3d and 11a) which change in response to the flow field of the permanent gyres and transient eddies and jets. Rhines (1979) has shown that the large-scale effect of this complex stretching and swirling of Q contours is equivalent to a Fickian diffusion of QS along isopycnic surfaces. That diffusion is believed to be important in shaping the permanent gyres (Rhines and Young 1982; Cox 1985).

11.7 Isopycnic Potential Vorticity Gradient (IPVG)

The gradient of Q, measured along the isopycnic surface, is known as the isopycnic potential vorticity gradient (IPVG),

$$IPVG := \nabla \rho Q \tag{16}$$

The existence of an IPVG between a pair of isopycnals at some location means that the upper isopycnal is inclined to the lower by more than can be explained by the gradient of absolute vorticity. It means that the vertical shear of the geostrophic current changes with depth, as in the beta-spiral (Stommel and Schott 1977).

11.8 Dynamical Effects of IPVG

There is a fundamental difference between the isopycnic gradients of temperature and potential vorticity. The former is dynamically passive: the latter is dynamically active. The geostrophic flow depends on the depth profile of baroclinicity, and therefore changes with IPVG. The redistribution of isotherms along isopycnals effected by the geostrophic flow produces no feedback to the flow (*i.e.* no acceleration). But when the flow redistributes Q contours, it changes the IPVG, and therefore the baroclinicity profile, altering the hydrostatic pressure gradient driving the flow, which therefore accelerates. The flow around transient eddies greatly increases IPVG when it draws out Q contours into streaks and winds them up, as in Figures 3d and 7. The secondary fronts formed in this way are particularly active and can produce vigorous vertical motion and the highest chlorophyll concentrations (Horch 1987).

11.9 The Spectrum of Q

As Q contours are drawn out into streaks by the geostrophic circulations around transient eddies, the spectrum of Q variance (called "potential enstrophy") extends to higher horizontal wavenumbers. This transfer of scalar variance from large scale to small is called a cascade in turbulence theory. Dynamical theories have been developed for the potential enstrophy cascade, making it possible to predict the statistics of Q at small scale from a knowledge of the source at large scale. Batchelor (1969) showed that the key variable in the dynamics of such cascades is the IPVG. Although potential vorticity is conserved in an active cascade, the variance of IPVG grows vigorously, causing local accelerations in the flow with the right phase to drive the variance of Q to small scale, where it can be dissipated by diapycnic processes. The existence of dynamical cascade theories (*e.g.* Rhines 1979) make it possible to predict the statistics of IPVG at small scale from the large scale maps of IPVG. Now that small scale IPVG can be related to patchiness in primary production, it will be possible to relate the average concentration of primary production "hot spots" to the climatological average IPVG.

11.10 Climatological Maps of Q

Climatological IPVG is computed by averaging spot values of Q (for water parcels bounded by a pair of isopycnals) over some area (*e.g.* a square with sides one degree of latitude and longitude) and/or some time interval. If there are enough independent samples in that box, then the random vortex stretching due to cyclonic and anti-cyclonic relative vorticity will cancel and the average spacing between the isopycnals will equal h_S. That is true regardless of the horizontal dimensions of the averaging area. Average Q equals average Q_S. If the only product required is the average potential vorticity, it is not necessary to measure the relative vorticity of the individual samples.

The choice of time interval for the sample depends on the location in the water column. Two factors must be considered in specifying the time interval. The first is that Q is defined in the context of specified isopycnals (Woods 1985), so it is necessary to avoid any period during which the defining isopycnals are no longer present in the water column. That means avoiding all-year sampling in the seasonal boundary layer, or all-day sampling in the diurnal boundary layer.

The second consideration is that the water parcels are mobile, and may move across the sampling area in the sampling time. Contours of Q_S are advected and diffused along isopycnic surfaces by the large scale circulation, just as contours of Q are advected by transient eddy motions. If the sampling time is too long, advective distorion in maps of Q_S may bias IPVG. This is most likely to be important in the seasonal thermocline where gyre-scale motion is strongest, and the proximity to the source of Q gives sharp horizontal gradients.

ON REGULATION OF PRIMARY PRODUCTION BY PHYSICAL PROCESSES THE OCEAN: TWO CASE STUDIES*

Terrence M. Joyce
Woods Hole Oceanographic Institution
Woods Hole, MA   02543
U.S.A.

ABSTRACT. Interactions between the physical environment of the ocean and biology have been recognized since the early days of oceanography. In fact, deep ocean expeditions used to be multidisciplinary in character, and it was recognized that biological provinces and the physical environment were related on the large scale. This association has been extended to the scales of fronts and isolated eddies. This effort, largely descriptive in nature, has only begun to touch on the interaction between the physical and biological dynamics. Two case studies will be presented which deal with coupled physical and biological dynamics of phytoplankton and primary production. The first deals with light-limiting conditions and the effect of vertical oscillatory motion. The second considers nutrient-limited conditions in a warm-core ring which is slowly decaying due to friction. Both case studies suggest that patchiness will develop on scales determined by the physical environment and illustrate the importance of internal gravity waves and oceanic eddies to phytoplankton dynamics.

1. INTRODUCTION

Two case studies are presented to illustrate the importance of physical processes to phytoplankton dynamics. In the first study, an extension will be made to earlier work by Holloway (1984) who considered the effects of small amplitude, high frequency vertical motion upon phytoplankton growth. Inasmuch that productivity is a nonlinear function of depth (being dependent on incident light levels), the vertical motions of water parcels about some mean depth, on average, can alter the dynamics of phytoplankton growth from the stationary "control" condition. This effect will be most easily presented in a coordinate frame moving with the local water parcels; a Lagrangian reference frame. Earlier results will be

*Woods Hole Oceanographic Institution Contribution No. 6409.

B. J. Rothschild (ed.), Toward a Theory on Biological-Physical Interactions in the World Ocean, 39–50.
© 1988 by Kluwer Academic Publishers.

extended to vertical motions of arbitrary amplitude and frequency. One result will be that the co-varying of the vertical motion and a time-dependent light field is capable of modulating phytoplankton growth to a greater extent than previously suspected.

In a second study an examination will be made of the interaction of the physical preconditioning of a Gulf Stream warm-core ring with the near-surface pool of nutrients, which could be responsible for episodic pulses of nutrients into the mixed layer. The preconditioning is such that the gravitational potential energy to vertically mix the surface layer down to the level of the first detectable nitrate is much smaller in the center of the ring than on its periphery. Moderate and observable pulses of wind-driven mixing can erode the stratification in the center of the ring and bring nutrients up from below into the depleted surface layer. This process of vertical mixing of a preconditioned physical environment could be responsible for ring scale enhancement of the vertically integrated primary production in a warm-core ring.

2. CASE 1: TIME DEPENDENT LIGHT AND VERTICAL MOTION

The physical effects of vertical motion and time dependent light will be presented following a more extensive treatment by Flierl and Joyce (in prep.). The biological dynamics are assumed to be light limited and no explicit dependence of growth upon nutrients is included. Following fluid particles, the model equation for the phytoplankton concentration ϕ_p, expressed in terms of mass of carbon or chlorophyll-a is:

$$\frac{\partial \phi_p}{\partial t} = \lambda_p \, \phi_p - \varepsilon \phi_p^2 \tag{1}$$

where the subscript p, henceforth dropped, denotes a fluid particle is being followed. It will be assumed that fluid particles and phytoplankton cells are equivalent. It is possible to include a slow sinking of the cells into the model, but since there is no mechanism for retaining these cells near the surface in this model, no stationary solution can be achieved. In (1) λ is the specific growth rate of the phytoplankton and ε is introduced to limit unbounded exponential increases of ϕ. Following Flierl and Joyce, the solution of (1) satisfying the initial condition at t = 0, $\phi = \phi_0$ is

$$\phi(t) = \phi_0 e^{\bar{\lambda} t} \, [1 + \varepsilon \, \phi_0 \int_0^t dt' e^{\bar{\lambda} t}]^{-1} \tag{2}$$

$$\bar{\lambda} = \frac{1}{t} \int_0^t \lambda \, dt' \tag{3}$$

That the growth rate λ is time dependent can be seen by examining a model by Platt et al. (1980),

$$\lambda = \lambda_s \, [1 - e^{-(\alpha I/\lambda_s)}] \, e^{-(\beta I/\lambda_s)} \tag{4}$$

expressing the relationship between light intensity I and specific growth rate. If light is assumed to decay exponentially with depth, z following

$$I = I_s(t)e^{\gamma z}$$

than a particle oscillating about a depth z_0

$$z = -z_0 + \eta(z_0, t)$$

with a time dependent amplitude η will see a time variable light field independent from the diurnal light changes at the ocean surface, $I_s(t)$. Because the vertical displacement of a fluid particle, $\eta(z_0, t)$, can have a broad spectrum and because the light field is a non-linear function of depth, the light history of an individual particle will be rather complicated. One result is that the average light seen by an oscillating particle will be greater than that of a stationary particle at the same mean depth. This is due to the exponential dependence of light upon depth. The non-linear dependence of the growth rate upon light intensity further complicates the phytoplankton dynamics.

If the characteristic amplitude of vertical displacement, a, is small compared to the optical scale depth γ^{-1}, then the growth rate can be simplified to

$$\lambda(z_0 + \eta) = \Lambda_0 + \Lambda_0' \eta + (1/2)\Lambda_0'' \eta^2 + O(\gamma a)^3, \quad (\gamma a) < 1 \qquad (5)$$

where

$$\Lambda_0 = \lambda(I(z_0))$$

$$\Lambda_0' = (\partial\lambda/\partial I)_0 (\partial I/\partial z)_0$$

$$\Lambda_0'' = (\partial^2\lambda/\partial I^2)_0 (\partial I/\partial z)_0^2 + (\partial\lambda/\partial I)_0 (\partial^2 I/\partial z^2)_0 \qquad (6)$$

and the growth rate is expanded in a Taylor series about the point z_0. Denoting the time average growth rate by an overbar, the average growth rate can be written

$$\bar{\lambda} = \Lambda_0 + \overline{\Lambda_0'\eta} + (1/2)\overline{\Lambda_0''\eta^2} + O(\gamma a)^3 \qquad (7)$$

2.1 Small Amplitude, High Frequency Limit

If the motion is due to <u>high frequency</u> internal waves whose frequencies are high compared to the diurnal cycle of the light, then (equation 7) simplifies to

$$\bar{\lambda} = \Lambda_0 + (1/2)\Lambda_0'' \overline{\eta^2} \qquad (8)$$

since

$$\overline{\Lambda_0' \, \eta} \simeq \Lambda_0' \, \bar{\eta} \simeq 0$$

the result (equation 8) was obtained first by Holloway (1984) and shows that the average growth rate following a rapidly moving particle differs from that of a stationary particle by an amount $(1/2)\Lambda_0'' \, \overline{\eta^2}$, the sign of which depends on Λ_0''. For low light conditions much below photoinhibition levels

$$\Lambda_0 \simeq \alpha I_s e^{-\gamma z_0}$$

$$\Lambda_0'' \simeq \alpha \gamma^2 I_s e^{-\gamma z_0}$$

letting $\eta = a \sin \omega t$, the average growth rate is enhanced over the non-moving case by a factor of $1 + 1/4(\gamma a)^2$. For shallower depths, z_0, or greater mean light levels, the effect of photoinhibition becomes important and greater mean light levels may reduce mean growth rates. In Figure 1 the Platt et al. (1982) relationship is plotted for arctic diatoms at the 1% light level (Sc-93 in Table 1, p. 1162). For light levels well below $60 w/m^2$, the $\lambda(I)$ relationship is linear and Λ_0'' is dominated by the second term in equation (6) as noted above. For increasing mean light levels, the first term in equation (6) becomes increasingly larger in magnitude and negative. Near the maximum in the $\lambda(I)$ curve the first term in equation (6) dominiates and any vertical motion will <u>reduce</u> net growth rate $\bar{\lambda}$.

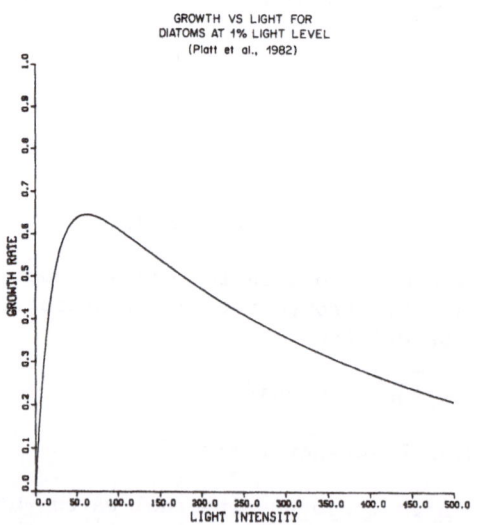

Figure 1. From Platt et al. (1982), a growth curve for an assemblage of arctic diatoms (see text) at the 1% light level. Specific growth rate is in $(hr)^{-1}$ with a scale factor conversion for mg Carbon/mg chlorophyll-a, assumed to be constant on the model. Light initensity units are watts m^{-2}.

The solution for the low light condition appropriate to this light frequency limit is given in Figure 2 using the full solution (equations 2,3). The daily light cycle is a "clipped" sinusoid shown by the solid line in the top panel. The time-varying light following a half-hour period internal wave with amplitudes of 5 (dashed) and 10 (chain dashed) meter amplitude about a mean depth of 46 m is indicated. The average growth rates for the two amplitudes and the stationary case are shown in the second panel. The enhanced growth rate due to the high frequency motion is evident and is reflected in the larger phytoplankton concentration, ϕ. In this example the steady-state solution is not reached until several days (only 25 hours are plotted).

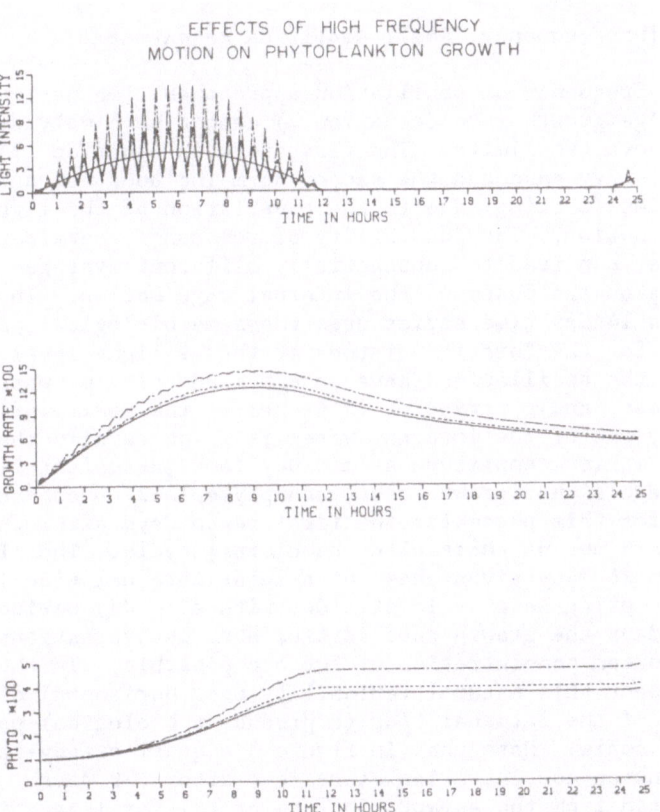

EFFECTS OF HIGH FREQUENCY
MOTION ON PHYTOPLANKTON GROWTH

Figure 2. Effects of high frequency internal waves on phytoplankton light intensity (upper panel), average growth rate, $\bar{\lambda}$ (middle), and phytoplankton concentration ϕ (lower) are shown for three cases. The solid, dashed, chain dashed curves are for half-hour period internal waves of amplitudes of 0, 5, 10m, respectively. Model parameters are for $\gamma = 0.1$ m^{-1}.

2.2 Low Frequency, Small Amplitude Limit

In the limit of very low frequency motion in which the period of oscillation is much larger than a day, the daily-averaged growth rate can be obtained from equation (7) to be

$$\bar{\lambda} \simeq \overline{\Lambda_0} + \overline{\Lambda_0'} \; \eta(t)$$

In this limit there will be greater or lesser growth depending upon whether $\eta(t)$ is positive or negative, respectively. This limit is appropriate to a model suggested by Kahru (1983) for oscillations with a 13 day period in the Baltic. It should be noted that this limit takes no account of photoadaptation (e.g. Perry, 1981) which may be significant over these longer time scales.

2.3 Medium Frequency, Small Amplitude Resonance

When the frequency of oscillation approaches time periods of a day, the average growth rate (equation 7) cannot be simply decomposed as in the above two limits. The first term in equation (7) represents the stationary case and the second term includes sum and difference frequencies (beating) due to the covariation of the light and the vertical motion. The possibility of resonance, considered by Flierl and Joyce, can lead to substantially different averaged growth rates depending on the phase of the internal wave motion. In Figure 3 a simulated 14 day time series uses the same biological parameters as Figure 2 for the "arctic" diatoms at the 1% light level. In this example, the oscillations have an amplitude of 5 m, phases of 90 and 270 degrees, and a period of 12.42 hrs.: the semi-diurnal tide. Initially one of the internal waves is in phase with the sun reaching maximum amplitude at mid-day (and just after midnight). The average light, growth rate, and phytoplankton concentration is largest for this phase for the first seven days after which vertical motions are out of phase with the diurnal cycle. The light variation for any given phase of a lunar internal tide is very much like the spring-leap cycle of tides with a 14 day period. After several days the growth rate settles down but variations in phytoplankton concentration of 50% are possible. In the ocean one would expect this natural variability over horizontal distances of 10-50 km of the internal tide to produce a biological patchiness in the same scale. Note that in Figure 3 a quasi-steady-state on the phytoplankton concentration is reached after 8-9 days. This can be anticipated from the asymptotic form of (2) for large times:

$$\phi \sim \bar{\lambda}/\varepsilon$$

3. CASE 2: WIND-INDUCED MIXING IN A WARM-CORE RING

A warm-core ring of the Gulf Stream was the object of a time series investigation in 1982. This ring, 82B, was formed in February of

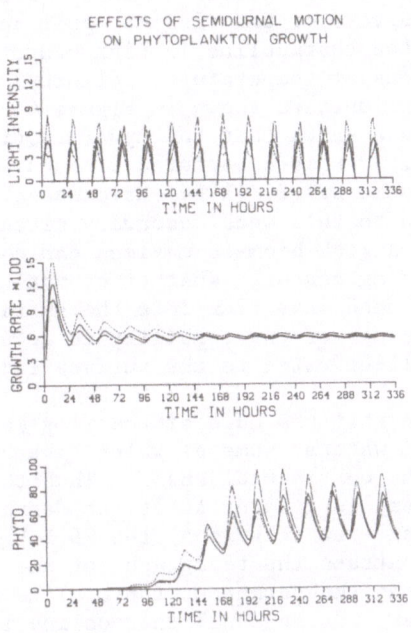

Figure 3. As in Figure 2 but for a 5m internal wave of period 12.42 hrs. Dashed and chain dashed curves are for two different internal waves having a phase difference of 180°.

1982 and was followed over most of its 6-7 month lifetime (Joyce and Wiebe, 1983); Evans *et al.*, 1985). Due to wintertime cooling its central core of 18°C Sargasso Sea Water was modified and the temperature reduced to 15.7°C. The initial cooling brought up nutrients from the thermocline into the surface layer (Fox and Kester, 1986); by June, however, the surface layer nutrients were depleted. A spring phytoplankton bloom in the ring core developed somewhat later than in the surrounding Slope Water (Brown *et al.*, 1985) and McCarthy and Nevins (1986) point out that prior to the bloom much of the primary production was 'new', not involving recycled nitrogen. The phytoplankton biomass maximum in the center of the ring observed in June by Nelson *et al.* (1985) was believed to have developed *in situ* rather than being the result of lateral mixing of the ring with its surroundings. Following the onset of stratification in the ring core in late April/early May, the observed rates of nitrate uptake by the phytoplankton should have resulted in a much greater net depletion of integrated nitrate in the upper 100m than observed. This prompted Nelson *et al.* (in preparation) to look for additional nitrate sources in the ring center.

One source of nitrate could be provided by slow upwelling in the

core of the ring expected due to its frictional delay (Flierl and
Mied, 1985). Hydrographic surveys of the ring in April and in June
indicated that the depth of the thermocline in ring center decreased
at about 1 meter/day. Sections of temperature, salinity, potential
density and oxygen for 82B in June are shown in Figure 4. Joyce and
McDougall (in preparation) have shown that the upward motion of the
main thermocline at depths of 500m is associated with outward
motion near the surface. Franks et al. (1986) have modelled the
response of the phytoplankton to this weak secondary circulation of
the ring and have shown that a weak biomass maximum can develop in
the seasonal thermocline at ring center. Whether or not this motion
is sufficient to provide the necessary flux into the surface layer
is difficult to determine because of the inability to accurately
extrapolate thermocline upwelling rates to the surface layer in the
presence of surface heating.

From Figure 4 one can see that the core of the ring is
surrounded by fresher surface waters, some of which have been
entrained by the ring from the continental shelf. Thus the
stratification on the periphery is greater as is the depth of the
seasonal thermocline. In plan view (Figure 5) the 26.2 kg/m^3 σ_θ
surface has been used to illustrate the topography of the seasonal
thermocline. Note that whereas the permanent thermocline of 82B is
300m deeper in the ring center, the seasonal thermocline is about
20m shallower than in the periphery. Nelson et al. (in prep.) have
shown that the surface layer is depleted in nitrate and that the
first detectable levels are found at σ_θ = 26.10 in ring center and
26.28 in the Slope Water. Given the observed topography of the
seasonal thermocline and a first detectable nitrate at σ_θ = 26.2
(approximately), the following question can be posed. How much
potential energy would be required to vertically mix the surface
layer down to σ_θ = 26.2? The relevant energy, PE, is defined by

$$PE = \int_{-z_c}^{0} g(\bar{\rho} - \rho)z\,dz$$

where g is the acceleration of gravity ($\approx$ 9.8 m/s^2), z is the
vertical coordinate, z_c is the depth of the 26.2σ_θ surface, and $\bar{\rho}$ is
the average density between 0 and -z_c. This energy, in units of
Joules/m^2, is also plotted in Figure 5. It can be seen that PE is
nearly two orders of magnitude smaller in the center of the ring than
on the ring periphery. This pre-conditioning of 82B is such that a
passing storm could perferentially mix the center of the ring,
bringing up nutrients into the surface layer.

In a mixed layer experiment, Davis et al. (1981) have related
the rate of change of potential energy to the rate of working by the
surface wind stress, τ. Though the definitions of potential energy
in equation (9) and in Davis et al. differ, prior to mixing down to
a depth z_c, the following relation should hold for net heating
conditions

$$\frac{\partial PE}{\partial t} = \rho\, m_0 u_*^3, \qquad u_* = (\tau/\rho)^{1/2} \tag{10}$$

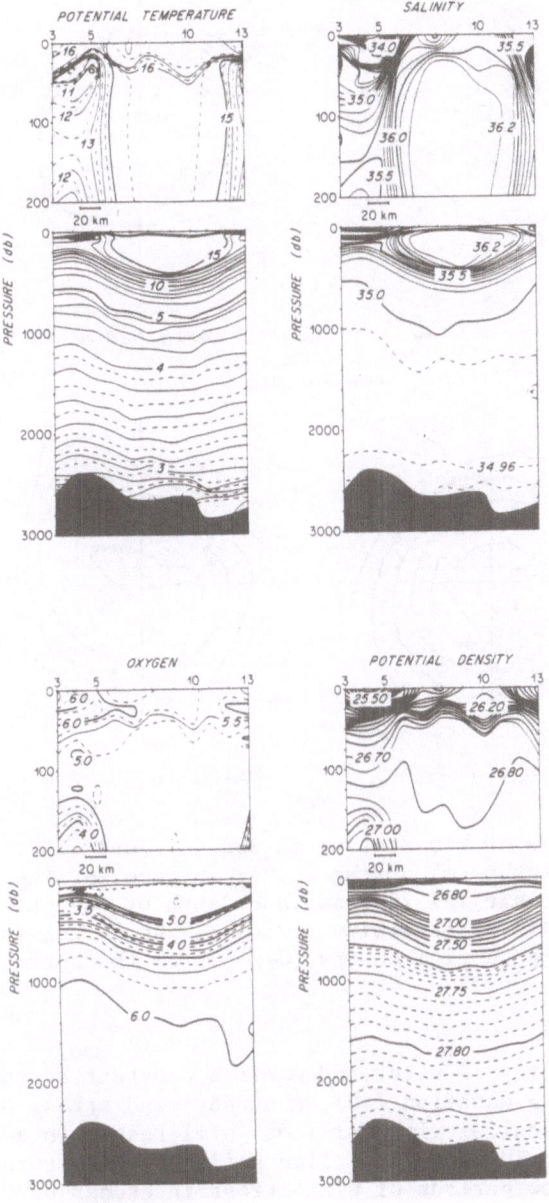

Figure 4. CTD section across warm core ring 82B showing potential temperature (°C), salinity (°/oo) (4a), oxygen (ml/l), and potential density (kg/m^3) (4b). For the pressure range shown, pressure (decibars) is approximately equal to depth in meters. Data are from the R/V ENDEAVOR 86 cruise June 1982.

48

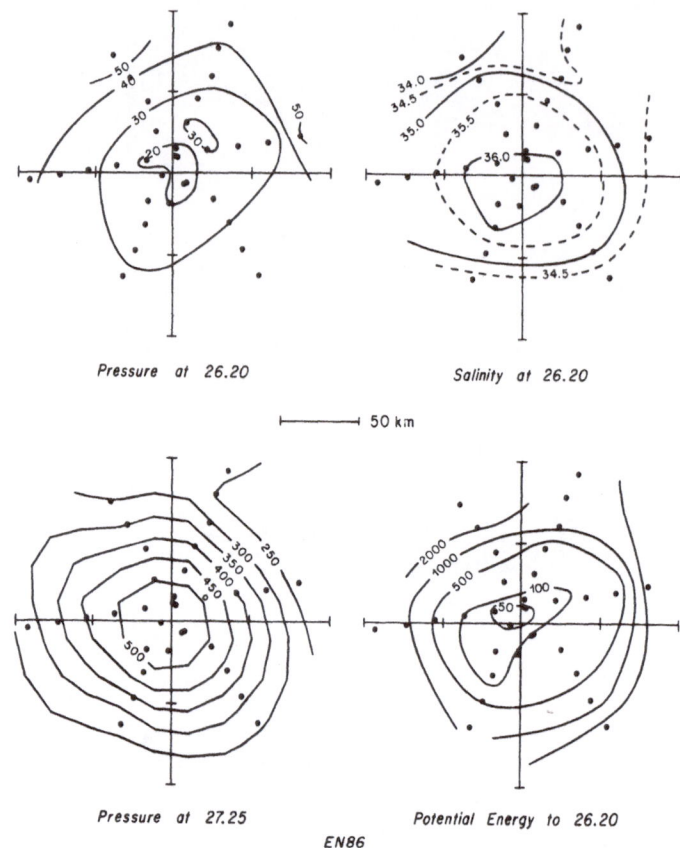

Pressure at 26.20   Salinity at 26.20

⊢————⊣ 50 km

Pressure at 27.25   Potential Energy to 26.20

EN86

Figure 5. Plan view of CTD station in 82B for June 1982. Shown are the pressures or depths (m) of the $26.2\sigma_\theta$ surface for the Seasonal thermocline (upper left) $27.25\sigma_\theta$ surface depth of the main thermocline (lower left), salinity ($^o/oo$) on the $26.2\sigma_\theta$ surface (upper right), and potential energy (J/m^2) from the surface to the $26.2\sigma_\theta$ surface.

where ρ is the density of water and m_o is a constant approximately equal to 0.5. Using equation (10), a steady wind stress of 1 dyne/cm^2 or 0.1 Pascal blowing for 1 day will result in a change in potential energy of 50 joule/m^2. Time series of wind stress from the June cruise show periods of wind stress in excess of 0.2 Pascal for several hours. It would thus appear possible that wind mixing could produce episodic injections of nitrate into the surface layer of 82B. This mixing would have been most likely soon after the onset of stratification when PE was small. As net heating builds up the stratification, the level of the first detectable nitrate becomes more difficult to approach by wind-induced mixing at the surface.

4. DISCUSSION

Two case studies have been presented to illustrate the importance of physical dynamics to the problem of primary production. In the first, low light conditions were assumed to be limiting for phytoplankton growth. A simple model equation was presented and solved following fluid particles as they underwent vertical oscillations with amplitude a. If incident light exponentially decays from the ocean surface with an e-folding distance γ^{-1}, then oscillating particles will, on average, see higher light levels than stationary ones. If (γa) is < 1, then the effects can be characterized by an $0(\gamma a)^2$ increase in the average growth rate for high frequency internal waves and an $0(\gamma a)$ modulation due to near-resonance of the more energetic low frequency internal waves with the daily-varying surface light field. This modulation will produce regions of enhanced and depressed growth depending on the dominant phase of the internal wave.

In the second study the preconditioning of a Gulf Stream warm-core ring was such that the seasonal thermocline and water with significant nitrate were closer to the sea surface in the ring center than on the periphery. The first detectable nitrate levels on the edge of the ring were deeper and the overlying fluid cap consisted of lighter, low salinity water of shelf/slope origin. As a result of this physical environment, the energy required to vertically mix from the surface down to levels where nutrients were found was more than an order of magnitude less in ring center. Passing storms of moderate strength and duration could episodically pulse nitrate into the nutrient depleted surface waters in the center of the ring, but not on the ring's periphery. The effect of wind forcing during the period of re-stratification could play a significant role in the dynamics of the spring phytoplankton bloom in the central core of warm-core ring 82B.

5. ACKNOWLEDGEMENTS

Travel support for this meeting was provided by NATO. The research support of the U.S. National Science Foundation Grant OCE8501176 is gratefully acknowledged.

6. REFERENCES

Brown, O.B., R.H. Evans, J.W. Brown, H.R. Gordon, R.C. Smith, and K.S. Baker: 1985, 'Phytoplankton blooming off the U.S. east coast: A satellite description', *Science* 229, 163-167.

Davis, R.E., R. DeSzoeke, and P. Niiler: 1981, 'Variability in the upper ocean during MILE. Part II: Modelling the mixed layer response', *Deep-Sea Research* 28A(12), 1453-1475.

Evans, R.H., K.S. Baker, O.B. Brown, and R. Smith: 1985,
'Chronology of warm-core ring 82B', *J. of Geophysical Research*
90(C5), 8803- 8811.

Flierl, G.R. and R.P. Mied: 1985, 'Frictionally induced
circulations and spin down of a warm-core ring', *J. of
Geophysical Research* **90(C5)**, 8917-8927.

Flierl, G.R. and T.M. Joyce: MS, (in prep.).

Fox, M.A. and D.R. Kester: 1986, 'Nutrient distributions in warm-
core ring 82B, April-August', *Deep-Sea Research* **33(11/12)**, 1761-
1772.

Holloway, G.: 1984, 'Effects of velocity fluctuations on vertical
distributions of phytoplankton', *J. of Marine Research* **42**, 559-
571.

Joyce, T.M., and P. Wiebe: 1983, 'Warm-core rings of the Gulf
stream', *Oceanus* **26(2)**, 34-44.

Joyce, T.M. and T.J. McDougall: MS, 'Physical structure and temporal
evolution of Gulf Stream Warm-Core Ring 82B', (in prep.).

Kathru, M.: 1983, 'Phytoplankton patchiness generated by long
internal waves: A model', *Mar. Ecol. Prog. Ser.* **10**, 111-117.

McCarthy, J.J. and J.L. Nevins: 1986, 'Utilization of nitrogen
and phosphorus by primary producers in warm-core ring 82B
following deep convection mixing', *Deep-Sea Research* **33(11/12)**,
1773-1788.

Nelson, D.M., H.W. Ducklow, G.L. Hitchcock, M.A. Brzenzinski,
T.J. Cowles, C. Garside, R.W. Gould, Jr., T.M. Joyce, C.
Langdon, J.J. McCarthy, and C.S. Yentsch: 1985, 'Distribution
and composition of biogenic particulate matter in a Gulf stream
warm core ring', *Deep-Sea Research* **32(11)**, 1347-1369.

Nelson, D.M., J.J. McCarthy, T.M. Joyce, and H.W. Ducklow: (in
prep.).

Perry, M.J., M.C. Talbot, and R.S. Alberte: 1981,
'Photoadaptation in marine phytoplankton: Response of the
photosynthetic unit', *Marine Biology* **62**, 91-101.

Platt, T., W.G. Harrison, B. Irwin, E.P. Horne and C.L. Gallegos:
1982, 'Photosynthesis and photoadaptation of marine
phytoplankton in the arctic', *Deep-Sea Research* **29(10A)**, 1159-
1170.

LAGRANGIAN SIMULATION OF PRIMARY PRODUCTION IN THE PHYSICAL ENVIRONMENT - THE DEEP CHLOROPHYLL MAXIMUM AND NUTRICLINE

K. U. Wolf
Institut fuer Meereskunde an der Universitaet Kiel
Duesternbrooker Weg 20
D-2300 Kiel 1
F.R. Germany

and J. D. Woods
Robert Hooke Institute
Oxford University
England

ABSTRACT.   A one-dimensional numerical model is used to simulate the transition from spring plankton bloom to oligotrophic profiles. The phytoplankton population is represented by a statistically significant number of individual particles. This Lagrangian-ensemble method permits explicit integration of non-linear equations for the interaction between physics, biochemistry and physiology. The model accurately simulates the form of observed nutrient and chlorophyll profiles. It shows how the subsurface chlorophyll maximum results from the physical environment and nutrient depletion in the upper levels. Nutrient uptake creates a nutricline which slowly descends during the late spring; the chlorophyll maximum intensifies and descends with the nutricline. An oligotrophic situation in the ocean is characterised by a space of a few decametres between the nutricline and the nocturnal turbocline. Slow upwelling does not change this result.

1.   INTRODUCTION

The subsurface chlorophyll maximum is the most striking feature of plankton profiles in the oligotrophic regime which forms each summer in the temperate ocean (Steele, 1964; Strickland, 1968; Anderson, 1969 and 1972; Cullen and Eppley, 1981).
    Nevertheless, we know little about the mechanism of its formation and its relation to the spring plankton bloom. Riley, Stommel and Bumpus (1949) were the first to simulate the deep chlorophyll maximum with a mathematical model. In 1960 Steele and Yentsch used a similar model and introduced a sinking rate of phytoplankton cells which was assumed to vary with light and nutrient concentration, and parametrized as varying with depth. The model simulated phytoplankton

B. J. Rothschild (ed.), Toward a Theory on Biological-Physical Interactions in the World Ocean, 51–70.
© *1988 by Kluwer Academic Publishers.*

maxima, even below the euphotic zone. Jamart *et al.* (1977) extended the latter model, parametrizing sinking speed in terms of nutrient concentration, and including: a vertically-varying eddy diffusivity, light varying with depth and time, respiration, nutrient limitation and grazing. Jamart *et al.* (1979) showed that nutrient-dependent sinking rate was not essential for the formation of a chlorophyll maximum in this model. Nevertheless many authors (*e.g.* Cullen and Eppley, 1981; Bienfang, 1983) still argue that it may be caused by a decrease in the sinking speed. On the other hand, Smetacek (1985) advocates for oligotrophic states of diatom blooms increased sinking rates caused by aggregation as a survival strategy of the population. One of the motivations of our investigation was to see whether the Lagrangian ensemble method of modelling plankton growth could help clarify this confusion in the literature.

The most important factor was the dependence of growth on nutrient concentration, which has been parameterized by Droop (1968, 1974, 1975 and 1982) and Maske (1982) by means of the 'Quota'-method, in which the growth function depends on an internal nutrient pool of the plankton.

All these studies are based on the Eulerian continuum method, which cannot treat explicitly the diversity in the properties of cells at the same depth. Woods and Onken (1982) developed a Lagrangian-ensemble method to simulate the variations of individual particle movement and the interaction to adaptation and production. We used an extension of their method to obtain the results presented in this paper. The Lagrangian-ensemble method enables the modeller to treat explicitly those aspects of the non-linear feedback between physics, chemistry and biology that depend on the independence of individual plankters. For example features like the correlation between the internal nutrient and energy pools of the algae with the growth function (Eppley and Strickland 1968) can be treated adequately only with the individual handling of the algae cells.

This paper confirms that the development of the oligotrophic plankton profiles is mainly controlled by the interaction between phytoplankton growth and the nutrient regime, as suggested by Venrick *et al.* (1973), and shows in detail how that is achieved.

2. METHOD

2.1. An Improved Version of the Lagrangian-Ensemble Method

Our investigation is based on the Lagrangian ensemble method introduced by Woods and Onken (1982), in which the biomass is divided between an ensemble of particles which move vertically according to rules that apply to individual plankters. The model equations are integrated separately for each particle; the attributes of every member of the ensemble are updated at each time step of the model run. The model yields two types of products: (1) the development of each particle from start to finish of the integration (50 days in this case), and (2) profiles of such properties as biomass and seawater turbidity derived from the ensemble statistics at each (one hour) time step of the model integration.

In reality, every plankter follows a different path through the sea giving it a unique developmental history: this gives diversity to the state of development of particles that happen at any instant to lie at the same depth. (In a one-dimensional description, the same depth means the same physical environment.) Furthermore, plankton physiology involves relative slow adaptation to its changing environment (in particular the illumination), so one of the varying attributes of a phytoplankter is its state of adaptation to the physical environment, which depends on its previous environmental history. Thus there is diversity in the state of adaptation of a crowd of plankters that happen at any instant to occupy the same depth: so each member of the crowd is responding differently to the physical environment at that depth. The long term development of plankton profiles is sensitive to these diversities in the state of development and state of adaptation to the environment; neglecting them in plankton models leads to qualitative and quantitative error. The Eulerian continuum method cannot take account of the diversities: the Lagrangian ensemble method provides an economical way to do so.

The constraint on the Lagrangian ensemble method is that no computer is fast enough to integrate the individual histories of every plankter in a bloom. The compromise is to divide the biomass into a limited number of particles which behave like individual plankters, but which are deemed to be carrying much more material than a single plankter. The particles can be thought of as representatives of the total community, each bearing an information tag recording its depth, how many cells (i.e. how much biomass) the particle represents, the state of light-adaptation and the nutrient and energy pools of each of the cells. By definition all the cells in one particle are clones with identical histories; they divide simultaneously. The values on each information tag are updated every time step of the integration. The plankton profile at each time step is computed by polling the information attached to the particles in each depth interval (which is typically one metre thick).

In order to obtain a statistically significant estimate it is necessary to ensure that there are sufficient particles in each depth interval; experience shows that 20 particles normally suffice. Thus a one-dimensional model extending to a maximum depth of 250 metres requires at least 5,000 particles, and more if they are unevenly distributed. That criterion was not satisfied by Woods and Onken (1982) who used only 100 particles; the mixed layer rapidly became depleted of particles. That was not so serious for their 5-day integration, but it would become unacceptable for 50-day integrations needed to simulate the transition to an oligotrophic regime. If their technique of keeping a fixed number of particles had been retained, we would have had to start with an initial injection of many millions of particles to ensure that the particle count never dropped below the threshold of 20 in any one-meter depth interval. We indroduced a more economical solution by designing the model to count the number of particles in each depth interval at a very time step and whenever the

number fell below 20 automatically to divide the biomass and all other
attributes of the biggest particles in that depth interval each into
two new particles. Thereafter, the new particles carry the divided
biomass along independent trajectories.

2.2. The Physical environment

Consider a water column in the open ocean with an area of one square
meter, a maximum depth of 250 metres, a rigid surface and no flow
through the side walls, located at 40°N somewhere in the central North
Atlantic. For simplicity the physical environment is reduced to
prescribed temporal variations of (1) the depth of the turbocline
(marking the base of the mixed layer) and (2) the radiation profile
$I(z,t)$; these were computed using the one-dimensional model of Woods
and Barkmann (1986). Thus we deliberately neglected the feedback
between biology and physics provided by the impact of plankton
concentration and optical properties of the water; our colleague
F. Dörre (1985) has shown how to include that feedback in our
Lagrangian ensemble model.

Mixing is often thought to lead to homogenization. That is the
case if mixing is steady and uniform. However, where there are spatial
and temporal variations in mixing on the relevant time scale, it can
be a powerful generator of heterogeneity. That is what happens in the
euphotic zone, which has two mixing regimes: the upper mixed layer,
where vertical homogenization occurs on time scales of less than an
hour, and the thermocline, where it takes weeks or months, so long
that it can be ignored. There is a sharp interface between these two
mixing regimes: the turbocline, which rises and falls through the
water column with the diurnal cycle of solar heating and with the
seasons (Figure 1). The diversity in plankter histories arises from
their different transition between the two regimes, brought about by a
combination of the diurnal cycle in turbocline depth and the random
mixing above the turbocline. That was the novel contribution of Woods
and Onken (1982).

2.3. Particle motion

The model considers three contributions to the motion of the particle,
which is constrained to one dimension, the vertical. The first
contribution is the motion of the particle through the water (*i.e.* the
particle behaviour), which we assumed, for simplicity, to be a steady
sinking of two metres per day. The second is the bulk
upwelling/downwelling of the water, which is assumed to have a
constant speed independent of depth in the seasonal thermocline and
reducing linearly in the diurnal thermocline and mixed layer to zero
at the surface. The third contribution (see Figure 2b and 2c) is
turbulent mixing, which is negligible below the turbocline according
to the measurements of Denman *et al.* (1983). There is some weak
turbulence in the diurnal thermocline, but it cannot displace
particles significantly because the vertical scale of the energy-

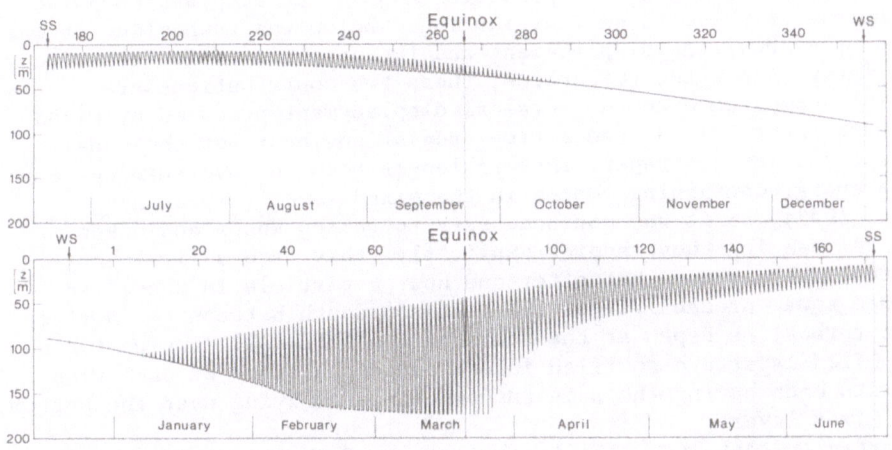

Figure 1. Seasonal variation of the simulated turbocline depth (from Woods and Barkmann, 1986).

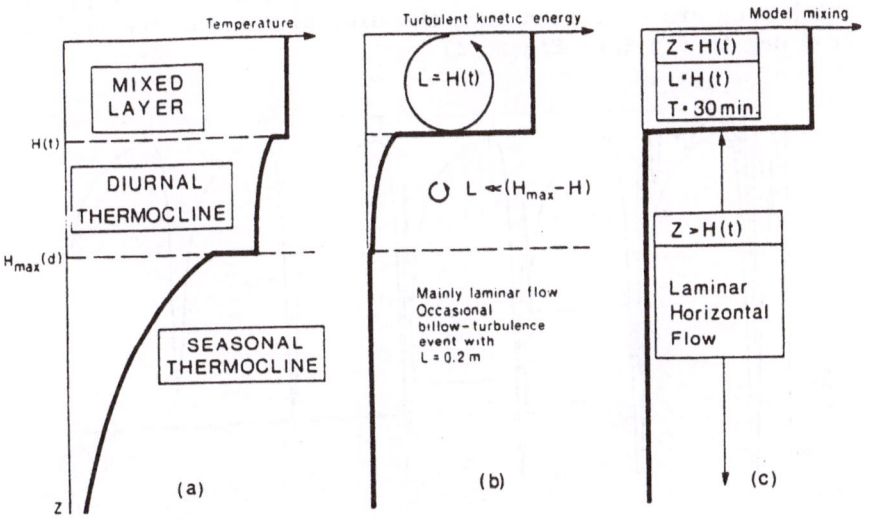

Figure 2. Schematic relation of (a) the mixed layer with diurnal and seasonal thermocline, (b) mixing scales and (c) the simplified model mixing (from Woods and Onken, 1982).

containing eddies is much less than the one-metre vertical resolution of the model (Woods and Strass, 1986; Ross, 1987).

The change of depth of a particle in one time step was computed from the sum of these three contributions. Below the turbocline it was based on the upwelling displacement and the particle motion relative to the water. Above the turbocline, these two contributions are negligible compared with the vertical displacement produced by mixed layer turbulence. We adopted a time step of one hour for the model runs reported in this paper. That is longer than the overturning time for the energy-containing eddies in the mixed layer (Denman and Gargett, 1983), so it was not necessary to follow Woods and Onken (1982) in treating those eddies explicitly (they used a five-minute time step). We assumed that after one hour a particle in the mixed layer had equal probability of lying at any depth between the surface and the turbocline depth at the start of the interval. Thus particles in the mixed layer are shuffled into a new depth order at each time step, with each having the same random chance of lying near the bottom of the mixed layer.

Particles that lie close to the bottom of the mixed layer are left behind in the diurnal thermocline when the turbocline rises in the morning. This is the crucial transition between the mixed layer and thermocline. Once in the thermocline, the particles move slowly down through the water. Most of them are re-entrained into the mixed layer when the turbocline descends next night. Because the depths of particles in the mixed layer are shuffled randomly each time step, the depth at which they are detrained into the diurnal thermocline varies randomly from day to day (see Figure 3).

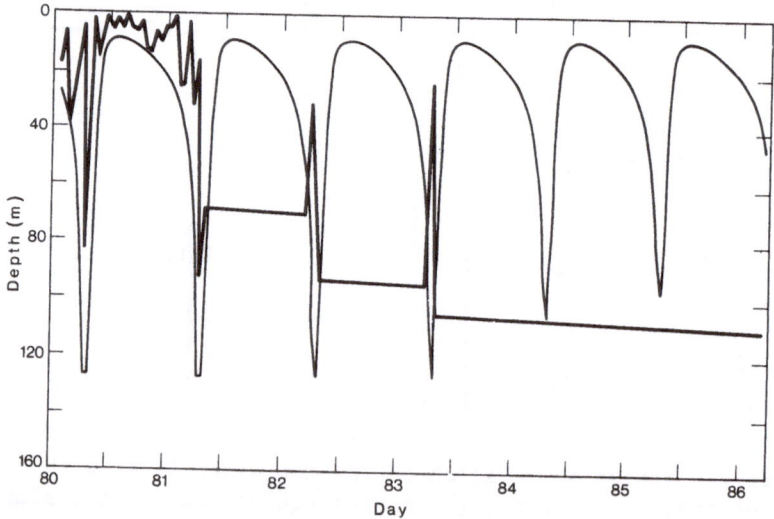

Figure 3. Path of one selected particle during the first six days of simulation.

Particles that happen to be detrained from mixed layer early in the morning when the turbocline is still deep are re-entrained last.

If the maximum depth of the turbocline attained on successive nights decreases (as it does after a storm, or climatologically in the spring) then the deepest detrained particles will not be re-entrained. They have made the transition from the diurnal thermocline into the seasonal thermocline, where they will remain until the autumn, unless a storm drives the turbocline down to their level. Once in the seasonal thermocline, the particles are immune from nightly shuffling. Those in a given depth band remain together as they sink through the water; they experience the same daily cycle of illumination, and lose difference in adaptation forced by earlier transitions through the mixed layer.

It will be shown later that the deep chlorophyll maximum comprises particles that have made that transition into seasonal thermocline and are therefore growing at nearly the same rate controlled by their depth.

2.4. Light

The Lagrangian description presented above leads to a new understanding of how to approach the question of light and plankton growth. It will be summarized here because it differs substantially from classical accounts based on the Eulerian approach (see, for example, Jamart et al., 1979). During a single day particles in the diurnal and seasonal thermoclines do not change their depths significantly compared with the e-folding depth of 400 nm light. During the hours of daylight, they experience a level of illumination that depends on their depth and the regular variation of solar elevation (which are included in our model) plus any fluctuations due to changing cloud cover (which is not). Particles that are detrained from the mixed layer early in the morning (and therefore deep) receive relatively little light that day: if they emerge late they receive much more. This erratic change in day-to-day light level must be one physical stress that has provoked the evolution of photo-adaptation in phytoplankton. The few that survive all day in the mixed layer suffer bright, fluctuating illumination, possibly at a level that leads to (or is interpreted as exhibiting) photo-inhibition.

Our model calculates the solar energy flux intercepted by a particle from its depth z, the prescribed irradiance profile $I(z,t)$ for that day and hour, the effective area F of one cell and the number of cells in the particle (shelf-shading is not included in these model runs). An average illumination I_m over the adaptation period (taken to be 5 hours in the present model runs) was computed along the particle trajectory. This is used to compute each particle's efficiency factor

$$\varepsilon = \exp(-I/I_m)$$

which controls the light energy absorption Eabs per cell and per time step

$$E_{abs} = F \cdot I(z,t) \cdot \varepsilon$$

This energy accumulates in the particle energy pool, until the conditions for cell division are satisfied.

2.5. Nutrients

The nutrient uptake U of a phytoplankton cell per time step depends on the nutrient concentration $N(z,t)$ in the surrounding seawater. It was computed for each cell using the Monod (1942) parametrization:

$$U = U_{max} \cdot N/(N+k_s)$$

where U_{max} is the maximum nutrient uptake per time step and k_s the half saturation constant. The nutrient uptake of each particle is credited to its nutrient pool, and removed from the nutrient content of the water column at its depth. This impact of plankton growth on the nutrient profile is essential for simulation of the deep chlorophyll maximum.

2.6. Energy and nutrient storage

The energy and nutrient pools of each particle have the inputs described above. The energy pool suffers a steady respiration loss E_{res}. The particle's compensation depth, at which energy gain equals respiration loss in 24 hours, depends on its efficiency factor, controlled by the erratic depth history of previous days. The vertical spread of particle compensation depths is a measure of the diversity in adaptation in the ensemble. If the respiration loss exceeds the photosynthesis gain for long enough to empty the energy pool, the particle is deemed to have died and it becomes detritus.

2.7. Cell division and production

Eppley and Strickland (1968) found that the production rate of a phytoplankton community correlates better with the intercellular nutrient content than with the nutrient concentration in sea water. The power of the Lagrangian ensemble method is that it permits the modeller to specify plankton growth in terms of the status of individual particles rather than the surrounding environment. We did this by specifying that all the cells in a particle divide when its energy pool and its nutrient pool both pass threshold values. After cell division those amounts are subtracted from the pools and the surplus is divided between mother and daughter cells.

We do not claim that this represents the optimal set of equations and parameters for modelling plankton growth: merely that they sufficed for the purpose of this study. It would be trivial to incorporate more exotic formulations for the energetics and nutrient intake (for example they might be coupled), and to test the sensitivity of the model output to various hypotheses (see Wolf, 1985).

3. RESULTS

We begin by presenting the results of a single run of the model without upwelling. The model run starts on 22 March, just before the end of the winter cooling season, and continuing for fifty days through the initial light-limited spring plankton bloom to the late spring oligotrophic regime.

3.1. The Fate of a Typical Particle

The internal working of the model is revealed in Figure 4 which shows the development of one fairly successful particle, in which the cells divide thirteen times during the first forty days, giving an eight thousand fold increase in particle biomass. The particle stays in the diurnal boundary layer until day 107 (17 April) when it is lost permanently to the seasonal thermocline. Up to then it had ten cell divisions. Examining Figures 4b and 4c it is seen that the first seven divisions were light-limited, and the next three were nutrient-limited. After 17 April, as the particle sinks into the seasonal thermocline below 35 metres, light becomes the limiting factor. The particle passed into the seasonal thermocline before the nutrients ran out in the mixed layer (on 23 April, see Figure 6e), but the descending nutricline catches it up on 29 April (day 120) sharply reducing the rate of nutrient uptake. But the descent of the nutricline then slows allowing the particle (which is falling through the water at 2 metres per day) to get below it on 2 May and resume rapid nutrient assimilation. Meanwhile, the particle has descended to nearly 80 metres, where the light is dimmer slowing the input to the energy pool. If the integration had been continued the particle might have achieved one more cell division before sinking below its compensation depth.

Even in this simple calculation, we see that the particle growth proceeds at a rate controlled by the presence of its fellow plankters. This environmental feedback was revealed even more dramatically in a later set of numerical experiments (not reported here), which included self-shading, which raised the compensation depth.

3.2. The Spring Plankton Bloom

Now we consider the total production of the plankton community represented by the ensemble of particles (Figure 5). The growth can be divided into three phases. Phase I starts in winter when the mixed layer is replete with nutrients (Figure 6a) and extends each night to a depth of 125 metres (Figure 3). Even in winter the daytime mixed layer is remarkably shallow. Some 'lucky' particles are detrained into the diurnal thermocline quite near the surface, where there is sufficient light (it is already the equinox) for rapid filling of their energy pools. A few cell divisions occur on the second day of the integration (24 March). Thereafter, the number of divisions occurring each day increases steadily (although perhaps not as dramatically as one might expect), and the stock of cells rises super-

60

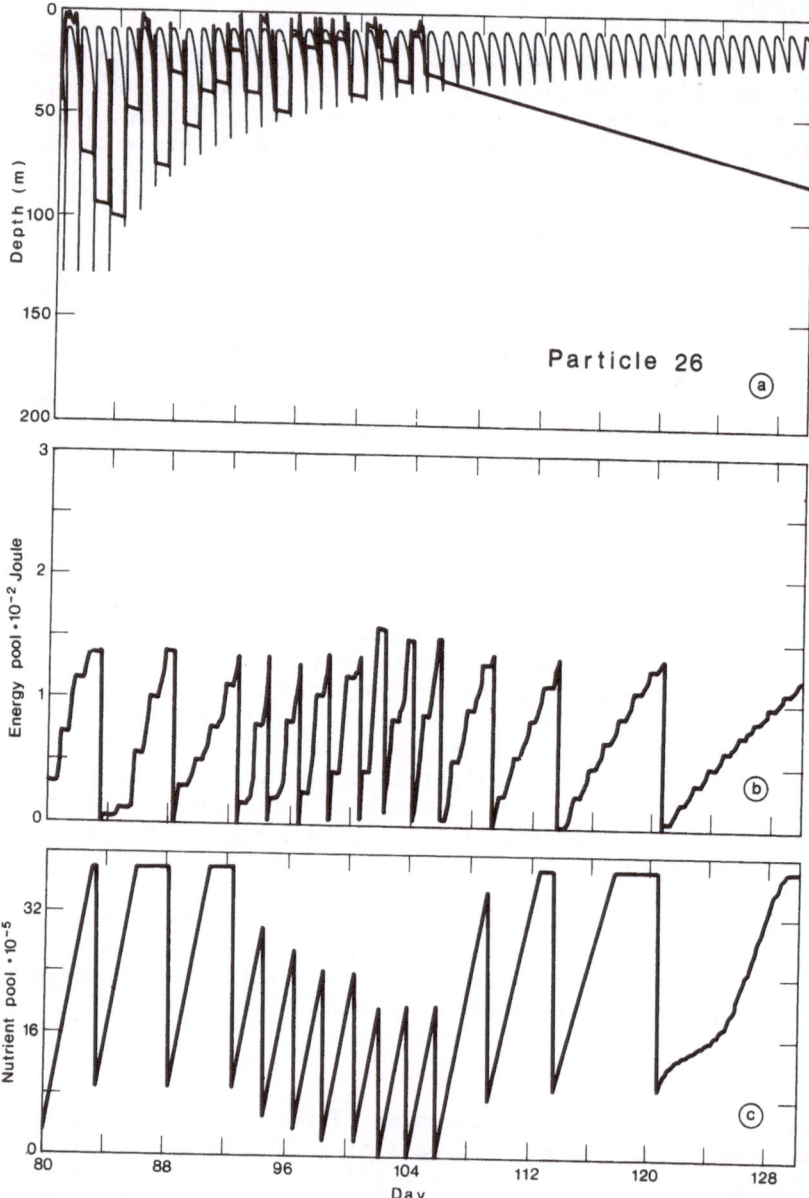

Figure 4. (a) The path of a particle in the physical environment (depth versus time); (b) and (c) are the corresponding contents of the nutrient and energy pool versus simulation time.

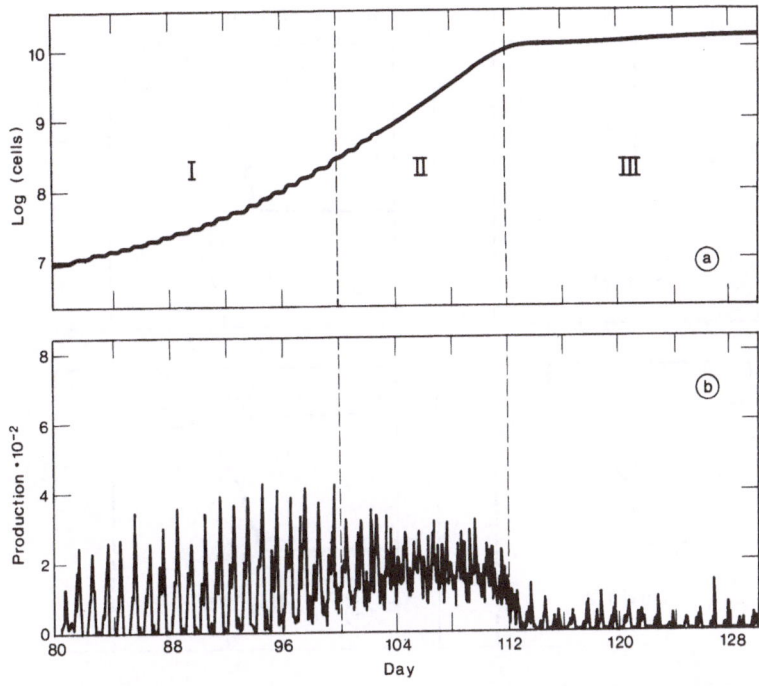

Figure 5. (a) The development of the total phytoplankton biomass of the water column over an integration period of 50 days; (b) the corresponding productivity expressed in average cell divisions per cell and time step.

exponentially from 10 million on 23 March (day 80) to 250 million at the end of Phase I on 13 April (day 100). During the first fortnight, the cell division occurs almost totally during the day; the exceptions must be particles which fill their energy pool just before sundown, leaving nutrient uptake at night to achieve the required double condition for cell division. Nocturnal cell division occurs increasingly until 12 April (day 99) when, for a fortnight there is no hour of the day in which cell division does not occur. Phase II is characterized by rapid filling of the particle energy pool, but a gradual slowing down of the rate of nutrient pool filling as the nutrient concentration in the mixed layer declines, with the overall result that the mid-day cell division rate decreases, and the diurnal steps in biomass disappear. Nevertheless, the total stock of cells continues to rise super-exponentially, reaching 10,000 million by the end of Phase II on 22 April (day 112). The start of Phase III coincides with the final stage of depletion of nutrients in the mixed layer (Figure 6).

62

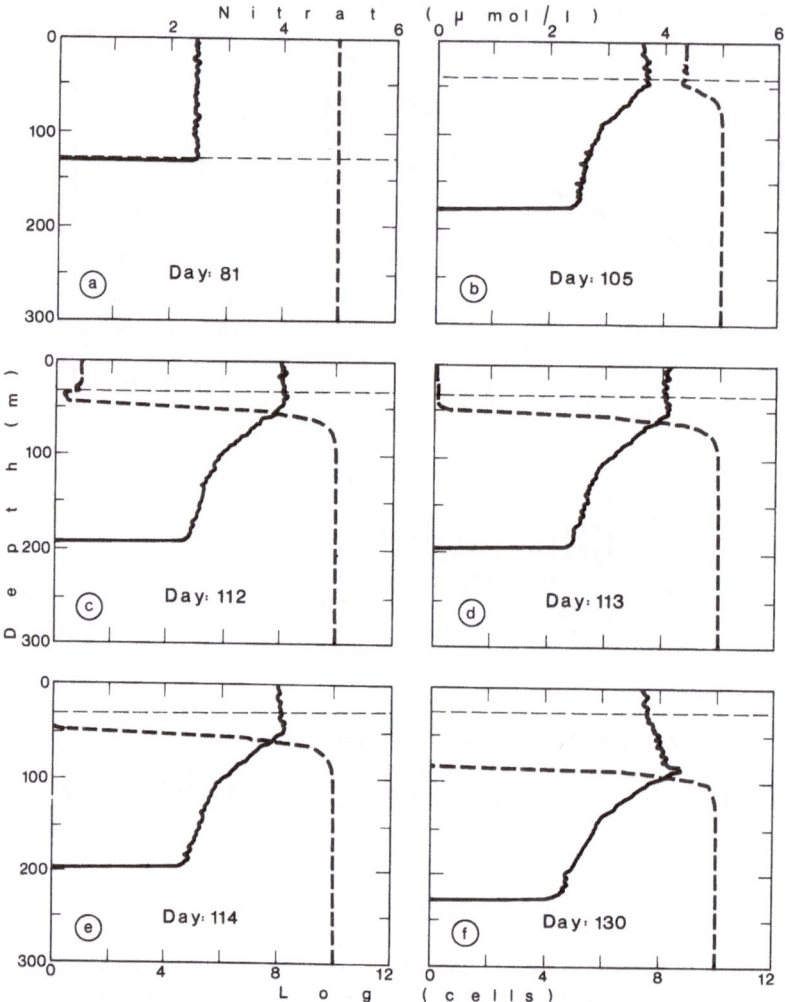

Figure 6a-f. Nitrate (dotted line) and phytoplankton profiles on selected simulation days (note the logarithmic axis for plankton cell concentration). The faint horizontal line indicates the maximum mixed layer depth within the last 24 hours.

The total cell concentration in this phase is not simulated realistically because grazing was not included. Nevertheless it is possible to make some qualitative statements about this phase of simulation. After 24 April (day 114) the diurnal boundary layer is exhausted of nutrients and further growth can only occur in the seasonal thermocline, where the light is much weaker. The oligotrophic regime has begun just one month after the start of the heating season. Thereafter production is much weaker and the stock of cells increases slowly, reaching 11,400 million by the end of the integration on 9 May (day 130).

3.3. The Nutricline

The nutricline is a sharp transition between the nutrient-free mixed layer and the nutrient-rich deep thermocline in the oligotrophic regime. Its development is illustrated in Figure 6. Observed June nitrate profiles for the North Atlantic (Figure 7a and 7b) have the same general form as Figures 6b/c and 6f.

If there is no upwelling, the nutricline in our model can only move through the water column as the result of consumption by phytoplankton. That leads to a steady descent from 60 metres to 100 metres during the last fortnight of the integration (Figure 6e to 6f); an average descent speed of 3 metres per day, initially faster and then slowing down. (Compare this with the steady 2 m/d sinking speed of the plankters; see section 3.1).

The nutrients are all 'pre-formed': the model does not include nutrient recycling, which is likely to be increasingly important in Phase III and later in the season.

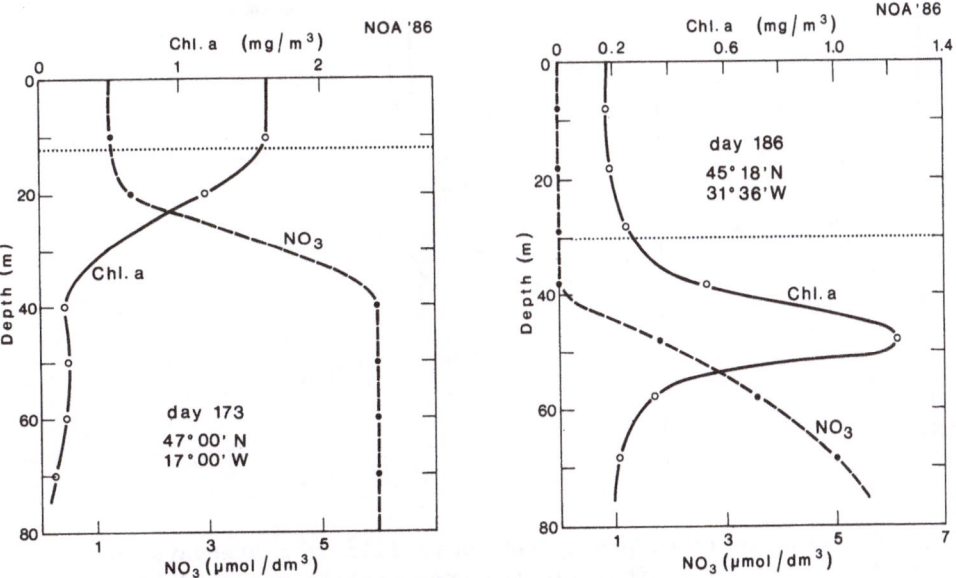

Figure 7. Two examples of nutrient- and chlorophyll-a bottle cast profiles measured in the northern North Atlantic. Note the different depths of the chlorophyll maximum and the corresponding nutrient profiles.

3.4. The Subsurface Chlorophyll Maximum

Now we consider the vertical distribution of the plankton. The computation was started with an initial injection of particles into the 125 metre deep mixed layer. The lower edge of that distribution descended at the fall speed of the particles reaching 200 metres on 27 April. The concentration in that deepest set decreased as the particles fell below their compensation depth. Growth continues throughout the integration at depths down to 100 metres (there is negligible nutrient loss below that in Figure 6f).

Figure 8a shows the mature oligotrophic profile, with a sharply defined deep chlorophyll maximum, predicted by the model for 5 May. Figure 8b shows a profile of chlorophyll fluorescence measured in the North Atlantic four months later but 450 km further north when the mixed layer depth was similar. The profiles have the same form, but the modelled chlorophyll maximum was much deeper. This is because the light was penetrating too deep into the ocean. Subsequent calculations with a model that included self-shading produced a much shallower chlorophyll maximum. See Strass and Woods (1987) for a detailed discussion of the meridional dependence of the plankton bloom and the subsurface chlorophyll maximum.

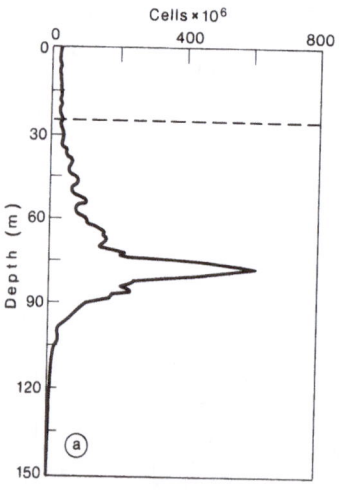
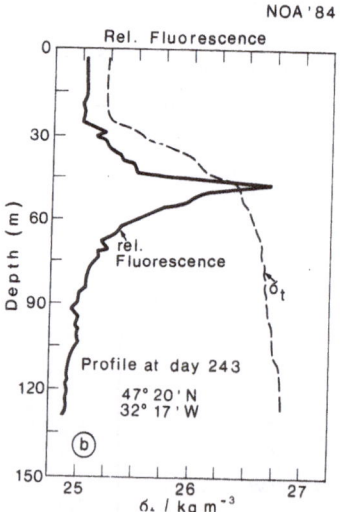

Figure 8. (a) Simulated phytoplankton profile of a standard model run on day 130. (b) Typical fluorescence profile measured in late summer. Data provided by V. Strass. Note the similarity in the form of the maxima.

Figure 6 shows how the deep chlorophyll maximum forms as a consequence of the surface exhaustion of nutrients. Growth is merely limited to plankters in and below the nutricline. Cell division is nutrient-limited in the upper reaches of the nutricline; light-limited below. The most rapid growth occurs where the daily input to energy and nutrient pools is in balance: that occurs in the middle of the nutricline, which is therefore the site of the chlorophyll maximum (as observed by Herbland and Voituriez, 1979 and Cullen and Eppley, 1981). Bottle cast measurements in the North Atlantic (see Figure 7a and b) made in June 1986 to test the model results show these typical chlorophyll/nutrient profiles which have been predicted by the model and are characteristics of the described primary production phases.

The model produces a deep chlorophyll maximum without invoking variation of particle sinking speed with depth, light or nutrient concentration. The chlorophyll maximum lies well below the mixed layer. It comprises particles that are no longer subject to the nocturnal mixing which forces inter-particle diversity in adaptation; all cells at the same depth in the chlorophyll maximum grow at nearly the same rate.

3.5. Upwelling

Slow steady upwelling brings particles closer to the surface where there is more light, so that their energy pool fills more rapidly. The model runs reported here ignored self-shading, so the irradiance of 400 nm light reaching the particles below the nutricline varies exponentially with depth. Consequently, the energy uptake by those particles and (because their growth is light-limited) their rate of cell division increases exponentially with upwelling speed. The increased growth leads to faster nutrient consumption, driving the nutricline down faster through the water column. A series of model runs with upwelling at different constant speeds reveal a weak (c = 0.07) exponential increase in production (Figure 9) and showed that the faster descent of the nutricline balanced the upwelling speed until the latter reaches the prescribed sinking rate of the plankters (2 metres per day).

If the upwelling speed is greater than the particle sinking speed, the particles move upwards into the mixed layer, accompanied by the nutricline. Once they pass the turbocline, particles and nutrients are both rapidly distributed vertically through the mixed layer mixing; then the chlorophyll maximum lies in the mixed layer. When that happens the production is very much more rapid than in the weak upwelling case. Such rapid upwelling is not sustained for long periods: it occurs in events lasting a few hours (Woods, this volume). Figure 10 shows chlorophyll and nutrient profiles on 24 April of two integrations of our model: (a) with no upwelling, and (b) with upwelling equal to 30 m/d on 23 April but otherwise zero. The chlorophyll maximum is in the seasonal thermocline at a depth of 55 metres in run (a) and within the mixed layer in run (b). The chlorophyll concentration is much higher in the latter case. This

result forms the basis for Woods's (this volume) interpretation of the
regional variation of chlorophyll interpretation of the regional
variation of chlorophyll concentration detected in satellite colour
images.

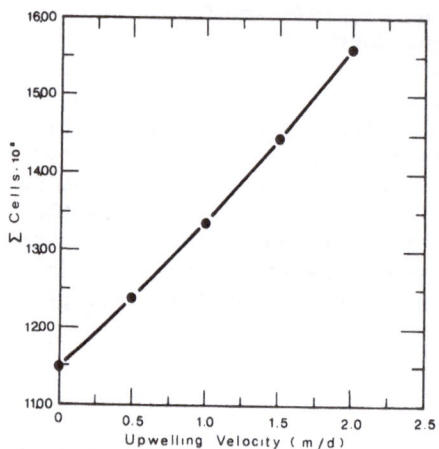

Figure 9.   Biomass as function of upwelling velocity; the dots each
represent the mean biomass from ten model runs after 50 days of
simulation for various constant upwelling velocities.

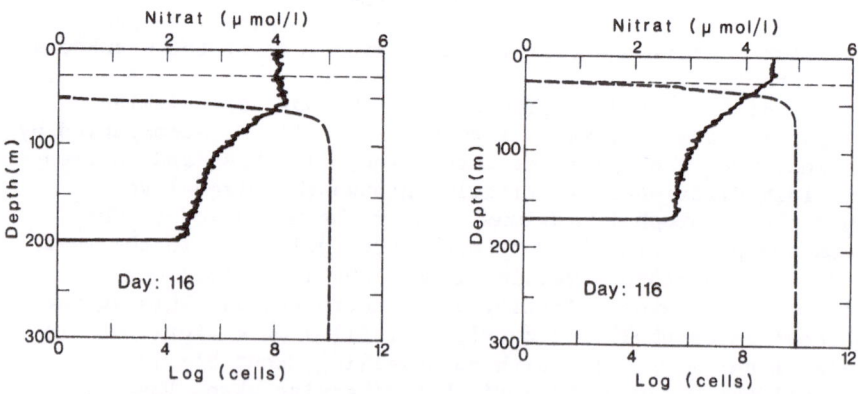

Figure 10.   Simulated nutrient and phytoplankton profiles (a)
without and (b) with a single upwelling event of 30 m/day on day
115.   Note the depths of phytoplankton maximum and nutricline.

4. DISCUSSION

The model used here to investigate the vertical vernal development of chlorophyll and nutrient profiles in the open ocean was deliberately simplified. The aim was to show that even with such elementary equations for phytoplankton growth it is possible to generate the essential features of summer profiles in the temperate ocean. The most serious omissions reported here are self-shading, grazing and nutrient recycling; the first two have been studied by Dorre (1985) and Burkert (1984) respectively. An improved model incorporating their techniques is currently been developed in Kiel to study the annual cycle. Preliminary results do not conflict with the conclusions reported here on the basis of the simple model.

The main advantage gained from using the Lagrangian ensemble method lies in the revealing histories of particles which behaved like individual plankters. The advantage of handling adaptation diversity explicitly was important when there are nutrients in the mixed layer; once the nutricline has descended into seasonal thermocline adaptation diversity as an important factor influencing depth dependent growth disappears. But adaptation diversity re-appears wherever rapid upwelling events drive the nutricline up to the mixed layer, and those sporadic events may control new production in oligotrophic regimes. Also, regenerated nutrients cause a low but steady production within the nutrient depleted mixed layer. So the computational overheads of the Lagrangian ensemble method are worthwhile all year round.

5. CONCLUSION

A simple model based on the Lagrangian ensemble method has been used to simulate the development of the spring plankton bloom starting from the end of winter, when the mixed layer is deepest, to the end of May, when the nutrients have run out in the mixed layer, leaving a nutricline and chlorophyll maximum in the seasonal thermocline. The distance between the bottom of the mixed layer and the nutricline increases with time, as the particles sink through the water.

The production in this oligotrophic regime is sensitive to upwelling. If the upwelling speed is slower than the fall speed of the plankters, the production increases exponentially ($c = 0.07$) with the upwelling speed, but the nutricline and chlorophyll maximum stay in the seasonal thermocline, well below the bottom of the mixed layer. This means that it is difficult to detect weak upwelling by examining mixed layer nutrient and chlorophyll concentrations. Strass and Woods (this volume) show, that these events can be detected by batfish-fluorescence measurements which have high vertical and horizontal resolution. The depth of the nutricline and therefore the deep chlorophyll maximum is nearly independent of upwelling speed, provided it is slower than the particle fall speed. Rapid upwelling events drive the nutricline and plankters upwards until it meets the turbocline and floods nutrients and plankton into the mixed layer where they multiply explosively, presumably giving the "hot spots" seen in satellite images of ocean colour. Sites of rapid upwelling are marked by a chlorophyll maximum in the mixed layer.

6. ACKNOWLEDGEMENTS

This work was funded by contract Wo 254/10-3 of the Deutsche Forschungsgemeinschaft. The authors acknowledge valuable discussions with their colleagues in the plankton group of the Kiel Institut fuer Meereskunde, Abteilung Regionale Ozeanographie: Volker Strass, Bernd Burkert, Frank Dörre and Reiner Onken.

7. REFERENCES

Anderson, G.C.: 1969, 'Subsurface chlorophyll maximum in the northeast Pacific Ocean', *Limnol. Oceanogr.* 14, 386-391.

Anderson, G.C.: 1972, 'Aspects of marine phytoplankton studied near the Columbia River, with special reference to a subsurface chlorophyll maximum', in A. T. Pruter and D. L. Alverson (eds.), *The Columbia River estuary and adjacent ocean waters; bioenviromental studies*, University of Washington Press, Seattle, 219-240.

Burkert, B.: 1984, 'Das Wachstum herbivorer Zooplankter unter dem Einfluss Physiologie, Verhalten und Umwelt Ein numerisches Modell', Diploma thesis, University of Kiel.

Bienfang, P.K., J. Szyper and E. Laws: 1983, 'Sinking rate and pigment responses to light-limitation of marine diatom: implications to dynamics of chlorophyll maximum layers', *Oceanol. Acta* 6, 55-62.

Cullen, J.J. and R.W. Eppley: 1981, 'Chlorophyll maximum layers of the southern California Bight and possible mechanisms of their formation and maintenance', *Oceanol. Acta* 4, 23-32.

Denman, K.L. and A.E. Gargett: 1983, 'Time and space scales of vertical mixing and advection of phytoplankton in the upper ocean', *Limnol. Oceanogr.* 28, 801-815.

Droop, M.R.: 1968, 'Vitamin BJ1J2 and marine ecology IV. The kinetics of uptake, growth and inhibition in Monochrisis lutheri', *J. Mar. Biol. Ass. U.K.* 48, 689-733.

Droop, M.R.: 1974, 'The nutrient status of algal cells in continuous culture', *J. Mar. Biol. Ass. U.K.* 54, 825-855.

Droop, M.R.: 1975, 'The nutrient status of algal cells in batch culture', *J. Mar. Biol. Ass. U.K.* 55, 541-555.

Droop, M.R., M.J. Michelson, J.M. Scott and M.F. Turner: 1982, 'Light and nutrient status of algal cells', *J. Mar. Biol. Ass. U.K.* 62, 403-434.

Dörre, F.: 1985, 'Selbstlimitierung des Phytoplanktonwachstums durch, Veraenderung der optischen Eigenschaften des Meerwassers - Ein numerisches Modell', Diploma thesis, University of Kiel.

Eppley, R.W. and J.D.H. Strickland: 1968, 'Kinetics of phytoplankton growth', Advances in Microbiology of the Sea, Vol. 1, Academic Press, London and N. Y., 23-62.

Herbland, A. and B. Voituriez: 1979, 'Hydrological structure analysis for estimating the primary production in the tropical Atlantic Ocean', *J. Mar. Res.* **37**, 87-101.

Jamart, B., D.F. Winter and K. Banse: 1979, 'Sensitivity analysis of a mathematical model of phytoplankton growth and nutrient distribution in the Pacific Ocean off the northwestern U.S. coast', *J. Plankton Res.* **1**, 267-290.

Jamart, B.M., D.F. Winter, K. Banse, G.C. Anderson and R.K. Lam: 1977, 'A theoretical study of phytoplankton growth and nutrient distribution in the Pacific Ocean off the northwestern U.S. - coast', *Deep-Sea Res.* **24**, 753-773.

Maske, H.: 1982, 'Ammonium - limited continuous cultures of Skeletonema Costatum in steady transitional state: Experimental results and model simulations', *J. Mar. Biol. Ass. U.K.* **62**, 919-943.

Monod, J.: 1942, *Recherches sur la Croissante des Cultures Bacteriennes*, Herman, Paris.

Riley, G.A., H. Stommel and D.A. Bumpus: 1949, 'Quantitative ecology of the plankton of the western North Atlantic', *Bull. Bingham Oceanogr. Coll.* **12**, 1-169.

Ross, H.: 1987, 'Der Einfluss der Scherung des Triftstromes auf die horizontale Dispersion in der planetarischen Grenzschicht', Diploma thesis, University of Kiel.

Steele, J.H.: 1964, 'A study of production in the Gulf of Mexico', *J. Mar. Res.* **22**, 211-222.

Steele, J.H. and C.S. Yentsch: 1960, 'The vertical distribution of chlorophyll', *J. Mar. Biol. Ass.* **39**, 217-226.

Strass, V. and J.D. Woods: 1988, 'Horizontal and seasonal variation of density and chlorophyll profiles between the Azores and Greenland', in B.J. Rothschild (ed.), *Toward a Theory on Biological-Physical Interactions in the World Ocean*, Kluwer Academic Publishers, Dordrecht, pp. 113-136.

Strickland, J.D.H.: 1968, 'A comparison of profiles of nutrient and chlorophyll concentrations taken from discrete depths and by continuous recording', *Limnol. Oceanogr.* **13**, 388-391.

Venrick, E.L., J.A. McGowan and A.W. Mantyla: 1973, 'Deep maxima of photosynthetic chlorophyll in the Pacific Ocean', *Fishery Bull.* **71**, 41-52.

Wolf, K.U.: 1985, 'Phytoplanktonwachstum unter Licht- und Naehrstoff-limitierung im Deckschichtmodell', Diploma thesis, University of Kiel.

Woods, J.D. and W. Barkmann: 1986, 'The response of the upper ocean to solar heating. I. The mixed layer', *Quart. J. Roy. Met. Soc.* **112**, 1-27.

Woods, J.D.: 1988, 'Scale upwelling and primary production', in B.J. Rothschild, ed., *Toward a Theory on the Biological-Physical Interactions in the World Ocean*, Kluwer Academic Publishers, Dordrecht, pp. 7-38.

Woods, J.D. and R. Onken: 1982, 'Diurnal variation and primary production in the ocean - preliminary results of a Lagrangian ensemble model', *J. Plankton Res.* 4, 735-756.

Woods, J.D. and V. Strass: 1986, 'The response of the upper ocean to solar heating. II. The wind-driven current', *Quart. J. Roy. Met. Soc.* 112, 29-42.

Louis Prieur
Laboratoire de physique et chimie marines
CEROV, B.P. 8
06230 Villefranche-sur-Mer
France

Louis Legendre
GIROQ, Département de biologie
Université Laval
Québec, Québec
Canada, G1K 7P4

ABSTRACT. New phytoplankton production in the oceans is primarily regulated by the local availability of photons and nitrate. Given average efficiencies of photon capture and conversion by phytoplankton, solar irradiance and light attenuation in the water column can be used to compute the shortest time $t_C(z)$ required by the cells to double their carbon biomass at any depth z. In a symmetrical fashion, concentration of nitrate in the deep reservoir and vertical eddy diffusion can be used to compute the shortest time $t_S(z)$ required by the cells to double their nitrogen biomass, given average characteristics of nutrient uptake. Computations must also take into account excursions around depths z resulting from vertical eddy diffusion. Maximum production per unit biomass would occur at the depth where $t_C(z) = t_S(z)$, and at times when both $t_C(z)$ and $t_S(z)$ exceed a critical doubling time above which there is no growth. Using this approach, surface irradiance, vertical light attenuation (or Secchi depth), concentration of nitrate at depth and vertical eddy diffusion (estimated from the density profile) are combined into numbers that set conditions for the timing and depth of new production.

1. INTRODUCTION

There are various ways to approach the analysis of ecological responses to environmental forcing. One of them is the statistical analysis of field data, which can be used to evidence relationships

B. J. Rothschild (ed.), *Toward a Theory on Biological-Physical Interactions in the World Ocean*, 71–112.
© 1988 by Kluwer Academic Publishers.

among measured variables, and to develop correlative (*i.e.* fore-casting) models. At the other end of the range, predictive model-ling is aimed at estimating the effects on some variables resulting from changes in others variables. These models are based on equa-tions, that relate response variables to causal input variables and from which it is possible to predict biological responses to physi-cal forcing. Alternatively, the same equations can be used to derive criteria to discriminate between different states of a response variable (*e.g.* growth or no-growth of a phytoplankton population), or to identify characteristic features of the system (*e.g.* the depth of maximum phytoplankton production per unit biomass). A well-known example of such criteria in oceanography is the stratification parameter of Simpson and Hunter (1974), whose values delimit well-mixed, frontal and well-stratified waters. This approach is less demanding than predictive modelling, since only characteristic points must be recognized; it can also be the first step leading to full predictive modelling. Accordingly, the present paper will be devoted to develop, from first principles, criteria for new phytoplankton production per unit biomass. These criteria will be based on environmental variables usually measured by oceanographers and that are therefore readily available for many oceanic basins, and on physiological characteristics for which values can be found in the literature. Such criteria can be used for identifying environmental variables to which new phytoplankton production is most sensitive, and also to eventually explore large-scale responses of phytoplankton to changes in phycical forcing.

Oceanographic criteria will be developed in this paper for two types of phytoplankton heterogeneities, namely the periods of significant phytoplankton growth and the subsurface maxima in production per unit biomass. Periods of significant phytoplankton growth generally start by a bloom. The now classical explanation for the spring phytoplankton bloom is the model of Riley (1942) and Sverdrup (1953), according to which a bloom occurs when carbon fixation by phytoplankton exceeds respiration losses per unit area. In this model, increased production at the time of the bloom results from the confinement of phytoplankton above a 'critical depth', caused by seasonal increases in both solar irradiance and vertical stratification. Considering the effects of vertical destabilization and stabilization on irradiance and nutrients, Legendre (1981) did generalize this approach to various time scales, including periodicities of tidal and of meteorological origin (*e.g.* summer blooms). The 'critical depth' model was designed to explain the initiation of productive periods, and not differences between productive and nonproductive periods. Conse-quently it only considers the effects of solar irradiance on phyto-plankton production, since nutrients are generally nonlimiting at the beginning of blooms. On the contrary, discrimination between productive and nonproductive periods must take into account both irradiance and nutrient effects.

The second type of heterogeneities discussed in the present paper concern the vertical distribution of phytoplankton production. There have been several explanations given for the presence of subsurface chlorophyll maxima in the water column. Cullen (1982) has rightly stressed that a chlorophyll maximum does not necessarily reflect a maximum in biomass, since chlorophyll per unit carbon is known to change with irradiance and nutrient conditions. From his review of the literature, he concluded that there exist different types of subsurface maxima, with different generating mechanisms. In addition to the mechanisms reviewed by Cullen (1982), hypotheses concerning subsurface maxima can be found for example in Kiefer and Kremer (1981), Holloway (1984; see also Joyce in this book) and Wolf and Woods (this book). Actual measurements of chlorophyll and production maxima have shown that natural situations may be highly variable. On the continental shelf at mid-latitudes for example, Holligan et al. (1984; Gulf of Maine) and Vandevelde et al. (1987; Gulf of St. Lawrence) found subsurface chlorophyll maxima associated with production maxima at the nitracline, as in the 'typical tropical structure' (Herbland and Voituriez, 1979). In a similar environment, Herman and Platt (1986; Scotian Shelf) found that much of the chlorophyll in the subsurface maximum was often photosynthetically inactive because of light limitation.

This shows that the mechanisms responsible for subsurface chlorophyll maxima are still an open question. Before tackling it, one must distinguish between the chlorophyll maximum and the production maximum, which may or may not coincide. Causes for chlorophyll maxima may be quite diverse, and they may include such factors as irradiance, nutrients, stratification, internal waves, cell behaviour, grazing, and production. Maxima in production per unit biomass, on the other hand, are expected to be governed by irradiance and nutrients, which are the proximal determinants of microalgal production (Legendre and Demers, 1984). The present paper will only consider subsurface maxima in production per unit biomass.

The first question to ask is then: what are the main environmental factors that control phytoplankton production in the oceans? Using 225 sets of data from various oceanic regions, de Lafontaine and Peters (1986) found that production per unit biomass is mainly influenced by the depth of the mixed layer, surface temperature, and the attenuation of light in the water column. Unfortunately, their statistical analysis did not explicitly include nutrient concentrations. Concerning more specifically the temporal changes that occur in water column production, the 'critical depth' model explains that, under conditions of nonlimiting nutrients, these changes are controlled by underwater irradiance and water column stratification (see above). This has been verified not only for the spring phytoplankton bloom, but also for intermittent summer blooms (e.g. Levasseur et al., 1984); in some cases, photosynthetic characteristics of the phytoplankton have been reported to respond to changes in 'critical depth' on very short time scales (*e.g.* for semidiurnal tides: Fortier and Legendre, 1979; Fréchette and Legendre, 1982). In environments where nutrient limitation occurs, 'new' (versus 'regenerated') production is driven by the replenishment of nitrate in upper waters (Dugdale and Goering, 1967; Eppley and Peterson, 1979). Replenishment of nitrate by vertical eddy diffusion is then a condition for sustained or inter-mittent new production (*e.g.* Takahashi *et al.*, 1977; Legendre *et al.*, 1982), and changes in uptake characteristics have been observ-ed by MacIsaac *et al.* (1985) in an upwelling area. On the other hand, the depth of subsurface production maxima has been related to underwater irradiance (*e.g.* Herman and Platt, 1986), vertical stability of the water column and vertical changes in photosynthet-ic characteristics (*e.g.* Holloway, 1984; Vandevelde *et al.*, 1987), nitrate availability (*e.g.* Holligan *et al.*, 1984), and vertical changes in nutrient uptake (*e.g.* Dugdale and MacIsaac, 1971).

New phytoplankton production, both temporally and on the ver-tical axis, therefore appears to respond to a set of five varia-bles. Two of them are related to the fixation of carbon; these are the underwater irradiance and the photosynthetic characteris-tics of the cells. Two others concern the uptake of nitrogen; they are the availability of nitrate and the uptake characteristics of the cells. Finally, vertical stability of the water column characterizes displacements of the cells through the vertical irra-diance and nutrient gradients. This is not to say that other envi-ronmental processes do not influence new production (*e.g.* air-sea interactions, internal waves, horizontal advection, etc.), but it is suggested that the timing and depth of new production per unit biomass are more directly controlled by the first five variables.

2. CRITERIA

2.1. General Conditions and Assumptions

To double their biomass, phytoplankton cells require definite amounts of photons and nutrients, supplied within a time frame compatible with the physiological characteristics of the cells. As phytoplankton continuously incur metabolic expenses, doubling their biomass can only happen if photons are captured and nutrients assimilated at rates that exceed those required for the maintenance of the cells. The time required by an average phytoplankton cell to double its biomass is given by a simple equation:

$$t_{net} = \frac{biomass}{rate\ of\ gains\ -\ rate\ of\ losses}$$

$$= (t_{gains}^{-1} - t_{losses}^{-1})^{-1} \tag{1}$$

When losses are independent from the gains, the two rates (and thus times t_{gains} and t_{losses}) can be estimated independently.

In principle, equations developed below could be applied to any substance taken by phytoplankton. In this paper however, only carbon and nitrogen will be considered. In aquatic environments, carbon uptake is generally not limited by the environmental concentration of HCO_3, but rather by underwater irradiance. It follows that t_C is a function of the rate of photon capture by phytoplankton. On the other hand, NO_3 will be the only environmental source of dissolved inorganic nitrogen considered in this paper. This is because NO_3 is the only nitrogenous nutrient which is not recycled *in situ* and of which, consequently, the supply is entirely under hydrodynamical control. It follows that new phytoplankton production will be discussed. Assuming that phytoplankton new production is regulated by the local availability of photons and NO_3, the following characteristic times can be defined (t_C and t_S are t_{gains}): t_C is the minimum time required by an average phytoplankton cell to double its carbon biomass, given an optimal rate of photon capture (Section 2.3); t_S is the minimum time required by an average phytoplankton cell to double its nitrogen biomass, given an optimal uptake rate of nitrate (S: substrate) (Section 2.4); t_{crit} is the critical mean doubling time, above which there is no growth of the population (Section 2.2.2).

Using the conditions that exist at the boundaries of the productive layer (irradiance at sea surface and nutrient concentration at depth), the downwards attenuation of light and the upwards diffusion of the limiting nutrient, and also photon capture and nutrient assimilation by the cells, it is possible to estimate t_C and t_S at any depth z. Minimum potential doubling time of phytoplankton at any depth z (t_{prod}) will be the longest of $t_C(z)$ or $t_S(z)$. Times $t_C(z)$ and $t_S(z)$ are monotonic functions of z; as depth z increases, $t_C(z)$ progressively lenghtens and $t_S(z)$ progressively shortens (Figure 1). Near the surface (Figure 1a), generally $t_S(z) > t_C(z)$ so that phytoplankton require time $t_S(z)$ to double their biomass; this indicates nutrient limitation. Deeper in the water column, $t_C(z) > t_S(z)$ so that doubling the biomass cannot be shorter than $t_C(z)$; this shows light limitation. Both $t_S(z)$ in the upper layer and $t_C(z)$ in the lower layer are longer than $t_C(z) = t_S(z)$ at mid depth. In this case, maximum production per unit biomass occurs at the depth where doubling time of the biomass is shortest, that is at the depth ($z_{C=S}$) where $t_C(z) = t_S(z)$ [$t_{C=S}$]. During the winter, it may happen that $t_C(z) > t_S(z)$ at all depths, which corresponds to light limitation over the whole water column (Figure 1c).

76

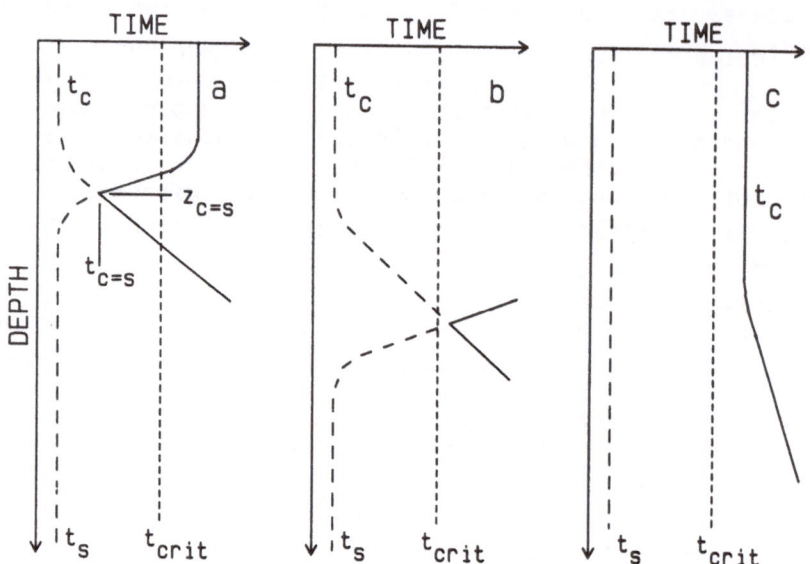

Figure 1.   Schematic changes with depth of minimum doubling times
for the carbon biomass [$t_C(z)$] and the nitrogen biomass [$t_S(z)$] of
an average phytoplankton cell. At any depth z, the minimum
doubling time (t_{prod}; solid line) is the longest of $t_C(z)$ or $t_S(z)$;
t_{crit} is the critical doubling time above which there is no growth
of the population. [a] Maximum new production per unit biomass at
depth $z_{C=S}$ where $t_C(z) = t_S(z)$. [b] No growth of the population
since $t_{C=S} > t_{crit}$ (*e.g.* in very oligotrophic waters). [c] No
growth of the population because $t_C(z) > t_{crit}$ at all depths (*e.g.*
during the winter at intermediate and high latitudes.

There are two sets of conditions resulting in no growth. The
first one (Figure 1b) would correspond to very oligotrophic waters,
where the upwards diffusion of nitrate may be too weak to sustain
steady new production; this would result in $t_{C=S} > t_{crit}$. The
second one (Figure 1c) corresponds to winter conditions at intermediate and high latitudes, where low irradiance combined with deep
vertical mixing leads to $t_C > t_{crit}$ at all depths.

The above considerations raise the question of the exact
meaning of times t_C and t_S. These times are defined for an average
phytoplankton cell. In the population, individual cells have characteristics and encounter environmental conditions that can depart
quite significantly from the mean. As a consequence, there will
almost always be some growing cells. When conditions are unfavour-

able however, a large number of cells deteriorate and eventually disappear, with the average result of a mean doubling time of the population longer than t_{crit}, and therefore no overall growth.

The next sections will be devoted to the estimations of $t_C(z)$ and $t_S(z)$ in the water column. These will require several simplifying assumptions, in order to provide oceanographic criteria that can be easily estimated from available data. The first assumption concerns environmental steady state. In the context of the present paper, it means that the physical characteristics and processes regulating phytoplankton production remain constant for periods of time that are long relative to $t_C(z)$ and $t_S(z)$. A corollary to this first assumption is that environmental conditions are not affected, on the short term, by phytoplankton production. According to the second assumption, the control of phytoplankton growth by irradiance and nitrate is alternative, not multiplicative. This is contrary to what is often supposed, but is in agreement with other authors (e.g. Kiefer and Kremer, 1981; Tett, 1981). Symbols used and units are listed in Table I.

2.2. Estimation of t_{net} and t_{crit}

2.2.1. Losses of C and S. As explained in Section 2.1, estimation of doubling times should take into account the fact that phytoplankton cells incur metabolic expenses. Concerning carbon, the two main sources of losses by phytoplankton are exudation of organic carbon and respiration. For example, carbon budgets in coastal waters often indicate significant dissolved carbon production (i.e. exudation), that can amount to $\approx$ 15-20% of net summer production (Larsson and Hagström, 1982), or $\approx$ 45% (Larsson and Hagström, 1979) and even $\approx$ 50% (Joiris et al., 1982) of net annual production. By contrast, Sharp (1977) concluded that excretion of photoassimilated carbon is low. Since minimum doubling times of phytoplankton are sought, no correction will be made in this paper for carbon exudation during period t_C. Concerning respiration, Blasco et al. (1982) related respiration rate (t_R^{-1}) to growth rate (t_{net}^{-1}) for diatoms: $t_R^{-1} \approx 0.1 t_{net}^{-1}$. Using this relationship in conjunction with (1) gives:

$$t_{C(net)} = 1.1 \, t_C \qquad (2)$$

Computed doubling time of the carbon biomass t_C will be multiplied by 1.1, to account for minimum respiration losses. Equation (1) offers the possibility for other corrections to t_C, if different hypotheses are made concerning exudation and respiration.

Nitrogen losses related to NO_3 uptake are of two types: exudation of nitrogenous organic matter and NO_2 excretion. According to

TABLE I. Symbols, units and dimensions of quantities in the text.

Symbol	Quantity	Units	Dimensions
a	Coef. of photon absorption per cell	m^{-1}	L^{-1}
$a*_{chl}$	Specific absorption coef. per chl.a	$mg^{-1}Chla\ m^2$	$M^{-1}L^2$
c_i	Intracellular concentration of chl.a	$mg\ m^{-3}$	ML^{-3}
E	Irradiance; photon fluence rate (PAR: photosynthetically available rad.)	quanta $m^{-2}\ d^{-1}$ Ein $m^{-2}\ d^{-1}$	$NL^{-2}T^{-1}$ $NL^{-2}T^{-1}$
E_k	E at the onset of saturation in P^B versus E curve	quanta $m^{-2}\ d^{-1}$	$NL^{-2}T^{-1}$
k	Coef. of vertical eddy diffusion	$m^2\ s^{-1}$	$L^2\ T^{-1}$
K_E	Irradiance E at which $\phi = 0.5\ \phi_{max}$	quanta $m^{-2}\ d^{-1}$	$NL^{-2}T^{-1}$
K_s	Half-saturation constant; substrate affinity (Michaelis-Menten equation)	mol m^{-3}	NL^{-3}
m	Depth at bottom of the mixed layer	m	L
M	Cellular biomass	mol	N
N	Brunt-Vaïsälä frequency	s^{-1}	T^{-1}
P^B	Production per unit pigment	mgC mg^{-1}Chl h^{-1}	T^{-1}
S	Concentration of substrate (nutrient)	mol m^{-3}	NL^{-3}
t_{crit}	Critical doubling time of population	day (d)	T
t_C	Minimum doubling t of carbon biomass	day (d)	T
t_{om}	Characteristic t of optics and mixing	day (d)	T
t_S	Minimum doubl. t of nitrogen biomass	day (d)	T
t_{sm}	Character. t of nutrients and mixing	day (d)	T
v	Cellular volume	m^3	L^3
V	Specific uptake rate of substrate S	d^{-1}	T^{-1}
z	Depth in the water column	m	L
ε	Rate of dissipation of turbulent kinetic energy	$m^2\ s^{-3}$	L^2T^{-3}
ρ	Water density	kg m^{-3}	ML^{-3}
σ_C	Absorption cross-section per unit C (also $a*_C$)	$mg^{-1}C\ m^2$	$M^{-1}L^2$
τ	Optical depth	(dimensionless)	
ϕ	Quantum yield	mol mol^{-1}	NN^{-1}
χ	Vertical attenuation coef. of PAR	m^{-1}	L^{-1}

Parsons and Takahashi (1973), exudation of nitrogenous substances by phytoplankton is rather small. On the contrary, significant release of NO_2 during NO_3 uptake has been documented for both cultures and natural populations (see the review of literature in Raimbault, 1986). According to Collos (1982) this fast excretion of NO_2 could lead to underestimates of NO_3 uptake if an isotopic method (e.g. ^{15}N) is used. This is contrary to the opinion of Harrison (1983), who thinks that such errors would be minimal because of the relatively slow recycling of NO_3. In fact, field measurements of uptake characteristics, which are generally not corrected for NO_2 excretion, provide values of net NO_3 uptake needed in this paper. Doubling times t_S computed for nitrogen will consequently be taken as equivalent to $t_{S(net)}$. As in the case of t_C above, (1) offers the possibility for other corrections to t_S, if different hypotheses are made concerning nitrogen losses.

It follows that t_C and t_S will be t_{gains}, corrected in this paper for only minimum losses. Estimated doubling times of the carbon and nitrogen biomasses will therefore be minimum.

2.2.2. <u>Maximum doubling time t_{crit}</u>. If computed doubling times $t_C(z)$ or $t_S(z)$ exceed some critical value (t_{crit}), cells at this depth will potentially never grow. Richardson et al. (1983) give average minimum irradiances (computed on a 24 h basis) for net photosynthesis (compensation) and for growth of various microalgal groups. The latter are 0.43 Ein m^{-2} d^{-1} for blue-green algae, 0.57 for dinoflagellates, and 0.55 for diatoms. According to equation (4b') (Section 2.3.2), these values correspond to doubling times of respectively 10.7, 8.2 and 8.4 d. In the present paper, t_{crit} is provisionally taken as 5 d.

2.3. Estimation of $t_C(z)$

2.3.1. <u>Equations</u>. Minimum time $t_C(z)$ required for doubling the carbon biomass of an average cell at depth z, given an optimal rate of photon capture, depends on a simple ratio (equation 1):

$$t_C(z) = \frac{\text{no. photons required to double the C biomass}}{\text{no. photons captured per unit time}} \quad (3)$$

To double its biomass, a cell must fix as many moles of carbon (M) as it already contains. Given the number of CO_2 molecules fixed per photon absorbed (ϕ: quantum yield), the number of photons required to double the biomass is $[M/\phi]$. The number of photons a cell captures per unit time at depth z is $[av\ E(z)]$, which is the product of irradiance at this depth $[E(z)$; photon fluence rate] with the absorption cross-section of the cell $[av]$. E is the photosynthetically available radiation (PAR: 400-700 nm), a is the coefficient of photon absorption by a cell and v is its volume. Replacing the numerator and the denominator of (3) by these expressions gives:

$$t_C(z) = \frac{M}{\phi\ av} \frac{1}{E(z)} = \frac{1}{\sigma_C\ \phi\ E(z)} \tag{4}$$

where $\sigma_C = [av/M]$ is the optical absorption cross-section [or specific absorption coefficient ($a{*}_C$) in Morel et al. (1987)] normalized to cellular carbon.

Quantum yield ϕ is known to monotonically decrease with increasing irradiance. In view of stressing the parallelism between irradiance and nutrients, an equation of the Michaelis-Menten type is used, where ϕ is a function of the average irradiation E, of the maximum quantum yield ϕ_{max}, and of the irradiance K_E at which $\phi = \phi_{max}/2$; K_E is equivalent to K_S in the Michaelis-Menten equation.

$$\phi = \phi_{max}\ \frac{1/E}{1/K_E + 1/E} = \phi_{max}\ \frac{K_E}{E + K_E} \tag{5}$$

Kiefer and Mitchell (1983) have shown that this equation adequately fits laboratory data. Using this relation, (4) becomes:

$$t_C(z) = \frac{1}{\sigma_C}\ \frac{E + K_E}{\phi_{max}\ K_E}\ \frac{1}{E(z)} \tag{4a}$$

Quantum yield ϕ is a physiological characteristic, of which the actual value depends on the recent light history of the cells. For cells found at depth z, E is the average irradiance they have experienced during a certain period prior to the determination of ϕ. The length of this period (a few hours to a few days) is fixed by the process of light/shade adaptation, which can also influence K_E (Jørgensen, 1969). By contrast, $E(z)$ is the irradiance at depth z during period $t_C(z)$ required for doubling the biomass. In general, $t_C(z)$ is longer than a few hours, so that $E(z)$ varies. The average irradiance at depth z during period $t_C(z)$ is then:

$$<E(z)> = [1/t_C(z)]\ _0\!\int^{t_C(z)} E(z,t)\ dt$$

Given the assumption of steady state, $<E(z)>$ is the average irradiance impinging on any cell, of which the average position in the water column is depth z. For such a cell, E and $<E(z)$ can practically be taken as equal, so that (4a) becomes:

$$t_C(z) = \frac{1}{\sigma_C \, \phi_{max} \, K_E} \left[1 + \frac{K_E}{<E(z)>} \right] . \qquad (4b)$$

Equation (4b) is an integral equation, since $<E(z)>$ includes an integral term which is itself a function of $t_C(z)$. This equation can be solved numerically. In general, irradiances at a depth z are computed from values measured at the sea surface $E(0)$ and from the vertical attenuation coefficient for PAR (χ) $[L^{-1}]$, thus defining the optical depth $\tau = z\chi$ (a dimensionless quantity):

$$<E(z)> = <E(0)> \, e^{-z\chi} = <E(0)> \, e^{-\tau} \qquad (6)$$

In this paper, χ rather than the usual k is used for the vertical attenuation coefficient, because the symbol for vertical eddy diffusion is also k. In (6), it could be possible to use χ variable with depth; however, for matter of simplicity, χ is assumed to be the same over the whole water column and to remain constant during period $t_C(z)$.

A final development in the estimation of $t_C(z)$ consists of incorporating vertical excursions of phytoplankton across irradiance gradients. Two possibilities will be considered, that modify (6); these are movements in the surface mixed layer, and displacements in a stratified layer.

The first case corresponds to vertical movements of phytoplankton in the surface mixed layer, which extends down to depth m. In the mixed layer, density stratification is taken as being null (*i.e.* density difference between surface and depth m <0.01 kg m^{-3}). During their vertical excursions, phytoplankton are exposed to the average irradiance in the mixed layer:

$$<E_m> = <E(0)>[1 - e^{-\tau_m}]/\tau_m \qquad (7)$$

where $\tau_m = m\chi$. This relationship is true only if the time it takes for one excursion across the mixed layer (t_m) is short relative to $t_C(z)$. Time t_m depends on the depth of the mixed layer (m) and on the coefficient of vertical eddy diffusion (k_m). The following formula is used to compute t_m (*e.g.* Lande and Wood, 1987), and compare it to $t_C(z)$:

$$t_m = m^2 / 2k_m \ll t_C(z)$$

Practical considerations concerning the estimation of k_m are deferred until the next section.

The second case concerns vertical displacements of phytoplankton in a stratified layer, at depths z > m. Excursions of phytoplankton through the water column are then estimated from the profile of vertical eddy diffusion (k). At the beginning of time interval t_C, there are phytoplankton cells at depth z. During time interval t_C, the magnitude of their vertical excursions on either sides of depth z is $\Delta z = (2kt_C)^{0.5}$. This assumes that k remains relatively constant over Δz. Then $\Delta \tau = \Delta z \chi = (2kt_C)^{0.5}\chi$, and the average irradiation to which the cells are exposed during their vertical excursions ($2\Delta z$) is:

$$<E(z)> = <E(0)> (1/2\Delta\tau) \int_{(\tau-\Delta\tau)}^{(\tau+\Delta\tau)} [e^{-\tau}] \, d\tau$$

$$<E(z)> = <E(0)> [1/\Delta\tau][e^{-\tau} \sinh(\Delta\tau)] \tag{8}$$

where sinh stands for hyperbolic sine. Equation (8) is an integral equation since it includes $\Delta\tau = (2kt_C)^{0.5}\chi$, where t_C is itself defined by integral equation (4b). This equation can be solved numerically.

In (8), it is assumed that k remains constant over Δz and thus that excursions of the cells are symmetrical above and below depth z. A more realistic approach in stratified waters (vertically changing k) would be to assume that the cells are found on the average half the time (0.5 t_C) above and half the time below depth z. Under this assumption, average irradiances during excursions above and below depth z are:

$$<E_{sup}> = <E(0)> (1/\Delta\tau_{sup}) \int_{\tau}^{\tau_{sup}} [e^{-\tau}] \, d\tau$$

$$= <E(0)> (1/\Delta\tau_{sup}) [e^{-\tau}(e^{\Delta\tau_{sup}} - 1)] \tag{8a}$$

$$<E_{inf}> = <E(0)> (1/\Delta\tau_{inf}) [e^{-\tau}(e^{\Delta\tau_{inf}} - 1)] \tag{8b}$$

where $\Delta\tau_{sup} = [\tau - \tau_{sup}] > 0$ and $\Delta\tau_{inf} = [\tau_{inf} - \tau] > 0$. The average irradiance experienced by the cells during t_C is the average of $<E_{sup}>$ and $<E_{inf}>$:

$$<E(z)> = 0.5<E(0)>e^{-\tau}[(e^{\Delta\tau_{sup}}-1)/\Delta\tau_{sup}+(e^{\Delta\tau_{inf}}-1)/\Delta\tau_{inf}] \tag{8c}$$

It is possible to numerically estimate t_C (equation 4b) using (8c), since τ_{sup} and τ_{inf} are defined (see equation 17) by:

$$[\Delta\tau_{sup}] \int_{\tau_{sup}}^{\tau} [1/2k\chi^2] \, d\tau = 0.5t_C$$

$$[\Delta\tau_{inf}] \;_\tau\!\!\int^{\tau inf} [1/2k\chi^2] \; d\tau = 0.5t_C$$

In the surface mixed layer (*i.e.* above depth m of the pycno-cline), all the cells receive the same average irradiance (when $t_m \ll t_C$). Vertical excursions are equivalent to displacements of total amplitude $2\Delta z = m$ around depth $z = m/2$; it follows that $\Delta\tau = \tau_m/2$ and $\tau = \tau_m/2$. Replacing these values in (8) gives back (7). Within and below the pycnocline, the role of $[\sinh(\Delta\tau)/\Delta\tau]$ in (8) depends on the magnitude of $\Delta\tau$. When $\Delta\tau < 1$, the vertical dis-placements do not significantly influence the average irradiance impinging on the cells since $<E(z)> \approx <E(0)> e^{-\tau}$, which is the same as in the case of no vertical displacement (equation 6). On the other hand, when $\Delta\tau \gg 1$, vertical excursions play a significant role, since $<E(z)>$ is larger than the average irradiance at fixed depth due to nonlinear change of irradiance with depth.

$\Delta\tau$ (or $\Delta\tau_{sup}$, $\Delta\tau_{inf}$) is therefore a very important character-istic for phytoplankton. However, it is not immediately available since $\Delta\tau$ is a function of time t_C, which is itself the end result of the computations. Independently from t_C, it is possible to define a time scale which combines the optical (χ) and mixing (k) characteristics (t_{om}) at any depth z:

$$t_{om} = 1/(2k\chi^2) \;:\; [L^2T^{-1}]^{-1}[L^{-1}]^{-2} = [T] \tag{9}$$

$\Delta\tau$ is a simple function of t_{om} and t_C: $\Delta\tau = (t_C/t_{om})^{0.5}$. Time scale t_{om} is a property of the environment, which can be used to compare water columns with different optical and mixing characteristics. For example, transparent waters (low χ) with strong vertical eddy diffusion (high k) may be equivalent, for phyto-plankton, to turbid waters (high χ) with weak diffusion (low k). Time scale t_{om} is therefore characteristic of the interaction between optical and mixing properties, and it specifies the time t_C above which verti-cal displacements of the cells result in shorter doubling time of the biomass. The same expression is also found in Lewis *et al.* (1986), where it characterizes the mixing over one optical depth. In the surface mixed layer $\tau_m = (t_m/t_{om})^{0.5}$, which makes τ_m inde-pendent from t_C and related instead to the time (t_m) it takes for cells to traverse the layer.

2.3.2. <u>Numerical estimates</u>. Irradiance measurements are reported in the literature with different units. Table II lists the factors used to convert various units into Ein m^{-2} d^{-1} of underwater PAR.

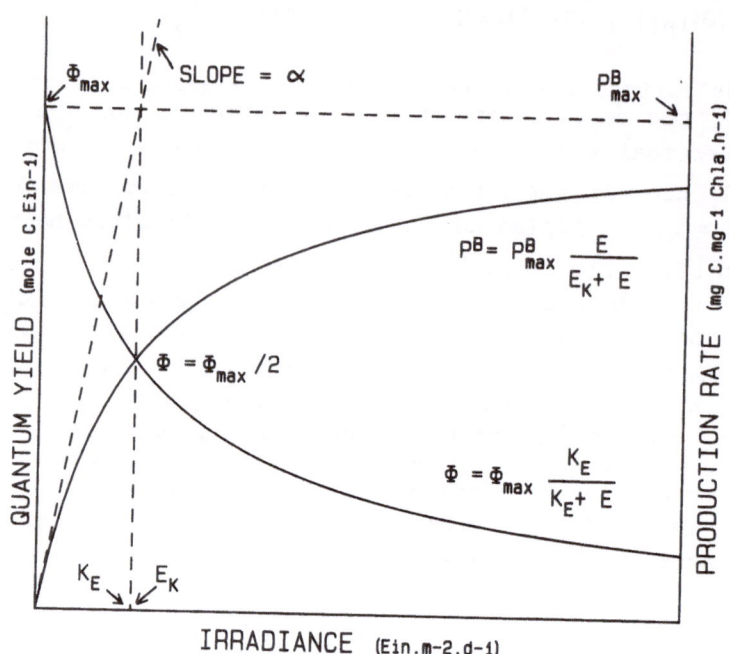

Figure 2.   Relationships between quantum yield (ϕ) and photo-synthesis per unit pigment (P^B), as a function of irradiance (E).

TABLE II.   Factors used to convert irradiance measurements into Ein m^{-2} d^{-1} of underwater PAR.

To convert from	To	Multiply by
μEin m^{-2} s^{-1}	Ein m^{-2} d^{-1}	0.0864
quanta m^{-2} d^{-1}	Ein m^{-2} d^{-1}	1.660×10^{-24}
quanta m^{-2} s^{-1}	Ein m^{-2} d^{-1}	0.1435×10^{-18}
J m^{-2} d^{-1} (of underwater PAR*)	Ein m^{-2} d^{-1}	4.15×10^{-6}**
W m^{-2} (of underwater PAR*)	Ein m^{-2} d^{-1}	0.3586**
g-cal cm^{-2} min^{-1} (of underwater PAR*)	Ein m^{-2} d^{-1}	250
[1 cal.g $\approx$ 4.18 J]		

* PAR constitutes only 43 % of the total irradiance (energy units) at the Earth's surface (Jerlov, 1974; Ivanoff, 1975; Jitts et al., 1976).

** Converting J into quanta depends on the spectral distribution of irradiance; for solar radiation in the 400-700 nm band (PAR*), 1 J $\approx 2.77 \times 10^{18}$ quanta above the water surface, and 1 J $\approx 2.5 \times 10^{18}$ quanta in marine waters (Morel and Smith, 1974).

The equation to compute t_C is (4b):

$$t_C(z) = \frac{1}{\sigma_C \, \phi_{max} \, K_E} \left[1 + \frac{K_E}{<E(z)>} \right]$$

Values can be found in the literature for several parameters of this equation. Falkowski et al. (1985) give $\sigma_C = 0.19$ m^2 g^{-1}C for the diatom *Thalassiosira weisflogii*, 0.21 for the chrysophyte *Isochrysis galbana* and 0.07 for the dinoflagellate *Prorocentrum micans*; these values decrease with increasing irradiances. At irradiances ≤ 2 Ein m^{-2} d^{-1}, Morel et al. (1987) report $\sigma_C \approx 0.3$. In view of estimating $t_{C(min)}$, maximum values must be used; taking $\sigma_C^{-1} = 0.29$ mol C m^{-2}, (4) becomes:

$$t_C(z) = \frac{0.29 \text{ mol C m}^{-2}}{\phi \, E(z,t)} \qquad (4')$$

Another approach could consist in taking $\sigma_C^{-1} = [M/av]$ as the product $[M/v][a]^{-1}$, which can be developed as:

$$[M/av] = [M/v][a^*_{chl} \, c_i]^{-1}$$

where a^*_{chl} is the specific absorption coefficient or optical absorption cross-section normalized to chlorophyll and c_i is the intracellular concentration of chlorophyll. Values can be found in the literature for $[M/v]$ (e.g. Mullin et al., 1966; Strathman, 1967; Takahashi et al., 1985), a^*_{chl} or k_C (e.g. Morel and Prieur, 1977; Kirk, 1983; Bannister and Weidemann, 1984; Falkowski et al., 1985; Sathyendranath et al., 1987) and c_i (e.g. Morel, 1987). Measurements of these parameters have generally not been made simultaneously, and a wide of values can be found for $[M/v]$ (0.05 to 0.4 pg C μm^{-3}), a^*_{chl} (0.004 to 0.027 m^2 mg^{-1}Chla) and c_i (0.3 to 15.2x10^6 mg Chla m^{-3}). Given the relationships between $[M/v]$, a^*_{chl} and c_i in phytoplankton cells, variations of $[M/v][a^*_{chl}c_i]^{-1}$ should be smaller than variations observed in individual parameters. Using central values 0.15 for $[M/v]$, 0.015 for a^*_{chl} and 4.5x10^6 for c_i results in $[M/av] \approx 0.2$ mol C m^{-2}. This value is of the same order of magnitude as those measured directly (see above).

To compute ϕ, both K_E and ϕ_{max} must be known (equation 5). Concerning K_E, it is useful to consider the photosynthesis (P^B) versus irradiance (E) relationship, since $\phi a^*_{chl} = P^B/E$ [P^B: mg C mg^{-1}Chl d^{-1}]. E_k is the irradiance at the onset of saturation in the P^B vs E curve, and it can be shown that $K_E = E_k$ (Figure 2). Kirk (1983; Table 10.1) reports that E_k (PAR) for marine phytoplankton varies between 100 and 500 μEin m^{-2} s^{-1}. Taking 12 h of daylight, these values correspond to $E_k = K_E = 4.3-21.6$ Ein m^{-2} d^{-1}. For cultures of the diatom *Thalassiosira weissflogii*, Kiefer and Mitchell (1983) give 10 Ein m^{-2} d^{-1}, and the value that best fits the field data of Morel (1978) is $\approx 5 \times 10^{24}$ quanta m^{-2} d$^{-1} \approx 8.5$ Ein m^{-2} d^{-1}. In the present paper K_E will be taken as 10 Ein m^{-2} d^{-1}. As to ϕ_{max}, various estimates can be found in the literature. Given the photosynthetic process, it takes not less than 8 Ein (moles) of light to reduce one mole of CO_2 to its carbohydrate equivalent, which sets the upper limit ϕ_{max} = 0.125 mol C Ein^{-1}. On the other hand, synthesis of protein, lipids and nucleic acids also require light quanta, so that the true minimum quantum requirement for gross assimilation of CO_2 does not exceed 10-12, which brings ϕ_{max} down to 0.08-0.10 mol C Ein^{-1} (Radmer and Kok, 1977). Kiefer and Mitchell (1983) found 0.06 mol C Ein^{-1}. Since the maximum photosynthetic rate is $P^B_{max} = \phi_{max} a^*_{chl} K_E$, taking $\phi_{max} = 0.08$ and $a^*_{chl} = 0.015$ results in P^B_{max} = 12 mg C mg^{-1}Chl h^{-1} (for 12 h of dayligth) which is a realistic maximum value for the natural environment.With 10 Ein m^{-2} d^{-1} for K_E and 0.08 mol C Ein^{-1} for ϕ_{max}, (5) becomes:

$$\phi \ (\text{mol C Ein}^{-1}) = 0.8 \ / \ (E + 10) \qquad (5')$$

Using (4') and (5'), (4b) gives:

$$t_C(z) = 1.1 \times 0.36 \ [1 + (10/<E>)] = 0.4 + [4/<E>] \qquad (4b')$$

where $<E>$ is in Ein m^{-2} d^{-1}, and 1.1 is the factor that accounts for respiration (equation 2). According to (4b'), the most rapid doubling time of phytoplankton biomass, at very high (but non-inhibitory) irradiance, would approach 0.4 d.

Very often, E(z) measurements are not available, so that irradiances in the water column must be estimated from surface values and the vertical attenuation coefficient (χ). The attenuation coefficient may be estimated from measurements with a submersible irradiance meter. Alternatively, χ may be approximated from Secchi depth (z_{SD}) using:

$$\chi = A \, / \, z_{SD}$$

Proposed values for A generally vary between 1.4 and 2.0 (Gordon and Wouters, 1978); values most often used are A = 1.7 in oceanic waters (Poole and Atkins, 1929) and A = 1.44 in coastal waters (Holmes, 1970). Using a water telescope, Højerslev (1986) found A = 3.3. Simonot and Le Treut (1986) give maps of Secchi depths and $\chi = 1.7 \, / \, z_{SD}$ for the world ocean. There is, however, no straightforward relationship between Secchi depth and χ (e.g.. Tyler, 1968; Preisendorfer, 1986). Even in clear waters, χ is normally higher near the surface due to strong absorption in the red part of the spectrum. In the surface layer (upper 5 m), χ generally varies between 0.15 m^{-1} (clear waters) and 0.5 m^{-1} (turbid oceanic waters). At depth, values are lower, typically between 0.05 and 0.3 m^{-1}. Mathematical relationships have been established between χ and the concentration of phytoplankton and other particulate matter (e.g. Smith and Baker, 1978; Prieur and Sathyendranath, 1981).

Formula $t_m = m^2/2k_m$ (Section 2.3.1) was found by other authors to provide good estimates of the time for one excursion across the surface mixed layer (e.g. Lande and Wood, 1987). However, using k_m to parametrize eddy diffusion is open to criticism, since it is not derived from the theory of fluid mechanics (e.g. Holloway, 1984). The value of k_m depends on the source(s) of turbulent kinetic energy in the surface mixed layer (i.e. wind stress, thermal convection, velocity shear, internal waves; Denman and Gargett, 1983; Gargett, 1984). Despite these problems and follow-ing Lande and Wood (1987), k_m will be taken as equal to 10^3 m^2 d^{-1} ($\approx 10^{-2}$ m^2 s^{-1}) which is a maximum value for the coefficient of vertical eddy diffusion. In a very turbulent mixed layer, formula $t_m = m^2/2k_m$ is no longer applicable, and hypotheses concerning turbulence must be explicited (Denman and Gargett, 1983). At sea, the effects of turbulence on phytoplankton production have been directly measured by Lewis et al. (1984).

In stratified conditions, a useful relationship to estimate the coefficient of vertical eddy diffusion (k) [L^2T^{-1}] from available field measurements is (Denman and Gargett, 1983; Gargett, 1984):

$$k = 0.25 \, \varepsilon \, N^{-2}$$

where ε $[L^2T^{-3}]$ is the rate of dissipation of turbulent kinetic energy and N is the Brunt-Vaïsälä frequency $[T^{-1}]$; $N^2 = (g/\rho)(d\rho/dz)$ where g $[LT^{-2}]$ is the acceleration of gravity and ρ is water density $[ML^{-3}]$. In this paper, fixed values of ε will be used, and k will be estimated from stratification (N) only. In the oceans however, ε and N are generally not independent, so that coupled values quoted by Denman and Gargett (1983) can be used for practical applications. Another approach could be to choose ε according to the type of forcing responsible for turbulence, and to enter its value in the above equation. As an alternative, Gargett (1984) proposed a relationship applicable when turbulence is caused by breaking internal waves: $k = a_0 N^{-1}$, where $a_0 = 10^{-7}$ m^2 s^{-2}.

In the upper layer, N varies between 3×10^{-3} s^{-1} for weakly stratified waters, and 3×10^{-2} s^{-1} for the pycnocline. Taking $\varepsilon = 2 \times 10^{-9}$ m^2 s^{-3}, which corresponds to low rates of transfer of turbulent energy, $5 \times 10^{-7} < k < 5 \times 10^{-5}$ m^2 s^{-1}. For high rates of transfer of turbulent energy on the other hand, $\varepsilon = 2 \times 10^{-7}$ m^2 s^{-3}, which gives $5 \times 10^{-5} < k < 5 \times 10^{-3}$ m^2 s^{-1}. Below the pycnocline, N progressively decreases from $\leq 3 \times 10^{-2}$ to 10^{-3} s^{-1}, and the above formula from Gargett (1984) gives $3 \times 10^{-6} < k < 10^{-4} m^2$ s^{-1}.

2.4. Estimation of $t_S(z)$

2.4.1. Equations. Minimum time $t_S(z)$ required for doubling the biomass of an average cell at depth z, given optimal nitrogen uptake, cannot exceed (equation 1):

$$t_S(z) = \frac{\text{amount of S required to double a cell}}{\text{amount of S taken per unit time}} \qquad (10)$$

Environmental nutrients are transported by phytoplankton at a given rate (mol S m^{-3} d^{-1}). To double its biomass, an average cell must fix at least as many moles of S as it already contains. In steady state, the minimum time required to do so is:

$$t_S(z) = \frac{\text{amount of S per cell}}{\text{transport rate of S}} \qquad (10a)$$

The inverse of (10a) is known as the specific uptake rate of a nutrient (V: d^{-1}), so that $t_S(z)$ is in fact the inverse of V. In principle, S could be any substance taken by the cells. As explained in Section 2.1, S stands here for nitrogen, and NO_3 will be the only environmental source of S considered . The Michaelis-

Menten equation (Dugdale, 1967; Eppley and Coatsworth, 1968) gives V as a function of substrate concentration (S: mol m^{-3}), of maximum uptake rate (V_{max}: d^{-1}) and of the concentration of S at which $V = 0.5\ V_{max}$ (K_S: mol m^{-3}), this half-saturation constant being also known as substrate affinity. The Michaelis-Menten equation is used to estimate $t_S(z) = V^{-1}$:

$$t_S(z) = \frac{1}{V} = \frac{K_S + S}{V_{max}S} = \frac{1}{V_{max}}\left[1 + \frac{K_S}{S} \right] \qquad (11)$$

V_{max} is taken here as the absolute maximum uptake rate, and not the maximum rate measured experimentally which can vary with the growth rate and is usually noted V'_{max}. It is well known that cell growth depends on a number of physiological processes, of which nutrient uptake is only one component. This is why the Michaelis-Menten equation cannot adequately describe growth, and why other equations have been developed that relate growth to substrate concentration (Monod, 1942) or to cellular nutrient concentration (cell quota; Droop, 1973). These equations are all equivalent when steady state is attained (e.g. Goldman and Glibert, 1983). In steady state, cells cannot increase their biomass faster than the rate at which they take substances from the environment, so that specific uptake rate V will be used in the present paper as the upper limit for growth.

In order to estimate S(z) it is assumed that variations in the vertical distribution of NO_3 are restricted to the upper layer of the ocean, between the surface and a depth where NO_3 is at maximum concentration (S_{max}). For the computation, it is assumed that there are initially (t = 0) no nutrients above a certain depth z_S, and that nutrient concentration is maximum (S_{max}) below z_S; depth z_S is used as the new origin $z' = 0$ (Figure 3a). Doubling time $t_S(z)$ is the time it would take for the concentration of S at depth z to reach such a level that the biomass of phytoplankton would have doubled (according to 11). It is also assumed that S uptake by the phytoplankton biomass is low relative to changes in the concentration of S, so that time changes in S at fixed depth z' can be described by the following differential equation:

$$\frac{\partial S(z,t)}{\partial t} = \frac{\partial}{\partial z'}\left[k\ \frac{\partial S(z',t)}{\partial z'} \right] \qquad (12)$$

Between times t = 0 and t = t_S, the (small) biomass M_0 will double. Assuming that the uptake is effected by M_0,

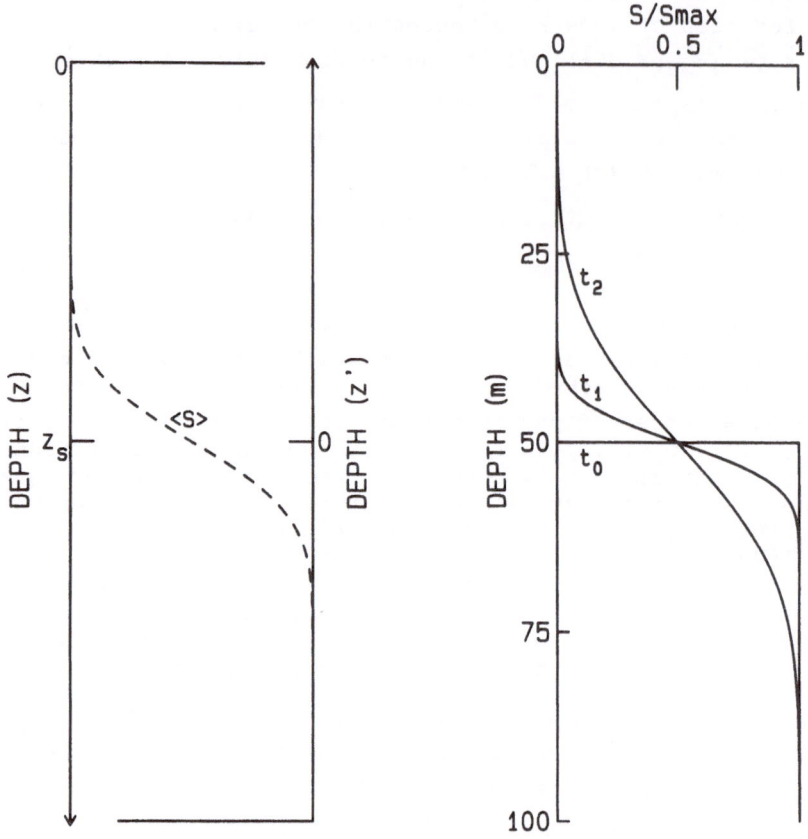

Figure 3.   Vertical nutrient distributions. [a] Profile of <S>
(average concentration) to show that depth z' = 0 (right-hand
ordinate) is defined to correspond to z = z_S (left-hand ordinate).
[b] Changes with time of S/S_{max}. At t_0 (t = 0), there are no
nutrients above z_S, and concentration is maximum below (Haeviside
function); as time progresses, nutrients diffuse upwards. With
k = 10 m^2 d^{-1}, t_1 = 1 and t_2 = 10 d; with k = 1, t_1 = 10 and t_2 =
100 d.

$$dM = M_0 \ V \ dt$$

$$_0\!\int^{t_S} dM = M_0 \ _0\!\int^{t_S} V \ dt = M_0$$

$$\frac{1}{V_{max}} = \ _0\!\int^{t_S} \frac{S(z',t)}{K_S + S(z',t)} \ dt \tag{13}$$

In view of computing t_S, (13) will be used rather than (11). This is because estimating t_S from (11) would require knowing the vertical profile of S, the concentration S being itself affected by MV. Computing equilibrium conditions in such a case would require a complex simulation, while using (13) makes it possible to estimate t_S for conditions where S limits growth and M does not significantly influence S.

In order to calculate t_S, differential equation (12) must first be solved for the initial conditions, at t_0, which are $S(z',0) = S_{max}H(-z')$ where $H(-z')$ is the Haeviside function. This function specifies the initial conditions explained above, that are $S(z',0) = 0$ for $z' > 0$, and $S(z',0) = S_{max}$ for $z' < 0$ (Figure 3b). The coefficient of vertical eddy diffusion k will first be taken as constant over the water column. The general solution for the heat equation (12) is (e.g. Landau and Lifchitz, 1971, p. 242):

$$S(z',t) =$$
$$[1/2(k\pi t)^{0.5}] \ _{-\infty}\!\int^{+\infty} S(\zeta,0) \ \exp -[(z'-\zeta)^2/4kt] \ d\zeta \tag{14}$$

where ζ is a depth variable used for the integration. Given the initial conditions $H(-z')$, $S(\zeta,0) = 0$ from 0 to $+\infty$, and $S(\zeta,0) = S_{max}$ from 0 to $-\infty$. It follows that:

$$S(z',t) = [S_{max}/2(k\pi t)^{0.5}] \ _{-\infty}\!\int^0 \exp[-(z'-\zeta)^2/4kt] \ d\zeta \tag{14a}$$

Using the complement of the error function (erfc = 1 - erf), this equation can be rewritten as:

$$S(z',t) = [S_{max}/2] \ \text{erfc} \ \{z'/[2(kt)^{0.5}]\} \tag{15}$$
$$\text{where erfc}(x) = 1 - (2/\pi^{0.5}) \ _0\!\int^x \exp[-\zeta^2] \ d\zeta$$

The complement of the error function (erfc) varies between 1 (for x = 0) and 0 (for x = $+\infty$); in addition erfc$(-x) = [2 - \text{erfc}(x)]$, which shows that $S(z',t) = S_{max}$ when $z' = -\infty$. Figure 3b gives changes with time in vertical profiles of $S(z',t)$, for k constant over the water column.

With k constant, $S(z',t)$ at depth z' is a simple function of time t. It is then possible to define a substrate mixing time scale

t_{sm}, analogous to t_{om} used in Section 2.3.1 for the optical and mixing characteristics (equation 9):

$$t_{sm}(z') = (z')^2/2k \qquad (16)$$

Combining (15) and (16), time changes in substrate concentration become a function of the sole ratio (t_{sm}/t):

$$S(z',t) =$$
$$[S_{max}/2]\ erfc\ [\pm(t_{sm}/2t)^{0.5}],\ where\ z'= \pm\ (2kt_{sm})^{0.5} \qquad (15a)$$

The sign $\pm$ is that of z'. It must be noted that $S(z',t)$ is estimated independently for each depth z'.

Up to now, k has been taken as constant over the water column. In most cases however, k varies with depth. In order to use (15) to compute $t_S(z)$ with (13), one must estimate a coefficient of vertical eddy diffusion k' equivalent to the depth varying k. This equivalent k' is computed in assuming that the average time for diffusion through a layer of thickness $n\Delta z$ (from depth z_S to depth z) is the average of times $t_{sm}(z)$:

$$(n\Delta z)^2/2k' = (1/n)\ (n\Delta z)^2 \sum_{i=1}^{n} (1/2k_i)$$
$$(1/k') = (1/n) \sum_{i=1}^{n} (1/k_i) \qquad (17)$$

In (17), all the layers Δz_i have the same thickness, and it is assumed that k_i does not vary within layer Δz_i. When k is constant over the whole layer $n\Delta z$, each $k_i = k$ and accordingly $k' = k$.

Equation (13) is an integral equation, where $S(z',t)$ is defined by (15). Numerical solution of (13) directly gives $t_S(z)$. Doubling time $t_S(z)$ at any depth can thus be estimated from the concentration of NO_3 in the deep reservoir (S_{max}) and the vertical density profile. Coefficients of vertical eddy diffusion (k_i) are computed from densities (through the Brunt-Vaïsälä frequencies) as explained above in Section 2.3.2; the choice of depth $z' = 0$ will be discussed in Section 2.4.2.

When computing $t_C(z)$, the equations were modified to take into account vertical excursions of the cells in the water column, across irradiance gradients. This was because the vertical distribution of underwater irradiance is independent from vertical eddy diffusion. In the case of nutrients, vertical eddy diffusion influences both the availability of nutrients and the vertical excursions of the cells. When vertical eddy diffusion is weak, movements of phytoplankton in the water column are minimal and they can therefore be neglected. On the other hand, when vertical eddy diffusion is strong, phytoplankton are subjected to significant vertical displacements but, simultaneously, the availability of nutrients tends to become homogenous. Again, vertical movements of the cells in the water column can be neglected, as phytoplankton traverse a weak nutrient gradient. Thus, in first approximation, vertical excursions of the cells in the water column can be neglected when computing $t_S(z)$.

2.4.2. <u>Numerical estimation</u>. As for $t_C(z)$ above (Section 2.3.2), several of the values in the equations for $t_S(z)$ can be found directly in the literature and considered, in first approximation, as constant. The most important variables for nitrogen uptake are V_{max} and K_S. For V_{max}, Kiefer and Kremer (1981) take 1.7 d^{-1}, and Dugdale and Wilkerson (1986) give values between 0.06 and 0.10 h^{-1} (1.4-2.5 d^{-1}). In the context of the present paper, V_{max} sets the maximum doubling rate (Section 2.4.1). Goldman et al. (1979) have compiled growth rates of phytoplankton for natural marine waters, which are in general ≤ 2 d^{-1}; Eppley (1981) explains that one set of exceptionally high values reported by Goldman et al. (1979) are in fact overestimates. As it is expected that the maximum absolute growth rate is seldom, if ever, measured at sea (Section 4.1), V_{max} is taken as 2.5 d^{-1}.

Equation (11) indicates that K_S scales the influence of S on doubling time t_S; for example, increasing K_S/S by one unit results in t_S longer by 0.4 d (for $V_{max} = 2.5$ d^{-1}). It is often stated (e.g. Parsons and Takahashi, 1973) that K_S values in eutrophic waters ($K_S \geq 1$ mmol m^{-3}) are well above those from oligotrophic environments (K_S down to < 0.1 mmol m^{-3}). Goldman and Glibert (1983) report $0.1 < K_S < 10.3$ mmol m^{-3} for cultured marine phyto-plankters, while values for eutrophic waters vary between 1.0 and 4.2 and those for oligotrophic conditions can reach 0.9 mmol m^{-3}. $K_S \leq 0.91$ have been observed by Eppley et al. (1977) in the upper eutrophic zone of the Central North Pacific Ocean. In the present paper, K_S will be taken as 10^{-3} mol m^{-3}; its influence on the computed value of t_S will be assessed in Section 2.5 by numerical simulation.

Calculation of t_S is not as straightforward as that of t_C. Using (15a) and (16), (13) can be transformed as follows:

$$\frac{1}{V_{max}\ t_{sm}} = {}_0\int^{t_S/t_{sm}} \frac{S(u)/S_{max}}{K_S/S_{max} + S/S_{max}}\ du,\ \text{where}\ u = t/t_{sm} \tag{13a}$$

For each depth z', t_{sm} is computed as a function of vertical eddy diffusion k (equations 16 and 17). Numerical solution of integral equation (13a) is achieved by increasing t_S until the right-hand side of the equation becomes equal to $[V_{max}\ t_{sm}]^{-1}$. In practice, t_S can be tabulated as a function of K_S/S_{max} and t_{sm} (Figure 4).

In addition, it is possible to compute the expected concentration of nitrate $<S(z')>$ at depth z' and time t_S:

$$<S(z')> = t_{sm}/t_S\ {}_0\int^{t_S/t_{sm}} S(u)\ du$$

When t_{sm} increases, t_S also increases but the ratio t_S/t_{sm} decreases; it follows that the average nutrient concentration at depth z' during period t_S decreases with higher t_{sm}, since $<S(z')>$ is the mean of a cumulative monotonic function. Higher t_{sm} corresponds (equation 16) to either depth z being farther from z_S (larger z') or vertical eddy diffusion being weaker (smaller k), both conditions which are conducive to lower nutrient concentration.

A practical problem when calculating t_S is the choice of reference depth z_S, above which there are no nutrients and below which $S = S_{max}$ at t_0 (Section 2.4.1). Given (14) and (15), nutrient concentration at depth z_S always remains equal to $0.5\ S_{max}$ (Figure 3); accordingly, z_S at an oceanographic station should be the depth where $S = 0.5\ S_{max}$. However, nitrate profiles are often influenced by phytoplankton uptake and short-term variability, so that actual profiles cannot be used (when available) to determine z_S. It is thus preferable to use as z_S the depth of the isopycnal where $S = 0.5\ S_{max}$; for a given area, this isopycnal is easily determined as shown in Section 3.2 (Figure 10).

2.5. Simulations

Numerical simulations were conducted in order to assess the influence of environmental conditions and of phytoplankton characteristics on depth $z_{C=S}$ and on minimum doubling time $t_{C=S}$. For such simulations, one must set: [1] depths z_S (or τ_S) and z_m (or τ_m), that are characteristic of stationary conditions; [2] boundary conditions for irradiance and nutrients ($<E(0)>$ and S_{max}, or alternatively $<E(0)>/K_E$ and S_{max}/K_S); and [3] coefficients that govern the downwards transfer of photons (χ) and the upwards transfer of

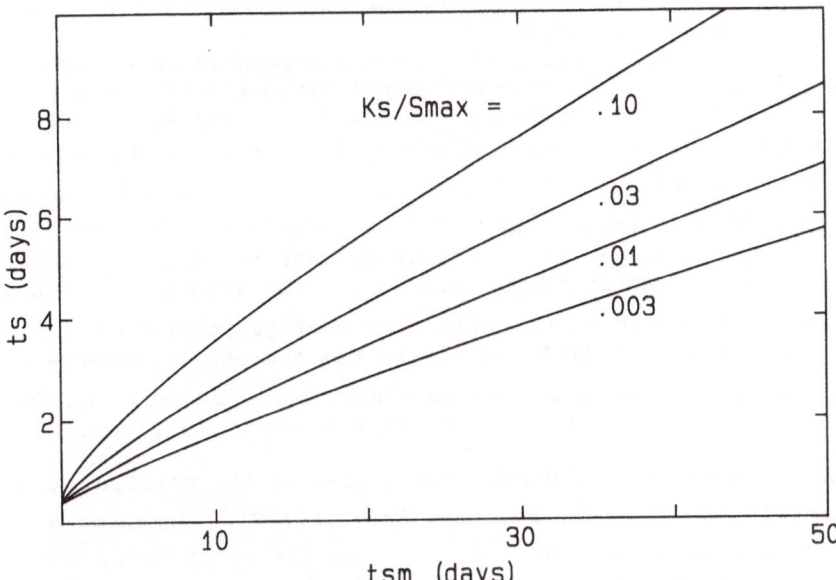

Figure 4.   Minimum doubling time of the nitrogen biomass (t_S) as a function of K_S/S_{max} and t_{sm}.

TABLE III. Values of parameters in the simulations. Half-saturation constants K_S and K_E, vertical attenuation coefficient χ, surface irradiance $<E(0)>$, optical depths of the surface mixed layer τ_m and of the nutrient source τ_S, parameters N_0 and α to calculate Brunt-Vaïsälä frequencies $N = N_0 \exp [-(\tau-\tau_S)/\alpha]$, and rate of dissipation of turbulent kinetic energy ε. In addition, computed $t_{C=S}$ and $\tau_{C=S}$.

Simul.	K_S/S_{max}	χ m^{-1}	$<E(0)>$ Ein m^{-2} d^{-1}	τ_S	N_0 s^{-1}	α	$t_{C=S}$ days	$\tau_{C=S}$
1	0.03	0.08	35	3.4	0.0090	3.92	2.3	2.8
2	0.03	0.08	35	4.6	0.0066	3.92	3.8	3.4
3	0.03	0.08	35	6.0	0.0046	3.92	6.4	3.8
4	0.003	0.08	35	4.6	0.0066	3.92	3.2	3.2
5	0.03	0.08	70	4.6	0.0066	3.92	2.7	3.6
6	0.03	0.08	17.5	4.6	0.0066	3.92	6.0	3.0
7	0.10	0.15	70	4.6	0.0050	8.00	1.8	3.2
8	0.10	0.15	35	4.6	0.0050	8.00	2.4	2.8
9	0.10	0.15	17.5	4.6	0.0050	8.00	3.0	2.6
GUIDOME	0.03	0.09	40	4.0	0.0224	2.70	2.0	2.8

$K_E = 10$ Ein m^{-2} d^{-1}; $\tau_m = 1.8$; $\varepsilon = 2 \times 10^{-8}$ m^2 s^{-3} [GUIDOME: $\varepsilon = 5 \times 10^{-7}$]

$t_{C(min)} = 0.4$ d, since $\phi_{max} = 0.08$ mol C Ein^{-1} and $s_C = 3.5$ m^{-2} mol^{-1}

$t_{S(min)} = 0.4$ d, since $V_{max} = 2.5$ d^{-1}

nutrients (k). Values for the simulations, together with some results, are given in Table III.

Results for each simulation are summarized in three panels (Figures 5 and 6). The left-hand panel gives the vertical profiles of E and N. The middle panel gives doubling times t_C and t_S; it must be remembered that maximum production occurs at depth $z_{C=S}$ where $t_C = t_S$, and that there is no growth of the population when $t_C > t_{crit}$ or $t_S > t_{crit}$ ($t_{crit} = 5$ d, Section 2.2.2). The right-hand panel gives (logarithmic scale) average <E> and <S>, normalized to boundary values <E(0)> and S_{max}. <E> (right-hand panel) is the irradiance impinging on cells with average position τ during period t_C, while E(z) (left panel) is the average irradiance at optical depth τ. <S> is the average nutrient concentration for small biomass; it also represents maximum potential biomass in steady state.

In Figures 5 and 6, depths are scaled by the vertical attenuation coefficient ($z\chi = \tau$), because <E> varies with depth as a function of τ (equations 6, 7 and 8). On the other hand, <S> is governed by the coefficient of vertical eddy diffusion, which is scaled by χ^2 [$k\chi^2 = (1/2t_{om})$; equation 9] when optical depths τ are used. In this case, χ and k play an equivalent role in the vertical changes of t_S. It follows that $t_{sm} = (\tau - \tau_S) t_{om}$, so that t_{om} is the physical time of reference to evaluate mixing time t_{sm}.

A first series of simulations (Figure 5a) compares three z_S (or τ_S) levels in the photic layer, set at respectively 3.5%, 1% and 0.25% of the surface irradiance. Doubling time t_S lengthens with deeper z_S, which results in longer $t_{C=S}$ and in production maxima deeper in the photic layer. Depths of maximum production are respectively 0.6, 1.2 and 2.2 units above τ_S, or 7.5, 15 and 27.5 m above z_S. Even when the source of nutrients is very deep, turbulence is thus strong enough to maintain the simulated production maximum in the photic layer, but the resulting mean doubling time may be too long for population growth. Depth z_S is therefore a critical factor in determining the minimum doubling time $t_{C=S}$.

A second series of simulations examines the influence of K_S (0.1 versus 1 mmol m^{-3}, for $S_{max} = 33$ mmol m^{-3}; Figure 5b). The resulting t_S profiles are almost identical for the two K_S values; on the other hand, nutrient profiles <S> seem to be quite influenced by K_S. Thus the choice of an exact value for K_S does not appear to be very critical for the calculation of t_S.

A third series of simulations (Figure 6a) considers the effect of surface irradiance <E(0)>. Increasing <E(0)> results in short-

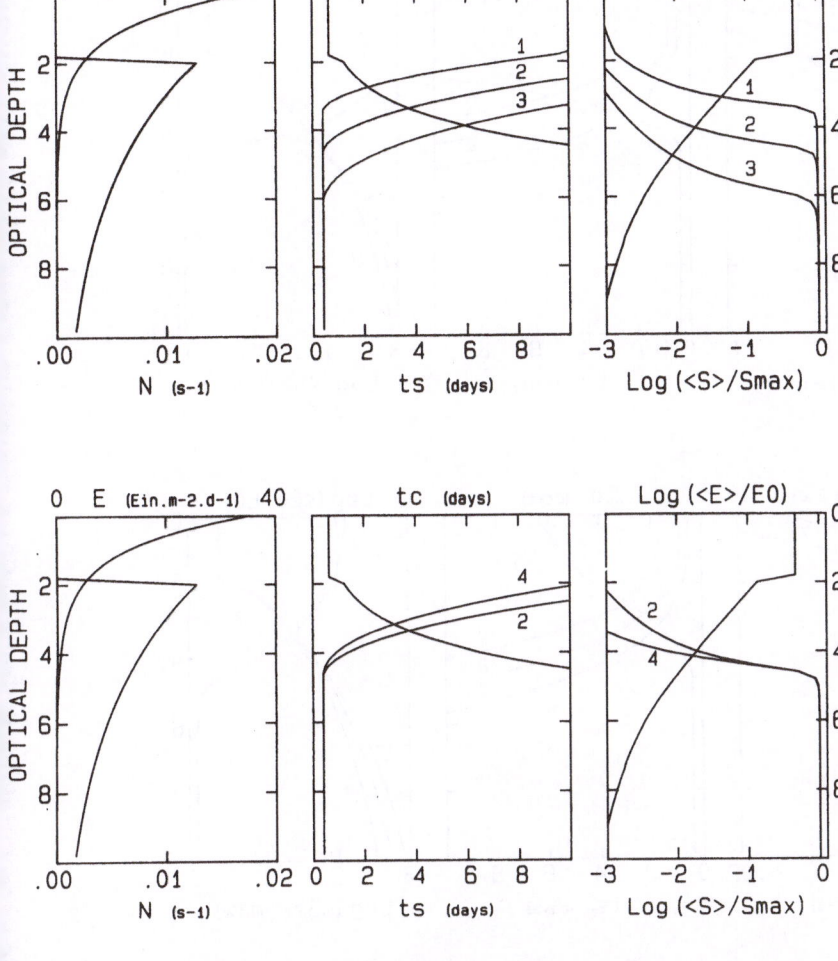

Figure 5.   Simulations. Vertical profiles of irradiance (E), Brunt-Väisälä frequency (N), minimum doubling times t_C and t_S, average irradiance <E> during period t_C, and average nitrate concentration <S> during period t_S. Comparisons of [a] three depths for τ_S (or z_S); [b] two values for K_S. Parameters and results are given in Table III.

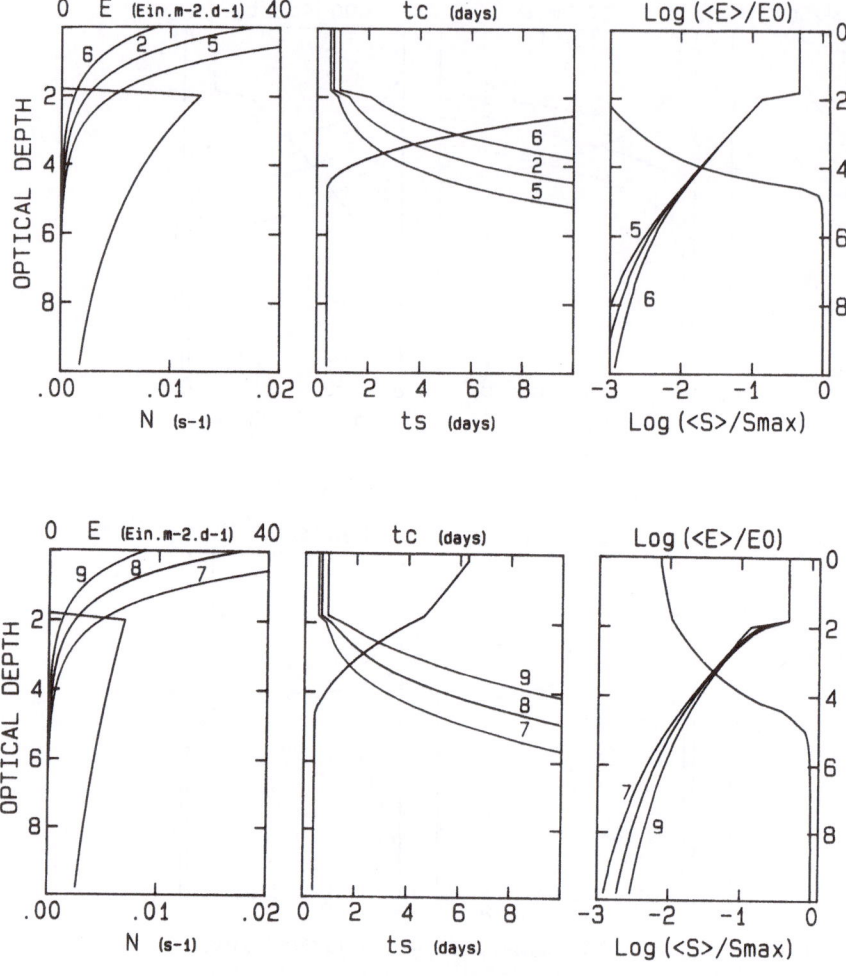

Figure 6.   Simulations (see Figure 5).   Comparisons of three
surface irradiances for [a] offshore and [b] inshore conditions.
Parameters and results are given in Table III.

ening $t_{C=S}$, with a slight deepening of the production maximum. The last series of simulations (Figure 6b) use the same three $<E(0)>$ as in Figure 6a, this time combined with lower S_{max}, weaker stratification and higher light attenuation. These conditions correspond to inshore (Figure 6b) versus offshore (Figure 6a) waters. Minimum doubling time $t_{C=S}$ is much shorter inshore and the production maximum is located higher in the photic layer.

3. FIELD EXAMPLES

Examples taken from the literature are presented in this section. These will show how to apply the equations of Section 3 to real data. Field examples will also demonstrate that the approach developed above leads to realistic oceanographic results.

3.1. Responses of Phytoplankton to Irradiance: Phytoplankton Blooms

When nutrients are nonlimiting, phytoplankton growth is governed by the availability of photons. This occurs, for example, at the beginning of the spring phytoplankton bloom, in upwelling areas, or in coastal environments with high nutrients (of natural or anthropogenic origin). In all these cases, surface irradiance is a major factor that determines doubling time t_C.

The critical depth model (Riley, 1942; Sverdrup, 1953) ascribes the spring bloom to seasonal increases in both solar irradiance and vertical stratification. These two factors, together with the vertical attenuation coefficient χ, set the average irradiance in the mixed layer (equation 7). Riley (1957) proposed from empirical evidence that $<E_m>$ must exceed 0.03 g-cal cm^{-2} min^{-1} (7.5 Ein m^{-2} d^{-1}) prior to any pronounced increased in growth rate. This average irradiance was calculated using values given by Sverdrup et al. (1942; Table 25) for total surface irradiance; PAR constitutes only 43% of it (see Table II), so that Riley's criterion is 3.25 Ein m^{-2} d^{-1} of PAR. This corresponds to a doubling time $t_C = 1.6$ d (equation 4b'); it must be noted that this value is much lower than the critical doubling time above which there is potentially no growth of the population ($t_{crit} = 5$ d; Section 2.2). Figure 7 shows how relatively small changes in surface irradiance can modify t_C; for example, with a mixed layer twice as deep as the photic layer ($\tau_m = 9.2$), increasing $<E(0)>$ from 14 to 31 Ein m^{-2} d^{-1} shortens t_C from 3.0 to 1.6 d.

In some instances of high nutrient supply, phytoplankton production can remain under the control of irradiance for the best part of the year ($t_C > t_S$). This was the case at a station in the Lower St. Lawrence estuary, where Levasseur et al. (1984) observed

100

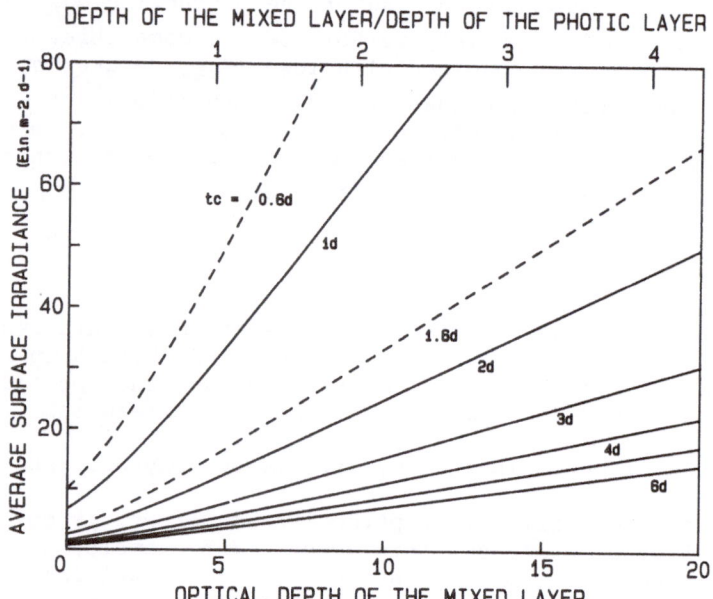

Figure 7.   Changes in minimum doubling time of the carbon biomass
(t_C) as a function of average surface irradiance and optical depth
of the mixed layer (τ_m).   Ratio [depth of the mixed layer/depth of
the photic layer] $= \tau_m/\log_e 100 = \tau_m/4.6$.

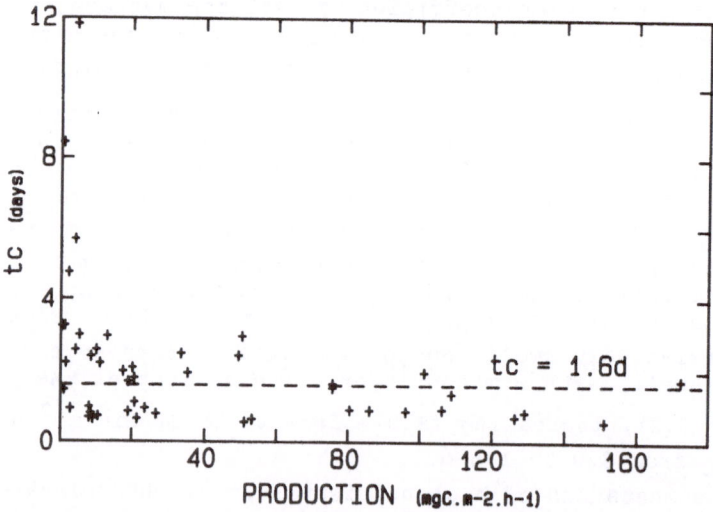

Figure 8.   Minimum doubling time of the carbon biomass (t_C) in the
mixed layer versus phytoplankton production in the photic layer.
Measurements done by Levasseur et al. (1984) in the St. Lawrence
Estuary, between June 1979 and July 1980.   The criterion for a
phytoplankton bloom is $t_C \leq 1.6$ d (dashed line).

generally high nutrient concentrations. They noted that instances
of high production occurred at times when the average irradiance in
the mixed layer was higher than the criterion of Riley (1957).
During a little more than one year, they made 47 measurements of
primary production at 5 depths in the photic layer; t_C values
calculated from their data are plotted in Figure 8 as a function of
measured production. It must be noted that in their paper irradiance values converted from g-cal cm^{-2} min^{-1} into W m^{-2} are too high
by a factor of 10; these must also be multiplied by 0.43 in order
to be transformed into PAR (see Table II). Figure 8 shows that
measured production was low when doubling times t_C exceeded the
criterion of 1.6 d, and that almost all the high production values
did correspond to $t_C \leq 1.6$ d. The wide scatter in production
measurements for $t_C \leq 1.6$ d comes from the fact that production
reflects both photosynthetic activity (P^B) and biomass. The criterion derived from Riley (1957) was thus consistent with observed
phytoplankton blooms, for over more than one year.

3.2. Control of Phytoplankton by Irradiance and Nutrients

In most instances, phytoplankton production is under the control of
both light and nutrients. As explained in Section 2.1, the timing
and depth of production in such a case is set by values t_C and t_S.
An example taken from tropical oligotrophic waters shows how irradiance and nutrients control the depth of the maximum in production
per unit biomass.
    Data were collected in the subtropical North Atlantic Ocean
(Coste, 1977). Figure 9a gives the observed vertical profiles of
density, nitrate, chlorophyll a, production and P^B. In the
sampling area, 0.5 S_{max} corresponded to 1025.86 kg m^{-3} (Figure 10);
at the station under study, this isopycnal was at $z_S = 42$ m.
Computed average NO_3 [<S>] during period t_S fit well the actual
measurements in the upper water column, even if no attempt was made
to adjust S_{max} (27 mmol m^{-3} in the general area, Figure 10) to the
locally observed concentration (25 mmol m^{-3}). Figure 9b gives the
results of a simulation with the same values (see GUIDOME in Table
III). In Figure 9a, t_{prod} was plotted as t_{prod}^{-1} to facilitate
comparison with measured P^B. There is good agreement between the
depths of maximum t_{prod}^{-1} and P^B. Below the maximum, there is a
factor $\approx$ 60 C Chl^{-1} between the values of t_{prod} and P^B, which is
similar to C:Chl ratios generally observed at sea; above the
maximum on the other hand, the difference between calculated and
observed values may be attributable to either increased C:Chl or

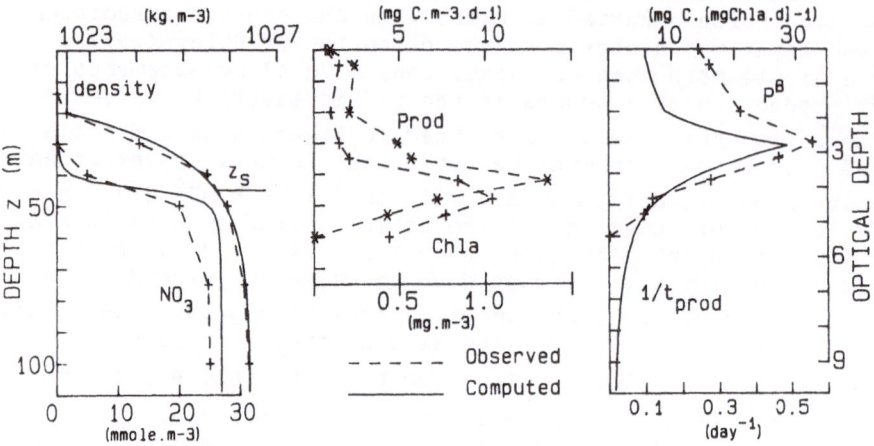

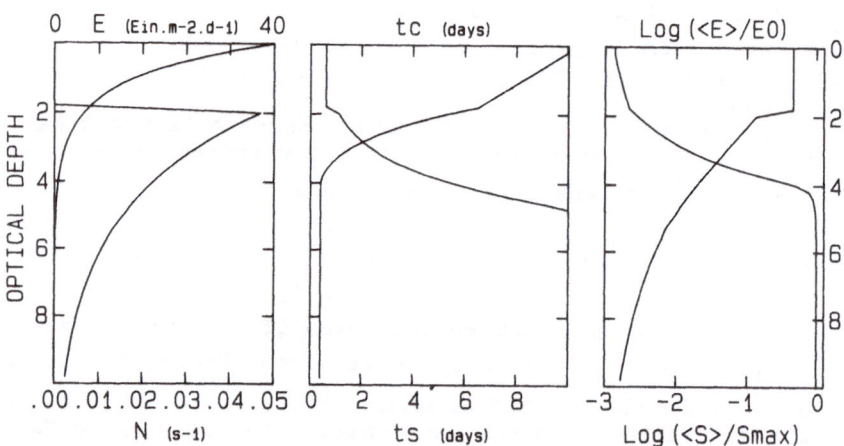

Figure 9.   Station GUIDOME-43 (22°10.3' W, 13°29.8' N), 4 October 1976. [a] Vertical profiles of density and nitrate (observed and computed), observed production, chlorophyll *a* and production per unit chlorophyll *a*, and inverse of computed minimum doubling time t_{prod}.   [b] Simulation (see Figures 5 and 6) conducted with observed values; parameters and results given in Table III.

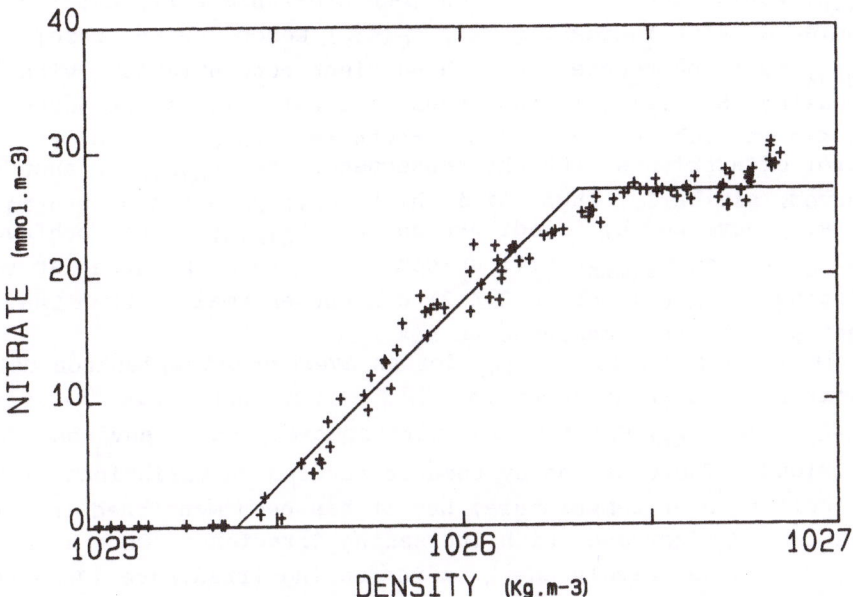

Figure 10.   Concentration of nitrate as a function of water density. Data collected in the general area of station GUIDOME-43 (Figure 9), from 5 to 7 October 1976 (Coste, 1977).

regenerated production. Overall, potential new production t_{prod} can account for most of the P^B maximum at this station, allowing for possible light adaptation and/or regenerated production near the surface.

4.   DISCUSSION

4.1. Control of Phytoplankton Growth by Irradiance and Nutrients

As explained in Section 2.1, it was assumed in the present paper that the control of phytoplankton growth by irradiance and nitrate is alternative, and not multiplicative.   This assumption made possible independent computation of t_C and t_S, which is to say that potential phytoplankton growth was considered to be limited by a single factor at a time. In addition, it follows from (4b) and (11) that t_C and t_S are the sums of two positive terms:

$$t_C = t_{C(min)} \ [1 + f_C(<E>)], \text{ where } f_C(<E>) = K_E/<E>$$
$$t_S = t_{S(min)} \ [1 + f_S(<S>)], \text{ where } f_S(<S>) \propto K_S/<S>$$

Minimum doubling times $t_{C(min)}$ and $t_{S(min)}$ are not directly governed by environmental irradiance or nutrients. In theory,

$t_{C(min)}$ would occur at very high (but non-inhibitory) average irradiance [with increasing $<E>$, $f_C(<E>)$ tends towards zero], and $t_{S(min)}$ would be reached at high nutrient concentration [with increasing $<S>$, $f_S(<S>)$ tends towards zero]. In nature however, when $<E>$ is high, phytoplankton growth is generally under the control of nutrients with the consequence that $t_{C(min)}$ cannot be observed; similarly, when $<S>$ is high, phytoplankton growth is generally governed by irradiance so that $t_{S(min)}$ is not achieved. Thus $t_{C(min)}$ or $t_{S(min)}$ will seldom be, if ever, measured at sea; accordingly, V_{max} (Section 2.4.2) was chosen smaller than the lowest growth rates measured at sea.

In Section 2.3.2, $t_{C(min)}$ for an average phytoplankton cell was taken as constant (equation 4b'). It is not known, however, if $t_{C(min)} = [\sigma_C \phi_{max} K_E]^{-1}$ varies with species and/or environmental conditions. There are no systematic studies on variations of σ_C with cell size or temperature, but it has been mentioned in Section 2.3.2 that σ_C decreases with increasing irradiance. On the other hand, K_E can positively adapt to increasing irradiance (Jørgensen, 1969) and temperature (Jewson, 1976), but its response to cell size is not documented. It is therefore possible that $[\sigma_C \phi_{max} K_E]$ stays relatively constant, except under transient irradiance or temperature conditions. This could be tested in the laboratory, since $[\sigma_C \phi_{max} K_E]$ is the maximum photosynthetic rate normalized to carbon P^C_{max}; estimates could also be derived from field measurements when phytoplankton samples are uncontaminated by detritus or other organisms, for example during blooms or after sorting the cells by flow cytometry.

Concerning $f_C(<E>) = [K_E/<E>]$ and $f_S(S) \propto [K_S/<S>]$, one must distinguish between short-term and long-term responses to environmental changes. On the short term K_E and K_S are fixed, so that lower $<E>$ results in longer t_C and similarly lower $<S>$ leads to longer t_S. On the long term, however, light/shade adaptation decreases K_E at low irradiance, thus shortening t_C (for $t_{C(min)}$ constant). The same is true for t_S, since K_S is generally lower in oligotrophic than in eutrophic waters (Section 2.4.2). Adaptive mechanisms may therefore result in maintaining doubling times t_C and t_S as short as possible.

In the multiplicative approach (*e.g.* Jamart *et al.*, 1977; Lewis *et al.*, 1986), phytoplankton is assumed to be simultaneously governed by irradiance and nutrients. This hypothesis fits the observation that nutrient uptake becomes independent from irradiance only at high light intensity (*e.g.* MacIsaac and Dugdale, 1972). The alternation hypothesis accounts for the same observation by shifting the control of phytoplankton from irradiance to nutrients as light intensity increases.

4.2. Interactions among Environmental and Biological Variables

In the above developments, boundary conditions were set independently. In the oceans however, average solar irradiance $<E(0)>$ and the depth z_S where $S = 0.5 \, S_{max}$ may be coupled over large time and space scales. For example, the vernal increase in solar irradiance simultaneously triggers a bloom (Section 3.1.1), which exhausts nutrients in the surface mixed layer, and increases the density stratification; this results in deepening z_S. As a consequence, $<E(0)>$ and z_S will be correlated at the seasonal scale (e.g. Wolf and Woods, this book). Since t_C and t_S are strongly influenced by respectively $<E(0)>$ and z_S, their seasonal trends will be somewhat coupled. If z_S remains associated with the same isopycnal throughout the year (Section 2.4.2), models that describe changes in vertical stratification as a function of solar irradiance (e.g. Gill and Turner, 1976; Jamart et al., 1977) can be used to describe seasonal changes in both $t_C(z)$ and $t_S(z)$. Similar approaches could also be used for spatial distributions.

A major assumption, up to this point, has been that environmental conditions were not affected, on the short term, by phytoplankton production. This is to say that changes in biomass during periods t_C and t_S were supposed to be small. On the other hand, as long as the computed doubling time remains shorter than t_{crit}, the biomass can double indefinitely. A stationary regime thus implies constant biomass, and therefore a balance between production and losses (e.g. grazing and sinking, which are not included in the present study). Loss terms, which are needed to close models, require additional assumptions and parameters; by contrast, criteria can be computed from environmental variables only.

Another effect of biomass may be its influence on the downwards attenuation of light and the upwards diffusion of nutrients. High concentrations of phytoplankton near the surface may reduce the availability of photons underneath (e.g. Herman and Platt, 1986); similarly, the presence of a strong subsurface maximum may reduce the availability of nutrients above (e.g. Jamart et al., 1977). This suggests that coincident subsurface maxima in chlorophyll and production per unit biomass may be the site of positive feedback, since the very presence of such a maximum would enhance the effects of light limitation below and nutrient limitation above, thus narrowly restricting the depth $z_{C=S}$. The effect of a strong maximum could be incorporated in the calculation of t_C and t_S by decreasing $<E(0)>$ and S_{max}.

4.3. Phytoplankton Production at Marine Ergoclines

Legendre *et al.* (1986) did observe that instances of high biological production, including temporal blooms and subsurface maxima, often occur at spatio-temporal transitions in the water column. They suggested that these transitions (called 'ergoclines' by Legendre and Demers, 1985) have the common characteristic of involving spatial and/or temporal gradients where physical processes can produce structures associated with enhanced biological production. They proposed that this occurs as the consequence of the matching or resonance of physical scales with biological scales.

It has been shown in Section 2 that minimum potential doubling time of phytoplankton (t_{prod}) is the longest of t_C or t_S. The values of t_C and t_S depend on $t_{C(min)}$ and $t_{S(min)}$, and also on f_C and f_S (Section 4.1). These functions can be written in such a way as to separate boundary conditions from transfer functions:

$$f_C = [K_E/<E(0)>] \; / \; T_C(\tau, \, t_m, \, t_{om}, \, t_C)$$
$$f_S = [K_S/S_{max}] \; / \; T_S(\tau_S, \, t_{sm}, \, t_{om}, \, t_S)$$

The adaptive roles of K_E and K_S have been discussed in Section 4.1. T_C and T_S are the transfer functions, and their values in the simulations are given in the right-hand panels of Figures 5, 6 and 9b [$<E>/<E(0)>$ and $<S>/<S_{max}>$].

During the spring bloom, $T_C = f(t_m, \, t_{om})$ in the mixed layer, so that:

$$t_C = t_{C(min)} \; \{1 \; + \; [K_E/<E(0)>] \; / \; T_C(t_m, \, t_{om})\}$$

In blooms that follow the replenishment of limiting nutrients in the surface mixed layer (e.g. summer or autumn blooms), $T_S = f(\tau_S, \, t_{sm}, \, t_{om}, \, t_S)$ and:

$$t_S = t_{S(min)} \; \{1 \; + \; [K_S/S_{max}] \; / \; T_S(\tau_S, \, t_{sm}, \, t_{om}, \, t_S)\}$$

Finally, at the depth $z_{C=S}$ of the subsurface maximum in production per unit biomass, $t_C = t_S$. When $t_{C(min)}$ and $t_{S(min)}$ have the same numerical value, as in the present paper, then:

$$[K_E/<E(0)>] \; / \; T_C(\tau, \, t_m, \, t_{om}, \, t_C) =$$
$$[K_S/S_{max}] \; / \; T_S(\tau_S, \, t_{sm}, \, t_{om}, \, t_S)$$

Given K_E, K_S, $<E(0)>$, S_{max}, τ and τ_S, the last three equations show that the growth rate of phytoplankton (t_C^{-1}, t_S^{-1}) at ergoclines is governed by physical rates (t_m^{-1}, t_{om}^{-1} and t_{sm}^{-1}).

This indicates that the hypothesis of Legendre *et al.* (1986) concerning biological production at aquatic ergoclines is not entirely correct, since rates and not scales in general are involved. The ergocline hypothesis should therefore be reformulated as:

> Enhanced biological production occurs at ergoclines as the consequence of the matching or resonance of physical rates with biological rates.

Resonance refers here to the fact that enhanced biological production can lead to development of a biomass maximum in the water column, which can in turn narrowly restrict the depth of the maximum (Section 4.2). The effects of light limitation below and nutrient limitation above the maximum can result in a positive feedback between physical rates and biological rates.

5. CONCLUSIONS

In this paper, potential maximum production per unit biomass was estimated as the shortest time t_{prod} required by an average phytoplankton cell to double its biomass, given the local availability of photons and nitrate. Values of $t_{prod}(z)$ depend on physical times, that are characteristic of environmental transfers, on physiological times $t_{C(min)}$ and $t_{S(min)}$, which may be constant, and on adaptive characteristics of the cells. Calculation of t_{prod} was based on boundary conditions $<E(0)>$ and S_{max}, on transfer functions T_C and T_S, and on parameters K_E and K_S of physiological adaptation. Data required for computing vertical profiles of t_{prod} are either derived from the literature or measured at sea; in most cases, K_E, K_S, ϕ_{max}, V_{max}, σ_C and ε will come from the literature, while $<E(0)>$, S_{max}, χ, τ_S or z_S, and $N(z)$ or $\rho(z)$ will be measured.

Criteria for new phytoplankton production per unit biomass have been derived from empirical evidence. Significant population growth does not seem to take place when t_{prod} exceeds 5 days. Furthermore, a spring bloom probably occurs only when $t_S < t_C \leq 1.6$ d; similarly, summer and autumn blooms likely correspond to $t_C < t_S \leq 1.6$ d. The spring bloom may be caused by increasing surface irradiance $<E(0)>$, or by decreasing optical depth of the surface mixed layer $\tau_m = (t_m/t_{om})^{0.5}$; the summer and autumn blooms, on the other hand, would occur as τ_S approaches τ_m, or when t_{om} decreases as the consequence of weaker stratification or increased turbulent kinetic energy (resulting for instance from wind events or heat losses). In cases when $t_C = t_S \leq 1.6$ d, a subsurface bloom could develop.

6. ACKNOWLEDGMENTS

Contribution to the programs of CNRS UA 353 and of the Groupe interuniversitaire de recherches océanographiques du Québec (GIROQ). The paper was written in Villefranche-sur-Mer, during a France-Canada scientific exchange. A grant from the Natural Sciences and Engineering Research Council of Canada to L. L. was instrumental in the completion of the work. The authors wish to thank Mr. M. Levasseur for access to unpublished data, as well as Drs. A. Bricaud, Y. Dandonneau, R. C. Dugdale, J. Gostan, D. A. Kiefer, A. Morel and F. P. Wilkerson for their suggestions.

7. REFERENCES

Bannister, T.T., and A.D. Weidemann: 1984, 'The maximum quantum yield of phytoplankton photosynthesis in situ' *J. Plankton Res.* **6**, 275-294.

Blasco, D., T. T. Packard, and P.C. Garfield: 1982, 'Size dependence of growth rate, respiratory electron transport system activity, and chemical composition in marine diatoms in the laboratory' *J. Phycol.* **18**, 58-63.

Collos, Y.: 1982, 'Transient situations in nitrate assimilation by marine diatoms. 2. Changes in nitrate and nitrite following a nitrate perturbation' *Limnol. Oceanogr.* **27**, 528-535.

Coste, B.: 1977, 'Résultats de la campagne GUIDOME (18 septembre - 13 octobre 1976)' Fasc. 1 *Résultats de campagnes à la mer* (13), CNEXO, Brest.

Cullen, J.J.: 1982, 'The deep chlorophyll maximum: comparing vertical profiles of chlorophylla ' *Can. J. Fish. aquat. Sci.* **39**, 791-803.

de Lafontaine, Y., and R.H. Peters: 1986, 'Empirical relationship for marine primary production: the effect of environmental variables' *Oceanol. Acta* **9**, 65-72.

Denmann, K.L., and A.E. Gargett: 1983, 'Time and space scales of vertical mixing and advection of phytoplankton in the upper ocean' *Limnol. Oceanogr.* **28**, 801-815.

Droop, M.R.: 1973, 'Some thoughts on nutrient limitation in algae' *J. Phycol.* **9**, 254-272.

Dugdale, R.C.: 1967, 'Nutrient limitation in the sea: dynamics, identification, and significance' *Limnol. Oceanogr.* **12**, 685-695.

Dugdale, R.C., and J.J. Goering: 1967, 'Uptake of new and regenerated forms of nitrogen in primary productivity' *Limnol. Oceanogr.* **12**, 196-206.

Dugdale, R.C., and J.J. MacIsaac: 1971, 'A computation model for the uptake of nitrate in the Peru upwelling region' *Inv. Pesq.* **35**, 299-308.

Dugdale, R.C., and F.P. Wilkerson: 1986, 'The use of ^{15}N to measure nitrogen uptake in eutrophic oceans; experimental considerations' *Limnol. Oceanogr.* **31**, 673-689.

Eppley, R.W.: 1981, 'Relations between nutrient assimilation and growth in phytoplankton with a brief review of estimates of growth rate in the ocean' In: T. Platt (Ed.) *Physiological bases of phytoplankton ecology*, Can. Bull. Fish. aquat. Sci. **210**, 251-263.

Eppley, R.W., and J.L. Coatsworth: 1968, 'Nitrate and nitrite uptake by *Ditylum brightwellii*. Kinetics and mechanisms' *J. Phycol.* **4**, 151-156.

Eppley, R.W., and B.J. Peterson: 1979, 'Particulate organic matter flux and planktonic new production in the deep ocean' *Nature (London)* **282**, 677-680.

Eppley, R.W., J.H. Sharp, E.H. Renger, M.J. Perry, and W.G. Harrison: 1977, 'Nitrogen assimilation by phytoplankton and other micro-organisms in the surface waters of the Central North Pacific Ocean' *Mar. Biol.* **39**, 111-120.

Falkowski, P.G., Z. Dubinsky, and K. Wyman: 1985, 'Growth-irradiance relationships in phytoplankton' *Limnol. Oceanogr.* **30**, 311-321.

Fortier, L., and L. Legendre: 1979, 'Le contrôle de la variabilité à court terme du phytoplancton estuarien: stabilité verticale et profondeur critique' *J. Fish. Res. Board Can.* **36**, 1325-1335.

Fréchette, M., and L. Legendre: 1982, 'Phytoplankton photosynthetic response to light in an internal tide dominated environment' *Estuaries* **5**, 287-293.

Gargett, A.E.: 1984, 'Vertical eddy diffusivity in the ocean interior' *J. mar. Res.* **42**, 359-393.

Gill, A.E., and J.S. Turner: 1976, 'A comparison of seasonal thermocline models with observations' *Deep-Sea Res.* **23**, 391-401.

Goldman, J.C., and P.M. Glibert: 1983, 'Kinetics of inorganic nitrogen uptake by phytoplankton' In: *Nitrogen in the marine environment*, Academic Press, New York, 233-274.

Goldman, J.C., J.J. McCarthy, and D.G. Peavey: 1979, 'Growth rate influence on the chemical composition of phytoplankton in oceanic waters' *Nature (London)* **279**, 210-215.

Gordon, H.R., and A.W. Wouters: 1978, 'Some relationships between Secchi depth and inherent optical properties of natural waters' *Appl. Opt.* **17**, 3341-3343.

Harrison, W.G.: 1983, 'Nitrogen in the marine environment: use of isotopes' In: E.J. Carpenter, and D.G. Capone (Eds.) *Nitrogen in the marine environment*, Academic Press, New York, 763-807.

Herbland, A., and B. Voituriez: 1979, 'Hydrological structure analysis for estimating the primary production in the tropical Atlantic' *J. mar. Res.* **37**, 87-101.

Herman, A.W., and T. Platt: 1986 'Primary production profiles in the ocean: estimation from a chlorophyll/light model' *Oceanol. Acta* **9**, 31-40.

110

Holligan, P.M., W.M. Balch, And C.M. Yentsch: 1984, 'The significance of subsurface chlorophyll, nitrite and ammonium maxima in relation to nitrogen for phytoplankton growth in stratified waters of Gulf of Maine' *J. mar. Res.* **42**, 1051-1073.

Holloway, G.: 1984, 'Effects of velocity fluctuations on vertical distributions of phytoplankton' *J. mar. Res.* **42**, 559-571.

Holmes, R.W.: 1970, 'The Secchi disk in turbid coastal waters' *Limnol. Oceanogr.* **15**, 688-694.

Højerslev, N.K.: 1986, 'Visibility of the sea with special reference to the Secchi disk' *Soc. photo-opt. Instrument. Engin.* **637** *Ocean Optics* **VIII**, 294-305.

Ivanoff, A.:1975, *Introduction à l'océanographie*, II, Vuibert, Paris.

Jamart, B.M., D.F. Winter, K. Banse, G.C. Andreson, and R.K. Lam: 1977, 'A theoretical study of phytoplankton growth and nutrient distribution in the Pacific Ocean off the northwestern U.S. coast' *Deep-Sea Res.* **24**, 753-773.

Jerlov, N.G.: 1974, 'A simple method for measuring quanta irradiance in the ocean' Kobenhavns Universitet, Inst. Fysik Oceanografi, Rep. 24.

Jewson, D.H.: 1976, 'The interactions of components controlling net phytoplankton photosynthesis in a well-mixed lake (Lough Neagh, Northern Ireland)' *Freshwater Biol.* **6**, 551-576.

Jitts, H.R., A. Morel, and Y. Saijo: 1976, 'The relation of oceanic primary production to available photosynthetic irradiance' *Aust. J. mar. Freshwater Res* . **27**, 441-454.

Joiris, C., G. Billen, C. Lancelot, H.M. Daro, J.P. Mommaerts, A. Bertels, M. Bossicart, J. Nijs, and J.H. Hecq: 1982, 'A budget of carbon cycling in the Belgian coastal zone: relative roles of zooplankton, bacterioplankton and benthos in the utilization of primary production' *Neth. J. Sea Res.* **16**, 260-275.

Jørgensen, E.G.: 1969, 'The adaptation of plankton algae. IV. Light adaptation in different algal species' *Physiol. Plant.* **22**, 1307-1315.

Kiefer, D.A., and J.N. Kremer: 1981, 'Origins of vertical patterns of phytoplankton and nutrients in the temperate, open ocean: a stratigraphic hypothesis' *Deep-Sea Res.* **28**, 1087-1105.

Kiefer, D.A., and B.G. Mitchell: 1983, 'A simple, steady state description of phytoplankton growth based on absorption cross section and quantum efficiency' *Limnol. Oceanogr.* **28**, 770-776.

Kirk, J.T.O.: 1983, *Light and Photosynthesis in Aquatic Ecosystems*, Cambridge Univ. Press, Cambridge.

Landau, L., and E. Lifchitz: 1971, *Mécanique des fluides*, Editions MIR, Moscou.

Lande, R, and A.M. Wood: 1987, 'Suspension times of particles in the upper ocean' *Deep Sea Res.* **34**, 61-72.

Larsson, U., and Å. Hagström: 1979, 'Phytoplankton exudate release as an energy source for the growth of pelagic bacteria' *Mar. Biol.* **52**, 199-206.

Larsson, U., and Å. Hagström: 1982, 'Fractionated phytoplankton primary production, exudate release and bacterial production in a Baltic eutrophication gradient' *Mar. Biol.* **67**, 57-70.

Legendre, L.: 1981, 'Hydrodynamic control of marine phytoplankton
    production: the paradox of stability' In: J.C.J. Nihoul (Ed.)
    Ecohydrodynamics, Elsevier, Amsterdam, 191-207.
Legendre, L., and S. Demers: 1984, 'Towards dynamic biological
    oceanography and limnology' *Can. J. Fish. aquat. Sci.* **41**, 2-19.
Legendre, L., and S. Demers: 1985, 'Auxiliary energy, ergoclines and
    aquatic biological production' *Naturaliste can.* **112**, 5-14.
Legendre, L., S. Demers, and D. Lefaivre: 1986, 'Biological
    production at marine ergoclines' In: J.C.J. Nihoul (Ed.) *Marine
    interfaces ecohydrodynamics*, Elsevier, Amsterdam, 1-29.
Legendre, L., R.G. Ingram, and Y. Simard: 1982, 'Aperiodic changes of
    water column stability and phytoplankton in an Arctic coastal
    embayment, Manitounuk Sound, Hudson Bay' *Naturaliste can.* **109**,
    775-786.
Levasseur, M., J.C. Therriault, and L. Legendre: 1984, 'Hierarchical
    control of phytoplankton succession by physical factors' *Mar.
    Ecol. Prog. Ser.* **19**, 211-222.
Lewis, M.R., W. G. Harrison, N.S. Oakey, D. Hebert, and T. Platt:
    1986, 'Vertical nitrate fluxes in the oligotrophic ocean' *Science*
    234, 870-872.
Lewis, M.R., E.P.W. Horne, J.J. Cullen, N.S. Oakey, N.S., and T.
    Platt: 1984, 'Turbulent motions may control phytoplankton
    photosynthesis in the upper ocean' *Nature (London)* **311**, 49-50.
MacIsaac, J.J., and R.C. Dugdale: 1972, 'Interaction of light and
    inorganic nitrogen in controlling nitrogen uptake in the sea'
    Deep-Sea Res. **19**, 209-323.
MacIsaac, J.J., R.C. Dugdale, R.T. Barber, D. Blasco, and T.T.
    Packard: 1985, 'Primary production cycle in an upwelling center'
    Deep-Sea Res. **32**, 503-529.
Monod, J.: 1942, *Recherches sur la croissance des cultures
    bactériennes*, 2nd ed. Hermann, Paris.
Morel, A.:1978, 'Available, usable, and stored radiant energy in
    relation to marine photosynthesis' *Deep-Sea Res.* **25**, 673-688.
Morel, A.:1987, 'Chlorophyll-specific scattering coefficient of
    phytoplankton. A simplified theoretical approach' *Deep-Sea Res.*
    (In press).
Morel, A., L. Lazzara, and J. Gostan: 1987, 'Growth rate and quantum
    yield time response for a diatom to changing irra-diances (energy
    and colour)' *Limnol. Oceanogr.* **32**, 1066-1084.
Morel, A., and L. Prieur: 1977, 'Analysis of variations in ocean
    color' *Limnol. Oceanogr.* **22**, 702-722.
Morel, A., and R.C.S. Smith: 1974, 'Relation between total quanta and
    total energy for aquatic photosynthesis' *Limnol. Oceanogr.* **19**,
    591-600.
Mullin, M.M., P.R. Sloan, R.W. Eppley: 1966, 'Relationship between
    carbon content, cell volume, and area in phytoplankton' *Limnol.
    Oceanogr.* **11**, 307-311.
Parsons, T.R., and M. Takahashi: 1973, *Biological Oceanographic
    Processes*, Pergamon Press, Oxford.
Poole, H.H., and W.R.G. Atkins: 1929, 'Photo-electric measurements of
    submarine illumination throughout the year' *J. mar. biol. Ass.
    U. K.* **16**, 297-324.

Preisendorfer, R.W.: 1986, 'Secchi disk science: visual optics of natural waters' *Limnol. Oceanogr.* 31, 909-926.

Prieur, L., and S. Sathyendranath: 1981, 'An optical classification of coastal and oceanic waters based on the specific spectral absorption curves of phytoplankton pigments, dissolved organic matter, and other particulate materials' *Limnol. Oceanogr.* 26, 671-689.

Radmer, R., and B. Kok: 1977, 'Photosynthesis: limited yields, unlimited dreams' *BioScience* 27, 599-605.

Raimbault, P.: 1986, 'Effect of temperature on nitrite excretion by three marine diatoms during nitrate uptake' *Mar.Biol.* 92, 149-155.

Richardson, K., J. Beardall, and J.A. Raven: 1983, 'Adaptation of unicellular algae to irradiance: an analysis of strategies' *New Phytol.* 93, 157-191.

Riley, G.A.: 1942, 'The relationship of vertical turbulence and spring diatom flowerings' *J. mar. Res.* 5, 67-87.

Riley, G.A.: 1957, 'Phytoplankton of the north central Sargasso Sea' *Limnol. Oceanogr.* 2, 252-270.

Sathyendranath, S., L. Lazzara, and L. Prieur: 1987, 'Variations in the spectral values of specific absorption of phytoplankton' *Limnol. Oceanogr.* 32, 403-415.

Sharp, J.H.: 1977, 'Excretion of organic matter by marine phyto-plankton: do healthy cells do it?' *Limnol. Oceanogr.* 22, 381-399.

Simonot, J.Y, and H. Le Treut: 1986, 'A climatological field of mean optical properties of the world ocean' *J. geophys. Res.* 91, 6642-6646.

Simpson, J.H., and J.R. Hunter: 1974, 'Fronts in the Irish Sea' *Nature (London)* 250, 404-406.

Smith, R.C., and K.S. Baker: 1978, 'The biooptical state of ocean waters and remote sensing' *Limnol. Oceanogr.* 23, 247-259.

Strathmann, R.R.: 1967, 'Estimating the organic carbon content of phytoplankton from cell volume or plasma volume' *Limnol. Oceanogr.* 12, 411-418.

Sverdrup, H.U.: 1953, 'On conditions for the vernal blooming of phytoplankton' *J. Cons. perm. int. Explor. Mer* 18, 287-295.

Sverdrup, H.U., M.W. Johnson, and R.,H. Fleming: 1942, *The Oceans. Their physics, Chemistry and General Biology,* Prentice Hall.

Takahashi, M., K. J. Kikuchi, and Y. Hara: 1985, 'Importance of picocyanobacteria biomass (unicellular, blue-green algae) in the picoplankton population of the coastal waters off Japan' *Mar. Biol.* 89, 63-69.

Takahashi, M., D.L. Siebert, and W.H. Thomas: 1977, 'Occasional blooms of phytoplankton during summer in Saanich Inlet, B. C., Canada' *Deep-Sea Res.* 24, 775-780.

Tett, P.: 1981, 'Modelling phytoplankton production at shelf-sea fronts' *Phil. Trans. R. Soc. Lond. A* 302, 605-615.

Tyler, J.E.: 1968, 'The Secchi disk' *Limnol. Oceanogr.* 13, 1-6.

Vandevelde, T., L. Legendre, J.C. Therriault, S. Demers, and A. Bah: 1987, 'Subsurface chlorophyll maximum and hydrodynamics of the water column' *J. mar. Res.* 45, 377-396.

HORIZONTAL AND SEASONAL VARIATION OF DENSITY AND CHLOROPHYLL PROFILES BETWEEN THE AZORES AND GREENLAND

V. Strass
Institut fuer Meereskunde an der Universitaet Kiel
Duesternbrooker Weg 20
D-2300 Kiel 1
F.R. Germany

and J.D. Woods
Robert Hooke Institute
Oxford University
England

ABSTRACT. The first results of a multi-year campaign designed to collect a data set suitable for testing theoretical models of the development of the plankton bloom and transition to the oligotrophic regime at mid-latitudes are reported. The data were collected with a towed 'batfish' undulating through the euphotic zone along a section running 2000 km north from the Azores through Ocean Weather Ship "C" and into the cyclonic sub-Arctic gyre. The data were analysed to give profiles of temperature, salinity, density, solar irradiance (at 500 nm) and chlorophyll concentration. They permit the first experimental tests of theoretical relationships between horizontal structure of the physics and biology, on both gyre-scale and mesoscale, during the heating season. It is shown that those results could not have been obtained by satellite remote sensing of ocean colour because the latter underestimates the chlorophyll concentration in the seasonal thermocline. The *in situ* measurements document the following features of the horizontal and vertical structures of the chlorophyll-a concentration, Chl:

(1)   A poleward migration of the near-surface Chl maximum during the heating season. This migration does not keep pace with the propagation of the mixed layer shallowing; rather, it follows the slow propagation of the 12°C-isotherm outcrop.

(2)   A poleward migration of the oligotrophic regime located south of the near-surface Chl maximum. The deep Chl maximum in the oligotrophic regime descends through the seasonal pycnocline during the heating season, and slopes down from north to south.

B. J. Rothschild (ed.), Toward a Theory on Biological-Physical Interactions in the World Ocean, 113–136.
© *1988 by Kluwer Academic Publishers.*

(3) Mesoscale variations of Chl are associated with (a) horizontal
differences of vertical stability in spring, (b) horizontal
gradients of isopycnal spacing and indicators of frontal
upwelling in the oligotrophic summer regime.

1. INTRODUCTION

The research reported in this paper was designed to identify the
physical processes influencing the horizontal variation of mid-
latitude open ocean phytoplankton concentration on scales of
kilometres to megametres. The motivation comes from the need to
incorporate phytoplankton primary production into computer models of
(1) climate change associated with variations of atmospheric carbon
dioxide concentration, and (2) the flow of carbon fixed in
phytoplankton to higher trophic levels.

Raymont (1980) has summarized the little that is known about the
regional variation of the seasonal development of plankton. It has not
yet been modelled and the accumulated data base of chlorophyll
profiles is too sparse to permit mapping. The data that come closest
to what is needed deal only with the chlorophyll in the mixed layer.
They include the monthly mean distributions of phytoplankton and
zooplankton derived from the Hardy plankton recorder towed at a
depth of 10 metres (Colebrook, 1982), and monthly-composite maps
of chlorophyll (Esaias et al., 1986) derived from the Coastal Zone
Colour Scanner (CZCS). The former suggest that the spring plankton
bloom in the open North-East Atlantic occurs in April off the Azores
and in June around OWS "C", and that the oligotrophic regime is
established by June off the Azores and by August at OWS "C". This
suggestion of a poleward migration in the seasonal cycle of primary
production has not yet been confirmed by analysis of the CZCS data.

Given a model of how primary production responds to seasonal
changes in the physical environment we might predict the regional
variation in the chlorophyll from the much better documented seasonal
cycle of the physical environment. Robinson et al. (1979) have
computed the monthly mean distribution of temperature in the top 150
metres of the North Atlantic from bathythermograph and hydrocast data
accumulated in the period 1948 to 1976. The mixed layer depth
diagnosed by Robinson et al. (1979) rises abruptly from May to June
between the Azores and Greenland, which is inconsistent with the idea
of a poleward migration of the spring bloom, if we rely solely on
Sverdrup's (1953) theory. However, the seasonal heat content does
migrate poleward at about the rate of the phytoplankton colour in
Colebrook's climatology. That is seen most clearly in the surface
outcrop of isopynals (Figure 1) computed from the Robinson et al.
(1979) data set by Bauer and Woods (1984) and Stammer and Woods
(1987). Does migration of the bloom depend on temperature-controlled
growth? Our inability to be more precise about the large-scale
regional variation of the seasonal cycle of primary production
reflects the lack of data and indicates the need for more sharply
focussed experiments.

Even less is known about the horizontal variability of primary production at smaller scales. The importance of fronts has been noted, especially in shelf seas (Le Fèvre, 1986; Le Fèvre and Frontier, 1988, this volume). Woods (1988, this volume) explains the local increase of primary production in terms of mesoscale upwelling due to vortex

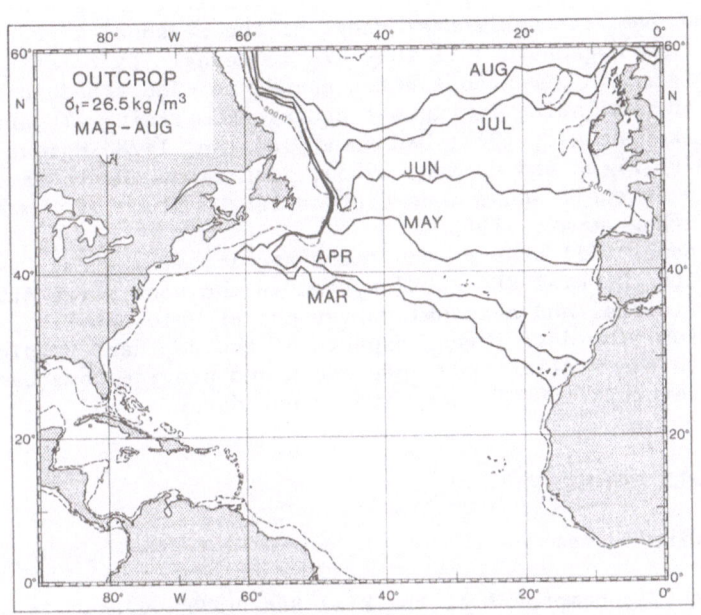

Figure 1. Climatological mean migration of surface outcrop of isopycnal σ_t = 26.5 during the heating season (from Stammer and Woods, 1987).

stretching at fronts, a mechanism which should be equally valid on the shelf and in the open ocean. He uses the physical variable IPVG (Isopycnic Potential Vorticity Gradient), which controls mesoscale jet speed and therefore upwelling rate, to relate small-scale patchiness of primary production to the large-scale circulation. Statistical

theories relating the spectrum of phytoplankton patch sizes to horizontal diffusion have been developed using the Kierstead and Slobodkin (1953) hypothesis (see, for example, Steele, 1974, and Wroblewski et al., 1975).

The goal of collecting a new data set to describe the horizontal variability of phytoplankton concentration as a function of season, and to clarify its relation to the physical environment poses considerable problems for the design of apparatus and experiment. One of the key requirements is that the new data should describe not only the initial bloom, with the chlorophyll maximum in the mixed layer, but also the subsequent oligotrophic phase, with the chlorophyll maximum in the seasonal thermocline, sometimes with a sharp peak which must be resolved (Dandonneau and Lemasson, 1987). Synoptic observations are needed if the data are to reveal the phase relationship predicted by mechanisms that connect physics to biology. High horizontal resolution is needed if the data are to reveal the mesoscale patches of phytoplankton concentration. The measurements must extend over the horizontal range of the seasonal migration of an isopycnal (two megametres) if they are to reveal its relationship to seasonal migration of phytoplankton abundance. The standard solution is to tow an undulating instrument package comprising CTD and optical sensors (Aiken et al., 1977; Denman and Herman, 1978; Herman and Denman, 1979; Platt and Herman, 1983). Such measurements have been made mainly in shelf seas; an exception is the study of the Azores front by Fasham et al., 1985.

This paper will present the first results from a data set obtained with the Kiel SEA ROVER system in the open North Atlantic, showing meridional and seasonal variations of chlorophyll concentration. The data reveal aspects of the dominant physical processes in the mid-latitude open ocean and provide some background for the interpretation of satellite colour data.

2. MATERIALS AND METHODS

2.1. Measuring System

The 'batfish', a component of the Kiel SEA ROVER system, is equipped with two thermometers and two conductivity cells, a pressure gauge, a fluorometer and a flat cosine radiometer sensitive to downwelling irradiance at 500 nm. Technical details of the Kiel version of the 'batfish' are described in Fischer et al. (1985). It typically covered the depth range from 10 m to a maximum depth of nearly 200 m. Within that depth range data were sampled with high vertical and horizontal resolution of about 10 cm and 0.7 km (half 'batfish' wave length), respectively. In addition, water samples pumped from below the ship's hull (4 m) were taken in regular 4-hourly intervals and occasionally with bottle casts. The timing of the regular near-surface samples was synchronized with the path of the fish to ensure that water samples and 'batfish' measurementswere taken from the same volume of water. Two subsamples were filtered from each water sample in parallel through Whatman GF/C glass fibre filters (to extract the plankton) and deep-frozen for the duration of the cruise; two other subsamples were preserved in bottles, and another two kept in bottles for post-cruise recalibration of the conductivity cells. Later in the laboratory the glass fibre filters were ground and dissolved in 90 % acetone, centrifuged and the pigment concentration determined spectrophotometrically (after Strickland and Parsons, 1968). Species composition and PPC (Particulate Phytoplankton Carbon) of the bottled samples were determined by microscopy. Nutrient data were taken during the 1986 and 1987 cruises.

2.2. Data Processing

The data received on board from the 'batfish' were stored on magnetic tape at full resolution. A desk computer was used to monitor the data in real time, using one data cycle every two seconds. Hourly profiles were plotted and stored on floppy disk. Figures 5, 7, 8 and 9 presented later are based on these hourly spot samples.

The magnetic tapes with the data were transferred to the shipboard computer, where salinity was calculated from temperature and conductivity. Salinity and (since 1986) fluorescence were despiked numerically with a median filter, density was calculated and the data were reduced by block-averaging over 5 cycles. In the last stage of on-board processing the 'batfish' data were merged with the navigation data.

Fluorescence was calibrated and salinity recalibrated and all data were interpolated to standard depth (spacing 1 m) and σ_T (spacing .025 kg/m^3) surfaces after each cruise. (Some of the later stages, including fluorescence calibration, had not been completed for the 1986 data set at the time of writing.)

2.3. Fluorescence Calibration

The calibration of the 'batfish' fluorescence signal (*i.e.* the conversion of Fl to Chl) is based on the equally-spaced samples and photometrically determined near-surface chlorophyll concentrations. Variations in time (or space) of fluorescence yield, $R = Fl/Chl$, are large, but are resolved by the time interval of 4 hours between adjacent determinations, as given by the autocorrelation function of R. The autocorrelation function is employed to smooth the variability of yield. This smoothed yield function is used to calibrate the fluorescence signal continuously, in which yield between supporting points is interpolated linearly. Statistically significant variations of R with solar irradiance measured at 500 nm appear to be negative and are eliminated during calibration.

The resultant error of calibrated fluorescence data relative to the spectro-photometrically determined Chl has been estimated to be on average about 10 % of the mean for near-surface values. The variation coefficient (standard error scaled by the mean) of R determined from parallel vertical profiles of fluorescence and Chl taken from bottle casts is less than 30 %; it gives an estimate of the relative calibration error of vertical Chl profiles.

Four large-scale surveys have been carried out at different phases of the heating season in the open North Atlantic. Each survey is designated NOA'nn, where nn denotes the year of the expedition. The timing of the cruises NOA'84, '85 and '86 and the ship tracks are shown in Figure 2. (NOA'87 took place during preparation of the manuscript, too late to be presented here.) The section extending northwards from the Azores passing through OWS "C" and ending at 55° N (SE of Cape Farewell, Greenland) has been measured during each of the cruises. The data from that section will be presented in this paper.

118

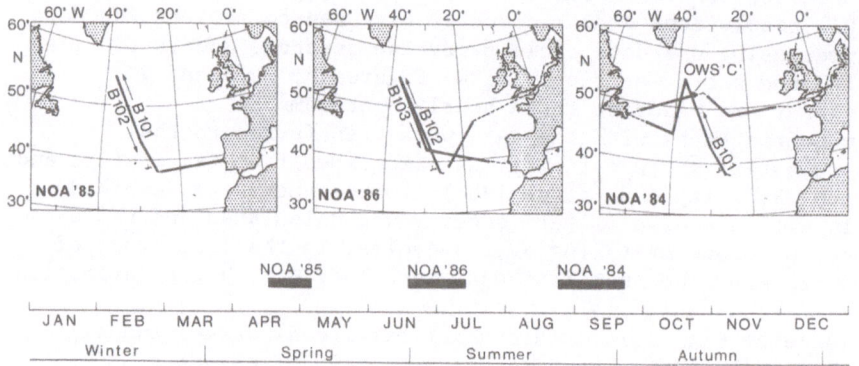

Figure 2.   Locations and seasonal timing of SEA ROVER surveys.

3.   RESULTS

3.1. Hydrographic situation

Figure 3 shows the general hydrographic situation along the standard
section, based on the density variation along selected isotherms.
Regions with small T-S variability are associated with the inner
regions of the sup-tropical anticyclonic gyre south of 43° N (km 800
in Figure 3) and of the sub-arctic cyclonic gyre north of 52°30' N (km
1700 in Figure 3). The North Atlantic current is marked by enhanced T-
S variability between 43° and 52°' N. The sharp T-S fronts and ribbons
of anomalously warm or cold water mark the sites where the section
crosses the various branches of the North Atlantic current (Krauss,
1986). The Polar Front is identified at 52° N (km 1650 in Figure 3) by
a temperature change of 3 K (within the resolved horizontal distance
of 14 km of that presentation) at $\sigma_t = 26.5$ kg m^{-3}.

3.2. Seasonal variation along the Azores-Greenland section

Each transit along the Azores-Greenland section yielded some 3000
profiles of density and chlorophyll concentration. Five transits in
April, June/July and August/September show the early stages of the
spring bloom, its transition to seasonal oligotrophy and document the
poleward migration of these two stages in the seasonal cycle of the
phytoplankton.

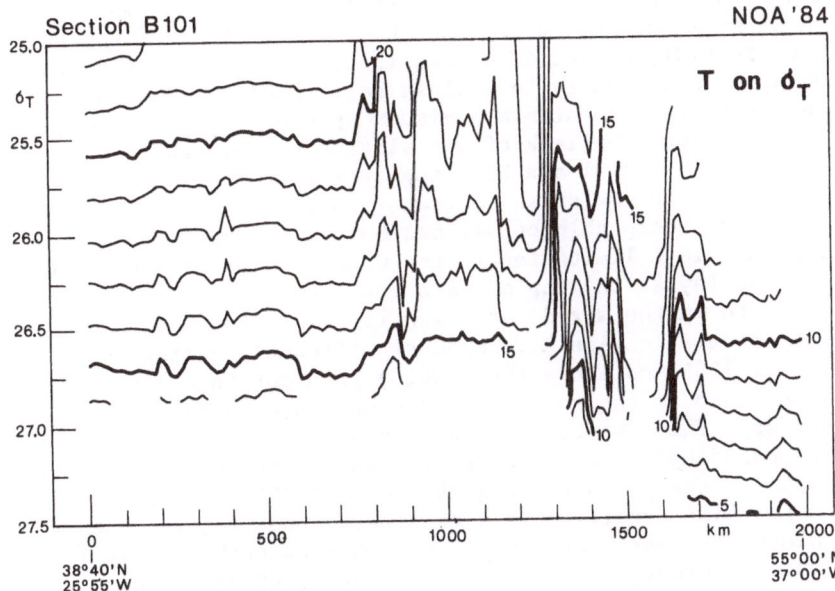

Figure 3. Isotherms plotted versus σ_t and distance along the section in late August (1984). The gross features can be regarded as illustrating the typical hydrographic situation along that quasimeridional transect: a quiet hydrographic region south of km 800, a broad band of high variability ending at the Polar Front (km 1650) and to the north again a quiet region. The data are averaged over 10 'batfish' wave lengths, giving a horizontal resolution of 14 km.

3.2.1. *The start of the spring bloom.* Figure 4 shows density and chlorophyll distributions from April 1985, collected as the ship proceeded north (B101: 18-24 April), then south (B102: 24-29 April) along the section during the start of the spring plankton bloom. South of km 1300, each location was sampled twice with an interval ranging from 3 to 11 days. The initial development of the bloom did not exhibit a steady migration to the north, but was patchy, reflecting the patchy development of the density stratification.

   Figure 4b shows that, in section B101, chlorophyll concentrations of greater than 1 mg/m^3 were encountered only at km 1000-1200, a region of high advection associated with the North Atlantic current. Southward of that early bloom, Chl exceeded 0.5 mg/m^3 only in isolated patches. Figure 4d shows that a few days later near-surface Chl remained less than 0.5 mg/m^3 at only a few locations. The lowest concentrations were found at the ends of the section. Near surface maxima exceeded 1.25 mg/m^3 in four patches distributed along the section. South of km 1300, the average depth of the 0.25 mg/m^3 contour lay at 50 metres in B101, and at 40 metres in B102 (a symptom that the bloom has developed sufficiently to exhibit self-shading); the contour sloped down to the north, reaching a maximum depth at about 100 km south of the Polar Front. The lack of correlation between the distribution of Chl along B101 and B102 suggests that there is comparable variability in the cross-section (along-current) direction. There is evidence of a deep chlorophyll maximum at the southern end of B102, suggesting nutrient limitation in the mixed layer; elsewhere the chlorophyll maximum lay at the top of each measured profile indicating that plankton growth in the mixed layer was light-limited.

The density stratification developed significantly during the period 18-29 April 1985: note how the density gradient in the top 50 metres range increased as the ship proceeded up and down the line (Figures 4a and 4c). The temperature range across the seasonal thermocline increased by 1 K in 10 days. In general terms this observation supports the paradigm (e.g. Raymont, 1980) that the increase in thermocline stability is a necessary condition for the start of the bloom. The horizontal scale of enhanced stability is less than 100 km, which indicates that during this early stage in the heating season the stratification of the seasonal thermocline is influenced as much by the internal motion of the upper ocean (which has that scale) as by increased surface heating (which has the much larger scale of weather systems in the atmosphere). This impact of oceanic motion on early spring stratification may be due to the permanent streams in the large-scale circulation, or to transient eddies and their associated mesoscale motions, or a combination of both. To a first approximation the location and horizontal scale of patches of early bloom follow those of early stratification, but closer examination shows that the correlation is not perfect. So while stratification is a necessary condition it is not sufficient, and cannot be used as the sole predictor of the early occurrence of the bloom.

3.2.2. *The transition to summer oligotrophy*. Figure 5 shows nutrient concentrations and phytoplankton fluorescence in the mixed layer, and the depths of the mixed layer and fluorescence maximum measured as the ship proceeded north along the standard section from 27 June to 2 July 1986. These measurements document the transition to summer oligotrophy, as nutrients run out in the mixed layer and the deep chlorophyll maximum becomes established in the seasonal thermocline.

The transition to oligotrophy occurs patchily, but if one had to choose a single latitude marking the northern boundary of the oligo-trophic regime it would be 46° N. There is still some residual nitrate in the mixed layer at 42°30′ N (Figure 5a) where the fluorescence maximum lies closely below the base of the exceptionally shallow mixed layer, suggesting a mesoscale upwelling event of the type described by Woods (1988, this volume). In all other samples from south of 46° N the mixed layer nutrients are exhausted and the fluorescence maximum lies at least 10 metres below the top of the

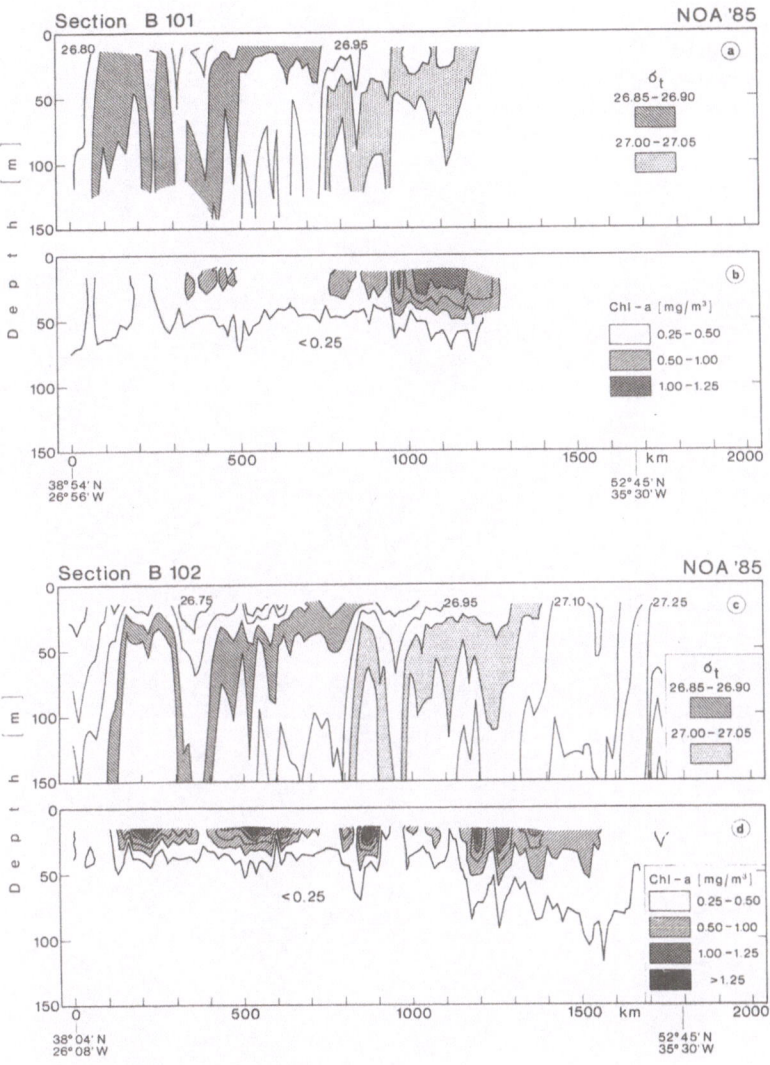

Figure 4. Isopleths of σ_t (a,c) and Chl (b,d) in April 1985. Section B101 heading north recorded from 18 - 24 April (data gap north of km 1300 due to instrument failure), B102 heading south from 24-29 April. Horizontal resolution is 15 km after along-track averaging.

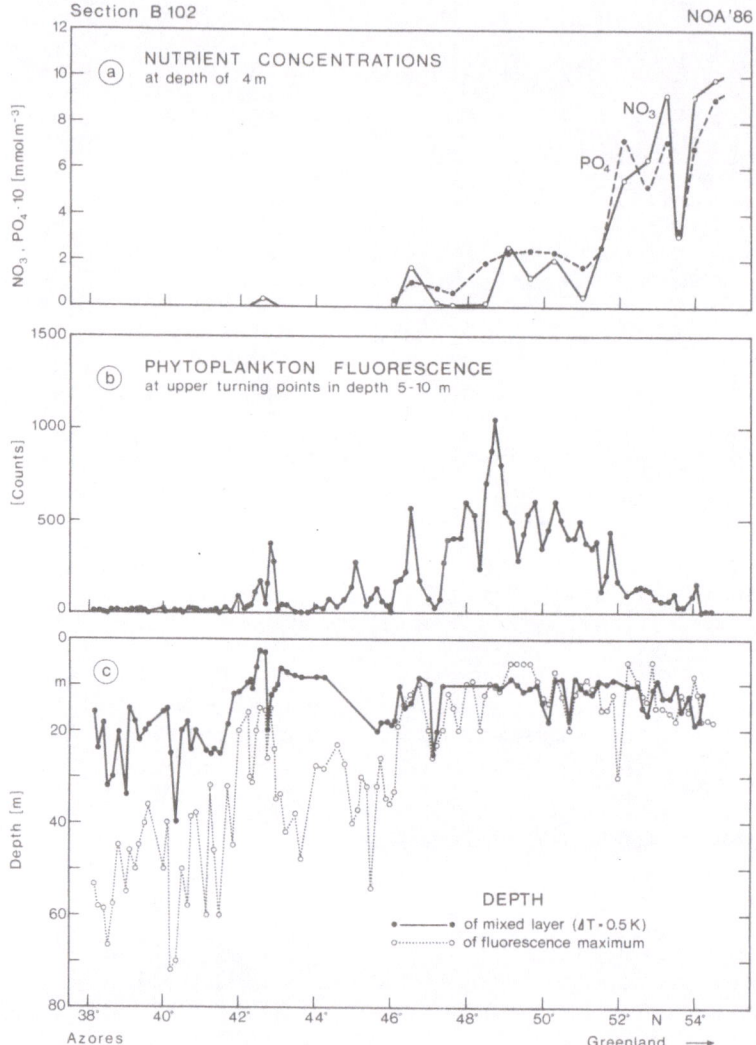

Figure 5.   Near-surface nutrient concentrations (a) and fluorescence (b), and depths of the top of the thermocline and fluorescence maximum (c) along the Azores-Greenland section, from 27 June to 2 July 1986. The maximum value of fluorescence corresponds to about 2 mg/m^3 of chlorophyll-a (the formal calibration will be available too late for inclusion in this paper). The data points represent spot samples from the data set, which will yield higher along track resolution when processing is complete.

seasonal thermocline and slopes downwards from north to south (Figure 5c). All samples north of 46° N showed some nitrate and phosphate in the mixed layer, and a fluorescence maximum above or close to the top of the seasonal thermocline, showing that the light-limited regime still prevailed in the mixed layer. The large-scale variation of mixed layer fluorescence (Figure 5b) is characterized by a broad maximum centred at 49° N. The decrease to the south is associated with nutrient depletion; the decrease to the north occurs abruptly at the Polar Front (52° N), with the sub-arctic gyre containing low chlorophyll and high nutrient concentrations at all depths. This broad variation is modulated by substantial fluctuation of mixed layer fluorescence at the limits of along-track resolution (60 km for nutrients, 15 km for spot samples of fluorescence and depths; the full resolution will be available when the 1986 cruise data are fully analysed).

3.2.3. *The late summer situation.* Figure 6 shows the distribution of density and chlorophyll concentration measured during a northward run along the Azores-Greenland section from 27 August to 2 September 1984; the temperature structure is shown in Figure 3. The vertical chlorophyll maximum lies in the mixed layer north of the Polar Front (km 1650) but to the south of the front it lies in the seasonal thermocline. This deep chlorophyll maximum slopes downwards towards the Azores.

The highest mixed-layer concentrations (up to 2 mg/m^3) were found north of the Polar Front in this section and others sampled in 1984 (see Figure 2). The maximum value lies just north of the Polar Front.

The highest chlorophyll concentrations measured in the oligotrophic regime south of the Polar Front ranged up to 5 mg/m^3.

They occurred around km 1180 and km 1280, in the density range 25.2 $25.2 \leq \sigma_t \leq 26.2$ kg/m^3. It is seen in Figure 6a that the Chl maxima occur where these isopycnals are displaced upwards, on either side of a ribbon of warmer water (compare with Figure 3). The vertical spacing between isopycnals within the warm water ribbon (between km 1150 and km 1300 in Figure 6a) is significantly different from that of the surrounding water, indicating a strong gradient of potential vorticity across the ribbon boundaries. The observation of high Chl values is consistent with the prediction of Wood's (1988) theory of enhanced primary production due to mesoscale upwelling at sites of high isopycnic gradient of potential vorticity.

3.2.4. *Seasonal migration of phytoplankton abundance.* Taken together, the measurements along the Azores-Greenland section in 1984, 1985 and 1986 document the migration of chlorophyll concentration during the heating season. (The present analysis ignores inter-annual variation, see section 4 below.)

124

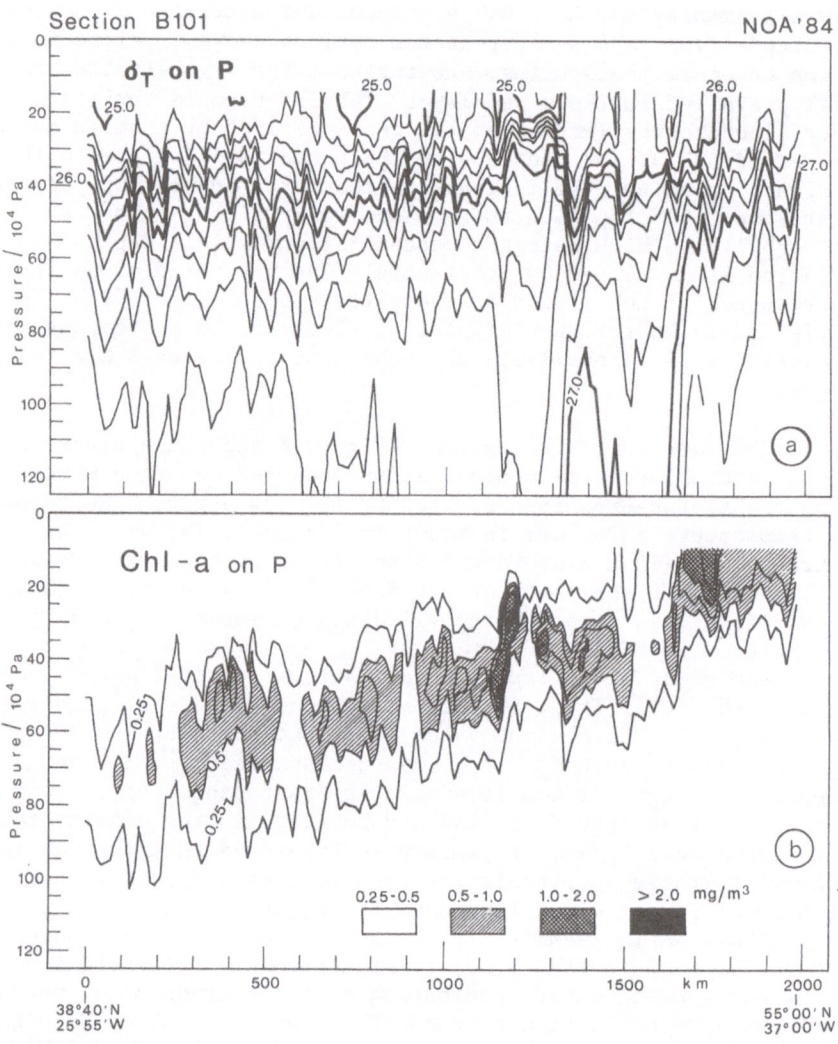

Figure 6. Vertical distribution of σ_t (a) and Chl (b) in late summer (27 Aug. to 2 Sept. 1984). Peak concentrations of chl-a (around km 1180 and km 1280) range up to 5 mg/m³. Horizontal resolution after averaging is 14 km.

Figure 7 shows the distributions of mixed layer chlorophyll
fluorescence along the section in late April, June/July and
August/September. They give the impression of a broad band of
phytoplankton with its peak concentration centred at: 45° N in April,
49° N at the end of June and 53° N at the end of August. Is it correct
to describe this as a poleward migration? If so it is necessary to
consider the migration of two lines, which mark the leading edges of:
(a) the onset of the spring bloom and (b) the transition to
oligotrophy.

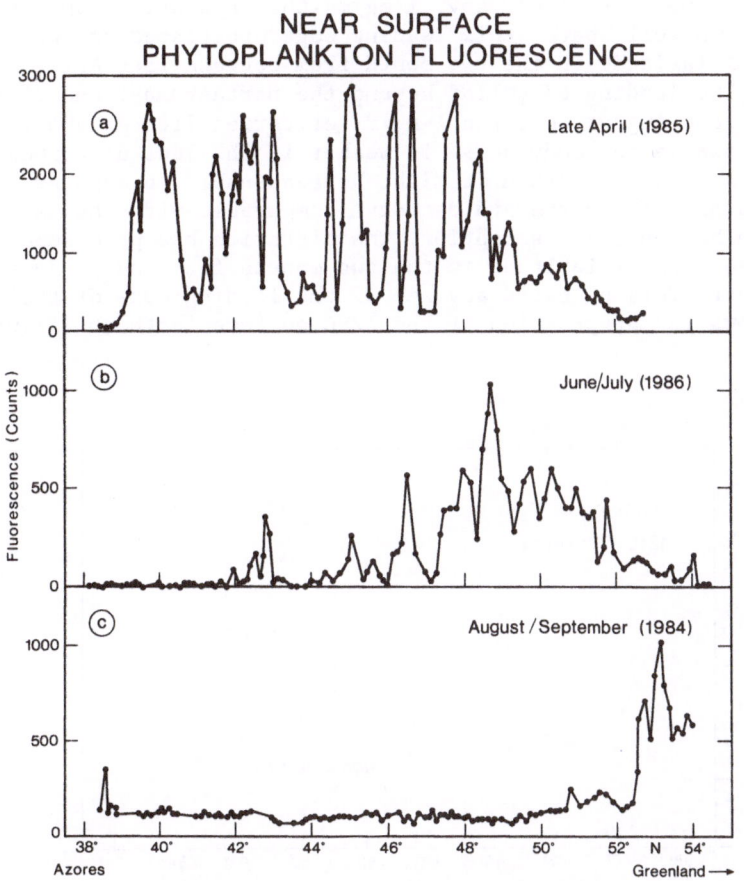

Figure 7. Development of the spatial distribution of chlorophyll
fluorescence in the mixed layer (upper turning point of the 'batfish')
during the heating season. Fluorescence peak values correspond to
approximately 2.0 mg/m^3 Chl in the 1986 data (which had not been
finally calibrated by the deadline for submission of this paper); 1984
and 1985 values have been calibrated, peaks correspond to
approximately 1.5 mg/m^3 Chl. The 1984 calibration had a zero offset of
150 counts, so the mixed layer Chl values are mostly zero south of 52° N.

It was shown earlier (Figure 4) that the spring bloom starts patchily in the anti-cyclonic gyre. There is no poleward migration in this portion of our section (south of 48° N). Along-section variation of fluorescence in the mixed layer (Figure 7a) is dominated by small scales which are not resolved by the resolution of the spot samples used for Figure 7.

There seems to be more resistance to the start of the bloom north of 48° N. The northern edge of the band of high fluorescence in Figures 7a and 7b is marked by an almost linear decrease from over 1 mg/m^3 Chl to near zero extending over about 5 degrees of latitude. Taking a threshold value of, say, 1 mg/m^3 Chl as a criterion for the bloom to be in full spate, this leading edge progressed poleward about 3 degrees of latitude in the two month from between late April and late June. The leading edge lies beyond the northernmost end of the section in late August, but the 1 mg/m^3 criterion lies at about 54° N, and, if we assume the eddy noise is weaker in the less-disturbed water of the sub-Arctic gyre, then it might be reasonable to suppose that it marks the start of a northward decrease comparable with that seen in earlier months. On that assumption, the migration has proceeded a further 3 degrees of latitude in the two months from end of June to end of August. This poleward advance of the leading edge of the spring bloom follows the progression of the 12°C surface isotherm (Figure 8).

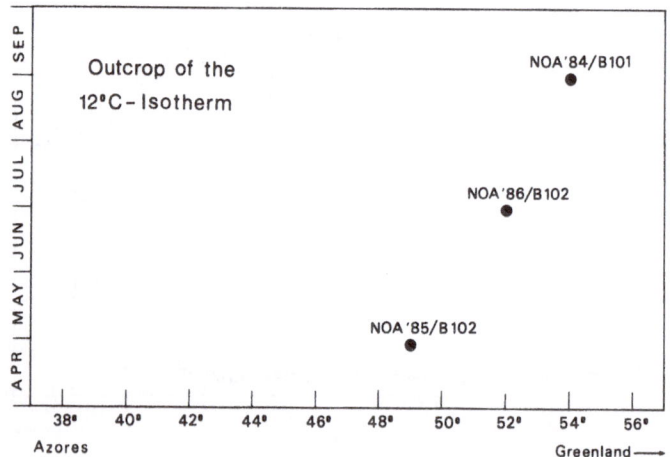

Figure 8.   Seasonal migration of surface outcrop of the 12°C-isotherm, taken from upper 'batfish' turning points.

The transition from bloom to oligotrophy migrates across the whole length of the section, being first encountered off the Azores in late April (Figure 4d) and in the sub-Arctic gyre just north of the

Polar Front in late August (Figure 6b). This transition is defined by a drop of near-surface fluorescence to near zero (or about 150 counts in 1984, when calibration revealed an instrumental offset) associated with the shaping of a deep fluorescence maximum. In late April this transition is found off the Azores (Figures 4d and 7a); in late June it occurs near 46° N, with one interruption of the horizontal extent of the oligotrophic regime at about 42°30' N (Figs. 5 and 7b). By the end of August, the oligotrophic regime extends from the Azores to the Polar Front at 52° N (Figs. 6b and 7c).

3.2.5. *New production in summer*. The deep chlorophyll maximum in the oligotrophic regime descends through the seasonal thermocline at approximately 10 metres per month (Figure 9). According to the model of Wolf and Woods (1988) that represents new production associated with the consumption of "preformed" nutrients brought up into the euphotic zone by deep mixing during the preceding winter.

The synoptic slope of the chlorophyll maximum noted in Figure 6b arose from a combination of the poleward migration of the transition to oligotrophy followed by this slow downward penetration due to new production.

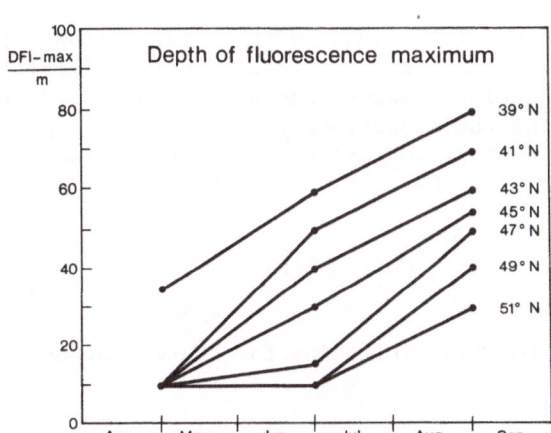

Figure 9. Seasonal and meridional variations of the depth of fluorescence maximum. Depth values are averaged by eye horizontally around given latitude. Maxima within the mixed layer are set to a standard depth of 10 m.

3.3. Vertically-integrated chlorophyll content

The chlorophyll content C_e of the euphotic zone can be estimated by vertical integration of our chlorophyll profiles:

$$C_e = \int_o^{Z_e} \text{Chl}(z) \cdot dz \qquad (1)$$

The limits of integration are set as follows: the upper turning point of the 'batfish' was normally in the mixed layer, so it was assumed that Chl(z) did not vary between it and the sea surface (z = 0). The lower turning point was normally deeper than 100 metres, at which depth Chl was observed to be close to zero (Figures 4b, d and 6b), and which depth lies below even the deepest observed Chl maxima (Figures 6b and 9); so the integration was terminated at z_e = 100 metres, which was designated the bottom of the euphotic zone.

The Azores-Greenland variation of chlorophyll content of the euphotic zone (mg/m^2) is shown for April 1985 in Figure 10b, and for August/September 1984 in Figure 11b. The corresponding mean volume concentrations $\text{Chl}_e = C_e/Z_e$ are shown in Figures 10c and 11c. The along-track mean chlorophyll content in the spring bloom was about 32 mg/m^2, with a standard deviation of 16 mg/m^2, and insignificant meridional variation. The corresponding values in late summer were along-track mean, 27 mg/m^2, standard deviation, 8 mg/m^2. The late summer section has two peaks of approximately 50 mg/m^2 at km 1180 (where the deep chlorophyll maximum is enhanced at indicators of high IPVG), and at km 1700 (where the surface chlorophyll maximum is enhanced just north of the Polar Front).

3.4. What would the CZCS have detected?

The vertical distribution of Chl must be taken into account in estimating chlorophyll content and concentration from satellite (including CZCS) ocean colour images, because the detected irradiance depends on the length of the subsurface path down to the plankton and back. The chlorophyll content (mg/m^2) of the euphotic zone estimated from CZCS (following e.g. Smith, 1981) is given by

$$C_s = \int_o^{Z_e} \text{Chl}(z) \cdot W(z) \cdot dz \qquad (2)$$

W(z) is the weighting function (allowing for two-way attenuation) given by

$$W(z) = \exp(-2k_d \cdot z) \qquad (3)$$

k_d is the diffuse attenuation coefficient of the light in the relevant spectral band. It is possible to compute C_s from our in situ measurements of Chl(z) from the fluorometer and k_d from the solar radiometer, which has a narrow band 500 nm filter providing a good estimate of k_d at 443 and 550 nm, CZCS wavelengths used for chlorophyll detection (Smith and Baker, 1982).

Satellite oceanographers seeking to estimate chlorophyll content from CZCS data do not normally have such in situ profiles to assist in

correcting for subsurface light attenuation. There is little advantage
to be gained from a few sample profiles, because their validity is
limited to within the horizontal correlation scale of a few
kilometres. So it is necessary to make assumptions about the form of
the profiles Chl(z) and W(z) based on climatology and model
predictions. The large number of samples in our data set, and their
spatial and seasonal coverage provide an opportunity to test and
calibrate proposed attenuation correction models derived for example
from Wolf and Woods (1988, this volume).

As an illustration of this application of our data we have
computed the chlorophyll content that would be estimated from CZCS
data using an attenuation model in which k_d is assumed to be
independent of depth, but to vary in time and place. For this purpose
we fitted a single exponential to each solar irradiance profile with
good measurements over a depth range of at least 50 metres. The
results (labelled C_s) are plotted in Figure 10b (for April 1985) and
Figure 11b (August/September 1984). Comparison with the chlorophyll
content measured *in situ* (C_e) shows that only a small fraction (about
12 % in April and about 4 % in August/September of the along-track
mean) is remotely sensed by satellite.

Scaling C_s with the effective satellite visibility depth

$$Z_s = \int_0^{Z_e} W(z) \cdot dz \qquad (4)$$

yields (*e.g.* Smith, 1981) the remotely sensed mean volume chlorophyll
concentration

$$\overline{Chl}_s = C_s/Z_s \qquad (5)$$

The along-track variation of Z_s is shown in Figure 10a (April 1985)
and Figure 11a (August/September 1984). It fluctuates about 8 m. The
mean attenuation length k_d^{-1} in April 1985 and August/September 1984
is about 15 m, *i.e.* about one seventh of the depth of the euphotic
zone ($Z_e = 100$ m, see above); and 99.99 % of the light that would be
detected by satellite comes from the depth range above Z_e, which is
covered by our simulation.

Figures 10c and 11c show to what degree $\overline{Chl}_e$ (the mean
concentration of chlorophyll in the euphotic zone) is related to $\overline{Chl}_s$
in April 1985 and August/September 1984, respectively. It is concluded
that Chl_e can be derived from Chls within an error of a factor of two.
The mean *in situ* chlorophyll concentration of the euphotic zone is
overestimated by a factor of nearly two in early spring, when the
plankton are concentrated near the surface, and underestimated in late
summer, when the plankton are concentrated in a layer deep in the
seasonal thermocline.

130

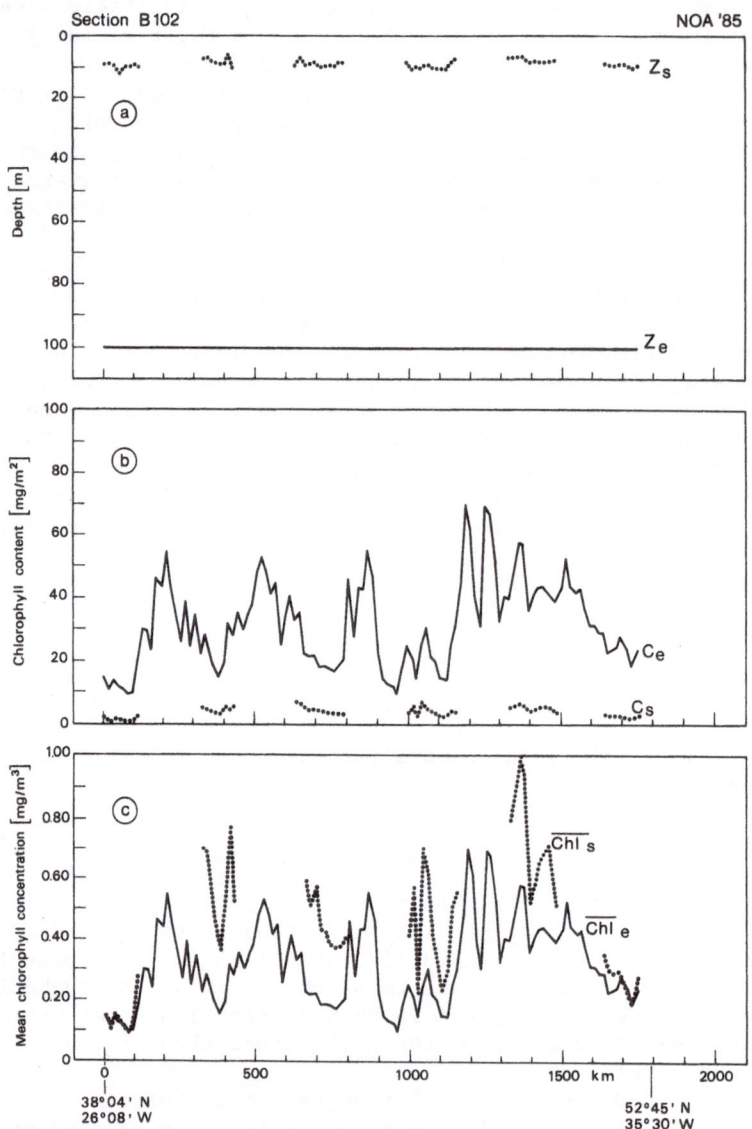

Figure 10. Water column integral quantities along spring section (24 April to 29 April 1985): a) depth of the euphotic zone, Z_e, and effective satellite visibility range, Z_s (see text); b) chlorophyll content measured *in situ*, C_e, and by satellite, C_s; c) mean chlorophyll concentration determined *in situ*, $\overline{Chl}_e$, and by satellite, $\overline{Chl}_s$.

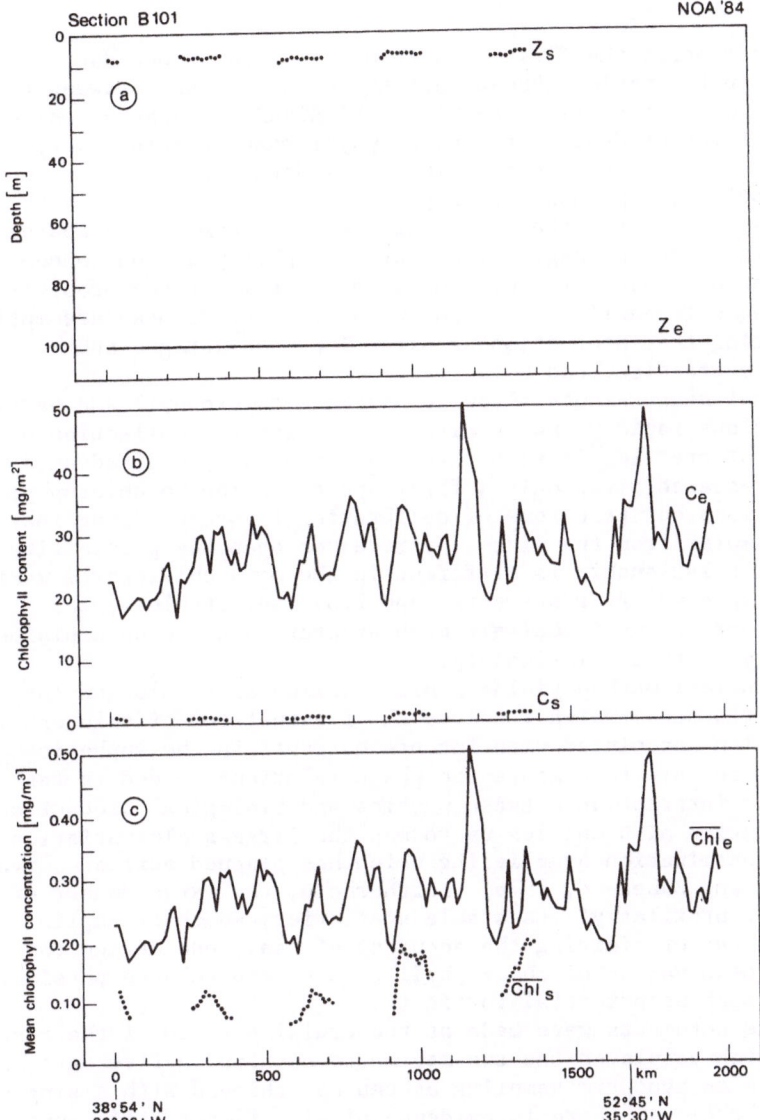

Figure 11. Water column integral quantities along late summer section (27 Aug. to 2 Sep. 1984): a) depth of the euphotic zone, Z_e, and effective satellite visibility depth, Z_s (see text); b) chlorophyll content measured *in situ*, C_e, and by satellite, C_s; c) mean chlorophyll concentration determined *in situ*, $\overline{Chl}_e$, and by satellite, $\overline{Chl}_s$.

4. DISCUSSION

This paper reports the first results derived from a new data set
collected in the period 1984-87 during annual cruises between the
Azores and Greenland using the Kiel "SEA ROVER" apparatus, which
yields profiles of density and chlorophyll concentration every 0.7
kilometre along the ship track. The total data set comprises over
20,000 profiles, together with supporting measurements of solar
irradiance, nutrients, etc.. The data set provides a rich source of
information about the regional patterns of phytoplankton concentration
in the heating season. It helps us to avoid some of the problems
encountered with earlier data, but we still have to make assumptions
in extracting the information we need. These advantages and
limitations are discussed below.

Optical measurements allow to estimate chlorophyll and particle
concentrations rapidly and *in situ*, permitting the collection of
thousands of profiles in each three-week cruise. The disadvantage is
that fluorescence gives only a first approximation to chlorophyll
concentration; our technique of calibrating fluorescence against
routine samples from the mixed layer leaves open the possibility that
the Fl:Chl relationship is different in the deep chlorophyll maximum
(for the expected error see paragraph 2.3). Nevertheless, the
presented variations in Chl are much greater than can be explained by
uncertainty in that relationship.

The optical method yields a high horizontal resolution (better
than one kilometre) which resolves the fluctuations of chlorophyll
concentration associated with intense currents in the euphotic zone,
making it possible to measure the phase relations needed to test
theories of interaction between physics and biological production. The
high resolution also enables us to map the large-scale variation
without contamination by aliasing which has plagued earlier surveys
(*e.g.* Krey and Babenerd, 1976). Furthermore, the large number of
independent profiles yields stable statistics, so that sampling error
is not the factor limiting the accuracy of, say, one-degree (of
latitude) mean values of chlorophyll content and related physical
variables such as potential vorticity.

The measurements were made at the cruising speed of the research
vessel, which completed the two megametre section in five days. That
is as close to synoptic sampling as can be achieved with a single
ship. Nevertheless, there is evidence of significant differences
between samples a week or so apart at the same location in the data
from 1985 and 1986, when the ship went up and down the line without
interruption. These differences are due partly to mesoscale patchiness
drifting through the section in the large-scale circulation, which is
orthogonal to the ship track. The non-synopticity does not impede
analysis of the large-scale regional distribution of the surface bloom
and deep chlorophyll maximum.

One of the principal goals of our experiment was to determine how
the large-scale regional variation changes during the heating season.
Ideally that would be extrated from a data set comprising a multi-year

time series of samples, each including sections at every month in the
heating season. Such a data set would allow us to compute a
climatological mean annual cycle, and the inter-annual deviations in
each year. Our data set comprises a four year (1984-87) time series
with a different phase of the annual cycle sampled each year. That
would contain sufficient information to construct the annual cycle if
the inter-annual variation were negligible; the results presented in
this paper have been discussed on that assumption. There is ample
evidence in the literature of inter-annual variation in the seasonal
thermocline (e.g. Colebrook and Taylor, 1979) and in primary
production (Raymont, 1980). There is also evidence of inter-annual
variability in our data set: for example, the chlorophyll distribution
in the May 1987 section does not lie exactly on the trend line in
Figure 9. We are investigating the possibility of using the more
extensive physical oceanographic data base to reduce the impact of
inter-annual variation on our description of the seasonal cycle.

5. CONCLUSION

Preliminary analysis of the Kiel "Sea Rover" sections between the
Azores and Greenland have revealed the following features of regional
variation of chlorophyll concentration during the heating season.

a.   The seasonal cycle follows that described by one-dimensional
     models (e.g. Wolf and Woods, 1988, this volume): the initial
     bloom has a maximum concentration of chlorophyll in the mixed
     layer; when the mixed layer nutrients run out production
     continues in the seasonal thermocline, with a deep chlorophyll
     maximum at the nutricline.
b.   From the Azores (38° N) to a latitude of about 48° N, the onset
     of the initial bloom occured within two weeks in April 1985, in
     phase with the decrease in mixed layer depth. Further north, the
     onset of the bloom migrated poleward at 1.5° latitude per month,
     in phase with the 12°C-isotherm outcrop which suggests
     temperature-limited growth. (Some evidences of a positive
     correlation between temperature and phytoplankton photosynthetic
     rate are summarized by Kirk, 1983.) The transition from bloom to
     oligotrophy migrates from the Azores (in late April) to the Polar
     Front at 52° N (in late August), a rate of 3.5° latitude per
     month.
c.   New production persists throughout the summer, causing the deep
     chlorophyll maximum to descend through the seasonal thermocline
     at a rate of 10 metres per month. The combination of this descent
     and the poleward migration of the time at which it starts causes
     the deep chlorophyll maximum to slope down towards south.
d.   Patches with the highest chlorophyll concentration in the
     seasonal thermocline were found where isopycnals were displaced
     upwards at indicators of strong isopycnic potential vorticity
     gradient, as in Woods's (1988) theory.

e. The vertically integrated chlorophyll content of the euphotic
    zone is dominated by mesoscale patchiness, with a significant
    peak at the site described above. The large-scale meridional
    variation is much smaller, both in the April bloom and in the
    September oligotrophic regime. The mean difference between the
    two regimes is less than the range of mesoscale variation.

f. Algorithms used to simulate CZCS estimates of chlorophyll content
    do not reproduce these results because they cannot correct for
    horizontal and seasonal variation in subsurface chlorophyll. The
    seasonal differences (between the spring bloom and oligotrophic
    regime) of both chlorophyll content and mean concentration in the
    euphotic zone are overestimated three to four times by CZCS
    simulations.

7. ACKNOWLEDGEMENTS

Thanks are due to all participants of the surveys, particularly C.
Meinke, V. Rehberg and Dr. J. Fischer for their technical support and
U. Wolf for determining the nutrient concentrations and his comments
on scientific aspects. Dr. H. Leach made valuable suggestions on
improvements to the manuscript. The work is financially supported by
the Deutsche Forschungsgemeinschaft (DFG) contracts Wo 254/10 and Le
550/2.

8. REFERENCES

Aiken, J., R.H. Bruce and J.A. Lindley: 1977, 'Ecological
    investigations with the undulating oceanographic recorder: the
    hydrography and plankton of the waters adjacent to the Orkney and
    Shetland Islands', Mar. Biol. 39, 77-91.
Bauer, J. and J.D. Woods: 1984, 'Isopycnic atlas of the North Atlantic
    Ocean - monthly mean maps and sections', Ber. Inst. Meeresk. 132,
    Univ. Kiel.
Colebrook, J.M.: 1982, 'Continuous plankton records: seasonal
    variations in the distribution and abundance of plankton in the
    North Atlantic Ocean and the North Sea', J. Plankt. Res. 4, 436-
    462.
Colebrook, J.M and A.H. Taylor: 1979, 'Year-to-year changes in sea
    surface temperature, North Atlantic 1948-1974', Deep-Sea Res.
    26A, 825-850.
Dandonneau, Y. and L. Lemasson: 1987, 'Water-column chlorophyll
    in an oligotrophic environment: correction for the sampling
    depths and variations of the vertical structure of density, and
    observation of a growth period', J. Plankt. Res. 9, 215-234.
Denman, K.L. and A.W. Herman: 1978, 'Space-time structure of a
    continental shelf ecosystem measured by a towed porpoising
    vehicle', J. Mar. Res. 36, 693-714.

Esaias, W.E., G.C. Feldman, C.R. McClain and J.A. Elrod: 1986, 'Monthly satellite-derived phytoplankton pigment distribution for the North Atlantic ocean basin', *EOS, American Geophys. Union* **67**, 835-837.

Fasham, M.J.R., T. Platt, B. Irwin and K. Jones: 1985, 'Factors affecting the spatial pattern of the deep chlorophyll maximum in the region of the Azores Front', *Prog. in Oceanogr.* **14**, 129-165.

Fischer, J., C. Meinke, P.J. Minnett, V. Rehberg and V. Strass: 1985, 'A Description of the IfM-Schleppfisch-System', Techn. Report, Inst. f. Meereskunde, Abt. Reg. Oz., Kiel, F.R.G., 2. edition.

Herman, A.W. and K.L. Denman: 1979, 'Intrusions and vertical mixing at the shelf/slope water front south of Nova Scotia', *J. Fish. Board Can.* **36**, 1445-1453.

Kierstead, H. and L.B. Slobodkin: 1953, 'The size of water masses containing plankton blooms', *J. Mar. Res.* **12**, 141-147.

Kirk, J.T.O.: 1983, *Light and photosynthesis in aquatic ecosystems*, Cambridge University Press.

Krauss, W.: 1986, 'The North Atlantic Current', *J. Geophys. Res.* **91**, 5061- 5074.

Krey, J. and B. Babenerd: 1976, 'Phytoplankton Production', Atlas of the International Indian Ocean Expedition. Inst. f. Meereskunde, Kiel University.

Le Fèvre, J.: 1986, 'Aspects of the biology of frontal systems', *Advances Mar. Biol.* **23**, 163-298.

Le Fèvre, J. and S. Frontier: 1988, 'Influence of temporal characteristics of physical phenomena on plankton dynamics, as shown by North-West European marine ecosystems', in B.J. Rothschild (ed.), *Toward a Theory on Biological-Physical Interactions in the World Ocean*, Kluwer Academic Publishers, Dordrecht, pp. 245-272.

Platt, T. and A.W. Herman: 1983, 'Remote sensing of phytoplankton in the sea: surface layer chlorophyll as an estimate of water column Chlorophyll and primary production', *Int. J. Remote Sensing* **4**, 343-351.

Raymont, J.E.G.: 1980, *Plankton and Productivity in the Ocean*, Vol. 1 Phytoplankton, Pergamon, Oxford 1980.

Robinson, M.K., R.A. Bauer and E.H. Schroeder: 1979, 'Atlas of North Atlantic - Indian Ocean monthly mean temperatures and mean salinities of the surface layer', Naval Oceanogr. Office, NSTL Station, Bay St. Louis, MS 39522.

Smith, R.C.: 1981, 'Remote sensing and depth distribution of ocean chlorophyll', *Mar. Ecol. Prog. Ser.* **5**, 359-361.

Smith, R.C. and K.S. Baker: 1982, 'Oceanic chlorophyll concentrations as determined by satellite (Nimbus 7 CZCS)', *Mar. Biol.* **66**, 269-279.

Stammer, D. and J.D. Woods: 1987, 'Isopycnic potential vorticity atlas of the North Atlantic Ocean - monthly mean maps', Ber. Inst. Meeresk. 165, Univ. Kiel.

Steele, J.H.: 1974, *The structure of marine ecosystems*, Harvard University Press, Cambridge, Massachusetts.

Strickland, J.D.H. and T.R. Parsons: 1968, 'A practical handbook of sea-water analysis', *Bull. Fish. Res. Bd. Can.* **167**, 1-311.

Sverdrup, H.U.: 1953, 'On conditions for the vernal blooming of phytoplankton', *J. Cons. Int. Explor. Mer.* **18**, 287-295.

Wolf, K.U. and J.D. Woods: 1988, 'Lagrangian simulation of primary production in the physical environment - The deep chlorophyll maximum and nutricline', in B.J. Rothschild (ed.), *Toward a Theory on Biological-Physical Interaction in the World Ocean*, Kluwer Academic Publishers, Dordrecht, pp. 51-70.

Woods, J.D.: 1988, 'Scale upwelling and primary production', in B.J. Rothschild (ed.), *Toward a Theory on Biological-Physical Interaction in the World Ocean*, Kluwer Academic Publishers, Dordrecht, pp. 7-38.

Wroblewski, J.S., J.J. O'Brien and T. Platt: 1975, 'On the physical and biological scales of phytoplankton patchiness in the ocean', *Mem. Soc. Roy. Sci. Liege*, 6th series, 7, 43-57.

SEASONAL OR APERIODIC CESSATION OF OLIGOTROPHY IN THE TROPICAL PACIFIC OCEAN

Y. Dandonneau
Centre ORSTOM
B. P. A5, Noumea
New Caledonia

ABSTRACT. Large regions of the tropical Pacific Ocean are probably in a permanent oligotrophic state. A monitoring of the sea surface chlorophyll concentration (SSCC) in cooperation with the crews of merchant ships shows however that the phytoplankton biomass increases in some of these regions during brief periods. Advection of nutrient-containing surface waters which come from the equatorial and Peruvian upwellings causes the zone between $5^{\circ}S$ and $15^{\circ}S$, at about $130^{\circ}W$, to be alternately in an oligotrophic or in a mesotrophic state; alternations occur at periods which are probably less than one month. In the southwestern tropical Pacific, the sea loses heat to the atmosphere each winter, and the depth of the mixed layer increases; the mixed layer then incorporates nitrate and new production is stimulated; the resulting increase of SSCC is observed every year from May to October, and extends northward to about $21^{\circ}S$. ENSO episodes are characterized by a rise of the thermocline in the western Pacific; during the 1982-83 ENSO, high SSCC was observed between the equator and $15^{\circ}N$, west of $170^{\circ}E$, from December 1982 to February 1983, caused by this rise of the thermocline and probably by vertical mixing. These observations show that the large scale coupling between physical processes and the phytoplankton can be mapped by very simple techniques, and suggest that other pulses of phytoplankton growth could be detected in oligotrophic areas using sea color satellite data.

1. INTRODUCTION

Oceanographic research vessels have permitted study of the structure of the photic layer and its consequences on the vertical distribution of chlorophyll (Riley *et al.*, 1949; Steele and Yentsch, 1960; Cullen, 1982). In situ sensors and rosette multi-samplers have increased our capability to resolve small scale features in the vertical. Small scale and mesoscale horizontal patterns can also be determined with records from sensors while the ship is moving, or with sea color data from satellites (Gordon *et al.*, 1981; Feldman *et al.*, 1984). There are, however, fewer data through which larger space scales can be addressed

137

B. J. Rothschild (ed.), Toward a Theory on Biological-Physical Interactions in the World Ocean, 137–156.
© 1988 by Kluwer Academic Publishers.

and still fewer data suitable for large time scale studies.
Replication of oceanographic cruises is generally inadequate because
these cruises are expensive, and only in a few regions where a
laboratory exists have long series of data been gathered (Menzel and
Ryther, 1961). Sea color data from satellites have seldom been
arranged in time series (Barale and Wittenberg Fay, 1986; Pelaez and
McGowan, 1986); up to now, small storage capacities for these data
have not permitted analysis of chronology over wide regions at short
time intervals. The time variations of oceanic phytoplankton biomass
have thus drawn less attention than horizontal or vertical variations,
and this gap in our perception is a major inhibitor of a better
understanding of the coupling between phytoplankton dynamics and
physical factors. This is especially true in oligotrophic areas where
fluxes of carbon or nitrogen result mostly from short events which are
often missed by observations too widely spaced in time (Jenkins and
Goldman, 1985; Platt and Harrison, 1985).

Horizontal variations at the ocean scale are also poorly known,
because wide oceanic zones have seldom or never been explored by
oceanographic cruises, and because standardization of techniques in
biology is generally not achieved. Comparison of the results obtained
by different laboratories is thus uncertain. The primary production
in the world ocean mapped by Koblentz-Mishke et al. (1970) is a fixed
and imperfect representation of the phytoplankton activity variations
from place to place. This representation can now be improved using sea
color satellite data. Measurements which can be made by ships of
opportunity on regular ship tracks are imprecise; however, they give
access to large scale studies, both in time and space (Colebrook,
1982). A programme based upon cooperation with merchant ships started
in 1969 in the Pacific ocean, supported by the ORSTOM Center in Noumea
(New Caledonia). At first, observations consisted only of sea surface
temperature and salinity; zooplankton sampling and filtrations for
chlorophyll measurements were added in 1978, and XBT probes have been
launched since 1979 as a part of the international TOGA programme. The
chlorophyll data collected through this programme are numerous, and
allow a monitoring of large scale events which occur in the tropical
Pacific Ocean (Figure 1). Increases in the phytoplankton biomass have
thus been observed in areas which were previously considered to be
permanently oligotrophic. Tropical oligotrophic zones are
characterized by a mixed layer (a) exhausted in nutrients, and (b)
thick, (c) maintained by a net heat flux from the atmosphere to the
ocean. If one out of these three conditions is not satisfied, the
light-nutrients status is modified, and new primary production is
stimulated, causing an increase in the chlorophyll concentration.
Such increases have been observed in the northeast of the Tuamotu
Islands, when waters drifting from the upwelling zones of the eastern
Pacific have a thick mixed layer - resulting from heat storage during
several thousands kilometers drift - which sometimes still contains
nutrients and a relatively high phytoplankton biomass (Dandonneau and
Eldin, in preparation). Each winter in the Coral sea, south of 20-
22°S, the mixed layer gets colder, deepens, and incorporates nutrient-
rich deeper layers (Dandonneau and Gohin, 1984). Finally, during the

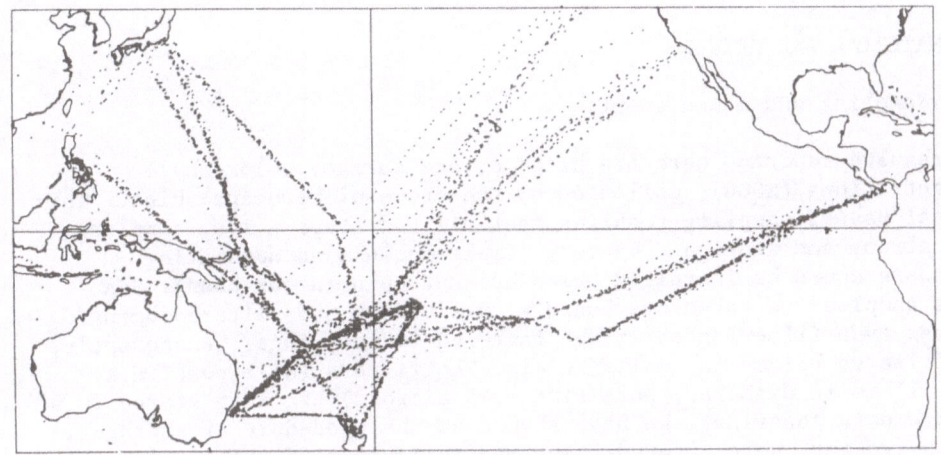

Figure 1. Position of the SSCC observations in 1983 (4645 data points).

1982-83 El-Niño-Southern Oscillation (ENSO), an increase of the sea surface chlorophyll concentration has been observed, centered at 7°N, in the western Pacific Ocean where the thermocline lifted by about 40 meters.

2. MATERIAL AND METHODS

2.1. Sampling and measurements

The results reported here are based on sea surface chlorophyll concentrations (SSCC), collected by the crews of merchant ships. This unusual way of sampling requires that all the steps - *i.e.* sampling, filtration, and storage - be very simple (unless an automatic procedure could be installed which has not been in our case). Sea water samples are taken overboard with a bucket and filtered using a Swinnex type filtering cartridge fastened to a syringe; consequently, the filtered volume is small (20 ml). The filters (Millipore, H.A. type, 13 mm in diameter, pore size 0.45 microns) are then stored in a dry and dark container; we have abandoned the procedure of storage in a deep freezer, which proved sometimes to cause difficulties for the crews and involved the risk of a breakage of the sequence of cold. The measurements are made in the laboratory at Noumea. The small filtered volume and long storage (one to three months) are not suited to classical fluorescence measurements after extraction of the photosynthetic pigments by acetone or methanol. The following technique was developed (Dandonneau, 1982): The fluorescence of the surface of the filters is measured in a fluorometer fitted with a specially adapted door and with two optical blue primary filters. The resulting value is linearly related to the chlorophyll concentration.
        The main difficulties in this technique result from the storage of the filters, during which some degradation of the pigments occurs. The fluorescence of the surface of the filters is produced by a mixture of pigments and degraded pigments; the results are expressed in milligrams of chlorophyll per cubic meter, but they must be considered only as an index of the chlorophyll concentration. For some voyages, poor conditions probably prevailed for the storage of the filters, while this cannot be ascertained. This problem has been considered in an earlier paper (Dandonneau, 1986); it cannot be solved easily because there exists no test for rejection of abnormally high or abnormally low chlorophyll concentrations (at least: no test with which I agree). These voyages reveal however the same horizontal patterns of chlorophyll distribution, which are an inherent information given by each transect, because all the SSCC values from these voyages, if biased, are biased in the same way by the same factor.

2.2. Defining the transition from upwelled to oligotrophic waters between Tahiti and Panama.

The ships sailing between Tahiti and Panama (Figure 1) cross a transition zone between (a) waters which originate in the equatorial and Peruvian upwellings and have a relatively high chlorophyll content, and (b) oligotrophic waters of the central south tropical Pacific. Each transect is then characterized by low SSCC values on the oligotrophic side, higher SSCC values to the equator, and a more or less abrupt step at the transition (Figure 2). The position of this step is not affected by possible biases caused by bad conditions of storage during some voyages. It was objectively determined for a transect consisting of observations SSCCi at longitudes Xi as the

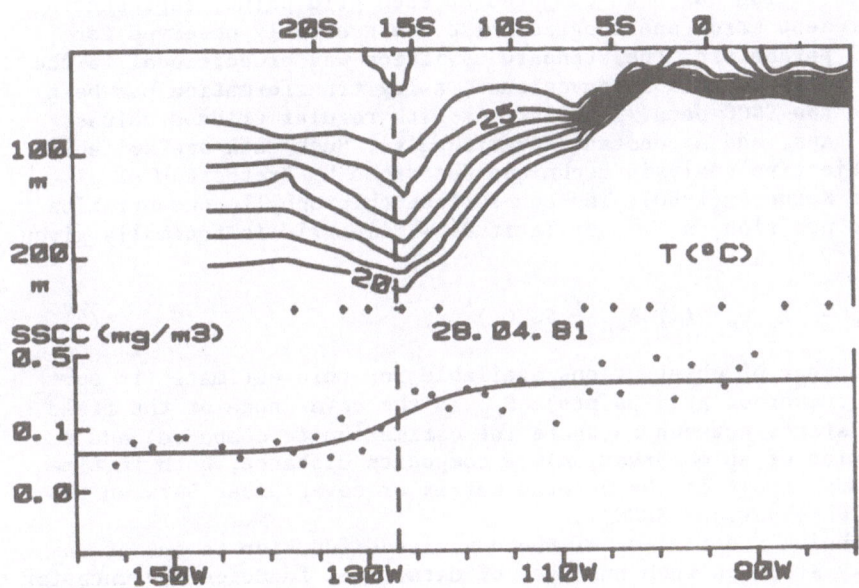

Figure 2. Above: Vertical section of temperature from Tahiti to Panama. The poleward slope of the isotherms between 3°S and 15°S corresponds to the South Equatorial Current; The equatorward slope south of 15°S corresponds to an eastward flow. Below: the points are SSCC values. The continuous line is the best fit of an hyperbolic tangent function. The transition between oligotrophic waters and richer waters coincides with the convergence. (Temperature and chlorophyll data from one out of the 117 Tahiti-Panama transects by cooperating merchant ships)

longitude X_0 of the point of inflexion of the function

$$\overline{SSCC} = A + B \tanh C (X - X_0) \tag{1}$$

where X is the longitude; A, B and X_0 are constants which minimize the sum of

$$(SSCC_i - \overline{SSCC}(X_i))^2$$

2.3. Objective analysis of SSCC data in the southwestern tropical Pacific Ocean

Six to eight ships call each month at Noumea (22°S, 166°E) so that the density of observations is high in this region. The effect of any particular voyage for which storage conditions for the filters may have been bad is then smoothed by the results from the other voyages. SSCC data have been averaged in a grid (one month, 2° in latitude, 15° in longitude - *i.e.* the whole width of the studied area -) and the standard deviation has been computed at each grid point, representing the measurement error and subgrid noise. As generally observed for biological parameters, the standard deviation was proportional to the mean. To palliate this inconvenience, a log transformation has been applied to the SSCC data. This left us with regular gridded values, with some gaps, and a constant subgrid noise. Such data are suitable for the objective analysis technique described by Bretherton *et al.* (1976) and Karweit (1980). The sea surface chlorophyll concentration at a given position, x, of our latitude - time grid is optimally given by:

$$SSCC_x = \sum_{r=1}^{N} C_{xr} \left(\sum_{s=1}^{N} A_{rs}^{-1} SSCC_s \right) \qquad (2)$$

N is the number of observations available for this estimate (in our case: the number of grid points). C_{xr} is the covariance of the field for the distance between x (where the estimation is computed) and r (the position of an observation), a composite distance, both in time and latitude. A_{rs}^{-1} is the inverse matrix of covariances between the available observations $SSCC_s$.

The choice and fitting of the covariance function is one of the most tricky steps in such analyses of data. This function accounts for the decrease of the information provided by a measurement at a place r when our estimate has to be made farther from r. We drew this function from the grid, after subtracting the seasonal variations, and estimated A, B and C in order to minimize

$$\sum_{D=1}^{15} \sum_{D=1}^{12} \sum_{i=1}^{N} (X_i \, X_{i+DL,Dt} - A \, B^{DL} \, C^{Dt})^2$$

where Dt is the time lag between two grid values, in months; DL is the difference in latitude, in degrees; N is the number of grid values; the X_i are the detrended grid values (LogSSCC minus seasonal variations). We obtained:

$$C(DL,Dt) = 0.108 \times 0.943^{DL} \times 0.682^{Dt} \qquad (3)$$

Equation (3) gives the covariances C_{xr} between the place where the estimation is made and each grid value, and C_{rs} between the grid values. Equation (2) gives then an optimal estimate of the field. SSCC values are recovered by adding the seasonal variations, and returning from the logarithms by an inverse transformation (Dandonneau and Gohin, 1984).

2.4. Mapping SSCC in the tropical Pacific Ocean

In the whole area covered by the merchant ships' network (Figure 1), the data are not dense enough to support an objective analysis. Most importantly, the covariance function which we adopted for the region around New Caledonia (equation 3), is valid only for that region and cannot be used everywhere in the tropical Pacific. It is not possible to define this function in regions where the data are scarce, because the small scale noise is high, and if we enlarge the grid up to three months and thousands of kilometers to smooth this noise, we shall not be able to explain the observed covariance of the SSCC field at such a large scale.

Thus, a simpler technique for smoothing and interpolation has been used. All the SSCC data collected since 1979 (excepting July 1982 to June 1983) have first been averaged on a long-term annual composite time-space grid (one month x $10°$ longitude x $2°$ latitude) in order to produce a standard year, which was then smoothed by a low pass filter (five months, $50°$ longitude, $10°$ latitude running mean, weighted by the number of observations). This provisional standard year was divided into twelve monthly files. Each year, from 1979 to 1985 was then averaged on a similar grid, and divided into twelve gappy raw monthly files. Processed final monthly files were obtained as follows:

- the chlorophyll concentration at a grid point with a rank i in longitude and j in latitude was first set to the value X_{ij} in the corresponding monthly file of the provisional standard year, with a weight equal to unity.
- the raw monthly file was then examined from rank $i'=i-5$ to $i+5$ in longitude and $j'=j-5$ to $j+5$ in latitude. Each value found, $X'_{i',j'}$ received a weight equal to $F E^{-d^2}$ where $d^2 = (i-i')^2 + (jj')^2$.
- $SSCC_{ij}$ in the processed final monthly file is the weighted mean of X_{ij} and all the $X'_{i',j'}$.

The constants E and F were respectively set to 1.17 and 4. For an estimation at a given grid point, these values give a weight equal to 4 if the value from the raw file comes from the same grid point, equal to 3.42 if it comes from an adjacent grid point, and equal to 0.24 if it comes from $i'=i\pm3$ and $j'=j\pm3$. This simple technique achieves both a) smoothing, b) interpolation, and c) considers the standard year with a relative weight which decreases when information from the year under study increases. The numerical values adopted for E and F have been chosen arbitrarily in such a way that these three roles be reasonably performed.

3. RESULTS

3.1. The southwestward limit of the waters drifting from the eastern Pacific Ocean upwelling zones

The area south of the cold water tongue which characterizes the equatorial upwelling in the eastern Pacific (Wyrtki, 1981) is mostly known owing to the EASTROPAC cruises (Blackburn et al.., 1970; Owen and Zeitschell, 1970) and to the research focused on the ENSO phenomenon (Barber and Chavez, 1983). The waters which drift westward in the South Equatorial Current originate in the Peruvian and equatorial upwellings, and the relative contributions of these two upwellings cannot be determined. The trajectories of drifting buoys in the eastern Pacific around 10°S alternately show southwestward branches, likely to transport water from the equatorial upwelling, and westward branches likely to transport waters from the Peruvian upwelling. These waters absorb heat during their drift, and the temperature increases to 26-27°C near 10-15°S, 130-140°W; well defined thermal fronts are not detected in this region (Newell, 1986; Legeckis, 1986). They have a high nutrient content when they are brought to the surface in the upwelling zone, and utilization of these nutrients occurs very slowly, so that they still contain nitrate at about 130°W, where the chlorophyll concentration is not as high as one would expect (Thomas, 1979). This has been explained by grazing pressure which limits the phytoplankton standing crop (Walsh, 1976) so that after a thousand kilometers drift, the phytoplankton have not yet consumed the nutrients in the mixed layer. The oligotrophic conditions which characterize the south Pacific central gyre are thus transferred to the west. The SSCC data collected between Tahiti and Panama show that the transition from the waters influenced by the upwellings to the oligotrophic waters where the chlorophyll content is very low, is marked by a more or less abrupt decrease of the SSCC values (Dandonneau and Eldin, in press). Temperature sections have been drawn, from XBTs launched from the same merchant ships. The transition between upwelled and oligotrophic waters generally corresponds to a convergence between the South Equatorial Current (characterized by a southward slope of the isotherms) and an eastward countercurrent (Eldin, 1983) which is shown by an equatorward slope of the isotherms (Figure 2). The position of this transition has been determined for 117 merchant ship transects. One of the most striking features is the high variability of the position of the transition, which moves between 5°S, 112°W and 17°S, 147°W on the Tahiti Panama direct track, and between 5°S, 104°W and 21°S, 138°W on the indirect track through the southern Tuamotu Islands (Figure 3). A linear transformation has been applied to the positions (latitude, longitude) of the transitions found on the southern track in such a way that the latitudes and the longitudes of the transition have the same mean and standard deviation on both tracks. The practical effect is a shifting of the positions

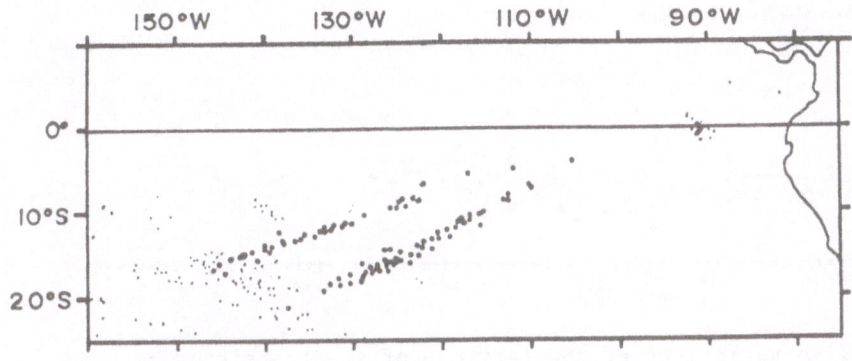

Figure 3. Positions of the transition between oligotrophic waters in the south and richer waters from the Peru and the equator (1980 to 1985).

from the southern track so that they statistically coincide with those on the northern track, achieving a unique series of data for the 117 observed transitions. The latitudes in this new series are an index of the width of the chlorophyll enriched area, having an improved definition in time (Figure 4). There are no clear seasonal variations: the southern limit of the chlorophyll enriched waters varies in latitude on a short time scale, with very little correspondence between consecutive observations of the transition. A low pass Fourier filter (cutoff period: six months) only explains 10% of the variance of this latitude index. The noise is thus unresolved by the frequency of our observations (117 transects for 6 years, mean period of observation = 19 days). This high variability might be related to equatorial long waves, which have a 20 to 30 day period (Legeckis, 1977), to short living meteorological events, or to instability of the eastward flows (Eldin, 1983) which bound the enriched waters to the south. One aspect is worthy of attention: there is a trend over the studied period for the transition latitudes to decrease, i.e. the transition tends to move towards the equator (Figure 4). Several hypotheses can be put forward:

- a decrease of the upwelling intensity, or a decrease of the nitrate concentration in the upwelled waters, so that the oligotrophic stage would be reached after a shorter time lag, and a shorter drift.

146

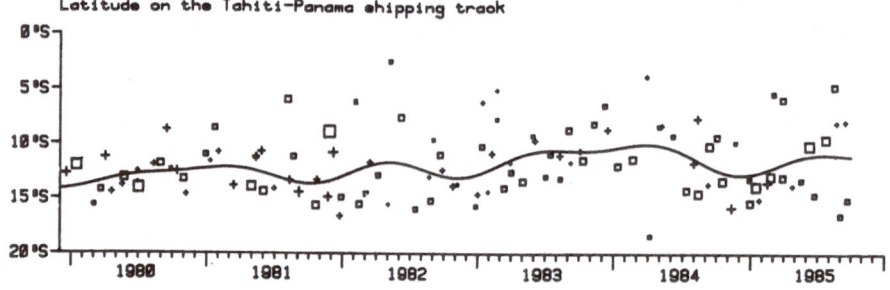

Figure 4. Time variations of the latitude of the transition between
oligotrophic waters in the south and richer waters from the Peru and
the equator. Squares: transitions found on the northern track.
Crosses: transitions found on the southern track and made compatible
with those found on the northern track (the two tracks are shown on
Figure 3). Continuous line: a low pass Fourier filter (cutoff period
:6 months) shows that most of the variations occurs at smaller scales.

- an equatorward shifting of the south Pacific central gyre,
  pushing back the southern limit of the South Equatorial
  Current.

No firm explanation can be given. This long term trend however is
probably not an artifact, because the transition between upwelled and
oligotrophic waters which is the base of this study is a structure
which does not depend upon the absolute values of SSCC, and is then
unaffected by possible biases in the measurements between transects.

3.2. Winter enrichment in the southwestern tropical Pacific

The annual cycle of phytoplankton in temperate regions is dominated by
a strong biomass increase after the vertical mixing in winter. At low
latitudes, seasonal variations have a small amplitude and are
uncertain, except in regions where there is a seasonal upwelling. In
the southwestern Pacific, a chlorophyll maximum is observed off Sydney
from August to October (Humphrey, 1963), and a secondary maximum in
February is caused by short intrusions of deep water on the
continental shelf. These maxima are governed respectively by the
winter vertical mixing of nitrate, and by coastal processes (Rochford,
1984). Farther north in Australia, at Townsville (19°S), Furnas and
Mitchell (1986) observe only summer chlorophyll maxima associated with
deep water intrusions through the Great Barrier Reef. These coastal
processes do not operate offshore, where winter mixing is the only

enrichment source. A SSCC maximum between 160°E and 175°E during the winter months (June to October) has been shown to decrease from 32°S toward lower latitudes, and to completely disappear north of 21°S (Dandonneau and Gohin, 1984). No seasonal variations of SSCC are observed between 14°S and 21°S. These results from merchant ships sampling covered the period from 1978 to 1982. They are confirmed by the observations collected later (Figure 5).

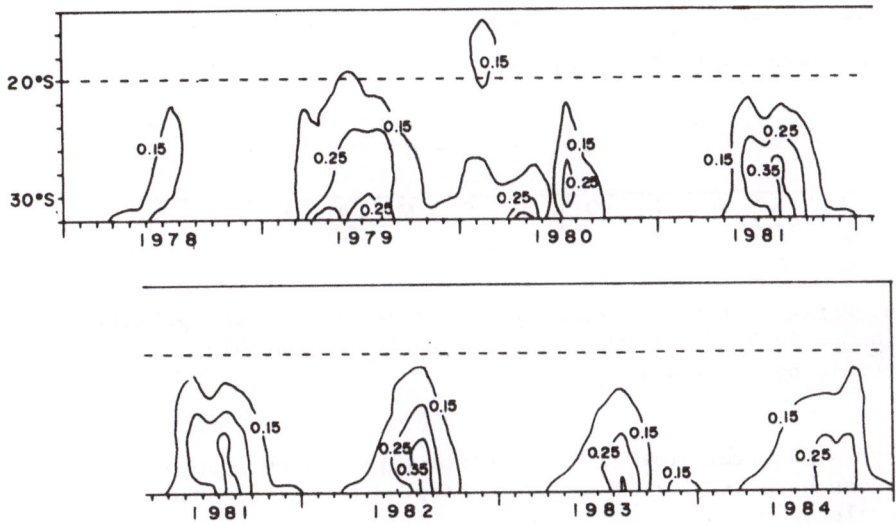

Figure 5. Time and latitude variations of SSCC between 14 and 32°S, 160 and 175°E (according to an objective analysis of about 7000 data points).

It is interesting to note that this winter increase is demonstrated by very simple measurements obtained in cooperation with voluntary observers, while earlier oceanographic cruises from Noumea (New Caledonia) during fifteen years have ignored it. A series of monthly oceanographic transects covering one year, with measurements at depth, was undertaken in 1983. The transect was located between the barrier reef of New Caledonia and an offshore station at 22°32'S, 165°42'E (Le Borgne et al.,1985); this position is close to the northern limit of the winter enrichment, and the annual signal has a low amplitude (Figure 5). The conclusions which can be drawn from the results appear uncertain (Figure 6). They would not, by themselves, establish the existence of a winter enrichment. This is not surprising since the signal to noise ratio at this latitude is 0.8 (Dandonneau and Gohin, 1984), and under such conditions, the description of an

148

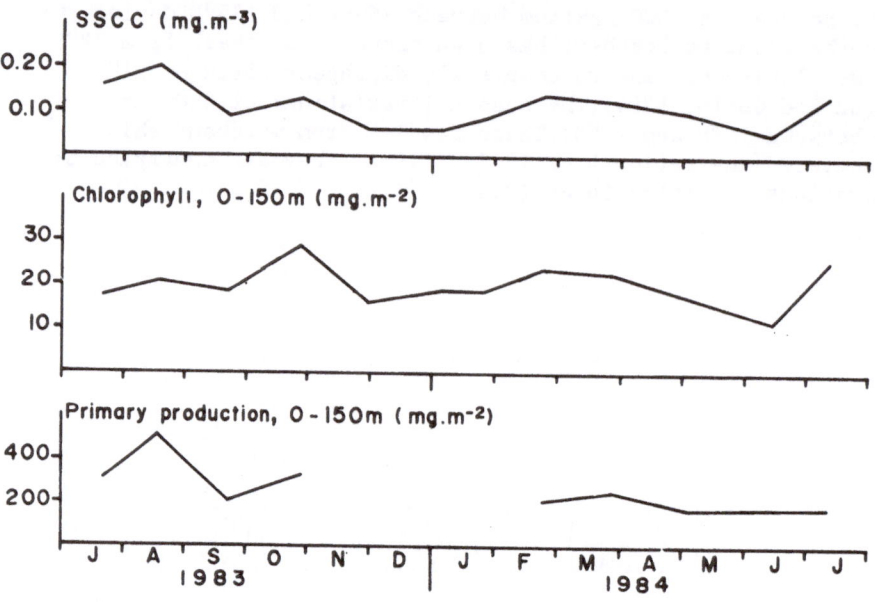

Figure 6. Variations of the chlorophyll concentration and primary production at 22°32'S, 165°42'E, from a series of monthly oceanographic observations.

annual cycle requires more than one station per month. Monthly observations have yet been a commonly adopted scheme for the study of annual cycles.

In this region, the mixed layer is permanently exhausted in nutrients; this layer however, around 22°S, is 50 to 80 m thick in summer and the nitracline is found 30 m deeper, while in winter, the mixed layer is about 125 m thick and nutrients are found immediately below. The sea surface temperature at 23°S in August is 22°C, and this temperature, in April, a few months before, corresponds to waters where nitrate concentration is 2.3 μmoles L^{-1}. The mixed layer thus thickens as it is cooled, and incorporates nitrate-rich waters, stimulating new production and causing an increase of SSCC. The cooling of the mixed layer is probably not caused by advection, because important cold water fluxes from the south are unlikely in this region (Rotschi and Lemasson, 1967). It is rather a consequence of heat loss from the ocean. The net heat flux to the atmosphere at 23°S, 165°E is estimated to be 120 W m^{-2} from May to July, 60 W m^{-2} in August, and reverses only from October (Weare et al., 1980). The duration, intensity, and northward extent of the winter enrichment vary between years (Figure 5). It would be interesting to relate these variations to anomalies of net heat flux, of horizontal advection, or to variations of the reservoir of nitrates below the thermocline.

3.3. Phytoplankton growth in the western tropical Pacific Ocean
     after the 1982-83 ENSO

The area bounded by $130^{o}E$ and $160^{o}E$, the equator, and $15^{o}N$ is
remarkable for important heat fluxes, through advection in the ocean,
or through exchanges with the atmosphere (Delcroix, 1987). The mixed
layer however is relatively constant, about 100 m thick, with a
temperature ranging up to $29^{o}C$ with small seasonal variations. Two
divergences between zonal currents are favourable to the growth of
phytoplankton: the equatorial divergence, where upwelling is
occasionally observed (Oudot and Wauthy, 1976), and a ridge of the
isotherms at the northern edge of the North Equatorial Countercurrent
at $9^{o}N$. The ridge usually does not produce any SSCC increase (Figure
7, upper). These two structures are deeply modified during ENSO
episodes (Meyers and Donguy, 1984): the equatorial divergence
disappears, and on the contrary, prevailing westerly winds tend to
induce a convergence at the equator, while the ridge is reinforced as
the North Equatorial Countercurrent speeds up. Westerly winds during
these episodes also drive eastward a large fraction of the warm water
stored in the western Pacific Ocean, causing the mixed layer thickness
to decrease by several tens of meters.
     During normal (non ENSO) conditions, the North Equatorial
Countercurrent has low SSCC, about 0.07 mg m^{-3}, while the equatorial
divergence is marked by SSCC values of about 0.13 mg m^{-3} (Figure 7,
above). During the 1982-83 ENSO, sampling was relatively dense in this
region (Figure 1), at a rate of one transect each month, permitting a
tentative monitoring of the SSCC variations on a quarterly basis
(Dandonneau, 1986). An outstanding SSCC increase centered at about $7^{o}N$
on the ridge of the North Equatorial Counter Current is the main
consequence of the ENSO in the entire investigated area (Figure 7,
lower). Some uncertainty results from possible low quality data which
would create an artifact. It must however be considered that a bias
would modify the data from a whole voyage by a multiplication factor,
but would not change their structure. In this respect, a) The
unchanged SSCC mean values during the ENSO in the western Pacific
between $20^{o}N$ and $30^{o}N$ and between $10^{o}S$ and $20^{o}S$, which result from
data collected during the same voyages, and b) the unquestionable
shift of the high SSCC spot which was centered at $8^{o}N$ and not at the
equator, are convincing arguments which favour an enrichment from the
North Equatorial Counter Current. When the transects are considered
one after the other (Figure 8), an evolution can be seen:

  - in September 1982, SSCC values are low, and present a weak
    maximum near the equator which disappears in October.
  - in November, an SSCC increase breaks out north of the equator,
    and grows up in intensity and width during December.
  - the SSCC maximum is centered around $10^{o}N$ in January 1983.
  - from February, the maximum shifts slowly toward the equator and
    lessens.
  - a maximum centered at the equator returns in April.

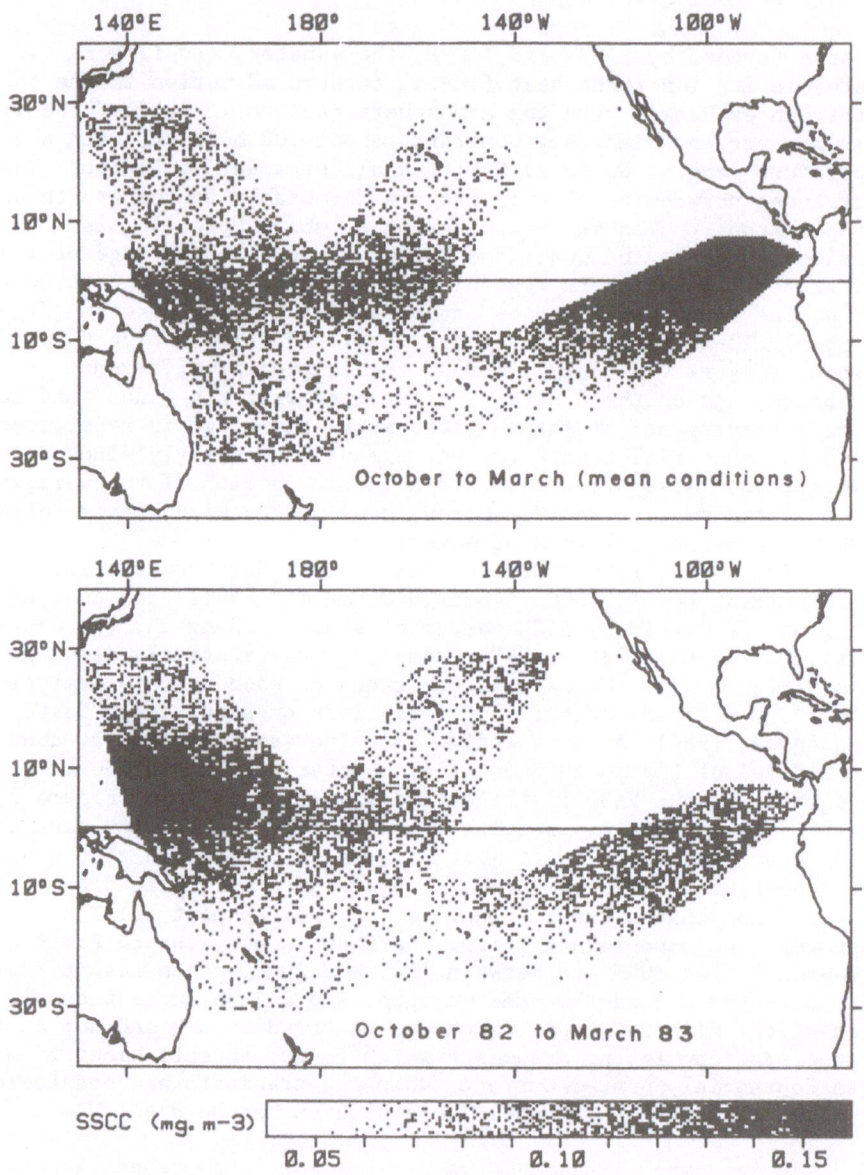

Figure 7. SSCC in the tropical Pacific Ocean: The 1982-83 ENSO caused a decrease of SSCC in the eastern Pacific, and an increase in the western Pacific, centered at the northern edge of the North Equatorial Countercurrent where the thermocline lifted by 40 m.

The rise of the thermocline in this region during the 1982-83 ENSO could account for the SSCC enrichment. At the latitude of the northern edge of the North Equatorial Countercurrent, the thermocline lifted by about 40 meters between January 1982 and January 1983 (Meyers and Donguy, 1984). Light penetration to nutrient rich layers was then improved, stimulating new production. Nitrate and phosphate measurements made during the biannual transects by R.V. Ryofu Maru along 137°E indeed show a consumption of nutrients from the deep reservoir during all the recent ENSO episodes (Dandonneau, 1986). It is however unlikely that improved light penetration below the thermocline cause a chlorophyll increase in the mixed layer, and a jump in surface concentrations from 0.07 to 0.40 mg m^{-3} in December and January. A 1°C decrease of the temperature of the mixed layer also occurred in this region during the 1982-83 ENSO; Meyers et al. (1986) relate this cooling at the equator mainly to intense evaporation after strengthening of the wind. They also compute a 13 W m^{-2} heat loss by vertical mixing after the thermocline had lifted. This vertical mixing probably concurred to the SSCC increase which we observed from December to March between 0° and 15°N. It is noteworthy that time-spaced oceanographic cruises again have not shown this increase: the results from the biannual cruises of R.V. Ryofu Maru show temperature and salinity anomalies at the equator related to ENSO episodes (Masuzawa and Nagasaka, 1975), but no indication emerges from the chlorophyll measurements made during these cruises. The relationship between temperature and nutrients (computed on data from the same cruises: Dandonneau, 1986) which shows an important consumption of nutrients during each ENSO episode indeed integrates the effects of pulses of new production over the six months which separate these cruises.

4. CONCLUSIONS

The three studies presented in this work deal with regions where the primary production is less than 150 mg C m^{-2}d^{-1} according to Koblentz-Mishke et al. (1970) and which are generally considered as permanently oligotrophic. The observing merchant ships network has permitted detection of phases of phytoplankton growth, during which these regions can be mentioned as 'mesotrophic', or even 'eutrophic'. Access to observational scales which are different from the scales which are imposed by use of oceanographic research vessels was decisive.

The transition between waters from the upwellings of the eastern Pacific and oligotrophic waters from the central south Pacific is not completely described by samples 100 km apart along a shipping track, even if contemporaneous temperature sections show that this transition generally coincides with a convergence between the South Equatorial Current and eastward flows. The advantage of sampling by merchant ships is the ability to cover long distances (about 6500 km between Tahiti and the Galapagos islands) nearly twice a month. This observational frequency is not high enough to resolve the rapid

152

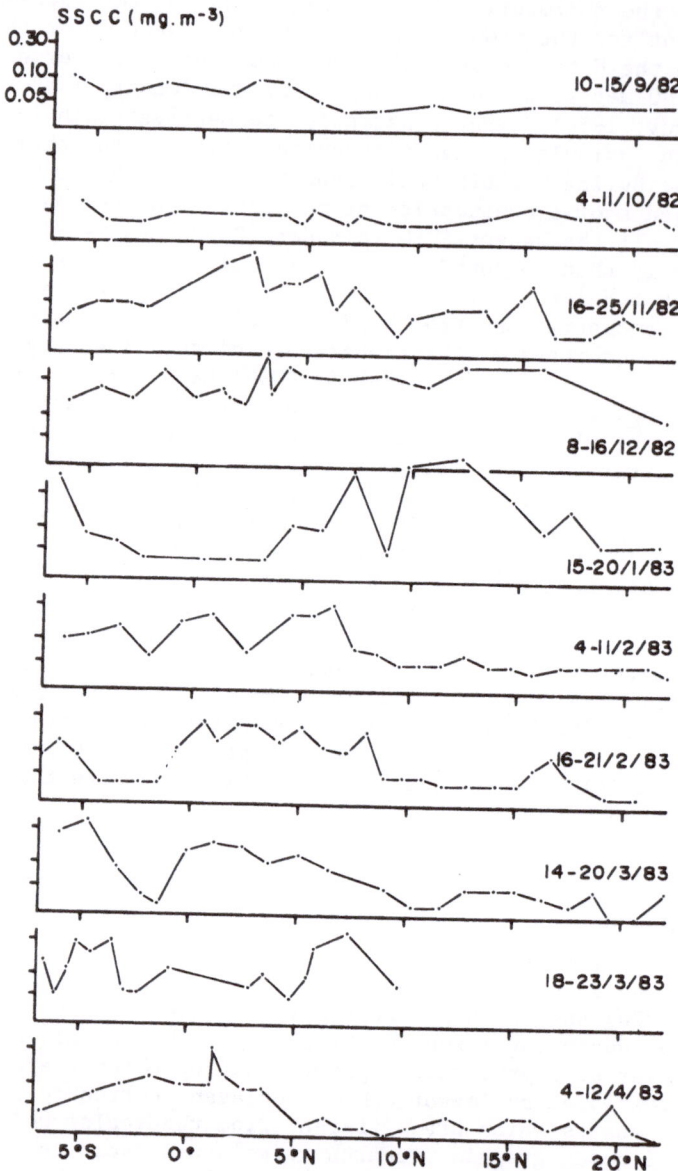

Figure 8. SSCC from individual merchant ships transects between New Caledonia and Japan.

shifting of the transition, but shows that periods less than one or two months dominate the shifting. The duration (six years) of the series of data also permits detection of a long term retreat of the transition toward the equator. No firm explanation can be proposed at the moment for this trend.

The high density of SSCC data collected around Noumea where the network converges provides a quasi permanent observation over the region. Here, it was possible to apply the principles of objective analysis. The range of the seasonal variations is small, even at 32°S, compared to other regions, and becomes negligible north of 22°S. On the contrary, the small scale noise is high. The low signal to noise ratio makes it difficult to detect these seasonal variations from oceanographic studies. The only report suggesting a winter SSCC increase in this region was based on statistics of Secchi disk transparency from measurements made by fishermen (Morita, 1973). The transparency of sea water is only an index of the phytoplankton biomass, like SSCC; these indexes can be obtained easily in great numbers. Efficiency of this strategy has been demonstrated also in the western Pacific between the equator and 15°N, where oceanographic cruises failed to detect the high primary productivity during the 1982-83 ENSO. This can be considered as a confirmation of the prediction by Munk (this volume) that large scale oversampling of biologically significant parameters will promote new approaches in biological oceanography, even if precision is low.

In these three examples, episodes of phytoplankton growth have been observed in oligotrophic regions. Oligotrophy in the tropical ocean is the result of a thick mixed layer, exhausted in nutrients, and stable. The last condition was reversed here by heat loss from the mixed layer which cools and incorporates deeper and deeper water, until it reaches the nutricline. Estimating the heat exchanges between the mixed layer and the atmosphere is difficult, because they result from several processes which generally cannot be known with precision. In the southwestern tropical Pacific, in winter, SSCC acts as an indicator of the heat losses, and could perhaps be used as a tool for the study of heat fluxes between the ocean and the atmosphere. The relationship between eastward fluxes, and the southern limit of waters coming from the upwellings of the eastern Pacific, could also help to understand the structure of the currents between 5°S and 15°S, east of Polynesia. This cannot be done with temperature alone, because the region shows no contrasting thermal fronts. The large time and space scale coupling between physical processes and SSCC suggests that sea color data measured from satellites would probably show many other unknown productivity features and document other poorly known ones.

5.   ACKNOWLEDGEMENTS

I would like to thank Drs. Andrew Bakun, Michael Mullin and Volker Strass for reading and annotating the manuscript. Several hundreds of officers of merchant ships have built the basis of this work by their patient and spontaneous aid in sampling. Probably none of them will read it. I would like however to pay homage to their kind cooperation.

6. REFERENCES

Barale, V., and R. Fay Wittenberg: 1986, 'Variability of the ocean surface color field in Central California near coastal waters as observed in a seasonal analysis of CZCS imagery', *J. Mar. Res.* **44**, 291-316.

Barber, R.T. and F.P. Chavez: 1983, 'Biological consequences of El-Niño', *Science* **222**, 1203-1210.

Blackburn, M., R.M. Laurs, R.W. Owen, and B. Zeitzschell: 1970, 'Seasonal and areal changes in standing stocks of phytoplankton, zooplankton, and micronecton in the eastern tropical Pacific', *Mar. Biol.* **7**, 14-31.

Bretherton, F.P., R.E. Davis, and C.B. Fandry: 1976, 'A technique for objective analysis and design ofoceanographic experiments applied to MODE-73', *Deep Sea Res.* **23**, 559-582.

Colebrook, J.M.: 1982, 'Continuous plankton records: phytoplankton, zooplankton and environment, North East Atlantic and North Sea, 1958-1980', *Oceanol. Acta* **5**, 473-480.

Cullen, J.J.: 1982, 'The deep chlorophyll maximum: comparing vertical profiles of chlorophyll a', *Can. J. Fish. Aquat. Sci.* **39**, 791-803.

Dandonneau, Y.: 1982, 'A method for the rapid determination of chlorophyll plus phaeopigments in samples collected by merchant ships', *Deep Sea Res.* **29**, 647-654.

Dandonneau, Y.: 1986, 'Monitoring the sea surface chlorophyll concentration in the tropical Pacific: consequences of the 1982-83 El-Niño', *U.S. Fish. Bull.* **84**, 687-695.

Dandonneau, Y. and G. Eldin: in press, 'The southwestward extent of chlorophyll enriched waters from the Peruvian and equatorial upwellings between Tahiti and Panama', *Mar. Ecol., Progr. Ser.*

Dandonneau, Y. and F. Gohin: 1984, 'Meridional and seasonal variations of the sea surface chlorophyll concentration in the southwestern tropical Pacific (14 to 32°S, 160 to 175°E)', *Deep Sea Res.* **31**, 1377-1393.

Delcroix, T.: 1987, 'Net heat gain of the tropical Pacific Ocean computed from subsurface ocean data and wind stress data', *Deep Sea Res.* **34**, 33-43.

Eldin, G.: 1983, 'Eastward flows of the south equatorial central Pacific', *J. Phys. Oceanogr.* **13**, 1461-1467.

Feldman, G., D. Clark, and D. Halpern: 1984, 'Satellite color observations of the phytoplankton distribution in the eastern equatorial Pacific during the 1982-83 El-Niño', *Science* **226**, 1069-1071.

Furnas, M.J. and A.W. Mitchell: 1986, 'Phytoplankton dynamics in the central Great Barrier Reef-I. Seasonal Changes in biomass and community structure and their relation to intrusive activity', *Continental Shelf Res.* **6**, 363-384.

Gordon, H.R., D.K. Clark, J.W. Brown, O.B. Brown, and R.H. Evans: 1981, 'Satellite measurement of the phytoplankton concentration in the surface waters of a warm core Gulf Stream ring', *J. Mar. Res.* **40**, 491-501.

Humphrey, G.F.: 1963, 'Seasonal variations in plankton pigments in waters off Sydney', *Aust. J. Mar. Freshw. Res.* **14**, 24-36.

Jenkins, W.J. and J.C. Goldman: 1985, 'Seasonal oxygen cycling and primary production in the Sargasso sea', *J. Mar. Res.* **43**, 465-481.

Karweit, M.: 1980, 'Optimal objective mapping: a technique for fitting surfaces to scattered data', in F. Diemer, J. Vernberg and D. Mirkes (eds.), *Advanced concepts in ocean measurements for marine biology*, University of South Carolina Press, Columbia, pp. 81-99.

Koblentz-Mishke, O.J., V.V. Volkovinsky, and J.G. Kabanova: 1970, 'Plankton primary production of the world ocean', in W.S. Wooster (ed.), *Scientific exploration of the southern Pacific*, Nat. Acad. Sci., Wash., D.C., pp. 183-193.

Le Borgne, R., Y. Dandonneau, and L. Lemasson: 1985, 'The problem of the Island mass effect on chlorophyll and zooplankton standing crops around Mare (Loyalty Islands) and New Caledonia', *Bull. Mar. Sci.* **37**, 450-459.

Legeckis, R.: 1977, 'Long waves in the eastern equatorial Pacific Ocean: a view from a geostationary satellite', *Science* **197**, 1179-1181.

Legeckis, R.: 1986, 'A satellite time series of sea surface temperatures in the eastern equatorial Pacific ocean', *J. Geophys. Res.* **91C**, 879-886.

Masuzawa, J. and K. Nagasaka: 1975, 'The 137°W oceanographic section', *J. Mar. Res. suppl.* **33**, 109-116.

Menzel, D.W. and J.H. Ryther: 1961, 'Annual variations in primary production of the Sargasso Sea off Bermuda', *Deep Sea Res.* **7**, 282-288.

Meyers, G. and J.R. Donguy: 1984, 'The North Equatorial Counter-current and heat storage in the western Pacific Ocean during 1982-83', *Nature* **312**, 258-260.

Meyers, G., J.R. Donguy, and R.K. Reeds: 1986, 'Evaporative cooling of the western equatorial Pacific Ocean by anomalous winds', *Nature* **323**, 523-526.

Morita, J.: 1973, 'Transparency observed by the Secchi disc in the western Pacific Ocean', *Far Seas Fish. Laboratory Bull.* **9**, 1-18.

Newell, R.E.: 1986, 'El-Niño: an approach toward equilibrium temperature in the tropical eastern Pacific', *J. Phys. Oceanogr.* **16**, 1338-1342.

Oudot, C. and B. Wauthy: 1976, 'Upwelling et dome dans le Pacifique tropical occidental', *Cah. ORSTOM ser. Oceanogr.* **14**, 27-48.

Owen, R.W. and B. Zeitzschell: 1970, 'Phytoplankton production: seasonal change in the oceanic eastern tropical Pacific', *Mar. Biol.* **7**, 32-36.

Pelaez, J. and J.A. McGowan: 1986, 'Phytoplankton pigment patterns in the California Current as determined by satellites', *Limnol. Oceanogr.* **31**, 927-950.

Platt, T. and W.G. Harrison: 1985, 'Biogenic fluxes of carbon and oxygen in the ocean', *Nature* **318**, 55-58.

Riley, G.A., H. Stommel, and D.F. Bumpus: 1949, 'Quantitative ecology of the plankton of the western north Atlantic', *Bull. Bingham Oceanogr. Coll.*, **12**, 1-169.

Rochford, D.J.: 1984, 'Nitrates in eastern Australian coastal waters', *Aust. J. Mar. Freshw. Res.* **35**, 385-397.

Rotschi, H. and L. Lemasson: 1967, 'Oceanography of the Coral and Tasman Seas', *Oceanogr. Mar. Biol. Ann. Rev.* **5**, 49-97.

Steele, J.H. and C.H. Yentsch: 1960, 'The vertical profile of chlorophyll', *J. Mar. Biol. Assoc. U.K.* **39**, 217-226.

Thomas, W.H.: 1979, 'Anomalous nutrient-chlorophyll interrelationships in the offshore eastern tropical Pacific ocean', *J. Mar. Res.* **37**, 327-335.

Walsh, J.J.: 1976, 'Herbivory as a factor in patterns of nutrient utilization in the sea', *Limnol. Oceanogr.* **21**, 1-13.

Weare, B.C., P.T. Strub, and M.D. Samuel: 1980, 'Marine climate atlas of the tropical Pacific ocean', *Contributions in Atmospheric Science, University of California,* Davis **20**, 147pp.

Wyrtki, K.: 1981, 'An estimate of equatorial upwelling in the Pacific', *J. Phys. Oceanogr.* **11**, 1205-1214.

THE DEEP PHAEOPIGMENTS MAXIMUM IN THE OCEAN: REALITY OR ILLUSION?

A. Herbland
IFREMER
BP 1049
Nantes
France

ABSTRACT. The vertical distribution of phaeopigments in the ocean is
traditionally interpreted as the result of the balance between two
processes: (1) Grazing by herbivores (= production) and (2) photo-
oxidation by excess of light (= degradation). If this interpretation
is still correct for coastal and surface waters (*e.g.* spring bloom of
phytoplankton and upwelled waters) there is today some evidence that
the permanent (or seasonal) deep phaeopigment maximum layer in the
bottom of the stratified euphotic zone (tropical and subtropical
oceans) is largely overestimated. The artifact would be due to the
presence of chlorophyll $\underline{b}$ which interferes with phaeopigments in the
widely and routinely used acid fluorometric method for determination
of $Chl\underline{a}$. The high concentration of $Chl\underline{b}$ relative to $Chl\underline{a}$ at depth (and
only at depth) suggests the existence, and probably the dominance, of
shade adapted green algae (eukaryotes) in the deep chlorophyll maximum
layer of the stratified euphotic zones. Recent direct observations and
counting of cells support this hypothesis. The example in this paper
shows that although they are irreplaceable, universal and simple
techniques must be used with due considerations of their properties
and limits when they are applied in structures where the conditions
are expected to be very different from those where the method has been
developed.

1.   INTRODUCTION

Measurements of photosynthetic pigments have played an important role
in studies of primary production in lakes and oceans. Among these
pigments, chlorophyll $\underline{a}$ ($Chl\underline{a}$) was specially investigated because it
is both the primary photosynthetic pigment in all oxygen evolving
photosynthetic organisms (the other algal chlorophylls are considered
as "accessory" or "secondary" photosynthetic pigments) and the most
abundant : although $Chl\underline{a}$ represents only about 1 % of the dry weight
of a phytoplankton cell, in quite variable proportion (Strickland,
1965; Shuter, 1979) the concentration of $Chl\underline{a}$ was and is still the
best and the most practical chemical indicator of phytoplankton

157

B. J. Rothschild (ed.), Toward a Theory on Biological-Physical Interactions in the World Ocean, 157–172.
© 1988 by Kluwer Academic Publishers.

biomass in natural samples (Cullen, 1982).

Yentsch (1965) and various authors cited by him have presented evidence that decomposition products of chlorophyll a can be present in natural-population extracts of phytoplankton, and one of the classical problems in the field of phytoplankton ecology has been to differentiate between "true" or "active" Chla and its degradation products in senescent phytoplankton and detritus.

The sequence of decomposition of Chla follows the pathways showed in Figure 1. Chla may lose the phytol (chlorophyllide a), only magnesium (Phaeophytin a or magnesium and phytol (phaeophorbide a). Prepared Chla extracts may easily be converted to magnesium-free products by the addition of acids and phytol may be removed by an enzyme, chlorophyllase. But since Chla and chlorophyllide a have identical visible absorption and fluorescence spectra, addition of acid only indicates whether the extract contains magnesium free products (phaeophorbide a and phaeophytin a) *i.e.* the phaeopigments.

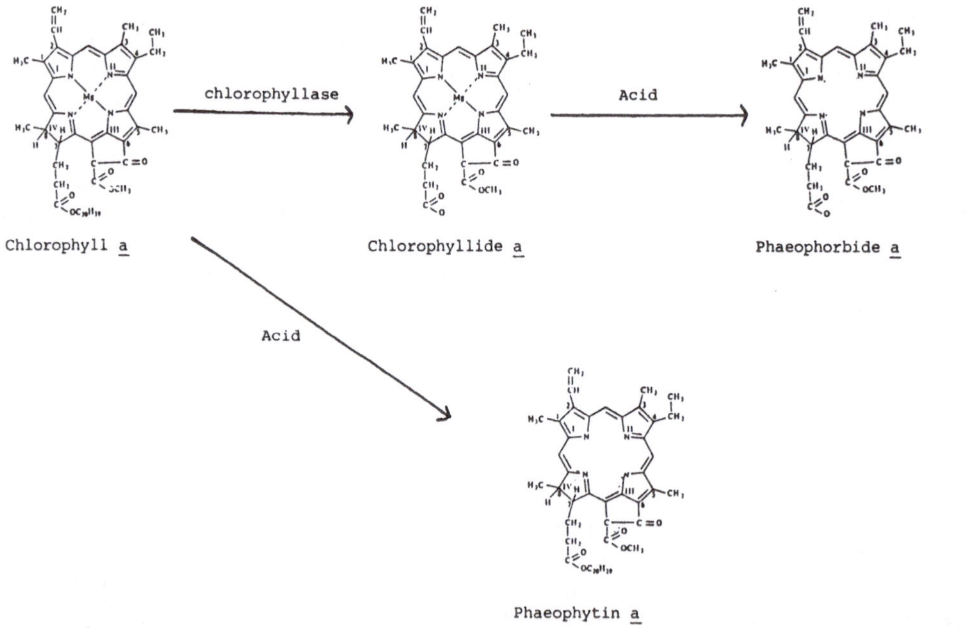

Figure 1. The principal steps in the decomposition of chlorophyll a. Phaeopigments = phaeophytin + phaeophorbide.

Accordingly a simple acidification step has been introduced to both the spectrophotometric (Lorenzen, 1967; Marker, 1972) and fluorometric (Yentsch and Menzel, 1963; Holm Hansen *et al*, 1965) techniques to help correct this source of error. In most works, it is the simple, rapid and sensitive fluorometric acidification technique which is used (Lorenzen and Jeffrey, 1980).

2. VERTICAL DISTRIBUTION OF PHAEOPIGMENTS AND THE CLASSICAL INTERPRETATION

A considerable amount of data exists on Chl$\underline{a}$ and phaeopigment vertical distribution. In the permanent thermally stratified waters of the tropical ocean and in the seasonally stratified waters of the subtropical and temperate ocean, the vertical distribution of Chl$\underline{a}$ and phaeopigments are closely related and show a typical pattern (Yentsch, 1965; Lorenzen, 1967; Soo-Hoo and Kiefer, 1982a and Figure 2):

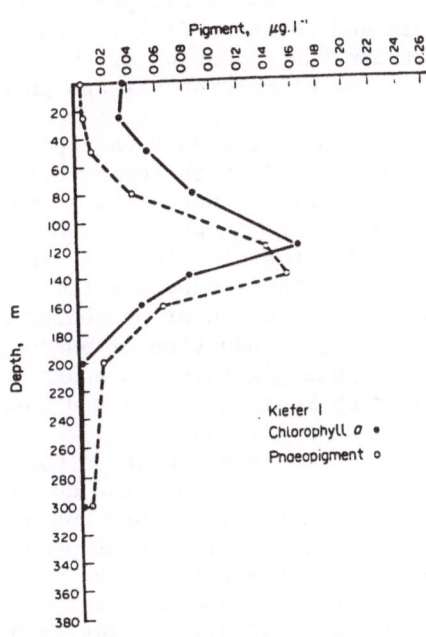

Figure 2. Vertical distribution, typical of open-ocean waters, of chlorophyll $\underline{a}$ and phaeopigments. Data are for 29°00'00"N and 122°31'12"W. (From Soo Hoo and Kiefer, 1982).

At the surface, and in the warm, nutrient-poor mixed layer, values of Chl$\underline{a}$ are low but phaeopigments are found at lower concentrations. Deeper down, in the deep chlorophyll maximum (DCM) generally located in the thermocline at the top of the nitracline (see the concept of Typical Tropical Structure by Herbland and Voituriez, 1979; Cullen, 1982) phaeopigments concentration increases, the maximum being reached systematically at a slightly deeper level than that of DCM. This phaeopigments decline but less than Chl$\underline{a}$. As a result the more or less constant percentage of phaeopigment in the mixed layer increases with depth.

2.1. How to explain such a wide-spread distribution?

Although the phaeopigments formation may result from a number pathways Chl**a** degradation (bacterial and viral degradation, prolonged darkness, according to Yentsch, 1965, and nutrient deficiency according to Wolken *et al*, 1955) it is generally recognized that the most frequent fate of phytoplankton is to be grazed by herbivores, so that the primary source of chlorophyll break-down products in natural waters is in fecal materials. Currie, (1962); Lorenzen, (1967); Nemoto, (1968, 1972); Daley, (1973) and Jeffrey, (1974), all demonstrated that phaeopigments are abundant in the guts and fecal pellets of herbivorous zooplankton. Shuman and Lorenzen (1975) found a stochiometric relationship between the ingestion of chlorophyll by grazing *Calanus* sp and the egestion of chlorophyll and phaeophorbide in fecal material.

It has also been known for a long time that phaeopigment quality is related to light : significant losses of phaeopigments were noted during photodegradation experiments by Lorenzen (1967), Moreth and Yentsch, 1970, Shuman (1978). The change in phaeopigment concentration follows a first-order decay kinetics that implies loss is proportional to the amount of radiation irrespective of exposure time.

The idea that the vertical distribution of phaeopigments represents a net result of grazing (= production of phaeopigments) and photooxidation (= destruction of phaeopigments) on short term scales in the water column was proposed by Yentsch (1965) and confirmed by Moreth and Yentsch (1970) and Lorenzen (1976).

In the former studies, most emphasis has been on copepods grazing. More recently, two papers have extended the concept to evaluate the role of microzooplankton grazing : Soo-Hoo and Kiefer (1982a, 1982b) considering the extreme small size of phaeopigment-containing particles suggested two pathways : a large particle pathway and small-particle pathway (see Figure 3a and its legend). Because feces of macrozooplankton are not quantitatively sampled by water bottles, patterns in vertical distribution of phaeopigments would result from processes affecting small particles.

In the same way, and more recently, Welschmeyer and Lorenzen (1985) constructed a model describing the dynamic budget of Chl**a** and phaeopigments within the euphotic zone of two different oceanic areas (Figure 3b). The model, based on field measurements of the vertical distribution of pigments, the vertical flux of solar radiation through the water column, experimental photodegradation experiments and downwards flux of pigments provided estimates of phytoplankton growth rates and microzooplankton grazing intensity. For the North and South central Pacific gyre phytoplankton growth rates averaged 0.2 d^{-1} and 95 % of daily grazing was due to microzooplankton herbivores, and the model suggests that grazing and growth are in balance.

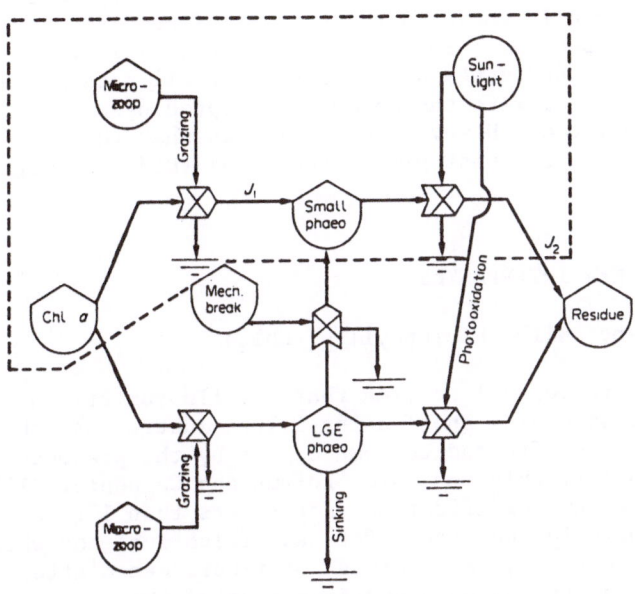

Figure 3a. Schematic representation of the major pathways of chlorophyll a degradation in the upper ocean. The part enclosed by the dashed lines describes the pathway attributed to microzooplankton grazing. The scheme is presented in the energy circuit language of Odum (1971).

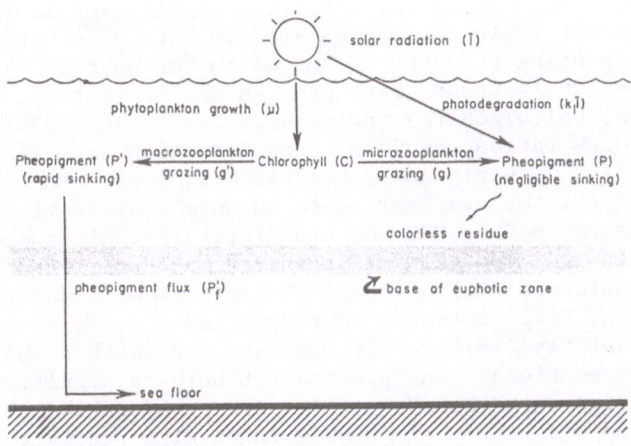

Figure 3b. Simplified diagram showing the dominant processes affecting the concentrations of chlorophyll and phaeopigments in the euphotic zone. (From Welschmeyer and Lorenzen, 1985).

To summarize, the vertical distributions of phaeopigments in the oceans and especially those in the stratified euphotic layer does not seem to raise a problem : *in situ* determinations, *in vitro* production and degradation experiments and models give, all together a coherent picture which agrees with the commonly accepted dynamics of the planktonic ecosystem. However, we shall see that there is to day some evidence that the deep phaeopigments maximum (DPM) is largely overestimated.

3. TOWARDS A NEW INTERPRETATION

3.1. Interference with chlorophyll b (Chlb)

It has been known for a long time that the fluorometric acidification techniques and specially those using filter instrument (the most commonly used) are affected to some degree by the presence of the other chlorophylls: Chlb and Chlc. Loftus and Carpenter (1971) noted that Chlb showed an acidification factor less than 1 (*i.e.* fluorescence of Chlb increases after acidification) and when Chlb is present in the extract, the decrease of fluorescence after acidification due to the phaeophytinisation of Chla is counteracted by the increase of fluorescence due to the formation of phaeophytin b. This results in a slight underestimation of Chla and calculation of significant quantities of phaeopigments even if none are present! The errors as a function of Chlc/Chla ratios are less severe than those associated with Chlb/Chla ratios.

In spite of the warning and the re-examination of the method by different authors (Holm-Hansen and Rieman, 1978; Gibbs 1979, Coveney 1982) the interference with Chlb was generally neglected because it was thought that Chlb concentrations are insignificant in the sea, particularly in the truly oceanic waters. In a study devoted to the presence of Chlb using the HPLC technique in the eastern North Pacific Ocean, Lorenzen (1981) found that although the pigment was detected in 72 % of samples, Chl(b)/Chl(a) ratios were less than 0.09 in 95 % samples. With such ratios the Chla underestimation is less than 5 % and the artificial phaeopigment /Chla ratio is less than 0.06. From the analysis of all the available data on Chlb distribution and relative abundance to Chla SOO-HOO and Kiefer (1982a) concluded : "until more data are reported as concentrations determined by separation techniques, evaluation of fluorometrically determined phaeopigments and Chla remains a consideration".

However, Jeffrey (1976) early demonstrated (with a thin layer chromatography technique), the presence of Chlb in significant proportion (ratios Chlb/Chla 0.5 at 100 m depths) in the Central North Pacific Gyre and Gieskes *et al.* (1978) found high ratios (0.2 to 0.56) for samples collected in the deep chlorophyll maximum layer of the Tropical North Atlantic. Today, more data exist and particularly on the vertical axis. Firstly, Neveux and De Billy (1986), using a modified spectrofluorometric method, have studied the distribution of

Chl<u>a</u>, Chl<u>b</u>, Chl<u>c</u> of the Indian Ocean where oligotrophic conditions characterized most of the area. Vertical profiles showed typical DCM located at depth of 1 % of surface light. These DCM often were associated with high Chl<u>b</u>/Chl<u>a</u> ratio. More over, *spectrofluorometry showed no significant increase in the percentage of phaeopigments <u>a</u> in the vicinity of the DCM, whereas such an increase was observed when simultaneously using the traditional fluorometric method* : overestimations as much as 300 % of the value computed by spectrofluorometry were observed. The artifact was partially attributed to the relative abundance of Chl<u>b</u> in the DCM.

To the contrary, in the upwelling region, an underestimation of phaeopigments concentration was observed, but always less than 50 %. This was related to the relative abundance of Chl<u>c</u> which tends to underestimate phaeopigments (Loftus and Carpenter, 1971; Neveux, 1976).

Secondly, using reverse phase HPLC, Gieskes and Kraay (1986) have made the same observations in the stratified waters of the tropical Atlantic Ocean where a permanent DCM occurs in the bottom of the euphotic zone : At the depth of the DCM layer, Chl<u>b</u> concentration was considerable (Chl<u>b</u>/Chl<u>a</u> ratios near 0.7) while near the surface only traces were found. They only detected traces of phaeoplytin a and phaeophorbide a with HPLC methodology in the deep samples and they concluded : "the abundance of phaeopigments measured in the conventional way near the bottom of the euphotic zone can therefore

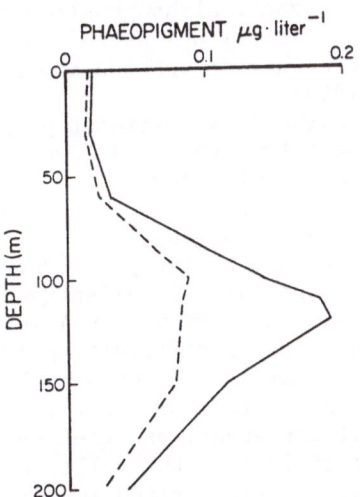

Figure 4. Vertical profile of phaeopigment concentration in tropical waters South of Hawaii. Composite profile for four stations ($n = 45$). Solid line: phaeopigments as estimated in the fluorometer; dashed lines: phaeopigment concentrations after correcting for chlorophyll <u>b</u> concentration in the sample. (From Vernet and Lorenzen, 1987).

not simply be used as an indicator of the presence of detritus or of grazing. *It may just be an artifact of the method used for its measurement reflecting the abundance of Chl b containing phytoplankton*".

Finally, and very recently Vernet and Lorenzen (1987), in a paper specially devoted to the problem of interference with Chlb, found consistently high concentrations of Chlb at the bottom of the euphotic zone in the tropical North Pacific Ocean (100-200 m) *but it was not detected in several samples of the mixed layer (0-60 m)*. The discrepancy between the phaeopigments concentration measured in the water column, by fluorometry and its Chlb corrected value (Figure 4) is most pronounced between 100 and 150 m, at the depth of phaeopigments maximum. If the Chlb correction, in that case, does not remove the feature, it decreases the overall concentration of phaeopigments suspended in the euphotic zone and flattens the deep phaeopigments maximum (DPM).

3.2. Field observations of Chlb containing algae

Now that there is some evidence that the DPM is largely overestimated by the presence of a Chlb rich layer, the question is : Are there field observations of vertical distribution of algae containing Chlb that support the pigment distribution? Chlb is an accessory pigment of algae belonging to the Chlorophyceae, Prasinophyceae and Euglenophycea and is not found in other algal groups. (Jeffrey, 1976). These classes form the commonly designated "green algae". Although Allen (1961) listed green algae in records of Pacific phytoplankton, they were often regarded as being coastal and estuarine species and not present in truly oceanic waters (Butcher, 1959).

In fact, the absence of green algae in significant proportion in the stratified euphotic layer may be attributed to a lack of appropriate methodology : most of the studies devoted to the vertical distribution of phytoplankton have used the Utermöhl *settling technique* with an inverted microscope on *preserved samples*.

Recently, it has been noted that much of the photosynthetic activity and Chla of water samples (especially in oligotrophic waters) is contained in very small cells (less than few microns) which have remained largely ignored because they are so difficult to identify and count (Fogg, 1986). Because of their fragility, they are also selectively destroyed by the use of preservatives (Booth, 1987) : for example, even with the best known of preservatives (glutaraldehyde), Murphy and Haugen (1985) found that 15-20 % of the cells in prasinophytes (= green algae) cultures disintegrated immediately and another 35-40 % were lost within 8 days! It is only with the general and recent use of the epifluorescence microscopy - initially designed to count stained bacteria (Hobbie *et al.*, 1977) that it became evident that these very small phytoplankters (*i.e.* picoplankton and small nannoplankton) make a significant contribution to total phytoplankton (see for example LI *et al.* 1983; Platt *et al.* 1983).

Although the studies of the *in situ* size distribution of "Phaeopigments"-containing particles (that is Chlb - containing algae)

165

are less numerous than those of Chla containing particles, there is
evidence that the small size categories dominate : Soo-Hoo and Kiefer
(1982a) reported that "phaeopigments" passed through the 5 μm filter
and were more usually found in particles smaller than those containing
Chla. Herbland *et al*. (1985) found the same result in the whole
equatorial Atlantic Ocean : the percentage of "Phaeopigments" was
larger in the < 2 and < 1 fm fraction and the < 1 μm "Phaeopigments"
showed a typical vertical pattern with a well defined maximum in the
nitracline, slightly deeper than the Chla maximum. Are there recent
reports of direct observations suggesting the presence, in significant
proportions of small "green algae" in the vicinity of the DPM?

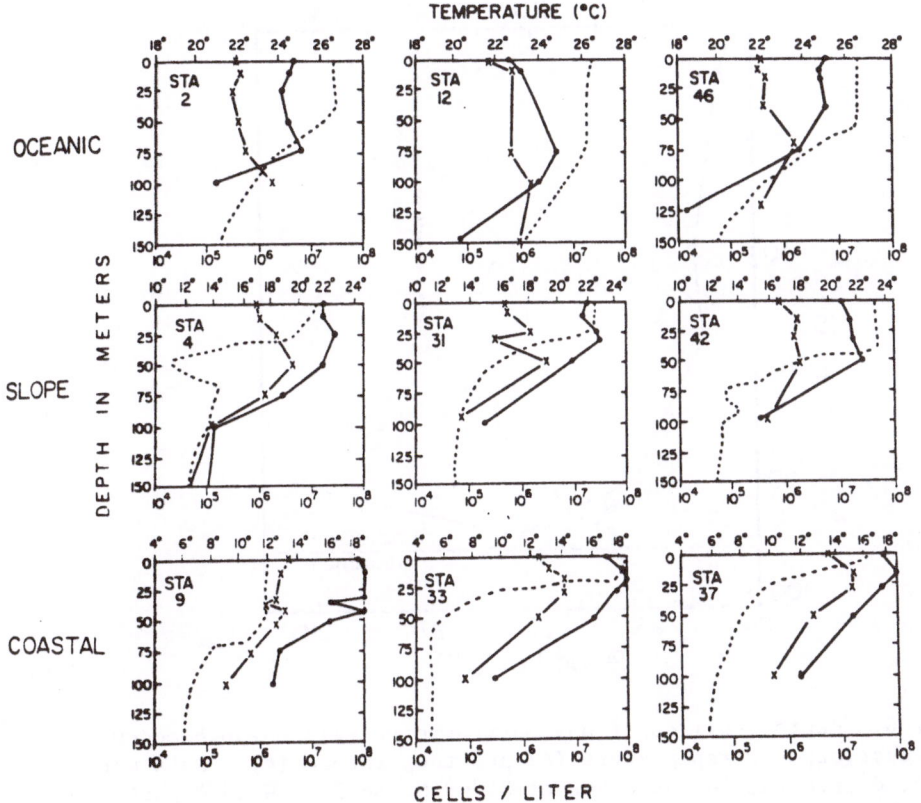

Figure 5. Comparison of the vertical distributions of cyanobacteria
(○) and eucaryotes (x) in vertically stable water columns. (From
Murphy and Haugen, 1985).

Murphy and Haugen (1985) compared the vertical distribution and
abundance of procaryotic chroococcoid cyanobacteria and eucaryotic
phototrophic ultraplankton at 50 stations in the North Atlantic. In
the isothermal layer of the vertically stable waters, cyanobacteria
outnumbered eucaryotes by roughly an order of magnitude (Figure 5).
They reached a maximum at the thermocline at about the 1 % light
level, and their abundance decreased rapidly below that. In contrast,
the eucaryotic assemblage continued to increase for another 25 m (the
maximum is at about the 0,5 % light level) and decreased less rapidly
thereafter. Thus in the thermocline and below the eucaryotes equaled
or exceeded the cyanobacteria. Glover *et al.* (1985) found similar
results for the Sargasso Sea (Figure 6). Moreover, Glover *et al.*

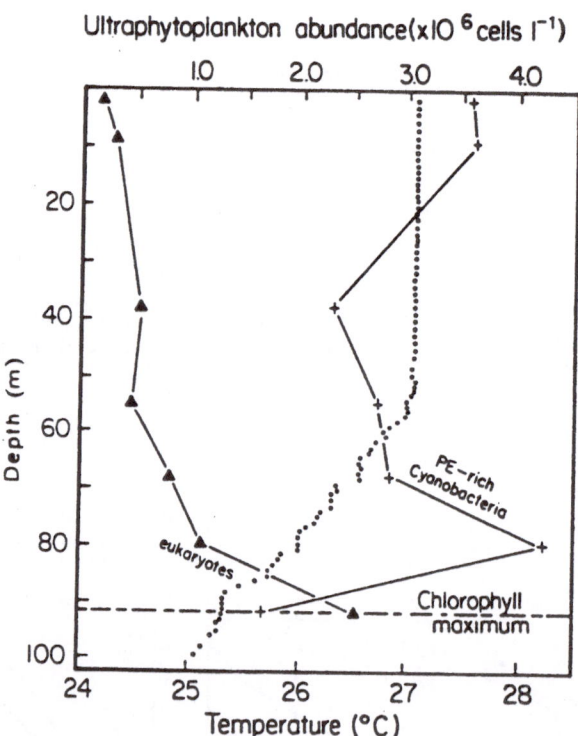

Figure 6. Depth profiles of phycoerythrin(PE)-rich *Synechococcus*
(+), eucaryotic ultraplankters (Δ) and temperature (∘). Data were
collected from the Sargasso Sea at 34°49'N, 66°20.6'W in August
1983. Pump water was immediately filtered at <125 mm Hg through a 3-
μm Nuclepore filter. The filtrate was passed through a 0.2-μm
Nuclepore filter and cells were classified as either *Synechococcus*
or eucaryotes. (From Glover *et al.*, 1985).

(1986), using photosynthesis and growth experiments with variable light quality and quantity found that eucaryotic ultraplankters have greater photosynthetic and growth efficiencies than *Synechococcus* in the dim-blue violet light occurring at the bottom of the euphotic layer (Figure 7).

Thus, the candidates for the validation of Chl_b_ distribution do exist, but they have been largely ignored because their smallness and fragility require new techniques of identification and counting. It remains to be unambiguously demonstrated that these small dominant eucaryotes are Chl_b_ containing algae.

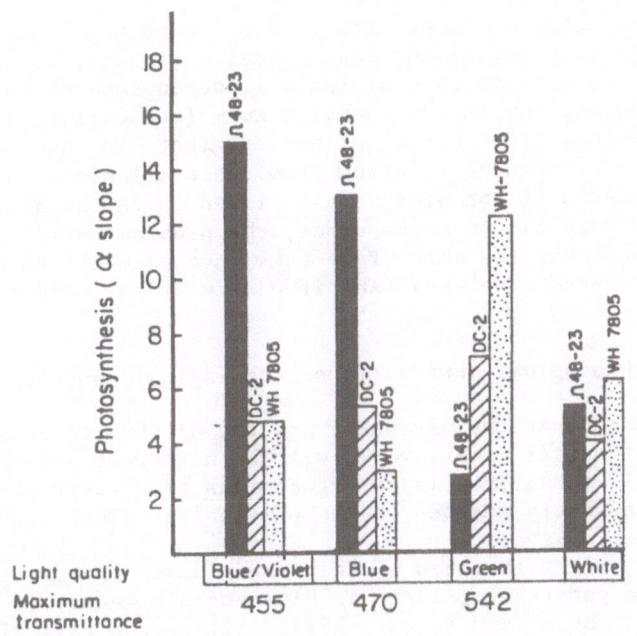

Figure 7. Photosynthetic efficiencies in different light qualities for an ultraplanktonic prasinophyte clone Ω48-23 (□). *Synechococcus* clone (DC-2 (□) and *Synechococcus* clone WH 7805 (□). Batch cultures were grown at 20°C, in 14/10 h light/dark cycles in blue-violet light at 1.6 x 10[15] quanta cm^{-2}s^{-1}. In 'log-phase', we determine chlorophyll a concentrations[24] and photosynthetic rates[25] at four rate-limiting fluxes of blue-violet, blue, green or white light. (From Glover *et al.*, 1986).

4. CONCLUSION

For the non-specialist, the "story of phaeopigments" may appear
somewhat tedious, of limited interest, and irrelevant to the problem
of biological-physical interactions in the ocean. If I concede that
the matter is an affair of specialist I think on the opposite that
this example concerns a general feature, with important ecological
implications, including physical-biological interactions.

4.1. DPM or Chl_b layer : a general feature in the ocean

Relevante and Gilmartin (1973) and Shulenberger (1978) showed that the
DCM crosses the Pacific Ocean at 28°N and that it is quasi-permanent
between 40°N and 25°S in the eastern part of the Pacific. Similar
observations have been made in the whole equatorial Atlantic and in
the eastern part of the tropical Atlantic (Gieskes and Kraay, 1986;
Herbland, 1983 and Herbland et al. 1985).
     Accordingly, there is large amount of evidence that the DCM is a
permanent feature in tropical and subtropical ocean and, if not fully
explained, it is considered as a biological consequence of the
presence of a thermocline in the euphotic zone (see Cullen, 1982).
Although the vertical distribution of phaeopigments is less documented
in the literature, it can be inferred from the available data that the
DCM and DPM are often if not always, associated in the stratified
euphotic zone (Figure 2). In consequence, the problem deals with about
40 % of the world ocean (or more if we take into account the seasonal
thermocline in temperate waters). And if we are wrong, we are wrong at
a large scale!

4.2. Physical - biological interactions : ecological consequences

It stands to reason that the vision of how the planktonic ecosystem in
tropical and subtropical waters works will be different according to
whether the DPM is actually a layer of detritus or a layer of active
ultraplankton, perfectly adapted to the physical-chemical conditions
that prevail at this level.
     For example, light absorbed by Chl_b can promote photosynthesis at
this depth : deep penetrating light in blue oceanic waters is near 480
nm (Jerlov, 1978 ; Morel and Prieur, 1977) ; it agrees well with in
vivo absorption (470 nm : Yentsch and Guillard, 1969) and corrected
excitation fluorescence peaks (486 nm : Neveux, 1982) of Chl_b.
     The recent discovery of a deep Chl_b layer reinforces the concept
of two superimposed layers of phytoplankton living in the stratified
euphotic zone (Venrick, 1982) with different dominating pigments,
different mean size and growth rates (Gieskes and Kraay, 1986) and
probably different nutrients requirements (Fogg, 1986). It gives a new
argument for the presence of a "shade flora" (Sournia, 1982), and the
existence of closed interactions between physical forcing, (light
spectrum) and biological response (pigments).
     On the other hand, it implies that recent estimates of
microzooplankton grazing from pigment budgets (Soo-Hoo and Kiefer,

1982a, Welschmeyer and Lorenzen, 1985) would be lower than previously believed for certain parts of the ocean, *i.e.* where a thermally stratified water column exists.)

Obviously, further investigations, with a new vision and new techniques, are necessary to complete the picture of a component of the ecosystem that may be responsible for most of the energy transformations and recycling of material taking place in the oceans (Fogg, 1986).

Three concluding remarks

1)   The interference with Chl_b was neglected during 20 years principally *because corrections found in surface samples were applied to deep samples*. If comparison and extrapolation are two techniques which are widely used in the scientific approach, this example shows that they must be applied with care. One would make today the same mistake if we should consider that the existence of phaeopigments is doubtful in the old upwelled waters or after a spring bloom of diatoms in temperate waters. The concepts of Lorenzen (1967) relating the amount of phaeopigments to zooplankton grazing are still valid. It is only the generalization to any environmental conditions which is questionable.

2)   Another cause of error is less the lack of a suitable method to detect the presence of Chl_b (as early as 1976 Jeffrey drew the attention upon the problem and wrote that thin layer chromatography was used in her laboratory as a routine method for field samples) - than the availability of a rapid, sensible and cheap method for measuring Chl_a. The attractive "easy to use" of the routine fluorometric technique has prevented the extension of more sophisticated and appropriate methods, more especially they give a smaller number of samples (quantity versus quality). However, a few profiles of precisely known different pigments in representative vertical structures far surpasses data obtained by some hundred profiles by the routine technique. Although, they are irreplaceable the *universal and simple methods should be used with due considerations of their properties and limits*. The controversy with the ^{14}C method is an other important example of the necessity of such a caution.

3)   Finally, for about 20 years (1965-1985) there was a kind of consensus (the word in vogue is "paradigm") on the phaeopigments status in the ocean: *in situ* observations, laboratory experiments and models agreed, even in case where it appears today that the presence of phaeopigments is doubtful! However, some observations (unpublished data of the author), were not, or hardly taken into account by the previous concepts : for example : how to explain the simultaneously short term increases of Chl_a and "phaeopigments" in certain low-lighted tamised cultures of natural phytoplankton of a deep chlorophyll maximum ? Why use photo-oxidation properties of *in vivo* phaeopigments as an explanation for their vertical *in situ*

distribution and forget the same properties of *in vivo* chlorophyll ? These two inconsistencies are well explained by the new interpretation. Let us bet that now the old paradigm is attacked, many other unpublished contradictions will appear in the literature.

5. REFERENCES

Allen, M.B.: 1961, 'Our knowledge of the kinds of organism in Pacific phytoplankton', *Proc. Pacif. Sci. Congr.* **10**, 58-60.

Booth, B.C.: 1987, 'The use of autofluorescence for analysing oceanic phytoplankton communities', *Bot. Marina* **30**, 101-108.

Butcher, R.W.: 1959, 'An introductory account of the smaller algae of British coastal waters. I. Introduction and chlorophycae', *Fishery Invest.*, London (**Ser. 4**), 1-74.

Coveney, M.: 1982, 'Elimination of chlorophyll b interference in the fluremetric determination of chlorophylle a and phaeopigment a. *Arch. Hydrobiol. Beih. Ergebn. Limnol.*, **16**, 77-90.

Cullen, J.J.: 1982, 'The Deep chlorophyll Maximum : comparing vertical profiles of chlorophyll a', *Can J. Fish Aquat. Sci.* **39**, 791-803.

Currie, R.: 1962, 'Pigments in zooplankton faeces', *Nature* **193**, 956-957.

Daley, R.J.: 1973, 'Experimental characterization of lacustrine chlorophyll diagenesis : II. Bacterial, viral and herbivore grazing effects', *Archiv fur Hydrobiologie* **72**, 402-439.

Fogg, G.E.: 1986, 'Light and ultraphytoplankton', *Nature* **319**, 96.

Gibbs, C.F.: 1979, 'Chlorophyll b interference in the fluoremetric determination of chlorophyll a and phaeopigments', *Aust. J. Mar. Freshwater Res.* **30**, 597-606.

Gieskes, W.W.C., G.W. Kraay and S.B. Tijssen: 1978, 'Chlorophylls and their degradation products in the deep pigment maximum layer of the tropical North Atlantic', *Neth. J. Sea Res.* **12**, 195-204.

Gieskes, W.W. and G.W. Kraay: 1986, 'Floristic and physiological differences between the shallow and the deep nanophytoplankton community in the euphotic zone of the open tropical Atlantic revealed by HPLC analysis of pigments', *Mar. Biol.* **91**, 567-576.

Glover, H.E., A.E. Smith and L. Shapiro: 1985, 'Diurnal variations in photosynthetic rates : comparisons of ultraphytoplankton with a larger phytoplankton size fraction', *J. Plank. Res.* **7**, 519-535.

Herbland, A. and B. Voituriez: 1979, 'Hydrological structure analysis for estimating the primary production in the tropical Atlantic', *J. Mar. Res.* **37**, 87-101.

Herbland, A.: 1983, 'Le maximum de chlorophylle dans l'Atlantique tropical oriental : description, Ecologie, interpretation', *Oceanogr. Trop.* **18**, 295-318.

Herbland, A., A. Lebouteiller and P. Raimbault: 1985, 'Size strucutre of phytoplankton biomass in the equatorial Atlantic Ocean', *Deep Sea Res.* **32**, 819-836.

Hobbie, J.E., R.J. Daley and S. Jasper: 1977, 'Use of nuclepore filters for counting bacteria by fluorescence microscopy', *Appl. Environ. Microbiol.* **33**, 1225-1228.

Holm-Hansen, O. and B. Riemann: 1978, 'Chlorophylle a determination : improvements in methodology', *OIKOS* **30**, 438-447.

Holm-Hansen, O., C.J. Lorenzen, R.W. Holmes and J.D.H. Strickland: 1965, 'Fluoremetric determination of chlorophyll', *J. Cons. perm. Int. Explor. Mer.* **30**, 3-15.

Jeffrey, S.W.: 1974, 'Profiles of photosynthetic pigments in the ocean using thin layer chromatography', *Mar. Biol.* **26**, 101-110.

Jeffrey, S.W.: 1976, A report of green algal pigments in the Central North Pacific ocean', *Mar. Biol.* **37**, 33-37.

Jerlov, N.G.: 1978, 'The optical classification of sea water in the euphotic zone', Rept nx 36 of the Inst. for fysik oceanografi, Kobenhauns Universitet.

Li, W.K. and others: 1983, 'Autotrophic picoplankton in the tropical ocean', *Science* **219**, 292-295.

Loftus, M.C. and J.H. Carpenter: 1971, 'A fluoremetric method for determining chlorophylls a, b and c', *J. Mar. Res.* **29**, 319-338.

Lorenzen, C.J. and S.W. Jeffrey: 1980, 'Determination of chlorophyll in sea water', UNESCO technical papers in marine science, no. 35, 20 p.

Lorenzen, C.J.: 1967, 'Determination of chlorophyll and phaeopigments: spectrophotometric equations', *Limnol. Oceanogr.* **12**, 343-346

Lorenzen, C.J.: 1967, 'Vertical distribution of chlorophyll and phaeopigments : Baja California', *Deep Sea Res.* **14**, 735-745.

Lorenzen, C.J.: 1976, 'Primary production in the sea', in D.H. Cushing and J.J. Walsh (eds.), *The ecology of the seas*, Saunders, London, 173-185.

Lorenzen, C.J.: 1981, 'Chlorophyll b in the eastern North Pacific Ocean', *Deep Sea Res.* **28**, 1049-1056.

Marker, A.F.H.: 1972, 'The use of acetone and methanol in the estimation of chlorophyll in the presence of phaeophytin', *Freshwat. Biol.* **2**, 361-385.

Morel, A. et L. Prieur: 1977, 'Energie radiative disponible pour la photosynthse (PAR). Mesures du flux de photons et analyse spectrale', in B. Coste (ed.), *Campagne GUIDOME 1976, fasc. 2, Resultats des campagnes a la mer*, no. 13 Publ. du CNEXO.

Moreth, C.M. and C.S. Yentsch: 1970, 'The role of chlorophyllase and light in the decomposition of chlorophyll from marine phytoplankton', *J. exp. mar. Biol. Ecol.* **4**, 238-249.

Murphy, L.S. and E.M. Haugen: 1985, 'The distribution and abundance of phototrophic ultraplankton in the North Atlantic', *Limnol. Oceanogr.* **30**, 47-58.

Nemoto, T.: 1972, 'Chlorophyll pigments in the stomach and gut of some macrozooplankton species', in A.Y. Takenouty (ed.), *Biological Oceanography of the northern North Pacific Ocean*, Indemitsu Shoten, Tokyo, 411-418.

Neveux, J. and G. de Billy: 1986, 'Spectrofluoremetric determination of chlorophylls and phaephytins. Their distribution in the western part of the Indian ocean (July to August 1979)', *Deep Sea Res.* **33**, 1-14.

Neveux, J.: 1976, 'Dosage de la chlorophylle a et de la phaeophytine a par fluorimetrie', *Annales de l'Institut Oceanographique* **52**, 165-174.

172

Neveux, J.: 1982, 'Pigments du phytoplancton : composition et activite photochimique des chlorophylles', Signification Ecologique de la fluorescence *in vivo* de la chlorophylle a. Thse de Doctorat d'Etat, Univ. P. et M. Curie, Paris 146 p.

Platt, T.D., V. Subbarao and B. Irwin: 1983, 'Photosynthesis of picoplankton in the oligotrophic ocean', *Nature* **301**, 702-704.

Relevante, N. and M. Gilmartin: 1973, 'Some observations on the chlorophyll maximum and primary production in the Eastern North Pacific', *Rev. ges. Hydrobiol.* **58**, 819-834.

Shulenberger, E.: 1981, 'The deep chlorophyll maximum and mesoscale environmental hererogeneity in the western half of the North Pacific Central Gyre', *Deep Sea Res.* **25**, 1193-1208.

Shuman, F.K. and C.J. Lorenzen: 1975, 'Quantitative degradation of chlorophyll by a marine herbivores', *Limnol. Oceanogr.* **2**, 580-586.

Shuman, F.K.: 1978, 'The fate of phytoplankton chlorophyll in the euphotic zone - Washington coastal waters', Ph. D. Dissertation, University of Washington, 125 pp.

Shutter, B.: 1979, 'A model of physiological adaptation in unicellular algae', *J. Theor Biol.* **78**, 519-552.

Soo-Hoo, J.B. and D.A. Kiefer: 1982a, 'Vertical distribution of phaeopigments - I. A simple grazing and photo-oxydative scheme for small particles', *Deep Sea Res.* **29**, 1539-1551.

Soo-Hoo, J.B. and D.A. Kiefer: 1982b, 'Vertical distribution of phaeopigments. II. Rates of production and kinetics photo-oxidation', *Deep Sea Res.* **29**, 1553-1563.

Sournia, A.: 1982, 'Is there a shade flora in the marine plankton?', *J. Plank. Res.* **4**, 391-399.

Strickland, J.D.H.: 1965, 'Production of organic matter in primary stages of the marine food chain', in J.P. Rikley and G. Skirrow (eds.), *Chemical Oceanography*, Academic Press, London, pp. 447-610.

Venrick, E.L.: 1982, 'Phytoplankton in an oligotrophic ocean : observations and questions', *Ecol. Monogr.* **52**, 129-154.

Vernet, M. And C.J. Lorenzen: 1987, 'The presence of chlorophyll b and the estimation of phaeopigments in marine phytoplankton', *J. Plank. Res.* **9**, 255-265.

Welschmeyer, N.A. and C.J. Lorenzen: 1985, 'Chlorophyll budgets : Zooplankton grazing and phytoplankton growth in a temperate fjord and the Central Pacific Gyres', *Limnol. Oceanogr.* **30**, 1-21.

Wolken, J.J., A.D. Mellon and C.L. Greenblatt: 1955, 'Environmental Factors affecting growth and chlorophyll synthesis', *J. Protozool.* **2**, 89-96.

Yentsch, C.S.: 1965, 'Distribution of chlorophyll and phaeophytin in the open ocean', *Deep Sea Res.* **12**, 653-666.

Yentsch, C.S. and D.W. Menzel: 1963, 'A method for the determination of phytoplankton chlorophyll and phaeophytin by fluorescence', *Deep Sea Res.* **10**, 221-231.

Yentsch, C.S. and R.R.L. Guillard: 1969, 'The absorption of chlorophyll b *in vivo*', *Photochemistry and photobiology*, **9**, 385-388.

FOOD CHAINS AND FISHERIES: AN ASSESSMENT AFTER 20 YEARS

REUBEN LASKER
Southwest Fisheries Center
NOAA/NMFS
La Jolla, CA 92038
U.S.A.

ABSTRACT. Oceanic food chain research has been notably unsuccessful in predicting fish yield, chiefly because of inadequate methods for determining primary production and for the lack of data on transfer efficiencies between steps in food chains. Methods have improved in recent years in assessing primary production (both recycled and new) but the state of the science does not yet permit grand means of primary production to be assigned to the world's oceans nor for them to be used to predict fisheries yield.

1. INTRODUCTION

Almost 20 years ago, at the 1968 symposium on marine food chains (Steele, 1970) I presented data to show that it was impossible to maintain the 1932-33 population of sardines of about 3,000,000 metric tons (mtons) in the California Current with the primary production measured for the California Current and inshore waters where the sardine resides (Lasker, 1970). My data source for primary production was the late John D.H. Strickland who provided me with his most up-to-date estimate ($1gC/m^2$ per day) of primary production for the area where sardines were found. These data were subsequently published (Eppley *et al.*, 1970). Sardines are zooplankton feeders and therefore they were at the top of a two step food chain. I used a conversion efficiency of 10% between steps in the food chain.

In the final analysis, there was no accounting for sustaining the large biomass with the primary production measured in the sardine's habitat. At that time I speculated that the sardine biomass may have been estimated at too high a level, or the estimates for primary and secondary production were too low.

Food chain research has contributed more to the knowledge of biogeochemical cycles in the ocean and to describing the driving forces behind the sinking flux of biogenic particles than to fisheries science in recent years. Primary production may not be the most important environmental variable for fish stocks. This is suggested by poor correlations between primary production and fish biomass and the fact that aggregations of food rather than total integrated food may

B. J. Rothschild (ed.), Toward a Theory on Biological-Physical Interactions in the World Ocean, 173–182.

be more important to larval fish survival (Lasker, 1975).
Nevertheless at some level there must be a reckoning between the
energy requirements of fish and the production of their food.

In this paper I reexamine this problem in light of the latest
data available, examine other biological systems and try to come to
some conclusion whether Food Chain Research has made a contribution to
Fisheries Science.

2. THE PRIMARY PRODUCTION CONTROVERSY

For some time there has been general unhappiness among biological
oceanographers with the Carbon-14 method of primary productivity
measurements (Peterson, 1980). Studies of oxygen distribution in the
north Pacific by Shulenberger and Reid (1981) have also cast doubt on
previous work done in the "deserts of the sea" on primary production.
They suggested that measurements of oxygen trapped below the surface
layers of nutrient-poor oligotrophic regions, were evidence of an
amount of primary production, heretofore not measured, and two to four
times as much as ever reported for these areas throughout the world.
Jenkins (1982) and Jenkins and Goldman (1985) came to similar
conclusions based upon oxygen distributions in the north Atlantic.
Recent measurements on the CO_2 system also seem to support the concept
of extra oxygen production and unmeasured primary production (Brewer,
et al. 1986, and Pers. Comm.).

Notwithstanding the argument by Platt (1984) and Platt and
Harrison (1986) who see errors in interpretation and magnitude in the
subsurface oxygen hypothesis, the ocean science community may have
been underestimating primary production by some multiple in the open
ocean (Kerr, 1983) and by some smaller fraction in inshore waters.

Toxicity of samplers and incubation bottles has been implicated in
depressing the amount of carbon incorporation in the carbon-14
technique (Fitzwater *et al.*, 1982). The trace metal "clean" methods of
these authors appear to have resolved much of the problem and have been
used in recent U.S. National Science Foundation programs aimed at
improving primary production measurements.

3. RYTHER'S 1966 PAPER ON FISH PRODUCTION

The paper by Ryther (1966) is a good example of the extent to which
primary production figures have been used in calculations on fish
yield. Ryther divided the world's oceans into open ocean, coastal, and
upwelling regions, then assigned 5, 3 and 1.5 trophic levels and 10,
15 and 20% efficiencies of transfer respectively, leading to the
production of fish.

Based on productivity figures compiled and published by Russian
scientists (Koblentz-Mishke, *et al.* 1970; this was In Press in 1969)
and Steemann Nielsen and Jensen, (1957) using the Steemann Nielsen
(1957) technique of Carbon-14 incorporation into phytoplankton, Ryther
deduced that primary production would be at its maximum at 20 x 10^9

mtons of carbon per year in the world and that the world fish production can be only about 24×10^7 mtons (wet weight) per year. From this figure he concluded that the world's commercial fish maximum sustainable yield could not exceed 10×10^7 mtons per year.

Fishery biologists were quick to dispute Ryther's conclusions but chiefly his low estimate of what tropical oceans can produce in fish. Alverson *et al.* (1970) argued with Ryther's conclusion that 90% of the ocean is a "biological desert." They pointed out that it did not square with the fact that the yield in tunas, billfishes and other open ocean pelagic fishes in the mid 1960s was 2.5 to 3×10^6 mtons annually, as against Ryther's estimate of only 1.6×10^6 mtons per year total fish production. The tuna and billfish catch alone exceeded 2.2×10^6 mtons in 1984, the last year for which statistics were compiled (FAO, 1986). To make their case, Alverson *et al.* (1970) included "tuna-like" fishes, many of which occur in coastal and upwelling regions, although they argued that it made no difference to their case.

In fairness to Ryther, he made the conservative assumption that there are most likely 5 steps in the food chain in the open ocean, as opposed to about 1.5 in upwelling systems, and used a reasonable 10% efficiency in energy transfer between steps in the open ocean. The four to five step food chain has been verified by Mearns *et al.*(1981) and Rau *et al.* (1983). The important point to be made however is that Ryther calculated a maximum sustainable yield for fisheries for all the world's oceans of 100 million mtons. This maximum was the major point of contention from fishery scientists. Alverson *et al.* (1980) were convinced this was too low. A recent paper by Wise (1984) makes the point that 100-120 million mtons limitation may only be due to fishing practices and consumer acceptance. He leaves his readers with the impression that all previous prognostications of status quo for world fisheries (*e.g.* United States, 1980) have been mistaken and will continue to underestimate harvests from the sea. We have very reliable statistics on the world marine fish catch which show that the tuna and billfish catch has been increasing since 1968 by 70000 mtons per year and for all marine fish by 1.5×10^6 mtons per year for the last 35 years (FAO, 1986) and this has been one of the causes for the optimism of fishery scientists.

A recent estimate of primary production for the global ocean is 51×10^9 mtons of carbon per year (Martin *et al.*, 1987). The mean productivity rates for the open ocean, coastal zone and upwelling areas of Martin *et al.* exceed those summarized by Ryther (1969) by factors of 2.6, 2.5 and 1.4 respectively. All but about 7×10^9 mtons of this production is recycled in the upper 100 m by processes including fish metabolism. The fish catch itself is an export from the productive system of the surface ocean and thus cannot exceed global "new production" (Dugdale and Goering, 1967). Global new production is probably about 20% of the total. A recent estimate of new production in the equatorial Pacific alone is 1×10^9 mtons per year (Chavez and Barber, 1987). Global new production is well in excess of the fish catch but the fate of most of this new production is to end up as biogenic particles in the depths of the sea (Eppley and Peterson, biogenic particles in the depths of the sea (Eppley and Peterson, 1979).

The uncertainties of Ryther's educated assumptions and the importance of correct figures in predicting the world fish catch have been an impetus to food chain scientists to make the pertinent measurements more accurate and therefore to obtain more reliable production values for the world's oceans. Should the fishing community view the future with unbridled optimism or is the leveling off of world fisheries just a matter of a few years' time?

4. WHAT CARBON INCORPORATION FIGURES TO USE?

For the coastal Pacific sardine, (*Sardinops sagax* = *S. caeruleus*), we have population figures from California and Baja California fishery data (Murphy, 1966 and MacCall, 1979). At its maximum, in 1932-33, the subpopulation centered in the Southern California Bight was 3.2 million mtons. The energy required for respiration by this population was 2.8×10^{12}Kcal/month, and the population required an intake of 2.2×10^{11}g C/month (Lasker, 1970).

The area covered by this population was 4.14×10^{10}m^2. At 1g/m^{2}C produced for a 30 day month, the primary production for this area is calculated to be 1.24×10^{12}gC/month.

Primary production.	1.24	x	10^{12}gC/month
Zooplankton production.	0.124		"
Sardine respiration.	0.22		"

These results are not substantially different from what I reported in 1970. Ryther used a 20% efficiency figure between phyto- and zooplankton while I used 10%. On the other hand I used 1gC/m^2 per day as a best estimate of coastal productivity and now the figure, derived from many more Carbon-14 measurements, is best put at 0.5g C/m^2 per day for the inner Southern California Bight (Smith and Eppley, 1982) while higher values are found offshore in the California Current (Hayward and Venrick, 1984). When the Ryther and Smith and Eppley figures are used, the calculation is:

Primary Production.	0.62	x	10^{12}gC/month
Zooplankton production.	0.12		"
Sardine Respiration.	0.22		"

Despite the 2X discrepancy between calculated zooplankton production and the amount of carbon needed to support the 3 million mtons of sardines, the figures are as good as we can expect given the present state of knowledge of primary production and transfer efficiency in the coastal zone, the lack of information on secondary production, and the errors in fish population estimation. For example, is estimating the sardine population, Murphy (1966) used an assumed natural mortality (M) figure of 0.5. Had he used 0.4 the estimated biomass would have been about 20% higher.

It also seems likely that to the north, off central and northern California, the highly energetic "jets and squirts" which move large quantities of cold, nutrient-enriched water offshore and become entrained into the southward moving California current (Lasker, et al., 1981; Treganza et al., 1987) add more primary production to the habitat of the sardine than has been included in the gross overall average. For the sardine and other fauna of this region it seems that we have achieved about the best input-output balance for the population if we agree that there are still uncertainties in our understanding of transfer efficiencies and in the methods and time scales of sampling used to measure primary production, and that there are probably nutrient inputs not yet accounted for. A promise of satellite technology is that we may acquire in the future much more comprehensive information on phytoplankton production on synoptic scales (Eppley et al. 1985).

5. OTHER UPWELLING SYSTEMS

While the northern and central parts of the California coast are areas of intense upwelling, with the production of jets and squirts bringing in as yet unmeasured quantities of nutrients, the Peruvian upwelling system by contrast supports much greater populations of fish, particularly the Peruvian anchoveta, Engraulis ringens and the Chilean sardine, Sardinops sagax. At its height the spawning biomass of the Peruvian anchoveta exceeded 20 million mtons (Csirke, 1980; Tsukayama, 1982) and the catch alone in 1968 was greater than 12 million mtons. The anchoveta is a phytoplankton feeder and the one-step food chain undoubtedly helped make this remarkable upwelling system more productive of fish. A 20 million ton anchoveta population has a carbon intake rate of approximately 6.8×10^{13} gC/yr based on the respiration measurements of Villavicencio (1981) where the Peruvian anchoveta has a total metabolic requirement of 87.2 cal./g per day. The conversion of calories to carbon is based on the combustion of glucose where 1g C yields 9350 calories. Chavez and Barber (1987) conclude from their study that 1409gC/m^2 per year are produced in the Peruvian upwelling zone (3.84gC/m^2 per day) and that the area of that zone is 1.82×1011 m^2. The total production was calculated to be 2.6×10^{14} gC per year, 40 times more than required by a 20 million ton anchoveta population. These exercises serve to illustrate that, at least in upwelling areas, the correspondence between carbon production measured and that needed by large fish populations is within reasonable bounds. Whether the discrepancies are due to primary production measurements or the area to which they are applied, other inaccuracies, e.g. estimating the biomass of large fish populations, the utilization of phytoplankton by other organisms, or extrapolation of laboratory oxygen consumption figures to caloric needs of field populations, cannot be stated at this time, and indeed these may be unanswerable questions.

6. FISH POPULATIONS IN "OLIGOTROPHIC" SEAS

Olson and Boggs (1986) have produced a significant work in their study of the apex predator, *Thunnus albacares*, the yellowfin tuna of the eastern tropical Pacific Ocean. This species is abundant throughout the world's tropical seas and may typify the abundances of the major apex predators in the open ocean. Sharp and Francis (1976) estimated an unexploited population of about 600,000 mtons was required to support the yellowfin surface fishery of 1966-71. Olson and Boggs used a 300,000 mton estimate for 1970-72 standing stocks, which was calculated by Inter-American Tropical Tuna Commission scientists (Anon, 1986) to calculate predation rates of the population. By using the more conservative figure of 300,000 mtons, an annual consumption of 4.5×10^9 kg of forage ($= 1.24 \times 10^{12}$ gC) and 5 steps in the food chain (Mearns *et al.*, 1981; Rau *et al.* 1983) with an efficiency between steps of 10%, we find that for the area of the yellowfin tuna habitat, 1.7×10^{13} m^2, 730gC/m^2 primary production per year are needed or approximately 2gC/m^2 per day over the whole area. This has to be an underestimate since no other top predators are included in the biomass. The most recent (1985) biomass estimate for yellowfin tuna in this area is 450,000 mtons, the result of a post-El Niño succession of three exceptional year classes (Anon. 1986).

Feldman (1986) has shown that the productive area of the eastern tropical Pacific has varied an order of magnitude between years based on satellite chlorophyll images. Dandonneau (this volume) using extensive ship-of-opportunity data, confirmed and extended these observations to the southwestern tropical Pacific and suggested a mechanism for this increased production. Cooling of the surface layer in the winter increases the depth of the mixed layer which overturns and brings up nitrate from the deep.

Owen and Zeitschel (1970) showed a seasonal change in primary production for this region using the Carbon-14 method, from 127 mg/m^2 per day to 318 mg/m^2 per day, or an average annual production of 75g/m^2, an order of magnitude lower than needed by this population of yellowfin tuna as calculated using the 10% transfer efficiency and the excellent data on forage requirements by Olson and Boggs (1986), but slightly more than the rate used by Ryther (50g C/m^2 per year) to make his calculations.

Between 90° and 180°W in the equatorial tropical Pacific is an oceanic upwelling zone which is unusually productive for open ocean areas. Ryther (1969) included areas like this in the estimate of 100gC/m^2 per year for the coastal zone (but not upwelling zone) production. Chavez and Barber (1987) suggest that the oceanographic upwelling zone of the tropical Pacific produces about 197gC/m^2 per year, about twice what Ryther estimated. While not an order of magnitude higher, the vast area over which this production occurs leads to the conclusion that populations in nearby areas, such as the yellowfin tuna in the eastern tropical Pacific, must benefit from this production.

Another possible mechanism for enhancing production in oligotrophic areas is nitrogen fixation. Blue-green algae in tropical seas have been estimated to fix 5×10^6 mtons annually (Carpenter, 1983). It is difficult to put this figure in perspective because of the uncertainty of the amount of total new nitrogen available in tropical seas.

7. CONCLUSIONS

Marine food chain research seems still to be in its infancy when it comes to energy budgets. The numbers being generated for primary production are constantly being upgraded and the trend has been toward substantial increases for all areas of the world's oceans. Other new findings also increase the values to be used in calculations of ocean productivity. These are 1) the determination of markedly increased production by equatorial upwelling areas as well as the size of those areas (Chavez and Barber, 1987), 2) nutrient enrichment by jets and squirts of the coastal zone, 3) nutrient enrichment by eddy entrainment (Simpson, 1986), 4) nutrient enrichment by deepening and overturning the mixed layer in tropical seas (Dandonneau, 1987), 5) the possibility of a utilizable pool of dissolved organic nitrogen in "oligotrophic" seas (Suzuki et al., 1985), 6) the continuing technical improvements in the methods for determining primary production, and 7) new methods for the more accurate determination of fish biomass (Lasker, 1985).

Other problems, not very different than those discussed 20 years ago, will also have to be resolved. For example, too few data on secondary production and percent transfer of energy through steps in the food chain prevent the determination of grand means for different ocean regimes. We lack too the information on feedback mechanisms in food chains which may influence transfer efficiencies.

All of these points bring us to the obvious conclusion that ocean scientists are not yet in a position to determine what the yield of food from the sea will be for the foreseeable future. The exercises to do so thus far have been important in pointing out the major problems in determining the starting point (primary production) and where we end up (commercial fisheries). A conference of this kind in 10 years may be able to resolve some of these issues, but we cannot do it today.

8. ACKNOWLEDGEMENTS

Thanks to Richard Eppley, Robert Olson and Alec MacCall who read the original manuscript and offered valuable suggestions.

9. REFERENCES

Alverson, D.L., A.R. Longhurst and J.A. Gulland: 1970, 'How much food from the sea?', Science 168, 503-505.

180

Anon: 1986, *Annual Report of the Inter-American Tropical* Tuna Commission, La Jolla, California, U.S.A., 248 p.

Brewer, P.G., K.W. Bruland, R.W. Eppley and J.J. McCarthy: 1986, 'The Global Ocean Flux Study (GOFS): Status of the U.S. GOFS Program', *Eos* 67, 827-832.

Carpenter, E.J.: 1983, 'Nitrogen fixation by marine Oscillatoria (Trichodesmium) in the world's oceans', in E.J. Carpenter and D.G. Capone (eds.), *Nitrogen in the Marine Environment*, Academic Press, New York, p.65-103.

Chavez, F.P. and R.T. Barber: 1967, 'An estimate of new production in the equatorial Pacific', *Deep-Sea Res*. In Press.

Dandonneau, Y.: 1988, 'Seasonal or aperiodic cessation of oligotrophy in the tropical Pacific Ocean', in B.J. Rothschild (ed.), *Toward a Theory on Biological-Physical Interaction in the World Ocean*, Kluwer Academic Publishers, Dordrecht, pp. 137–156.

Dugdale, R.C. and J.J. Goering: 1967, 'Uptake of new and regenerate forms of nitrogen in primary productivity', *Limnol. Oceanogr.* 12, 196-206.

Eppley, R.W. and B.J. Peterson: 1979, 'Particulate organic matter flux and planktonic new production in the deep ocean', *Nature* 282, 677-680.

Eppley, R.W., F.M.H. Reid and J.D.H. Strickland: 1970, 'Estimates of phytoplankton crop size, growth rate, and primary production', *Bull. Scripps Inst. Oceanogr.* 17, 33-42.

Eppley, R.W., E. Stewart, M.R. Abbott and U. Heyman: 1985, 'Estimating ocean primary production from satellite chlorophyll. Introduction to regional differences and statistics for the Southern California Bight', *J. Plankton Res.* 7, 57-70.

FAO: 1986, *Yearbook of fishery statistics: catches and landings*, 1984. 58, 451 p.

Feldman, G.C.: 1986, 'Variability of the productive habitat in the eastern equatorial Pacific', *Eos* 67, 106-108.

Fitzwater, S.E., G.A. Knauer and J.H. Martin: 1982, 'Metal contamination and its effect on primary production measurements', *Limnol. Oceanogr.* 27, 544-551.

Hayward, T.L. and E.L. Venrick: 1982, 'Relation between surface chlorophyll, integrated chlorophyll and integrated primary production', *Mar. Biol.* 69, 247-252.

Jenkins, W.J. and J.C. Goldman: 1985, 'Seasonal oxygen cycling and primary production in the Sargasso Sea', *J. Mar. Res.* 43, 465-491.

Kerr, R.A.: 1983, 'Are the ocean's deserts blooming?', *Science* 220, 397-398.

Koblentz-Mishke, O.I., V.V. Volkovinsky and J.G. Kabanova: 1970, 'Plankton primary production of the world ocean', in *Scientific Exploration of the South Pacific*, National Academy of Sciences, Washington, D.C. p.183-193.

Lasker, R.: 1970, 'Utilization of zooplankton energy by a Pacific sardine population in the California current', in J.H. Steele (ed.), *Marine Food Chains*, Oliver and Boyd, Edinburgh, p. 265-284.

Lasker, R.: 1975, 'Field criteria for survival of anchovy larvae: the relation between inshore chlorophyll maximum layers and successful first feeding', *Fish. Bull. (U.S.)* **73**, 453-462.

Lasker, R., J. Pelaez and R.M. Laurs: 1981, 'The use of satellite infrared imagery for describing ocean processes in relation to spawning of the northern anchovy (Engraulis mordax)', *Remote Sensing of Environment* **11**, 439-453.

Lasker, R. (ed.): 1985, 'An egg production method for estimating spawning biomass of pelagic fish: application to the northern anchovy, *Engraulis mordax*', *NOAA Technical Rep. NMFS* **36**, 99p.

MacCall, A.D.: 1979, 'Population estimates for the waning years of the Pacific sardine fishery', *Calif. Coop. Oceanic. Fish. Invest. Rep.* **20**, 72-82.

Martin, J.H., G.A. Knauer, D.M. Karl and W.W. Broenkow: 1987, 'VERTEX: Carbon cycling in the northeast Pacific', *Deep-Sea Res.* **34**, 267-285.

Mearns, A.J., D.R. Young, R.J. Olson, and H.A. Schafer: 1981, 'Trophic structure and the cesium-potassium ratio in pelagic ecosystems', *Calif. Coop. Oceanic Fish. Invest. Rep.* **22**, 99-110.

Murphy, G.I.: 1966, 'Population biology of the Pacific sardine (*Sardinops caerulea*)', *Proceedings Calif. Acad. Sci.* **34**, 1-84.

Olson, R.J. and C.H. Boggs: 1986, 'Apex predation by yellowfin tuna (Thunnus albacares): independent estimates from gastric evacuation and stomach contents, bioenergetics, and cesium concentrations', *Can. J. Fish. Aquat. Sci.* **43**, 1760-1775.

Owen, R.W. and B. Zeitzschel: 1970, 'Phytoplankton production: seasonal change in the oceanic eastern tropical Pacific, *Mar. Biol.* **7**, 32-36.

Peterson, B.J.: 1980, 'Aquatic primary productivity and the ^{14}C-CO_2 method: a history of the productivity problem', *Ann. Rev. Ecol. System* **11**, 359-385.

Platt, T.: 1984, 'Primary productivity in the central North Pacific: comparison to oxygen and carbon fluxes', *Deep-Sea Res.* **31**, 1311-1319.

Platt, T. and W.G. Harrison: 1986, 'Reconciliation of carbon and oxygen fluxes in the upper ocean', *Deep-Sea Res.* **33**, 273-276.

Rau, G.H., A.J. Mearns, D.R. Young, R.J. Olson, H.A. Schafer, and I.R. Kaplan: 1983, 'Animal $^{13}C/^{12}$ correlates with trophic level in pelagic food webs', *Ecology* **64**, 1314-1318.

Ryther, J.H.: 1969, 'Photosynthesis and fish production in the sea.' *Science* **166**, 72-76.

Shulenberger, E. and J.L. Reid: 1981, 'The Pacific shallow oxygen maximum, deep chlorophyll maximum, and primary productivity, reconsidered', *Deep-Sea Res.* **28**, 901-919.

Simpson, J.J.: 1986, 'Processes affecting upper ocean chemical structure in an eastern boundary current', in J.D. Burton, P.G. Brewer and R. Chesselet (eds.), *Dynamic Processes in the Chemistry of the Upper Ocean*, Plenum Publ. Corp, p.53-77.

Smith, P.E. and R.W. Eppley: 1982, 'Primary production and the anchovy population in the Southern California Bight: Comparison of time series', *Limnol. Oceanogr.* **27**, 1-17.

Steele, J.H. (ed.): 1970, *Marine Food Chains*, Oliver and Boyd, Edinburgh, 552 p.

Steemann Nielsen, E.: 1952, 'The use of radio-active carbon (C^{14}) for measuring organic production in the sea', *J. Cons. Perm. Int. Explor. Mer* **18**, 117-140.

Steemann Nielsen, E. and E.A. Jensen: 1957, 'Primary oceanic production. The autotrophic production of organic matter in the oceans', *Galathea Report* **1**, 49-136.

Suzuki, Y., Y. Sugimura and T. Itoh: 1985, 'A catalytic oxidation method for the determination of total nitrogen dissolved in seawater', *Mar. Chem.* **16**, 83-97.

Thomas, W.H., E.H. Renger and A.N. Dotson: 1971, 'Near-surface organic nitrogen in the eastern tropical Pacific ocean', *Deep-Sea Res.* **18**, 63-71.

Treganza, E.D., D.G. Redalje and R.W. Garwood: 1987, 'Chemical flux, mixed layer entrainment and phytoplankton blooms at upwelling fronts in the California coastal zone', *Continental Shelf Res.* **7**, 89-105.

United States. Council on Environmental Quality: 1980, *The Global 2000 Report to the President: Entering the Twenty-First Century*, Vol. 2, Technical Report, G.O. Barney, Study Director, Washington, D.C., Government Printing Office, 766p.

Villavicencio R.Z.: 1981, 'Investigacion preliminar de los requerimientos energeticos de anchoveta adulta (metabolismo estandar y actividad)', *Bol. Inst. del Mar del Peru*, Volumen Extraordinario, p. 193-205.

Wise, John P.: 1984, 'The future of food from the sea', in J.L. Simon and H. Kahn (eds.), *The Resourceful Earth, a response to Global 2000*, Basil Blackwell Publ. Oxford and New York, p. 113-127.

RECRUITMENT DEPENDENCE ON PLANKTONIC TRANSPORT IN COASTAL WATERS

William C. Boicourt
Horn Point Environmental Laboratories
University of Maryland Center for
Environmental and Estuarine Studies
Cambridge, Maryland 21613
U.S.A.

ABSTRACT. In the attack on the complexities of the recruitment variability problem, knowledge of the physical transport mechanisms serves to reduce the dimensionality and sharpens the questions driving field research. In coastal regions, topographic and oceanographic boundary constraints contribute to this structuring by limiting the transport domain and clarifying the transport pathways.

A measure of the state of our understanding of planktonic transport is the ongoing controversy concerning active versus passive mechanisms. The tendency has been to regard them as mutually exclusive, whereas observations support their acting together, with advection often dominating. Innovative field and laboratory studies on planktonic behaviour are needed to allow extention of numerical transport experiments beyond the point where an intractable number of possibilities are presented for testing via field experiment. In addition, an Eulerian description of water motion will be required on the small scales necessary for the estimation of Lagrangian mean velocities and for the incorporation of spatial variations of diffusion into transport formulations.

1. INTRODUCTION

Theoretical relationships between the physics of the ocean and the living organisms contained therein would greatly enhance our ability to attack the fundamental and perplexing problem of the marked variability in recruitment to marine populations. An indication of the importance of this problem is the fact that the converse is also true--a firm empirical understanding of the recruitment process would substantially aid the development and testing of a theoretical framework for the biology of the ocean.

In spite of the elusiveness of the details of recruitment mechanisms, some of the reasons for the complexity and the large variabilities are clear. There are multiple influences on recruitment, each one of which is a candidate for controlling factor and of which our knowledge is meager. Large ratios between eggs, larvae, juveniles and recruits allow small fluctuations in mortality

B. J. Rothschild (ed.), Toward a Theory on Biological-Physical Interactions in the World Ocean, 183–202.

rates at each stage to generate large fluctuations in the numbers of organisms surviving to recruitment. In the presence of this complexity and uncertainty, a welcome reduction in the dimensionality can be achieved by an appeal to physical transport. Among the many simplifications that transport knowledge can bring to the problem, perhaps the most significant is an estimate of the advective and diffusive losses of eggs and larvae, which have often been masked within an integrated mortality. In addition, when transport pathways are known, the processes of reproduction, growth, nutrition, predation and other mortality can be addressed in a Lagrangian frame. Temporal variability in this frame is predominantly the result of interactions among organisms and their immediate environment rather than the result of patches of organisms advecting past a point in space. This transformation from the traditional Eulerian view of water motion is essential if we are to make significant progress on the recruitment problem. Moreover, the merits of this approach can only be realized if the circulation and distribution of organisms are described with a high spatial and temporal resolution over the planktonic transport distance.

Although the recruitment problem is not unique to the coastal ocean, it is there that we can find substantial encouragement in our efforts to develop testable models of the controlling interactions. The strong signals and relative ease of obtaining observations on the short time and space scales appropriate to the problem make it an attractive laboratory in which to seek guidance for the construction of these models. In addition, the topographic and oceanographic boundaries in the coastal ocean provide a structuring of the problem, thereby reducing the dimensionality and increasing the clarity of the tests. A third reason that the coastal ocean is the likely site for achieving progress in recruitment is that three-dimensional numerical modeling of this region is now at the point where not only the water motion can be described on the appropriate scales, but also transport processes (which are inherently Lagrangian) can be directly simulated. Another important aspect of the shallow-water environment concerns motivation--the benefits of a predictive understanding of the recruitment process to the management of the world's fisheries (which are predominantly coastal) are obvious. Fisheries-motivated research has produced a large body of observational information to which we can appeal for guidance in both experimental design and theoretical development (Cushing, 1975, 1982).

While the virtues of the coastal ocean laboratory for relating the biological variability to physical processes are clear, so are the attendant disadvantages. The small time and space scales and the vulnerability to the often voiced complaint of "site specificity" or lack of generality of process are potential pitfalls that must be addressed directly by researchers working on the continental shelf or in the estuary. An overview of recent work in the coastal ocean reveals not only that modern observational techniques can achieve resolution on the appropriate scales (Allen *et al.*, 1983), but also that the understanding of individual coastal circulations is developing to the point where physical oceanographers are extracting

the general principles (Brink, 1987). A particular coastal topography
and sequence of driving forces can then be seen, not so much as
introducing an unique and intractable complexity, but rather as
selecting the balance among a catalog of possible circulations.

Coastal and estuarine circulation acts to transport planktonic
stage organisms from spawning grounds to nurseries. In some cases,
this transport process has been shown to influence, or even control,
the recruitment variability. Whether recruitment is controlled by a
physical or biological process during the transport interval, these
cases can provide lessons by which to sharpen the attack on the
problem. In the following paper, such lessons will be sought from
selected observational case histories of recruitment in the estuary
and on the continental shelf. Of specific interest is the problem of
how motile or marginally motile planktonic organisms are transported
in these environments.

2. THE ESTUARY

While recruitment has commanded increased attention in recent years,
it is a recruitment problem that provided the motivation for the early
physical oceanographic studies leading to the development of our
fundamental ideas on the circulation of estuaries. In the James River
estuary, a tributary to Chesapeake Bay, larvae of the oyster
Crassostrea virginica were found to set in large numbers far upriver
from the beds of adult oysters in the lower reaches of the estuary.
That larvae were even retained within an estuary, much less
transported upriver, seemed paradoxical to those who expected that
river flow would act to flush the larvae from the system. The classic
Chesapeake Bay Institute James River Study of 1949-1950 addressed the
mechanisms of this transport. The results led Pritchard (1951) to
suggest that the newly discovered two-layer estuarine circulation
could provide the means for the upriver transport of oyster larvae.
From these results came the basic understanding of what is now
referred to as the classical estuarine circulation (Pritchard, 1953,
1954; Beardsley and Boicourt, 1981). Pritchard was careful to point
out, however, that the observed larval distributions do not
necessarily reflect the action of physical transport and dispersion
alone; the distribution may also be influenced by any ability of
larvae to aggregate, swarm, sink, or swim in a deterministic manner.
In the James River Study, the technical inability to sample oyster
larvae with sufficient resolution and coverage prevented a clear
separation between the effects of passive transport and controlled
sinking or swimming by the planktonic larvae.

The combination of this inability to describe the distributions
of organisms with the inability to describe the details of the water
motions fuels the controversy between proponents of active versus
passive mechanisms of planktonic transport. One school of thought is
that the observed distribution patterns are the result of active
mechanisms such as the phasing of vertical swimming activity with the
tide (Carriker, 1951; Nelson, 1957; Haskin, 1964; Wood and Hargis,

1971)   If, for instance, negatively buoyant organisms increased their
swimming activity on flood tide, then they could selectively move
toward the head of an estuary.   The opposing school of thought
suggests that passive transport alone is sufficient to explain larval
retention (Korringa, 1952; de Wolf, 1974; Andrews, 1979,1983; Banse,
1986; Hannan, 1984).

Behavioural mechanisms that might effect the transport of larvae
in an estuarine environment have been demonstrated.   There is evidence
from laboratory studies (Haskin, 1964; Hidu and Haskin, 1978; Kennedy
and van Heukelem, pers. comm.) that larvae of the oyster *Crassostrea
virginica* can increase their swimming activity upon detection of an
increase in salinity.   Kennedy and van Heukelem found that the
detection sensitivity was high, with the larvae responding to changes
in salinity of less than 3 psu.   The argument for an active transport
mechanism is that the larvae would sink to the bottom during the ebb
cycle and swim during the beginning of flood tide, when salinity is
increasing.   This behaviour would seem to aid the upestuary transport
of only those larvae in proximity to the bottom because differential
salinities are the cues to activate the motility modes.   If larvae
were in the benthic boundary layer, then favourable salinity changes
could be produced by flood tidal currents.   Above the benthic boundary
layer, more intricate mechanisms would be required to produce a change
of salinity surrounding the planktonic organism moving with a water
mass.   Such mechanisms as time-limited swimming activity or circadean
tidal rhythms could operate without direct dependence upon
differential salinities.   If differential salinities are necessary to
trigger active sinking or swimming, however, then the vertical
salinity gradients and vertical mixing would have to be sufficiently
large to produce tidal variations in the water-mass salinities.
Salinity variations have been observed on tidal time scales in the
estuary, but from an Eulerian perspective.   Observations following the
planktonic trajectories would be necessary to test this hypothesis.
There is some evidence that late-stage larvae of the eastern oyster
are in concentration near the bottom (Carriker, 1951), where they
would be in a position to take advantage of flood tidal currents.
Whether such a mechanism could produce significant transports can be
decided from straightforward modeling, applying laboratory swimming
results to a tidal benthic-boundary layer model.   To determine if such
a mechanism controls the transport of larvae in the environment,
however, will require some detailed and probably elaborate
experimentation in the field.

Cronin (1982) and Cronin and Forward (1982) found evidence that
larvae of the xanthid crab *Rhithropanopeus harrisii* are retained in
the upper reaches of the Newport River estuary, North Carolina via a
timed vertical migration, synchronized with the tide, and in the
presence of a two-layer gravitational circulation.   The estuary is
sufficiently shallow in these reaches (4 m) that both salinity and
light (controlled by tidal height) cues can be invoked as triggers to
swimming and sinking behaviour (see Sulkin, 1984, for review).
Individual cues are not necessary if a variety of cues develop a
circadian rhythm over time.   Such a rhythm was demonstrated by Cronin

and Forward for *R. harrisii* in the laboratory. While the
demonstration of a tidal synchrony in vertical migration is clear in
these and later (Cronin and Forward, 1986) field experiments, the
relative roles of this active migration versus passive transport is
not clear. Cross spectra between larval mean-depth time series and
tidal currents show significant coherence at the 6.2-hour period, the
semidiurnal current speed cycle. In some experiments, the coherence
between the mean-depth and the salinity was significant at the
semidiurnal (12.4-h) period. The coherence between larval mean-depth
and current speed at 6.2 hours is highly suggestive of a passive
vertical mixing via bottom-generated turbulence, especially for these
shallow depths. The coherence with salinity at the 12.4-hour period
may be an artifact (due to the strong tidal signal), but it also could
be the result of salinity cues. Again, if the behavioural mechanism
is to increase swimming activity with an increase in salinity, then it
is still difficult to conceive of a mechanism to produce a decrease in
salinity surrounding larvae in the upper portions of the water column,
a decrease to cue the larvae to cease swimming or begin active
sinking. A mechanism that would produce active transport, and that
would seem to make sense energetically, is one where the swimming
activity upon detection of a salinity increase is transient, in a
burst, such that the swimming occurs for only a portion of the tidal
cycle. If, over the remainder of the tidal cycle, the larvae sink,
then no decreasing salinity cue would be necessary for the larvae to
be in a position to take advantage of the following flood tide.
Cronin and Forward's data are not inconsistent with such a mechanism,
if one allows for the effects of turbulence generated by tidal
currents. If, however, the *R. harrisii* larvae are negatively buoyant
and swim for only a minor part of the tidal cycle, then the data could
also support a strictly passive transport interpretation. Negative
buoyancy would act to force the distribution of larvae downward in the
water column. Tidally generated mixing in the presence of the two-
layer gravitational circulation of the Newport River, would then act
both to disperse the larvae and to favor the upestuary transport of
the majority. The demonstration of a tidal vertical migration in the
laboratory argues for at least a component of the transport being
active, working in concert with the circulation processes.
Quantitatively distinguishing between these mechanisms will require
modeling and experimentation efforts similar to the case of the
oyster.
     The present state of sampling and counting techniques for oyster
larvae hampers, or even precludes, obtaining distributional
information on the time and space scales required for the transport
problem. For this reason, investigators have resorted to alternative
field approaches. Boicourt, Kennedy, and Rives (1987) employed
indirect measures of larval transport in an attempt to explain the
marked and puzzling disparity in settlement success between two
adjacent and apparently biologically similar tributary estuaries
(Broad Creek and Tred Avon River) in the Chesapeake Bay.   Working
from larval settlement measures, they found clear circulation
differences that could explain differences in larval supply, given

that the primary sources of larvae were the productive oyster beds in
the larger body of water connecting the two tributaries. Both
Eulerian (moored current meters) and Lagrangian (dye tracer)
techniques were necessary to delineate the small-scale circulation
processes. Rhodamine-WT dye tracers were used, not as direct
surrogates for larval transport (because the timing and location of
the specific larval sources were not known), but as detailed measures
of transport. The widespread distribution of productive oyster bars,
both within and without the two tributaries, and the demonstration
(Kennedy and Krantz, 1982) of gametogenesis in oysters sampled from a
variety of bars, led to the conclusion that the larval sources were
also widespread. Conceptually, the problem was to sum the
contributions from the various sources and see if a statistically
greater number of larvae were retained within or delivered to one
tributary over the other. While Boicourt et al. demonstrated that
the differences in meteorological responses between the two
tributaries were sufficient to account for the observed differences in
larval settlement, they also could infer that active tidal transport
mechanisms do not contribute strongly to this difference. The similar
depths of Broad Creek and Tred Avon River and the similarity in tidal
amplitude appear to eliminate tidally phased vertical swimming and
advection with flood tidal currents as the dominant transport mode.
Active mechanisms are by no means ruled out, however. If late-stage
larvae are concentrated near the bottom during a meteorologically
driven, lower-layer inflow, they could detect the associated increase
in salinity, increase their swimming activity, and thereby enhance
their upestuary transport. Although the circulation observations in
this study were sufficiently intensive to reveal new details of both
the gravitational and meteorological circulations in such shallow
estuaries, they were still inadequate to document specific transport
paths or to locate the sources of larvae settling in a particular
location. Moreover, the lack of both spatial and temporal details of
the larval distributions prevent satisfyingly conclusive statements on
the transport issue.

A problem with the indirect approach such as employed by Boicourt
et al. (1987) is that another process, (such as settlement in benthic
invertebrates) unrelated to larger-scale transport may control what is
taken to be the indirect measure of transport. Recruitment control
may thereby be incorrectly ascribed to these transport processes, when
it may very well be the details of the small-scale transport during
settlement, or the temporal and spatial characteristics of growth and
mortality during the planktonic drift that are the primary
determinants of the number of recruits. Boicourt et al. appealed to
the apparent biological similarity of the two tributaries (i.e.
similar depths, substrates, food availability, predation) to reduce
these uncertainties, but many possibilities remain. In the midst of
these indeterminacies, we can find both encouragement and a model for
quantitative experimentation in the work of Butman (1986a, 1986b,
1986c) on the benthic invertebrate settlement process. Butman
combined careful laboratory and field experimentation on polychaete
larvae with analyses of the interaction of their motility with the

small-scale flows in the benthic boundary layer. The goal was to examine the alternative hypotheses of active habitat selection versus passive deposition for creating the observed patterns of settlement. The spatial scales (tens of centimeters) for which active habitat selection had been demonstrated in the laboratory were one to six orders of magnitude smaller than the species/sediment-composition correlation scales observed in the field. Butman (1986a) demonstrated that, although there are active transport mechanisms for polychaete larvae in the benthic boundary layer, the scale of these transports are limited to the order of centimeters to meters for the flows addressed in the study. As in the case of oyster larvae, the active and passive mechanisms are not seen as mutually exclusive, but acting together on different time and space scales. There is the possibility that each benthic species exhibits a different settlement behaviour. If this is the case, then Woodin's (1986) call for extensive further investigation into the settlement process should be heeded. Whereas the ratio between passive and active transport scales may be large for polychaete larvae, for crustacean and bivalve larvae, which have significantly greater vertical motility (Sulkin, 1984), these transports may be of the same order. Experimental attempts to test this hypothesis would be well served by Butman's innovative and quantitative approach.

Attacking the transport problem for planktonic organisms having greater motility or greater transport distances presents a different host of sampling difficulties and a wider scale of physical processes to address. Where the expanses of the continental shelf make the location and definition of a planktonic population (or even a local patch) difficult to define, the side boundaries of the estuary help to confine the population to a more two-dimensional distribution. Tyler and Seliger (1978) took advantage of this structuring to describe the role of physical transport in the annual cycle of a dinoflagellate (*Prorocentrum mariae-lebouriae*) in the Chesapeake Bay. They present evidence that this organism is present only in the lower portions of the Chesapeake Bay in winter and that it employs the two-layer estuarine circulation to move in late spring to the upper reaches of the Bay, where it is seen in bloom concentrations at the surface. The extensive (40 cruises over two years) distributional information is convincing that P. mariae-lebouriae undergoes large transport distances within the Bay, and that the gravitational circulation plays an important role. The details of this transport process, however, and the details of the distributions of organisms, are not clear. The tendency to view the dinoflagellate population and the water density as quasiconservative rather than a dynamic result of many processes may hamper the attempt to gain insight. Over a limited time scale, the assumption of an approximately conservative behaviour of the organism distribution probably holds. However, the inferred transport times are of the order months, during which the population (and hence, the distribution) is also under the influences of growth, reproduction, grazing, vertical mixing, longitudinal mixing (via the interaction of vertical motion and vertical mixing with the shear of the gravitational circulation), phototaxis, and inhomogeneities in the

distribution of light and nutrients.   Any motility that would enable
the dinoflagellate to undergo vertical migration would also be likely
to affect the longitudinal transport and distribution of the
population.   The observed complexity of the vertical distribution of
Prororcentrum (Tyler and Seliger, 1978, Figure 12) suggests an
interaction between vertical motility (both swimming, directed by
positive phototaxis, and sinking) and the fine scale density
structure.   The variety of generation mechanisms for this fine scale
structure and the inherent time variability contribute an additional
complexity.   Tyler and Seliger's presentation of the dinoflagellate
distributions suggests that the signals are sufficiently strong to
develop a transport picture in the presence of the identified noise.
In general, the transport process possibilities range from a
"pipeline" extreme, where large numbers of organisms are continuously
advected up the estuary in a confined layer below the pycnocline (and
with reproduction approximately balancing mortality), to a more
stochastic extreme, where a comparatively few organisms survive the
pitfalls of mortality, grazing, and advective-diffusive losses (to the
upper layer) to arrive sporadically in the upper estuary via the
statistically favourable lower layer flow.   These organisms would then
serve as a seed population to trigger a bloom under conducive light,
nutrient, vertical transport and mixing conditions.   While both
extremes may suffice to produce the seasonal transport of Prorocentrum
from lower Chesapeake Bay to the Upper Bay, an experimental design to
distinguish between them would be at the limit of modern technique.
To point out this multidimensionality of the transport process is not
to quibble about the contribution of Tyler and Seliger, who showed
that longitudinal advection of *Prororcentrum* from the lower reaches of
Chesapeake Bay toward the head of the Bay via the lower-layer
estuarine flow is an important component of its life cycle.   The
existence of this multidimensionality, however, means that moving
beyond the present picture, even in the presence of strong signals,
will require careful, quantitative estimates of these other processes
and their associated effects on the transport and distribution of
organisms.

     Topographic and flow constraints can provide a structuring of the
recruitment problem, not only for estuarine species, but also for
organisms that move between the estuary and the continental shelf
during different stages of their life cycle.   An example of such
control has been demonstrated by Pietrafesa *et al*. (1986) for spot
(*Leiostomus xanthurus*) in Pamlico Sound, North Carolina.   These fish
spawn offshore, near the Gulf Stream, and the larvae of the recruits
drift passively toward the barrier-island inlets.   In the Sound, the
nursery areas for the juveniles are located along the western
boundary, 40-50 km from their point of entrance.   Pietrafesa *et al*.
showed that the variability of the meteorologically driven
circulation of the Sound acting over these transport distances can
help explain the observed variability of juvenile spot in the nursery
regions.   To provide a proper representation of the circulation,
Pietrafesa *et al*. had to resort to a three-dimensional numerical
model.   A two-dimensional, vertically integrated model will not

suffice for the wind-driven circulation because the bottom stress is incorrectly specified. In the model, westerly winds were effective in creating near-bottom currents transporting water from the barrier island inlets toward the nursery grounds (and oppositely directed from the wind). These lower-layer currents would favour the transport of the demersal spot toward the nurseries. Conversely, easterly or northeasterly winds should drive the postlarval menhaden (*Brevoortia tyrannus*), located in the upper portion of the water column, toward the nurseries. No clear relationship was found between the menhaden abundance and the winds, but this lack of connection could be ascribed to the sampling technique. Pietrafesa *et al*. caution that, even with the satisfying transport story for the juvenile spot, passive transport by the water is not likely to be the only influence on the delivery of juveniles to the nursery; temperature and salinity cues are likely to be active. They state, as has been stated for many of the cases cited herein, that "Existing data sets provide inadequate detail...to understand fish migrations across the Sound on the short physical and biological time scales that appear to govern the process."

3. THE CONTINENTAL SHELF

The increase in the range of space scales and the decrease in the number of side boundaries upon moving from the estuary to the continental shelf might lead us to expect a substantial increase in the difficulty of tracking planktonic transport. On the other hand, fisheries-motivated investigations have produced a body of knowledge on the life cycles of a variety of organisms (see Cushing, 1982, for examples) that provides a framework by which to focus the search. In addition, bottom topography and oceanographic features such as upwelling centers, coastal fronts, shelf-break fronts, jets, squirts, eddies, and river plumes provide a wealth of boundaries and structure. Recent interdisciplinary research (OPUS, MECCAS, *etc*.) on the continental shelf or continental slope (WARM-CORE RINGS) has exploited this ordering in the attempt to define a semi-enclosed ecosystem and thereby attack a limited domain.

It is perhaps appropriate to begin the continental shelf examples with a case study involving both the estuary and the shelf--the larval transport of the blue crab (*Callinectes sapidus*). In the Chesapeake Bay, there has been a longstanding uncertainty as to the dispersal of bue-crab larvae and the mechanisms of recruitment. The abundance of blue crab larvae in coastal waters indicated dispersal, from the adult population in the estuary, to the continental shelf (Strathmann, 1982, suggests that the move to the continental shelf may not be for dispersal, as such, but a migration to areas more favourable to larval development). Sulkin *et al*. (1980) and Sulkin and van Heukelem (1982) conducted behavioural experiments in the laboratory, demonstrating patterns of vertical motion that would favor export of early stage larvae from the estuary and the maintenance of later zoeal stages nearshore, where they would be available for subsequent

return. The primary uncertainties have been: how are the larvae
exported to the continental shelf returned to the vicinity of the
estuary mouth, and how are they then transported back into the
estuary? Two similar answers have been offered for the first
question. Boicourt (1982) suggested that the wind-driven inner shelf
band of water moving northward prevents the wholesale transport of
blue crab larvae to the south (with the long-term mean flow) during
the July and August drift on the shelf. The larvae are then kept
within a range where the shelf currents can return the annual
recruitment to the Bay. Johnson, Hester, and McConougha (1984)
proposed a wind-driven transport (in the absence of other continental
shelf currents) that would return the larvae to the mouth of
Chesapeake Bay in September and October, when the recruits enter the
Bay. Both of these suggestions are postulates, consistent with the
data, but do not constitute a proven mechanism for recruitment.
Moreover, the correlation that Johnson *et al*. show between wind-
driven transport and later catch of blue crabs in Chesapeake Bay is
far from strong. A high correlation should not be expected, perhaps,
since catch data can be imprecise measures of recruitment. Another
inference, however, is that the determinants of recruitment are not
singular, but manifold. The present interest in the precise
mechanisms transporting larvae into the Chesapeake Bay from the
approaches to the Bay is an encouraging sign that simple correlative
connections no longer suffice as answers to the recruitment problem.
The structure and variability of the flow through the entrance to the
Bay is comparatively well understood. The primary need in the
attempt to decipher the blue crab recruitment is information on the
distribution of larvae, not only in the entrance to the Bay, but
throughout their planktonic drift on the shelf. Epifanio and Dittel
(1982), Provenzano *et al*. (1983), and Epifanio, Valenti, and Pembroke
(1984) presented distributional information supporting Sulkin *et
al*.'s (1980) export-return scenario. Distinguishing among the
possible return transport mechanisms will require more spatial and
temporal detail in the larval distributions (in conjunction with some
monitoring of the circulation), especially in the Bay entrance
region.

The blue crab recruitment problem in Chesapeake Bay stands in
comparatively good stead for achieving progress in our understanding.
The previous work on both larval distributions and circulation has
developed a body of knowledge upon which further studies can build.
In addition, zooplankton sampling technique has improved to the point
where sufficiently detailed vertical profiles can be obtained within
these shallow depths (<30 m). If this increased sampling capability
were combined with automated counting techniques, then larval
distributions could be mapped with resolution and coverage appropriate
to the transport problem.

On the continental shelf, as in the estuary, an integrated
modeling approach may be the only avenue for near term improvement in
our understanding of the interaction of the circulation with vertical
motility modes. In comparison to modeling the circulation, obtaining
a mathematical description of even the better understood vertical

migration behaviour of zooplankton is difficult. The state of our
understanding is such, however, that knowlingly simplistic models of
the vertical distribution of larvae can be used to gain insight, if
only to scale the magnitude of the contributory processes. A combined
circulation--motility model approach was employed by Rothlisberg,
Church, and Forbes (1983) to address the recruitment of the shrimp
Penaeus merguiensis in the Gulf of Carpentaria, Australia. Four
subregions of the Gulf have been identified with markedly different
patterns of recruitment from the two annual spawning periods.
Rothlisberg *et al.* argued that this structure can not be produced by
random diffusion and that circulation must be invoked to explain the
timing and spatial patterns. To examine the effect of vertical
migration on the larval transport, schematic migration modes were
extracted from field measurements. These migration patterns (with
additional extreme cases for sensitivity testing) were then combined
with circulation information from a numerical model to produce larval
transport paths which were qualitatively in agreement with the
transports inferred from the recruitment. Rothlisberg *et al.* conclude
that the phasing of the diurnal vertical migration and the dominant
diurnal tidal component (K1) control the larval transport. This
phasing has a slow progression through an annual cycle. They also
conclude that the effects of wind driven motion are negligible. While
there are serious shortcomings (which may affect these conclusions) in
the treatment of the circulation modes, the approach of Rothlisberg *et
al.* has yielded the fundamental demonstration that vertical migration
in the presence of vertical shear in the currents produces planktonic
transport distances that are significantly different from the case of
nonmotile organisms. A similar calculation for the copepod *Calanus
marshallae* in the coastal upwelling zone off Oregon led Wroblewski
(1982) to the same quantitative conclusion. The primary shortcomings
in the Gulf of Carpentaria circulation model are the use of seasonal
mean winds and the uncertainty in the baroclinic effects. For the
circulation predictions, a three-dimensional model was constructed by
superposing an Ekman model on a vertically integrated, two-dimensional
model. By employing the seasonal mean winds to drive the vertical
shear, however, the implicit assumptions are that the mean currents
over the larval drift interval are driven by the mean winds and that
the mean interaction of the time-varying wind drift and vertical
migration is small. Neither assumption can be accepted *a priori*. In
addition, the neglect of the baroclinic effects appears unjustifiable,
especially given the evidence from previous surveys that these effects
may be important. These shortcomings could be addressed with further
circulation studies in the Gulf of Carpentaria and should not
unnecessarily cloud the contribution of Rothlisberg, Clark and Forbes.
If these studies were combined with extensive sampling of the penaeid
shrimp larvae, then fewer simplifications would be necessary in the
modeling of the both the circulation and the vertical migration.

    While the larval transport process and its influence on
recruitment is a focus herein, any quantitative treatment of the
recruitment process must obviously include the biological sources and
sinks, which have been central to fisheries models. Spawning and

growth, and predation and mortality have often proved difficult to measure to order of magnitude accuracies, much less to accuracies acceptable for quantitative modeling. One of the benefits of the larval transport and the modeling thereof is the prospect for sharpening the direct or indirect measure of these processes and thereby reducing the overall dimensionality of the problem. A specific example of a model that includes both physical and biological losses and that provides and an ordering and structuring to the recruitment problem is Flierl and Wroblewski's (1985) treatment of variations in larval fish distributions on Georges Bank and the Middle Atlantic Bight regions of the northeastern United States continental shelf. Although many of the modeled processes are by admission oversimplified parameterizations of complex interactions, Flierl and Wroblewski address the relatively strong signals of the influence of warm-core rings on larval abundance on the shelf. For reasonable values of the onshore and offshore ring-induced velocities at the shelf break, markedly different larval loss rates (due to advection, diffusion, and mortality) are produced, depending upon the relative velocities of the alongshelf flow and of the ring translation along the continental slope. The greatest larval loss occurred when the shelf flow was slightly greater than the ring translation velocity. The observational support for Flierl and Wroblewski's predictions is somewhat tenuous, since the evidence is correlative occurrences of low recruitment with ring activity. In addition, the phasing and location of spawning, which influence the location of the possible ring interaction, are unknown. Flierl and Wroblewski, in spite of the explicit simplifications of their model, produce larval losses consistent with the co-occurrence of low recruitment and support the conclusion that advective losses of larvae can influence, or even control, the recruitment process. They add quantitative backing, as do Parrish, Nelson, and Bakun (1981), to the qualitative evidence developed by a long line of researchers beginning with Hjort (1914). Flierl and Wroblewski suggest that their model is a first step. Future efforts should address the effects of non-uniform distribution of larvae on the shelf, larval motility modes such as vertical migration, the spatial and temporal structure of ring-shelf interactions, and the decomposition of the biological gain and loss term into the component mechanisms. The recognition that recruitment may not be determined entirely during the larval stages underscores our lack of knowledge of the details of the life history or the larval distributions on the continental shelf. Sissenwine, Cohen, and Grosslein (1983), for instance, show that predation in the postlarval stages may control the passage to recruitment.

The role of physical transport in the delivery or return of larvae to the nursery region has been demonstrated for many estuarine and continental shelf environments. When the transition between the coast and the deep ocean is abrupt, then simple retention nearshore would seem the only likely mechanism for recruitment. This retention would not appear to serve the evolutionary need for gene dispersal. A growing body of evidence indicates that for the case of oceanic islands, the retention concept should be extended to include deep-

195

water flow features such as stationary eddies or von Karman vortex streets, which provide a mechanism for both dispersal and for return of larvae to the nursery. Lobel and Robinson (1986) studied the distribution of reef-fish larvae and the circulation in the lee of the island of Hawaii. Employing radio-tracked drifters, expendable bathythermographs, and objective analysis of the dynamic height fields (with supporting evidence from satellite radiometry), they described a geostrophic mesoscale eddy of diameter 50-60 km which was nearly stationary off the west coast of Hawaii for 2 months. Concurrently, plankton tows were made in a rough transect from the center of the eddy to the high-velocity edge. Larval and postlarval distributions suggested to Lobel and Robinson that the eddy formed the nursery for the fish and that a portion of the population was subsequently returned to the reefs of origin as recruits. With the ichthyoplankton sampling necessarily limited by the size of the research vessel, the spatial and temporal coverage was inadequate to confirm the details of Lobel and Robinson's eddy-nursery and return hypothesis. The concentrations and the space scales of the larvae appear to be sufficiently small that, even with an appeal to the extensive previous zooplankton studies in the region, testing this hypothesis will require an intensive program of plankton and circulation studies. Lobel and Robinson's clear documentation of a slowly varying eddy in the vicinity of the larval dispersal region, however, is attractive because it provides offshore dispersal, a developmental nursery, and a mechanism for the delivery of recruits.

4. DISCUSSION

The criteria for selecting the larval transport studies in the previous sections were that they serve as examples of our present understanding and provide lessons for future attacks on the recruitment problem. Toward these ends, many additional works would serve as well. In all cases (including the retention-transport of *Rhithropanopeus*; Cronin and Forward, 1982) transport is seen to act as an influence or a control on the recruitment process. The control can be a transport-path distance from the spawning location or dispersal region, or an oceanographic gate such as a warm-core ring shunt offshore (Flierl and Wroblewski, 1983), or a topographic gate such as an entrance to an estuary or a barrier-island inlet. The transport-path control process also has a temporal aspect, which can be seen as an extention of Cushing's (1967) match-mismatch hypothesis. If the transport is too slow or too fast, or it misses the gates when they happen to be open, then the larvae may arrive in the nursery grounds under less than optimum conditions of nutrition and predation, or in less than optimum densities.

While larval transport studies address abiotic, physical influences, which Legendre and Demers (1984) argue to be crucial to our understanding of ecosystems, the degree of evidencial support varies substantially. The best can be characterized as exemplary beginnings toward obtaining a predictive capability in recruitment.

The support for some others can be characterized as loosely correlative. There is a tendency on this end of the spectrum toward assertions of transport processes that correspond to the investigator's preconceived notions as to the dominant mechanisms. One of the primary lessons to be learned from recruitment studies is that correlative support alone will not suffice to prove a deterministic connection. To say this is not to imply that the problem is straightforward, or that researchers have somehow been remiss in overlooking the need for unequivocal documentation. The problem is obviously complex (having many dimensions, as Rothschild (1988) reminds us) and at the limit of our combined physical-biological art. If we are to determine connections from an observational perspective alone, then the state of larval sampling and counting techniques and the limited ability to describe the circulation on scales appropriate to larval transport will not yet support definitive quantitative statements on cause and effect.

In the attempt to reduce the dimensionality of the recruitment problem to a tractable simplicity, separation of the component processes may appear to be an attractive method of attack--the "divide and conquer" approach. Such an attack may prove fortuitously fruitful in a system with a dominant control mechanism and strong signals. Among the dangers in this separation is the tendency is to forget the obvious, that recruitment has been shown to be, not a linear, single-input process, but a product of many inputs, which often covary or interact. With physical transport of larvae being the focus here, the selection process for the example studies carries an inherent bias toward successful demonstrations wherein transport appears as a strong signal. The danger of responding to the lure of strong signals is the possibility that the controlling processes will be active at a lower level (perhaps below the level of detection or interest) and over a broader spatial scale than addressed in a particular study. In addition, transport distances may be large (see Scheltema, 1986a, 1986b, for oceanic scale transport), but relatively invariant interannually when compared with the effective variation in other physical influences (such as temperature, light, or upwelling) that may determine nutrition and predation, and ultimately, recruitment. In spite of the possible difficulties associated with strong and weak signals, the separation approach still has merit and will probably be crucial in attaining the next stage where the integrating capability of complex models can be exploited. In the meantime, Rothschild's (1988) admonition to restrict our attention to "minimal-length causal chains" may prove wise.

Knowledge of larval, postlarval, and juvenile transport paths serves to reduce the dimensions of the recruitment problem in a variety of ways. An obvious, but perhaps trivial case is where transport variability is the overwhelming determinant to recruitment success. In this situation, the subtle biological and physical interactions are dominated by the transport signal and thereby reduced below the threshold of consideration when a measure or predictability of recruitment is being formulated. In a more realistic situation, knowledge of the transport affords a reduction of the biological sampling problem from an spatial Eulerian search to a Lagrangian time series, following a population of larvae. In the process, the topographic and oceanographic features of the coastal ocean can provide an additional structuring of the problem, delimiting the population of larvae. An estuarine plume on the continental shelf, for instance, forms a defined ecosystem wherein population changes can be deduced from time-series measurements (Boicourt et al., 1987).

Clearly, the sampling problem would not be magically eliminated, even if the circulation details were known to scales beyond those presently achievable by coastal primitive-equation models. While multiple nets and large-volume pumping have increased the vertical resolution attainable in zooplankton sampling, the sampling time scales for these techniques limit the horizontal resolution and coverage. Modern variations of the Longhurst-Hardy plankton recorder, with electronic particle counting or photographic or video imaging and controllable depth profiling, promise to increase both horizontal and vertical resolution (Ortner, Pieper, and Mackas, 1982). Automatic counting or image analysis will be necessary in order to achieve the desired sampling scales in the coastal ocean. Acoustic techniques continue to show promise, (with the inherent limitations of sound in providing definition), but it is easy to share Mullin's (1988) frustration that their potential has yet to be realized.

Mullin (1988) provides examples of zooplankton distributions and behaviour from the California Current system that reveal the time and space scales challenging sampling techniques. The complexity of the population dynamics and the rich spatial and temporal structure of the distributions might lead to the conclusion that, even if the advanced sampling techniques were available, the level of required sampling effort would be substantially greater than the typical one-vessel, single sampling cruise of order tens of days. A glimpse of the possibilities that may be achievable via the new techniques can be found in an intensive program in the relatively stationary and confined Ligurian Sea front. Boucher (1984, 1988) and Boucher, Ibanez, and Prieur employed continous pumping and vertical profiling across the front to document the seasonal progression in zooplankton populations. Their sampling intensity was sufficiently matched to the scales of variability that they could begin the separation between biological and physical effects. As has been shown time and again, Boucher et al. (1987) demonstrated that concurrent monitoring of the physical structure is crucial to the interpretation of the biological structure.

The understanding of coastal and estuarine circulation processes has advanced to the point where models of vertical motility modes can be superposed on existing high-resolution circulation models. The transport of planktonic organisms can be simulated directly to examine the role of transport in recruitment success. At present, the major limitation to this effort is the inadequate knowledge of the details of vertical motility. This knowledge is sufficiently rudimentary that laboratory behaviour experiments have proved valuable despite arguments that the stratification, turbulence, or other conditions may not be representative of the field environment. These experiments have revealed quantitative motility measures and behavioural modes before they have been revealed through field investigation. What are needed are the specific responses to the stimuli of changes in light, salinity, temperature, hydrostatic pressure, and the energetics of these responses. From these building blocks, the vertical distribution of organisms can be developed for given conditions of stratification and mixing, and for given cues to the swimming or sinking activity. Establishing these building blocks will require painstaking experimentation in the laboratory and the field.

While a divide-and-conquer approach to larval transport is advocated herein, a strong skepticism of apparent deterministic connections should be maintained until the quantitative details are well in hand. Artifactual connections arise all too easily from situations where recruitment appears geographically separate from spawning. With information on the less tractable biological processes lacking, then the dominant signal becomes an inferred transport from spawning to nursery. Unfortunately, the time scale of this transport is seasonal, the same time scale for the variation of the other important controls. Thus, with the answer given that transport is active, then many qualitative explanations will work. The tendency is to examine the transport process first, rather than examine the more difficult processes such as predation (Hunter, 1987) or larval nutrition. Houde's (1987) investigation in to the mortality of larvae at various stages in development should serve as a warning that small perturbations in any of the processes can have profound effects on recruitment. In spite of this sublety and complexity in the recruitment process, however, the existing studies should not be criticized for failing to encompass all possible influences on the recruitment process. On the contrary, the evidence that the field has moved beyond the first correlative stage toward an explicit recognition of these complexities is heartening.

5. REFERENCES

Allen, J.A., R.C. Beardsley, J.O. Blanton, W.C. Boicourt, B. Butman, L.K. Coachman, A. Huyer, T.H. Kinder, T.C. Royer, J.D. Schumacher, R. L. Smith, W. Sturges, and C.D. Winant: 1983, 'Physical oceanography of continental shelves', *Reviews of Geophysics and Space Physics*, **21**, 1149-1181.

Andrews, J.D.: 1979, 'Pelecypoda: Ostreidae', in A.C. Giese and J.S. Pearse (eds.), *Reproduction of Marine Invertebrates, Vol. 5. Molluscs: Pelecypoda and Lesser Classes*, Academic Press, New York.

Andrews, J.D.: 1983, 'Transport of bivalve larvae in James River, Virginia', *J. Shellfish Res.* **3**, 29-40.

Banse, K.: 1986, 'Vertical distribution and horizontal transport of planktonic larvae of echinoderms and benthic polychaetes in an open coastal sea', *Bulletin of Marine Science* **39**, 162-175.

Beardsley, R.C. and W.C. Boicourt: 1981, 'On estuarine and continental shelf circulation in the Middle Atlantic Bight,' in B.A. Warren and C. Wunsch, eds., *Evolution of Physical Oceanography*, MIT Press, Cambridge, MA. pp. 198-233.

Boicourt, W.C.: 1982, 'Estuarine larval retention mechanisms on two scales', in V.S. Kennedy, ed., *Estuarine Comparisons*, Academic Press, New York, pp. 445-457.

Boicourt, W.C., S.-Y. Chao, H.W. Ducklow, P.M. Glibert, T.C. Malone, M.R. Roman, L.P. Sanford, J.A. Fuhrman, C. Garside, and R.W. Garvine: 1987, 'Physics and microbial ecology of a buoyant estuarine plume on the contintntal shelf', *EOS* **68**, 666-668.

Boicourt, W.C., V.S. Kennedy and S.R. Rives: 1988, 'Oyster larvae settlement success in adjacent Chesapeake Bay tributaries: the case for transport control', (unpublished manuscript).

Boucher, J.: 1984, 'Localization of zooplankton populations in the Ligurian marine front: role of ontogenic migration', *Deep-Sea Res.* **31**, 469-484.

Boucher, J.: 1988, 'Space-time aspects in the dynamics of planktonic stages', in B.J. Rothschild, ed., *Toward a Theory on Biological-Physical Interactions in the World Ocean*, Kluwer Academic Publishers, Dordrecht, pp. 203-214.

Boucher, J., F. Ibanez and L. Prieur: 1987, 'Daily and seasonal variations in the spatial distribution of zooplankton populations in relation to the physical structure in the Ligurian Sea Front', *J. Mar. Res.* **45**, 133-173.

Brink, K.H.: 1987, 'Coastal ocean physical processes', *Reviews of Geophysics* **25**, 204-216.

Butman, C.A.: 1986, 'Larval settlement of soft-sediment invertebrates: some predictions based on an analysis of near-bottom velocity profiles', in J.C.J. Nihoul, ed., *Marine Interfaces Ecohydrodynamics*, Elsevier Oceanography Series 42, Elsevier, Amsterdam, pp. 487-513.

Butman, C.A.: 1987, 'Larval settlement of soft-sediment invertebrates: the spatial scales of pattern explained by active habitat selection and the emerging role of hydrodynamical processes', *Ocean. Mar. Biol. Ann. Rev.* **24**.

Butman, C.A., J.P. Grassle, and E.J. Buskey: 1988, 'Horizontal swimming and gravitational sinking of *Capitalla* Sp. larvae: implications for settlement', In press: *Second International Polychaete Conference Proceedings*, 18-22 August, 1986, Copenhagen.

Carriker, M.R.: 1951, 'Ecological observations on the distribution of oyster larvae in New Jersey estuaries', *Ecol. Monographs* **21**, 19-38.

Cronin, T.W.: 1982, 'Estuarine retention of larvae of the crab *Rhithropanopeus harrisii*', *Estuar. Coast. Shelf Sci.* **15**, 207-220.

Cronin, T.W., and R.B. Forward: 1982, 'Tidally timed behaviour: effects on larval distributions in estuaries', in V.S. Kennedy, ed., *Estuarine Comparisons*, Academic Press, New York, pp. 505-520.

Cronin, T.W. and R.B. Forward: 1986, 'Vertical migration cycles of crab larvae and their role in larval dispersal', *Bull. Mar. Sci.* **39**, 192-201.

Cushing, D.H.: 1967, 'The grouping of herring populations', *J. Mar. Biol. Assoc. United Kingdom, N.S.* **47**, 193-208.

Cushing, D.H.: 1975, Marine Ecology and Fisheries, Cambridge University Press, Cambridge.

Cushing, D.H.: 1982, *Climate and Fisheries*, Academic Press, London.

deWolf, P.: 1974, On the retention of marine larvae in estuaries', *Thalassia Jugoslavica* **10**, 415-424.

Epifanio, C.E. and A.I. Dittel: 1982, 'Comparison of dispersal of crab larvae in Delaware Bay, USA, and the Gulf of Nicoya, Central America', in V.S. Kennedy, ed., *Estuarine Comparisons*, Academic Press, New York, pp. 477-487.

Epifanio, C.E., C.C. Valenti and A.E. Pembroke: 1984, Dispersal and recruitment of blue crab larvae in Delaware Bay, U.S.A.', *Est. Coast. Shelf Sci.* **18**, 1-12.

Flierl, G.R. and J.S. Wroblewski: 1985, 'The possible influence of warm core Gulf Stream rings upon shelf water larval fish distribution', *Fish. Bull.* **83**, 313-330.

Hannan, C.A.: 1984, 'Planktonic larvae may act like passive particles in turbulent near-bottom flows', *Limnol. Oceanogr.* **29**, 1108-1116.

Haskin, H.H.: 1964, 'The distribution of oyster larvae in Delaware Bay', in N. Marshall (ed.), *Proceedings of a Symposium on Experimental Marine Ecology*, University of Rhode Island Occasional Publication No. 2., pp. 76-80.

Hidu, H. and H.H. Haskin: 1978, 'Swimming speeds of oyster larvae *Crassostrea virginica* in different salinities and temperatures', *Estuaries* **1**, 252-255.

Hjort, J.: 1914, 'Fluctuations in the great fisheries of northern Europe viewed in the light of biological research', *Rapp. P.-v. Reun. Cons. int. Explor. Mer* **20**, 1-228.

Houde, E.D.: 1987, 'Fish early life dynamics and recruitment variabilty', In press, American Fisheries Society Symposium No. 2.

Hunter, J.R.: 1982, 'Predation and recruitment', in B.J. Rothschild and C.G.H. Rooth, convenors, *Fish Ecology III A foundation for REX--A recruitment experiment*, University of Miami Technical Report No. 82008, pp 173-209.

Johnson, D.R., B.S. Hester, and J.R. McConaugha: 1984, 'Studies of a wind mechanism influencing the recruitment of blue crabs in the Middle Atlantic Bight', *Cont. Shelf Res.* **3**, 425-437.

Kennedy, V.S.: 1980, 'Comparison of recent and past patterns of oyster settlement and seasonal fouling in Broad Creek and Tred Avon River', *Proc. Nat. Shellfish Assoc.* **70**, 36-46.

Kennedy, V.S., and L.B. Krantz: 1982, 'Comparative gametogenic and spawning patterns of the oyster *Crassostrea virginica* (Gmelin) in central Chesapeake Bay', *J. Shellfish Res.* **2**, 133-40.

Korringa, P.: 1952, 'Recent advances in oyster biology', *Quart. Rev. Biology* **27**, 339-365.

Legendre, L. and S. Demers: 1984, 'Towards dynamic biological oceanography and limnology', *Can. J. Fish. Aquat. Sci.* **41**, 2-19

Lobel, P.S. and A.R. Robinson: 1986, 'Transport and entrapment of fish
    larvae by ocean mesoscale eddies and currents in Hawaiian waters',
    Deep-Sea Res. 33, 483-500.
Mullin, M.M.: 1988, 'Production and distribution of nauplii and
    recruitment variability--putting the pieces together', in B.J.
    Rothschild, ed., *Toward a Theory on Biological-Physical
    Interactions in the World Ocean*, Kluwer Academic Publishers,
    Dordrecht, pp. 297-320.
Nelson, T.C.: 1953, 'Some observations on the migrations and setting
    of oyster larvae', *Proc. Nat. Shellfish Assoc.* 43, 99-104.
Nelson, T.C.: 1955, 'Observations of the behaviour and distribution of
    oyster larvae', *Proc. Nat. Shellfish Assoc.* 45, 23-28.
Ortner, P.B., R.E. Pieper and D.L. Mackas: 1982, 'Advances in
    zooplankton sampling, in B.J. Rothschild and C.G.H. Rooth,
    convenors, *Fish Ecology III A foundation for REX--a recruitment
    experiment*, University of Miami Technical Report No. 82008, pp.
    355-380.
Parrish, R.H., C.S. Nelson and A. Bakun: 1981, 'Transport mechanisms
    and reproductive success of fishes in the California Current',
    Biol. Oceanogr. 1, 175-203.
Pietrafesa, L.J., G.S. Janowitz, J.M. Miller, E.B. Noble, S.W. Ross
    and S.P. Epperly: 1986, 'Abiotic factors influencing the spatial
    and temporal variability of juvenile fish in Pamlico Sound, North
    Carolina', in D.A. Wolfe, ed., *Estuarine Variability*, Academic
    Press, New York, pp. 341-353.
Pritchard, D.W.: 1951, 'The physical hydrography of estuaries and some
    applications to biological problems', *Trans. 16th North American
    Wildlife Conf.*, March 5-7, 1951, 368-376.
Pritchard, D.W.: 1952, 'Salinity distribution and circulation in the
    Chesapeake Bay estuaries system', *J. Mar. Res.* 11, 106-123.
Pritchard, D.W.: 1953, 'Distribution of oyster larvae in relation to
    hydrographic conditions', *Proc. Gulf Caribbean Fisheries
    Institute* 1952, 123-132.
Pritchard, D.W.: 1954, 'A study of the salt balance in a coastal plain
    estuary', *J. Mar. Res.* 13, 133-144.
Pritchard, D.W.: 1956, 'The dynamic structure of a coastal plain
    estuary', *J. Mar. Res.* 15, 33-42.
Provenzano, A.J., J.R. McConaugha, K.B. Philips, D.F. Johnson and J.
    Clark: 1983, 'Diurnal vertical distribution of first stage larvae
    of the blue crab, *Callinectes sapidus*, at the mouth of Chesapeake
    Bay', *J. Estuar. Coast. Shelf Sci.* 16, 489-499.
Rothlisberg, P.C., J.A. Church and A.M.G. Forbes: 1983, 'Modelling the
    advection of vertically migrating shrimp larvae', *J. Mar. Res.*
    41, 511-538.
Rothschild, B.J.: 1986, *Dynamics of Marine Fish Populations*, Harvard
    University Press, Cambridge, 277pp.
Rothschild, B.J.: 1988, 'Biodynamics of the sea: the ecology of high
    dimensionality systems', in B.J. Rothschild, ed., *Toward a Theory
    on Biological-Physical Interactions in the World Ocean*, Kluwer
    Academic Publishers, Dordrecht, pp. 527-548.

Scheltema, R.S.: 1986a, Long-distance dispersal by planktonic larvae of shoal-water benthic invertebrates among central Pacific islands', *Bull. Mar. Sci.* **39**, 241-256.

Scheltema, R.S.: 1986b, 'On dispersal and planktonic larvae of benthic invertebrates: an eclectic overview and summary of problems', *Bull. Mar. Sci.* **39**, 290-322.

Sissenwine, M.P., E.B. Cohen, and M.D. Grosslein: 1984, 'Structure of the Georges Bank ecosystem', *Rapp. P.-v. Reun. Cons. int. Explor. Mer* **183**, 243-254.

Strathmann, R.R.: 1982, 'Selection for retention or export of larvae of estuaries', in V.S. Kennedy, ed., *Estuarine Comparisons*, Academic Press, New York, pp. 521-536.

Sulkin, S.D.: 1984, 'Behavioural basis of depth regulation in the larvae of brachyuran crabs', *Mar. Ecol. Prog. Ser.* **15**, 181-205.

Sulkin, S.D., W. Van Heukelem, P. Kelly and L. Van Heukelem: 1980, 'The behavioural basis of larval recruitment in the crab, *Callinectes sapidus* Rathbun: A laboratory investigation of ontogenetic changes in geotaxis and barokinesis', *Biol. Bull.* **159**, 402-417.

Sulkin, S.D. and W. Van Heukelem: 1982, 'Larval recruitment in the crab *Callinectes sapidus* Rathbun: an amendment to the concept of larval retention in estuaries', in V.S. Kennedy, ed., *Estuarine Comparisons*, Academic Press, New York, pp. 459-475.

Tyler, M.A. and H.H. Seliger: 1978, 'Annual subsurface transport of a red tide dinoflagellate to its bloom area: water circulation patterns and organism distributions in the Chesapeake Bay', *Limnol. Oceanogr.* **23**, 227-246.

Wood, L. and W.J. Hargis: 1971, 'Transport of bivalve larvae in a tidal estuary', in D.J. Crisp, ed., *Fourth Marine Biology Symposium*, Cambridge University Press, Cambridge, pp. 29-44.

Woodin, S.A.: 1986, 'Settlement of infauna: larval choice?', *Bull. Mar. Sci.* **39**, 401-407.

Wroblewski, J.S.: 1982, 'Interaction of currents and vertical migration in maintaining *Calanus marshallae* in the Oregon upwelling zone--a simulation', *Deep-Sea Res.* **29**, 665-686.

SPACE-TIME ASPECTS IN THE DYNAMICS OF PLANKTONIC STAGES

J. BOUCHER
IFREMER
Laboratoire Pêche
BP 337
29273 BREST CEDEX
FRANCE

ABSTRACT. Factors determining the space-time distribution of planktonic stages of benthic and pelagic populations are analyzed to define the impact of the hydrodynamic variability on their dynamics. Vertical behavioural patterns relative to hydrodynamic structures generate the observed spatial distributions. Ontogenetic time series of behavioural change contribute to the spatial or geographical persistence of the populations. The behavioural changes programmed within the life cycle interact with the hydrodynamics in modifying spatial distribution. The consequences of both timing and the space-time distribution on the variability of the dynamics of the zooplanktonic stages are analyzed.

1. INTRODUCTION

For the past ten years, analyses of plankton patchiness has related observed zooplankton spatial patterns to the influence of hydrodynamics. A review of these works leads Legendre and Demers (1984) to conclude that hydrodynamics could be recognized as "the driving force of aquatic ecosystems". The biotic and abiotic environmental factors act as proximate agents in transmitting the hydrodynamic variability to the organisms. These authors point out that the description of biomass heterogeneities or abundance variability permits only a deductive interpretation of the biological processes involved. Thus in determining the spatial distribution of the planktonic organisms, hydrodynamics could act directly by transport or dispersal, or indirectly by determining the dynamics of the environmental factors organisms must cope with or exploit. A first step might be a hierarchical classification of the hydrodynamic impacts.

Despite the high mortality rates of the planktonic stages, the overall benefit for the species or the adaptative value, is related to population dispersal in the broad sense : genetic flux (between populations), genetic "stirring" (within populations) and

B. J. Rothschild (ed.), Toward a Theory on Biological-Physical Interactions in the World Ocean, 203–214.
© 1988 by Kluwer Academic Publishers.

colonization of new territories (Thorson, 1950; Mileikovski, 1971; Crisp, 1976; Sastry, 1983). The inherent constraint of such a gain is a limited spatial dispersal in order to maintain population (as the reproductive unit) coherency. Sinclair (in press) emphasizes this constraining aspect of the sexual reproduction with regard to the larval spatial dispersal as a main cause of the fluctuations in the marine populations. He gives a review of the observed cases of geographical persistence in planktonic and pelagic populations. The persistence is obvious for the sessile bottom fauna.

Nevertheless the processes allowing such a persistence, despite the dispersal in the planktonic stages, the influence of the behaviour relative to the hydrodynamics, and their impact in the recruitment fluctuations are still uncompletely understood (Sastry, 1983).

This paper deals with the determination of the space-time distribution of the plankton, particularly with regard to the influence of ontogenetic behavioural change interacting with hydrodynamic factors to modify the exposure of organisms to varying environmental conditions. Consonant with Sinclair's theory (in press) that the population coherency is the ultimate biological factor, hydrodynamics will be considered as the driving force in the physical sense of the transport vector.

The role of behaviour relative to spatial hydrodynamic structures in determining geographic distribution, the event sequences through the ontogenesis of planktonic stages, and then the time-space cycle of an holoplanktonic population will be successively analyzed.

2. DETERMINATION OF THE SPACE-TIME DISTRIBUTION IN THE PLANKTONIC PHASE

Zooplankton have a great diversity of swimming abilities (e.g escape, hunting, swarming, cyclic vertical migration, spawning migrations. All of them swim (except egg stages) and move in the vertical plane. Thus there is a great diversity of vertical distribution patterns varying with the species and the developmental stages (cf. Vinogradov, 1968). Superposition of these vertically oriented motions and behaviours on horizontal water motions therefore contribute greatly to spatial and temporal distribution.

2.1. Spatial Distribution as Determined by Behaviour and Water Motion

Plankton behaviour can be classified according to whether it results in an apparent constant depth (requiring displacement or energy to be maintained) or a variable depth (i.e. vertical migrations).

In the first, "constant-depth" organisms are advected by the current flowing at the depth they occupy. Their geographical distribution is mainly defined by residual circulation. For example, gastropod teleplanic veligers for drift long-distances in the Pacific Ocean (Scheltema and Williams, 1983), while in contrast, (Figure 1a) nauplii and early copepodite stages of *Calanus marshallae* are constrained in the coastal convection cell of the coastal Oregon upwelling (Peterson *et al.*, 1979) resulting in geographic persistence.

In the second, where organisms actually migrate vertically, geographic transport is an integration of stage duration, of horizontal transport resulting from the current velocity and direction, and of residence time at each depth. Diel vertical migrations in an area where current and countercurrent are associated contribute to a geographically localized persistence by compensation of the horizontal transports. The process is modeled for *Calanus marshallae* in the Oregon upwelling (Wroblewski, 1982). Such process allowing geographical persistence was observed *in situ* in the Mauritanian upwelling for *Euchaeta* sp. and *Pleuromamma* sp. (Figure 1b) by cross comparisons of sampling referred to fixed geographical station and water drift (Boucher, 1982). This last example also illustrates the impact of sampling strategy on the significance of obtained information : the temporal variability of the specific compositions of the samples is due to the varying spatial distribution of the taxa with regard to the sampling spatial reference.

The same specific behavioural pattern in different hydrodynamic structures generates different geographical distributions. In the vertically well-mixed Bay of St. Brieuc, the bivalve and pectinid larvae are transported according to the tidal residual circulation despite their diel vertical migrations. In contrast, vertical migrations allow upper water estuary transport with limited drift by tidal current of such larvae in the two layered James River Estuary, Virginia (Wood and Hargis, 1971). The geographical distribution of the sessile barnacles is explained in the same way by two processes : i) passive transport and retention by hydrodynamic processes (de Wolf, 1973) or ii) vertical distribution and behaviour relative to water circulation in the estuaries (Bousfield, 1955).

These examples emphasize the role of the behaviour in the determinism of the spatial distribution. Other examples of the behaviours acting with passive transport are analyzed and discussed by Boicourt (this volume). In some of these examples this process is sufficient to explain the geographical persistence of a given stage. What happens when a planktonic drifts with the current ?

2.2. Series of Spatial Distribution Through the Ontogenesis

The changes in the behavioural patterns through the ontogenesis relative to the current structure or its evolution result in a coherent distribution series within a geographical area.

On the west coast of Australia, the early phyllosoma larvae of the lobster *Panulirus cygnus* are carried offshore by wind transport in the surface layer whereas the mid- and late-stage phyllosoma return to the coastal area with the underlying circulation features (Phillips *et al.*, 1978, 1979, 1981). In the same area the larvae of *Scyllarus bicuspidatus* released at the same time, have different

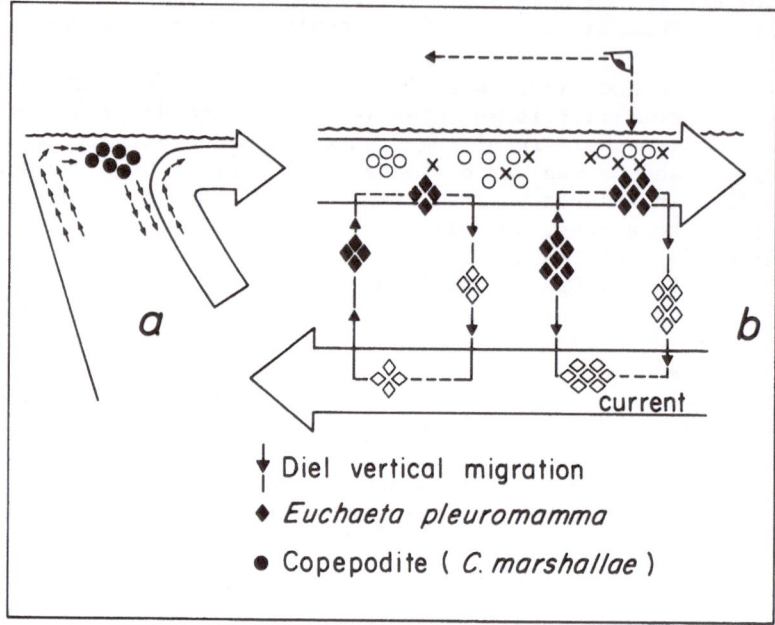

Figure 1. (a) Persistence of the copepodite stages of *Calanus marshallae* maintaining constant depth in a coastal upwelling convection cell. (b) Drift of planktonic stages (cross and circle) maintaining constant depth in an offshore upper layer. Differential advection by upper current during the night and deeper countercurrent during the day allows geographical persistence of other species performing diel vertical migrations (diamond-shaped) as observed outside the mauritanian upwelling (from Boucher, 1982).

vertical distributions resulting in different horizontal transport features (Phillips *et al.*, 1981).

On the Oregon coast similar interactions are involved in the geographic persistence of the *Cancer magister* crab larvae (Lough, 1976). The early larvae are found in near shore areas, constrained from offshore transport by the northwest flowing Davidson current. Late zoeae and megalops occupy deeper layers in the water column and are transported towards the coast by bottom flow (Lough, 1976). An interesting feature here is the time coincidence of the late stage metamorphosis with the seasonal change in March from the Davidson current to a south-southwest flowing current. Close patterns of relations between larval behaviour and hydrodynamics in the determinism of the space-time distribution have been shown for the blue crab *Callinectes sapidus* in the Chesapeake Bay (Provenzano *et al.*, 1983).

The spatial separation of the atlantic herring (*Clupea harengus*) and the Capelin (*Mallotus villosus*) populations are explained by the selection of the spawning grounds, the behaviour of the early larvae, their changes during the development and the two layered circulation in the middle St. Lawrence Estuary (Fortier and Leggett, 1982, 1983). The herring larvae hatched near the bottom (fixed eggs) are transported landward whereas the Capelin larvae spawned and hatched in the surface layer are transported seaward. The diel vertical migrations coincident with the tidal cycle retained the later herring larvae in the upper part of the middle estuary. The later Capelin larvae performed small scale diel vertical migrations and are transported towards the lower estuary (see also an analysis in Legendre and Demers, 1984). All of these examples explain the georaphical distribution and its persistence for a given population. They introduce the idea of a series of event programmed in the life cycle : time and site of spawning, different buoyancy or attachment of the eggs, behavioural patterns and changes through the ontogenesis. In particular, vertical behaviours are triggered by external factors with scalar properties (temperature, salinity, hydrostatic pressure, chemical concentrations) and oriented (or triggered and-) by vectorial ones, light and gravity (Rice, 1967; Crisp, 1974). But the behavioural responses to these factors change between species and ontogenic stages occupying a same water mass at the same or almost the same time. External information is the same for all the organisms but their response patterns are coded in the life cycle.

The time-space cycle of Calanus helgolandicus illustrates both aspects of the complex behavioural sequence and the relative independence of the life-cycle events relative to the dynamics of the hydrological and biological environment. This will add the case of holoplanktonic oceanic populations to the neritic ones (benthic and pelagic) analyzed above.

2.3. Time-Space Cycle of the *Calanus helgolandicus* Populations

The vertical swimming behaviour (depth of occurence, diel vertical migrations or absence, migration amplitude, spawning migration...) of *C. helgolandicus* in the Celtic Sea depends on the ontogenic stage and the seasons within a given stage (Williams and Conway, 1984). The sibling species *C. finmarchicus* exhibits a different behaviour resulting in a spatial separation. A similar series of behaviour changes through ontogenesis and seasons was deduced from the space-time distributions observed for *C. helgolandicus* in the mediterranean Liguro-Provençal Basin (Boucher, 1984; Boucher *et al.*, 1987; Ibanez et Boucher, 1987). The behaviours generating the successive spatial distributions and their persistence are very close to the previously analyzed cases. They are not detailed here. The hydrodynamic structure in the Liguro-Provencal area is characterized by the cyclonic circulation of the Coastal Ligurian Current bound by a highly dynamical thermo-haline front where divergent and convergent cells are associated within some kilometres. The distribution of the individuals in the so-defined hydrological boxes, change through the ontogenesis and the annual

cycle of the population (Figure 2). The population has two phases :
one phase generation occurs in the upper layer (100-0m) from spring
to early summer. The second phase is an "overwintering phase"
occupying deeper layers (> 300 m) until the following winter. At the
end of the winter the later type migrates towards the upper layer
where it sexually matures and reproduces. The copepodite (III, IV, V)
cohorts age and persist in the frontal zone (Boucher, 1984). The
distribution of the resulting adults is independent of the
hydrological structure. At the beginning of the summer these
individuals migrate towards deeper layers until the next winter. A
similar ontogenic migration was described for *Calanoides carinatus* in
the Gulf of Guinea whose reproductive generation migrates towards the
seasonal coastel upwelling (Binet and de Sainte Claire, 1976).

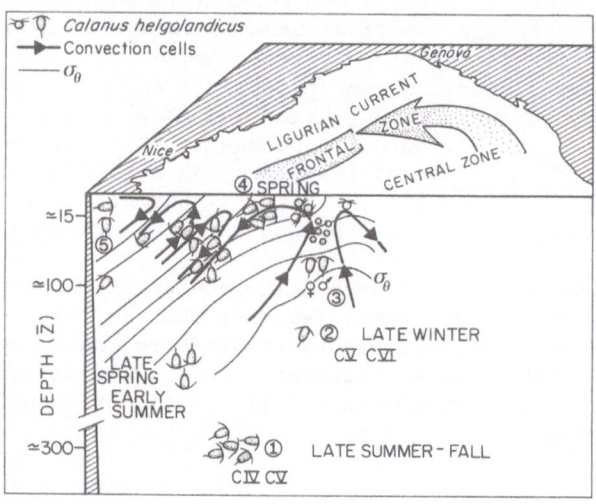

Figure 2. Time-space cycle of *Calanus helgolandicus* in the Ligurian
marine front
   1) overwintering phase in the lower layer
   2) and in late winter
   3) sexual maturation and spawning in the main divergence
      hypothesized form Peterson *et al.*, (1979) and Binet and
      De Sainte Claire (1976)
   4) maintained localization in the frontal zone of ageing cohorts
      (Boucher, 1984)
   5) dispersed individuals and adults with independent distribution.

The temporal dynamics of these ontogenic migrations is not
synchronized by the environmental changes. The onset of overwintering
and reproductive phases of the Celtic populations appear to occur
slightly later (respectively in October and April) than those of the
Mediterranean Sea. But the Copepods peaks are not always co ncident
with phytoplankton peaks (Williams and Conway, 1984, for the Celtic
Sea). The frontal structure in the Ligurian Sea is permanent all over

the year but more robust (isopycnal slopes) in winter with a higher phytoplankton biomass.

The last point of interest is the different metabolic orientation between spring and winter generations. The former synthesize and store lipid reserves whereas the latter use available external energy for sexual maturation and reproduction. The process involved in the determinism of the metabolic orientation is still an unsolved question. Hirche (1983) hypothesized an "endocrine process acting as the "biological clock" that controls termination of the overwintering stage" in his analysis of higher latitude individuals of the species. Whatever the internal process is it acts as an internal time sense determining the generation behaviour in order to match more or less future environmental features, from the Gulf of Guinea to the northern high lattitudes. Parallel process is established in the triggering of resting egg production in neritic copepods by increasing photoperiod (cold water acclimated populations) or decreasing ones (warm water acclimated populations) (Uye, 1985).

3. GENERALIZATION, INTEREST AND CONSEQUENCES IN THE DYNAMICS OF THE PLANKTONIC PHASE

Our primary purpose is to define the main processes involved in determining the spatial distribution of planktonic phases of marine populations in order to estimate the influence of the hydrodynamic variability on population fluctuations. This lies in the interaction of ontogenetically changing behavioural patterns with the hydrodynamic structure. It is a platitude to write that the distribution of individuals is a space-time continuum from the egg to the adult. But following such an approach the hydrodynamics appear as a the transport mechanism acting in the determination of the spatial distribution as an "effector" of the vertical behavioural pattern coded or programmed in the life cycle. Abiotic factors act mainly as regulators, triggering and orienting the vertical displacements. The biotic and abiotic factors secondarily enhance or disturb the process. Williams and Conway (1984) have outlined this innate aspect "Within this complex (vertical) distribution of *Calanus helgolandicus* there is an inherent behavioural pattern which always expresses itself dependent on the maturity of the copepod".

A main evidence supporting the generalization of such an analysis is the considerable amount of literature describing the vertical distribution of zooplanktonic organisms, emphasizing varying patterns according to species and ontogenic stages, but generating very confusing discussions on the role of the biological processes (feeding, predation, exclusion...) which could determine these distribution patterns (*cf.* for instance, the discussion in Sameoto, 1985). In the same way Bougis's analysis (1974) of zooplanktonic dynamics outlined that not only are quantitative relationships (food and feeders) involved but also space-time sequences of convenient events.

Our analysis of space-time distribution of zooplanktonic stages emphasizes the innate characteristic of the behavioural pattern coded or programmed in the life cycle generating the geographical persistence of the population. Therefore these innate behaviours are the determinants of the space-time interactions with the environment. Such an approach modifies the importance of the role of the trophic relationships according to the food-web paradigm.

To take advantage of these observations, an *in situ* experimental approach should be defined from the life cycle date, duration of life in the stage and the amplitude of the displacements allowed by the behaviour. As other mixed possibilities the definition of the population, the adaptative value of the behaviour in the space partition and in the persistence of the population, the probability of prey and predator time-space interaction could be considered. The discussion here will be limited to the main implications for the larval dynamics considering separately the timing of the events and the time-space analysis of the dynamics.

3.1. Time Aspects

Whereas spatial aspects will appears quite well defined and quantifiable, time aspects are poorly identified. A main implication, the timing of the life cycle in relation to the environmental events can be identified with the basic principle of the Match-Mismatch hypothesis : an independent timing of the process and the conditions allowing its realization.

The principal manifestation of the timing problem is the sexual cycle well known in the temperate neretic regions, but which also occurs in buffered environments as tropical and abyssal areas. For instance co-occurrying species of abyssal echinoderms have different sexual cycles (Tyler *et al.*, 1985). The regulation of these behaviours in a broad sense implies not only exogeneous factors but also endogeneous ones timed by external chronological cues. The complexity of the regulatory system decreases with the phyllogenetic evolution level of the species so that the dynamics of the lower appear more timed with the dynamics of their environment than that of the higher ones. But in all cases external chronological cues are involved in the timing of the life cycle events and therefore in the variability of the dynamics.

For example, the external timing of resting egg production limits the duration of the reproductive generations independent of phytoplankton abundance. The timing of the reproductive and the overwintering phases in *Calanus* species could contribute to the decreasing trend in the population abundance shown by Colebrook (1985) in the North Atlantic and the North Sea. These examples imply trophic relationship limitation. But there are other behaviours regulated by celestial cues. Different cyclic activities, spawning, hatching of larval, rhythmic physiological activities and swimming behaviour are synchronized by solar, lunar and tidal cycles at various time scales ranging from hours to the year (*cf.* De Coursey, 1983, a review of the principles with numerous cases for marine crustaceans). There is here a source of variability in the population dynamics following the match of the behaviour determination by such a biological timing and environmental variability. The "spawning" migrations of immature pelagic fishes are other examples : the following decrease of the population fecundity creates variability in the larval dynamics. The timing and the synchronization of the spawning or the hatching is an important factor in the determination of the spatial distribution and dispersal. The match or mismatch with physical features (*cf.* the blue crab case) will determine the dispersal of the larval cohorts.

More systematic observations need to be conducted to study the adaptative value of the timing in the life and the sexual cycles and its impact on the larval dynamics.

3.2. Time-Space Analysis of the Planktonic Larval Dynamics

The hydrodynamic mechanisms which result in part from climatic variability and which determine spatial distribution can be classified according to their potential impact.

i) direct : changes in organisms transport resulting from variability in kinetic structure, (*e.g* wind stress acting on the entrainment or by mixing, stratification of well-mixed water column, swimming or behaviour inhibition by turbulence).

ii) indirect : changes in the dynamics of the abiotic and biotic factors regulating the behaviour (triggering cues, life duration, swimming ability).

iii) interactions.

The analysis of the hydrodynamic variability is developped in some of the examples analyzed. Two approaches are available depending on the emphasis on long or short time scales.

i) time series analysis of the relation between variability of the climate processes which modify the hydrodynamic transport features and the fluctuation of the populations. Such are the cases of the wind-stress actions on the blue crab fluctuations (Hester in Johnson *et al.*, 1984) or the zooplankton dynamics (Taggart and Leggett, 1987).

ii) given the spatial reference defined by the habitat extent, *in situ* analysis of the cohort time-fluctuations (Capelin case, Fortier and Leggett, 1985).

Such approaches imply considerable logistic activity in the definition of the habitat space-time extent. But such a knowledge allows the definition of a reduced sampling grid (Taggart and Leggett, 1987). Another consequence of these hydrodynamic variability analysis is the need for a more sophisticated description of the physics than a mean or a short term model of the residual circulation.

212

4. REFERENCES

Binet, D. and Suisse de Sainte Claire, E.: 1975, 'Le copepode planctonique Calanoides carinatus, Repartition et cycle biologique au large de la Cote d'Ivoire', Cahiers ORSTOM Ser. Oceanogr., XIII, 15-30.

Boicourt, W.D.: 1988, 'Recruitment dependence on planktonic transport in coastal waters', in B.J. Rothschild (ed.), Toward a theory on biological and physical interactions in the world ocean, Kluwer Academic Publishers, Dordrecht, pp. 183-202.

Boucher, J.: 1982, 'Peuplement de copepodes des upwelling cotiers nord-ouest africain. II-Maintien de la localisation spatiale', Oceanol. Acta, 5, 199-208.

Boucher, J.: 1984, 'Localization of zooplankton populations in the Ligurian marine front: role of ontogenic migration', Deep Sea Res., 31, 469-478.

Boucher, J., F. Ibanez, and L. Prieur: 1987, 'Daily and seasonal variations in the spatial distribution of zooplankton populations in relation to the physical structure of the ligurian sea front', J. Mar. Res. 45(1), 133-173.

Bougis, P.: 1974, 'Ecologie du plancton marin. II Le zooplancton', Masson, Paris, 200pp.

Bousfield, E.L.: 1955, 'Ecological control of the occurrence of barnacles in the Miramichi estuary', Bull. Nat. Mus. Can., 137, 1-69.

Colebrook, J.M.: 1985, 'Continuous plankton records: overwintering and annual fluctuations in the abundance of zooplankton', Mar. Biol., 84, 261-265.

Crisp, D.J.: 1974, 'Factors influencing the settlement of marine invertebrate larvae', in P.T. Grant and A.M. Mackie (eds.) Chemoreception in Marine Organisms, Academic Press, New York, pp. 177-265.

Crisp, D.J.: 1976, 'The role of a pelagic larvae', in P.S. Davies (ed.), Perspectives in Experimental Biology, I-Zoology, Pergamon Press, New York, pp. 145-155.

Cushing, D.H.: 1982, Climate and Fisheries, Academic Press, New York, 373pp.

DeCoursey, P.J.: 1983, 'Biological Timing', in F.J. Vernberg and W.B. Vernberg (eds.), Biology of Crustaceans, Academic Press, New York, 7, 107-162.

DeWolf, P.: 1973, 'Ecological observations on the mechanisms of dispersal of barnacle larvae during planktonic life and settling', Neth. J. Sea. Res., 6, 1-129.

Fortier, L. and W.C. Leggett: 1982, 'Fickian transport and the dispersal of fish larvae in estuaries', Can. J. Fish. Aquat. Sci., 39, 1150-1163.

Fortier, L. and W.C. Leggett: 1983, 'Vertical migrations and transport of larval fish in a partially mixed estuary', Can. J. Fish. Sci., 40, 1543-1555.

Fortier, L. and W.C. Leggett: 1985, 'A drift study of larval fish survival', Mar. Ecol. Progr. Ser., 25, 245-257

Herman, A.W., D.D. Sameoto and A.R. Longhurst: 1981, 'Vertical and horizontal patterns of copepods near the shelf break south of Nova Scotia', *Can. J. Fish. Aquat. Sci*, **38**, 1065-1076.

Hirche, J.J.: 1983, 'Overwintering of *Calanus finmarchicus* and *C. helgolandicus*', *Mar. Ecol. Progress. Ser.*, **11**, 281-290.

Ibanez, F. and J. Boucher: 1987, 'Anisotropie des populations zooplanctoniques dans la zone frontale de Mer Ligure', *Oceanol. Acta*, **10(2)**, sous presse.

Iles, T.D. and W. Sinclair: 1982, 'Atlantic herring: Stock discreteness and abundance', *Science* 215, 627-633.

Johnson, D.R., B.S. Hester and J.R. McConaugha: 1984, 'Studies of a wind mechanism influencing the recruitment of blue crabs in the Middle Atlantic Bight', *Cont. Shelf. Res*, 3, 425-437.

Kullenberg, G.: 1985, 'Views on the recruitment problem. In relation to the ICES ad hoc study group on Recruitment', ICES, IREP problems, Session Q, 8pp.

Legendre, L. and S. Demers: 1984, 'Towards dynamic biological oceanography and limnology', *Can. J. Fish. Aquat. Sci*, 41, 2-19.

Lough, R.G.: 1976, "Larval dynamic of the Dungeness crab, *Cancer magister* off the Central Oregon coast, 1970-1971', *Fish. Bull.*, 74, 353-376.

Mileikovsky, S.A.: 1971, 'Types of larval development in marine bottom invertebrates: their distribution and ecology significance, a re-evaluation', *Mar. Biol.*, 10, 193-212.

Peterson, W.T., C.B. Miller, and A. Hutchinson: 1979, 'Zonation and maintenance of copepod populations in the Oregon upwelling zone', *Deep Sea Res.*, 26, 467-494.

Phillips, B.D.F., D.W. Rimmer, and D.D. Reid: 1978, 'Ecological investigations of the late stage phyllosoma and puerulus larvae of the western rock lobster *Panulirus longipes cygnus*', *Mar. Biol.*, 45, 347-357.

Phillips, B.D.F., P.A. Brown, D.W. Rimmer, and D.D. Reid: 1979, 'Distribution and dispersal of the phyllosoma larvae of the western rock lobster, *Panulirus cygnus* in the Southeastern Indian Ocean', *Aust. J. Mar. Freshwater Res.*, 30, 773-783.

Phillips, B.D.F., P.A. Brown, D.W. Rimmer, and S.J. Braine: 1981, 'Distribution and abundance of late larval stages of the Scyllaridae (slipper lobsters) in the Southeastern Indian Ocean', *Austr. J. Mar. Freshwater Res*, 32, 417-437.

Provenzano, A.J., J.R. McConaugha, K.B. Phillips, D.F. Johnson, and J. Clark: 1983, 'Diurnal vertical distribution of first stage larvae of the blue crab, *Callinectes sapidus* at the mouth of the Chesapeake Bay', *J. Est. Coast and Shelf Sci.*, 16, 489-499.

Rice, A.L.: 1967, 'The orientation of the pressure responses of some marine crustacea', *Proc. Symp. Crustacea, Ser. 2, Mar. Biol. Assoc. India*, III, 1124-1131.

Sameoto, D.D.: 1985, 'Environmental factors influencing diurnal distribution of zooplankton and ichthyoplankton', *J. Plank. Res.*, 6, 767-792.

Sastry, A.N.: 1983, 'Pelagic larval ecology and development', in F.J. Wernberg and W.B. Wernberg (eds.), *The Biology of Crustacea*, Academic Press, New York, 7, 213-280.

Scheltema, R.S. and I.P. Williams: 1983, 'Long distance dispersal of planktonic larvae and the biogeography and evolution of some polynesian and western Pacific Mollusks', *Bull. Mar. Sci.*, 33, 545-565.

Sinclair, M.: 1987, 'Marine populations. An essay on population regulation and speciation in the oceans', (in press).

Taggart, T.C. and W.C. Leggett: 1987, 'Wind-forced hydrodynamics and their interaction with larval fish and plankton abundance: A time-series analysis of physical-biological data', *Can. J. Fish. Aquat. Sci.*, 44, 438-451.

Thorson, G.: 1950, 'Reproductive and larval ecology of marine bottom invertebrates', *Biol. Rev.*, 25, 1-45.

Tyler, P.A., A. Muirhead, J.D. Gage, and D.S.M. Billet: 1985, 'Gametogenic strategies in deep-sea echinoids and holothurians from the N.E. Atlantic', in B.F. Keegan and D.B.S. O'Connor (eds.), *Proc. Vth Intern. Echinodermata Conf.*, Galway 24-29 September 1984, A.M. Balkema, Rotterdam, Boston, pp. 135-140.

Uye, S.: 1985: 'Resting egg production as a life history strategy of marine planktonic copepods', *Bull. Mar. Sci.*, 37, 440-449.

Vinogradov, M.E.: 1968, 'Vertical distribution of the oceanic zooplankton', Nauka publications, traduction francaise, Israel Program for scientific translation, Jerusalem, 1970, 339pp.

Williams, R. and O.V.P. Conway: 1984, 'Vertical distribution, and seasonal and diurnal migration of *Calanus helgolandicus* in the Celtic Sea', *Mar. Biol.* 79, 63-73.

Wood, L. and W.J. Hargis, Jr.: 1971, 'Transport of bivalva larvae in a tidal estuary', in J.D. Crisp (ed.), *4th Europ. Mar. Biol. Symp.*, Cambridge, 21-44.

Wroblewski, J.S.: 1982, 'Interaction of currents and vertical migration in maintaining *Calanus marshallae* in the Oregon upwelling zone - a simulation', *Deep Sea Res.*, 29, 665-686.

REVIEW OF OCEANIC TURBULENCE: IMPLICATIONS FOR BIODYNAMICS

Hidekatsu Yamazaki and Thomas R. Osborn
Chesapeake Bay Institute
The Johns Hopkins University
Suite 315/The Rotunda, 711 W 40th Street
Baltimore, Maryland 21211
U.S.A.

ABSTRACT. This review focuses on the nature of oceanic turbulence.
It describes some of the existing ideas and measurements to bring
attention to the possible effects of turbulence on biodynamics. Much
of the turbulence occurs in the upper part of the ocean where the
biological productivity is greatest. Reported intensities extend from
the noise levels of the instrumentation (about 10^{-7} W/m^3) to 1 W/m^3.
Sources of turbulent energy are the current shear, surface waves,
internal waves, tides, intrusions, free convection and forced
convection, and topographic features. In many instances the
probability distribution of turbulent dissipation values can be
approximated by a lognormal distribution. Models relating the
intensity of the turbulence to the mean fields--*e.g.* wind, currents,
and surface waves--are developing and appear useful to support future
biological modelling.

1. INTRODUCTION

The physical environment affects marine organisms in a variety of
ways. Temperature, salinity, and dissolved oxygen can affect growth
rates, metabolism and reproduction, while currents can transport
larvae into favorable or unfavorable environments. The relationships
between the physical environment and the temporal and spatial
variability of marine organisms must be investigated to enable further
modelling of marine production systems. As with any highly non-linear
and multifaceted system, it is difficult to relate the temporal and
spatial variability of the result (*e.g.* marine production) to the
temporal and spatial variability of the forcing functions (light,
wind, evaporation and precipitation). Non-linear systems inherently
transfer information (signal) across wavenumber and frequency bands.
Due to the wide range of scales, the simpler problems of oceanic
circulation and weather prediction are almost intractable, in spite of
well known equations of motion and state. Computer memories and
operating speeds are insufficient to solve the full equations.

B. J. Rothschild (ed.), Toward a Theory on Biological-Physical Interactions in the World Ocean, 215–234.
© 1988 by Kluwer Academic Publishers.

Simplified equations are derived with a truncated range of scales. Sometimes one parameterizes the small scales by derivatives of larger scale parameters, or one imbeds a small-scale problem into a predetermined, large-scale, mean flow.

Many pelagic creature have aggregated distribution of abundance in a wide variety of space and time scales. Haury *et al.* (1978) discuss time-space scales of plankton. The smallest scale, the micro-scale, is between 1 cm and 10 m. The largest scale, the mega-scale, has been observed in the order of 1000 km. Large scales are relatively easy to identify and correlate with the surrounding hydrographic environment. On the other hand, the role of small scale physics has not been studied in conjunction with the population dynamics. An essential step in biological modelling is to ascertain the effect of the smallest scales of motion on biological primary-secondary productivity. Contact and ingestion are fundamental activities in secondary production. Rothschild and Osborn (In Press) suggest that the turbulent velocities (Figure 1) modify the contact rates between small organisms.

During the last 20 years there has been a dramatic increase in our ability to sample and describe oceanic turbulence. Direct measurements of the 1 cm to 1 m scale velocity and temperature fluctuations give quantitative information about turbulence; the intensity, the spatial and temporal scales, the effects of stratification, and the generating mechanisms. Physical oceanographers are now using these observations to understand details of the global dynamics. Improved knowledge of oceanic turbulence can reveal links between the physical environment and biological productivity. Fortunately, the region of biological interest, the upper ocean, is the most heavily sampled. In the next section we briefly review the nature of turbulence and discuss the mechanisms generating oceanic turbulence. Section 3 summarizes the observations, while some speculations about the effects of turbulence on "biodynamics" are in section 4.

2. THE NATURE OF OCEANIC TURBULENCE

Turbulence is a random, three-dimensional motion with the velocity and vorticity irregularly distributed in time and space. It is characterized by an energy transfer from large to small scales where dissipation of kinetic energy is taking place. The stirring motion of the turbulence leads to enhanced mixing.

The familiar patchiness of passive materials in the ocean is the result of turbulent diffusion, a direct consequence of the turbulent motion. Suppose the density $C(x,t)$ of particles follows Fick's law.

$$\partial C/\partial t + \mathbf{U} \cdot \nabla C = D\nabla^2 C \qquad (1)$$

where $\mathbf{U}(x,t)$ is an instantaneous velocity and D is the molecular diffusivity. Turbulence effect can be realized by separating C and U into the mean and the fluctuating parts.

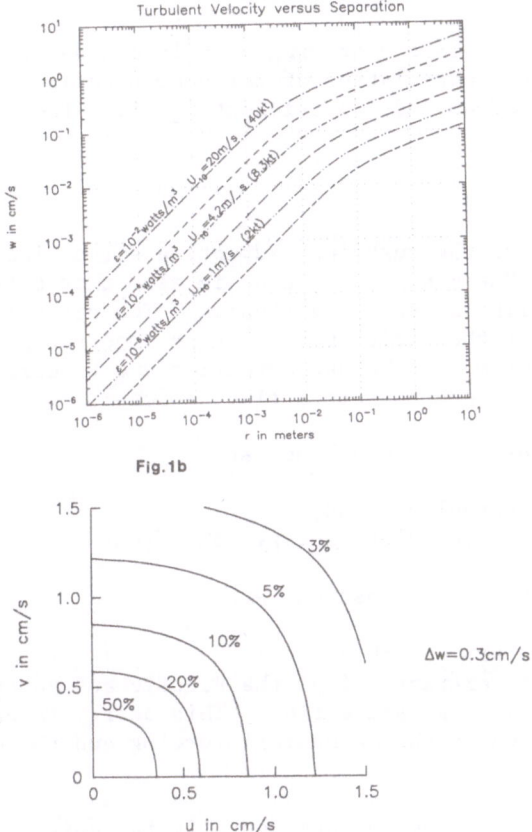

Fig.1b

Figure 1. (a) The root-mean-square turbulent velocity, w, that is uncorrelated with the turbulent velocity fluctuations at a point a distance r away. The curves represent increasing values of ε, the turbulent dissipation rate. Wind velocities that might generate particular dissipation rates are indicated. While wind speeds of 22m/s are relatively large, dissipation rates of 10^{-2} or 10^{-1} watts/m^3 would not be unusual in tidally mixed areas. (b) Contours showing the percentage increase in contact rate as a function of predator speed (v) and prey speed (u) for an uncorrelated rms turbulent velocity of 0.3 cm/s (Rothschild and Osborn, In Press).

$$C = C_0 + c \qquad (2.1)$$
$$U = U_0 + u \qquad (2.2)$$

Substitution and ensemble averaging yield;

$$\partial C_o/\partial t + U_o \cdot \nabla C_o - D\nabla^2 C_o - \nabla \cdot <uc> \qquad (3)$$

where $< >$ represents ensemble averaging. It is customary to recast the last term which is a correlation of the concentration and velocity fluctuations in terms of an eddy coefficient K_d times the mean gradient.

$$- <uc> - K_d \nabla C_o \qquad (4)$$

This relation means that the turbulent advective effect looks like molecular diffusion. One must always keep in mind that diffusion due to turbulence is not like molecular diffusion. The diffusion is not an inherent property of the fluid, but it is an inherent part of the flow. Turbulent diffusion is the net advection due to correlation between the velocity and concentration fluctuations.

Two important questions about turbulence are:

1. What causes turbulence and,
2. Where does the turbulent energy come from?

The simplest answers to these questions are:

1. Smooth flows are unstable.
2. The energy is extracted from the kinetic and potential energy of the large scale flow. This energy is provided at the boundaries by the mechanical forcing and the buoyancy flux.

To answer these questions in more depth it is convenient to introduce the equation for the kinetic energy of the turbulent velocity fluctuations. The vertical coordinate, z, is positive upward and the mean flow, U_o, is parallel to the x-axis. The rate of change of the turbulence kinetic energy, $1/2 u \cdot u - q - 1/2 \{u^2 + v^2 + w^2\}$, is, (Phillips, 1977)

$$\partial q/\partial t - - \partial\{ <wp/\rho_o> + <wq> \}/\partial z - <uw>\partial U_o/\partial z - g<\rho w>/\rho_o - \varepsilon. \qquad (5)$$
$$\quad\quad\quad\quad P \quad\quad\quad\quad\quad Q \quad\quad\quad\quad\quad S \quad\quad\quad\quad B \quad\quad Z$$

Where ρ_o and ρ are mean and fluctuating part of density, p is the fluctuating pressure, and g is the acceleration of gravity.

The first two terms, P and Q, on the right hand side of equation (5) represent the redistribution of the kinetic energy in space. These terms are divergences of vectors so the energy that arrives in one volume has come from another place. Q includes the turbulent diffusion of turbulent energy, e.g. the spreading of turbulence from a breaking wave. The third term, S, is usually a source term for the turbulent kinetic energy, providing energy from the mean flow. Small-scale mixing in the presence of a velocity shear extracts energy from the larger scales. The fourth term, B, is the buoyancy flux term. Turbulent mixing of a stratified fluid produces a positive buoyancy flux, which reduces q while increasing the potential energy of the system. However, a negative buoyancy flux, for example night time convection in the upper layer, acts as a major source term for the kinetic energy. The dissipation rate, ε, is always positive as it extracts energy from the kinetic energy field. It is a fundamental parameter of the flow and determines the smallest scales of turbulent motion. The Kolmogorov wavenumber, k_s, is determined by the dissipation rate and the kinematic viscosity, ν, $k_s = (\varepsilon/\nu^3)^{1/4}$. The associated wavelength, $\lambda_s = 2\pi/k_s$, is the lower boundary of the turbulent velocity fluctuations. At smaller scales only the straining motion from the turbulence is left. The dissipation can be estimated from the shear spectrum or inferred from the large scale parameters of the flow. Suppose U_L is the velocity for the energy containing eddies of scale L. Then $\varepsilon \approx U_L^3/L$. The turbulent velocity spectrum extends from $k = 2\pi/L$ to k_s, and if the Reynold's number is sufficiently large, there is also a universal shape to the spectrum. Table 1 summarizes time scales and spatial scales for various terms in equation (5).

The cartoon in Figure 2 (copied and compiled from Thorpe 1985 and 1987) shows many of the prevalent ideas about the sources for turbulent energy and mixing (see also Turner 1981, Gregg 1987 and

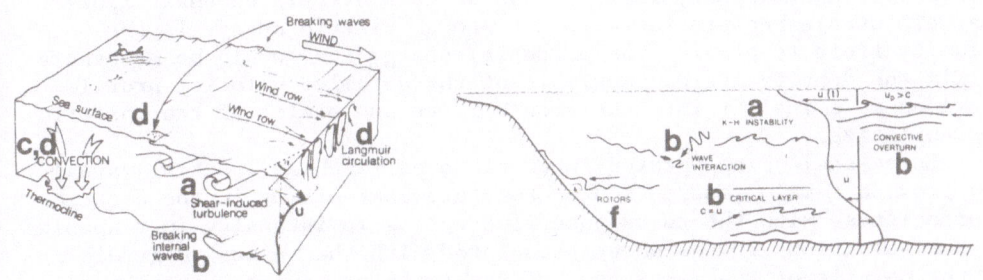

Figure 2. Cartoon drawings of turbulent upper layers with the mechanisms labelled following equation 5. (The top Figure from Thorpe, 1985 and the bottom Thorpe, 1987.)

Thorpe 1987 for nice reviews). Breaking waves, both surface and internal gravity waves, are felt to be responsible for much of the turbulence near the surface and in the stratified part of the ocean. The energy is taken from the wave field and dissipated with turbulence, at least in the case of internal waves. Turbulence also uses a small amount of the energy to cause local mixing and hence to raise the center of mass (positive buoyancy flux). Further mixing

in the upper layer is driven by "free convection" due to surface
cooling and evaporation. These effects produce denser water which
falls through the water column, converting potential energy into
turbulent kinetic energy. "Forced convection" is the mixing driven by
the wind stress applied to the surface. Here the energy is supplied
by the work done by the wind stress on the sea surface.

Kelvin-Helmholtz instabilities are important mechanisms for
producing turbulence. These instabilities (Lamb, 1932; Woods, 1968;
Thorpe, 1973) grow, roll up, and become gravitationally unstable then
collapse into a turbulent patch. Large internal waves can develop
shear instabilities. Once the turbulence is started it can grow, if
it extracts more energy from the mean flow (term S in equation 5) than
it converts to other types of energy, e.g. ε and B. The stirring leads
to collisions between parcels. The stirring conserves momentum but not
kinetic energy - remember the 'plastic' collisions in our early
physics courses where two objects stick together. This phenomenon
occurs when you stir a sheared flow and mix it to a uniform flow. The
process conserves momentum but removes energy from the mean flow by
dissipating kinetic energy through turbulence. The energy source is
the mean flow, but the energy is only available if the flow is
sheared. The presence of stratification complicates the picture a
bit. If the velocity field is mixed, the density field becomes mixed.
This raises the center of mass and increases potential energy. To go
from one state to another, without putting energy into the system, the
decrease in kinetic energy must exceed the increase in potential
energy. This result can be expressed as a Richardson Number =
$g(\Delta\rho/\Delta z)/(\Delta U/\nabla z)^2 > 0.25$ being a sufficient requirement for
stability (Chandrasekar, 1961). Thus we have stratified shear flows
that are stable because there is not enough kinetic energy in the
velocity field to provide the potential energy that would be necessary
to mix the density field. Analysis of the detailed velocity profile
provides more insight into the growth rates and scales of the specific
instabilities.

Large-scale, long-period internal waves (such as inertial waves)
may provide a major energy source for turbulent mixing in the oceanic
thermoclines, with the turbulence fluctuating in intensity in response
to the variations in the shear associated with the waves. Regardless
of the details of the mechanisms of the transfer, current shear is
important aspect of the oceanic turbulence problem. The shear not
only distort and stretch water parcels but also serves to bring
distant parcels of water together. Eckart (1948) noted the two aspects
of stirring are 1) to increase surface area of a parcel and 2) to
increase the property gradients at the boundary. Both of these
effects increase the rate of molecular transport which reduces
inhomogeneities and is called mixing.

Recent developments in acoustic doppler instrumentations made
possible to profile the ocean currents remotely. An advanced system
can sample over 100's of meters with 10 m spatial resolution resolving
the current speed up to a centimeter per second in a few minutes
(Pinkel, 1984). Figure 3 shows measurements from a submarine using a
1.2 mHz system with 1 m vertical bins, 1 cm/s rms noise in velocity,

and one minute sample period. The large shear feature seen at 30m depth extends horizontally over several hundred meters with a maximum shear of $2.10^{-2}s^{-1}$, a rather large value. Such horizontal features are good sources for turbulence as well as sites of substantial lateral differential displacement.

Topography also generates turbulence; directly due to blockage and flow separation as well as indirectly by generating lee waves, hydraulic jumps, and other instabilities in the flow Gregg and Sanford, 1980). The large variations in water depth and bottom material over the continental shelf and coastal regions interact with the tide. It is amplified by the shallower water to provide very intense stirring of the water column. As well, many waves (Kelvin waves and shelf waves) and upwelling enrich the turbulence.

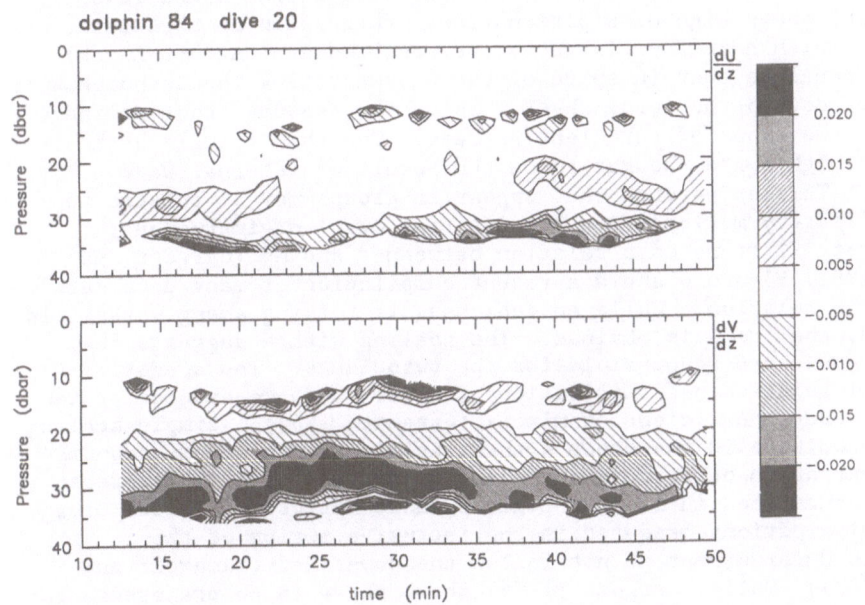

Figure 3. Contours of shear derived from a submarine mounted, doppler-acoustic, current profiler. The vessel is travelling at 1.4 m/s. V is the axial component of the current and U is athwartship.

3. OBSERVATIONS OF OCEANIC TURBULENCE

The world map in Figure 4 shows the distribution of reported turbulence measurements in the ocean. Obviously the sampling is skewed towards certain regions while neglecting others. However, we can still infer general natures of turbulence from the available results. We begin with two vertical profiles of dissipation from the subtropical front region of the central North Pacific (approximately 30°N 133°W) in late January 1980 (Lueck, 1987). There are thick, turbulent patches in the top of the thermocline (0 to 100 m in Figure 5a). Although the turbulent patches show highly intermittent, the top 100 meter is well mixed and active mixing events are taking place. The surface wind speed was not noticeably high, it was only 5 m/sec, but it had started picking up high wind speed. Considering the time of the profile obtained "free convection" is not a likely cause of the mixing event. The energy source is probably from the wind stress. Uniformly mixed layers are often found in temperature profiles, however, mixing events are not always associated with the mixed layers. A middle portion of mixed layer (Figure 5b) shows such a example. Shay and Gregg (1986) discuss this problem of differentiating between "mixed layers" and "mixing layers". A mixed layer does not necessary contain active mixing events. The density profile can remain long after the mixing event has ended. The thermocline often contains relatively high value of ε even though it is highly stratified (Osborn, 1978). The base of the mixed layer (Figure 5b) shows high dissipation rates. This patch of turbulence was associated with near-inertia waves (personal communication, Lueck). One must conclude that in spite of the buoyancy flux the turbulence is successful due to the local shear. Below the seasonal thermocline there are two types of turbulent patches. The thin (single estimates) layers of high ε are thought to be the result of internal wave breaking. Thicker patches that appear in groups may be related to some oscillatory motions (*e.g.* an inertial wave) or intrusions.

Gargett first noted a relation between ε and N^2 (Gargett and Osborn, 1981) Figure 6 shows a recent compilation of many data sets to examine the relation. While no consensus is forming among workers in the field, the trend is obvious. The scaling with N suggests that internal waves are responsible for the turbulence. The specific functional relation has implications for the depth dependence of the turbulent eddy coefficient of mixing (Gargett, 1984). Simple scaling of the turbulence in the ocean with stratification neither accounts for the variation of inertial wave energy nor explains the increase in microstructure seen in frontal regions where intrusions occur (Gregg, 1980). Dissipations measured in the intrusive region of the California Undercurrent do not follow the power law (Yamazaki and Lueck, 1987). While, Gargett (1976) shows there is no preference for the microstructure to occur on density interfaces, that does not preclude the intrusive regions from having greater turbulence because of the frontal dynamics (Kunze and Sanford, 1984; Kunze and Lueck, 1986). In general higher dissipation values are associated with regimes of intrusion.

In experiments focused on the upper layer, Oakey and Elliott (1982) and Oakey (1985) find a fairly direct relationship between wind speed and turbulence intensity. The integrated dissipation in the upper layer is about 1% of the downward energy flux which is proportional to the wind speed cubed (U_{10}^3, the wind at 10 m height).

Strong diurnal cycling of the upper layer is reported by Moum and Caldwell (1985) in the equatorial Pacific. The turbulence cuts off

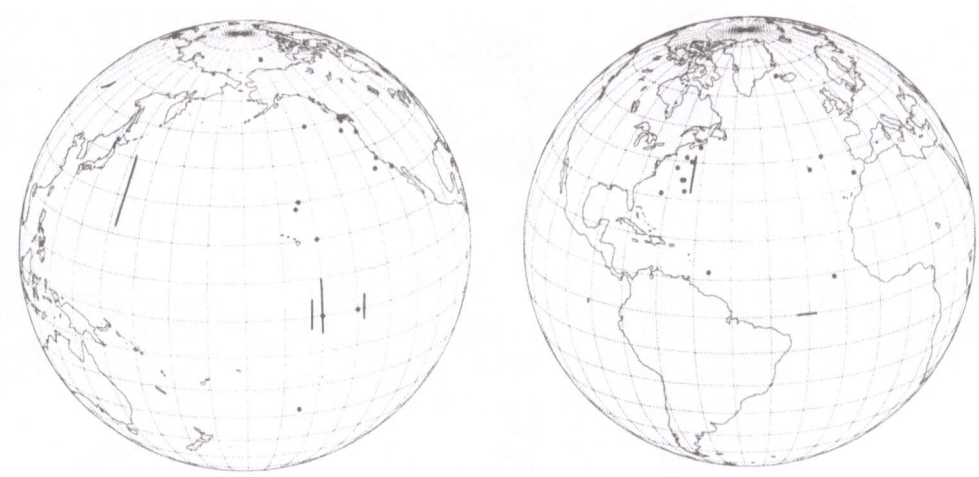

Figure 4. Locations of major observations of oceanic turbulence. A
single solid circle shows each independent observations. Solid lines
are major legs. The left part of Figure shows observations in the
Pacific ocean and the right Figure shows observations in the Atlantic
ocean.

with solar heating with minimum values of dissipation in the late
afternoon. These authors report (personal communication) a similar
diurnal cycle at mid-latitudes. The dissipation averaged over 10 days
is still about 1% the energy flux from the wind field.

Shay and Gregg (1984,1986) examine the turbulence in a convective
mixed layer. Below the level of wind forcing, the dissipation is 45%
of the buoyancy flux as seen in atmospheric studies. The diurnal
cycle of heating and cooling thus produces a diurnal cycle of
turbulence in the mixed layer. This diurnal cycle can be broken by
storms or outbreaks of cold air from continental areas, but the
results of this work suggest that modelling for the turbulence levels
from buoyantly driven convection is possible.

The processes in the top of the mixed layer include significant
contributions from wave breaking and the shear production associated
with the stress transfer through the ocean surface. Direct
measurements in high sea states are difficult. Dillon *et al.* (1981)
suggest the turbulence near the surface scales like the "law of the

224

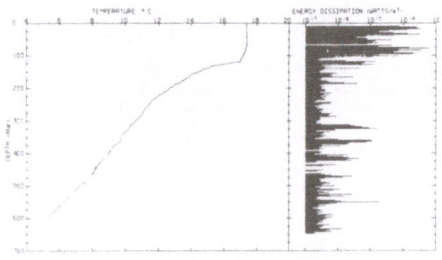

Fig.5a

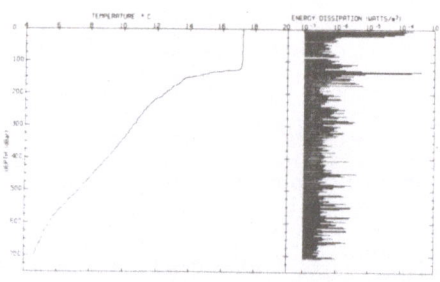

Fig.5b

Figure 5. Two dissipation profiles from January 1980 (Lueck, 1987).
Figure 5a is from the northern side of the subtropical front and
Figure 5b is in the surface manifestation of the front.

wall." Kitaigorodskii *et al*. (1983) show a peak in the turbulence
spectrum at the wave frequency associated with the advection of
turbulent fluctuations by the wave orbital motion.   The turbulence
very near the interface drops below the level predicted from the log
law in the presence of stratification (Kitaigorodskii, 1987).   Thorpe
et al. (1982) uses acoustic measurements of bubbles to infer the
turbulent diffusion rates from a model which balances diffusion,
absorption, compression, and buoyancy of the bubbles. It still remains
to clarify the processes and their magnitudes.
     Other upper layer processes are also important. Langmuir
circulation is being observed in detail now (Weller *et al*. 1985). The
vertical scales are larger than previously thought.   The circulation
frequently reaches the bottom of the mixed layer. Other coherent
structures in the upper layer, such as wave breaking and air bubbles,
are related to the turbulence (Thorpe 1984,1985) but their nature is
still poorly understood and their relative role in mixing is not
known.

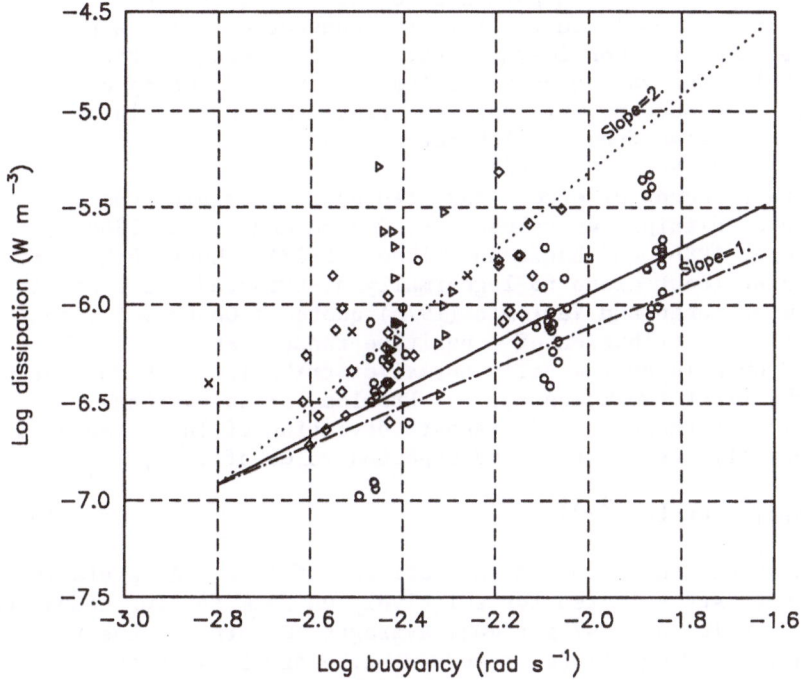

Figure 6. N versus ε for four different data sets. The x's are from
Moum and Osborn (1986), the squares are from Gregg *et al.* (1986), the
diamonds are from Lueck *et al.* (1983), the circles are from Yamazaki
and Lueck (1987) outside of the intrusive region, and the triangles
are from the same paper representing the intrusive regions.

    The presence of a bottom or other fixed solid boundary requires
the velocity to vanish. The tangential velocity component usually
meets the boundary conditions through a turbulent boundary layer and a
viscous sublayer. Observations of the bottom boundary layer at 200m
depth off the Oregon coast (Caldwell and Chriss, 1979) reveal a 0.6
cm viscous sublayer where the shear is 12 s^{-1}. In the turbulent
boundary layer, the estimated dissipation is 0.2 W/m^3, a substantial
value. Tidal energy propagating into estuarine environments can
easily produce even higher numbers. Swift and Brown (1983) report
depth integrated values of the dissipation of tidal energy that
correspond to between 0.1 and 1 W/m^3. The tidal rectification and
frontal circulation on Georges bank discussed by Loder and Wright
(1985) show such a region to be a site of strong turbulent mixing,
generated by the interaction of the large scale circulation, wind,
tides, and wave processes with the shallow topographic feature.
    No discussion of mixing mechanisms in the ocean is complete
without mentioning salt fingers. It is a form of convection that
removes potential energy from the salinity field dissipating about 50%

while using the rest to transport heat downward. This process is a subset of phenomena call doubly-diffusive convection. These processes were thought to play a considerable role in the mixing of the ocean, (Schmitt, 1981). However, recent results (Lueck 1987, Gregg and Sanford 1987, and Osborn 1987) show the fluxes are not as large as previously calculated from the laboratory results extrapolated to the ocean (Gargett and Schmitt, 1982).

A moderate amount of effort has been spent examining the distribution of dissipation values(e.g. Osborn and Lueck, 1985 a,b; Shay and Gregg, 1986: and Baker and Gibson, 1987). Most authors find the data are at least close to lognormally distributed. The result is consistent with Monin and Yaglom's (1975) quote of Obukhov "...that any probability distribution of a positive random variable can be approximated by a logarithmically normal distribution with correct values of the first two moments,..." Problems arise over the interpretation of the mean and standard deviation of ln ε. Suppose ε is log-normally distributed, the expected value of ε is,

$$E[\varepsilon] = \exp\{\mu+\sigma^2/2\}, \tag{6}$$

where μ and σ^2 are the mean and the variance of ln ε. A carefully restricted data set selected for relatively uniform forcing conditions shows a small value of $\sigma^2=2$ for 40cm averages (Osborn and Lueck, 1985a). Shay and Gregg (1986) report $\sigma^2=1.55$ and 1.44 in the convecting part of the upper layer (the former value corresponds to a data set they found data were not lognormal at the 95% confidence level. These latter data sets included data from several different days and intensity of forcing. The value of σ^2 decreases as the averaging increases. Osborn and Lueck (1985b) report $\sigma^2=0.5$ for 10m averages of data from a towed body in a single patch in the seasonal thermocline. Baker and Gibson (1987), on the other hand, report values of σ^2 between 3 and 7. They combine extensive portions of the water column together including both active and inactive regions. It is necessary to separate discussion of dissipation statistics depending upon generation mechanisms of turbulence. Suppose we observe a random variable which is drawn from two identical distributions with different means and variances. The sample variance is mostly larger than the population variances, and we should not apply a simple distribution to fit the samples. There are difficulties in using parametric density estimation for multimodal distribution. Uncertainty remains as to the best approach to sample and represent the statistical variation of ε.

Strong surface forcing can create a 50 m to 100 m or more turbulent mixing layer (Figure 5). Twenty-five to thirty meter patches are found in the thermocline. Otherwise the patches are thinner with an exponential distribution (Yamazaki and Lueck, 1987). Patch can be as thin as 5 centimeters, (Schoeberlein, 1985). "Patch" does not mean a single overturning event from the top to the bottom, but rather, it means a section of dissipation exceeding a fixed threshold. Thick patches are usually associated with stronger turbulence and last longer than thin ones. Gregg et al. (1986)

classify them into two category; short-lived *puffs* and *persistent patches*. Internal gravity waves cause *puffs*. The spatial scales are a few meters for thickness and less than an internal wave length horizontally. Gregg *et al.* (1986) report *puffs* disappear from sequential dissipation profiles a few hundreds meters apart. They also show *persistent patches* coincide with inertial currents and can last for several hours. The *persistent patches* are also found at interfaces of intrusions (Yamazaki and Lueck, 1987).

Two dimensional pictures (Figure 7) of oceanic mixing processes come from towed thermistor chains (Dugan *et al.*, 1985). These data are from 300 thermistors spaced along a 100m chain, temperature and its variance are plotted with time versus depth. This two dimensional cut through the ocean shows the range of vertical scales of turbulence from a meter to many tens of meters and the horizontal scales of kilometers that occur.

4. THE EFFECTS OF TURBULENCE ON BIODYNAMICS

Turbulent diffusion has a pronounced kinematic effect on small organisms. Large scale turbulent diffusion can transport larvae away from a spawning region, whence they may reach favorable environments, but the majority are swept into fatal environments. Reef fish off Hawaii (Loebel and Robinson, 1986) and blue crabs in the Chesapeake Bay (Cronin and Mansueti, 1971) have oceanic life stages that are advected about by the currents and must be adapted to survive in the existing circulation pattern. The patches and spatial distributions of many planktonic creatures are partially a result of the spatial characteristics of the turbulent mixing (Denman and Platt, 1976). Phytoplankton spectra follow the turbulent spectra at high wavenumbers but the low wavenumber part of the spectra is proportional to k^{-1}. The critical wavenumber is $k_c = (\lambda^3/\varepsilon)^{1/2}$ growth rate of the plankton. Despite argent needs of understanding the roles of oceanic turbulence in recruitment of marine organisms many questions need to be answered; How are creatures affected by the mean and fluctuating portions of the circulation? Have they responded to the temporal and spatial distribution of turbulence? Are they adapted to the 'variability of the ocean' rather than the 'mean?' Although increasing evidences of high production sites where different types of water come across, there still remain uncertain the roles of lateral mixing process at the boundaries, *e.g.* fronts and eddies, in abundance of organisms.

Another effect of turbulent diffusion is the transport of nutrients across the thermocline into the upper layer. McGowan and Hayward (1978) report the doubling of the primary production in 1969 in the North Pacific is due to a series of mixing events. There is no doubt about the direct relation between the vertical transport of nutrients and the biological productivity. Vertical stirring and mixing are necessary to sustain high concentrations of plants and animals.

Woods and Onken (1982) show that as the mixing layer deepens in a storm, the cloud of plankton that has sunk into the seasonal

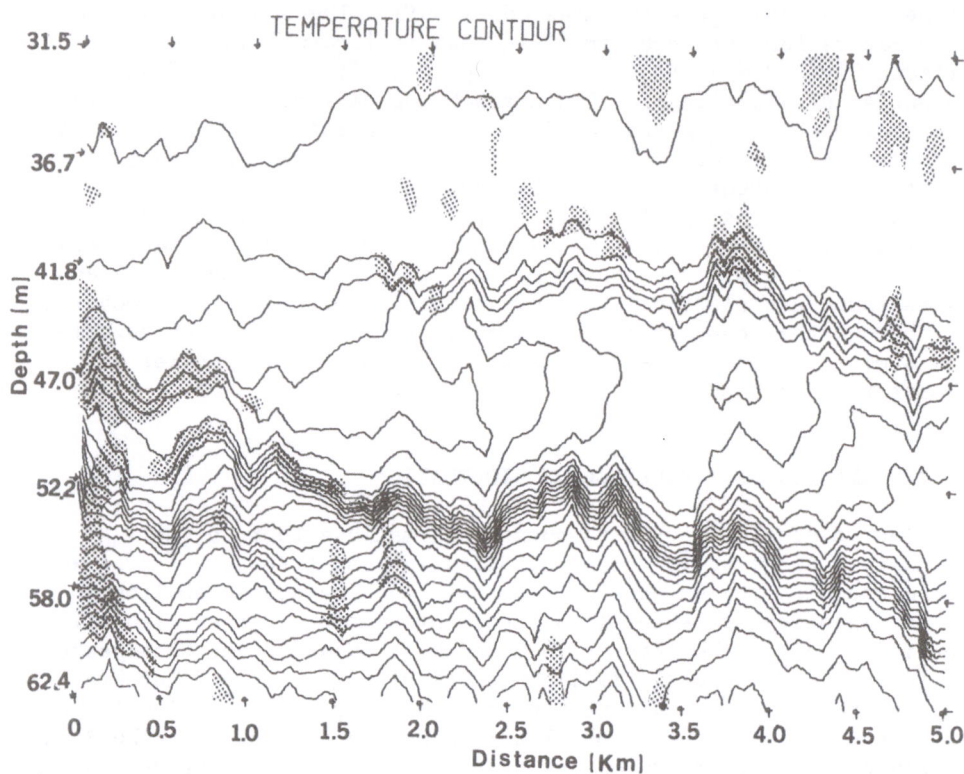

Figure 7. Temperature and its variance from Dugan *et al.* (1985).
Shaded areas are regions of high temperature variance. Note the
horizontal and vertical scales of the patches and how they follow the
oscillations of the isotherms.

thermocline is entrained back into the mixed layer. Recent
observations, (Price *et al.*, 1986), show that the depth of mixed
layer is changed diurnally by the surface heating and cooling.
Turbulent mixing of the upper ocean at the equator, (Moum and
Caldwell, 1985), is intensified right after the sunset and the
turbulent dissipation continuously increases until the sunrise.
Diurnal change in mixing layer depth may be playing a significant
role in diurnal migration of marine organisms.

Many zooplankton show diurnal vertical migration. Some *Calanus*
stay near the surface at night and move as deep as at 400 m during
daytime, (Raymont, 1983). Do the zooplankton benefit from the
nighttime turbulence peak? Similarly, do phytoplankton benefit from
the lack of turbulence during the day? Are they better off at a
relatively constant level, or would they grow faster with more depth

and light cycling, (Lande and Wood, 1987)? The production of
phytoplankton tends to increase by the alteration of mixing and
quiescent periods (Legendre, 1981). The seasonal variation of the
depth of mixed layer may be sensed by certain marine organisms. Some
Calanus finmarchicus aggregate at about 10 m in April, go down to 20
m in June, and then stay near the surface throughout August and
September, (Marushige, 1977). Time scale of mixing events relative
to biological reproduction time scale must be an important factor of
population dynamics (Tett and Edwards, 1984).

A laboratory study of Harder (1968) shows many zooplankters tend
to aggregate at density interfaces. Some field observations support
his results. Turbulence measurements often show a high level of
kinetic energy dissipation at density interfaces. Figure 8, Haury
and Yamazaki (1987), shows simultaneous observations of zooplankton
abundance and dissipation rate sampled from a submarine at the base
of mixed layer (20 meters in depth). The data show no obvious
correlation between the plankton counts and dissipation values.
However, many species show higher abundances in the same region where
dissipation rates are slightly higher than the preceding part of leg
(see distance between 20 and 24 km in Figure 8). The swimming
abilities of most copepods would permit them to select their
preferred depth.

Many taxa of plankters use either mechanoreception or
chemoreception to detect prey (Okubo, 1986). Mechanoreception is
based on changes in pressure field by the presence of prey and
chemoreception primarily depends on a chemical leak from prey.
Pressure field can be altered by turbulence. Smallest turbulent eddy
can be as small as few millimeters. So that turbulence might mask the
approach of the predator, and thus delay the prey's detection of the
predator until too late to escape. On the other hand, turbulence may
hide prey from predator by disturbing pressure field. Similar
speculations can be drawn for chemoreception. Since turbulence
affects the contact rates between small organism (Rothschild and
Osborn, In Press), it would influence the energy expended in
searching for and capturing prey. Feeding rate, ultimately growth
and reproduction rates are therefore affected by turbulence.

Lastly, shear layers are commonly associated with active
turbulence layers, so it may be a beneficial mechanism to transport
plankton one place to another. Turbulence may help plankton getting
out and in the shear layers, so they can move horizontally wide area
without swimming by themselves.

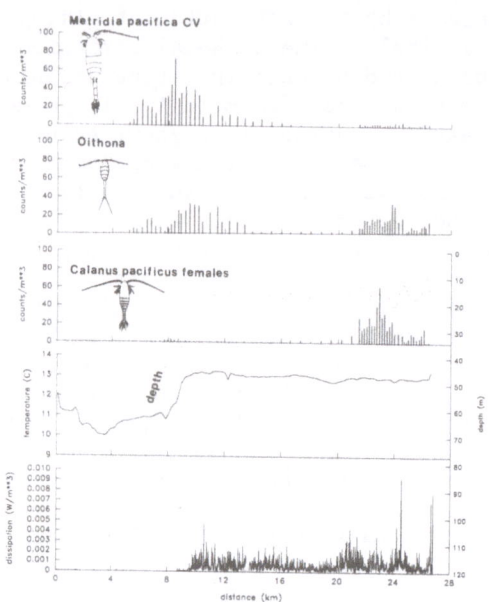

Figure 8. Time series of turbulent dissipation rates, temperature and copepods (Haury and Yamazaki, 1987). The depth of submarine is appeared as a dotted line. Many species show abundance either at 8 km or 22 km on distance axis. The dissipation rates at 22 km is slightly higher than preceding sections.

Table 1. The source of turbulent energy and participating terms in equation (5). The scales are typical sizes. The buoyancy period, $2\pi/N$ is about 10 minutes in the seasonal thermocline. Unknown scales are shown with "?".

Code	Energy source	Term in ( )	Time scales	Vertical scales(m)	Horizontal scales(m)
a	Shear instability	S	minutes to day	1-30	10^3
b	Internal wave	S also B	N	1-30	10^2
c	Free convection	B	hours	10-150	10^3
d	Forced convection	S	hours to day	10-150	10^3
e	Double diffusion	B	?	1	10^3
f	Solid boundary effect	S	n/a	n/a	n/a
g	Advection of q	Q	?	?	?

5. REFERENCES

Baker, M.A. and C.H. Gibson: 1987, 'Sampling turbulence in the stratified ocean: Statistical consequences of strong intermittency', *J. Phys. Oceanogr.*, in press.

Caldwell, D.R. and T.M. Criss: 1979, 'Viscous sublayer at the sea floor', Science 205, 1131-1132.

Chandrasekar, S.: 1961, *Hydrodynamic and Hydromagnetic stability*, Oxford at the Clarendon Press, 654pp.

Chriss, T.M. and D.R. Caldwell: 1984, 'Turbulence spectra from the viscous sublayer and buffer layer at the ocean floor', *J. Fluid Mech.* **142**, 39-55.

Cronin, L.E. and A.J. Mansueti: 1971, 'The biology of the estuary, A symposium on the biological significance of estuaries', Sport Fishing Institute, Washington D.C., 14-39.

Denman, K.L. and T. Platt: 1976, 'The variance spectrum of phytoplankton in a turbulent ocean', *J. Mar. Res.* **34**, 593-601.

Dillon, T.M., J.G. Richman, C.G. Hansen and M.D. Pearson: 1981, 'Near-surface turbulence measurements in a lake', *Nature* 290, 390-392.

Dugan, J.P., B.S. Okawa and G.O. Marmorino: 1985, 'Patchiness of small-scale temperature fluctuations in the seasonal thermocline: Results using a preliminary patch processor algorithm', Naval Research Laboratory Memorandum Report 5649, Washington D.C., 48pp.

Eckart, C.: 1948, 'An analysis of stirring and mixing of incompressible fluids', *J. Mar. Res.* 7, 265-275.

Gargett, A.E.: 1976, 'An investigation of the occurrence of the oceanic turbulence with respect to finestructure', *J. Phys. Oceanogr.* **6**, 139-156.

Gargett, A.E. and T.R. Osborn: 1981, 'Small-scale shear measurements during the Fine and Microstructure Experiment (Fame)', *J. Geophys. Res.* **86**, 1929-1944.

Gargett, A.E. and R.W. Schmitt: 1982, 'Observations of salt fingers in the central waters of the eastern north Pacific', *J. Geophys. Res.* **87**, 8017-8029.

Gargett, A.E.: 1984, 'Vertical eddy diffusivity in the ocean interior', *J. Mar. Res.* **42**, 359-393.

Gregg, M.C. and M.G. Briscoe: 1979, 'Internal waves, finestructure, microstructure, and mixing in the ocean', *Rev. Geophy. Space Phys.* 17, 1524-1548.

Gregg, M.C. and T.B. Sanford: 1980, 'Signatures of mixing from the Bermuda slope, the Sargasso sea, and the Gulf stream', *J. Phys. Oceanogr.* **10**, 105-127.

Gregg, M.C.: 1980, 'Microstructure patches in the thermocline', *J. Phys. Oceanogr.* **10**, 915-943.

Gregg, M.C., E.A. D'Asaro, T.J. Shay and N. Larson: 1986, 'Observations of persistent mixing and near-inertial internal waves', *J. Phys. Oceanogr.* **16**, 856-885.

Gregg, M.C. and T.B. Sanford: 1987, 'Shear and Turbulence in Thermohaline Staircases', *Deep-Sea Res.*, in press.

Gregg, M.C.: 1987, 'Diapycnal mixing in the thermocline: a review', *J. Geophys. Res.* **92**, 5249-5286.

Harder, W.: 1968, 'Reactions of plankton organisms to water stratification', *Limnol. Oceanogr.* **13**, 156-168.

Haury, L.R., J.A. McGowan and P.H. Wiebe: 1978, 'Patterns and processes in the time-space scales of plankton distributions', in J.H. Steele (ed.), *Spatial patten in plankton communities*, Proc. NATO Conf. on Mar. Biol., Erice, Italy, Plenum press, 277-327.

Haury, L.R. and H. Yamazaki: 1987, 'Zooplankton and Oceanic Turbulence: Measurements from a Submarine', in progress.

Kitaigorodskii, S.A., M.A. Donnelan, J.L. Lumley and E.A. Terray: 1983, 'Wave-turbulence interactions in the upper ocean. Part II statistical characteristics of wave and turbulent components of the random velocity field in the marine surface layer', *J. Phys. Oceanogr.* **13**, 1988-1999.

Kitaigorodskii, S.A.: 1987, 'Notes on similarity theory for atmospheric boundary layers in presence of background stable stratification', submitted to Tellus.

Kunze, E. and T.B. Sanford: 1984, 'Observations of near-inertial waves in a front', *J. Phys. Oceanogr.* **14**, 566-581.

Kunze, E. and R.G. Lueck: 1986, 'Velocity profiles in a warm-core ring', *J. Phys. Oceanogr.* **86**, 991-995.

Lamb, H.: 1932, *Hydrodynamics*, 6th ed., Cambridge at the University Press, 738pp.

Lande R. and A.M. Wood: 1987, 'Suspension time of particles in the upper ocean', *Deep-Sea Res.* **34**, 61-72.

Legendre, L.: 1981, Hydrodynamic control of marine phytoplankton production: the paradox of stability, in J.C.J. Nihoul (ed.), *Ecohydrodynamics*, Elsevier, 191-207.

Loebel, P.S. and A.R. Robinson: 1986, 'Transport and entrapment of fish larvae by ocean mesoscale eddies and currents in Hawaiian waters', *Deep-Sea Res.* **33**, 483-500.

Loder, J.W. and D.G. Wright: 1985, 'Tidal rectification and frontal circulation on the sides of Georges Bank', *J. Mar. Res.* **43**, 581-604.

Lueck, R.G., W.R. Crawford and T.R. Osborn: 1983, 'Turbulent dissipation over the continental slope off Vancouver Island', *J. Phys. Oceanogr.* **13**, 1809-1818.

Lueck, R.G.: 1987, 'Microstructure measurements in thermohaline staircase', *Deep-Sea Res.*, in press.

Lueck, R.G.: 1987, 'Turbulence mixing at the Pacific subtropical front, submitted to *J. Phys. Oceanogr.*

Marushige, R.: 1977, *Plankton ecology, Science of Ocean Environment*, in S. Horibe (ed.), 182, University of Tokyo Press, pp. 151-182 (in Japanese).

McGowan, J.A. and T.L. Hayward: 1978, 'Mixing and ocean productivity', *Deep-Sea Res.* **25**, 771-793.

Monin, A.S. and A.M. Yaglom: 1975, 'Statistical fluid mechanics: Mechanics of turbulence vol.2', ed. J.L. Lumley, The MIT Press, 874pp.

Moum, J.N., and D.R. Caldwell: 1985, 'Local influences on shear-flow turbulence in the equatorial ocean', *Science* **230**, 315-316.

Moum, J.N. and T.R. Osborn: 1986, 'Mixing in the main thermocline', *J. Phys. Oceanogr.* **16**, 1250-1259.

Oakey, N.S. and J.A. Elliott: 1982, 'Dissipation within the surface mixed layer', *J. Phys. Oceanogr.* **12**, 171-185.

Oakey, N.S.: 1985, 'Statistics of mixing parameters in the upper ocean during JASIN phase 2', *J. Phys. Oceanogr.* **15**, 1662-1675.

Okubo, A.: 1986, 'Fantastic voyage into the deep: marine biofluid mechanics', in E. Teramoto and M. Yamaguchi (eds.), *Lecture notes in Biology*, Proc. International symposium on mathematical biology, Springer-Verlog.

Osborn, T.R.: 1978, 'Measurements of energy dissipation adjacent to an island', *J. Geophys. Res.* **83**, 2939-2957.

Osborn, T.R. and R.G. Lueck; 1985a, 'Turbulence measurements with a submarine', *J. Phys. Oceanogr.* **15**, 1502-1520.

Osborn, T.R. and R.G. Lueck: 1985b, 'Turbulence measurements from a towed body', *J. Atmos. Oceanic Technol.* **2**, 517-527.

Osborn, T.R.: 1987, 'Signatures of double-diffusive convection and turbulence in an intrusive regime', submitted to *J. Phys. Oceanogr.*

Phillips, O.M.: 1977, *The dynamics of the upper ocean 2nd edition*, Cambridge University Press, 336pp.

Pinkel, R.: 1984, 'The wavenumber frequency spectrum of the internal wavefield', in P. Muller and R. Pujalet (eds.), *Internal gravity waves and small-scale turbulence*, Proc. 'Aha Huliko'a Hawaiian winter workshop, 113-128.

Price, J.F., R.A. Weller and R. Pinkel: 1986, 'Diurnal cycling: Observations and models of the upper ocean response to diurnal heating, cooling and wind mixing', *J. Geophys. Res.* **91**, 8411-8427.

Raymont, J.E.G.: 1983, *Plankton and productivity in the oceans 2nd ed., Vol. 2, Zooplankton*, Pergamon Press, 824pp.

Rothschild, B.J. and T.R. Osborn: In Press, 'Small-scale turbulence and plankton contact rates', *J. Plankton Res.*

Schmitt, R.W.: 1981, 'Form of the temperature-salinity relationship in the Central Water: evidence for double-diffusive mixing', *J. Phys. Oceanogr.* **11**, 1015-1026.

Schoeberlein, H.C.: 1985, 'A statistical analysis of patches of ocean small-scale activity', Johns Hopkins APL Tech. Dig. 6.

Shay, T.J. and M.C. Gregg: 1984, 'Turbulence in an oceanic convective mixed layer', *Nature* **310**, 282-285.

Shay, T.J. and M.C. Gregg: 1986, 'Convectively-driven turbulent mixing in the upper ocean', *J. Phys. Oceanogr.* **16**, 1777-1798.

Swift, M.R. and W.S. Brown: 1983, 'Distribution of bottom stress and tidal energy dissipation in a well-mixed estuary', *Estuarine Coastal and Shelf Science*, **17**, 297-317.

Tett, P. and A. Edwards: 1984, 'Mixing and plankton: an interdisciplinary theme in oceanography', *Oceanogr. Mar. Biol. Ann. Rev.* **22**, 99-123.

Thorpe, S.A.: 1973, 'Experiments on instability and turbulence in a stratified shear flow', *J. Fluid Mech.* **61**, 731-751.

Thorpe, S.A.: 1975, 'The Excitation, Dissipation and Interaction of Internal Waves in the Deep Ocean', J. Geophys. Res. 80, 328-338.

Thorpe, S.A., A.R. Stubbs, A.J. Hall, and R.J. Turner: 1982, 'Wave-produced bubbles observed by side-scan sonar', *Nature* **296**, 636-638.

Thorpe, S.A.: 1984, 'The effect of Langmuir circulation on the distribution of submerged bubbles caused by breaking wind waves', *J. Fluid Mech.* **142**, 151-170.

Thorpe, S.A.: 1985, 'Small-scale processes in the upper ocean boundary layer', *Nature* **318**, 519-523.

Thorpe, S.A.: 1987, 'Transitional phenomena and the development of turbulence in stratified fluids: a review', *J. Geophys. Res.* **92**, 5231-5248.

Turner, J.S.: 1981, 'Small-scale mixing processes', in B.A. Warren and C. Wunsch (ed.), *Evolution of Physical Oceanography*, The MIT Press, pp. 236-263.

Vlymen, W.J. III: 1974, 'Swimming energetics of the larval anchovy, *Engraulis Mordax*', *Fish. Bull.* 77, 885-899.

Weller, R.A., J.P. Dean, J. Marra, J.F. Price, E.A. Francis, and D.C. Boardman: 1985, 'Three-Dimensional Flow in the Upper Ocean', *Science* **227**, 1552-1556.

Woods, J.D.: 1968, 'Wave-induced shear instability in the summer thermocline', *J. Fluid Mech.* 32, 791-800.

Woods, J.D. and R. Onken: 1982, 'Diurnal variation and primary production in the ocean - preliminary results of a Lagrangian ensemble model', *J. Plank. Res.* 4., 735-756.

Yamazaki, H. and R.G. Lueck: 1987, 'Turbulence in the California Undercurrent', *J. Phys. Oceanogr.*, in press.

THE NORTHERLY WIND

David H. Cushing
198 Yarmouth Road
Lowestoft
Suffolk NR32 4AB
ENGLAND

ABSTRACT. Under the stress of northerly winds off East Greenland in winter during the sixties the Great Salinity Anomaly of the Seventies was formed and it drifted across the North Atlantic for nearly twenty years. It crossed the spawning grounds of a number of "deep water" stocks and the recruitments to eleven out of fifteen were significantly reduced during the years of passage of the Great Slug. North of Iceland and on the Grand Bank both primary and secondary production were reduced, which suggests that the reduced recruitment was linked to lack of food for the larvae.

1. INTRODUCTION

Northerly winds over periods of years have had profound biological effects over the North Atlantic. There were two distinct phenomena, first, between the fifties and the seventies a pressure anomaly ridge became established between Iceland and Morocco and it generated increased northerly winds over the North Sea; the consequence was a decline in the production of zooplankton between the fifties and the seventies, which reversed in the early eighties (Dickson *et al.*, 1986). The second event was a sharp increase in northerly winds in the region east of Greenland between November and March during the sixties. The result was that a large mass of low salinity water drifted round the North Atlantic between 1962 and 1982. It was the Great Salinity Anomaly of the Seventies (the "Great Slug"), the drift of which is shown in Figure 1. This shows the average drift of water in the top 1000m of the North Atlantic (Dietrich *et al.*, 1975) on which is superimposed the dates at which the Great Slug was recorded at different positions in the most pronounced way.

The ridge pressure anomaly appeared in the early fifties and comparing the periods 1956-1965 and 1966-70, it became more intense during the later period. Northerlies increased over the Norwegian-Greenland Sea and temperature declined over the North Atlantic. The proportion of polar water increased in the East Greenland and East Icelandic currents and they became cooler and fresher; indeed they

235

B. J. Rothschild (ed.), Toward a Theory on Biological-Physical Interactions in the World Ocean, 235–244.
© 1988 by Kluwer Academic Publishers.

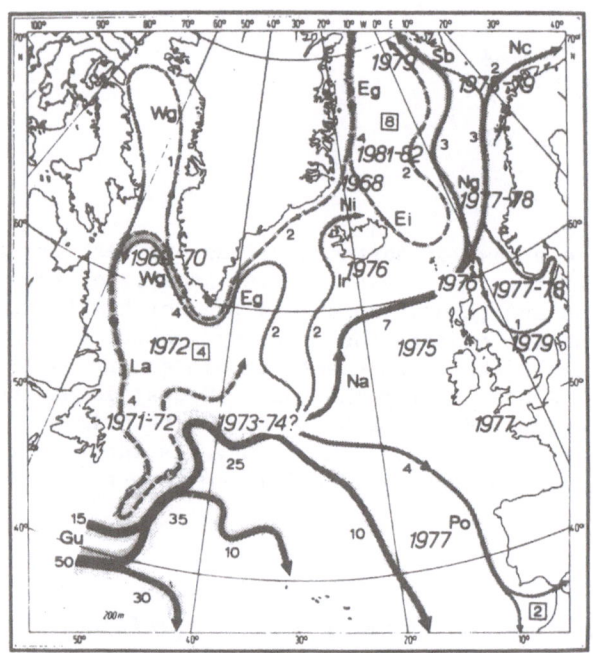

Figure 1. The draft of the Great Salinity Anomaly around the North
East Atlantic, based on the chart of currents of Dietrich *et al.*
(1975); the dates give the positions by years from Dickson *et al.*
(1984). On the chart are superimposed the positions of the Ocean
Weather Ships, B, C and M.

became polar currents carrying ice in the period between 1964 and
1971. Dickson *et al* (1975) showed that in June the water was very
cold and dilute between Iceland and Jan Mayen. They described the
temperature and salinity anomalies in June in the area 67°N to 69°N,
11°W to 15°W.

The same anomaly was found in the Arctic between 1976 and 1981
(Dickson and Blindheim, 1974) and they called it the Great Salinity
anomaly of the Seventies. Between 1964 and 1971 the salinity of the
East Icelandic and East Greenland Currents fell below 34.70‰ and the
surface layers were not dense enough to mix with the more saline
below and ice could form and remain (Malmberg, 1969). Between 1964
and 1971 this slug of cool fresh water north of Iceland extended
down to 200m. From Figure 1 the passage of the Great Slug was as
follows as peak dates: North Iceland in 1968, West Greenland in
1969 to 1970, the Grand Bank in 1971 to 1972, in the North Atlantic
Drift in 1973 to 1974, in the Irminger Current in 1976, in the Faroe-
Shetland Channel in 1976, in the Norwegian Sea in 1977 to 1978, off
the North Cape in 1978 to 1979, and off West Spitzbergen in the same

years. It took seven years to pass north of Iceland and six years
to cross the waters off the North Cape. Off the North coast of
Iceland for five years the water was more than 2°C cooler and for
five years off the North Cape it was 0.7°C cooler. In salinity the
anomaly was -0.5 °/oo off Iceland and 0.11 °/oo in the North Cape
Current. Dickson *et al.* (1984) showed that an earlier salinity
anomaly had occurred between 1907 and 1924 in the Faroe Shetland
Channel and north of Iceland.

2. EFFECTS UPON PRODUCTION AND ON YEAR CLASS STRENGTH

The biological events during the passage of the Great Slug were
well described north of Iceland. Primary production was reduced to
less than a third (Astthorsson *et al.*, 1983). In May 1962 the drift
ice lay close to the north west coast of Iceland and blocked the
passage of the Irminger Current to the north of Iceland and the
herring larvae from their spawning grounds (Jakobbsson, 1978).
Between 1966 and 1968 herring shoals lay between 100 to 400 miles
north and east of Iceland (Malmberg *et al.*, 1967; Malmberg and
Vilhgamsson, 1969; Vilhjamsson and Stefansson, 1968; Malmberg *et
al.*, 1968); before the passage of the "Great Slug" the shoals had
been found relatively close to the north coast (Jakobsson, 1978).
On the Grand Banks in 1972, during the passage of the "Great Slug"
the plankton recorder network revealed that the quantity of
phytoplankton, total copepods and euphausiids was reduced to less
than one third of the average from samples from the previous
thirteen years (Robinson *et al.*, 1975). Cod and haddock
distributions (from echo survey) in the Barents Sea were shifted to
the westward (Middtun *et al.*, 1981) and echo patches of capelin
moved to the south and southwest (Loeng *et al.*, 1983). Blacker
(1981) and Southward and Mattacola (1980) recorded blue whiting and
Norway pout in the English Channel in 1976 to 1979. The reduction
of primary production and in zooplankton north of Iceland and of
zooplankton on the Grand Bank are the most important of these
observations because they show how profoundly the ecosystem was
affected.

Fifteen "deep water" stocks were examined and the recruitments
spawned during the passage of the "Great Slug" were compared with
the non-Slug year classes. The years of "Slug" passage were
identified in Dickson *et al.*, (1984) from the hydrographic evidence.
Table 1 gives the results by stocks, years of anomaly and
difference in year class strength as tested by a Wilcoxon rank test.
The "Slug" years were significantly different in eleven out of the
fifteen stocks; the exceptions were the Faroe Plateau cod and
haddock, the cod in areas 2J, 3K and 3L on the Grand Bank and the
North east Arctic Saithe. The time series of recruitment for all
fifteen stocks are shown in Figure 2 and in all cases the anomalous
year classes are low. The stocks in the tidal waters round the
British Isles and on the continental shelf south of St. Pierre et
Miquelon were excluded because the effect of the Great Slug would

have been diluted.

Of the four stocks which did not respond to the "Great Slug", in three (2J 3K 3L cod, Faroe Plateau haddock and North east Arctic Saithe) there are periods of three or four low year classes which do not correspond exactly with those indicated by the hydrographic evidence. There is no such group in the Faroe Plateau cod stock. For the 2J 3K 3L stock the year classes were low in 1969-71 and not in 1971-73 as expected from the hydrographic evidence; for Faroe Plateau haddock they were low in 1977-79 and not in 1975-77; for the North East Arctic Saithe they were low in 1979-82 and not in 1978-81 (see Figure 2j; the other three stocks are not illustrated). It is possible that the larval drifts of the three stocks are not sampled by the hydrographic stations so far used; research is continuing on this point.

For many years the Russians have conducted 0 group surveys in the Barents Sea (including material on the North East Arctic cod and haddock. The results of these surveys show in both stocks that the year classes of 1982-84 recovered sharply after the low "Slug" year classes of 1978-81 (Figure 2, 1, n). In the material so far published the recovery cannot be seen in the four year old recruits (Figure 2,k,m).

The Icelandic spring spawning herring failed to recover after 1962. Normally their larvae were drifted from their southern spawning ground to the rich grounds north of the island in the Irminger current. The year classes from 1962 onward were progressively reduced never to recover. The evidence is based on catches and it would be interesting to know whether a stock survives at low level.

3. POSSIBLE PHYSICAL EFFECTS

The "Great Slug" was cool and fresh. Of itself there is no reason to believe that reduced salinity should affect recruitment. Reductions in salinity and temperature tend to compensate each other within the ranges observed (see Figure 183, a T-S diagram in Sverdrup et al., 1942); the effect of low salinity on ice formation referred to above was limited to the East Greenland Current. Shepherd et al. (1984) show that the links between recruitment and temperature are extensive and pervasive; indeed of all the many stocks studied all were affected.

The effect of reduced temperature should be examined in the region of larval drift and good estimates of the seasonal distribution of temperature are needed before the period of the "Great Slug". It is unlikely that such seasonal distributions are well known for the area of larval drift for all stocks, some of which live in inhospital seas.

At the Ocean Weather Ship Stations seasonal distributions of sea surface temperature are available (Rodewald, 1972). Five year running means were fitted to the material for the decade, 1951-61, from O.W.S. stations, B, C and M, the positions of which are shown

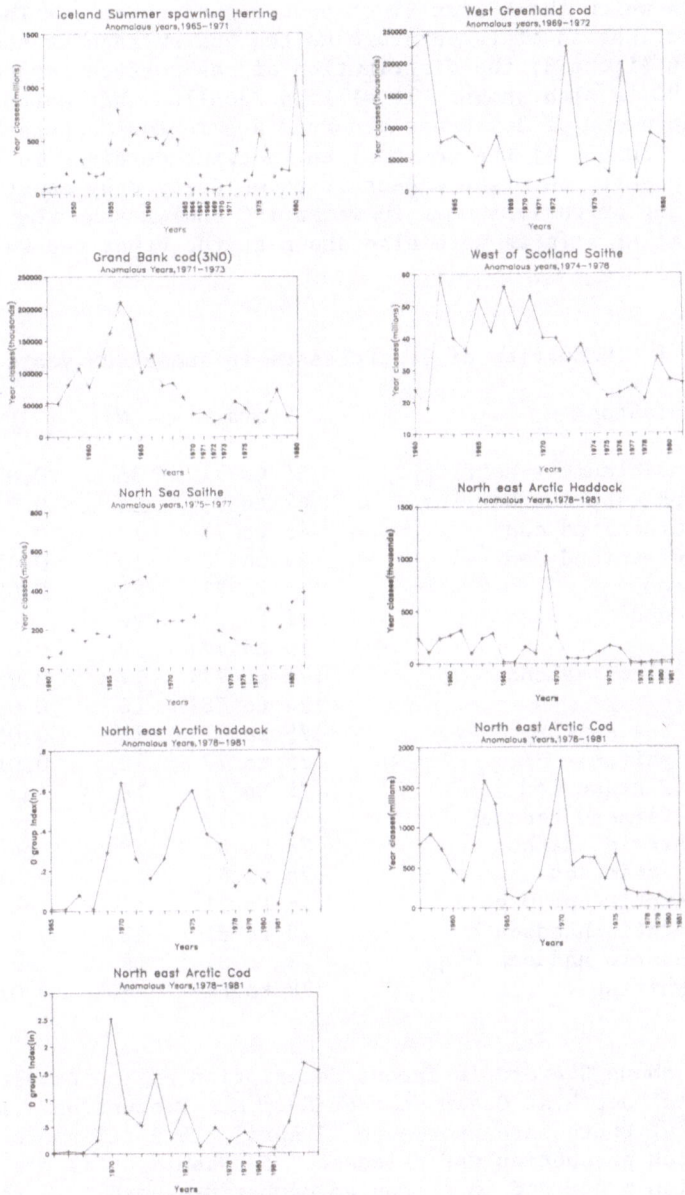

Figure 2. The distribution of year classes in time series for the "deep water" stocks with the years of the Great Salinity Anomaly shown: a) Iceland summer spawning herring (p0.01); b) West Greenland cod (p0.01); c) 3NO cod (p0.05); d) West of Scotland saithe (p0.05; p0.01); e) North Sea saithe (p0.05); f) North east Arctic haddock (p0.01); g) North east Arctic haddock 0 group (n.s.); h) North east Arctic cod (p0.01); i) North east Arctic cod 0 group (n.s.).

on Figure 1; it is assumed that they represent an average of the region across which the "Great Slug" passed subsequently. The seasonal distribution of temperature at the sea surface at stations B is shown in Figure 3; the distribution of sea surface temperature reduced by $1^{\circ}C$ is also shown. It will be recalled that north of Iceland a decrement of $2.5^{\circ}C$ was observed and in the Barents Sea one of $0.7^{\circ}C$ was found. If the seasonal thermocline develops in the same way each year, we might expect it to be delayed by about a month in spring or early summer in water $1^{\circ}C$ cooler than the average. Similar effects were also shown at the other two Weather Ships.

Table 1. Reduction of year classes in anomalous years.

Stock	Years	N	P
Icelandic summer herring	65 to 71	36	0.01
Icelandic spring herring	62 to 71	45	0.01
East Greenland cod	65 to 71	13	0.05
West Greenland cod	69 to 72	15	0.01
3 NO cod	71 to 73	22	0.05
2J3k1 cod	71 to 73	20	ns
	[75 to 77]	18	0.05
W. Scotland saithe	[74 to 77]	17	0.05
	[74 to 78]	16	0.01
North Sea saithe	75 to 77	18	0.05
Faroe saithe	75 to 77	18	0.01
Faroe Plateau cod	75 to 77	18	ns
Faroe Plateau haddock	75 to 77	18	ns
N.E. Arctic saithe	78 to 81	19	ns
N.E. Arctic cod	78 to 81	21	0.01
N.E. Arctic cod 0 gp	78 to 81	15	ns
N.E. Arctic haddock	78 to 81	22	0.01
N.E. Arctic haddock 0 gp	78 to 81	16	ns
Blue whiting	78 to 81	6	0.01

Figure 4 shows Sverdrup's famous description of the development of the critical depth at Ocean Weather Ship M. Production started on 4 April, but there were storms on 15 April, 28 April and on 4-5 May after which production was released. Between 4 April and 4-5 May, production proceeded in a stop-go manner presumably in the period of transient thermoclines. It is this period which may be of interest. Colebrook (1982) showed that the spring outburst was fully developed before stratification was fully established.

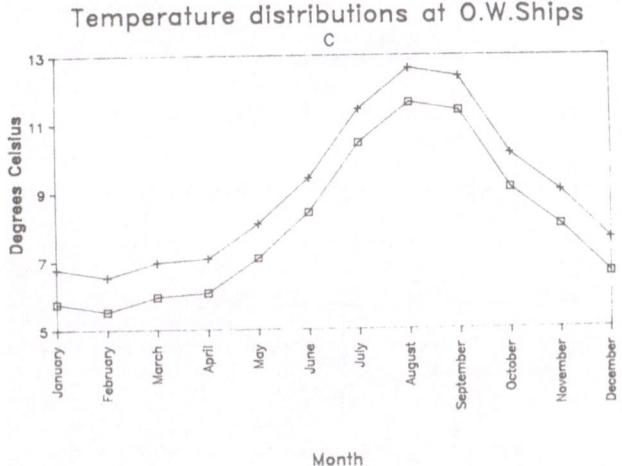

Figure 3. The seasonal distribution of temperature at Ocean Weather
Ship, C, for the decade 1951 to 1961; a decrement of 1°C throughout
the season is shown to demonstrate the delay to the productive
season.

4. CONCLUSION

The production of algae and of zooplankton was reduced
considerably north of Iceland and on the Grand Banks during the
passage of the "Great Slug" then the lack of food may well have
extended the period of larval development and that of adventitious
predation. The development of eggs and larvae is an inverse
function of temperature. Riley and Thompson (1980) show that a
reduction of 1°C extends the development of cod eggs by 2 days in 23
at spring temperatures in the northern North Atlantic; we might
expect the development of larvae to be extended by ten days in a
hundred, assuming that they were well fed. If the larvae were not
well fed, as is likely, during the passage of the "Great Slug",
larval development will be delayed, but by how much is not known.
    Primary production was reduced either directly by temperature or
indirectly by a lengthening of the period of transient thermoclines.
 I suspect that the effect of temperature is less important than
that of irradiance. Is it possible that the period of transient
thermoclines was extended?
    Recruitment to the "deep water" stocks was reduced during the
passage of the "Great Slug". Production of larval food was probably
reduced, larval development extended and larval mortality increased.
 Just after metamorphosis the herring seek nursery grounds inshore
and cod and haddock find their way to the sea bed some time after
metamorphosis. In Iceland the herring nursery grounds are inshore

242

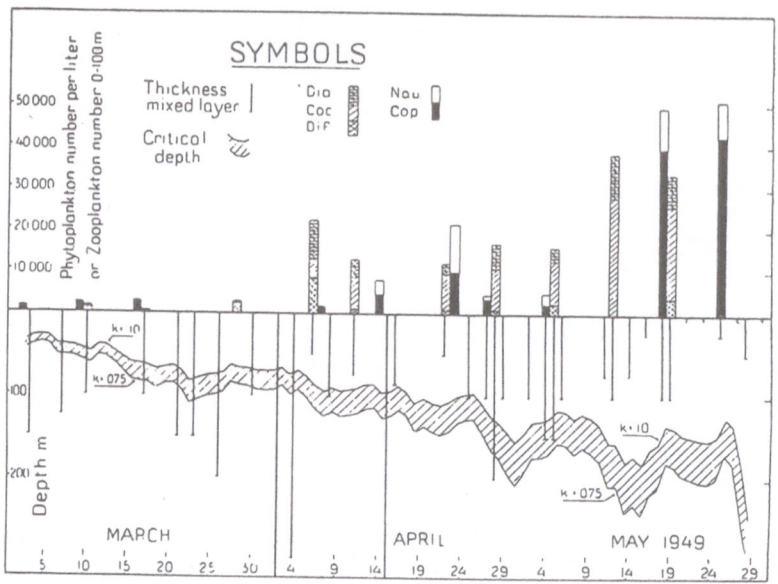

Figure 4. Sverdrup's (1953) figure which describes the relationship
between critical depth (the diagonally hatched area for extinction
coefficients, k = 0.10 and k = 0.075), the depth of mixing (shown as
vertical bars) and the production of diatoms coccolithophorids,
dinoflagellates and nauplii. Note that the storms of 2-4 April and
15 April reduced production for a week or ten days.

in the north coast fjords (Jakobsson, 1978) and those of the cod in
the Barents Sea lie on the Svalbard Shelf (Maslov, 1942); both
regions are out of the path of the "Great Slug", although both may
be subject to sluggish mixture. Hence there is some suggestion that
recruitment was at least largely modified before the animals reach
the nursery. Further, this suggests that recruitment is affected by
the factors that modulate primary production.

5. REFERENCES

Anon: 1985, 'Report of the saithe (coalfish) working group', ICES CM
    1985 Assess 18, 108pp.
Anon: 1985, 'Report of the blue whiting assessment group', ICES CM
    1985 3, 40pp.
Anon: 1985, 'Report of the working group on the cod stocks off East
    Greenland', ICES CM 1985 Assessment 6, 26pp.
Anon: 1985, 'Report of the arctic fisheries working group', ICES CM
    1985 Assessment 12, 52pp.

Anon: 1985, 'Herring assessment working group for the area south of 62 N', ICES CM 1985 Assessment 12, 148pp.

Astthorsson, O.S., I. Hallgrimsson and G.S. Jonsson: 1983, 'Variations in zooplankton in Icelandic waters in spring during the years 1961-82', *Rit. Fiskideildar* 7, 73-113.

Baird, J.W. and C.A. Bishop: 1985, 'Assessment of the cod stock in NAFO Division 2j and 3k1', *North Atl. Fish. Org. S.C.R.* **85/39**, 38pp.

Bishop, C.A. and J.W. Baird: 1985, 'Assessment of the cod stock in NAFO Division 3 NO', *North Atl. Fish. Org. S.C.R.* **85/39**, 38pp.

Blacker, R.W.: 1981, 'Recent occurrences of blue whiting, *Micromesistius poutassou* and Norway pout, *Trisopterus esmarkii* in the English Channel and Southern North Sea', *J. Mar. Biol. Assn. N.S.* **61(2)**, 307-314.

Colebrook, J.M.: 1982, 'Continuous plankton records: Seasonal variations in the distribution and abundance of plankton in the North east Atlantic and the North Sea', *J. Plankt. Res.* 4, 435-462.

Dickson, R.R., H.H. Lamb, S-A. Malmberg and J.M. Colebrook: 1975, 'Climatic reversal in the northern North Atlantic', *Nature* 256, 479-482.

Dickson, R.R. and J. Blindheim: 1984, 'On the abnormal hydrographic conditions in the European Arctic during the 1970s', *Rapp. Proces.-Verb. Cons. int. Explor. Mer* **185**, 201-213.

Dickson, R.R., S.A. Malmberg, S.R. Jones and A.J. Lee: 1984, 'An investigation of the earlier Great Salinity Anomaly of 1910-1914 in waters west of the British Isles', ICES CM 1984 GEN 4, 15pp.

Dickson, R.R., P.M. Kelly, J.M. Colebrook, W.S. Wooster, and D.H. Cushing: 1986, 'North winds and production in the eastern North Atlantic', ICES CM C37, 10pp.

Dietrich, G., K. Kalle, W. Krauss and G. Siedler: 1975, 'General Oceanography', 2nd edn. John Wiley, New York, 626pp.

Horsted, S.A.: 1978, 'Subarea 1 Cod: data for 1976-77 and first months of 1978 and estimate of biomass and yield, 1978-80', *Int. Commn. North West Atl. Fish. Res. Doc.* 78/VI/44, 38pp.

Horsted, S.A., J.M. Jensen, B.W. Jones, J. Messtorf and A. Schumacher: 1980, 'An assessment of the Greenland cod stock', ICES CM 1980 G49, 7pp.

Jakobsson, J.: 1978, 'The North Icelandic herring fishery and environmental conditions', *Symp. Biol. Basis. Pel. Fish Stock Management*, ICES Mimeo No. 30, 101pp.

Jakobsson, J.: 1980, 'Exploitation of the Icelandic spring and summer spawning herring in relation to fisheries management, 1947-77', *Rapp. Proces-Verb. Cons. int. Explor. Mer* **177**, 23-42.

Loeng, H., Odd Nakken and Askjell Raknes: 1983, 'Loddas utbredelse i Barentshavet i forhold til temperaturfeltet i perioden 1974-82', *Fisken Hav.* 1983(1), 1-17.

Malmberg, S.A.: 1969, 'Hydrographic changes in the waters between Iceland and Jan Mayen in the last decade', *Jokull* **19(1969)**, 30-43.

Malmberg, S.A., I. Hallgrimsson and J. Jakobsson: 1967, 'Report
of the joint meeting on the distribution of herring in relation
to hydrography and plankton', Seydisfjordur, 20-23 June 1965,
Ann Biol. Cons. Int. Explor. Mer 22, 188-195.

Malmberg, S.A., T. Thoradottir and H. Vilhjalmsson: 1968, 'Report
on the joint meeting on Atlanto-Scandian herring distribution
held at Akureyri, 18-19 June, 1966, *Ann Biol. Cons. Int. Explor.
Mer* 23, 215- 220.

Malmberg, S.A. and H. Vilhjalmsson: 1969, 'Report of the joint
meeting of Icelandic, Norwegian and Soviet investigators on the
Atlanto-Scandian herring distribution in relation to
oceanographic conditions held at Seydisfjordur, 5-6 June 1968',
Ann. Biol. Cons. int. Explor. Mer 25, 260-265.

Maslov, N.A.: 1942, 'Bottom living fishes of the Barents Sea',
Trans. Knipovich Inst. Murmansk 8, 3-186.

Midttun, L., O. Nakken, and A. Raknes: 1981, 'Variasjoner i
utbredelsen av Torsk i Barentshavet i perioden 197-81', *Fisken
og Havet* 1981, 4, 16pp.

Rodewald, M.: 1972, 'Long term variations of the sea temperature
in the area of nine North Atlantic Ocean Weather Stations during
the period, 1951-1968', *Rapp. Proces-Verb. Cons. Int. Explor.
Mer* 162, 139-153.

Southward, A.J., and A.D. Mattacola: 1980, 'Occurrence of Norway
pout *Trisopterus esmarkii* (Nilsson) and blue whiting
Micromesistius poutassou (Risso) in the western English Channel
off Plymouth', *J. Mar. Biol. Assn. N.S.* 60(1), 39-44.

Sverdrup, H.U.: 1953, 'On conditions for the vernal blooming of
phytoplankton', *J. Cons. int. Explor. Mer* 18, 287-295.

Sverdrup, H.U., M.W. Johnson, and R.H. Fleming: 1942, *The oceans,
their physics, chemistry and general biology*, Prentice-Hall, New
York, 1085pp.

Thompson, B.M. and J.D. Riley: 1981, 'Egg and larval development
studies in the North Sea cod (*Gadus morhua* L.), *Rapp.
Proces-Verb. Cons. Intern. Explor. Mer* 178, 553-559.

Thoradottir, T.: 1977, 'Primary production in North Icelandic
waters in relation to recent climatic changes', in M.J. Dunbar
(ed.), *Polar Oceans*, Arctic Inst. North America, pp. 655-665.

Influence of temporal characteristics of physical phenomena on plankton dynamics, as shown by North-West European marine ecosystems.

Jacques Le Fèvre* and Serge Frontier**

* *Laboratoire d'Océanographie Biologique*
Université de Bretagne Occidentale
29287 BREST CEDEX, France

** *Laboratoire d'Écologie Numérique*
Université des Sciences et Techniques de Lille (SN3)
59655 VILLENEUVE D'ASCQ CEDEX, France

Abstract: This contribution shows that food chain structures, not only general productivity levels, are governed by temporal characteristics of nutrient dynamics at physical interfaces, or, in other words, by the temporal modulation of auxiliary energy transfers at ergoclines. The examples used, based upon both literature reviews and preliminary original results from a recent cruise, are taken from pelagic ecosystems in north-west European seas, where nutrient enrichment processes are periodic and tide-dependent. The period is that of the M_2 tide at the Celtic Sea shelf-break, where time-integration in the ecosystem appears to result in enhanced general productivity in the form of a classical herbivore food chain. The period is fortnightly (neaps-springs alternation) at shelf-sea tidal fronts, where herbivores apparently cannot adapt themselves to short-lived phytoplankton outgrowths every second week; phytoplankton biomass is accordingly allowed to accumulate in hydrodynamic traps and recycled first by microheterotrophs, which are preyed upon, in turn, by larger-sized microphagous zooplankters.

1. Introduction

A widespread view that biological productivity should be enhanced at marine physical interfaces is probably best illustrated in the case of fronts. For example, increased primary production has been repeatedly postulated, especially since the mid-1970s, on north-west European shelf-sea tidal fronts (e.g. Holligan, 1981). Enhanced productivity at a variety of interfaces, including ice-water and sediment-water, has also recently become the idea underlying the ergocline theory, to which an International Liège Colloquium on Ocean Hydrodynamics was recently

245

B. J. Rothschild (ed.), Toward a Theory on Biological-Physical Interactions in the World Ocean, 245–272.
© *1988 by Kluwer Academic Publishers.*

devoted (*cf.* Legendre *et al.*, 1986). There are, however, important differences in the approaches to the question.

The high-productivity theory relevant to north-west European tidal fronts was founded on the observation of high (sometimes very high) phytoplankton standing stocks in summer at tidal fronts, especially near the island of Ushant at the western end of the English Channel. As first described by Pingree *et al.* (1975), these high standing stocks most often consist of dense patches of the dinoflagellate *Gyrodinium aureolum* Hulburt, extending for some distance at the pycnocline level towards the stratified area on one side of the front. The high-productivity theory, as it has become well-known through a number of papers (e.g. Pingree *et al.*, 1975, 1976, 1978; Simpson *et al.*, 1979; Holligan, 1978, 1981; Tett, 1981; Holligan *et al.*, 1983, among many others) largely considered high biomass in itself as sufficient evidence for high production. Two main factors were invoked, namely light and nutrient (nitrogen) availability. Productivity would be high at the surface on the front because only there would phytoplankton be freed from light limitation prevailing in the tidally mixed area, due to mixing extending below Sverdrup's (1953) critical depth, and from nutrient limitation prevailing in the stratified area, due to summer exhaustion of the euphotic layer. Productivity would also be high at the thermocline level in the stratified area because only there would phytoplankton cells both receive a sufficient amount of illumination and have access to the nutrient reserve just below the pycnocline. Taking into account only vertical exchanges of matter and energy, the theory is typically one-dimensional, as explicitly emphasized in some papers by its supporters (Tett, 1981; Holligan *et al.*, 1984a). In a sense, the theory is also somewhat static, since it postulates a kind of permanent régime, with little consideration of time-fluctuations other than the annual cycle, which is much too long to be relevant to the proposed mechanisms. This, of course, involves some degree of oversimplification, since only a detailed review (*cf.* Le Fèvre, 1986) can discuss the full variety of points of views involved in high frontal productivity theories. Departure from strict one-dimensionality is found, for instance, in the work of Savidge (1976), showing that the waters on the two sides of a front may be complementary with respect to more subtle phytoplankton needs than light or mass supply of major nutrients; hence the need for horizontal water exchanges to account for increased productivity, as shown in this case by ^{14}C uptake experiments. It is also found in the work of Pingree (1978) showing how turbulent frontal dynamics can provide mechanisms for such exchanges. Some attention has also been paid in the earlier papers to non-annual rhythms, especially to the alternation between spring and neap tides (Pingree *et al.*, 1975, 1977). The interest in the fortnightly tidal rhythm, however, largely faded away as the permanent-régime picture emerged. Indeed, apart from a remark by Simpson and Bowers (1979) that neaps-springs oscillations of the frontal position, although restricted in their range due to a physical feedback mechanism, could still be of "considerable biological importance", no mention of this cycle has been made in the past decade by the proponents of what Loder and Platt (1985) in their critical paper call the "classical" theory of frontal productivity. The latter can still, therefore, be considered one-dimensional and based upon a relatively static form of permanent-régime assumption.

On the other hand, the theory of ergoclines, as summed up by Legendre *et al.* (1986), is basically a dynamic one, stating that ecosystems are largely shaped by inputs of non-photosynthetic energy ("auxiliary energy") at well-defined marine energetic interfaces, in space or in time. An important point in this respect is made by Legendre (1981), who emphasizes that neither stabilization nor destabilization of the water column favours phytoplankton production, a rôle rather played by the alternation between stabilization and destabilization which can take place at various frequencies. More generally, Legendre and Demers (1984) argue that hydrodynamics is the driving force of aquatic ecosystems, with such factors as light and nutrients acting only as "proximal agents". This, of course, was already apparent in Sverdrup's (1953) critical-depth theory, where the partial stabilization of the water column which results in the onset of the spring bloom can be viewed as a temporal ergocline. Hydrodynamic control is also evident in the annual phytoplankton cycle and its species successions. For temperate waters, the "classical" cycle, as described by Cushing (1959), consists of two diatom outbursts, a major one in spring and a minor one in autumn, with a less productive summer dinoflagellate maximum in between. The impressive wealth of data accumulated by the surveys with the Continuous Plankton Recorder, as summarized for instance in Cushing (1975), showed that in the North Atlantic and adjacent seas important variations are found in the sharpness and relative size of the two peaks. The very existence of a summer trough between the two peaks, and of a corresponding dinoflagellate maximum, is actually dependent on the establishment of a seasonal thermocline. In permanently well-mixed tidal areas, such as the western English Channel, the tendency is rather for a single primary production maximum in (early) summer, for which diatoms are responsible (Grall, 1972a, 1972b; Boalch *et al*, 1978). A partial generalization is found in Holligan *et al.* (1980), showing that around the British Isles different dinoflagellate assemblages are characteristic of different hydrodynamic régimes. The point is emphasized with respect to more general phytoplankton communities in Le Fèvre *et al.* (1983a) and the matter is discussed in more detail in Le Fèvre (1986). Intermediate cases between year-round tidal mixing and summer-long stratification are found in some areas where the hydrodynamic equilibrium is easily offset. Le Fèvre *et al.* (1981a) report an example on the southern coast of Brittany, where wind-induced mixing events result in more or less abortive summer diatom outgrowths. Legendre (1981) also quotes a number of papers reporting cases where non-periodic (weather-induced) or periodic (e.g. neaps-springs alternation) changes in water column stability result in production events, with a more or less pronounced influence on production budgets. This can be used to introduce the idea that the response of ecosystems to the input of auxiliary energy at ergoclines will largely depend on how this input is modulated in time, or, as put in a correction by Prieur and Legendre (this volume) to a statement by Legendre *et al.*

(1986), that "Enhanced biological production occurs at ergoclines as the consequence of the matching or resonance of physical rates with biological rates". Indeed, following suggestions by Le Fèvre *et al.* (1983b) and Le Fèvre (1986) based upon cases of pelagic ecosystems in north-west European waters, Frontier (1986) emphasized that the very structure of food chains, especially at such ergoclines as fronts, is dependent on the time-match or tuning between physical periodicities and time-constants in the response of living organisms. This is the view to which the present contribution is devoted, with, in addition to a review of the literature, some recently obtained original results given as supportive evidence.

2. Time constants and alternative food chains

There is little question that time constants and variations in time-match play an important role in marine life. One of the best-known examples is Cushing's match-mismatch hypothesis (*cf.* Cushing and Dickson, 1976) showing how the effect of global climatic variations can be mediated in the ecosystem by the lesser or greater degree of synchronization between the hatching of fish larvae and the abundance of their prey, resulting in large, sometimes dramatic, changes in the abundance and species composition of stocks available to commercial fisheries. Living organisms, however, can also adapt themselves to physical conditions by developing periodic behaviour patterns. An example is given by Binet (1977) with respect to pelagic copepods off the Ivory Coast. In this area, the eastwards Guinea current flows at the surface, roughly parallel to the coast, while an undercurrent in the opposite direction is found near the bottom on the continental shelf. Several copepod species undertake "ontogenic migrations", through which the copepodite stages migrate to deeper waters as they grow older, until they reach the undercurrent, which may bring them back in, or near, their place of origin. The continued existence of the population in a definite area is accordingly dependent on proper duration and timing of the migrations, which may have evolved through selective pressure. Tyler and Seliger (1978, 1981) also give fine examples of how proper time-match between circulation régimes and the life-history of an organism can select dominant species, in this case red-tide dinoflagellates on the Atlantic coast of the USA.

Red-tide conditions are precisely a good introduction to the concept of alternative food chains, which will be linked later to time-constants. High biomass values, as found in red tides, and as found in frontal patches of *Gyrodinium aureolum* on north-west European tidal fronts (which often reach red-tide conditions) do not necessarily result from enhanced growth at the actual place where they are found. This was pointed out by several authors in the 1950s (e.g. Bary, 1953; Ryther, 1955) who suggested that high cell densities resulted instead from accumulation by appropriate circulation patterns. A typical example, and probably the most commonly found, is that of a dinoflagellate red tide occurring at a front, where mechanical accumulation results from surface convergence and from upwards migration (case of motile species) or buoyancy (e.g. *Noctiluca scintillans*) that would prevent the organisms from following the downwelling water at the front. About two decades after this kind of explanation was first proposed in an elaborate way, Wyatt and Horwood (1973) summarized the approaches to the question by saying that all hypotheses put forward to account for red tides could be ascribed to either a nutrient theory (which will not be discussed here) or a hydrographic theory (involving some form of accumulation process). They also suggested that red-tide patches accumulated at high densities can be viewed as the final phase of a senescent culture, experiencing self-shading and nutrient depletion, and whose ultimate fate is putrefaction. In the particular case of some dinoflagellates, this condition is, at least in part, brought about by the inhibition of grazing by herbivores due to the presence of toxins. Instead of being passed on to a classical food chain, the accumulated phytoplankton biomass will be recycled by bacteria, which, in turn, can be grazed upon by microheterotrophs, especially flagellates and ciliates. It has actually been known for a long time (e.g. Margalef, 1956) that ciliates often proliferate at the ultimate stage when a red tide collapses.

The work of Fenchel (1982a-d) shows that such microheterotrophic food chains, with bacteria as their first link and flagellates and ciliates as subsequent consumers, are rather widespread and not restricted to cases of red tide. Newell and Linley (1984) also suggest that such food chains make up a "decomposer pathway", as an alternative to the more classical "herbivore pathway", and that the former may dominate over the latter in various situations. As suggested by Le Fèvre (1986) and Frontier (1986), intermediate cases can be found, since a number of larger-sized microphagous zooplankters (e.g. appendicularians, salps, doliolids or pteropods such as *Limacina*) may feed upon microheterotrophs and be, in turn, prey to larger animals (e.g. fish larvae preying upon appendicularians or hyperiid amphipods eating salps and being in turn eaten by tuna fish), thus redirecting microheterotrophic production to more classical pathways. Accordingly, Le Fèvre (1986) put forward the view that in such places as fronts an "accumulation biotope" will often develop, whose functional structure may include a range of possibilities. One of them is fast cycling of organic matter through a pure microheterotrophic food chain; another might be the development of a specific herbivore pathway (e.g. through the simultaneous accumulation of neuston animals and suitable phytoplanktonic food); yet another one is that large predators or scavengers be attracted to the accumulated biomass, with the exploitation of microheterotrophic production by microphagous zooplankton possibly playing a rôle in the process. In the latter case, biomass exportation out of the accumulation biotope will be likely to take place, e.g. through migration if the final link consists of birds or large fish. Frontier (1986) views such a situation as a case of "exploitation of an ecosystem by another one", of which examples are also known in terrestrial environments, and the energy involved in biomass transport as "secondary auxiliary energy". The above shows that there is a kind of competition between the decomposer pathway and the herbivore pathway as regards the first links in the food chains. The decomposer pathway is especially likely to dominate if a herbivore community cannot establish itself, which can result from the presence of toxins in the phytoplankton cells, but can also result, as will be shown below with more precise examples, from improper time-match between phytoplankton growth and zooplankton response. This does not seem at first sight to be necessarily related to the accumulation biotope, but it should be borne in mind that biomass being allowed to become senescent through lack of grazing will always tend to accumulate in some hydrodynamic trap, while any biomass accumulation both implies some previous dominance of ageing over grazing and further favours physiological senescence (e.g. through self-shading due to high phytoplankton cell concentrations), which is likely to make the accumulated material more palatable for decomposers than for herbivores.

The accumulation biotope and microheterotrophic food chain should also be given consideration with respect to the subsurface chlorophyll maximum which is a quasi-universal feature in the world ocean (e.g. Steele and Yentsch, 1960). This maximum has most often been ascribed to *in situ* phytoplankton growth, with some form of escape from both nutrient-limitation and light-limitation being postulated, as recalled above in the case of the stratified area close to European tidal fronts. A frequent characteristic of the subsurface chlorophyll maximum, however, is to be coincident with the pycnocline. The physical discontinuity (or steep gradient) is by itself an accumulating system for any sinking particulate material, due to the variations in density and viscosity involved. This was clearly shown through appropriate calculations by Yamamoto (1983, 1984), who did find a subsurface maximum at the pycnocline level in the western Pacific for passive tracers, namely quartz grains of comparable size to phytoplankton cells. Any phytoplankton with a tendency to sink, as is generally the case of ageing cells (*cf*. Smayda, 1970), will accordingly accumulate at the pycnocline, even though the rate of the physical process will not be the same as in the case of quartz grains. A distinct possibility for the subsurface chlorophyll maximum is therefore its consisting of accumulated senescent cells (whose physiological condition could, however, somewhat improve on reaching the nitracline, hence further reduction of the sinking rate), which would not be very attractive for herbivores and would rather be recycled through a food chain starting with microheterotrophs. There is, however, no reason why this should be the general case; the subsurface chlorophyll maximum may also originate from *in situ* enhanced growth (together with some increase in chlorophyll content per cell due to shade adaptation) in situations where escape from both light-limitation and nutrient-limitation does take place at some depth (and where, in addition, the grazing pressure is low enough for a high phytoplankton biomass to be maintained). Falkowski (1983) considers both possibilities and gives the difference in taxonomic composition between the surface layer and the subsurface chlorophyll maximum as a measure of the latter's originating from *in situ* growth; examples of the different possible situations in this respect can actually be found in the literature (for references, see Le Fèvre, 1986). A number of other clues are also available. Accumulation on the pycnocline can be suspected for instance when, as is often the case (e.g. Lorenzen, 1967; Longhurst, 1976), the primary production maximum is found higher in the water column than the chlorophyll maximum. This, however, can involve some uncertainty, since it is very difficult to obtain truly representative measurements of primary production in the chlorophyll maximum, even with *in situ* techniques, where the pycnocline oscillates at high frequencies (for more detail, see Le Fèvre, 1986). Although also beset with uncertainties in the interpretation, the existence of a maximum in dissolved oxygen concentration (percentage saturation) above the chlorophyll maximum, as reported by Le Fèvre *et al.* (1983b) for the stratified area west of the Ushant front in summer, will provide another indication that the phytoplankton maximum at the thermocline may result from accumulation there rather than from *in situ* growth, the maximum in photosynthesis being rather likely to be associated with the oxygen maximum.

The key point, however, is whether the phytoplankton biomass at the maximum in the thermocline is grazed upon by herbivores or recycled by microheterotrophs. As pointed out by Legendre and Demers (1984), the question of the level at which the zooplankton maximum is found in the water column, and of a possible coincidence with the level of the maximum of either phytoplankton biomass or phytoplankton production, is highly debatable, due to the rarity of high-resolution data on zooplankton vertical distribution (e.g. the controversy between Ortner *et al.*, 1980, and Longhurst and Herman, 1981). Some papers, however, do report the existence of a zooplankton maximum coincident with a production maximum and located somewhat higher in the water column than the chlorophyll maximum (e.g. Longhurst, 1976; Longhurst and Williams, 1979). The work of King *et al.* (1987) may help cast further light on the situation. These authors investigated, on two cruises carried out in June and September 1982 in the Gulf of Maine, the vertical distribution of microzooplankton and macrozooplankton biomass, together with the amount of glutamate dehydrogenase in the two fractions, in relationship with the depth of the subsurface chlorophyll maximum. Glutamate dehydrogenase (GDH) is an enzyme involved in ammonium regeneration, and its concentration may be taken as an indication of the intensity of zooplankton metabolism. In September, the GDH maximum is found, with the exception of one station, higher in the water column than the subsurface chlorophyll maximum and coincident with the primary production maximum, as is the maximum of zooplankton (> 153 μm) biomass; in June, the GDH maximum is rather coincident with the subsurface chlorophyll maximum, but it is also closely associated with the biomass peak of the tintinnid ciliates *Parafavella* spp. Despite the authors' statement in their abstract that "GDH maxima were generally observed to correspond to the depth of the chlorophyll maximum", their results do support, therefore, the view that the herbivore food chain tends to be best developed at the production maximum in the upper layer, while the microheterotrophic food chain often tends to take advantage of accumulated phytoplanktonic biomass at the pycnocline level.

For more detailed examples, it is appropriate to consider again the summer situation in west European waters, such as the stratified areas in the Celtic Sea and in the English Channel west of the Ushant front. In such areas, the phytoplankton in the subsurface chlorophyll maximum generally consists of small flagellates where the thermocline is deepest and of *Gyrodinium aureolum*, i.e. the same dinoflagellate as found near the surface, in the vicinity of the front. According to Falkowski's (1983) criterion, therefore, accumulation would be likely to be dominant close to the front and *in situ* growth farther away, which would be contrary to the assumptions of some versions of the high-productivity theory. However, although variations are found in the situation depending on the place and time of the year (for more detail see Le Fèvre, 1986), convergent indications suggest that, whether dominated by microflagellates or *Gyrodinium*, the biomass maximum at the thermocline level would most often be recycled by microheterotrophs. Among them are the observations reported by Le Fèvre *et al.* (1983b) according to which, on a pluriannual basis, the correlation is poor in summer between oxygen saturation and chlorophyll concentration, with an oxygen maximum generally higher in the water column than the chlorophyll maximum. Holligan *et al.* (1984b) also showed that both bacteria and protozoa had their maximal biomass in the same depth-range as the chlorophyll maximum, while larger zooplankton was more abundant in the upper layer. The vertical resolution of the sampling performed by these authors is too coarse for a certainty to be obtained, but Jacq (1986) and Jacq and Prieur (1986) did find at the thermocline level a maximum in the activity of particle-bound bacteria (as opposed to free-living bacteria), whose peculiarity is that they are dependent on dead or ageing material. Finally, Southward and Barrett (1983) report a situation where the only large zooplankters definitely associated with the subsurface chlorophyll maximum were the appendicularians and *Limacina*, i.e. microphagous forms very likely to be indicative of a food chain starting with microheterotrophs.

Similar indications are also available with respect to the surface environment in the frontal region itself. Direct evidence of frontal accumulation of various organisms (*Noctiluca*, cladocerans, fish eggs, neustonic copepods, etc.), together with scum lines or patches of flotsam, has been reported many times in the region (e.g. Le Fèvre and Grall, 1970; Grall *et al.*, 1971; Pingree *et al.*, 1974, 1975; Grall *et al.*, 1980; Le Fèvre *et al.*, 1981b). Some of these observations (e.g. Pingree *et al.*, 1975) involved *Gyrodinium aureolum*, which is moreover reported by Holligan *et al.* (1983) to be grazed upon by *Noctiluca* at the heart of frontal high-density patches, a situation suggestive of a food chain peculiar to the accumulation biotope. Some of the densest *Gyrodinium* patches were reported not to be growing at all (Holligan *et al.*, 1984a; Jordan and Joint, 1984) and to be associated with high bacterial activity rather than high zooplankton biomass (Holligan *et al.*, 1984b; Newell and Linley, 1984). Holligan (1981) pointed out the lack of a clear relationship between phytoplankton and zooplankton abundance in regions where tidal fronts occur, a characteristic which was again reported, on a vertically integrated basis and from pluri-annual data series, by Le Fèvre *et al.* (1983b). Some zooplankters are, however, associated with fronts: Le Fèvre (1986) reported frontal maxima in the abundance of *Limacina* and an unidentified doliolid (probably *Doliolum nationalis*), i.e. again two microphagous forms.

If, as suggested above, biomass accumulation takes place at European tidal fronts and gives way to a microheterotrophic food chain rather than to a more classical herbivore food chain, fertilization processes and their temporal characteristics may provide an explanation. A cold bottom water mass (< 12°C), segregated early in the annual cycle and retaining to a large extent the physical conditions of the winter time when it was formed, is found until late autumn over most of the Celtic Sea and the Armorican area on the shelf south of Brittany. This structure was described in detail by Vincent and Kurc (1969), who quote early observations by Le Danois in the 1920s and give it the name "bourrelet froid" (which may be translated as "cushion-shaped mass of cold water"); it is similar to the "cold pool" decribed from the Mid-Atlantic Bight of the USA by Bigelow in the 1930s (*cf.* for instance Houghton *et al.*, 1982). Although the "classical" high-productivity theory of frontal systems (in the sense of Loder and Platt, 1985) has taken for granted that the main nutrient reserve in frontal regions was the permanently well-mixed area, it has been clearly shown (Morin, 1984; Morin *et al.*, 1985) that this rôle is actually played by the bourrelet froid. The bourrelet not only stores inorganic nutrients, but actually accumulates them: nitrate-nitrogen concentrations in excess of 7 μmole l^{-1} are, for instance, common there in summer, while the well-mixed area seldom reaches 4 μmole l^{-1}. Accordingly, Le Fèvre *et al.* (1983b) postulated that periodic release of nutrients from the bourrelet on the neaps-to-springs increase in tidal amplitude was the dominant fertilization process in the area and that the lack of a herbivore community on the fronts despite the high phytoplankton standing stock might be due to "the zooplankton's being unable to benefit from a phytoplankton bloom lasting for a few days every second week". A demonstration of the fortnightly fertilization process is actually given by Morin *et al.* (1985). With a phase-lag of 2-3 days, consistent with the observations of Simpson and Bowers (1979) on the response of the physical structure to the variation in tidal range, the edge of the bourrelet, which is usually found close to the front position, is eroded, resulting in the release of water from the bourrelet in the well-mixed areas and local nutrient enrichment there. Cross-frontal exchanges, e.g. through frontal eddies as suggested by Pingree (1978), is the most likely way through which the process can result in enhanced growth of phytoplankton at the stratified side of the front. This timing of events would be contrary to some versions of the high-productivity theory, where the growth-stimulating phase of the fortnightly cycle was assumed to be the stabilization of (supposedly nutrient-rich) well-mixed water (see again Loder and Platt, 1985), but reviewing available reports actually shows that higher frontal chlorophyll concentrations occur on spring tides or slightly after them. Accordingly, the picture drawn by Le Fèvre (1986), is that growth of phytoplankton, namely *Gyrodinium aureolum*, would be triggered or enhanced by the nutrient input on spring tides, which, concurrently with mechanical accumulation, would lead to an increasing standing stock in the frontal area. As the tide slackens towards neaps, the nutrient supply from the bourrelet would be severed and growth become nutrient-limited. Cell density in the frontal patches could, however, go on increasing for some time through accumulation by convergent circulation until the final stage of a senescent population is reached, possibly in the form of a red tide. Some of the cells would tend to sink along the isopycnals for some distance, which would account for the origin of the subsurface *Gyrodinium* maximum. Uptake of nutrients at the thermocline level would allow at least the survival of a sufficient population there to serve the seeding function when a new cycle is next initiated. Given their generation time, herbivores would indeed be unable to adjust their population levels in response to short-lived fortnightly phytoplankton blooms, while senescent biomass is the natural food for microheterotrophs.

A counter-example in the same region is provided by Le Fèvre *et al.* (1983b) in the case of the Rade de Brest, a semi-enclosed bay where fertilization is mainly dependent on land runoff, which fluctuates essentially on an annual basis. A pluri-annual survey at a fixed location in this bay showed oxygen saturation to be highly correlated with chlorophyll concentration and zooplankton biomass to be fairly well correlated with chlorophyll. A noteworthy point is that zooplankton biomass was even better correlated with oxygen saturation than it was with phytoplankton (in the form of chlorophyll); this is interpreted as showing that zooplankton is more dependent on phytoplankton production than biomass and the general picture is that of a classical herbivore food chain taking advantage of active primary production under a kind of steady fertilization régime allowing proper time for the responses of zooplankton to phytoplankton changes. This may be debatable, since the sampling was carried out at fortnightly intervals (constant tidal range), which would have masked any fortnightly fluctuation, and since the environment is a very coastal one, allowing a quick response to an increase in phytoplankton standing stock in the form of a release of larvae of benthonic animals. Upwelling ecosystems, however, do provide classical examples of herbivore food chains thriving under rather stable fertilization régimes.

Another, and more original, counter-example is provided by the shelf-break area bordering the Celtic Sea. Use of infrared satellite-borne remote sensing showed in the late 1970s that the surface waters in this area are colder in late summer by about 2°C than over both the continental shelf and the abyssal plain of the Bay of Biscay (e.g. Pingree, 1979). A variety of hypotheses were put forward to account for the phenomenon, including, given the existence of some indications on a local biological richness, that of a classical wind-induced upwelling (e.g. Dickson *et al.*, 1980; Heaps, 1980). A model study carried out by Mazé (1980) suggested, however, that the significant phenomenon over the continental slope was the generation of internal waves due to the interaction between bottom topography and the propagation of the barotropic tidal wave towards the continental shelf, consistent with previous knowledge that internal waves were important in such areas (see Pingree and Mardell, 1981, and Le Fèvre, 1986, for more detailed references). The model approach was refined several times and field observations have been carried out (Mazé, 1983; Pingree et al., 1984; Mazé et al., 1986; Pingree et al., 1986; Mazé, 1987), so that a reasonably safe picture of what is taking place has now emerged. Large oscillations of the thermocline (with a range up to 60 m or more) are actually generated over the continental slope, with the same period as the incoming tidal wave (hence the name "internal tide" given to the phenomenon); their amplitude is maximal at their place of generation, near the inshore edge of the slope, and they progressively dampen out as they propagate towards both the shelf and the ocean. The propagation is strongly non-linear towards the shelf (due to the interaction with barotropic tidal currents), but the internal tide tends to become sinusoidal after it has propagated for some distance towards the abyssal plain. Complex circulation patterns, such as vertical shear and periodic spatial variations of residual currents, are associated with the phenomenon. At high internal tide, the thermocline is shallower than the depth of equilibrium consistent with wind-mixing conditions; this results in enhanced vertical mixing, which explains the cooling of the surface layer. The process has been modelled, showing that once an appropriate wind régime is established the temperature in the surface layer will decrease in steps, in phase with the M_2 tidal cycle, and this has been confirmed by field observations (Mazé et al., 1986). Hydrographic and weather conditions are most suitable for the process in late summer, which explains why the shelf-break area shows up as a band of cooler water in most infrared satellite images recorded at this time of the year. Enhanced vertical mixing of course also results in the input of nutrients from deeper waters into the euphotic layer. Due to the non-linear character of the internal tide, this enrichment may take the form of brief pulses (Holligan et al., 1985); on a longer-term basis, however, the key point is that fertilization is periodical, and that the period is that of the M_2 tide (i.e about 12 h). This period is likely to be time-integrated by the ecosystem, which will respond in the same way as in the case of a permanent régime. The preliminary results given by Le Tareau et al. (1983) actually show a high degree of consistency between vertically integrated distributions of phytoplankton and zooplankton biomass in the shelf-break area, which may be taken as an indication of the dominance of a herbivore food chain.

It is finally worth noticing that this kind of fertilization process is not unique to the Celtic Sea shelf-break. Similar non-linear internal tides are known for instance off both the Atlantic and Pacific coasts of North-America and have been suggested or shown to induce nutrient pulses (e.g. Haury et al., 1979, 1983; Shea and Broenkow, 1982; Sandström and Elliott, 1984). Phenomena so far recognized as "secondary upwellings", i.e. shelf-break surface cooling and nutrient enrichment in regions where a classical upwelling is found closer to the coast (e.g. Bang, 1971), may even turn out in some cases to involve an internal tide instead.

3. A check from a recent cruise

Cruise ONDINE 85 (from the French "ONDes INternEs": internal waves) was carried out between the beginning of September and mid-November 1985, under the responsibility of the Service Hydrographique et Océanographique de la Marine (SHOM, Hydrographic and Oceanographic Department of the French Navy) with the cooperation of the Physical, Chemical and Biological Oceanography laboratories in the Université de Bretagne Occidentale, Brest, and several other institutions. The area investigated was the southern Celtic Sea, from the west coast of Brittany to about 10°W, the north of the Bay of Biscay and the shelf-break zone in between; the main purpose was to obtain further information on the processes taking place over the shelf-break and their physical, chemical and biological characteristics and consequences. The programme involved two research vessels and included a very large complement of methods and sampling techniques, including instrument moorings, infrared remote-sensing, continuous measurement of surface temperature and *in vivo* phytoplankton fluorescence, CTD and bottle casts for measuring physical and chemical parameters, as well as chlorophyll concentration, and zooplankton sampling. Among the various operations, the point of interest here is that a network of 31 stations (see Fig. 1 below) was repeatedly sampled in the course of the cruise, yielding significant information on the time-variations of the pelagic ecosystem over the whole area investigated.

3.1. Material and methods

Network stations typically consisted of a hydrographic cast (CTD or bottles, depending on the technical capabilities of the vessel), which provided data on temperature, salinity, dissolved oxygen, inorganic nutrients and chlorophyll concentration, and of a zooplankton net haul. Zooplankton data will be given prime consideration here because, among the partial set of data as yet available, they provide a very convenient check for the alternative food chains hypothesis as described above. The sampling scheme was the result of a compromise, due to the inavailability of sophisticated equipment and to the need to minimize the time spent at a given station, so that the whole network could be occupied as quickly as possible (about 5 days). Accordingly, no direct information could be collected on zooplankton vertical distribution, integrated water column sampling being performed instead, by means of vertical hauls with the 200 µm mesh-sized WP$_2$ net (UNESCO, 1968), from the vicinity of the bottom to the surface on the continental shelf, from 200 m to the surface in deeper areas. Three identical nets were mounted on a common frame and operated simultaneously. One of the samples was immediately filtered on a numbered disc of pre-weighed gauze identical to the filtering material of the net, quickly rinsed with distilled water, to remove the salt content of interstitial water without loosing too much organic matter due to the osmotic shock, and immediately frozen at about -20°C. Back in the laboratory, the discs and their plankton load were dried for 3 days at 60-70°C, allowed to cool and weighed to obtain the mesozooplankton dryweight biomass (as total weight minus initial disc weight). The second sample was concentrated onboard and preserved with formalin (about 4% final formaldehyde concentration), for subsequent identification and counting of species or higher-order taxonomic categories. The use made of the third sample is not relevant here.

Enumeration of the taxonomic samples has so far been completed only for the first coverage of the network (5-10 September 1985). Subsampling was performed by adjusting the volume of any given sample to 250 ml and withdrawing an aliquot with a 5 ml Stempel pipette (subsampling factor F=50). After enumeration was performed on the aliquot, counting of those organisms whose numbers reached a threshold considered statistically significant was discontinued. A second fraction was withdrawn for counting rarer organisms, then enumeration was discontinued for those organisms whose cumulated numbers in the first two fractions (F=25) reached the same threshold as above. The process was repeated until the whole sample (F=1) was screened for the enumeration of the rarer organisms. In order to save some of the time devoted to the task, the statistical threshold for discontinuing the enumeration of any taxon was set at 31 specimens, i.e. half an order of magnitude lower than recommended by Frontier (1972).

In the absence of a flow-meter, the volume of water filtered by the net was calculated as the product of mouth area (0.25 m^2) by height hauled. The hydrodynamic design of the WP$_2$ net was originally adjusted for this procedure to be legitimate (UNESCO, 1968), which was subsequently confirmed by independent checks by various authors (e.g. Razouls and Thiriot, 1972; Bhaud et al., 1974; Evans, 1977). The final data, both for biomass and taxonomical analysis, have been calculated as values per unit volume (1 m^3 for biomass, 10 m^3 for animal numbers) and values per unit area (1 m^2, for the whole water column sampled). Only values per unit area will, however, be considered here. Biomass values were checked for a possible systematic difference between day and night measurements by calculating whole-cruise averages, separately for the continental shelf and for the areas where the depth was 200 m or more. Night values were found to be significantly higher than day values in both cases, with a night-to-day ratio of about 1.3 on the shelf and about 1.4 elsewhere. For the purpose of geographic comparisons (e.g. distribution charting as below), day values were accordingly multiplied by the factor relevant to the area where they were obtained.

3.2. Results

In west-European seas, the summer of 1985 was a cold and rainy one, while very fine weather prevailed from the beginning of September to mid-November. This probably explains why plankton abundance was found to increase in the course of the cruise. The variation is shown here in the case of zooplankton dryweight biomass (vertically integrated for the 200 upper metres), for which distribution charts are given for the two first parts of the cruise, 5-10 September (Fig. 1a) and 23-28 September (Fig. 1b). Leaving aside an isolated high value at the northernmost station, Fig. 1a shows a coherent pattern, where the most conspicuous features are a maximum probably related to the Ushant front (see the infrared satellite image Fig. 2 for the hydrographic situation) and a nucleus of values above 4000 mg m^{-2} over the shelf-break, which happens to be coincident with a cold spot on Fig. 2. The maximum on the Ushant front is in apparent contradiction with the theory proposed above, but it has disappeared by the time of the second survey (Fig. 1b), while the nucleus over the shelf-break has grown considerably and a second such nucleus has appeared to the west, also over the shelf-break. Very different plankton dynamics are therefore found on the Ushant front and over the shelf-break.

A model study recently carried out by Serpette (1987) within the same framework as the cruise is helpful in accounting for plankton dynamics in the shelf-break area. The model assumes a two-layer ocean and simulates the generation and propagation of internal waves in the whole of the Bay of Biscay. Various characteristics (amplitude of the oscillation of the interface for the different tidal constituents, barotropic and baroclinic currents, vertical shear, etc.) are computed at the points of a 5×5 km grid (while all earlier models were one-dimensional) giving a reasonably realistic view of the influence of bottom topography on the physical phenomena. Figure 3 shows some of the results, in the form of a chart of the range of the M$_2$ oscillation of the interface, i.e. the characteristic most directly associated with the fertilization process postulated above. The internal tide is not uniform over the shelf-break, three nuclei of maximal range being found instead over the 200 m isobath, centered at about 5°50', 6°30' and 7°W. Other maxima are found over the shelf, most noticeably a large one at about 47°35'N, 6°05'W; no corresponding maximum is found, however, in the barotropic forcing term and Serpette's (1987) conclusion is that the feature results from constructive interferences between the internal waves generated at the two easternmost nuclei over the 200 m isobath. Taken together, these two primary nuclei and the secondary one on the shelf make up a single, complex, area where internal tide energy is maximal. This area is coincident with a conspicuous cold spot over the shelf-break on Fig. 2 and with the area where a nucleus of high zooplankton biomass is found on Fig. 1a and has grown on Fig. 1b; the core (highest values) of the second biomass maximum on Fig. 1b also corresponds to a maximum in internal tide amplitude on Fig. 3.

These results are, therefore, highly consistent with the view that the main fertilization process in the area closely depends on the M$_2$ internal tide, and that it results in enhancement of general plankton productivity, showing up as progressive increase in zooplankton standing stock under favourable climatic conditions.

Figure 1 (facing page) - Distribution of zooplankton dryweight biomass down to 200 m (mg m^{-2}) on the two first surveys of cruise ONDINE 85: a, 5-10 September 1985; b, 23-28 September 1985. The values are corrected for diel variations (see text); solid dots correspond to station positions; the background is a bathymetric chart. The darkest spots enclosed by isolines bearing no indications correspond to values above 6000 mg m^{-2}.

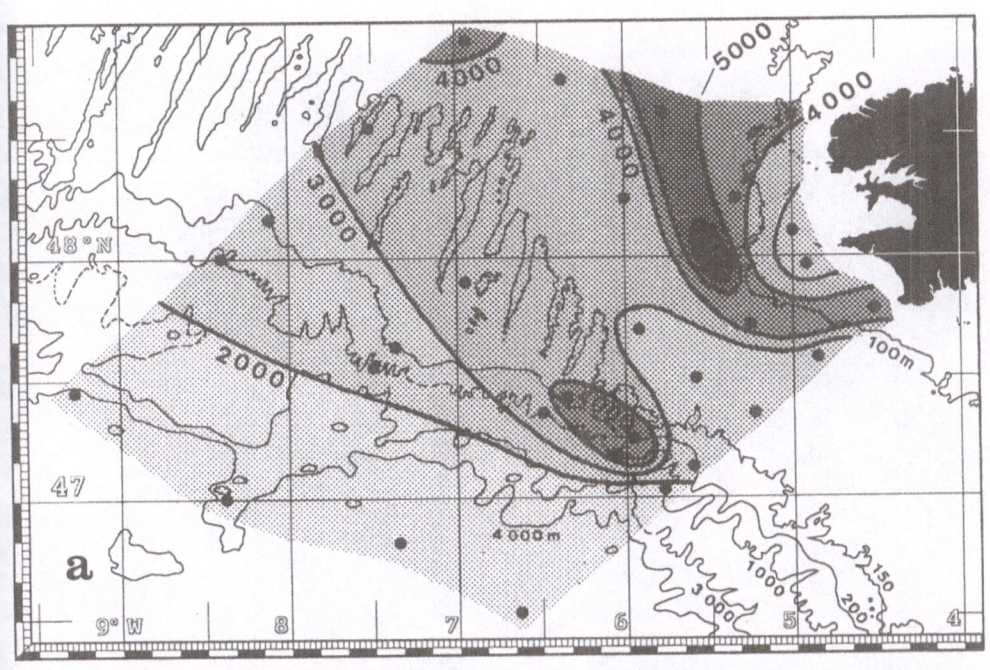

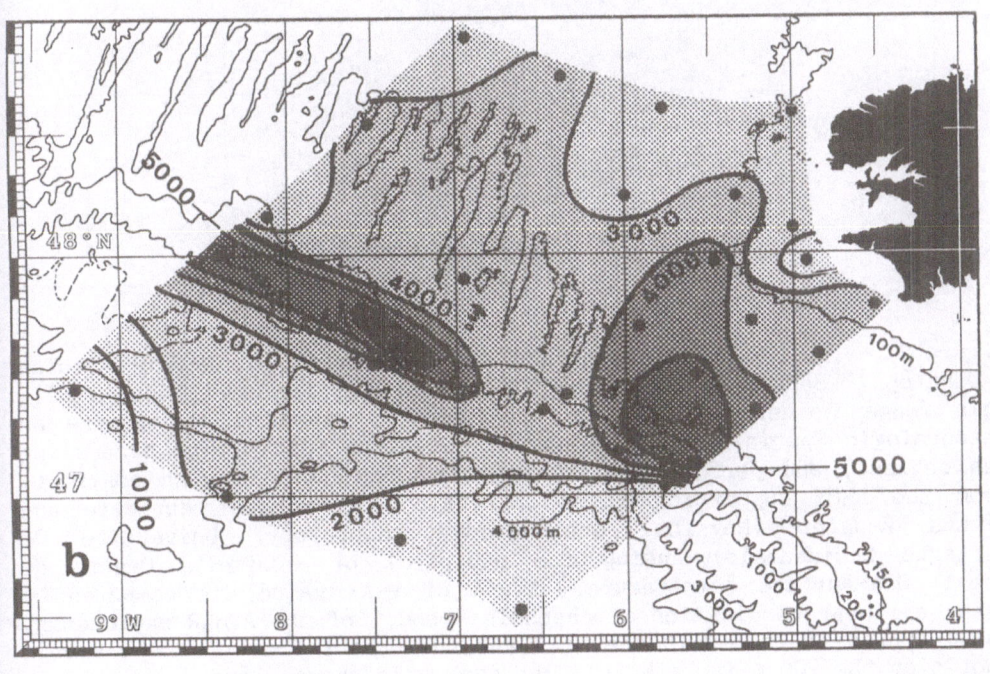

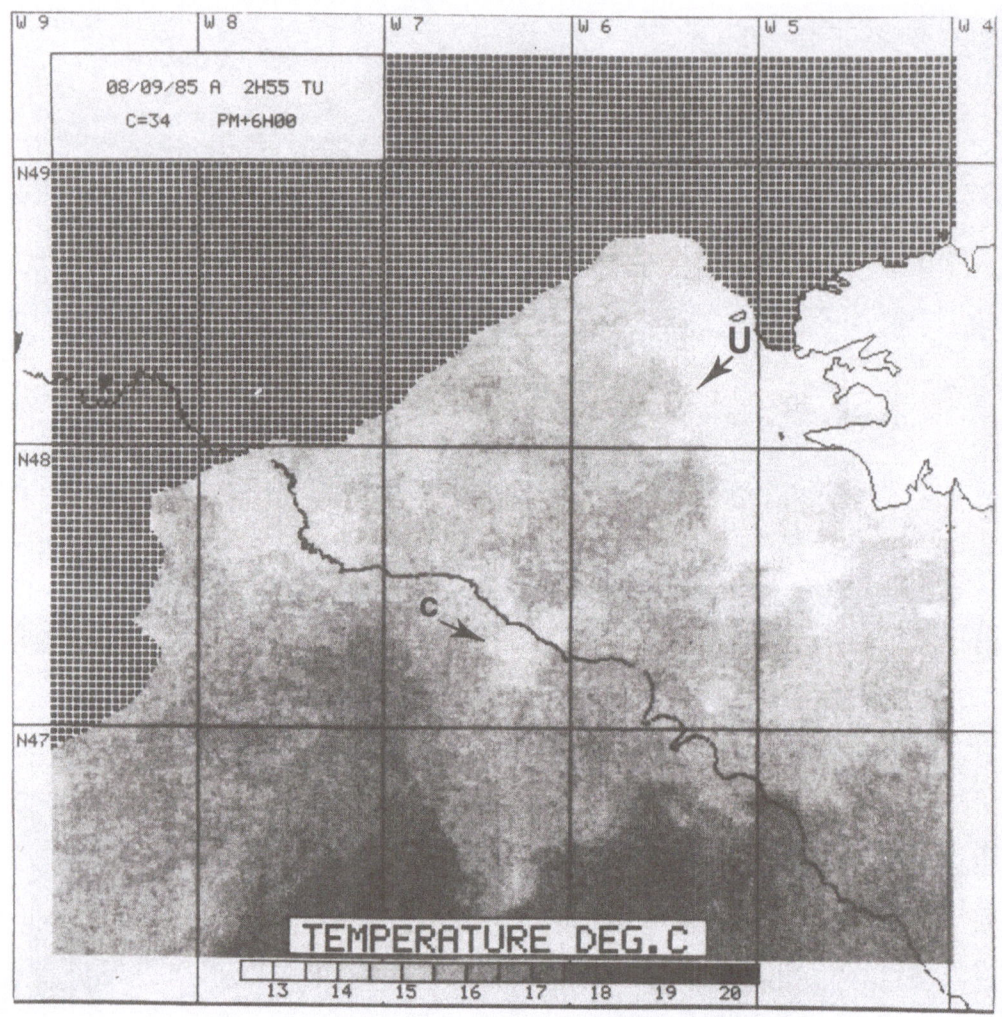

Figure 2 - Hydrographic situation on the first survey of cruise ONDINE 85 (8 September 1985, 2h55 GMT, low-water in Brest, neap tide). This image is derived from infrared data received at the Centre de Météorologie Spatiale in Lannion from satellite NOAA 9. The processing was devised, and performed within the framework of research contracts from the SHOM supporting cruise ONDINE 85, by Vincent Mariette and Gilles Rougier (Physical Oceanography Laboratory, Université de Bretagne Occidentale), using the equipment of IFREMER's Centre de Brest. Sea-surface temperature (±0.2°C of real value) is computed by using a linear combination of channels 4 and 5 of the AVHRR radiometer. The features mentioned in the text are the Ushant front (U) and a cold spot over the 200 m isobath (c). The dark grid shows cloud cover.

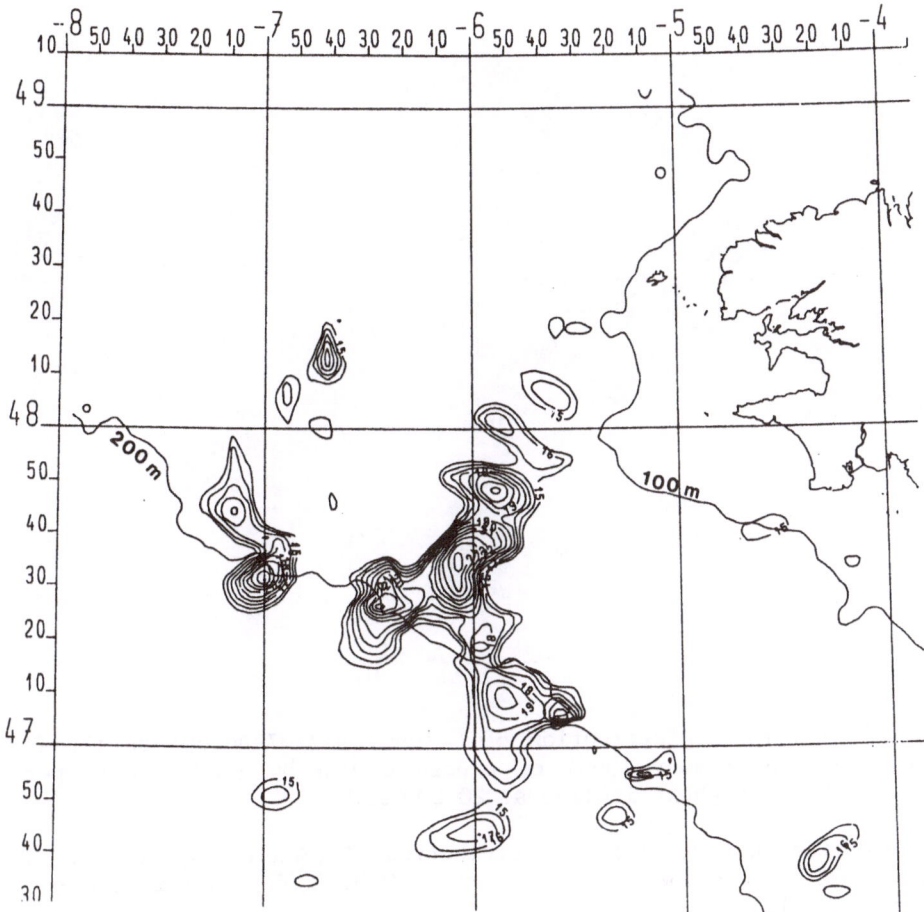

Figure 3 - Predicted range (m) of the M_2 internal tide. This map shows only part of the results of a numerical model covering the whole of the Bay of Biscay. The range of the M_2 internal tide is only one of several quantities (e.g. S_2 range, baroclinic currents, vertical shear) calculated at the points of a 5×5 km grid. The assumptions on which the model rests include a two-layer ocean and fixed values for various parameters, such as the mean depth of the thermocline (40 m) and the reduced gravity corresponding to buoyancy forces (g'=0.0153 m s^{-2} in the case shown here). Different values for these parameters would produce some differences in the patterns, especially on the continental shelf. In addition, the range given here is almost certainly overestimated on the inner part of the shelf, since Pingree *et al.* (1986) have shown the amplitude of the internal waves to be reduced by a factor 3 after they have propagated onshelf for about 70 km (from the 200 m isobath); this is not taken into account in the way the propagation is modelled. [From Serpette, 1987].

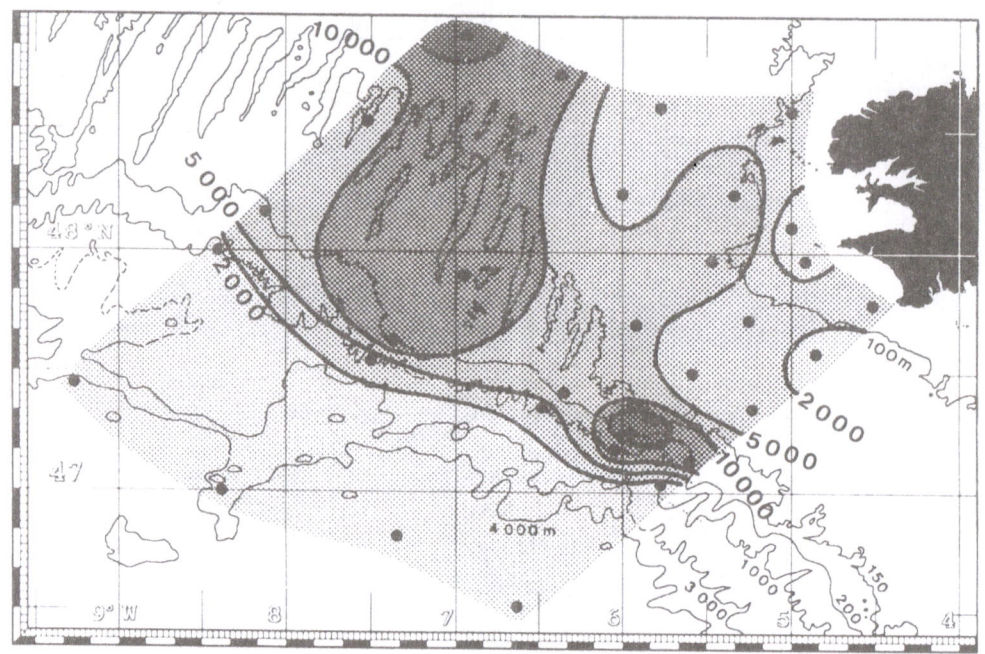

Figure 4 (above) - Distribution of *Calanus helgolandicus* on the first survey (5-10 September 1985) of cruise ONDINE 85 (individuals per m² down to 200 m). Highest isoline at 20 000 ind. m⁻².

Figure 5 (facing page, top) - Distribution of *Limacina* sp. on the first survey (5-10 September 1985) of cruise ONDINE 85 (individuals per m² down to 200 m).

Figure 6 (facing page, bottom) - Position of Vincent and Kurc's (1969) "bourrelet froid" on the continental shelf (compiled from pluri-annual data series). The 11.7°C bottom isotherm delineates the maximal extension of the bourrelet in spring; in autumn, the structure tends to break up into two bourrelets (hatched areas), one in the Celtic Sea (**C**) and one on the Armorican shelf (**A**). The transect on the chart is not relevant here. [From Le Magueresse, 1974].

If the theoretical views above are correct, the enhanced plankton productivity should also correspond to a predominantly herbivore food chain; the high-biomass nuclei at or near the shelf-break should therefore correspond to a maximum in the abundance of some herbivores, which should not be the case of the biomass maximum near the Ushant front at the beginning of September (Fig. 1a).

Given the data so far available, the relevant check is performed here by charting the abundance (individuals per m²) on the first survey (5-10 September 1985) of the calanoid copepod *Calanus helgolandicus*

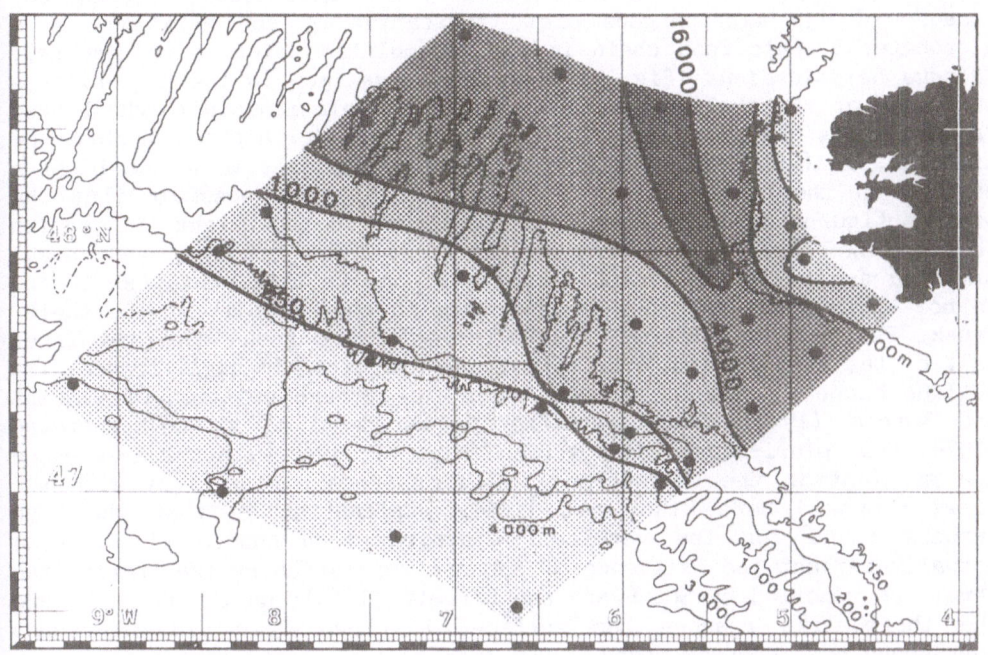

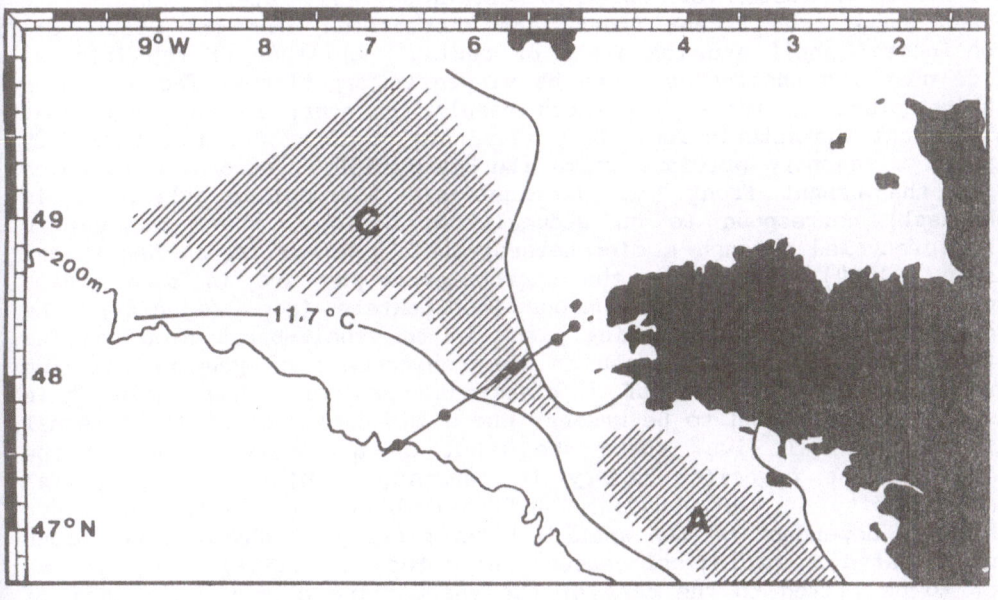

(Claus), considered representative of the herbivore pathway, and of the pteropod *Limacina* sp. (probably *L. retroversa*), considered, because of its microphagous diet, indicative of the existence of a microheterotrophic food chain in the accumulation biotope (*cf. supra*). *Calanus helgolandicus* (Fig. 4) does show a maximum in the high-biomass nucleus over the shelf-break and no apparent relationship to the Ushant front in its distribution. On the contrary, the highest numbers of *Limacina*, > 16 000 ind. m^{-2} (Fig. 5), are coincident with the biomass maximum on the Ushant front (Fig. 1a); it is even possible that the biomass figures there be biased in excess due to the weight of *Limacina* shells (the samples were not decalcified). Out of this peak area, *Limacina* are well-represented essentially on the continental shelf and do not exhibit any increase in abundance in the vicinity of the shelf-break. Fairly high numbers (> 4000 ind. m^{-2}) are found over the inner half of the shelf, in an area of precise hydrographic significance: its outline happens to be coincident with the average position of Vincent and Kurc's (1969) "bourrelet froid", as compiled by Le Magueresse (1974) from pluri-annual data series (Fig. 6), i.e. with the area where the gradient in the thermocline is the sharpest. Consistent with the views proposed here, the most likely explanation is that the high numbers found for the vertically integrated abundance of *Limacina* actually reflect the existence of a dense population at the thermocline level, as reported by Southward and Barrett (1983); on the front, where the thermocline reaches the surface, high-density would of course correspond to a surface patch.

4. Conclusion

All three systems for which observations and interpretations are given above, the Ushant front, the thermocline over the Celtic Sea continental shelf and the areas of maximal amplitude of the internal tide near the shelf-break, can be viewed as ergoclines. The processes taking place at these interfaces result, however, in very different biological characteristics. The preliminary results of cruise ONDINE 85 appear to support previous views (Le Fèvre, 1986), according to which both the Ushant front and the shelf-sea thermocline (where it is sharpest) correspond to an accumulation biotope, where biomass is first recycled through a microheterotrophic pathway (*sensu* Newell and Linley, 1984), and where the microheterotrophs are in turn grazed upon by larger-sized microphagous zooplankters (e.g. *Limacina*). The fertilization processes taking place in the shelf-break area, on the other hand, appear to result in the enhancement of general plankton production in the form of a classical herbivore food chain. This enhancement is found to be maximal where the amplitude of the internal tide is maximal, i.e. where the input of auxiliary energy in the sense of the ergocline theory is largest, resulting in a spatial heterogeneity in the abundance and distribution of living organisms which is dependent on the spatial heterogeneity of physical phenomena but mediated by biological causes. While the theoretical framework has yet to be fitted to the case of the thermocline over a large part of the shelf, the contrast between the frontal region and the shelf-break is clearly accounted for by the difference in time-match between the fertilization processes and the response of living organisms. In both cases the fertilization processes are periodical and tide-dependent. In the shelf-break area, the dominant periodicity is that of the M_2 tide; this is time-integrated in the response of the living organisms and the outcome is similar to what is found in the case of a more or less permanent régime such as an upwelling. In the frontal region, the dominant periodicity is that of the neaps-springs alternation, apparently resulting in short outbursts of phytoplankton growth every second week; herbivores cannot adjust themselves to such a situation, which allows the biomass to be accumulated and recycled by microheterotrophs. This is again consistent with previous views (Frontier, 1986) according to which not only the general level of plankton production, but the very structure of the food chains as well, may be shaped by the temporal characteristics of the transfer of auxiliary energy into the ecosystem at ergoclines.

A number of points, however, still require clarification. One of them is the case of the thermocline, where, according to the literature, various situations can be found, ranging from accumulation of ageing phytoplankton cells to an actively growing local population. Temporal characteristics in the variations of physical factors may also turn out to play an important rôle there, e.g. through the period of internal waves: a common high-frequency phenomenon is the Brunt-Väisälä oscillation of the pycnocline, whose rhythm is dependent on the vertical density gradient and will affect the light-history, and therefore the physiological condition, of the phytoplankton cells in the subsurface chlorophyll maximum. The significance, for the general biological production, of the particular food chain link where microheterotrophs are grazed upon by microphagous zooplankters also still has to be ascertained in a variety of situations.

5. Acknowledgements

The first author's participation in cruise ONDINE 85 and in the analysis of the data was supported by research contracts from the Direction des Recherches et Études Techniques (DRET), Ministry of Defence, Paris, and the Service Hydrographique et Océanographique de la Marine (SHOM). Permission to use the results in the present paper was also obtained by courtesy of DRET and SHOM. Miss Christine Voisin greatly assisted in the field and laboratory work, and in particular performed most of the tedious task of counting plankton samples. Alain Serpette kindly communicated a copy of his latest model study, as yet unpublished, on the generation and propagation of the internal tide. The sources of the satellite image Fig. 2 are acknowledged in the legend to the figure.

6. References

Bang N.D., 1971. 'The southern Benguela Current region in February 1966. Part II. Bathythermography and air-sea interactions'. *Deep-Sea Research*, **18**, 209-224.

Bary B.M., 1953. 'Sea-water discoloration by living organisms'. *New-Zealand Journal of Science and Technology*, **B 34**, 393-407.

Bhaud M., Bourdillon A., Castelbon C. and Bastiani F., 1974. 'Étude de la répartition verticale du zooplancton de deux secteurs de la Méditerranée: la campagne "Médiplanct 1972". II. Analyse des méthodes d'échantillonnage'. *Annales de l'Institut Océanographique, Paris*, **59**, 41-50.

Binet D., 1977. 'Cycles biologiques et migrations ontogéniques chez quelques copépodes pélagiques des eaux ivoiriennes'. *Cahiers de l'ORSTOM, Série Océanographie*, **15**, 111-138.

Boalch G.T., Harbour D.S. and Butler E.I., 1978. 'Seasonal phytoplankton production in the western English Channel 1964-1974'. *Journal of the Marine Biological Association of the United Kingdom*, **58**, 943-953.

Cushing D.H., 1959. 'The seasonal variation in oceanic production as a problem in population dynamics'. *Journal du Conseil International pour l'Exploration de la Mer*, **24**, 455-464.

Cushing D.H., 1975. *Marine Ecology and Fisheries*. Cambridge University Press, Cambridge.

Cushing D.H. and Dickson R., 1976. 'The biological response in the sea to climatic changes'. *Advances in Marine Biology*, **14**, 1-122.

Dickson R.R., Gurbutt P.A. and Narayana Pillai V., 1980. 'Satellite evidence of enhanced upwelling along the European continental slope'. *Journal of Physical Oceanography*, **10**, 813-819.

Evans F., 1977. 'Seasonal density and production estimates of the common planktonic copepods of Northumberland coastal waters'. *Estuarine and Coastal Marine Science*, **5**, 223-241.

Falkowski P.G., 1983. 'Light-shade adaptation and vertical mixing of marine phytoplankton: a comparative field study'. *Journal of Marine Research*, **41**, 215-237.

Fenchel T., 1982a. 'Ecology of heterotrophic microflagellates. I. Some important forms and their functional morphology'. *Marine Ecology - Progress Series*, **8**, 211-223.

Fenchel T., 1982b. 'Ecology of heterotrophic microflagellates. II. Bioenergetics and growth'. *Marine Ecology - Progress Series*, **8**, 225-231.

Fenchel T., 1982c. 'Ecology of heterotrophic microflagellates. III. Adaptations to heterogeneous environments'. *Marine Ecology - Progress Series*, **9**, 25-33.

Fenchel T., 1982d. 'Ecology of heterotrophic microflagellates. IV. Quantitative occurrence and importance as bacterial consumers'. *Marine Ecology - Progress Series*, **9**, 35-42.

Frontier S., 1972. 'Calcul de l'erreur sur un comptage de zooplancton'. *Journal of Experimental Marine Biology and Ecology*, **8**, 121-132.

Frontier S., 1986. 'Studying fronts as contact ecosystems'. In: Nihoul J.C.J. (ed.), *Marine Interfaces Ecohydrodynamics*, Proceedings of the 17th International Liège Colloquium on Ocean Hydrodynamics, Elsevier Oceanography Series, 42, 55-66.

Grall J.R., 1972a. '*Recherches quantitatives sur la production primaire du phytoplancton dans les parages de Roscoff*'. Thèse de Doctorat ès Sciences Naturelles, Université de Paris 6.

Grall J.R., 1972b. 'Développement "printanier" de la diatomée *Rhizosolenia delicatula* près de Roscoff'. *Marine Biology*, 16, 41-48.

Grall J.R., Le Fèvre-Lehoërff G. and Le Fèvre J., 1971. 'Observations sur la distribution du plancton à proximité d'Ouessant en juin 1969 et ses relations avec le milieu physique'. *Cahiers Océanographiques*, 23, 145-170.

Grall J.R., Le Corre P., Le Fèvre J., Marty Y. and Tournier B., 1980. 'Caractéristiques de la couche d'eau superficielle dans la zone des fronts thermiques Ouest-Bretagne'. *Oceanis*, 6, 235-249.

Haury L.R., Briscoe M.G. and Orr M.H., 1979. 'Tidally generated internal wave packets in Massachusetts Bay'. *Nature, London*, 278, 312-317.

Haury L.R., Wiebe P.H., Orr M.H. and Briscoe M.G., 1983. 'Tidally generated high-frequency internal wave packets and their effects on plankton in Massachusetts Bay'. *Journal of Marine Research*, 41, 65-112.

Heaps N.S., 1980. 'A mechanism for local upwelling along the European continental slope'. *Oceanologica Acta*, 3, 449-454.

Holligan P.M., 1978. 'Patchiness in subsurface phytoplankton populations on the northwest European continental shelf'. In: Steele J.H., (ed.), *Spatial Patterns in Plankton Communities*, Plenum Press, New York, London, 221-238.

Holligan P.M., 1981. 'Biological implications of fronts on the northwest European continental shelf'. *Philosophical Transactions of the Royal Society of London*, A 302, 547-562.

Holligan P.M., Maddock L. and Dodge J.D., 1980. 'The distribution of dinoflagellates around the British Isles in 1977: a multivariate analysis'. *Journal of the Marine Biological Association of the United Kingdom*, 60, 851-867.

Holligan P.M., Viollier M., Dupouy C. and Aiken J., 1983. 'Satellite studies on the distributions of chlorophyll and dinoflagellate blooms in the western English Channel'. *Continental Shelf Research*, 2, 81-96.

Holligan P.M., Williams P.J. le B., Purdie D. and Harris R.P., 1984a. 'Photosynthesis, respiration and nitrogen supply in stratified, frontal and tidally mixed shelf waters'. *Marine Ecology - Progress Series*, 17, 201-213.

Holligan P.M., Harris R.P., Newell R.C., Harbour D.S., Head R.N., Linley E.A.S., Lucas M.I., Tranter P.R.G. and Weekley C.M., 1984b. 'Vertical distribution and partitioning of organic carbon in mixed, frontal and stratified waters of the English Channel'. *Marine Ecology - Progress Series*, 14, 111-127.

Holligan P.M., Pingree R.D. and Mardell G.T., 1985. 'Oceanic solitons, nutrient pulses and phytoplankton growth'. *Nature, London*, **314**, 348-350.

Houghton R.W., Schiltz R., Beardsley R.C., Butman B. and Lockwood Chamberlain J., 1982. 'The Middle Atlantic Bight cold pool: evolution of the temperature structure during summer 1979'. *Journal of Physical Oceanography*, **12**, 1019-1029.

Jacq E., 1986. '*Étude des peuplements bactériens planctoniques dans deux systèmes côtiers de Bretagne: la rade de Brest et la zone frontale d'Ouessant*'. Thèse de Doctorat (Océanographie: Aquaculture et Pêche), Université de Bretagne Occidentale, Brest.

Jacq E. and Prieur D., 1986. 'Les associations bactéries-matière particulaire en milieu pélagique côtier: exemples de variations spatiales et temporelles'. *In: GERBAM - Deuxième Colloque International de Bactériologie Marine - CNRS, Brest, 1-5 octobre 1984*. Actes de Colloques, IFREMER, Brest, **3**, 229-236.

Jordan M.B. and Joint I.R., 1984. 'Studies on phytoplankton distribution and primary production in the western English Channel in 1980 and 1981'. *Continental Shelf Research*, **3**, 25-34.

King F.D., Cucci T.L. and Townsend D.W., 1987. 'Microzooplankton and macrozooplankton glutamate dehydrogenase as indices of the relative contribution of these fractions to ammonium regeneration in the Gulf of Maine'. *Journal of Plankton Research*, **9**, 277-289.

Le Fèvre J., 1986. 'Aspects of the biology of frontal systems'. *Advances in Marine Biology*, **23**, 163-299.

Le Fèvre J. and Grall J.R., 1970. 'On the relationships of *Noctiluca* swarming off the western coast of Brittany with hydrological features and plankton characteristics of the environment'. *Journal of Experimental Marine Biology and Ecology*, **4**, 287-306.

Le Fèvre J., Cochard J.C. and Grall J.R., 1981a. 'Physical characteristics of an inshore area on the Atlantic coast of Brittany and their influence on the pelagic ecosystem: the case of the "Rivière d'Étel".' *Estuarine, Coastal and Shelf Science*, **13**, 131-144.

Le Fèvre J., Quiniou-Le Mot F. and Tournier B., 1981b. 'Structures thermiques et distribution de certains organismes planctoniques: nouvelles méthodes d'approche à partir de l'exemple du site de Plogoff'. *In: Deuxièmes Journées de la Thermo-Écologie: Influence des Rejets Thermiques sur le Milieu Vivant en Mer et en Estuaire*, Electricité de France, Paris, 229-244.

Le Fèvre J., Viollier M., Le Corre P., Dupouy C. and Grall J.R., 1983a. 'Remote sensing observations of biological material by LANDSAT along a tidal thermal front and their relevancy to the available field data'. *Estuarine, Coastal and Shelf Science*, **16**, 37-50.

Le Fèvre J., Le Corre P., Morin P. and Birrien J.L., 1983b. 'The pelagic ecosystem in frontal zones and other environments off the west coast of Brittany'. *Oceanologica Acta*, Proceedings of the 17th European Marine Biology Symposium, special issue, 125-129.

Legendre L., 1981. 'Hydrodynamic control of marine phytoplankton production: the paradox of stability'. *In*: Nihoul J.C.J. (ed.), *Ecohydrodynamics*. Elsevier, Amsterdam, 191-207.

Legendre L. and Demers S., 1984. 'Towards dynamic biological oceanography and limnology'. *Canadian Journal of Fisheries and Aquatic Sciences*, **41**, 2-19.

Legendre L., Demers S. and Lefaivre D., 1986. 'Biological production at marine ergoclines'. *In*: Nihoul J.C.J. (ed.), *Marine Interfaces Ecohydrodynamics*, Proceedings of the 17th International Liège Colloquium on Ocean Hydrodynamics, Elsevier Oceanography Series, **42**, 1-29.

Le Magueresse A., 1974. '*La structure thermique sur le plateau continental dans le secteur Ouest-Bretagne, son évolution annuelle et quelques aspects de sa variabilité dans une zone frontale*'. Thèse de Doctorat de Spécialité (Océanographie Physique), Université de Bretagne Occidentale, Brest.

Le Tareau J.Y., Mazé R., Le Fèvre J., Billard C. and Camus Y., 1983. 'Envat 81, campagne multidisciplinaire en Atlantique. Aspects météorologiques, chimiques, biologiques, hydrologiques et thermodynamiques'. *Metmar*, **18**, 6-25.

Loder J.W. and Platt T., 1985. 'Physical controls on phytoplankton production at tidal fronts'. *Proceedings of the 19th European Marine Biology Symposium*, 3-21.

Longhurst A.R., 1976. 'Interactions between zooplankton and phytoplankton profiles in the eastern tropical pacific ocean'. *Deep-Sea Research*, **23**, 729-754

Longhurst A.R. and Herman A.W., 1981. 'Do oceanic zooplankton aggregate at, or near, the deep chlorophyll maximum ?' *Journal of Marine Research*, **39**, 353-356.

Longhurst A.R. and Williams R., 1979. 'Materials for plankton modelling: vertical distribution of Atlantic zooplankton in summer'. *Journal of Plankton Research*, **1**, 1-28.

Lorenzen C.J., 1967. 'Vertical distribution of chlorophyll and phaeopigments: Baja California'. *Deep-Sea Research*, **14**, 735-745.

Margalef R., 1956. 'Estructura y dinámica de la "purga de mar" en la Ría de Vigo'. *Investigación Pesquera*, **5**, 113-134.

Mazé R., 1980. 'Formation d'ondes internes stationnaires sur le talus continental. Application au Golfe de Gascogne'. *Annales Hydrographiques*, **8**, 45-58.

Mazé R., 1983. '*Mouvements internes induits dans un golfe par le passage d'une dépression et par la marée. Applications au Golfe de Gascogne*'. Thèse de Doctorat ès Sciences Physiques, Université de Bretagne Occidentale, Brest.

Mazé R., 1987 'Generation and propagation of non-linear internal waves induced by the tide over a continental slope'. *Continental Shelf Research*, **7**, 1079-1104.

Mazé R., Camus Y. and Le Tareau J.Y., 1986. 'Formation de gradients thermiques à la surface de l'océan, au-dessus d'un talus, par interaction entre les ondes internes et le mélange dû au vent'. *Journal du Conseil International pour l'Exploration de la Mer*, **42**, 221-240.

Morin P., 1984. '*Évolution des éléments nutritifs dans les systèmes frontaux de l'Iroise: assimilation et régénération, relation avec les structures hydrologiques et les cycles de développement du phytoplancton*'. Thèse de Doctorat de Spécialité (Chimie Appliquée - Chimie Marine), Université de Bretagne Occidentale, Brest.

Morin P., Le Corre P. and Le Fèvre J., 1985. 'Assimilation and regeneration of nutrients off the west coast of Brittany'. *Journal of the Marine Biological Association of the United Kingdom*, **65**, 677-695.

Newell R.C. and Linley E.A.S., 1984. 'Significance of microheterotrophs in the decomposition of phytoplankton: estimates of carbon and nitrogen flow based on the biomass of plankton communities'. *Marine Ecology - Progress Series*, **16**, 105-119.

Ortner P.B., Wiebe P.H. and Cox J.L., 1980. 'Relationships between oceanic epizooplankton distributions and the seasonal deep chlorophyll maximum in the Northwestern Atlantic Ocean'. *Journal of Marine Research*, **38**, 507-531.

Pingree R.D., 1978. 'Cyclonic eddies and cross-frontal mixing'. *Journal of the Marine Biological Association of the United Kingdom*, **58**, 955-963.

Pingree R.D., 1979. 'Baroclinic eddies bordering the Celtic Sea in late summer'. *Journal of the Marine Biological Association of the United Kingdom*, **59**, 689-698.

Pingree R.D. and Mardell G.T., 1981. 'Slope turbulence, internal waves and phytoplankton growth at the Celtic Sea shelf-break'. *Philosophical Transactions of the Royal Society of London*, A **302**, 663-682.

Pingree R.D., Forster G.R. and Morrison G.K., 1974. 'Turbulent convergent tidal fronts'. *Journal of the Marine Biological Association of the United Kingdom*, **54**, 469-479.

Pingree R.D., Pugh P.R., Holligan P.M. and Forster G.R., 1975. 'Summer phytoplankton blooms and red tides along tidal fronts in the approaches to the English Channel'. *Nature, London*, **258**, 672-677.

Pingree R.D., Holligan P.M., Mardell G.T. and Head R.N., 1976. 'The influence of physical stability on spring, summer and autumn phytoplankton blooms in the Celtic Sea'. *Journal of the Marine Biological Association of the United Kingdom*, **56**, 845-873.

Pingree R.D., Holligan P.M. and Head R.N., 1977. 'Survival of dinoflagellate blooms in the western English Channel'. *Nature, London*, **265**, 266-269.

Pingree R.D., Holligan P.M. and Mardell G.T., 1978. 'The effects of vertical stability on phytoplankton distributions in the summer on the northwest European Shelf'. *Deep-Sea Research*, **25**, 1011-1028.

Pingree R.D., Griffiths D.K. and Mardell G.T., 1984. 'The structure of the internal tide at the Celtic Sea shelf break'. *Journal of the Marine Biological Association of the United Kingdom*, **64**, 99-113.

Pingree R.D., Mardell G.T. and New A.L., 1986. 'Propagation of the internal tides from the upper slopes of the Bay of Biscay'. *Nature, London*, **321**, 154-158.

Prieur L. and Legendre L. 'Oceanographic criteria for new plankton production'. [*This volume*].

Razouls C. and Thiriot A., 1972. 'Données quantitatives du mésoplancton en Méditerranée occidentale (saisons hivernales, 1969-70)'. *Vie et Milieu, Série B,* **23,** 209-241.

Ryther J.H., 1955. 'Ecology of autotrophic dinoflagellates with reference to red water conditions'. *In:* Johnson F.H. (ed.), *The Luminescence of Biological Systems,* American Association for the Advancement of Science, Washington, publication n° 41, 387-414.

Sandström H. and Elliott J.A., 1984. 'Internal tides and solitons on the Scotian shelf: a nutrient pump at work'. *Journal of Geophysical Research,* **89,** 6415-6426.

Savidge G., 1976. 'A preliminary study of the distribution of chlorophyll *a* in the vicinity of fronts in the Celtic and western Irish Seas'. *Estuarine and Coastal Marine Science,* **4,** 617-625.

Serpette A., 1987. *'Présentation d'un modèle numérique simulant la génération puis la propagation des marées internes semi-diurnes dans le Golfe de Gascogne'.* Scientific Report, Laboratoire d'Océanographie Physique, Université de Bretagne Occidentale, Brest, for the Établissement Principal du Service Hydrographique et Océanographique de la Marine (EPSHOM), Contract n° 87-473, Convention n° 4/85.

Shea R.E. and Broenkow W.W., 1982. 'The role of internal tides in the nutrient enrichment of Monterey Bay, California'. *Estuarine, Coastal and Shelf Science,* **15,** 57-66.

Simpson J.H. and Bowers D., 1979. 'Shelf sea fronts' adjustments revealed by satellite IR imagery'. *Nature, London,* **280,** 648-651.

Simpson J.H., Edelstein D.J., Edwards A., Morris N.C.G. and Tett P.B., 1979. 'The Islay front: physical structure and phytoplankton distribution'. *Estuarine and Coastal Marine Science,* **9,** 713-726.

Smayda T.J., 1970. 'The suspension and sinking of phytoplankton in the sea'. *Oceanography and Marine Biology, an Annual Review,* **8,** 353-414.

Southward A.J. and Barrett R.L., 1983. 'Observations on the vertical distribution of zooplankton, including post-larval teleosts, off Plymouth in the presence of a chlorophyll-dense layer'. *Journal of Plankton Research,* **5,** 599-618.

Steele J.H. and Yentsch C.S., 1960. 'The vertical distribution of chlorophyll'. *Journal of the Marine Biological Association of the United Kingdom,* **39,** 217-226.

Sverdrup H.U., 1953. 'On conditions for the vernal blooming of phytoplankton'. *Journal du Conseil International pour l'Exploration de la Mer,* **18,** 287-295.

Tett P., 1981. 'Modelling phytoplankton production at shelf-sea fronts'. *Philosophical Transactions of the Royal Society of London,* A 302, 605-615.

Tyler M.A. and Seliger H.H., 1978. 'Annual subsurface transport of a red tide dinoflagellate to its bloom area: water circulation patterns and organism distributions in the Chesapeake Bay'. *Limnology and Oceanography,* **23,** 227-246.

Tyler M.A. and Seliger H.H., 1981. 'Selection for a red tide organism: physiological responses to the physical environment'. *Limnology and Oceanography,* **26,** 310-324.

UNESCO, 1968. *'Zooplankton Sampling'*. Monographs on Oceanographic Methodology, 2, UNESCO Press, Paris.

Vincent A. and Kurc G., 1969. 'Hydrologie. Variations saisonnières de la situation thermique du Golfe de Gascogne en 1967'. *Revue des Travaux de l'Institut des Pêches Maritimes*, 33, 79-96.

Wyatt T. and Horwood J., 1973. 'Model which generates red tides'. *Nature, London*, 244, 238-240.

Yamamoto S., 1983. 'Settling velocity and accumulation of silt-sized quartz grains in water column of deep sea: a computational approach'. *Bulletin of the College of Science, University of the Ryukyus*, 36, 117-128.

Yamamoto S., 1984. 'Concentration and behavior of detrital mineral grains in the water column of the open sea in the western Pacific'. *Journal of the Oceanographic Society of Japan*, 40, 80-89.

SPATIAL AND TEMPORAL DISCONTINUITIES OF BIOLOGICAL PROCESSES IN PELAGIC SURFACE WATERS

Joel C. Goldman
Woods Hole Oceanographic Institution
Woods Hole, MA 02543
U.S.A.

ABSTRACT. The classical paradigm of an unproductive, nutrient-poor pelagic zone where primary production is fueled almost exclusively by nutrient regeneration processes, appears at odds with the contemporary view that new primary production, supported by a stoichiometric input of oxidized nutrients into the euphotic zone, is considerably higher than previously thought. One way to accomodate both scenarios is to invoke the two layer concept in which the bulk of new primary production occurs at or near the base of the euphotic zone in response to pulsed injections of NO_3^- and PO_4^{3-}. Productivity in the upper euphotic zone where nutrients and biomass are trapped would be regulated almost exclusively by regenerative and degradative processes that occur within the microbial food loop. Since the microbial food loop which consists of a tightly-knit assemblage of phototrophic and heterotrophic nanno- and picoplankton persists throughout the euphotic zone, most of the energy and carbon processed by these small microbes would be lost through respiration and thus would not contribute to new production exiting to deeper waters. This raises the perplexing question of how biological processes are coupled to the input of new nutrients which, in turn, is controlled by physical events that occur on greatly varying temporal and spatial scales. Possibly, short-lived, local mixing events provide the right combination of light and new nutrients to allow rapid and undetected bursts of growth of larger phytoplankton species, in effect, creating ephemeral eutrophication zones. The resulting food chain may be short and simple so that newly fixed carbon can exit the euphotic zone rapidly while leaving behind an oxygen signal.

1. INTRODUCTION

It has been suggested recently by a number of researchers that marine primary production is far greater than has been measured by the classical ^{14}C incubation technique (Johnson *et al.*, 1981; Schulenberger and Reid, 1981; Jenkins, 1982; Jenkins and Goldman, 1985). Some of these newer estimates, based on integration of water

B. J. Rothschild (ed.), *Toward a Theory on Biological-Physical Interactions in the World Ocean*, 273–296.
© *1988 by Kluwer Academic Publishers.*

column oxygen measurements over long time and space scales, have not gone unchallanged, however (Platt, 1984; Platt and Harrison, 1985, 1986). At the core of the resulting controversy is the question of how the old ^{14}C measurements, which often are presumed to represent estimates of total primary production (the sum of new and regenerated production) (e.g. Eppley and Peterson, 1979; Platt and Harrison, 1985) and are performed on short temporal and spatial scales, can be reconciled with the newer oxygen measurements that only are of new production.

In trying to sort out the distinguishing characteristics of the two types of measurements and what each represents it is instructive to define first some important terms. Dugdale and Goering (1967) in their now classical paper distinguished new (P_N) from regenerated primary production (P_R) on the basis of the relative uptake of NO_3^- and NH_4^+ by phytoplankton populations. By a simple mass balance calculation, and using nitrogen as an example, new production was defined as photosynthetically derived particulate organic nitrogen that exited the euphotic zone to deeper waters. This new production was balanced over long time scales by the quantity of dissolved nitrogen (primarily NO_3^-) that was transported into the euphotic zone by physical processes and was taken up by phytoplankton during growth. Regenerated production, in contrast, was fueled solely by reduced nitrogen sources (primarily NH_4^+ but also dissolved organic N including urea) that originated from excretion and degradation processes of heterotrophs that grazed primary producers within the euphotic zone. Thus the sum of these forms of production equalled total production (P_T).

The early view that the pelagic ocean was a biological desert (Ryther, 1969) gained considerable support during the 1970's and early 1980's from numerous researchers who used the ^{15}N tracer incubation technique in field studies. Two major conclusions were that the percentage of total production derived from regeneration processes [R = $P_R/P_T(100)$] frequently was as high as 80-100% in surface waters of the open ocean (Eppley and Peterson, 1979) and that the very small (<10 μm) size class of microbes played a major role in the recycling process (Glibert, 1982). Thus the contemporary view that new production is considerably higher than previously thought appears at odds with the classical paradigm of an unproductive, nutrient-poor pelagic zone in which primary production is fueled almost exclusively by nutrient regenerative processes.

A major difficulty in accepting the possibility of a highly productive open ocean is that our knowledge of how nutrients are being supplied to the euphotic zone and how the biology responds to this input is limited. Based on tradional sampling practices, there is little evidence for either the enhanced nutrient input or the higher phytoplankton biomass required to support the quantity of oxygen that accumulates seasonally in some oceanic locales. Hence we are either dealing with a major artifact in the oxygen numbers or else we have failed to characterize adequately all the important physical and biological processes that are occurring at the interface between the aphotic and euphotic zones of the open ocean. In support of the latter possibility, Knauer *et al.*, (1984) and Jenkins and Goldman (1985) suggested that the euphotic zone could be divided into an upper and lower layer. In the upper layer where new nutrients were scarce regeneration would be high and new production low. The lower layer, in contrast, would be a zone of high new production and low regeneration, supported by episodic inputs of new nutrients that were difficult to identify with conventional sampling techniques. In effect, nutrient regeneration, when integrated over the entire euphotic zone, would be considerably lower than the values extrapolated from ^{15}N studies. In this paper I will expand on this concept and suggest that very different biological processes are responsible for regeneration and new production. In dealing with this question I will emphasize that the appropriate temporal and spatial scales upon which each of the processes function may be very different.

2. RELATIONSHIP BETWEEN NEW PRODUCTION AND REGENERATION

2.1. The $\underline{f}$ ratio and Available NO_3^-

Following the lead of Dugdale and Goering (1967), many researchers have attempted to determine the regeneration efficency R by measuring the relative rates of NH_4^+ and NO_3^- uptake by phytoplankton with ^{15}N techniques (Eppley *et al.*, 1979; Glibert, 1982; Glibert and McCarthy, 1984; Harrison *et al.*, 1983; Harrison *et al.*, 1987). Eppley and Peterson (1979) used the term $\underline{f}$ ratio" which is $(1 - R/100)$ to describe the ratio of new to total production $(P_N:P_T)$. In practice,

$$\underline{f} = \partial NO_3^- / (\partial NH_4^+ + \partial NO_3^-) \tag{1}$$

in which is NO_3^- and NH_4^+ are, respectively, the $^{15}NO_3^-$ and $^{15}NH_4^+$ uptake rates of the phytoplankton populations during the course of shipboard incubations on discrete and confined samples. Adjustments to Eq. 1 to account for urea uptake by phytoplankton and isotope dilution of $^{15}NH_4^+$ by grazer excretion of unlabelled NH_4^+ during the course of the incubation are necessary but difficult to make (Glibert *et al.*, 1982; Harrison *et al.*, 1987). A far more serious technical problem is that ambient NO_3^- and NH_4^+ concentrations in surface waters of the open ocean frequently are below detection limits (~ 0.03 μg atoms/liter) (McCarthy and Goldman, 1979). Under such conditions ^{15}N is not a true tracer and addition of even the smallest possible quantity of ^{15}N-labelled substrate (typically 0.03-0.1 μg atoms/liter) increases the substrate concentration greatly, leading to anomolously high uptake rates. Past difficulties in making accurate measurements of NO_3^- and NH_4^+ at these low concentrations have compounded this problem.

The recent attempts by Platt and Harrison (Platt and Harrison, 1985; Harrison *et al.*, 1987) to show a general relationship between the $\underline{f}$ ratio and ambient NO_3^- concentration provide a clear example of the difficulties in using the ^{15}N tracer method to determine regeneration efficiencies. As seen in Figure 1 from Harrison *et al.*

(1987), plots of $\underline{f}$ versus ambient NO_3^- for individual data sets from 8 field studies (mostly those from productive coastal or near-shore waters) appear to give reasonably good exponential fits of the data. On close scrutiny, however, it is evident that the bulk of the data in most of the plots fall into two distinct groups, one for ambient NO_3^- concentrations below or near the detection limit where $\underline{f}$ varies greatly from zero to close to the maximum values attained, and the other where $\underline{f}$ is maximum at concentrations of NO_3^- above several μg atoms/liter. The variablity of the data in the first group presumably occurs because of enhanced uptake due to the pulsing effect of the tracer addition.

The problem of variable $\underline{f}$ at very low NO_3^- concentration is seen clearly in one of the data sets from the Middle Atlantic Bight study of Harrison et al. (1983). By replotting the data on appropriate scales, it is evident that the absolute uptake rates for NO_3^- (Figure 1A) and NH_4^+ (Figure 1B) at ambient concentrations below 0.2 μg atoms/liter were highly variable and greatly influenced by the addition of ^{15}N tracer (0.1 μg atoms/liter in this case): NO_3^- varied from 0.1 to 6.9 μg atoms/m3/h and NH_4^+ from 0.3 to 11.8 μg atoms/m3/h when ambient NO_3^- or NH_4^+ was undetectable. The resulting $\underline{f}$ ratios ranged from 0.10 to 0.79 at concentrations of ambient NO_3^- between undetectable and 0.25 μg atoms/liter (Figure 1C). For most of the other data sets analyzed by Harrison et al. (1987) there was a similar variability in $\underline{f}$ when ambient NO_3^- concentrations were at the detection limit. Glibert and McCarthy (1984) experienced the same difficulty in their diurnal study at a station in the Sargasso Sea: $\underline{f}$ varied from 0.07 to 0.36 when ambient NO_3^- and NH_4^+ concentrations which were <0.05 μg atoms/liter were increased greatly by addition of 0.03 μg atoms/liter of ^{15}N tracer.

Thus, while plots of $\underline{f}$ versus NO_3^- may give the appearance that a quantitative relationship exists between the two variables (e.g. Figure 1D), a potentially serious bias is introduced when the resulting curve is fit from data sets representing the two extreme portions of the curve where, in fact, no real relationships exist between $\underline{f}$ and NO_3^-. When NO_3^- concentrations are very low $\underline{f}$ is indeterminant and when NO_3^- concentrations are very high $\underline{f}$ plateaus at a saturating value. For this and other reasons (Jenkins and Goldman, 1985) the ^{15}N technique simply may be unsuitable for determining regeneration efficiencies in oceanic waters when nutrient concentrations are at or near undetectable levels.

2.2. The Accuracy of Regeneration Measurements

The importance of having an accurate estimate of R is seen in Figure 2, which is simply a plot of the ratio $P_T:P_N$ (= $\underline{1/f}$) and R (= 1 - $\underline{f}$). When R is less than 60% the ratio of total to new production is relatively insensitive even to large errors in R. However, when R is in the range 80-90+% then the difference between total and new production becomes highly sensitive to small experimental errors in the determination of R. Given that the prevailing view has been that regeneration efficiencies in the open ocean are >90% (Eppley and

Peterson, 1979), ironically determined by ^{15}N methodologies, and that the ratio $P_T:P_N$ really is the ratio of production measured by the ^{14}C and water column O2 techniques, it is easy to see why great emphasis has been put on having an accurate estimate of R.

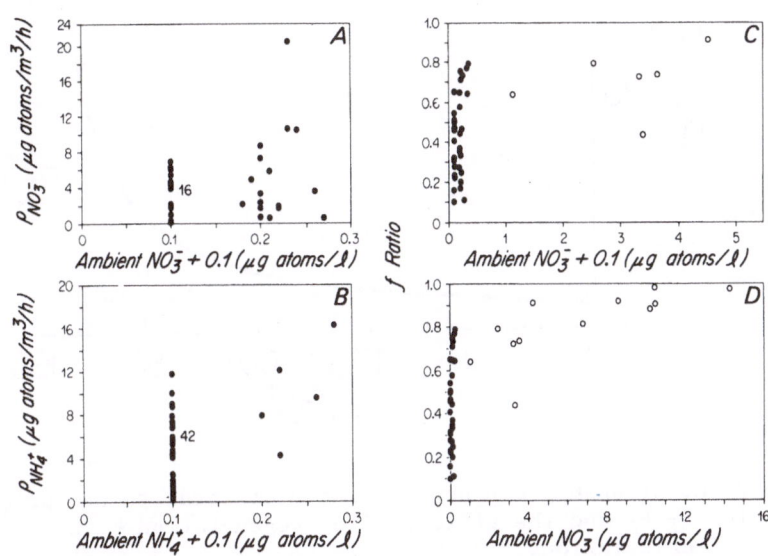

Figure 1. Absolute uptake rates for NO_3^- (A) and NH_4^+ (B) and the resulting f ratio (C,D) as a function of NO_3^- concentration from the study of Harrison *et al.* (1983). Panel C is the expanded version (NO_3^- concentration between 0-5 µg atoms/l) of the full data set shown in Panel D (NO_3^- concentration between 0-16 µg atoms/l). Open symbols represent data for which ambient NO_3^- concentrations were >0.25 µg atoms/l. 15N tracer addition of 0.1 µg atoms/l were added to ambient NO_3^- or NH_4^+ concentration in Panels A, B, and C, whereas only ambient NO_3^- concentration is presented in Panel D. Numbers in Panels A and B indicate number of data points for experiments in which ambient NO_3^- or NH_4^+.

Jenkins and Goldman (1985), in advocating the two layer system, actually were suggesting that the biological processes of regeneration and new production were distinctly different, the former involving a highly dynamic microbial food web where energy was wasted and the latter a simple and efficient food chain where energy was conserved. In the balance of this paper I will compare these two different systems and show that there is no incompatibility in the two systems existing simultaneously. In fact, the primary producers involved in new production may simply go unnoticed because of their small numbers but large size and because their growth is tightly coupled to episodic nutrient input into the euphotic zone.

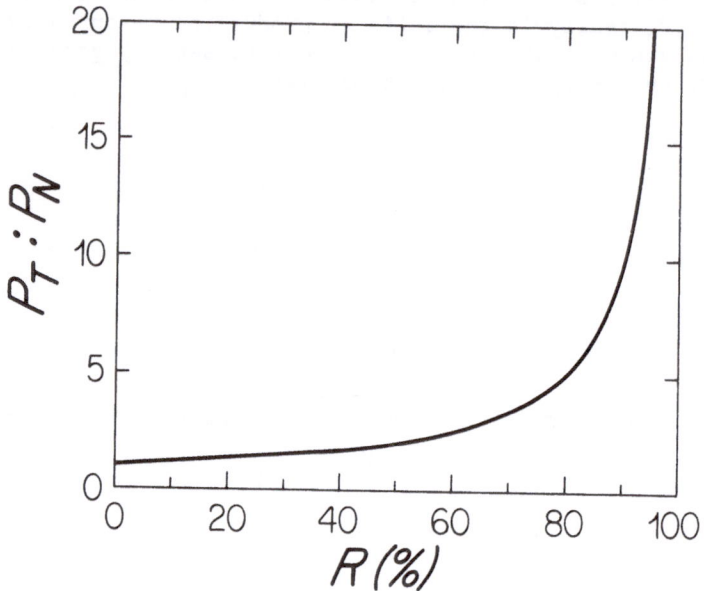

Figure 2. Relationship between the ratio of total to new primary production $P_T:P_N$ and the efficiency of regeneration R. See text for details of calculations.

3. THE MICROBIAL FOOD LOOP

3.1. Size Structure in the Microbial Loop

Pomeroy (1974) provided the first clues that the classical marine food chain consisting of diatoms, copopods, and fish was not the only major pathway by which energy and materials were channelled in the ocean. He viewed the complicated microbial food web consisting of small phototrophs, bacteria, and protozoa as playing a vital, if not major role in these processes. Several discoveries in recent years have added further fuel to this argument. First, there is little doubt now that photosynthetic picoplankton (<2 μm in size) are ubiqitous in the world's oceans in very large numbers (Johnson and Sieburth, 1979; Waterbury et al., 1979). On the basis of size fractionation studies with ^{14}C, these phototrophs appear to contribute greatly to total primary production (Li et al., 1983; Platt et al., 1983). And second, evidence is mounting that the bulk of nutrient regeneration in pelagic surface waters occurs within the highly dynamic and complicated microbial food web first described by Pomeroy (the "microbial food loop") (Azam et al., 1983; Goldman and Caron, 1985). The actual structure of this food web and the types and sizes of the microbes involved have been exceedingly difficult to characterize. Small

protozoa are believed to play important roles in grazing the
phototrophs, and, along with bacteria, in processing nutrients and
dissipating energy. Fenchel (1982), in fact, has argued that the
dominant predators of bacteria (and presumably of photosynthetic
ultraplankton) in the microbial food loop are phagotrophic
flagellates, microbes that generally are smaller than 10-20 μm in
size. Sieburth et al. (1978) distinguished two size classes of
microbes among the smaller organisms, the picoplankton (0.5-2 μm) and
the nanoplankton (2-20 μm), whereas Murphy and Haugen (1985), in
finding that small (2-5 μm) eucaryotic phototrophs were widely
distributed in large numbers in the North Atlantic along with the
picoplankton, enlarged the smaller size group to include all organisms
<5 μm in size (the ultraplankton).

3.2. The Spinning Wheel Concept

Based on evidence that phytoplankton in the open ocean may be growing
at their maximum possible growth rates (Goldman et al., 1979),
together with the view that regeneration efficiencies in this system
are very high, the microbial food loop has been likened to a "spinning
wheel" that turns at a speed controlled by the maximum possible growth
rate of the phototroph component (Goldman, 1984a,b). I have previously
argued that the relative growth rate $\mu:\mu(I_0)$ of the phytoplankton in
this wheel approaches 1 in which μ is the specific growth rate and
$\mu(IO)$ is the maximum specific growth rate for a particular light
intensity (Goldman, 1980; 1986). Thus because light intensity varies
with depth, the absolute growth rate of phytoplankton is expected to
decrease with depth, but without a change in the physiological state
of the cell population. The size of the wheel (i.e. the biomass of
both phototrophs and heterotrophs), however, while a function of the
total nutrients added to the system, should be independent of the rate
at which the wheel spins (Goldman, 1984a,b).

3.3. New Production in the Microbial Loop

As the pendulum has swung from paradigm of the diatom-copopod-fish
linkage to that of the microbial loop, there is a growing perception
that the bulk of primary production in the open ocean occurs among the
ultraplankton. Little distinction is made between the sources of new
and regenerated production. Yet the only data in support of this
argument comes from ^{14}C incubation studies on size-fractionated
samples (Li et al., 1983; Platt et al., 1983; Takahashi and Bienfang,
1983; Glover et al., 1985; Iturriaga and Mitchell, 1986). It must be
recognized, however, that such results provide no information on how
phototrophs of the microbial loop may contribute to new production.
Because of the extremely small size of the microbes involved, sinking
rates of individual untraplankton cells are very low (<0.1-0.2 m/d)
(Smayda, 1970; Bienfang, 1985); hence, we would expect these small
phototrophs to contribute negligible quantities of organic material
directly to new production. Results from sediment trap studies tend to
confirm this conclusion (Silver et al., 1986). Thus there are only two

alternative ways in which biomass within the loop could possibly sink out of the euphotic zone, either by efficient grazing by metazoa and release of large fecal material, or by physical aggregation of the microbial populations comprising the "loop" into larger particles.

Direct metazoan grazing on ultraplankton probably is not a major pathway by which biomass is passed up the marine food chain. Retention efficiencies for particles less than a few microns in size by most suspension-feeding metazoa generally is very low (Jorgensen, 1984). It is now generally believed that protozoa, particularly phagotrophic microflagellates because they are the main grazers of the ultraplankton, are a link between the very small components of the microbial loop and the larger zooplankton (Conover, 1982; Sherr et al., 1986). It must be recognized, however, that as the number of grazing steps between the ultraplankton and the metazoa increases, the total food chain efficiency decreases. Even considering that protozoa have gross growth efficiencies as high as 50-60% (Fenchel, 1982; Sherr et al., 1983; Caron et al., 1985), with only two grazing steps in a microbial food chain (e.g. small phototroph - flagellate - ciliate) only 25% of the initial biomass remains. To account for the very high nutrient regeneration efficiencies found in the microbial loop at least 3 grazing steps are required if at each step the gross growth efficiency is 50% (Goldman and Caron, 1985; Goldman et al., 1985). Such a food chain would leave even less prey biomass for metazoan grazers. Clearly, new production, at the levels suggested, does not occur by way of a grazing link between the microbial loop and larger marine animals.

Physical aggregation of particles into flocculent masses occurs in the ocean at many scales, ranging from the formation of microaggregates that are microns to tens of micron in size (Riley, 1970), to macroaggregates ("marine snow) that are hundreds to thousands of microns in size (Trent et al., 1978), and up to floating mats of large phytoplankton that are meters to tens of meters in size (Carpenter et al., 1977; Alldredge and Silver, 1982). These organic aggregates, formed by complex physical-chemical-biological interactions (McCave, 1975; 1984), often are enriched by orders of magnitude over the bulk fluid with nutrients and a diverse flora of bacteria, photosynthetic ultraplankton, and protozoa (Trent et al., 1978; Silver et al., 1978; Shanks and Trent, 1979; Caron et al., 1982). I have suggested in the past that aggregates of all sizes act as oases in the desert where the bulk of photosynthesis, grazing, and nutrient regeneration within the microbial loop may occur (Goldman, 1984a,b). Alldredge and Cohen (1987) recently showed that marine snow particles, indeed, had a very different chemical environment than the bulk fluid and could be important sites of microbial processes in the ocean.

Sinking rates of marine snow particles can vary from 1-350 m/d and frequently are over 100 m/d (Silver, 1986). These rates are substantial compared to sinking rates of most phytoplankton species in the ocean (<10-20 m/d) (Smayda, 1970); hence the microbial loop potentially could be an important source of new production if the photosynthetic ultraplankton were a significant component of the

flocculent mass sinking out of the euphotic zone. This does not seem
to be the case, however, as only a few percent of the total population
of photosynthetic ultraplankton in the euphotic zone have been found
as intact cells within marine snow material (Silver *et al.*, 1986).

4. SOURCES OF NEW PRODUCTION

4.1. Large Phytoplankton Species and New Production

A very distorted picture of global production may occur if the rare,
but potentially important biological events in the ocean are ignored.
The ^{14}C and ^{15}N incubation methods, by relying on a few discrete grab
samples to represent the entire euphotic zone, simply can not provide
the necessary temporal and spatial coverage for measuring annual
production and nutrient cycling accurately over large areas of the
ocean. Results from size fractionation studies with both techniques,
although instrumental in furthering our understanding of the microbial
loop, still are of limited value because only relative information is
obtained on the captured populations.

   Although the classical view that primary production was
restricted to diatoms left little room for the contribution of the
ultraplankton, neither does the contemporary view, emphasizing the the
importance of the ultraplankton, leave any room for a role for the
diatoms. Yet if the arguments presented above are correct, that is,
phototrophs of the microbial loop are not major contributors to new
production, then the only other source of this production must be the
larger phytoplankton species. It is my own contention that the great
flurry of recent interest over the ultraplankton and the microbial
loop has caused us to lose sight of two simple but crucials points.
First, one very large phytoplankton cell can equal the biomass of
ultraplankton orders of magnitude greater in number. And second,
biological events in the ocean are patchy on temporal and spatial
scales that are not amenable to quantitative measurement. For example,
the number of 1 μm cells required to equal the carbon content of one
diatom cell increases almost by three orders of magnitude for every
order of magnitude increase in the diameter of the diatom (Figure 3).
Thus in one liter of oceanic water it would take either a single 1000
um diatom or 100 diatoms 100 μm in size to equal the carbon content of
the entire population of photosynthetic picoplankton (~106-107 cells
according to Waterbury *et al.*, 1986). Clearly, the chances of
adequately representing primary production in the ocean by dividing a
captured population into size fractions and incubating each fraction
in a bottle would be proportional to the chances of capturing all the
contributors to that production.

   To dramatize the potential importance of larger phytoplankton
cells to new production and the difficulties that might be encountered
in recognizing and measuring this contribution, consider the extreme
scenario in which the microbial loop is a completely closed system:
the wheel spins at its maximum possible rate and regeneration is 100%
efficient. The loop operates throughout the euphotic zone and is

282

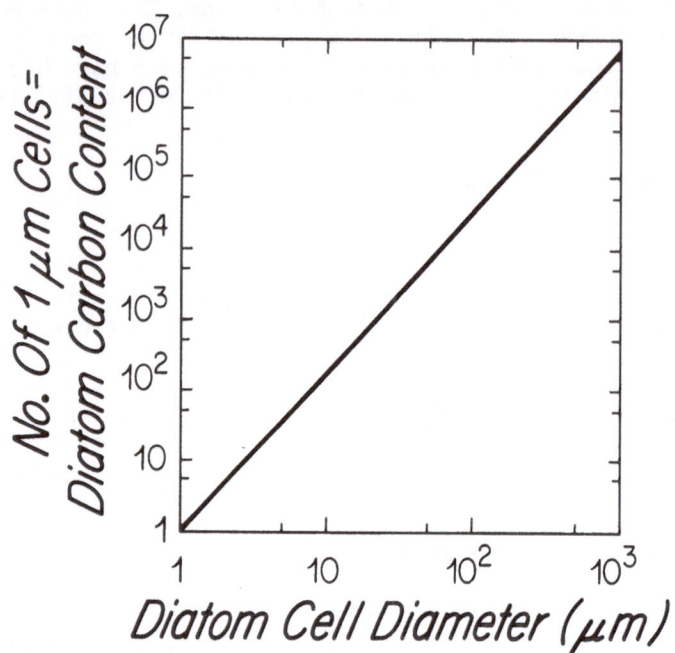

Figure 3. Relationship for the number of 1 μm picoplankton cells that equal the carbon content of one diatom cell of a varying diameter. The Strathmann equation (Strathmann, 1967) (log C (pg carbon) = 0.758 log V (μm^3) - 0.422) was used to convert cellvolume to cell carbon.

essentially a steady state system. Superimposed on this zero production system is an independent production system comprised of large phytoplankton cells that grow rapidly in response to an episodic input of new nutrients and then sink out of the euphotic zone, leaving behind an integrated oxygen signal. Hence, the microbial loop and the system responsible for new production are entirely separated, both spatially and temporally (Figure 4).

For this new production system it can be shown that only a small number of larger phytoplankton cells are necessary to account for all the new production and that, with current sampling practices, this production easily could go unnoticed. As an example, consider that this new production occurs in a typical oligotrophic water such as the Sargasso Sea at a rate of 5 M O_2/m^2/yr, as reported by Jenkins and Goldman (1985) (equivalent to 2.8 M C/m^2/yr, assuming a photosynthetic quotient of 1.8). Next, consider that all of this production is limited either to the bottom 25 m or bottom 50 m of the euphotic zone and occurs continuously (steady state) or in one of several episodic sequences.

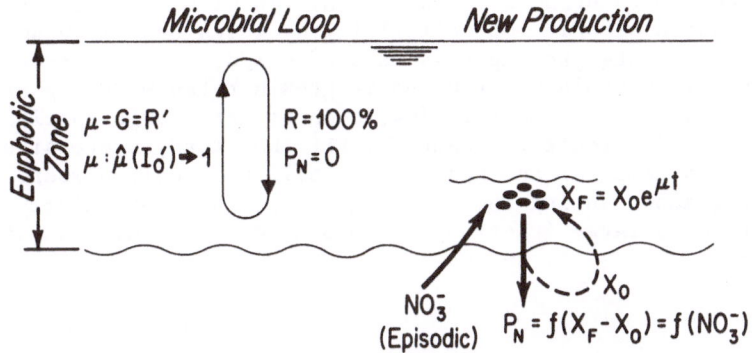

Figure 4. Schematic views of the extremes of the microbial loop and the system for new production. The microbial loop is portrayed as a spinning wheel where the rates of phytoplankton growth (μ), protozoan grazing (G), and nutrient regeneration (R') are equal and the relative phytoplankton growth rate as a function of light intensity [$\mu:\hat{\mu}(I_0)$] approaches 1 and where regeneration efficiency (R) is 100% so that new production (P_N) is zero. This system is dominated by ultraplankton and small protozoa and occurs throughout the euphotic zone. The new production system is fueled by episodic inputs of NO_3^- and an inoculum cell population (X_0) into the lower portion of the euphotic that leads to blooms of large phytoplankton (X_F). The large phytoplankton are not readily grazed and thus sink to deeper waters.

New production in the steady state system simply is:

$$P_N = \mu X \tag{2}$$

in which μ and X respectively are the steady state growth rate (1/d) and cell concentration (cells/l) of the phytoplankton population. It is assumed that there are no losses and all cells sink out of the euphotic zone. Then, by converting cell carbon to cell volume (for diatoms in this example) (Strathmann, 1967), it is possible to determine the required steady state population of phytoplankton cells of different diameters necessary to produce all of the new carbon as a function of specific growth rate (Figure 5). Considering a realistic range of specific growth rates for phytoplankton in the water column of 0.1 to 1.5/d (Eppley, 1972; Goldman et al., 1979; Goldman, 1986), new production by 1 μm size picoplankton at the levels suggested could be maintained only with steady state populations of about 3×10^6 to 4×10^7 cells/l if production occurred only in the bottom 25 m (Figure 5A) and between 10^6 and 2×10^7 cell/l if it occurred in the bottom 50 m (Figure 5B). While these picoplankton concentrations are about equal to those found commonly in oligotrophic waters (Waterbury et al., 1986), it must be recognized that much higher cell concentrations would be required to sustain total production.

The picture changes drastically, however, if very large cells are responsible for new production. For example, all of the new production in the bottom 25 or 50 m could be achieved with a steady state cell concentration of only 305 cells/l if the cells were 1000 μm in diameter and were growing at a rate of 0.1/d; even lower cell concentrations would be required if growth rates were higher (Figure 5). Cell concentrations this low, even of very large species, may be exceedingly difficult to quantify. For the steady state situation there is no temporal and spatial varibility in cell number, but capturing and counting the few large cells that potentially could account for a large fraction of new production becomes a major challenge.

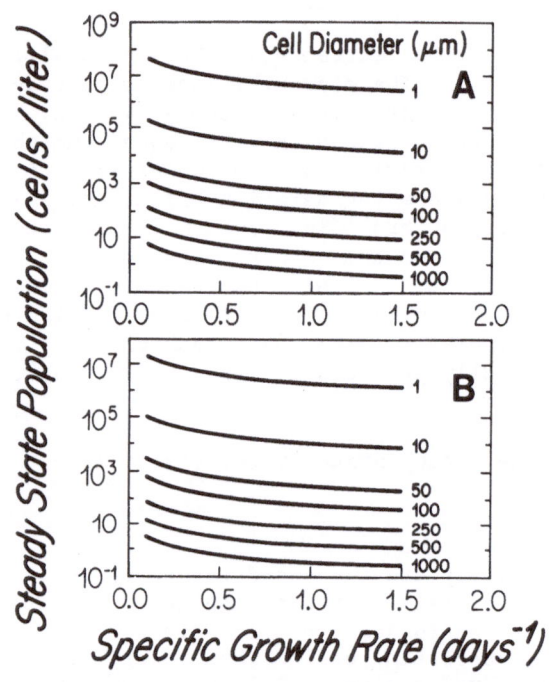

Figure 5. Nomographs of the steady state phytoplankton populations required for diatom cells of varying diameters as a function of specific growth rate to support new production at a rate of 5 M O_2/m^2/yr if production occurred only in the bottom 25 m (A) or 50 m (B) of the euphotic zone. See text for explanantion of calculations.

The problem of identifying the contribution of large phytoplankton cells to new production becomes more evident if production is tightly connected to episodic inputs of new nutrients. For example, consider four arbitrarily chosen episodic situations, the first two in which all of total yearly new production (2.8 M C/m²/yr) occurs during one continuous bloom period, one for 3.7 days duration (1% of the time) and the other for 18.3 days (5% of the time). Production in the remaining two situations occurs during repetitive bloom periods, one involving five 10-day periods, and the other ten 5-day periods. For this analysis, it is assumed that an inoculum (X_0) of only 1 cell/l for each bloom of phytoplankton is recycled back into the euphotic zone with the new nutrient water (Figure 4). Also, new production only occurs in the bottom 25 m and all of the cells produced during the bloom (X_F) sink out of the euphotic zone immediately after each bloom ends. Annual new production thus is proportional to $X_F - X_0$ or $X_0(e^{ut} - 1)$ and can be expressed as:

$$P_N = 25 \text{ n C } X_0(e^{ut} - 1)(10^3) \tag{3}$$

in which n is the number of bloom periods per year, t is the duration of each bloom period, and C is the conversion factor from cell number to cell carbon for cells of different diameters (once again, using the Strathmann equation for diatoms). Regeneration in this new production system simply is $(X_0/X_F)100$.

From the resulting nomograph for the four situations (Figure 6), it is evident that the required growth rates to support yearly new production vary tremendously from one scenario to the other. Both the one 3.7 day bloom and the ten 5-day bloom scenarios require unrealistically high growth rates (>3/d) for small cells (10 μm) and approach the limits of attainability (<2/d) only when cells >50-250 μm in size are involved. Growth rates for all size classes considered (10-1000 μm) in the other two scenarios, however, are well within the attainable range (0.3-1.4/d) even at the reduced light levels expected in the bottom 25 m of the euphotic zone (Chan, 1978; Brand and Guillard, 1981). The cell populations attained at the end of each bloom period likewise varies greatly among the different scenarios, ranging from 7.4 x 10^6 cells/l in the single growth period of 3.7 days and involving 10 μm size cells to 20 cells/l for each of the ten 5-day periods when 1000 μm size cells are involved (Table 1). Regeneration in each of these cases is essentially zero.

The above examples were meant to be gross exaggerations of how new production might occur. My purpose in presenting them solely is to highlight two main points:

(1) under both steady state and transient conditions the contribution of large phytoplankton cells to new production can be disproportionately large relative to that of the smaller cells and that this production could easily go unnoticed without intense sampling;

(2) the required growth rates during these blooms are well within the known physiological limits of the phytoplankton.

4.2 Rare Events and Variability in Food Chain Structure

Ryther (1969) in his classical study of the factors governing fish production in the ocean pointed out that the size of the primary producers was the key factor in determining the number of links in the marine food chain. Coastal and upwelling regions where nutrients are in greatest supply have long been recognized as major sites of global

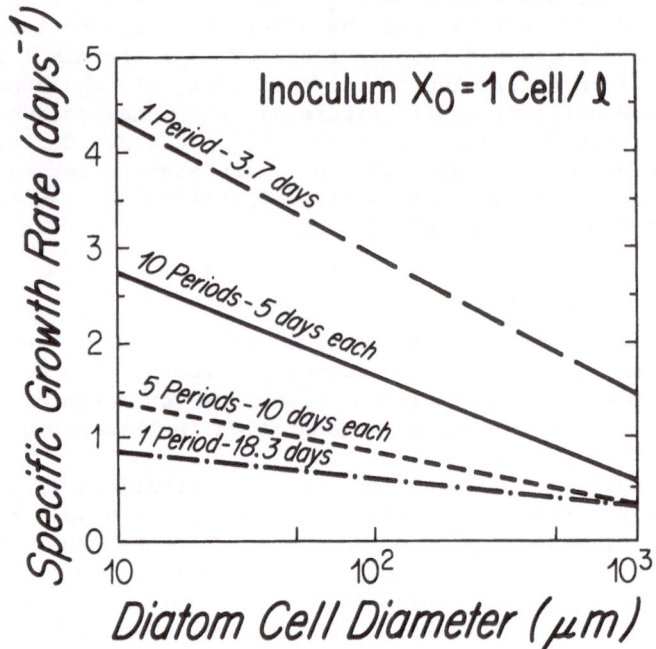

Figure 6. Nomograph of the required specific growth rate of diatoms as a function of cell diameter for four different hypothetical situations where 5 M $O_2/m^2/yr$ are produced through episodic events in the bottom 25 m of the euphotic zone and the inoculum for each bloom period is 1 cell/l. See text and Table 1 for details of calculations and description of episodic events.

Table 1. Final phytoplankton cell concentration resulting from episodic blooms of different durations and yearly frequencies and for cells of varying diameters when the inoculum (X_0) is 1 cell/l. See text and Figure 6 for details of analysis.

Yearly Frequency	Bloom Duration (days)	Final Cell Concentration (cells/1)			
		10 μm	100 μm	500 μm	1000μm
1	3.7	7.4×10^6	3.9×10^4	9.8×10^2	2.0×10^2
1	18.3	7.4×10^6	3.9×10^4	9.8×10^2	2.0×10^2
5	10.0	1.5×10^6	7.7×10^3	2.0×10^2	4.1×10
10	5.0	7.4×10^5	3.9×10^3	9.8×10	2.0×10

fish production because they sustain the growth of larger phytoplankton species leading to short and relatively efficient food chains. The notion that large phytoplankton species and species that

aggregate into large colonies by forming gelatinous or filamentous
masses are relatively more abundant in nutrient enriched regions than
they are in impoverished waters often is advocated (Ryther, 1969;
Malone, 1971). Some factors which seem to favor the production of the
larger species, *e.g.* heterogeneous nutrient supply (Harrison and
Turpin, 1982) and high turbulence (Eppley *et al.*, 1978; Margalef,
1978), are characteristic of nutrient enriched environments.
Additional factors that might favor the larger species include reduced
respiration rates (Laws, 1975), large storage vacuoles (Grenny *et al.*,
1973), and temporal lags in the growth of grazers (Steele and Frost,
1977).

Large phytoplankton species are not restricted to seemingly
productive coastal and upwelling waters. In fact, the extremely large
species often are cosmopolitan species. It is well established that a
diverse array of large diatom and dinoflagellate species are
ubiquitous in oceanic waters in greatly varying background numbers
($<1/m^3$ to $>100/1$) (Belyayeva, 1972; Semina, 1972; Guillard and Kilham,
1977). However, recognition of the importance of the large (>60-100
um) species to food chain dynamics and new production in oceanic
waters has been hampered because of the difficulty in identifying and
enumerating these plankton; quantitative analyses of these
phytoplankton species can only be made after collection in net hauls
(Belyayeva, 1972) or in sediment traps (Takahashi, 1986).

Truly huge species such as the diatom *Ethmodiscus rex* (up to 2000
μm in size) and the dinoflagellate *Pyrocystis noctiluca* (350-450 μm in
diameter) generally are found in very low numbers (<1-$200/m3$), but
they are distributed over large areas of the pelagic environment.
(McHugh, 1954; Belyayeva, 1970; Swift *et al.*, 1976). Somewhat smaller
(50-100 μm) diatoms such as *Hemiiaulus hauckii*, *Mastogloia*
(*Stigmaphora*) *rostrata* and a variety of *Rhizosolenia* species (which
vary greatly in size) frequently are the dominant diatoms found in
oligotrophic waters (Guillard and Kilham, 1977). Semina (1972), on the
basis of numerous net hauls in the Pacific Ocean, found that the
average diameter of phytoplankton cells captured was >80 μm in an
extensive area between about 200S and 200N; species of *Pyrocystis* and
Rhizosolenia were the dominant forms. Blooms of many of these large
diatom species have been observed on occasion with numbers reaching up
to 103-104/1 (Clemons and Miller, 1984; Riley, 1957; Venrick, 1974).

Why large phytoplankton species are ever present in in oceanic
waters in background numbers and why they bloom on occasion when
nutrients seem not to be available are unanswered questions that have
plagued biological oceanographers for a long-time (Guillard and
Kilham, 1977; Clemons and Miller, 1984). Thus, because of the
difficulties in collecting the large species and the sparcity of
sampling, the few anecdotal accounts of such blooms really provide no
information on the actual frequency and extent of their occurance in
the open ocean. From a few long term studies on phytoplankton
distributions and particulate flux at specific oceanic stations,
however, we can gain some perspective of the episodic nature of these
blooms and their potential significance to new production. For
example, from the two year study of Riley (1957) on weekly

phytoplankton distributions at a station in the North Central Sargasso Sea (Figure 7), it is evident that, aside from the expected spring diatom bloom which extended vertically throughout the euphotic zone, there were several periods when intense but short-lived blooms occurred at about 100 m resulting in cell concentrations between 104-4 x 10^4/1. The bloom period during early September, 1950 is of particular interest. During this period when the water was still thermally stratified the diatom *Rhizosolenia hebetata* f semispina (~150 μm long and 10 μm wide) reached a cell concentration of 2.1 x 10^4/1, accounting for 65% of the total water column diatom population sampled on that date. The short-lived nature of the bloom and its occurance in proximity to the base of the euphotic zone is suggestive of the episodic nutrient pulse-large phytoplankton species bloom scenario described earlier. Given that only 400 ml samples were obtained from 5 discrete depths for each sampling, the extent and magnitude of all the blooms indicated in Figure 7 are conservative, at best.

The fate of large phytoplankton species in oceanic waters is of crucial importance to their role in new production. It is becoming increasingly evident that sinking rates of bloom species often are much greater than would be predicted by the simple Stokes settling equation. Takahashi (1986), for example, estimated the sinking rates of a wide assortment of diatoms to be about 175 m/d in the subarctic Pacific. Similarly, the presence of photosynthetically-active phytoplankton cells at thousands of meters can be accounted for only by invoking accelerated sinking rates (Smadya, 1971; Platt *et al.*, 1983). These accelerated sinking rates can only be accounted for by cell aggregation. Smetacek (1985), in fact, has suggested that rapid sinking of diatoms through aggregation is a survival strategy whereby cells switch from a growth phase when nutrients and light both are available to a resting dark stage and also to obtain shelter from grazing when the nutrient supply is exhausted; in this scenario cells not only avoid stress brought on by nutrient limitation, but also act as the seed population when nutrients are brought into the euphotic zone. This scenario dove-tails exactly with the scenario I have proposed earlier (Figure 4). For short-lived oceanic blooms triggered by episodic nutrient inputs, grazing, chiefly by copopods, simply can not keep pace with phytoplankton growth because the resident copopod population is too small and the growth rates of these grazers are low compared to those of the phytoplankton growth rates. If aggregation is a major mechanism by which bloom cells form large masses that can sink rapidly out of the euphotic zone then it is not necessary that the bloom species be large. For example, small coccolithophores which often are the dominant species in oligotrophic waters (Hulbert *et al.*, 1960, Smayda, 1980) frequently appear in sediment traps as aggregated masses of intact cells (Honjo, 1982; Cadee, 1985).

The most dramatic evidence for direct and rapid transport of intact cells to great depth directly following a bloom and the uncoupling between primary and secondary production comes from time series photographs of massive accumulations of diatomaceous material

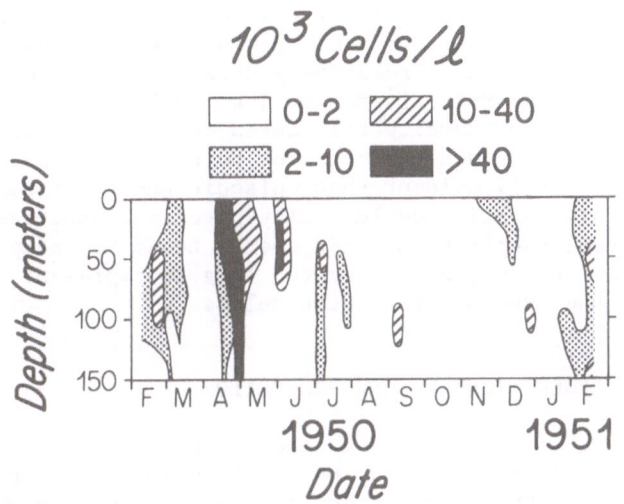

Figure 7. Seasonal and depth distribution of diatoms numbers determined in time series analyses of phytoplankton distributions in the North Central Sargasso Sea during 1950-1952. Modified from Riley (1957). Reproduced in a modified form here by kind permission of the American Society of Limnology and Oceanography, Inc.

at depths between 1300 and 4500 m in the northeast Atlantic over periods of a few days (Billett et al., 1983; Lampitt, 1985; Rice et al., 1986). Accelerated sinking rates of 100 m/d were estimated from the time interval between the bloom and the deposition of the material at the sea floor (Billett et al., 1983). Sinking thus must be a common and major way in which new production exits the euphotic zone. The resulting food chain is short and simple and the $\underline{f}$ ratio during these periods is very low.

Just how important relatively rare events may be to new production can been seen in the long-term (eight year), time-series study by Deuser (in press) of total particulate flux at 3200 m at Station S in the Sargasso Sea. Based on the first two to three years of bi-monthly measurements of particulates collected at the site, Deuser and Ross (1980) and Deuser et al. (1981) found a seasonal cycle in flux that matched closely (with an approximate one month lag) the seasonal trend in ^{14}C-based primary productivity measured earlier by Menzel and Ryther (1961) at the same location. Subsequently, when eight years of data were analyzed the same yearly trends in flux were evident, but there were three sampling periods (one each in 1981, 1982, and 1983) in which unusually high particulate fluxes were recorded. These fluxes were in excess of the averaged data for the entire study (with the three periods excluded). In fact, the average daily flux for the period April 8-May 26, 1981 (117 mg dry wt/m^2/d) was higher by a factor of 2-3 than for any similar time of year during the eight years of measurements. By including the three bi-monthly periods, the eight year average flux (based on 43 individual measurements of accumulated sediment trap material over two month periods) was increased by 8.2%. Since organic material was a relatively constant fraction of total particulates (Deuser, 1986), it can be assumed that the observed trends in the fluxes of total particulates were paralleled by those of biogenic material. Although the cause of these "rare events" has yet to be determined, it is evident that episodic and significant new production at this location could easily have gone unnoticed without adequate long-term sampling and that even a two-month sampling interval in this case may have been too long to represent quantitatively the true nature of the pulses.

5. SCALES OF REGENERATION AND NEW PRODUCTION

A major objective of this paper has been to show that two rather separate biological processes may be involved in new and regenerated production and that the temporal and spatial scales upon which each process functions are very different. The microbial loop dominates when nutrients are scare, which is the condition in any parcel of water in the euphotic zone of pelagic environments most of the time. The new production system, in contrast, may be dominated by large phytoplankton species which are not readily grazed and which occurs only when ephemeral patches of nutrients are introduced into the euphotic zone. The two systems operate side by side and are not, as proposed by Platt and Harrsion (1985), at variance with one another.

The microenvironment is the major domain of the microbial loop (Goldman, 1984a,b; Sieburth, 1984). The microbes in this system most likely lead a feast or famine existence where the bulk fluid is basically void of nutrients and prey. To survive, an individual ultraplankter or protozoan must search for food on temporal scales of seconds to days and over spatial scales of microns to thousands of microns. If this search occurs only in the bulk fluid where all microbes are randomly dispersed then molecular diffusion and viscosity become major physical barriers to success. For example, using a standard molecular diffusion coefficient of 10^3 μm^2/sec and a seawater kinematic viscosity value of 10^6 μm^2/sec, and by assuming random distributions of organisms in the water column, it is easy to demonstrate the limitations of the proposal of McCarthy and Goldman (1979) regarding the importance of rapid nutrient uptake by phytoplankton in the presence of ephemeral point sources of nutrients (Jackson, 1980; Williams and Muir, 1981; Currie, 1984a,b). Yet, if we consider that organic aggregates of varying sizes (from tens to thousands of microns in size) are sites of nutrient enrichment because of their unique surface and bulk chemical properties (more viscous environments and hence lower molecular diffusion rates), and that organisms are attracted to these sites by chemical cues, then the microenvironment becomes an oasis in the desert where much of microbial activity may occur (Goldman, 1984b). Bulk fluid properties under such conditions provide no real clues as to the important processes of nutrient acquisition, predation, and nutrient regeneration.

The "needle in the haystack" scales of the new production system are in marked contrast to the pertinant micro scales of the microbial loop. In this case we are dealing with physical and biological events on spatial scales of meters to hundreds of meters (and greater) and temporal scale of days to weeks, both well within our conceptual frame of reference and both easily measurable. But now the limiting factor is that we must be at the right place at the right time, in essence, we must how know to locate the needle in the haystack. The vastness of the ocean has, in effect, made it impossible until now to sample the rare events that may be a major source of nutrient enrichment of the euphotic zone and the new production that follows (Jenkins and Goldman, 1985). Clearly, Ryther (1969) was correct in assessing the need for short simple food chains to sustain fish production in the oceans. The application of this principle to reconciling the microbial loop contraints on new production with the actual measurements based on integrated O_2 in the water column becomes evident only when the appropriate scales of the two systems depicted in Figure 4 are considered.

6. ACKNOWLEDGEMENTS

This work was supported by Grant No. OCE-8511283 from the National Science Foundation. Contribution No. 6545 from the Woods Hole Oceanographic Institution.

7. REFERENCES

Alldredge, A.L. and Y. Cohen: 1987, 'Can microscale chemical patches persist in the sea? Microelectrode study of marine snow, fecal pellets', Science 235, 689-691.

Alldredge, A.L. and M.W. Silver: 1982, 'Abundance and production rates of floating diatom mats (Rhizosolenia castacanei and R. imbricata var. shrubsolei) in the Eastern Pacific Ocean', Mar. Biol. 66, 83-88.

Azam, F.T., T. Fenchel, J.G. Field, L.A. Meyer-Reil, and F. Thingstad: 1983, 'The ecological role of water-column microbes in the sea', Mar. Ecol. Prog. Ser. 10, 257-263.

Belyayeva, T.V.: 1970, 'Abundance of Ethmodiscus in Pacific plankton', Oceanology 10, 672-675.

Belyayeva, T.V.: 1972, 'Distribution of large diatoms in the southeastern Pacific', Oceanology 12, 400-407.

Bienfang, P.K.: 1985, 'Size structure and sinking rates of various microparticulate constituents in oligotrophic Hawaiian waters', Mar. Ecol. Prog. Ser. 23, 143-151.

Billett, D.S.M., R.S. Lampitt, A.L. Rice, and R.F.C. Mantoura: 1983, 'Seasonal sedimentation of phytoplankton to the deep-sea benthos', Nature 302, 520-522.

Brand, L.E. and R.R.L. Guillard: 1981, 'The effects of continuous light and light intensity on the reproduction rates of twenty-two species of marine phytoplankton', *J. exp. mar. Biol. Ecol.* 50, 119-132.

Cadee, G.C.: 1985, 'Macroaggregates of *Emiliana huxleyi* in sediment traps', *Mar. Ecol. Prog. Ser.* 24, 193-196.

Caron, D.A., P.G. Davis, L.P. Madin, and J. McN. Sieburth: 1982, 'Heterotrophic bacteria and bacterivorous protozoa in oceanic macroparticulates', *Science* 218, 795-797.

Caron, D.A., J.C. Goldman, O. K. Andersen, and M.R. Dennett: 1985, 'Nutrient cycling in a microflagellate food chain: II. Population dynamics and carbon cycling', *Mar. Ecol. Prog. Ser.* 24, 243-254.

Carpenter, E.J., G.R. Harbison, L.P. Madin, N.R. Swanberg, D.C. Biggs, E.M. Hulburt, V.L. McAlister, and J.J. McCarthy: 1977, '*Rhizosolenia* mats', *Limnol. Oceanogr.* 22, 739-741.

Chan, A.T.: 1978, 'Comparative physiological study of marine diatoms and dinoflagellates in relation to irradiance and cell size. I. Growth under continuous light', *J. Phycol.* 14, 396-402.

Clemons, M.J. and C.B. Miller: 1984, 'Blooms of large diatoms in the oceanic, subarctic Pacific', *Deep-Sea Res.* 31, 85-95.

Conover, R.J.: 1982, 'Interrelations between microzooplankton and other plankton organisms', *Ann. Inst. oceanogr. Paris* 58, 31-46.

Currie, D.J.: 1984a, 'Microscale nutrient patches: do they matter to the phytoplankton?', *Limnol. Oceanogr.* 29, 211-214.

Currie, D.J.: 1984b, 'Phytoplankton growth and the microscales nutrient patch hypothesis', *J. Plankton. Res.* 6, 591-599.

Deuser, W.G.: 1986, 'Seasonal and interannual variations in deep-water particle fluxes in the Sargasso Sea and their relation to surface hydrography', *Deep-Sea Res.* 33, 225-246.

Deuser, W.G.: in press, 'Variability of hydrography and particle flux: transient and long-term relationships', *Mitt. Geol. -Palaent. Inst. Univ. Hamburg* 62.

Deuser, W.G. and E.H. Ross: 1980, 'Seasonal change in the flux of organic carbon to the deep Sargasso Sea', *Nature* 283, 364-365.

Deuser, W.G., E.H. Ross, and R.F. Anderson: 1981, 'Seasonality in the supply of sediment to the deep Sargasso Sea and implications for the rapid transfer of matter to the deep ocean', *Deep-Sea Res.* 28, 495-505.

Dugdale, R.C. and J.J. Goering: 1967, 'Uptake of new and regenerated forms of nitrogen in primary production', *Limnol. Oceanogr.* 12, 196-206.

Eppley, R.W.: 1972, 'Temperature and phytoplankton growth in the sea', *Fish. Bull.* 70, 1063-1085.

Eppley, R.W. and B.J. Peterson: 1979, 'Particulate organic matter flux and planktonic new production in the deep ocean', *Nature* 282, 677-680.

Eppley, R.W., P. Koeller, and G.T. Wallace Jr.: 1978, 'Stirring influences the phytoplankton species composition within enclosed columns of coastal sea water', *J. exp. mar. Biol. Ecol.* 32, 219-239.

Eppley, R.W., E.H. Renger, and W.G. Harrison: 1979, 'Nitrate and phytoplankton production in southern California coastal waters', *Limnol. Oceanogr.* 24, 483-494.

Fenchel, T.: 1982, 'Ecology of heterotrophic microflagellates. II. Bioenergetics and growth', *Mar. Ecol. Prog. Ser.* 8, 225-231.

Glibert, P.M.: 1982, 'Regional studies of daily, seasonal and size fraction variability in ammonium remineralization', *Mar. Biol.* 70, 209-222.

Glibert, P.M. and J.J. McCarthy: 1984, 'Uptake and assimilation of ammonium and nitrate by phytoplankton: indices of nutritional status for natural populations', *J. Plankton Res.* 6, 677-697.

Glibert, P.M., J.C. Goldman, and E.J. Carpenter: 1982, 'Seasonal variations in the utilization of ammonium and nitrate by phytoplankton in Vineyard Sound, Massachusetts, USA', *Mar. Biol.* 70, 237-249.

Glover, H.E., A.E. Smith, and L. Shapiro: 1985, 'Diurnal variations in photosynthetic rates: comparison of ultraplankton with a larger size fraction', *J. Plankton Res.* 7, 519-535.

Goldman, J.C.: 1980, 'Physiological processes, nutrient availability, and the concept of relative growth rate in marine phytoplankton ecology', in P.G. Falkowski (ed.), *Primary productivity in the sea*, Plenum Press, N.Y., pp. 179-194.

Goldman, J.C.: 1984a, 'Oceanic nutrient cycles', in M.J. Fasham (ed.), *Flows of energy and materials in marine ecosystems: theory and practice*, Plenum press, N.Y., pp. 137-170.

Goldman, J.C.: 1984b, 'Conceptual role for microaggregates in pelagic surface waters', *Bull. Mar. Sci.* 35, 462-476.

Goldman, J.C.: 1986, 'On phytoplankton growth rates and particulate C:N:P ratios at low light', *Limnol. Oceanogr.* 31, 47-55.

Goldman, J.C. and D.A. Caron: 1985, 'Experimental studies on an omnivorous microflagellate: implications for grazing and nutrient regeneration in the marine microbial food chain', *Deep-Sea Res.* 32, 899-915.

Goldman, J.C., J.J. McCarthy, and D.G. Peavey: 1979, 'Growth rate influence on the chemical composition of phytoplankton in oceanic waters', *Nature* 279, 210-215.

Goldman, J.C., D.A. Caron, O.K. Andersen, and M.R. Dennett: 1985, 'Nutrient cycling in a microflagellate food chains: I. Nitrogen dynamics', *Mar. Ecol. Prog. Ser.* 24, 231-242.

Grenny, W.J., D.A. Bella, and H.C. Curl: 1973, 'A theoretical approach to interspecific competition in phytoplankton communities', *Am. Nat.* 107, 405-425.

Guillard, R.R.L. and P. Kilham: 1977, 'The ecology of marine planktonic diatoms', in D. Werner (ed.), *The biology of diatoms*, University of California Press, Berkeley, pp. 372-469.

Harrison, W.G., D. Douglas, P. Falkowski, G. Rowe, and J. Vidal: 1983, 'Summer nutrient dynamics of the middle Atlantic Bight: nitrogen uptake and regeneration', *J. Plankton Res.* 5, 539-556.

Harrison, W.G., T. Platt, and M.R. Lewis: 1987, '*f*-Ratio and its relationship to ambient nitrate concentration in coastal waters', *J. Plankton Res.* 9, 235-248.

Honjo, S.: 1982, 'Seasonality and interaction of biogenic and lithogenic particulate flux at the Panama Basin', *Science* **218**, 883-884.

Hulburt, E.M., Ryther, J.H., and R.R.L. Guillard: 1960, 'The phytoplankton of the Sargasso Sea off Bermuda', *J. Cons. perm. int. Explor. Mer.* **25**, 115-127.

Iturriaga, R. and B.G. Mitchell: 1986, 'Chroococcoid cynaobacteria: a significant component in the food web dynamics of the open ocean', *Mar. Ecol. Prog. Ser.* **28**, 291-297.

Jackson, G.A.: 1980, 'Phytoplankton growth and zooplankton grazing in oligotrophic oceans', *Nature* **284**, 439-441.

Jenkins, W.J.: 1982, 'Oxygen utilization rates in the North Atlantic Subtropical Gyre and primary production in oligotrophic waters', *Nature* **300**, 246-248.

Jenkins, W.J. and J.C. Goldman: 1985, 'Seasonal oxygen cycling and primary production in the Sargasso Sea', *J. Mar. Res.* **43**, 465-491.

Johnson, K.M., C.M. Burney, and J. McN. Sieburth: 1981, 'Enigmatic marine ecosystem metabolism measured by direct CO_2 and O_2 flux in conjunction with DOC release and uptake'. *Mar. Biol.* **65**. 49-60.

Johnson, P.W. and J. McN. Sieburth: 1979, 'Chroococcoid cynaobacteria in the sea; A ubiquitous and diverse phototropic biomass', *Limnol. Oceanogr.* **24**, 928-935.

Jorgensen, C.B.: 1984, 'Effect of grazing: metazoan suspension feeders', in J.E. Hobbie and P.J.leB. Williams (eds.), *Heterotrophic activity in the sea*, Plenum Press, N.Y., pp. 445-464.

Knauer, G.A., J.H. Martin, and D.M. Karl: 1984, 'Further evidence of a two-layered euphotic zone in oceanic waters', *EOS, Trans. Am. Geophys. Soc.* **65**, 923 (abstr.).

Lampitt, R.S.: 1985, 'Evidence for the seasonal deposition of detritus to the deep-sea floor and its subsequent resuspension', *Deep-Sea Res.* **32**, 885-897.

Laws, E.A.: 1975, 'The importance of respiration losses in controlling the size distribution of marine phytoplankton', *Ecology* **56**, 419-426.

Li, WK.W., D.V. Subba Rao, W.G. Harrison, J.C. Smith, J.J. Cullen, B. Irwin, and T. Platt: 1983, 'Autotrophic picoplankton in the tropical ocean', *Science* **219**, 292-294.

Malone, T.C.: 1971, 'The relative importance of nanoplankton and netplankton as primary producers in tropical oceanic and neritic phytoplankton communities', *Limnol. Oceanogr.* **16**, 633-639.

Margalef, R.: 1978, 'Life-forms of phytoplankton as survival alternatives in an unstable environment', *Oceanologica Acta* **1**, 493-509.

McCarthy, J.J. and J.C. Goldman: 1979, 'Nitrogenous nutrition of marine phytoplankton in nutrient-depleted waters', *Science* **203**, 670-672.

McCave, I.N.: 1984, 'Size spectra and aggregation of suspended particles in the deep ocean', *Deep-Sea Res.* **31**, 329-352.

McHugh, J.L.: 1954, 'Distribution and abundance of the diatom
    Ethmodiscus rex off the west coast of North America', *Deep-Sea
    Res.* 1, 216-223.
Murphy, L.S. and E.M. Haugen: 1985, 'The distribution and abundance of
    phototrophic ultraplankton in the North Atlantic', *Limnol.
    Oceanogr.* 30, 47-58.
Platt, T.: 1984, 'Primary productivity in the central North Pacific:
    comparison of oxygen and carbon fluxes', *Deep-Sea Res.* 31, 1311-
    1319.
Platt, T. and W.G. Harrison: 1985, 'Biogenic fluxes of carbon and
    oxygen in the ocean', *Nature* 318, 55-58.
Platt, T. and W.G. Harrison: 1986, 'Reconciliation of carbon and
    oxygen fluxes in the upper ocean', *Deep-Sea Res.* 33, 273-276.
Platt, T., D.V. Subba Rao, J.C. Smith, W.K. Li, B. Irwin, E.P.W.
    Horne, and D.D. Sameoto: 1983, 'Photosynthetically-competent
    phytoplankton from the aphotic zone of the deep ocean', *Mar.
    Ecol. Prog. Ser.* 10, 105-110.
Pomeroy, L.R.: 1974, 'The ocean's food web, a changing paradigm',
    BioScience 24, 499-504.
Riley, G.A.: 1957, 'Phytoplankton of the North Central Sargasso Sea',
    Limnol. Oceanogr. 2, 252-269.
Riley, G.A.: 1970, 'Particulate organic matter in sea water', *Adv.
    Mar. Biol.* 8, 1-118.
Rice, A.L., D.S.M. Billett, J. Fry, A.W.G. John, R.S. Lampitt, R.F.C.
    Mantoura, and R.J. Morris: 1986, 'Seasonal deposition of
    phytodetritus to the deep-sea floor', *Proc. Roy. Soc. Endinburgh*
    88B, 265-279.
Ryther, J.H.: 1969, Photosynthesis and fish production in the sea',
    Science 166, 72-76.
Semina, H.J.: 1972, 'The size of phytoplankton cells in the Pacific
    Ocean', *Int. Revue ges. Hydrobiol.* 57, 177-205.
Schulenberger, E. and J.L. Reid: 1981, 'Oxygen saturation and carbon
    uptake near 28°N, 155°W', *Deep-Sea Res.* 33, 267-271.
Shanks, A.L., and J.D. Trent: 1979, 'Marine snow: microscale nutrient
    patches', *Limnol. Oceanogr.* 24, 850-854.
Sherr, E.B., B.F. Sheer, and G. Paffenhofer: 1986, 'Phagotrophic
    protozoa as food for metazoans: a "missing" trophic link in
    marine pelagic food webs', *Mar. Microb. Food Webs* 1, 61-80.
Sherr, B.F., E.B. Sherr, and T. Berman: 1983, 'Grazing, growth, and
    ammonium excretion rates of a heterotrophic microflagellate fed
    with four species of bacteria', *Appl. Environ. Microbiol.* 45,
    1196-1201.
Sieburth, J. McN.: 1984, 'Protozoan bacterivory in pelagic marine
    waters', in J.E. Hobbie and P.J.leB. Williams (eds.),
    Heterotrophic activity in the sea, Plenum Press, N.Y., pp. 405-
    444.
Sieburth, J. McN., V. Smetacek, and J. Lenz: 1978, 'Pelagic ecosystem
    structure: heterotrophic compartments and their relationship to
    plankton size fractions', *Limnol. Oceanogr.* 23, 1256-1263.

Silver, M.W., A.L. Shanks, and J.D. Trent: 1978, 'Marine snow: microplankton habitat and source of small-scale patchiness in pelagic populations', *Science* 201, 371-373.

Silver, M.W., M.M. Gowing, and P. Davoll: 1986, 'The association of photosynthetic picoplankton and ultraplankton with pelagic detritus through the water column (0-2000 m)', in T. Platt and W.K.W. Li (eds.), *Photosynthetic picoplankton, Can. Bull. Aquat. Sci.* 214, pp. 311-341.

Smayda, T.J.: 1970, 'The suspension and sinking of phytoplankton in the sea', *Oceanogr. Mar. Biol. Ann. Rev.* 8, 353-414.

Smayda, T.J.: 1971, 'Normal and accelerated sinking of phytoplankton in the sea', *Mar. Geol.* 11, 105-122.

Smayda, T.J.: 1980, 'Phytoplankton species succession', in I. Morris (ed.), *The physiological ecology of phytoplankton*, University of California Press, Berkeley, pp. 493-570.

Smetacek, V.S.: 1985, 'Role of sinking in diatom life-history cycles: ecological, evoluntionary, and geoloical significance', *Mar. Biol.* 84, 239-251.

Steele, J.H. and B.W. Frost: 1977, 'The structure of plankton communities', *Phil. Trans Soc. Lond. B* 280, 485-534.

Strathmann, R.R.: 1967, 'Estimating the organic carbon content of phytoplankton from cell volume or plasma volume', *Limnol. Oceanogr.* 12, 411-418.

Swift, E., M. Stuart, and V. Meunier: 1976, 'The *in situ* growth rates of some deep-living dinoflagellates: *Pyrocystis fusiformis* and *Pyrocystis noctiluca*', *Limnol. Oceanogr.* 21, 418-426.

Takahashi, K.: 1986, 'Seasonal fluxes of pelagic diatoms in the subarctic Pacific, 1982-1983', *Deep-Sea Res.* 33, 1225-1251.

Takahashi, M. and P.K. Bienfang: 1983, 'Size structure of phytoplankton biomass and photosynthesis in subtropical Hawaiian waters', *Mar. Biol.* 76, 203-211.

Trent, J.D., A.L. Shanks, and M.W. Silver: 1978, '*In situ* laboratory measurments on macroscopic aggregates in Monterey Bay, California', *Limnol. Oceanogr.* 23, 626-635.

Venrick, E.L.: 1974, 'The distribution and significance of *Richelia intracellularia* Schmidt in the North Pacific Central Gyre', *Limnol. Oceanogr.* 19, 437-445.

Waterbury, J.B., S.W. Watson, R.R.L. Guillard, and L.E. Brand: 1979, 'Widespread occurance of a unicellular, marine, planktonic, cyanobacterium', *Nature* 277, 293-294.

Waterbury, J.B., S.W. Watson, F.W. Valois, and D.G. Franks: 1986, 'Biological and ecological characterization of the marine unicellular cyanobacterium *Synechococcus*', in T. Platt and W.K.W. Li (eds.), *Photosynthetic picoplankton, Can. Bull. Fish. Aquat. Sci.* 214, pp. 71-120.

Williams, P.J. leB. and L.R. Muir: 1984, 'Diffusion as a constraint on the biological importance of microzones in the sea', in J.C.J. Nihoul (ed.), *Ecohydrodynamics*, Elsevier Publishing Co., Amsterdam, pp.209-218.

PRODUCTION AND DISTRIBUTION OF NAUPLII AND RECRUITMENT
VARIABILITY -- PUTTING THE PIECES TOGETHER

M. M. Mullin
Institute of Marine Resources, A-018
Scripps Institution of Oceanography
University of California, San Diego
La Jolla, California 92093
U.S.A.

ABSTRACT. The natural rate of production of copepod nauplii depends
on adult fecundity and abundance. The former appears to be frequently
food-limited, and this limitation can affect both potential fecundity
(stored reserves of the female) and realized rate of egg production
(reflecting quantity and quality of present food). Naupliar growth
and mortality rates are known for only a few natural marine
populations, and hence there are few direct measurements of secondary
production due to nauplii. Examples from the California Current
illustrate scales of variability in horizontal, vertical, and temporal
distributions of nauplii (particularly *Calanus pacificus*) relative to
the more frequently sampled adults. A significant problem, whose
solution will depend partly on advances in the technology for
identifying age categories of particular species, is to link causally
the variability of populations on several scales of space and time
with the important influences at each scale, be these behavioral or
due to the physical, chemical, or biotic environment, and to connect
variability on small scales to that on large.

1. INTRODUCTION

I start from the proposition that the comparative study of
recruitment, by which I mean the relations between the reproductive
and larval (or early juvenile) ecology of various species and the
sizes and compositions of their adult populations, can facilitate the
understanding of the variations in time and space of these
populations, and therefore of the assemblages, communities, and food
webs of which they are part. The fundamental units of concern in this
perspective are species or populations (not, say, trophic levels), and
the emphasis is on reproduction and the population dynamics of youth.
This proposition has long been a major, utilitarian tenet of fisheries
ecology; it is recognized conceptually by intertidal and benthic
ecologists, but (with significant exceptions) not studied

B. J. Rothschild (ed.), Toward a Theory on Biological-Physical Interactions in the World Ocean, 297–320.
© 1988 by Kluwer Academic Publishers.

quantitatively except at its beginning (fecundity) and end
(metamorphosis and settlement); and it has played a relatively minor
role in studies of marine zooplankton.

In spite of this curmudgeonly final assertion, I will illustrate
the functional and distributional relations between adults and young
in holoplanktonic copepods, particularly *Calanus* off California. I do
so because I am personally familiar with this example (or collection
of examples), and because such data are relatively rare, but I point
out that naupliar copepods are an important---perhaps the most
important---food source for many larval fishes, and so my examples
pertain tangentially to recruitment in commercially important species.
Indeed, several of the examples were originally studied for that
reason.

These illustrations or examples of functional relations and
patterns of distribution represent the bits and pieces of knowledge
which together ought to form an explanation of population variation,
ideally both general enough to have some predictive or extrapolative
power and also specific enough that the mechanisms or causal relations
are explicitly identified. Note, however, the words "ought to"; my
anticlimactic conclusion will be that the tantalizing bits and pieces
remain disturbingly heterogeneous---the task of obtaining in one
situation detailed distributional information on several scales,
measurements of the important rates (reproduction, growth, and
mortality), and the quantitative relations between the distributions
and rates and the physical and biotic environment has proven
insurmountable in practical terms. Our present picture is more
montage than seamless whole. Perhaps, however, the montage is good
enough to guide future research.

2. FUNCTIONAL RELATIONS OF REPRODUCTION, AND SECONDARY PRODUCTIVITY OF NAUPLII

The question of whether (or how often) the rates of production of
marine copepods are food-limited was brought into focus by McLaren
(1978) and others, and has been the subject of considerable
experimental study as our ability to work with living zooplankters has
improved. This question is actually a twofold one, since reproductive
output can be affected both by potential fecundity, reflecting the
females' nutritional history and stored reserves, and realized
fecundity, responding to current availability of food.

The first example, from Hakanson (in press), shows that the
weight and lipid contents of *Calanus pacificus* Copepodite V (the
stage immediately preceeding sexual maturity) in the California
Current vary on a large scale, and that the pattern of variation bears
some relation to a measure of availability of food, the concentration
of chlorophyll (Figure 1). Scatter plots show that although some fat,
heavy copepods were found where the chlorophyll concentration at the
time was <100 mg·m^{-2}, all the animals were fat and heavy where the
concentration was >200 mg·m^{-2} (Figure 2). This result can be linked to
fecundity through, for example, the demonstration by Runge (1984) that

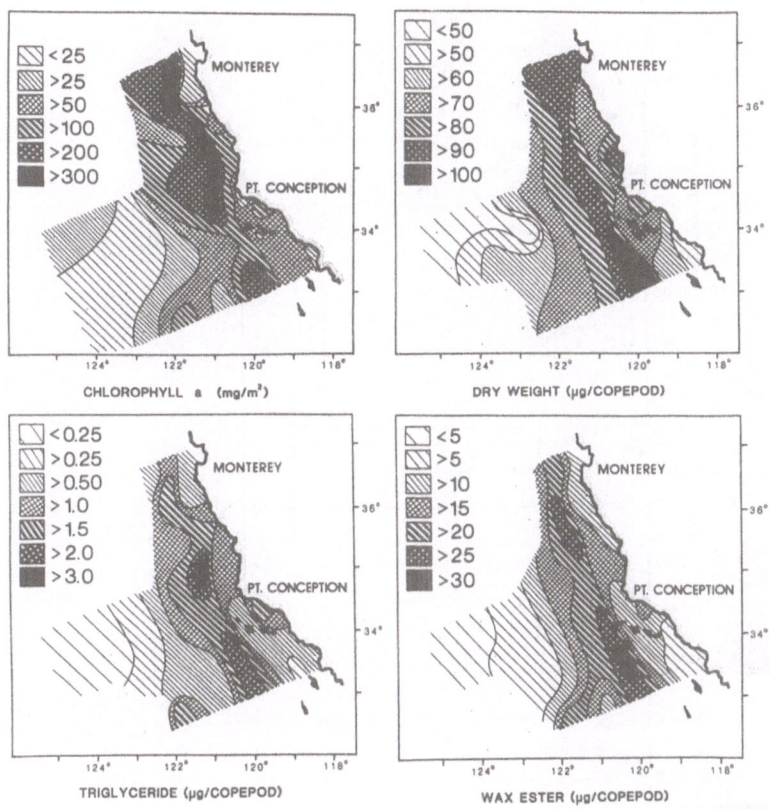

Figure 1. Spatial distributions of chlorophyll and of dry weight
and lipid contents of *Calanus pacificus* CV in the California
Current, April, 1984. From Hakanson (in press).

larger (here, longer) female *Calanus* produce more eggs per clutch in
the presence of abundant food (Figure 3). It may also be the case that
these eggs are of high "quality" (*i.e.* stored reserves per egg), but
this has not been demonstrated in zooplankton to my knowledge.

The immediate effect of food, which Dagg (1977) showed to be more
important for some species than for others, is summarized in Figure 4,
which is an extension of a summary in Runge (1984). The Figure
suggests the expected saturation kinetics (Figure 4A), but also
indicates important interspecific (and even perhaps intraspecific)

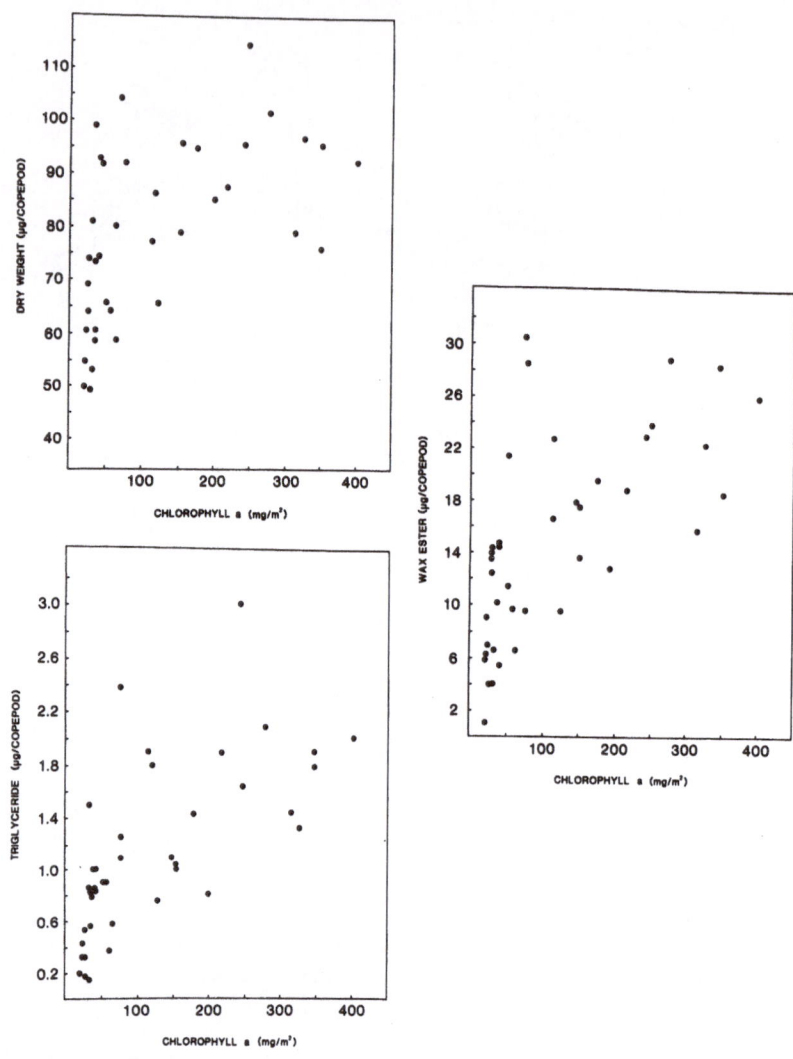

Figure 2. Relations of dry weight and lipid contents of *Calanus pacificus* CV to chlorophyll in the California Current, April, 1984. From Hakanson (in press).

differences and the signficance of food quality (cultured phytoplankton versus natural seston). Other sources of natural variability are related to season---temperature, type of phytoplankton, and vertical distribution of females relative to that of food (Runge, 1985; Peterson, 1985; Ambler, 1986).

The issue of food quality was also addressed by Checkley (1980a), who demonstrated, in addition to the saturation kinetics summarized in Figure 4, the importance of the carbon-to-nitrogen ratio of the food source. As shown in Figure 5A, female *Paracalanus* convert ingested nitrogen to nitrogen in eggs with constant efficiency, but as the relative concentration of nitrogen (and, presumably, protein) in the food decreases, the animals are increasingly wasteful of ingested carbon (Figure 5B). A C:N ratio of about 5 (g:g) seems optimal. Thus, the distribution of nitrogen in phytoplankton can be turned into a map of the degree of food limitation of reproduction in the Southern

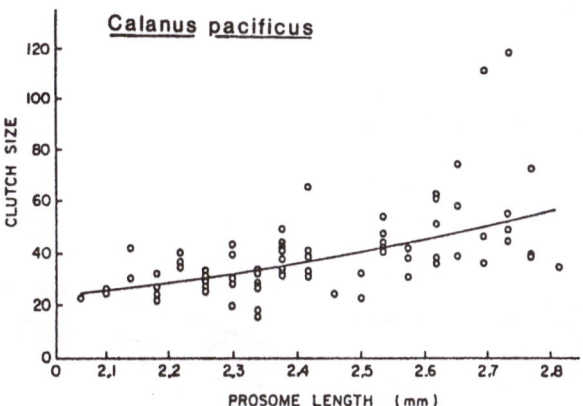

Figure 3. Clutch size as a function of prosomal length of female *Calanus pacificus* in the presence of excess food. From Runge (1984).

California Bight (Figure 6). Ambler's (1986) multifactorial analysis also pointed to the importance of the C:N ratio of the food of *Acartia* in a small lagoon. The study of *Temora stylifera* by Abou Debs (1984) indicated quite low efficiencies of conversion of nitrogen, however (Figure 5A).

An example which impresses me by its thoroughness is from Durbin *et al.* (1983) for *Acartia tonsa*. Figure 7 shows the details (and the usual experimental scatter) of the kinetics summarized in Figure 4, and Figure 8 demonstrates clearly that even at concentrations of chlorophyll exceeding 20 mg·m^{-3}, egg production can be enhanced by the supplementation of natural seston with cultured phytoplankton.

I, at least, am convinced that food limitation of reproduction by marine copepods is quite general, if not universal, and this probably means that secondary production also is so limited. The examples show further that it is important to measure the right property or quality of the food to demonstrate this limitation.

Once naupliar copepods hatch, they pass through six stages before metamorphosing to juvenile copepodids. Most species feed and grow in weight as well as size during this larval ontogeny, though a few species use only stored reserves for growth so there is a net loss of

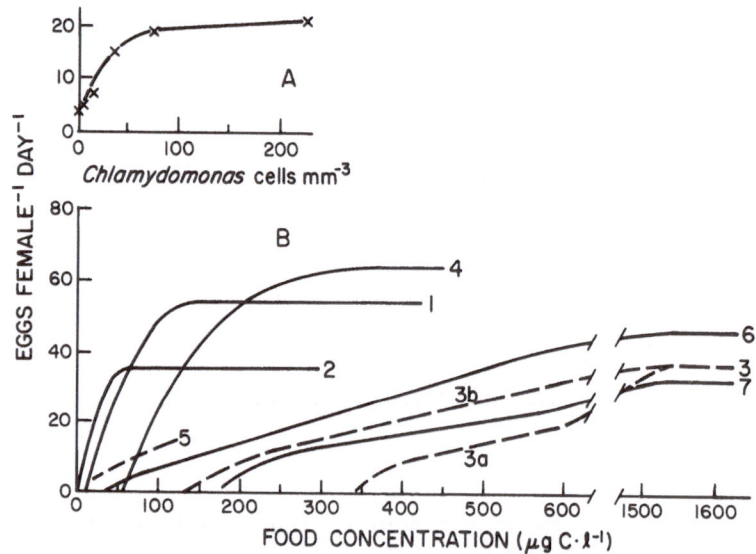

Figure 4. Functional reproductive response of marine copepods to concentration of food at various temperatures. A) Classical result for *Calanus finmarchicus* from Marshall and Orr (1955). B) Recent results for several species (modified from Runge, 1984). Solid lines indicate cultured phytoplankton as food; dashed lines indicate natural seston. 1) *Paracalanus parvus* (Checkley, 1980a); 2) *Acartia clausii* (Uye, 1981); 3) *Acartia tonsa*, a. total seston, b. chlorophyll x 60 (Durbin *et al.*, 1983); 4) *Calanus pacificus* (Runge, 1984); 5) *Acrocalanus inermis* (Kimmerer, 1984); 6) *Acartia tonsa* (Kiørboe *et al.*, 1985); 7) *Centropages typicus* (Smith and Lane, 1985)--Dagg (1978) found that rates up to 200 eggs·female^{-1}·day^{-1} were possible for this species. Peterson's (1985) results for *Temora longicornis* could be fit by line 3b, using the same conversion factor.

organic matter.   Secondary production, when calculated over a finite
time period, is a function of the mean biomass (individual weight
times abundance), the rate of individual growth, and the rate of
mortality (since some individuals disappear before the end of the
period).

There are, to my knowledge, fewer than a dozen published studies
in which the secondary production of nauplii in a natural population
of marine copepods was analyzed carefully and over at least a few
generations (though, of course, if one is willing to assume a
production/biomass ratio from an empirical generalization, one can
then turn any estimate of naupliar biomass into an estimate of
production).   Table I summarizes this set; the studies are categorized
by temperature regime (temperate or tropical) and bodily size of the
species (large or small).   I have expressed naupliar production as a
fraction of the total secondary production of the population so that
populations of different biomasses can be compared, and because there
are a number of other studies where total secondary production, or
production of copepodite stages, has been estimated, and naupliar
production could therefore be derived if a credible generalization
emerged.   Finally, since the eggs produced by the adults represent a
supply of food for larval fish in addition to the production resulting
from the somatic growth of the nauplii, I have tabulated two ratios
where possible.

Table I. Naupliar production, and egg + naupliar production, as
ratios of the population's secondary production in planktonic,
marine copepods.

POPULATION	LOCATION	NAUPLIAR SOMATIC GROWTH / TOTAL POPULATION PRODUCTION	EGG PRODUCTION + NAUPLIAR GROWTH / TOTAL POPULATION PRODUCTION	REFERENCE
Acartia tonsa (small)	Patuxent Estuary, Chesapeake Bay (temperate)	0.71[2]	n.a.	Heinle, 1966
Acartia clausi (small)	Jakle's Lagoon, San Juan Island, Washington (temperate)	0.15	0.53	Landry, 1978
Acartia clausi (small)	Onagawa Bay, Japan (temperate)	0.24	0.55	Uye, pers. comm. based on data in Uye, 1982
Pseudodiaptomus marinus (small)	Inland Sea of Japan (temperate)	0.02	0.52	Uye, pers. comm. based on data in Uye et al, 1983
Acrocalanus inermis (small)	Kaneohe Bay, Hawaii (tropical)	0.20	0.52[1]	Newberry & Bartholomew, 1976
Acrocalanus inermis (small)	Kaneohe Bay, Hawaii (tropical)	n.a.	0.01-0.42	Kimmerer, 1983
Calanus pacificus (large)	Southern California Bight (temperate)	0.01-0.04	0.28-0.57[1]	Mullin & Brooks, 1970
Calanus finmarchicus (large)	Fladen Ground, North Sea (temperate)	<0.10	<0.20	Fransz & Diel, 1985

1) assuming growth rate of copepodites is continued as egg production by females

2) too high, since naupliar biomass was considerably overestimated - Miller, Johnson, and Heinle, 1977

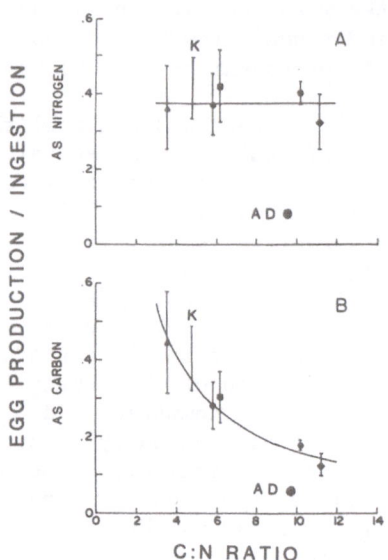

Figure 5. Gross efficiency of egg production (egg mass/ingested mass) as A) nitrogen or B) carbon realtive to C:N ratio of the food. From Checkley (1980a) for *Paracalanus parvus* (vertical lines are 95% confidence limits), Kiørboe *et al.* (1985) (=K) for *Acartia tonsa*, and Abou Debs (1984) (=AD) for *Temora stylifera*. Foods are cultured phytoplankton.

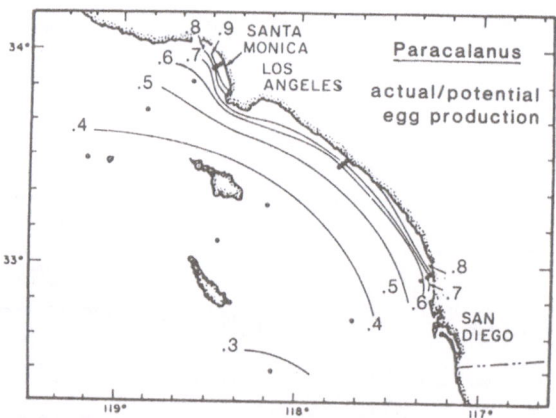

Figure 6. Ratio of actual to potential fecundity of *Paracalanus parvus* in the Southern California Bight (4-year average). From Checkley (1980b).

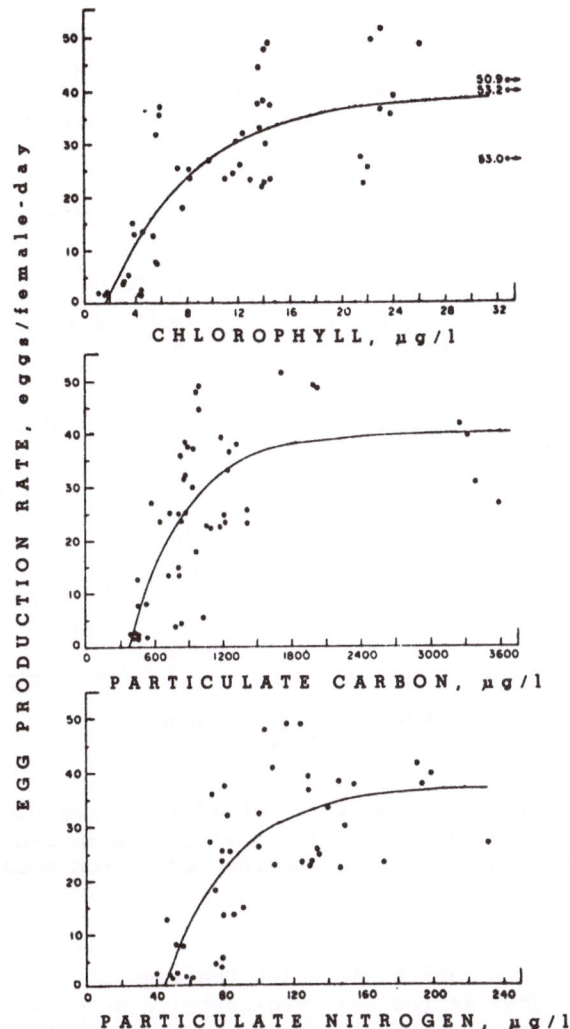

Figure 7. Egg production by female *Acartia tonsa* as functions of various measures of the biomass of natural food in Narragansett Bay. From Durbin *et al.* (1983).

This set of data is small (but hard-won!); <u>none</u> of the examples are oceanic, and most are from semi-enclosed basins. Nevertheless, I conclude that 25% is the likely upper limit to the fraction of total secondary production in a natural, marine population of copepods which is attributable to naupliar somatic growth, and for larger, long-lived species, 10% might be a more reasonable upper limit. Naupliar plus egg production is probably in the range 20-60% of total production.

The ratio of naupliar to total secondary production, though not the absolute magnitudes, is probably most strongly influenced by the age-frequency distribution of the population, which in turn reflects age-specific growth and mortality. It would therefore be useful to have data of the same quality from populations differing in their age-frequency distributions. One might make comparisons between species in which the eggs are carried by the female in a sac and species of

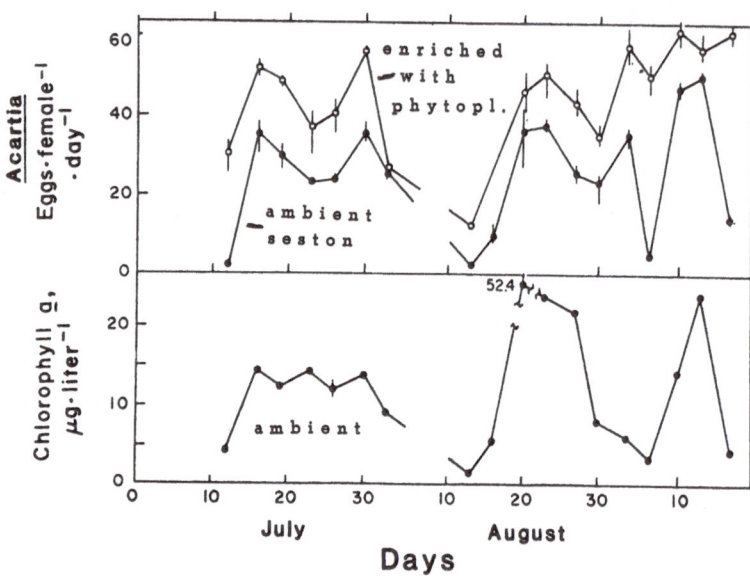

Figure 8. Ambient concentration of chlorophyll in Narragansett Bay for two months, and fecundity of *Acartia tonsa* in the ambient seston or in seston enriched with cultured phytoplankton. From Durbin *et al*. (1983).

similar size from the same habitat in which the eggs are freely broadcast (*Pseudodiaptomus* is the only sac-carrying species tabulated). I would also like to see at least one value for an oceanic population from oligotrophic waters (though I admit that the range of ratios is already so large that it is unlikely an oceanic population would fall outside it).

3. SPATIAL AND TEMPORAL DISTRIBUTIONS ON SEVERAL SCALES

I will now turn to a set of examples concerning the spatial and temporal distributions of nauplii in the California Current and Southern California Bight. Where possible, I will use *Calanus pacificus* as the example, and will compare the distribution of its nauplii to those of juveniles and adults, so that variability (or pattern) in recruitment can be visualized. Again, however, I will piece together studies done at different times, so it is difficult to

determine how patterns on one spatial scale or dimension relate to
patterns on another (but see Figure 14).

The distributional patterns affect the functional relations
discussed above, as implied by Figures 1 though 4---location relative
to food affects fecundity---and the scale is important because spatial
and temporal scales are related. Conditions on the scale traversed by
an individual zooplankter's daily ambit affect its "success" for that
day but are generally ephemeral, both with respect to the individual's
occupancy and to dispersion by turbulence. Conditions on the scale of
a population's range last much longer. The correct way to measure
time, is, of course, in units appropriate to the organisms---time to
starvation, generation time, lifespan, etc.

On a large scale, Figure 9 shows the locations of local "maxima"
(subjectively defined) in abundances of large nauplii of all species
in the upper 70 m, relative to 10-m isotherms and to the distributions
of nearshore water (which contained echinoderm plutei) and oceanic
water (which contained the copepod, *Mecynocera*). The maxima, with one
exception, were in a cool tongue of water near the interface between

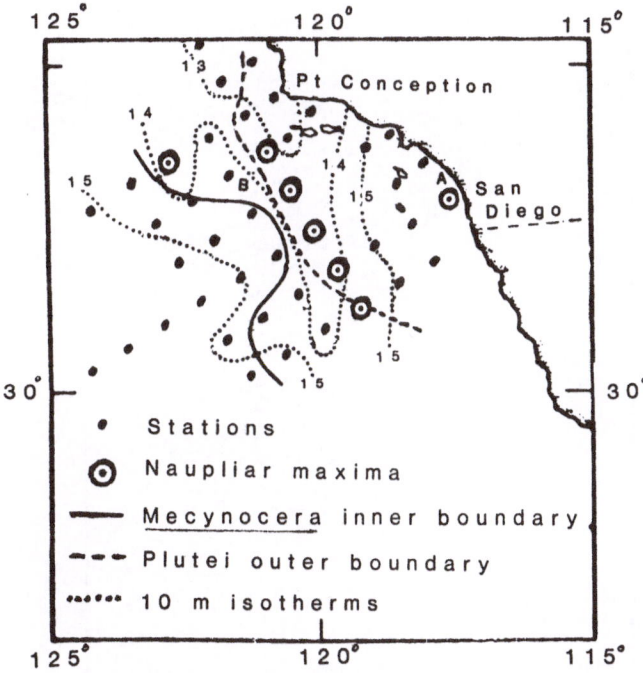

Figure 9. Relative maxima in abundances of naupliar copepods (those
wider than 143 μm isotherms, offshore limit of echinoderm plutei,
and inshore limit of *Mecynocera*. From Arthur (1977). Points A and
B indicate locations of transects shown in Figure 12.

308

the nearshore and oceanic biological indicators. Figure 10 shows that maxima frequently occur well offshore, especially north of Point Conception, resembling somewhat the distribution of the heaviest CV *Calanus* (Figure 1).

Calanus is not equally abundant throughout the year in the Southern California Bight. Figure 11 shows a spring and summer off La Jolla when the entire water columns at two stations were intensively sampled. On overall declines were superimposed a series of peaks of naupliar (NIII-NVI) and female abundances (perhaps causally related through maturation, though the implied development time is quite short and the possibility of advective changes cannot be ruled out). The ratio of nauplii to co-occuring females (a possible measure of recruitment) was also quite variable, but the results at the end of August suggest that a small adult population gave rise to recruitment which was relatively as successful as that in the spring. The concentration of chlorophyll at the stations was greater in late April

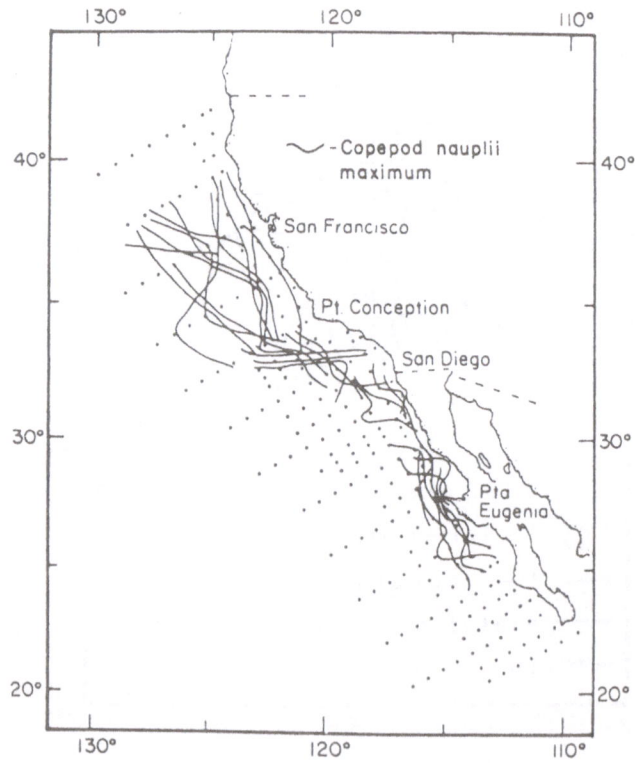

Figure 10. Positions of maxima in naupliar copepod abundances for individual cruises, June, 1949, to July, 1951. From Arthur (1977).

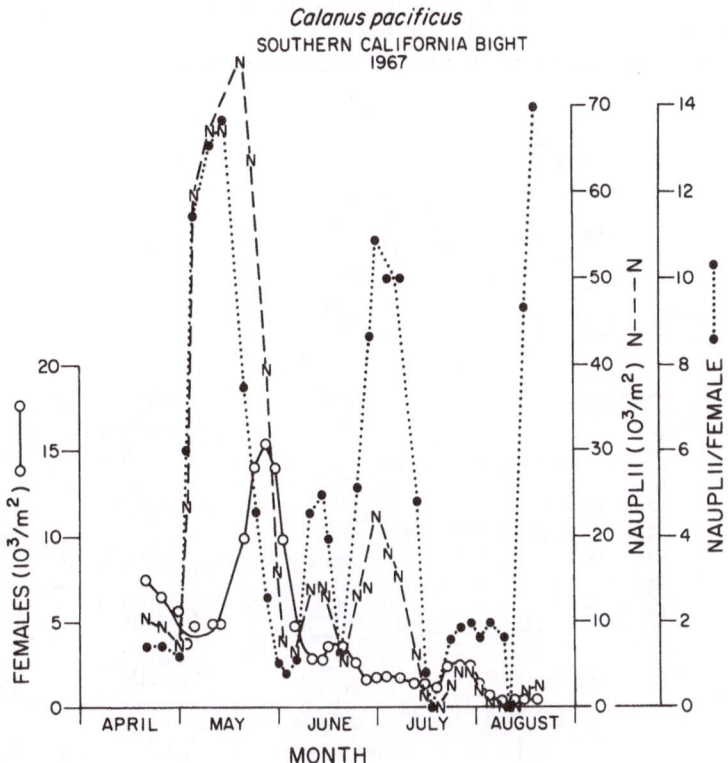

Figure 11. Abundances of naupliar and female *Calanus pacificus* in
the water column off La Jolla, California, in spring-summer, 1967,
based on two stations each sampled twice per date. data points are
3-date running means, and each represents 4-24 values. Data from
Mullin and Brooks (1970).

through mid May than in summer (Eppley *et al.*, 1970), but there was no
indication of increased concentrations in early August which might
have fueled an autumnal burst of recruitment.

On the horizontal scale of tens of km, *Calanus* nauplii do not
necessarily show patterns similar to those of the adults. Figure 12
shows abundances along transects sampled continuously at 35 m by pump,
one transect near the coast in the Southern California Bight and one
in the open California Current (locations A and B, respectively, in
Figure 9). If the data in Figures 11 and 13 were relevant to July of
1977 (the montage problem again!), the transects were sampled after
the maximal abundances of spring and slightly deeper than the optimal
strata for NIII-NVI. In this instance, the young stages were much
more abundant near the coast than offshore (cf. Figure 10).
Significant variation on the scale of a few tens of km is generally
apparent in Figure 12, especially for nauplii and early copepodite

stages, in spite of the sample-to-sample variability on the scale of a
few hundred meters. Indeed, analysis of variance showed that *Calanus*
nauplii differed significantly in abundance between successive nights

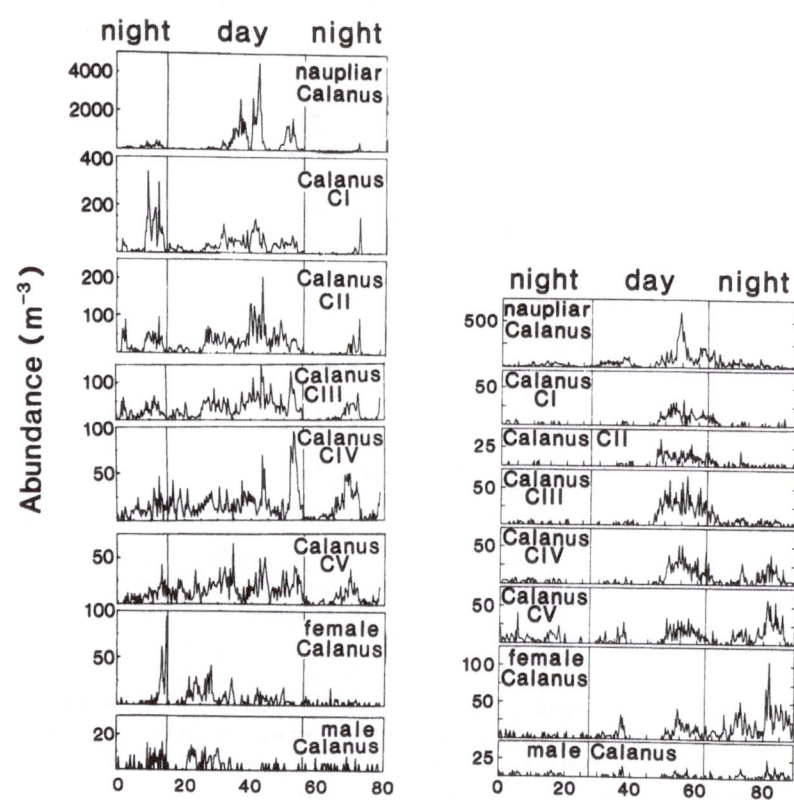

Figure 12. Abundances of developmental stages of *Calanus pacificus*
at 35 m on 80-km transects in late July, 1977. Vertical lines
separate night from day. Data from Star and Mullin (1981). Left,
Nearshore Southern California Bight, location A in Figure 9; right,
California Current, location B in Figure 9.

as well as between night and day, and the autocovariance functions
indicated that characteristic length scales of variability (=
"patches") were generally in excess of 1 km (Star and Mullin, 1981;
see Figure 16). Nearshore, the young stages (nauplii and early
copepodites) were positively correlated in abundance with the
concentration of chlorophyll at scales >1 km, but this was not true
offshore; nor were there correlations with temperature in either
transect (Star and Mullin, 1981).

The impression given by Figure 12 (confirmed by an objective cluster analysis---Star and Mullin, 1981) is of a gradation of patterns from the youngest to the oldest stages, rather than similarity of adults and larvae. This impression is in large part due to the increasing tendency for animals as they mature to migrate into the sampled stratum at night and live in deeper strata by day, rather than inhabiting the surface waters continuously (see Figure 13). However, even at night, the disparity between patterns of females and nauplii is striking.

The vertical distributions of the developmental stages of *Calanus* in the Southern California Bight in spring (Figure 13) conform to expectation for the genus (Marshall and Orr, 1955). The first two naupliar stages, which are non-feeding, occur deeper than later nauplii because eggs sink after being laid near the surface. The feeding naupliar stages (NIII-NVI) occur mainly in the upper 25 m, while the early copepodite stages occur slightly deeper. Diel vertical migration is apparent (and statistically significant) for CVs and adults. Total abundances (m^{-2}) were not significantly different between night and day for any stage except NI-NII, so avoidance of the sampler near the surface by day did not affect the results. There is indirect evidence (Koslow and Ota, 1981) that the migratory pattern of *Calanus* in the Southern California Bight changes seasonally; further, the sampling which led to Figure 13 was not deep enough to capture any deep-living, diapausing late copepodite stages (Alldredge *et al.*, 1984).

Most properties are more variable vertically than horizontally in the ocean, and naupliar *Calanus* are no exception. Within the upper 35 m, a vertical separation of 15 m is approximately equivalent, in terms of the difference in abundances which is typically encountered, to a horizontal separation of 10 km (Figure 14). Even so, quite large differences in abundances can sometimes occur horizontally within a hundred meters or so. Although the dissimilarity of absolute abundances tended to increase as separation between two samples at the same depth increased, the relative shapes of vertical profiles (as measured by percent similarity indices) could not be shown to change linearly with increasing separation.

Even on the scale of cm vertically, there can be differences in concentrations of nauplii which might be important for a predator (Owen, 1981), although nutrients showed little vertical structure (Figure 15). The real significance depends on the persistence of such heterogeneity relative to the searching behavior of the predators. On this scale, persistence probably depends in part on the degree to which nauplii are associated with macroscopic organic aggregates.

As stated earlier, it is difficult to extrapolate the scales and intensities of variability shown in Figures 11, 12 and 13 to the spatial scale of the population of *Calanus pacificus*, and to interannual or interdecadal time scales. If *Calanus* is anomalously rare in a large area or for a few years on average, does this mean

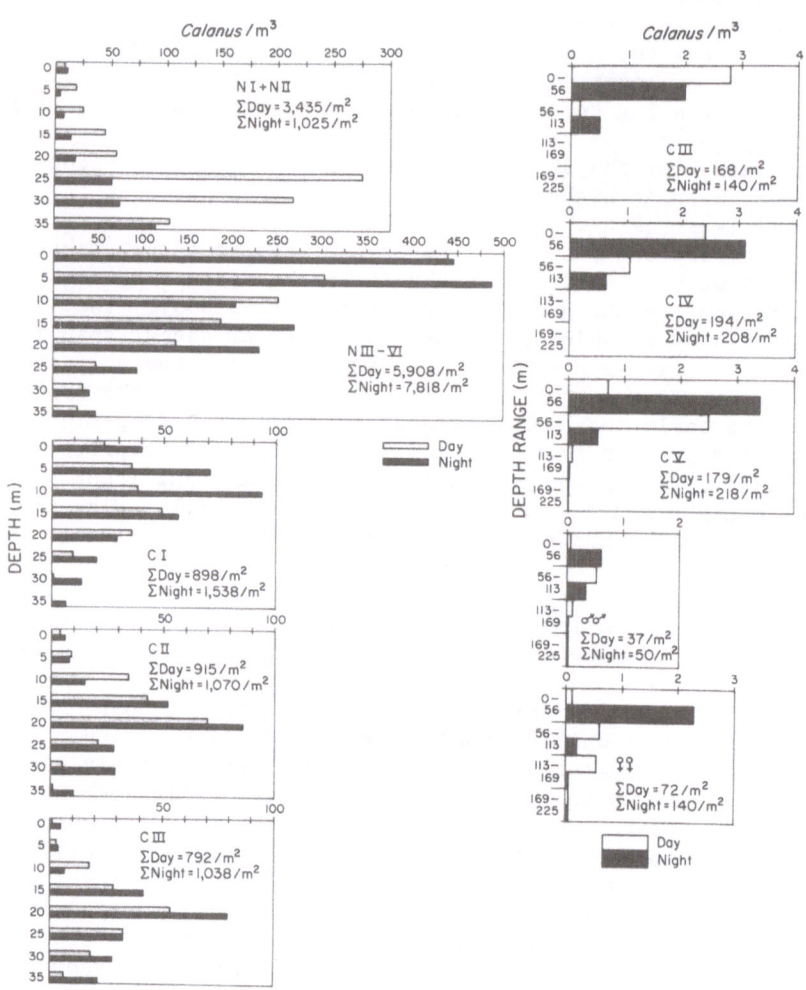

Figure 13. Vertical distributions of developmental stages of *Calanus pacificus* in the Southern California Bight in spring. Left, samples taken by pump at discrete depths in March, 1976; right, samples from 333 μm mesh, opening/closing nets fished obliquely within the indicated depth strata in April, 1965. Note that distributions of Copepodite III are shown twice to connect the 2 data sets. There were at least 8 samples per depth and time. From Mullin (1986).

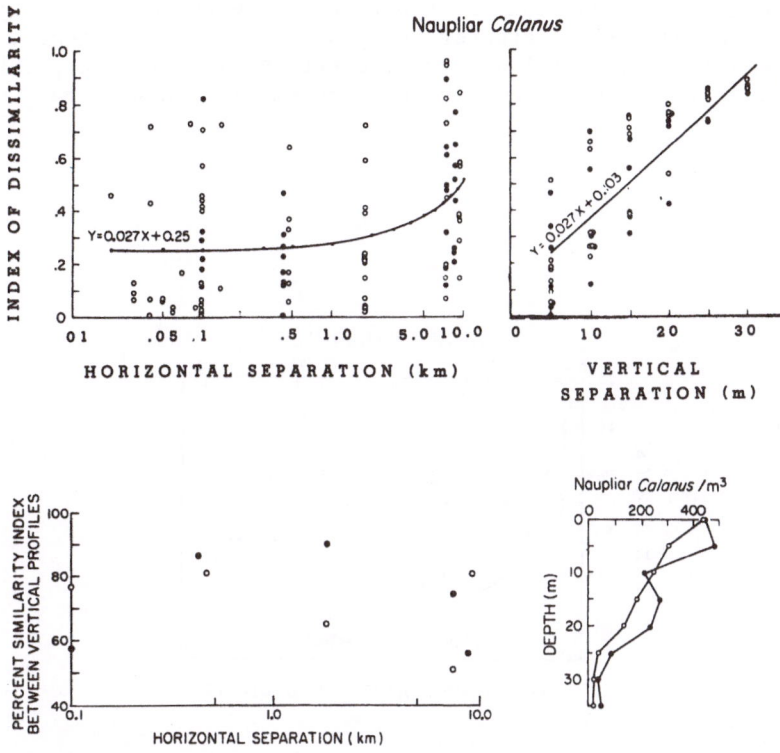

Figure 14. Horizontal and vertical variability in abundance of
naupliar *Calanus pacificus* in March, 1976, by date (open circles)
and at night (filled circles). The index of dissimilarity is
|A - B|/(A + B). Each horizontal separation (note logarithmic
abcissa) represents 2 ships sampling simultaneously at each of 8
depths by day or at night. The vertical profiles of abundance are
means of 10 diurnal and 10 nocturnal profiles over the same 8 depths
(see Figure 13), and give rise to the plot of dissimilarity versus
vertical separation. The percent similarity index is a measure of
the similarity in shape of any two vertical profiles taken
simultaneously, 8 depths per profile. From Mullin (1986).

that the anomaly applies equally to all seasons, or that only the
spring maximum or the dense patches are suppressed? A change in the
physical environment is likely to be an ultimate cause of such an
anomaly---is this also the direct proximate cause, through advection
of individuals, or has a new "carrying capacity" been established
through alteration of the balance of birth, growth, and death?

On the scale of tens of km, variability increases as mean
abundance increases (Figure 16A). Does this also occur when the

abundance of *Calanus* changes significantly throughout its range? If so, variability will be greater in "good" than in "bad" years, though this seems counterintuitive. From the perspective of a zooplanktivore, it is also important whether the characteristic spatial scale of

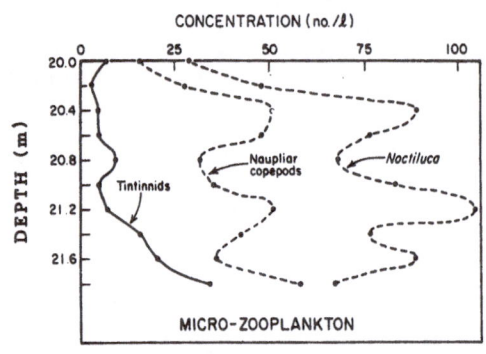

Figure 15. Vertical distributions of naupliar copepods, other micro-zooplankton, and nutrients in 2 locations sampled by an array of 10 simultaneously closing water bottles. From Owen (1981). Depth dimension is 2.0 m.

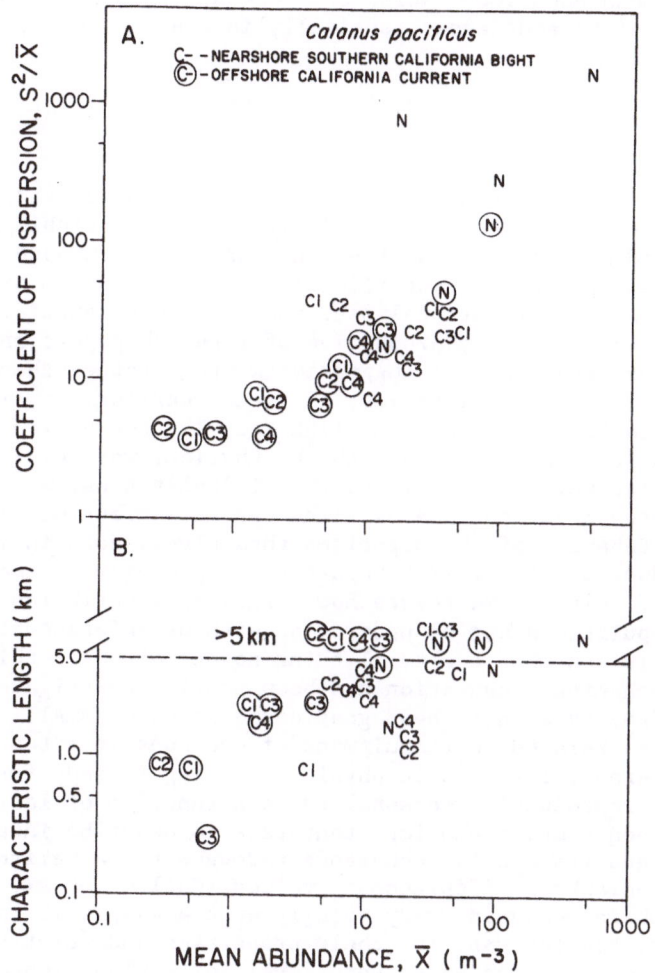

Figure 16. Coefficients of dispersion (variance/mean) and characteristic length scales (km) of *Calanus pacificus* developmental stages (N = nauplii; C1-4 = copepodite stages (I-IV) as functions of mean abundances. Data from Star and Mullin (1981), shown in Figure 12, with nights and days treated separately. The exact values for length scales exceeding 5 km could not be reliably calculated.

variability changes with overall abundance, since the size of patches determines (in part) their persistence, and therefore might affect the searching behavior of predators and variability in nutrition within the predatory population. Figure 16B suggests an increase in characteristic length scale (first zero crossing of the autocovariance function) as mean abundance increases. Whether this would be true for temporal changes in abundance, especially interannual ones, is unknown.

4. PROSPECTS

I have tried to draw together published information to exemplify the attempt to understand (rather than simply measure) variability in recruitment and secondary production in zooplankton through knowledge of the component processes (specifically, food limitation of reproduction) and distributional patterns on several spatial scales. This approach has been very productive of research papers and Ph.D. theses, but has resulted in a suggestive montage rather than a set of completely analyzed case histories, as stated earlier.

It is essential to attempt to link causally the variability in distributions in space and time with the physical variability of the environment, for which (at least to the biologist's naive eye) the connection between various scales---the variance spectrum---is better understood. Behavior of the organisms themselves, both in motility and in reproduction, is clearly important, especially on small scales, but it is difficult for me to see how large-scale variation in planktonic populations can be understood without reference to physical forcing on that scale. I think no one disputes this, but I argue that mechanistic causation has been poorly investigated. Nor, to my knowledge, have there been good comparisons of small- and meso-scale processes related to recruitment of copepods in situations of markedly different, large-scale physical forcing (except for assessment of reproductive seasonality as a function of latitude).

It has been a major and important task to describe interannual variability and its spatial coherence throughout the California Current in zooplankton displacement volume (Chelton *et al.*, 1982)---a property which is measured very quickly once a sample is removed from the sea. Yet, species vary in their properties and relations to environmental change, and hence some level of explanation or hypothesis-testing is not possible without information on species. Further, questions concerning rates of growth and mortality usually require age- or stage-specific information.

These considerations lead to a frustrating dilemma---my conviction that knowledge on a species by species basis is essential (though this does not mean that all species must be equally well understood, nor that only species-specific information is valuable), and my concern that technology has not greatly improved the speed or scale at which we gather such information. Put simply, I am not much more efficient in counting *Calanus* nauplii in a complex, natural sample than were Marshall and Orr (1955). I have been told since I was a first-year graduate student that acoustics was going to solve my problems, but the particular dilemma I have expressed has not, in my opinion, been resolved yet. *In situ* optical devices based on holography or lensless photography are promising, especially for revealing small-scale distributions, but have yet to generate a large body of data. The more extensive data from electronic particle counters can only be interpreted as species where communities are very simple. I see more hope in a combination of species-specific stains or tags (in the broadest sense) and optical devices analogous to cell sorters or image analyzers for processing samples once they have been removed from the sea. (See Dickey (this volume) for a more optimistic view of the utility of *in situ* instruments.)

My own interests have centered recently on indirect estimation of ecologically relevant rates by biochemical techniques---an extension of the "condition factor" measurements used by fisheries biologists in recognition that not all animals of a particular species and age are physiologically equivalent or have the same demographic future. Hakanson's study (Figures 1 and 2) represents a start in this direction, as does work by Cox *et al.* (1983) and Willason *et al.* (1986). I specifically look for a suite of techniques by which to estimate metabolic state or health integrated over different time intervals so that I can see how (or whether) spatial and temporal scales of environmental variation are reflected in those physiological parameters related to demography and production. Further, enough samples must be analyzed so that the variances can be determined. I believe it would be significant, for example, to determine whether the variance in a "long-term" property related to production (*e.g.* lifetime fecundity or lipid contents) is greater than or less than the variance in a "short-term" measure (*e.g.* daily fecundity or ATP or energy charge) and the variance in food supply (*e.g.* chlorophyll or primary production) on the same spatial scales, as well as determining whether or not the means are correlated. One could then judge whether short-term variability was mollified or exacerbated ("the rich get richer") over time.

5. REFERENCES CITED

Abou Debs, C.: 1984, 'Carbon and nitrogen budget of the calanoid copepod *Temora stylifera*: Effect of concentration and composition of food', *Mar. Ecol. Prog. Ser* 15, 213-223.

Alldredge, A. L., B. H. Robison, A. Fleminger, J. J. Torres, J. M. King, and W. M. Hamner: 1984, 'Direct sampling and *in situ* observation of a persistent copepod aggregation in the mesopelagic zone of the Santa Barbara Basin', *Mar. Biol.* 80, 75-81.

Ambler, J.: 1986, 'Effect of food quantity and quality on egg production of *Acartia tonsa* Dana from East Lagoon, Galveston, Texas', *Estuar. Coastal Shelf Sci.* 23, 183-196.

Arthur, D. K.: 1977, 'Distribution, size, and abundance of microcopepods in the California Current system and their possible influence on survival of marine teleost larvae', *Fish. Bull.* 75, 601-612.

318

Checkley, D. M., Jr.: 1980a, 'The egg production of a marine planktonic copepod in relation to its food supply: Laboratory studies', *Limnol. Oceanogr.* **25**, 430-446.

Checkley, D.M., Jr.: 1980b, 'Food limitation of egg production by a marine, planktonic copepod in the sea off Southern California', *Limnol. Oceanogr.* **25**, 991-998.

Chelton, D. B., P. A. Bernal, and J. A. McGowan: 1982, 'Large-scale interannual physical and biological interaction in the California Current', *J. Mar. Res.* **40**, 1095-1125.

Cox, J. L., S. Willason, and L. Harding: 1983, 'Consequences of distributional heterogeneity of *Calanus pacificus* grazing', *Bull. Mar. Sci.* **33**, 213-226.

Dagg, M.: 1977, 'Some effects of patchy food environments on copepods', *Limnol. Oceanogr.* **22**, 99-107.

Dagg, M.: 1978, °Estimated, *in situ*, rates of egg production for the copepod Centropages typicus (Kroyer) in the New York Bight', *J. Exp. Mar. Biol. Ecol.* **34**, 183-196.

Dickey, T.D.: 1988, '
in B.J. Rothschild, ed., *Toward a Theory on Biological-Physical Interactions in the World Ocean*, Kluwer Academic Publishers, Dordrecht, pp.

Durbin, E. G., A. G. Durbin, T. J. Smayda, and P. G. Verity: 1983, 'Food limitation of production by adult *Acartia tonsa* in Narragansett Bay, Rhode Island', *Limnol. Oceanogr.* **28**, 1199-1213.

Eppley, R. W., F. M. H. Reid, and J. D. H. Strickland: 1970, 'Estimates of phytoplankton crop size, growth rate, and primary production', in J. D. H. Strickland (ed.), *The Ecology of the Plankton off La Jolla, California, in the Period April through September, 1967*, Bull. Scripps Inst. Oceanogr. **17**, 33-72.

Fransz, H. G., and S. Diel: 1985, 'Secondary production of *Calanus finmarchicus* (Copepoda: Calanoida) in a transitional system of the Fladen Ground area (northern North Sea) during the spring of 1983', in P. E. Gibbs (Ed.), *Proc. 19th European Mar. Biol. Symp.* Cambridge Univ., pp. 123-133

Hakanson, J. L.: 1987, 'The feeding condition of *Calanus pacificus* and other zooplankton in relation to phytoplankton pigments in the California Current', *Limnol. Oceanogr.* (in press)

Kimmerer, W. J.: 1983, °Direct measurement of the production:biomass ratio of the subtropical calanoid copepod *Acrocalanus inermis*', *J. Plankton Res.* **5**, 1-14.

Kimmerer, W. J.: 1984, 'Spatial and temporal variability in egg production rates of the calanoid copepod *Acrocalanus inermis*', *Mar. Biol.* **78**, 165-170.

Kiørboe, T., F. Mohlenberg, and K. Hamburger: 1985, 'Bioenergetics of the planktonic copepod *Acartia tonsa*: Relation between feeding, egg production and respiration, and composition of specific dynamic action', *Mar. Ecol. Prog. Ser.* **26**, 85-97.

Koslow, J. A., and A. Ota: 1981, 'The ecology of vertical migration in three common zooplankters in the La Jolla Bight, April-August, 1967', *Biol. Oceanogr.* **1**, 107-134.

Landry, M. R.: 1978, 'Population dynamics and production of a planktonic marine copepod, *Acartia clausii*, in a small temperate lagoon on San Juan Island, Washington', *Intern. Rev. ges Hydrobiol.* **63**, 77-119.

Marshall, S. M., and A. P. Orr: 1955, *The Biology of a Marine Copepod*, Oliver & Boyd, Edinburgh.

McLaren, I. A.: 1978, 'Generation lengths of some temperate marine copepods: Estimation, prediction, and implications', *J. Fish. Res. Bd. Canada* **35**, 1330-1342.

Mullin, M. M.: 1986, 'Spatial and temporal patterns', in R. W. Eppley (ed.), *Plankton Dynamics of the Southern California Bight*, Springer-Verlag, Berlin, pp. 216-273

Mullin, M.M.: and E. R. Brooks: 1970, 'Production of the planktonic copepod, *Calanus helgolandicus*', in J. D. H. Strickland (ed.), *The Ecology of the Plankton off La Jolla, California, in the period April through September, 1967, Bull. Scripps Inst. Oceanography* **17**, 89-103.

Newbury, T. K., and E. F. Bartholomew: 1976, 'Secondary production of microcopepods in the southern, eutrophic basin of Kaneohe Bay, Oahu, Hawaiian Islands', *Pacific Sci.* **30**, 373-384.

Owen, R. W.: 1981, 'Patterning of flow and organisms in the larval anchovy environment', in G. D. Sharp (ed.), *Workshop on the Effects of Environmental Variation on the Survival of Larval Pelagic Fishes*, Intergov. Oceanogr. Comm. Workshop Rept. No. 28., pp. 167-200.

Peterson, W. T.: 1985, 'Abundance, age structure, and *in situ* egg production rates of the copepod *Temora longicornis* in Long Island Sound, New York', *Bull. Mar. Sci.* **37**, 726-738.

Runge, J. A.: 1984, 'Egg production of the marine, planktonic copepod, *Calanus pacificus* Brodsky: Laboratory observations', *J. Exp. Mar. Biol. Ecol.* **74**, 53-66.

Runge, J. A.: 1985, 'Relationship of egg production of *Calanus pacificus* to seasonal changes in phytoplankton availability in Puget Sound, Washington', *Limnol. Oceanogr.* **30**, 382-396.

Smith, S. L., and P. V. Z. Lane: 1985, 'Laboratory studies of the marine copepod Centropages typicus: Egg production and development rates', *Mar. Biol.* **85**, 153-162.

Star, J. L., and M. M. Mullin: 1981, 'Zooplanktonic assemblages in three areas of the North Pacific as revealed by continuous horizontal transects', *Deep-Sea Res.* **28A**, 1303-1322.

Uye, S.: 1981, 'Fecundity studies of neritic copepods *Acartia clausi* Giesbrecht and *A. steuri* Smirnov: A simple model of daily egg production', *J. Exp. Mar. Biol. Ecol.* **50**, 255-271.

Uye, S.: 1982, 'Population dynamics and production of *Acartia clausi* Giesdbrecht (Copepoda: Calanoida) in inlet waters', *J. Exp. Mar. Biol. Ecol.* **57**, 55-83.

Uye, S.: Y. Iwai, and S. Kasahara: 1983, 'Growth and production of the inshore marine copepod *Pseudodiaptomus marinus* in the central part of the Inland Sea of Japan', *Mar. Biol.* 73, 91-98.

Willason, S. W., J. Favuzzi, and J. L. Cox: 1986, 'Patchiness and nutritional condition of zooplankton in the California Current', *Fish. Bull.* 84, 157-176.

MODELLING OF THE RECRUITMENT OF MARINE SPECIES

P. Nival, F. Carlotti and A. Sciandra
Station Zoologique
Villefranche sur mer
France

ABSTRACT. The flux of young adults to the parent stock is controlled by various biological processes. The processes include the trophodynamics of each individual (feeding, excretion, metabolism, growth), as well as phenomena at the population level (predation, death rate). All these processes acting in combination determine the probability of mortality of a given individual in a cohort. The various forcing functions to these processes need to be considered to understand the evolution over time of the probability of mortality and its cumulative effects on the growth rate of the species.
Mathematical models permit one to evaluate the part played by each variable in the total mortality of a cohort, as soon as the basic information on which the theoretical structure is built is pertinent and informative enough.
A model of larval dynamics which takes into account larval age-and-mass budget to estimate the transfer rate from one stage to the other, can simulate some experimental results. It is used to estimate the effect of perturbations in the food concentration. Starvation can reduce the recruitment of larval stage 4 but the phase between hatching of larvae and the perturbation is very important if predation by a selective predators occurs.
The match between hatching, food abundance, and the mismatch with the predators are the conditions which determine the success of a cohort development.

1. INTRODUCTION

Many factors influence the dynamics of the diverse components of the pelagic ecosystem. Some of them are endogenous, others, outside the system, are usually considered as forcing variables.

Models of the system can be simple, complex or detailed. Most models that include physical, chemical and biological processes are based on a very simple picture of the biological variables of the ecosystem. Some models with a rather complicated structure can stimulate observations made at sea (Radach, 1982; Kremer and Nixon, 1978; Wroblewski, 1980; Platt *et al.*, 1981). Others take into account

B. J. Rothschild (ed.), Toward a Theory on Biological-Physical Interactions in the World Ocean, 321–342.
© 1988 by Kluwer Academic Publishers.

322

as many biological details as possible at the present state of our
knowledge of the processes. They are designed to simulate a
restricted part of the planktonic system (Lehman, 1976; Steele and
Frost, 1977).

The mathematical structure, which translates the picture into
mathematical relationships, implies that all the relations suspected
between variables are known, or, at least, that it is possible to
suggest the type of function which gives the known properties.

For different reasons, many models give a very crude
representation of biological variables. First, it is not necessary to
design a complicated structure for the biological part of a model
which is simple in its chemical or physical parts; second, the time-
and-space scales of the phenomena which are to be simulated do not
need to take into account complicated mechanisms if the input to these
mechanisms are constant; and third, the measurements of the variables
with which the output of the model are to be compared are highly
aggregated variables.

The first and second points can be generally accepted, but the
third must be considered carefully. As an example, consider the
variable "phytoplankton." Usually the phytoplankton density is
measured by chlorophyll concentration or nitrogen content. This
easily measured variable integrates a great variety of organisms with
different characteristics and behaviors. Depending on the "filter"
used to collect particulate matter, a selection of sizes is made.
This observation is also valid for such a variable as "zooplankton",
which may be based on measurements of total dry weight of organisms
collected by a plankton tow. A single relation between a variable
"phytoplankton" and a variable "herbivorous zooplankton" can be
misleading because the trophic relations are obscured (Figure 1). The
importance of the underlying structure of these variables must be
carefully evaluated when the relevant structure of a model is set (see
also Andersen, 1985). In order to take into account the difference in
growth-and-mortality between larval stages and adults of each species,
it is necessary to consider the developmental stages of each species
(cf. Rothschild 1986:Chapter 8).

As another example, the input of larvae of benthic organisms into
the water column often results in a sudden change in the total
planktonic population. These larvae, representing new herbivores and
also new prey for carnivores, disappear from the pelagic community at
the end of their larval development with a suddenness comparable to
their appearance. An oversimplified model could not simulate the
correct transient behavior of the natural system.

Another reason to take into account the complexity level of
developmental stages in models of pelagic ecosystems is the need to
estimate the variations in the recruitment of adults in marine species
exploitable by man. Moreover the understanding of stability of prey-
predator relations which shape the behavior of the whole pelagic
ecosystem, might depend strongly on the larval stage dynamic. The
variations in time and space of a given species might be influenced by
the relation not only of the adult, but also of the different
developmental stages to food, predators or physical conditions

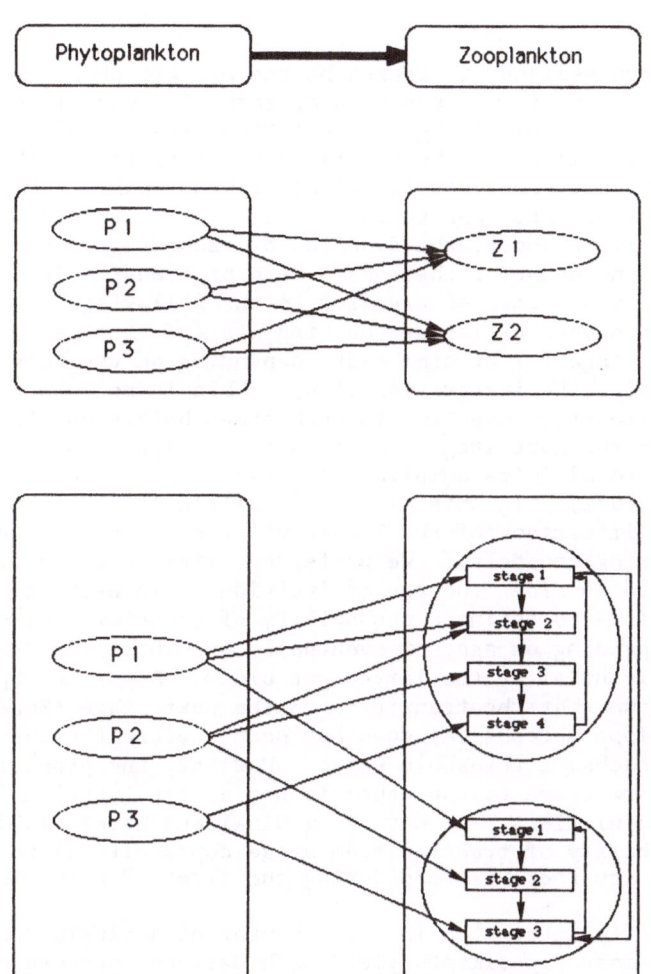

Figure 1. Different representations of two trophic levels in
models. (A) Species of phytoplankton and zooplankton are not
considered, but only biomass. This is the simplest representation
which ignores modifications of the transfer related to the species
characteristics. (B) The major species are considered as different
state variables. (C) A more complex picture of trophic relations
results from introduction of population dynamics of animal species.
This last picture is necessary to understand the variation in the
recruitment of one species. Some relations between species can be
specified only if this level of complexity is taken into account.

(Boucher, 1984; 1988). As it is easier to identify and count adults
of marine species than larvae, the data on larval distribution at sea
are scarce, in spite of the great number of experimental works done on
larvae.

2.   THE MODEL

The model we have designed to represent the pelagic phase, in the life
cycle of a marine species is based on differential equations (at least
one for each life stage) giving the variation rate of individuals.
The structure is similar to the models of Sciandra (1982, 1986) and
Davis (1984) who have improved the model of Wroblewski (1980).   Some
of the processes are represented in the same way as Andersen and Nival
(1986).   In a simple model, the dynamics of each stage would be
represented by one equation that has a term of transfer from the
preceding stage and a term of transfer to the following.   In this
case, a certain number of larvae reaching stage I at time t, would be
transferred to stage I+1 at time t+dt, depending on the rates of
transfer (dt being the integration step).   This is not what is usually
observed.   Larvae stay some time in each stage before moulting and
transferring to the next stage.   Some models designed for insects
include delays to minimize complicated formulations (Manly, 1974;
Blythe *et al.*, 1982).   In this case it is difficult to take into
account the modifications of the larvae related to changes in the
environment during the delay.   We preferred, like Sciandra (1982) and
Davis (1984), to consider the age of individuals in each stage.   It is
therefore possible to define a probability of transfer from one stage
to the next depending on age, or eventually on other processes which
may become influential as the larvae got older.   Anger and Spindler
(1987) have shown that the transition to the next stage (Zoea I to
Zoea II of *Menippe mercenaria*) does not occur before the concentration
of a hormone reaches a threshold value.   However, the probability of
transfer from one stage to the other is not a step function in terms
of the whole population.   The data from Nival and Nival (1983) show
that the probability of transfer from stage copepodite C4 to C5 of the
copepod *Temora stylifera* is zero during the first 12 h and increases
afterwards.

   We are mostly interested in the behavior of a single cohort of
larvae because most experiments yield such data and because many
marine species are semelparous so that their larvae belong
approximately to the same cohort.

   Blythe *et al.* (1985) and Gurney and Nisbet (1985) give examples
of sophisticated models of population dynamics used to study the
stability and dynamics of discrete generations.   Neither the
simplifications they adopt, nor their use of lag time are convenient
for our purpose, which is to estimate the effect of variable
environment on recruitment.

   The model we have designed, has two types of differential
equations.   One set gives the variation per unit time of the numbers
of larvae in each age class and each stage.   The other one gives the
variation per unit time of the weight of the larvae.   Each stage is
divided into 11 age classes (Figure 2).   The 11th age class
corresponds to larvae which do not proceed to a next stage.   The
numerical integration of the equations using the method of Runge Kutta

Stage i   Stage i+1

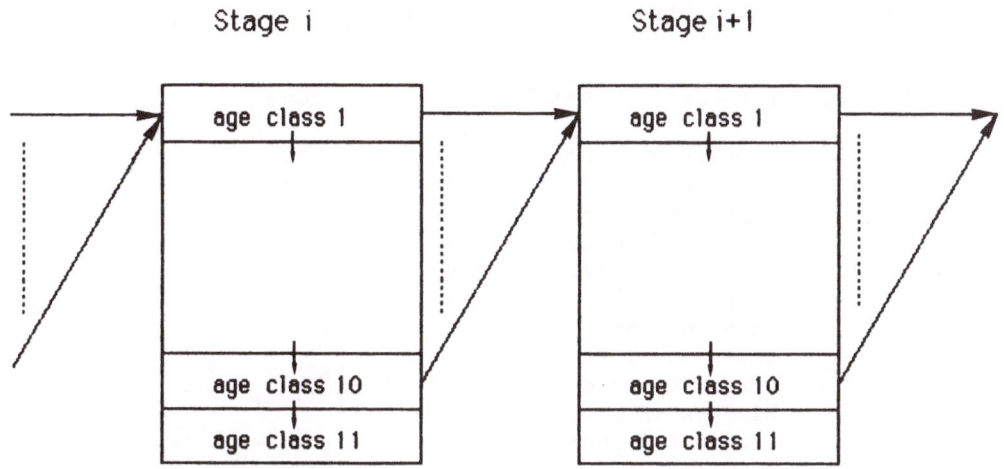

Figure 2. Conceptual framework of the model: two of the four
stages of the model are represented. Early stage is divided into 11
age classes. Every day, the larvae in one age class are shifted
into the next. At any time a proportion of larvae of an age class
can reach the age class of next stage. All the larvae of age class
which have not been transferred to the next larval stage are assumed
to be in bad condition and are stored in the 11th age class.

gives the time variation of weight and numbers for the different
stages.
        Tables 1 and 2 respectively, give the processes and equations of
the model. Details of the model structure can be found in Carlotti
(1986, 1987). Figure 3 shows the connection of the different
processes which modify transfer rate and mortality rate. We consider
that the number of individuals transfered from one stage to the other
depends on food, predators, mortality and growth rate. We consider
that an individual at any stage of its development has some
probability of dying, changing stage or being eaten by a predator.

1.   The probability of dying (M) depends on the amount of food
        ingested and available for growth. From a high value when
        the animal is starved, it declines to relatively low
        constant value when it is well fed.
2.   The probability of being eaten by a predator (P) depends on
        the abundance of predators and on the efficiency with which
        such predators catch each larval stage. For instance,
        Pennington and Chia (1984) show that the ability of the
        trochophore larvae of *Sabellaria cementarium* to expand
        larval setae can reduce the predation of some species on
        this development stage, compared to the previous one which
        has no setae.

Table 1. Mathematical formulation of the biological processes considered in the model.

Process Represented in Model		*Construct*
L	Biomass of larvae	
B	*Mass budget*	$B = I - E_g - E_x$
I	Ingestion rate	$I = b \cdot B_p \cdot W^{0.75}$
b	Filtrated volume	
B_p	Biomass of phytoplankton	
W	Weight	
E_g	Egestion rate	$E_g = 0.3 \cdot I$
E_x	Excretion rate	$E_x = 0.1 \cdot W$
P	*Predation rate*	$P = Pr \cdot Ip \cdot E/L$
Pr	Biomass of predators	
I_p	Predator ingestion rate	$I_p = I_{pmax} \cdot (L - L_{min}) / (K_1 + (L - L_{min}))$
		$I_p = 0$ if $L < L_{min}$
I_{pmax}	Maximum ingestion rate	
L_{min}	Threshold of larval biomass	
K_1	Half saturation constant	
E_c	Capture efficacity	
M	*Mortality rate*	$M = M_{max}$ $B < B_{min}$
		$M = M_{min} + a_m / (B - b_m)$ $B > B_{min}$
M_{max}	Maximum mortality rate	
M_{min}	Minimum mortality rate	
B_{min}	Threshold of mass budget	
a_m, b_m	Shape factors	
T	*Transfer rate*	$T = T_m \cdot f(S_b) \cdot g(W)$
Tm	Maximum transfer rate	
f(Sb)	Function of mass budget	$f(S_b) = S_b / (S_b + C_b)$
Sb	Mean mass budget over the last 5 hours	
Cb	Half saturation constant	
g(W)	Function of weight	$g(W) = W^k / (W^k + c_w k)$
W	Weight	
k	Shape factor	
C_w	Specific weight for half maximum transfer rate	

3. The probability of changing stage (transfer rate T) depends first on the mass budget of the animal, that is, to some extent, on the ingestion rate but also on the respiration or excretion rate, and on the weight which depends on the growth. Figure 4 shows how varies T in the space of mass budget and weight. We assume that this probability varies at two different time scales:
  -    the scale of hours (effect of short term variations in food abundance). This effect should be damped to take

Table 2. Equations giving the rates of variation in number (L) and weight (W). M: mortality rate; T: transfer rate; P: predation rate; Bp: phytoplankton concentration.

Equations for abundance:

$$\frac{dL(1,1)}{dt} = (-T(1,1)-M(1,1)-P(1,1)\cdot L(1,1)$$

$$\frac{dL(1,J)}{dt} = (-T(1,J)-M(1,J)-P(1,J))\cdot L(1,J) + \sum_{i=1}^{10} T(1,J-1)\cdot L(I,J-1)$$

$$\frac{dL(I,J)}{dt} = (-T(I,J)-M(I,J))-P(I,J)\cdot L(I,J)$$

$$\frac{dL(11,J)}{dt} = (-M(I,J)-P(I,J))\cdot L(I,J)$$

Equations for weight:

$$\frac{dW(I,J)}{dt} = ((0.7)\cdot b\cdot B_p\cdot W(I,J)^{0.75}) - ((0.1)\cdot W(I,J))$$

I:age classes (1 to 11)
J:stages (1 to 4)

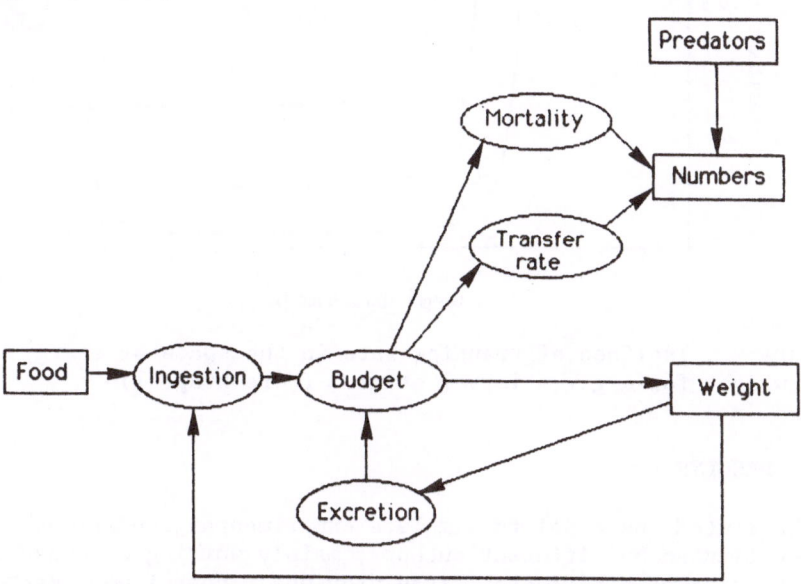

Figure 3. Relationship between processes. Food concentration affects numbers of organisms and their weight through series of steps, each one of them depending on others.

into account food storage in larvae that can smooth the
variability of resources.
the scale of a few days to a week to take into account
the cumulative effect of growth, or the building up of
some substances (hormones) that trigger morphological
changes.

Table 3 gives the coefficient used in the reference simulation
(Figure 5.A). Figure 5.B, 5.C and 5.D show some simulations of cohort
evolution for different values of the coefficients Cb, Cw and k which
shape the transfer rate T. When Cb increases (Figure 5.B1 to 5.B2) the
duration of stages is longer; k affects the shape of the abundance
curve in its early part: when k is high the slope is steeper
(Figure 5.C1 to 5.C2). Cw translates the position of the curves
(Figure 5.D1 to 5.D2).

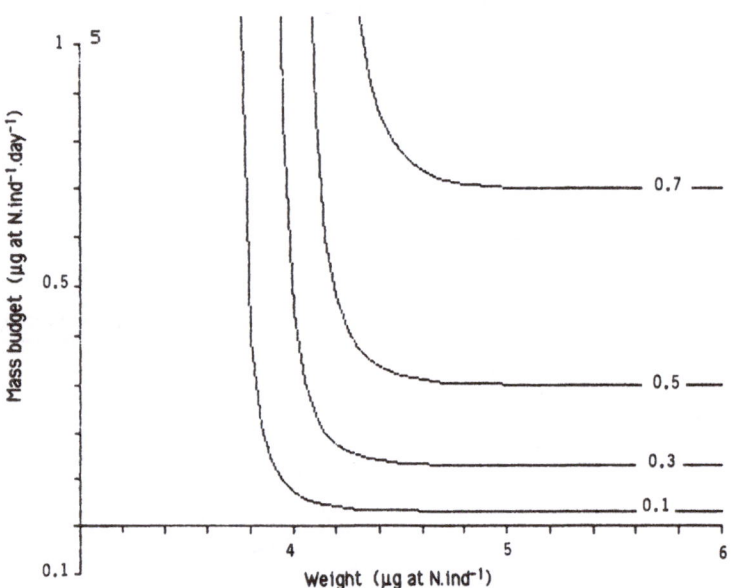

Figure 4. Isolines of transfer rate in the space of energy budget
and weight for a given larval stage. (Here stage 1)

3.   RESULTS

We have used the model to simulate experimental conditions
reconstituted by different authors, mainly working on crustaceans.
The molting of crustacean larvae provides a conspicuous marker for
changes in stage and permits simple experiments.

Table 3. Set of coefficient values used for reference simulation (Figure 5a).

Symbol of Coefficient	Units	General Process	Specific Process 1	2	3	4
B_p	μg at N$\cdot$1^{-1}	8				
P_r	ind$\cdot$1^{-1}	50				
b	1$\cdot\mu$g at N$^{-1}\cdot$j^{-1}		0.035	0.042	0.050	0.065
E_c			0.	1.	0.	0.
T_m	j^{-1}		1.	1.	1.	1.
C_b	μg at N$\cdot$j^{-1}		0.3	0.5	0.5	-
C_w	μg at N	30	4.	5.5	8.	-
I_{pmax}	j^{-1}	0.3				
L_{min}	ind$\cdot$1^{-1}	5				
K_1	ind$\cdot$1^{-1}	10				
B_{min}	μg at N$\cdot$1^{-1}	0				
M_{max}	j^{-1}	0.08				
M_{min}	j^{-1}	0.04				
a_m	μg at N$\cdot$j$^{-1}\cdot$ind^{-1}	0.06				
b_m	μg at N$\cdot$j$^{-1}\cdot$ind^{-1}	-1.5				

3.1 Summation of experimental results.

3.1.1. *Test of the transfer function*. It is possible to simulate for one larval stage the input of individuals coming from the previous stage and the output to the next stage. Figure 6 gives an example of the evolution with time of the number of larvae arriving at and leaving stage 2 during 7 hours. The maximum of these curves occurs at the inflexion points on the abundance curve, and output is equal to input at the maximum of the abundance curve. The difference in the maximum of the input and output curves depends on the mortality rate and on the difference in the coefficients of the transfer function from one stage to the other. The input curve influences the shape of the beginning of abundance curve, and the output one shapes the end of abundance curve. The data of Sciandra (1982) show such curves with a tail of late individuals that his model is unable to simulate because it has a too simple transfer function. Parslow *et al*. (1979) assume that the transfer function is a Gaussian curve.

Sulkin and Van Heukelem (1986) give results of this kind for the blue crab (*Callinectes sapidus*). Each day, they collected the megalopae which had appeared in a culture of zoeae and gave the evolution in time of the input in the megalopa stage. Figure 7 gives an example of the fit one can get by adjusting the duration of stage 2 in the model to the 30 days series of the experiment. In this case,

330

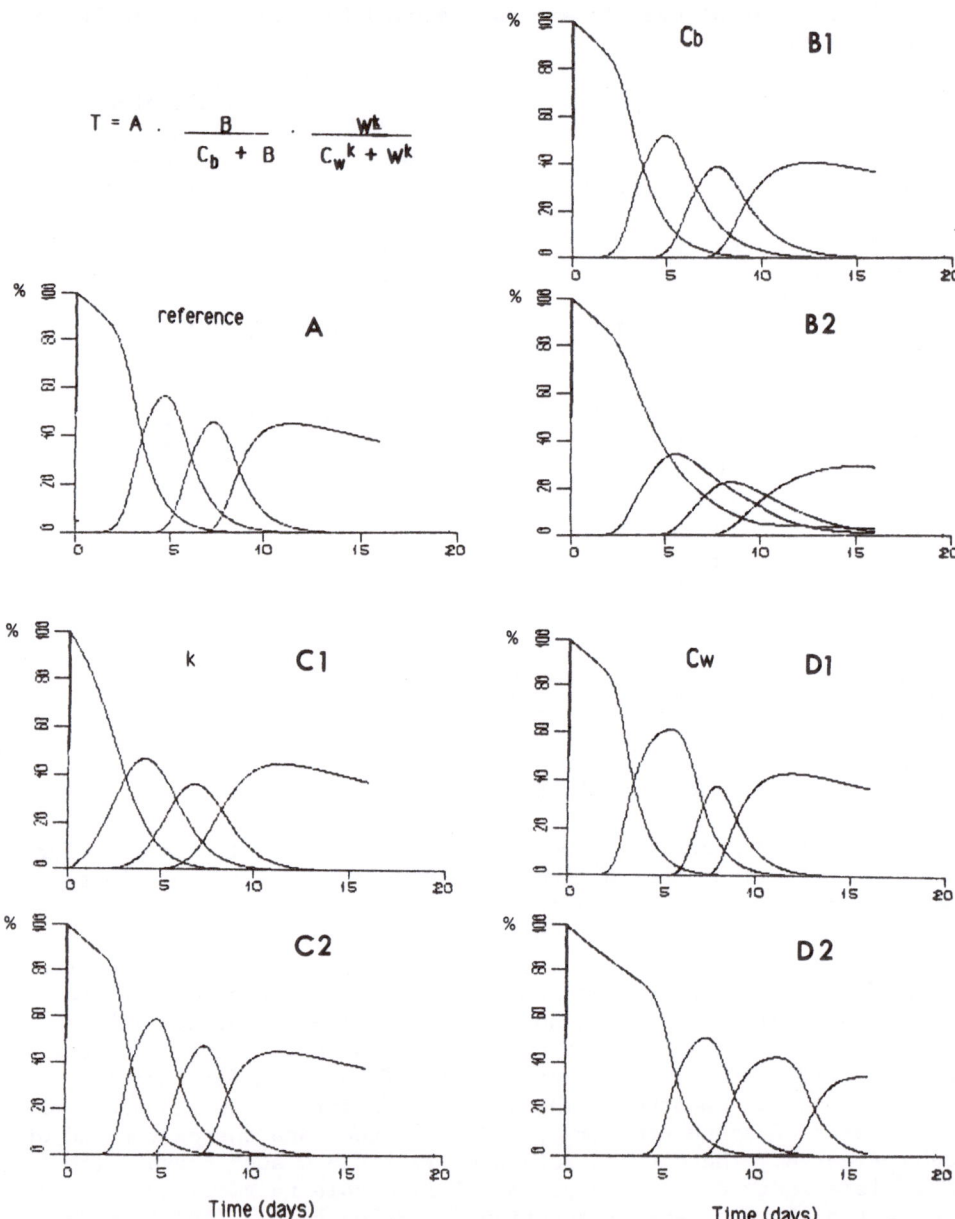

Figure 5. Effect of changes on the coefficient of the transfer rate formulation on the shape of abundance curve of each larval stage. (A) Reference simulation. Each simulation starts with 100 larvae, each one weighing 3 μg at N. (B) Effect of Cb [B1: Cb(1)=1, Cb(2)=2, Cb(3)=3; B2; Cb(1)=5, Cb(2)=6.5, Cb(3)=8]. (C) Effect of k [C1: k=10; C2: k=40]. (D) Effect of Cw [D1: Cw(1)=4, Cw(2)=6.2, Cw(3)=8; d2: Cw(1)=5, Cw(2)=7, Cw(3)=12].

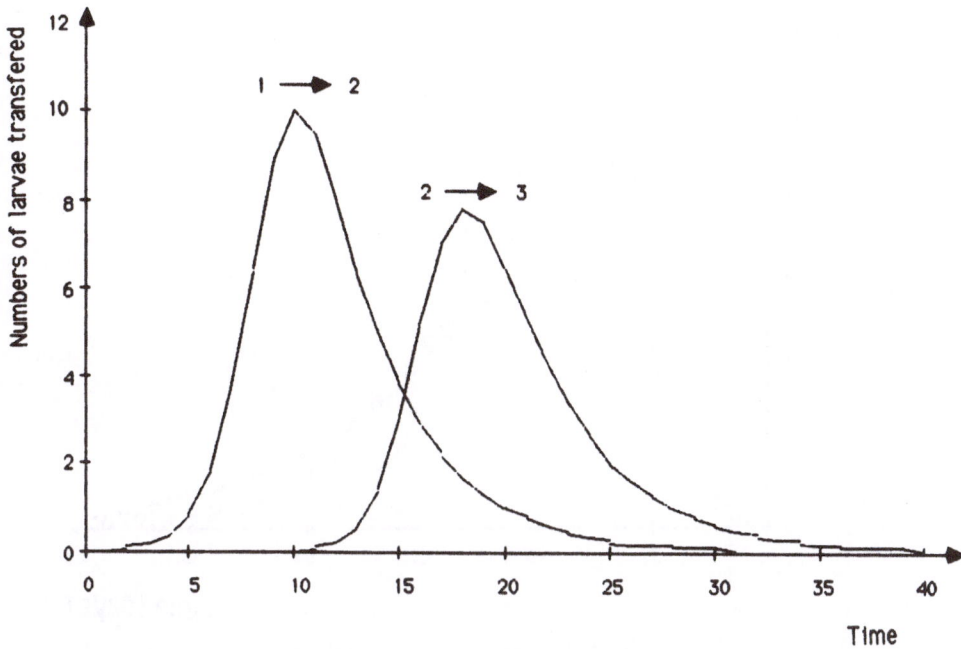

Figure 6. Variation in time of the transfer rate from 1st to 2nd stage (input to stage 2) and from 2nd to 3rd stage (output from stage 2). We have considered the number of larvae transferred during successive 7 hour intervals. The unit of time is 7 hours. These curves are skewed to the left. The leading tail simulate the late individual which grow slowly.

the model can reproduce the shape of transfer rate evolution. (The unit of time is 1/29th of the stage duration taken when input of larvae 2 is 1% of larvae initially present.)

3.1.2. *Test of the effect of starvation.* An experiment from Anger et al. (1981) provides an opportunity to test the model output. These authors wanted to estimate the effect of a starvation period during the stage Zoae I of a crab (*Menippe mercenaria*) on the following stage Zoae II. If the starvation period occurs early in the stage, the mortality is high and the duration of the stage is long; if the starvation period occurs later on, the effect on the mortality of starved stage is the same but the following stage is less affected (Figure 8).

We simulated the same experiment with the model which can reproduce the shape of the experimental results but the mean duration of the first and second stage are larger in the model than for *Menippe* (Figure 9). Some exceptions appear in Figure 9.A. The first two feeding regimes give larger values for development duration than measured in the experiment. This discrepancy is related to the

332

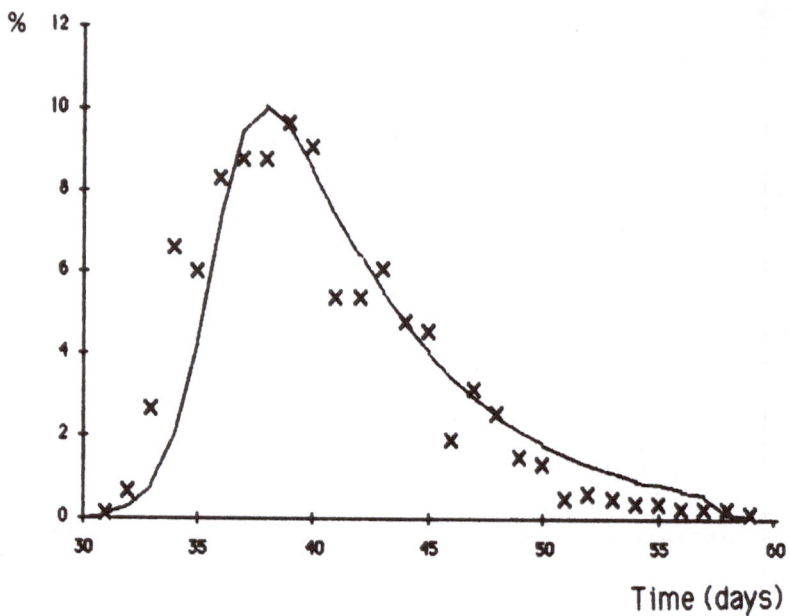

Time (days)

Figure 7. Comparison of the time variation of the transfer rate
(numbers of individuals per day during 29 days) from one stage to the
other (continuous line) with data from Sulkin and Van Heukelem
(1986): transfer rate from the zoea stage to the megalopa stage in
the *Callinectes sapidus* (+: % of total number of megalopae collected
during the experiment, data from Sulkin and Van Heukelem (1986); --:
model simulation].

criterion adopted to estimate the development duration from a
population which shows individual variability. The determination of
the duration of a stage is complicated if the larvae develop
asynchronously. We first assumed that the duration is the difference
in time of the maximum of abundance curve (Figure 9.A), second that it
is the difference in time of the 50% of abundance of the same stage
(Figure 9.B). We can see that the results of the model with the
second criterion are better fitted to the results of Anger *et al*.
(1981) than with the first criterion.

3.1.3. *Test of a time series*. The data from Blaszkowski and Moreira
(1986) can be used to compare the shape of the simulated curves and
data from a culture. Figure 10 shows that the timing and the
amplitude of abundance curves are well reproduced. This result is
obtained after minor changes from the reference set of coefficients: a
decrease in the mortality rate and a modification in the constant Cb
in the transfer function.
        This set of data obtained in constant food conditions cannot
allow a complete test of the ability of the model to simulate

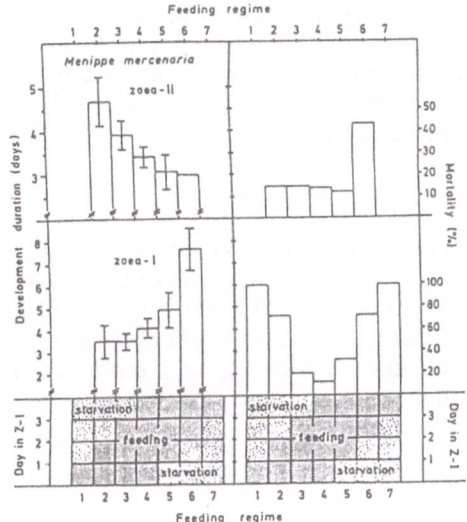

Figure 8. Effect of different feeding regimes in the development time and mortality rate for two successive stages of the larvae of the crab *Mennipe mercenaria* from Anger *et al.* (1981).

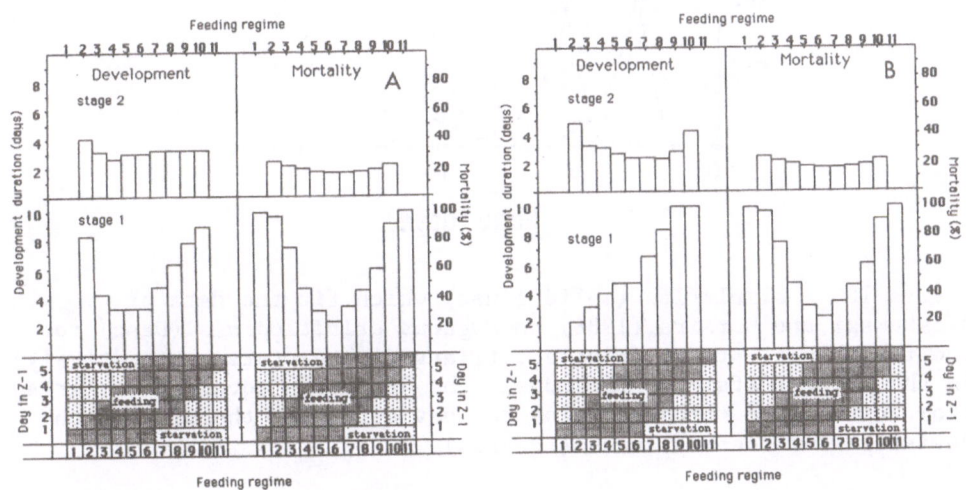

Figure 9. Results of simulations from the model for similar patterns of feeding regime. (A) Development time calculated from time interval between the maximums of abundance of stages 1 and 2, (B) Development time calculated from time interval between 50% of maximum abundance for stage 2. In this case, the output of the model is similar to the experimental results depicted by Figure 8.

334

population dynamics. It is necessary to rear larvae in variable food
conditions. We should compare the effect of a period of starvation on
the recruitment of a species reared in the laboratory and simulated by
the model. Because of the complicated output which is expected, such
experimental design are rarely set up.

3.2. Simulation of perturbations on larvae environment.

The preceding tests show that the model is able to reproduce some
properties of larval growth, but specific experiments should be
designed to verify the usefulness of all of its functions.
Nevertheless, we can use this theoretical structure to investigate
some situations relevant to the effect of the environment of the
larvae in simple situations.

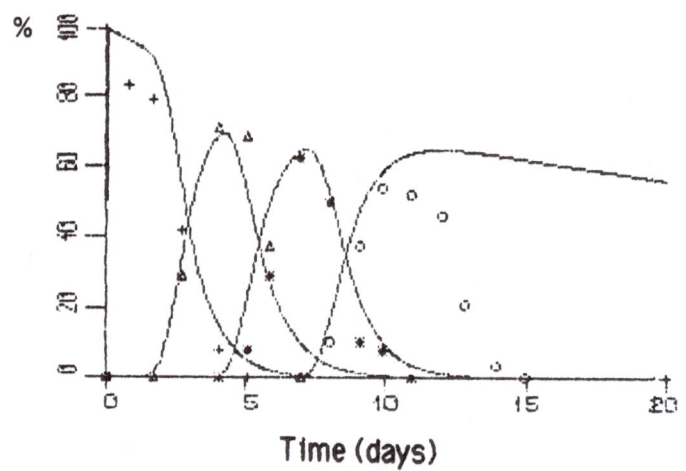

Figure 10. A simulation (solid lines) which fit the data of
Blaskowski and Moreira (1986) on *Pagurus criniticornis* (Dana) for
the first three stages (crosses, triangles and stars, respectively).
In the present state of the model, the 4th stage accumulates larvae
from third stage which explains the deviation why the simulation
deviates from the real data (squares).

3.2.1. *Effect of food on recruitment.* Food affects the cohort
evolution in a complicated way. It affects the ingestion rate which
determines the budget and consequently the transfer rate. It is a
three stage process, and each step is controlled by other processes.
If we simulate a laboratory experiment with constant food, the food
must exceed a threshold to allow the transfer from stage 1 to stage 2
(Table 4).

Table 4.   Effect of food level on recruitment of stage larvae.

Food	5	6	7	8	μg at $N \cdot l^{-1}$
Recruits	0.05	25.6	37.2	44.7	%
Time to reach stage 4	>20	17.7	13.62	11.42	days

Figure 11.A shows the modifications of the cohort evolution when a period of two days of starvation is translated from day 1 after hatching to day 7. Starvation reduces the transfer between two stages, so the stage following the one which is affected by the starvation appears later. The lengthening of development duration increases the loss of individuals by mortality, so the recruitment declines, but it does not change much with the period of starvation (Table 5). The main feature is the increase of the period of presence of the affected stage, allowing a predator selecting stage 2, for instance, to catch a large amount of them.

3.2.2. *Effect of predation mortality on the recruitment.* If the predator grows slowly (abundance nearly constant) and is able to collect all stages with the same efficiency, the result is an increase of the overall mortality and a reduction of the recruits.

We can first investigate the effect of a predator which is able to collect only stage 2. The stage specific predation is known for copepods (Mullin, 1979; Bailey and Yen, 1983) and for a variety of other planktonic invertebrate larvae (Rumrill et al., 1985; Pennington et al., 1986).

If we consider a predator which captures larvae when their concentration is over five per liter and has a maximum predation rate of 10 preys per day, the potential amount of larvae captured is maximal when the starvation begins on day 5 (Table 5). This increase of predation on stage 2 has an effect on the recruitment of stage 4.

Simulations considering the combination of a starvation period and the presence of a predator selecting stage 2 give results shown on Figure 11.B. The effect of the predation is more important when the concentration of larvae tends to be high during a long period of time (Figure 11.A. and 11.B. at 5-7).

The effect of starvation and predation on the recruitment depends on the duration of the starvation period and on the abundance of predators, but this simulation shows that the outcome is not simple.

3.2.3. *Effect of timing of maximum food concentration and maximum predator concentration.* There is considerable literature on the value of different food species to the rearing success of larval stages for marine animals (Hines, 1986). Sulkin (1975) shows that the survival of *Callinectes sapidus* larvae depends strongly on food species. Measurement of available food by particulate carbon or chlorophyll is certainly not correct if a specific item is needed. The species composition changes when the spring bloom develops (Robinson et al., 1986) so the short advantageous food species and the most efficient

Table 5. Effect of a starvation period of two days on the recruitment of larvae stage 4 (L4). Potential predation is the number of larvae which can be eaten by a predator selecting stage 2 during the total development time. Starvation influences the recruitment, but its timing has only a slight effect. On the contrary, when a selected predation on stage 2 is added, the number of recruits becomes highly variable.

	Without Predation			With Predation	
	Recruitment (%)	Time for max L4 (days)	Potential Predation (ind/pred)	Recruitment (%)	Time for max L4 (days)
Without starvation	44.7	11.4	35	19.7	11.4
Starvation Period (days)					
dates					
1-3	32.2	15.1	28	12.5	15.0
3-5	32.4	15.0	53	8.7	15.0
5-7	33.5	14.7	69	6.7	14.9
7-9	34.9	14.5	43	15.2	14.5

predator can be occurring in a sequence, relative to the hatching of larvae, which might not correspond to the sequence of the total phytoplankton and total zooplankton.

If we assume that the predator's maximum and the food maximum are not linked, it could be possible to find the occurrence of predators before the occurrence of food. The model can be used to determine the variation in recruitment for different lags of predator and food maximum relative to the hatching date.

Figure 12 shows that when the food species occurs early, the recruitment depends very much on the lag of the predator bloom. It confirms the intuition that when the predator bloom is late much of the development of the larvae has been completed and the recruitment is successful. If the food species blooms late, the development is lengthened and the natural mortality is higher. The recruitment is reduced, but the predators are less effective. This simulation put the emphasis on the fact that a successful recruitment requires an abundance of food with larval development and a mismatch in the advent of the predator.

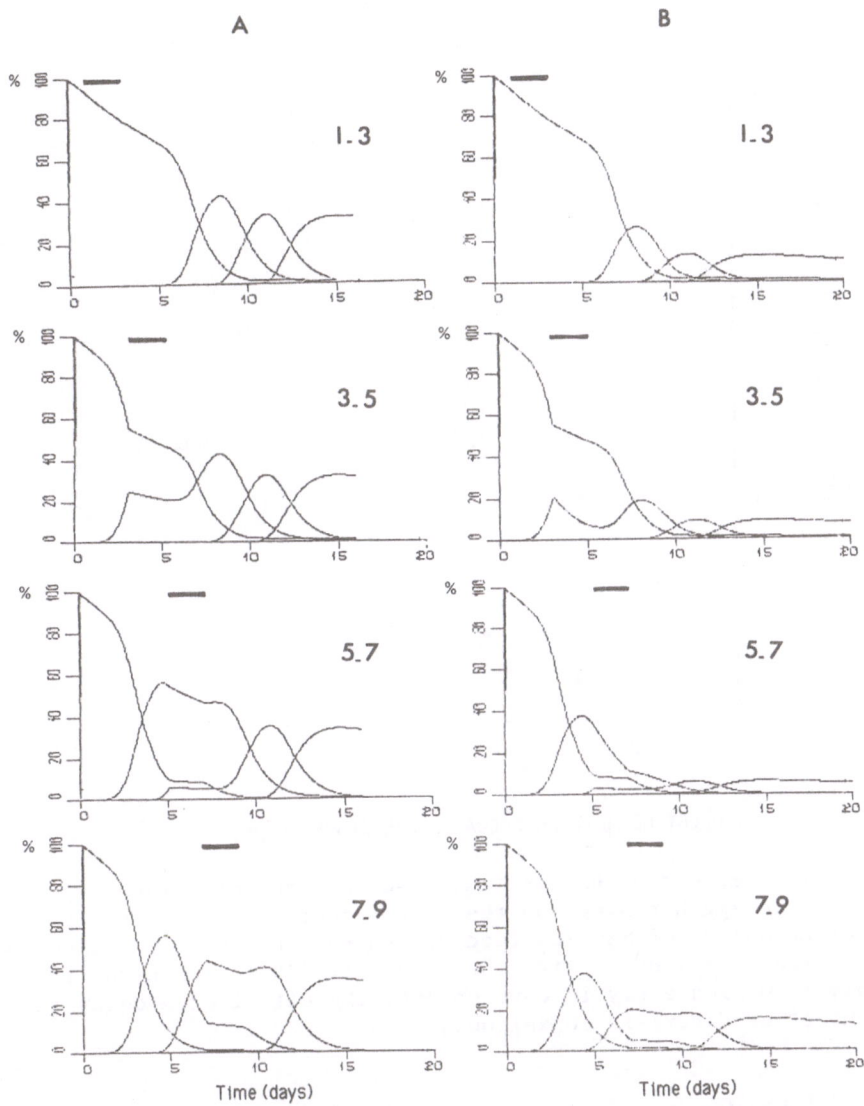

Figure 11. Effect of a starvation period of two days assumed to originate from a physical perturbation (mixing) of the food distribution on larval development and recruitment of stage 4. Food at a constant concentration except during the perturbation where it declines to zero. (A) No predator. Depending on the timing of the perturbation (---), stage 2 is delayed or extended. Table 5 gives the recruitment efficiency for the four situations. (B) Presence of a predator which feeds specifically on the stage 2 (1-3; 3-5; etc.: post hatching days subjected to the perturbation).

338

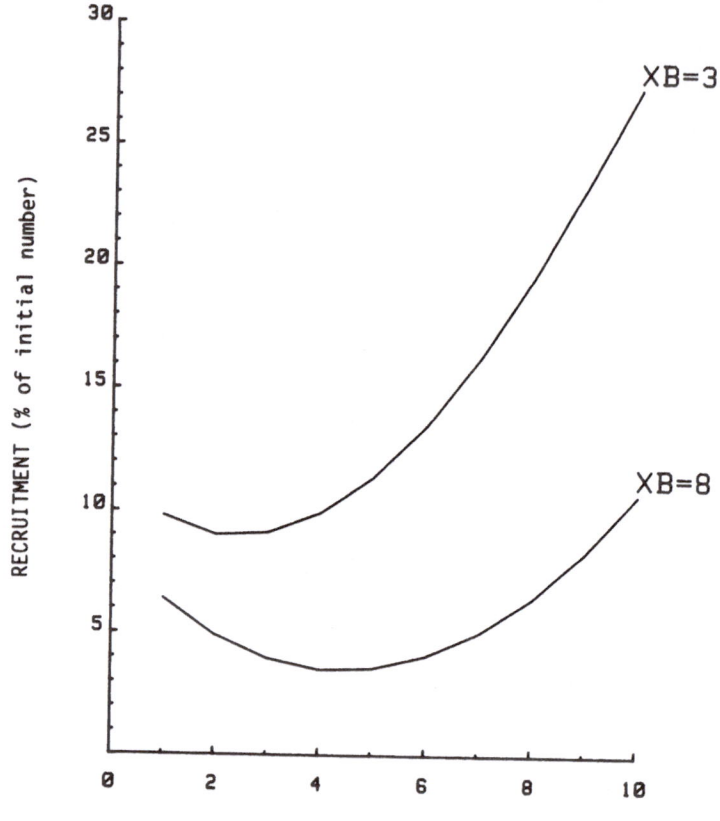

DELAY OF MAXIMUM PREDATOR ABUNDANCE (days)

Figure 12. Effect of the maximum predator concentration (XP) relative to egg hatching, on the recruitment of stage 4 for two different delays of maximum food concentration (XB). The predator can collect every stage with the same efficiency. When the food supply is maxima early in development, the effect of a delay in the advent of predators is conspicuous.

4. DISCUSSION

This model includes two kinds of processes that control the transfer rate and ultimately the amount of larvae reaching a defined stage in their life cycle. The first type is related to the physiology: ingestion, excretion, etc. It is sensitive to rapid variations of available food through mass budget. The second, based on the larval weight, is cumulative and takes into account the different events affecting the mass budget since hatching. We have kept our representation of the mass budget very simple, and we have assumed that the temperature is constant; this holds for species with a rapid development. Ingestion does not take into account the regulation by digestion rate and stomach volume (Lehman, 1976). It also ignores different capture efficiencies for the different food species the larvae can encounter, but there is less information in the literature for larvae than for copepods and it seems more complicated (Strathmann, 1971). Capture and growth efficiencies are important parameters which determine the fit of the larval development to the available food. They must be considered but we need better experimental evidence. The present model assumes that the food does not change in type during the larval development.

Another important control, which is not included in the model, is related to storage which damps the fluctuations of the food. This control is necessary to simulate the evolution of larvae hatching from telolecith eggs, or larvae which can accumulate stored reserves (e.g. crustaceans).

Models connecting biological and physical process usually simplify the biological coupling (Walsh, 1976; Klein and Steele, 1985). At the present time it is certainly the only thing to do because the knowledge of the connections or interactions of the processes at different time scales is not sufficient. Nevertheless, it is urgent to construct some models with processes working at different time scales because the structures which exist at different space scales should be the consequence of these processes. Jackson (1986) estimated the influence of physical process like transport on the settlement of planktonic larvae. Its conclusions depend largely on the biological process considered. The length of the pelagic period and the ability for larvae to attain competence for settling are not simple functions of time.

The quantitative estimation of the recruitment at a defined stage in the life cycle of a species must be based on a good understanding of the process controlling the different causes of mortality. However, a model will not be realistic if there is not enough knowledge of the basic processes and of the way they interact. An experimental approach well connected with a theoretical approach is necessary to focus the design of experiments on badly known aspects of biology or on processes to which the model is most sensitive. The model presented here shows that laboratory experiments on feeding and growth of specific larvae, are necessary to know the functions and values of their coefficients; however, experiments made in mesocosms are also very important because some processes are not easy to isolate from others and we only observe the results of their interactions. This is the situation when interaction of the biology and the environment is to be studied.

Models are also important for understanding the behavior of the biological variables to perturbation of their interactions by physical events. Wind stress, inducing a mixing with, for instance, changes the food concentration in the vicinity of the larvae, is not constant but has a characteristic frequency spectrum. This model shows that the effect of a perturbation in the food concentration on the outcome of the larval development is not simple because it modifies the relations of the larvae not only with their food but with their predators. This shows that it is important to identify the trophic structure appropriate for the type of larvae studied. A second important fact shown by the model is the phase between hatching (start of the chain of events in the growth of the larvae) and perturbation.

Models can allow one to test hypotheses about recruitment success if their structure takes into account the most important biological processes and the food web to which the larva belongs. A model is necessary to understand the behavior of complex systems, especially to answer the question: Is the recruitment depending mainly on food or on predators? We suggest here that the question is not so simple and cannot be answered by observations at sea only. More experimental studies of biological process and their modifications by physical processes are needed in order to determine the shapes of each element of the puzzle of interactions which determine the recruitment success. Such studies should help to improve the model structure.

5.   ACKNOWLEDGMENTS

We thank Nick Holland for his comments and kind help in improving the English. This paper is a contribution to UA716 CNRS-TOAE programs.

6.   REFERENCES

Andersen, V.: 1985, 'Modelisation de l'ecosysteme pelagique', Thèse de doctorat, Paris VI., 135 pp.

Andersen, V. and P. Nival: 1986, 'A model of the population dynamics of salps in coastal waters of the Ligurian sea', J. Plankton Res. 8, 1091-1110.

Anger, K., R.R. Dawir, V. Anger and J.D. Costlow: 1981, 'Effect of early starvation periods of zoea development of brachyuran crabs', Biol. Bull. 161, 199-212.

Anger, K. and K.D. Spinkler: 1987, 'Energetics, moult cycle and ecdysteroid titers in spider crab (Hyas araneus) larvae starved after the D0 threshold', Mar. Biol. 94, 367-375.

Bailey, K.M. and J. Yen: 1983, 'Predation by a carnivorous marine copepod, Euchaeta elongata Esterly, on eggs and larvae of the Pacific Hake, Merlucius productus', J. Plankton Res. 5, 71-82.

Blaszkowki, C. and G.S. Moreira: 1986, 'Combined effects of temperature and salinity on the survival and duration of larval stages of Pagurus criniticornis (Dana) (Crustacea, Paguridae)', J. Exp. Mar. Biol. Ecol. 103, 77-88.

Blythe, S.P., R.M. Nisbet and W.S.C. Gurney: 1982, 'Instability and complex dynamic behaviour in population models with long time delays', Theor. Pop. Biol. 22, 147-176.

Blythe, S.P., R.M. Nisbet and W.S.C. Gurney: 1985, 'Stability switches in distributed delay models', J. Math. Anal. Apll. 109, 388-396.

Boucher, J.: 1984, 'Localization of zooplankton populations in the Ligurian marine front: role of ontogenic migration', Deep Sea Res. 31, 469-484.

Boucher, J.: 1988, 'Space-time aspects in the dynamics of planktonic stages', in B.J. Rothschild (ed.), *Toward a Theory on Biological-Physical Interaction in the World Ocean*, Kluwer Academic Publishers, Dordrecht, pp. 203-214.

Carlotti, F.: 1986, Modele de recrutement de larves d'organismes marins. Influence de la nourriture et des predateurs', D.E.A. d'Oceanographie Biologique, Universite Paris VI., Paris, 34 p.

Carlotti, F.: 1987, 'Modele de recrutement de larves d'organismes marins', *J. Rech. Oceanogr.*, in press.

Davis, C.S.: 1984, 'Interaction of a copepod population with the mean circulation on Georges Bank', *J. Mar. Res.* 42, 573-590.

Gurney, W.S.C. and R.M. Nisbet: 1985, 'Fluctuation periodicity, generation separation and the expression of larval competition', *Theor. Pop. Biol.* 28, 150-180.

Hines, A.H.: 1986, 'Larval problems and perspectives in life histories of marine invertebrates', *Bull. Mar. Sci.* 39, 506-525.

Jackson, G.A.: 1986, 'Interaction of physical and biological process in the settlement of planktonic larvae', *Bull. Mar. Sci.* 39, 202-212.

Klein, P. and J.H. Steele: 1985, 'Some physical factors affecting ecosystems', *J. Mar. Res.* 43, 337-350.

Kremer, J.N. and S.W. Nixon: 1978, *A coastal marine ecosystem - simulation and analysis*, Springer Verlag, New York, 217 p.

Lehman, J.T.: 1976, 'The filter feeder as an optimal forager and the predicted shapes of feeding curves', *Limnol. Oceanogr.* 21, 501-516.

Manly, B.F.J.: 1974, 'Estimation of stage-specific survival rates and other parameters for insect populations developing through several stages', *Oecologia* 15, 277-285.

Mullin, M.M.: 1979, 'Differential predation by the carnivorous marine copepod *Tortanus discaudatus*', *Limnol. Oceanogr.* 24, 774-777.

Nival, P. and S. Nival: 1983, 'La variabilite individuelle au cours du developpement du copepode *Temora stylifera* Dana', *Rapp. Comm. Int. Mer Medit.* 28, 163-164.

Parslow, J., N.C. Sonntag and J.B.L. Matthews: 1979, 'Technique of system identification applied to estimating copepod population parameters', *J. Plankton Res.* 2, 137-151.

Pennington, J.T. and F.S. Chia: 1984, 'Morphological and behavioral defense of trochophore larvae of *Sabellaria cementarium* (Polychaeta) against four planktonic predators', *Biol. Bull.* 167, 168-175.

Pennington, J.T., S.R. Rumrill and F.S. Chia: 1986, 'Stage specific predation upon embryos and larvae of the Pacific sand dollar *Dendraster excentricus*, by 11 species of common zooplanktonic predators', *Bull. Mar. Sci.* 39, 234-240.

Platt, T., K.H. Mann and R.E. Ulanowicz: 1981, *Mathematical models in biological oceanography*, Monographs on oceanographic methodology 7, UNESCO Press, 156 p.

Radach, G.: 1982, 'Dynamic interactions between the lower trophic levels of the marine food web in relation to the physical environment during the Fladen Ground experiment', *Neth. J. Sea Res.* **16**, 231-246.

Robinson, G.A., J. Aiken and H.G. Hunt: 1986, 'Synoptic surveys of the western English Channel. The relationship between plankton and hydrography', *J. Mar. Biol. Ass. U.K.* **66**, 201-218.

Rothschild, B.J.: 1986, *Dynamics of Marine Fish Populations*, Harvard University Press, Cambridge, Mass., 277pp.

Rumrill, S.S., J.T. Pennington and F.S. Chia: 1985, 'Differential susceptibility of marine invertebrate larvae: laboratory predation of sand dollar, *Dendraster excentricus* (Eschscholtz) embryos and larvae by zoea of the red crab, *Cancer productus* (Randall)', *J. Exp. Mar. Biol. Ecol.* **90**, 193-208.

Sciandra, A.: 1982, 'Etude d'un système marin artificiel. Construction d'un modèle mathématique et application à l'exploitation d'une population de copépodes pélagiques *Euterpina acutifrons* (Dana)', Thèse de 3ème cycle. Univ. Paris VI., 108 pp.

Sciandra, A.: 1986, 'Study and modelling of the development of *Euterpina acutifrons* (Copepoda: harpacticoida)', *J. Plankton Res.* **8**, 1149-1162.

Steele, J.H. and B.W. Frost: 1977, 'The structure of plankton communities', *Phil. Trans. R. Soc. Lond. (B: Biol. Sci.)*, **280**, 485-534.

Strathmann, R.R.: 1971, 'The feeding behavior of planktotrophic echinoderm larvae: mechanisms, regulation, and rates of suspension-feeding', *J. Exp. Mar. Biol. Ecol.* **6**, 109-160.

Sulkin, S.D.: 1975, 'The significance of diet in the growth and development of larvae of the blue crab *Callinectes sapidus* Rathbun, under laboratory conditions', *J. Exp. Mar. Biol. Ecol.* **20**, 119-135.

Sulkin, S.D. and W.F. Van Heukelem: 1986, 'Variability in the length of the megalopal stage and its consequence to dispersal and recruitment in the portunid crab *Callinectes sapidus* Rathbun', *Bull. Mar. Sci.* **39**, 269-278.

Walsh, J.J.: 1976, 'Models of the sea', in D.H. Cushing and J.J. Walsh (eds.), *The ecology of the seas*, pp. 388-407.

Wroblewski, J.S.: 1980, 'A simulation of the distribution of *Acartia clausi* during Oregon upwelling, August 1973', *J. Plankton Res.* **2**, 43-68.

SIMULATION STUDIES OF FISH LARVAL SURVIVAL

R. Jones and E.W. Henderson
Department of Agriculture and Fisheries for Scotland
Marine Laboratory
P.O. Box 101, Victoria Road
Aberdeen, Scotland

ABSTRACT. The results of simulation studies of the fish larval stage
within a closed ecosystem model using input parameters appropriate to
a hypothetical, but largely "haddock-like" species are described. In
a seasonally varying world, a fish larva feeding on rapidly growing
food organisms, is likely to experience a relatively high
concentration of food for only a relatively short period of time. The
implication is that the probability of larval survival could be
relatively high provided two main criteria are met: (1) a larva
should be in the right place at the right time, to start feeding,
just as the biomass of suitable food organisms is beginning to
increase, and (2) a larva does not happen to be part of a larval
cohort so numerous that its overall grazing power is likely to
shorten the time during which the concentration food organism exceeds
a critical value. To survive, it is essential that while food is
abundant, a larva is able to grow large enough to be able to begin
feeding on other and larger food organisms. In practice, some larvae
are likely to grow well with a high probability of continued survival
whilst others are likely to grow more slowly with a much lower
probability of ultimate survival. Individuals that grow slowly may
be eliminated by predators before they have had time to exhibit
obvious signs of resource limitation, much less signs of starvation.
The main conclusion however is that whether larvae die of predation
or starvation, the actual number dying is likely to be largely a
function of resource limitation, and not something that can be
predicted simply from a knowledge of the number of larvae and of the
number of predators. Simulation studies suggest that there should be
an optimum period for first feeding of larvae but that this is likely
to depend on the number of first feeding larvae. In practice
therefore, the optimum period for first feeding (and hence spawning)
is not likely to to be precisely defined and could easily be spread
out over several weeks as is normally observed. The highest levels of
simulated larval production were obtained using relatively small
biomasses of first feeding larvae. Theoretical results suggested a
biomass of first feeding larvae of around 2-5 mg N/m^2 for optimum
larval production.

B. J. Rothschild (ed.), Toward a Theory on Biological-Physical Interactions in the World Ocean, 343–372.
© *1988 by Kluwer Academic Publishers.*

1.   INTRODUCTION

This paper is concerned with simulation studies of the fish larval
stage within a closed ecosystem model. It complements earlier
simulation studies of the larval stage (Jones, 1973, 1981 and Jones
and Hall, 1973) that were based on "non-closed" sets of equations and
extends simulation studies based on a closed ecosystem described by
Jones and Henderson (in press).

2.   THE POTENTIAL FOR RESOURCE LIMITATION AT THE LARVAL STAGE

It seems reasonable to suppose that, although resource limitation is
liable to operate at any life-history stage, it is likely
to operate most strongly at the larval stage. There are several
reasons for thinking this, all associated with the relatively high
growth rates of the fish larvae and of their food organisms.

2.1 Potential Food Requirements of a Larval Cohort of Haddock

As far as potential food requirements are concerned, it is a simple
matter to estimate the food that could be eaten by the larvae of an
average year class. An average North Sea haddock year class for
example, produces about 0.017 Kcal/m^2 of eggs (equivalent to about
0.0017 gC/m^2) (Jones, 1982). On average, each surviving larva will
increase its body weight by a factor of about 2500 during its first 80
days. If all the individuals from an average spawning were to survive
for this period, their biomass would increase to 0.0017 x 2500 = 4.25
gC/m^2. Assuming a food conversion efficiency of 30%, they would
therefore require about 14 gCm^2 of food energy. This number can be
put into perspective by realising that the total secondary production
for the entire North Sea is only about 15-20 gC/m^2 per year (Steele,
1974).
  If the same group of individuals were to survive for 120 days,
each individual would increase its body weight by a factor of 17500,
and the entire cohort in the absence of mortality would need about 100
gC/m^2. For the additional 40 days, much of the food would be
dependent on young fish (Robb and Hislop, 1980) and since this is
equivalent to tertiary production, the potential for resource
limitation is that much greater.
  Finally the food requirements for haddock represent those for
only one of a relatively large number of fish species. It is quite
clear therefore that the potential for resource limitation at the
larval and early juvenile stages must be enormous.

2.2 Necessity for Growth Rate to be Rapid

Although larval growth rates can vary in practice, there is a
necessity for rapid growth at some stage, simply because there is
such a large size difference between a newly hatched larvae and a
mature individual. A newly hatched haddock larvae weighs about 0.4

mg whereas a mature individual weighs at least 200 g. Before it can
reproduce an individual haddock therefore, has to increase its body
weight by a factor of at least 500,000. A period of extremely rapid
growth, at some stage in the life history, would therefore appear to
be unavoidable, and in practice, it is at the larval stage that much
of this rapid growth occurs. It seems reasonable to suppose
therefore that an overriding biological requirement at the larval
stage is to accomplish some necessary minimum amount of growth.

The fact that larval growth rates may slow down in winter, does
not necessarily conflict with the requirement that within the few
months when primary production is relatively high, body weight must
increase by at least several orders of magnitude. Autumn-spawned
herring larvae for example, are no exception. The period of very
rapid growth is simply interrupted until the following productive
season.

2.3 Growth Rates of Larvae and Food Organisms

One of the consequences of rapid growth is a constraint on the food
available for fish larvae. Many gadoid fish larvae grow at about 10%
per day in body weight, whilst the young stages of copepods on which
they mainly feed grow even faster. According to Mullin and Brooks
(1970) for example, *Calanus* grow exponentially from Nauplius I to
copepodite IV in about 28 days at a temperature of 10C during which
time, the weight change is from 0.1 to 16 μg C. This is equivalent to
a growth rate of about 20% per day. Just what this means to a rapidly
growing larva is illustrated diagrammatically in Figure 1. This shows
a fish larva in relation to copepod growth curves, on a logarithmic
scale, with the copepod body weights increasing at a rate of 20% per
day. Two copepod growth curves are shown, positioned in such a way
that the range of food particle sizes available to the fish larva is
just one order of magnitude (one order of magnitude being chosen
simply to illustrate the argument). When positioned in this way, it
can be seen from the Figure that the two copepod growth curves are
separated by just 13 days along the time axis. Similarly a larva able
to eat particle sizes over two orders of magnitude would only have 26
days food production to depend upon. The conclusion is that for so
long as the food organisms are growing at 20% per day, (*i.e.* for about
30 days in the case of the young stages of *Calanus* in the North Sea),
a larva constrained to feeding on particle sizes over a range even as
large as two orders of magnitude, would be unlikely to have more than
about four weeks production to feed on, at any one moment.

2.4 Food Potential Towards the End of the Larval Stage

Only when a larva is large enough to eat slowly growing food
organisms, will it have access to a range of particle sizes that could
have been produced over a relatively much longer time period.

The conclusion is that the potential for resource limitation is
likely to be greatest while the larvae and their food organisms are
growing rapidly, and diminished once the growth rates slow down.

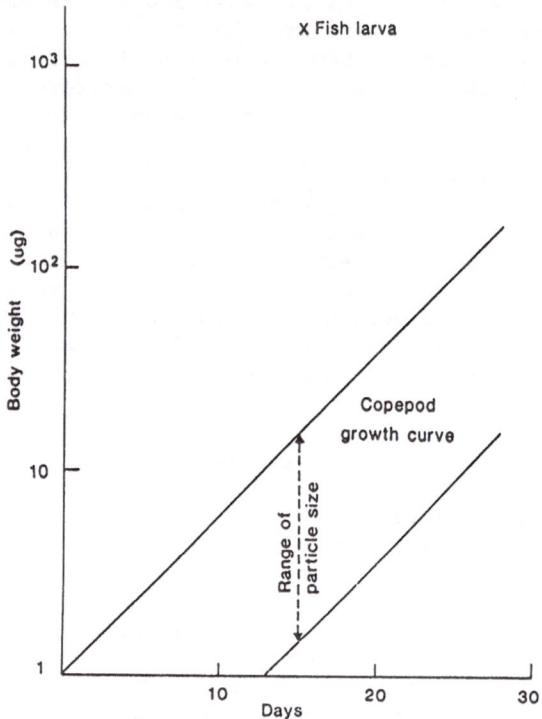

Figure 1. Showing the range of particle sizes available to a fish larva feeding on copepod nauplii growing at 20% per day.

3. SIMULATION STUDIES

Simulation studies using a simple model of the pelagic ecosystem (Jones and Henderson, in press) facilitated more rigorous investigation of the likely effect of seasonal variation on fish larvae/copepod nauplii inter-relationships. Figure 2 shows the inter-relationships between phytoplankton, zooplankton, and primary carnivores used for this study. Phytoplankton is represented by one pool. Zooplankton and primary carnivores are each represented by three pools, representing three different life stages (*i.e.* nauplii, copepodites and adults in the case of zooplankton, and larvae, juveniles and adult fish in the case of primary carnivores). In the simulations, the fish larval "pool" was further sub-divided into a number of smaller pools, each representing one day's growth for a non-food limited larva (see Figure 3). Each pool represents biomass, and is in nitrogen units. There are therefore flows both up the page and

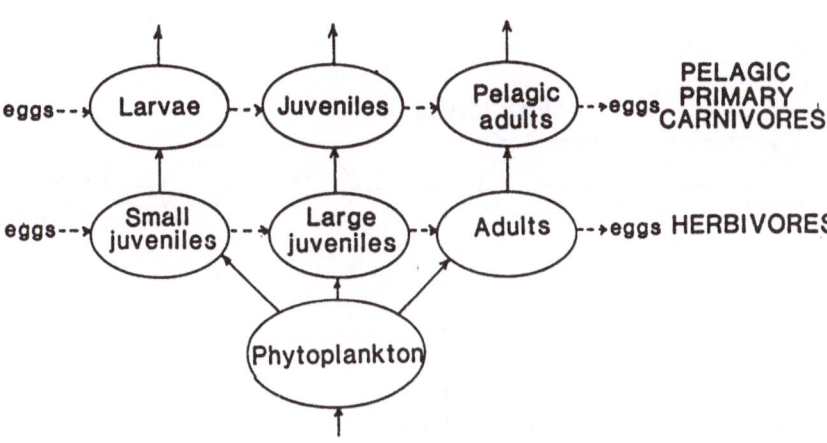

Figure 2. Showing the interrelationships between phytoplankton, three life stages of herbivores and three life stages of pelagic primary carnivores used in the simulation model.

across the page. Flows up the page, are due to grazing by zooplankton on plants and by fish on zooplankton. Flows from left to right across the page are due to the promotion of young life stages to older life stages as a result of growth. There are also inputs of eggs to the youngest stages of zooplankton and fish, calculated from adult biomass in the case of zooplankton, but introduced as an input number in the case of fish.

To avoid having to make assumptions about the rate of input of nitrogen into the phytoplankton pool, the system shown in Figures 2 and 3 was incorporated into the closed system depicted in Figure 4. This depicts a closed ecosystem comprising animals, plants, inorganic nutrient and dead organic matter. Details of this model are given by Jones and Henderson (in press) where it is shown that a system such as this, when seasonally perturbed, is capable of simulating an annual nutrient cycle with a reasonable degree of realism. References relevant to the model and to much of this paper are given in Jones and Henderson (in press), and details are also given in the Appendices to this paper.

348

Each run of the model was allowed to run for 10 years or more, to
find combinations of the input parameters for which the same seasonal
pattern of variation in each of the pools was reproduced year after
year without change. Runs were rejected if this did not happen, or if
the pattern of seasonal variations in any of the pools did not appear
reasonable. The results described in this paper are based on short
runs of the summer period under various conditions, starting out from
a post-bloom situation based on the results of the longer runs
referred to above.

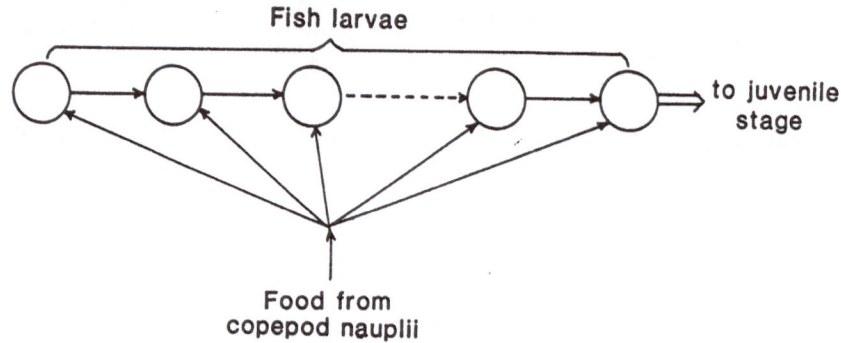

Figure 3. Showing the further subdivision of the fish larval stage
used in the simulations in this paper.

3.1 Zooplankton and Fish Larval Interactions at the End of the Spring Bloom

One of the more consistent results to emerge from the simulation
studies, involved the relationship between predator biomass and prey
biomass at the end of the spring bloom, when both consisted of young
and relatively very rapidly growing individuals. Figure 5 shows a
diagrammatic example of the kind of changes in the biomasses of
zooplankton nauplii and fish larvae typically observed at the end of
the spring bloom. Characteristic features are that both fish larvae
and zooplankton nauplii experience short periods of food abundance,
which permit their biomasses to increase relatively rapidly to
individual peaks. Each peak is typically followed by a sharp decline
in each case.

As far as the copepod nauplii are concerned, the decline in
biomass is influenced by food limitation, predation and promotion of
individuals to the next life stage.

From the point of view of a cohort of fish larvae, there is a
period of relatively high food concentration (period $t_1...t_2$ in
Figure 5) followed by a period of relatively low food concentration

(period $t_2...t_3$). Larvae that become large enough, quickly enough, can be promoted to the juvenile stage. Larvae that fail to make this transition in time, stop growing (in the model), and their combined biomass declines.

3.2 Effects of Varying Copepod and Fish Larval Growth Rates

Examples of the effect of varying the growth rates of the copepod nauplii and the fish larvae is shown in Figures 6a and 6b. In Figure 6a, potential naupliar growth rate (*i.e.* growth rate when food is unlimiting) is made equal to 30% per day. The potential fish larval

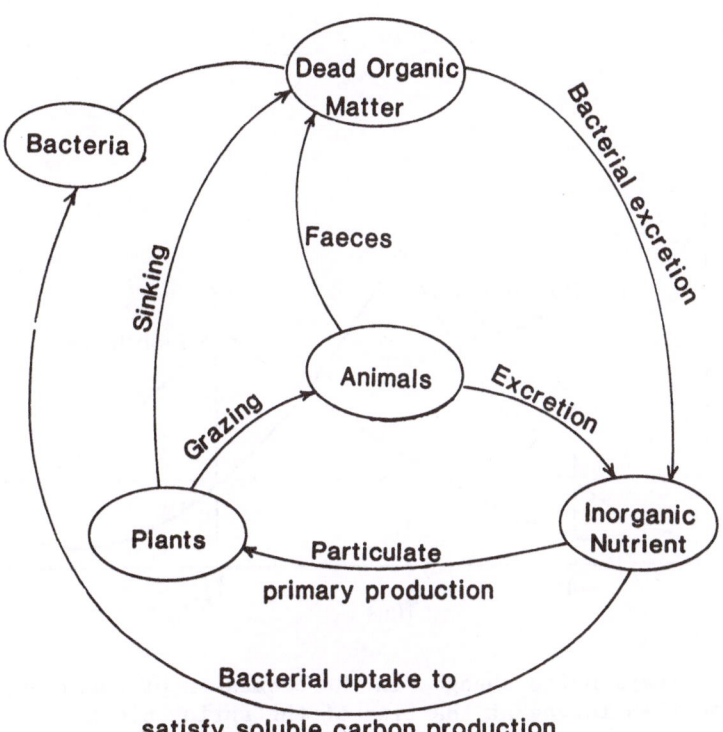

Figure 4. The simple closed ecosystem model within which the animal stages illustrated in Figure 3 were constrained.

growth rate is made equal to 10% per day. In Figure 6b, the corresponding growth rates are 15% and 5% per day respectively.

Comparison shows that peak biomasses are greater when potential growth rates are high. This is presumably advantageous. On the other hand, the duration of relatively high naupliar biomass is shorter when the growth rates are high and this is presumably not advantageous. The implication is that in nature, there is likely to be selection for an optimum growth rate in each case (for copepod nauplii in respect of the spring bloom and for fish larvae in respect of copepod nauplii).

As far as larval production is concerned, the effects of different growth rates are shown in Table 1.

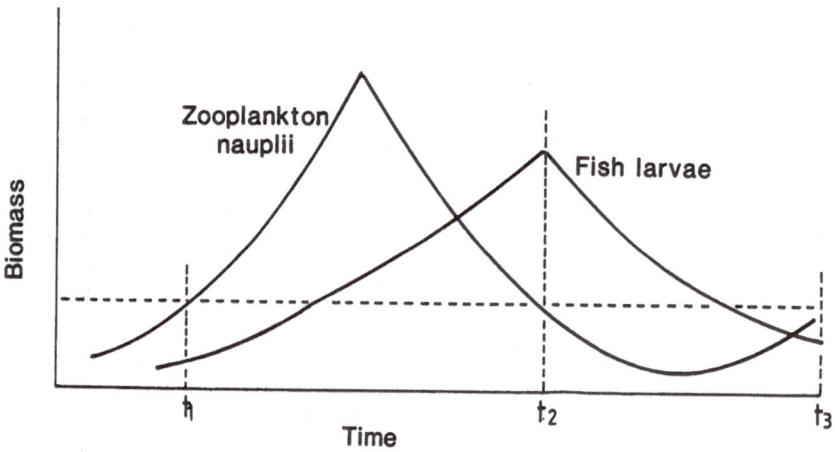

Figure 5. Diagrammatic changes in the biomasses of zooplankton nauplii and fish larvae at the time of the spring bloom.

3.3 The Effect of Varying the Duration of the Larval Stage

Increasing the minimum period of larval duration (i.e. the number of days (= steps) of unlimited growth needed for a larva to become classed as a juvenile) was found to be advantageous up to a point. If the duration was too long however, the larvae were more likely to run out of food before they were large enough to be promoted to the juvenile stage in the model.

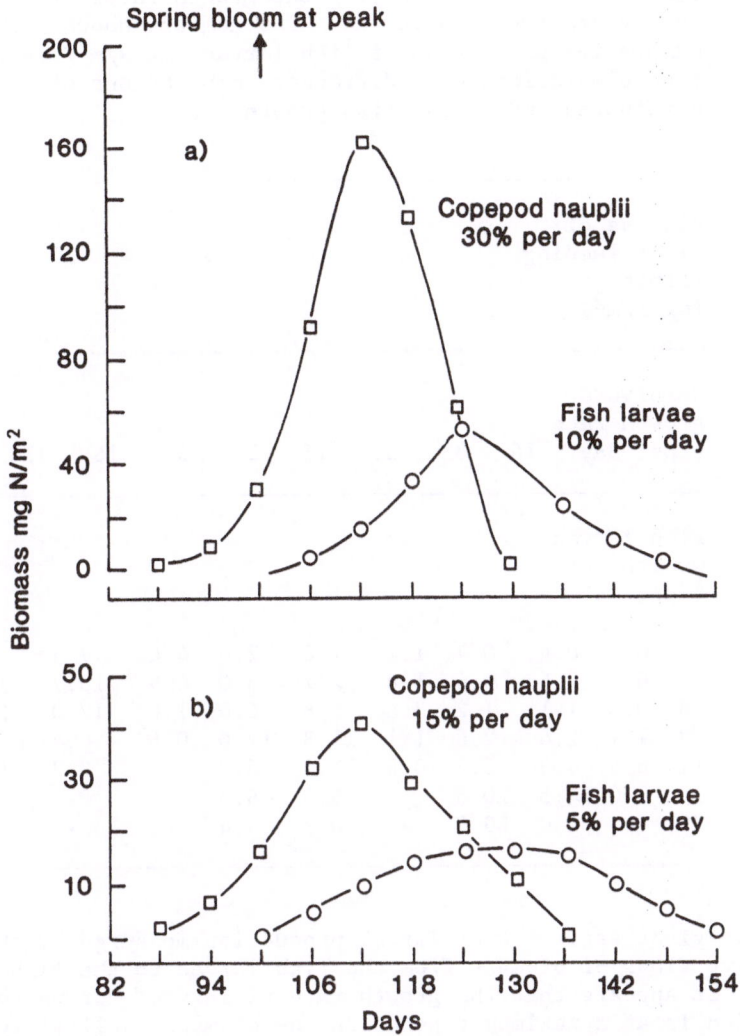

Figure 6. Particular relationships between the biomasses of copepod nauplii and fish larvae at the time of the spring bloom from the simulation model.

a) results when the potential growth rates of copepod nauplii and fish larvae were 30% per day and 10% per day respectively.
b) results when these potential growth rates were 15% per day and 5% per day respectively.

Table 1. Effect of varying the growth rates of herbivores and fish larvae. The numbers shown are cumulative production of fish larvae (mg N/m^2/season) from simulations with different combinations of herbivores and larval fish growth rates.

Biomass of first feeding larvae (mg N/m^2)	1			5			20		
Herbivore growth rate % per day	16	33	55	16	33	55	16	33	55
Fish larvae Growth rate % per day									
1	0.2	0.4	0.4	1.2	1.8	2.0	4.4	7.4	8.0
3	0.4	0.6	0.6	1.8	2.7	3.0	4.9	10.7	11.5
6	0.8	1.1	1.2	3.6	5.8	6.0	1.0	17.0	24.4
8	1.7	2.4	2.6	1.1	11.8	12.6	0.6	0.4	50.7
11	3.4	4.9	5.2	0.8	11.0	26.0	-	0.2	41.4
16	1.0	14.5	20.8	-	5.0	46.3	-	-	1.4
22	-	2.0	59.2	-	0.3	4.4	-	-	0.9

The Table gives estimates of larval production measured by the cumulative flows of biomass from the fish larvae to the fish juvenile stages. It appears that the growth rate of larvae, for which larval production is at a maximum depends on the biomass of first feeding larvae as well as on the growth rate of their food organisms.

3.4 Effect of Increasing the Biomass of First Feeding Larvae

Figures 7a-c show the effects of increasing the biomass of first feeding larvae. In general, it was found that as the biomass of first feeding larvae was increased, fish larval biomass tended to reach higher levels, but that the period of relatively high naupliar biomass tended to decrease. The conclusion is that as the initial number of first feeding larvae is increased, the time period of relatively high food concentration is shortened. As a result, larvae are more likely to stop growing before they have had time to become large enough to be promoted to the next feeding stage.

3.5 Effect of Spawning Period and Biomass of First Feeding Larvae

Estimates of larval production for different first feeding periods and different biomasses of first feeding larvae are given in Table 2.

Table 2. Effect of varying the biomass of first feeding larvae and the periods over which first feeding occurs. The cumulative production of fish larvae (mg N/m^2/season) from simulations with different combinations of larval biomass and feeding period.

First feeding period (day number)	Biomass of first feeding larvae (mg N/m^2)							
	0.5	1	2	5	8	14	20	40
60-80	0.1	0.2	0.4	0.9	1.4	2.3	3.0	5.2
70-90	0.4	0.8	1.5	3.3	4.9	7.1	8.8	9.9
80-100	1.0	1.9	3.8	8.2	11.8	3.8	1.1	0.7
90-110	1.8	3.6	6.8	16.4	3.3	0.5	N	N
100-120	2.5	4.9	9.9	11.0	3.6	N	N	N
110-130	3.0	6.0	12.0	8.5	2.2	N	N	N
130-150	4.1	8.2	16.4	5.2	0.1	N	N	N
140-160	4.8	10.1	19.2	3.8	0.01	N	N	N

N = Negligible

Figures 8a and 8b show larval production plotted against the biomass of first feeding larvae, for a range of first feeding periods.
Principal conclusions are:

a)   that, irrespective of spawning period, larval production reaches a maximum at some intermediate level of biomass of first feeding larvae, and

b)   that the optimum first feeding period depends on the biomass of first feeding larvae.

In practice there would presumably be selection for an optimum first feeding period (and hence spawning period) but these results indicate that this period would not necessarily be sharply defined, and could quite easily extend over as much as several weeks, as is observed in practice.

An important feature of the results in Figures 8a and 8b is that larval production tends to be relatively high at low levels of the biomass of first feeding larvae, over a wide range of spawning times.

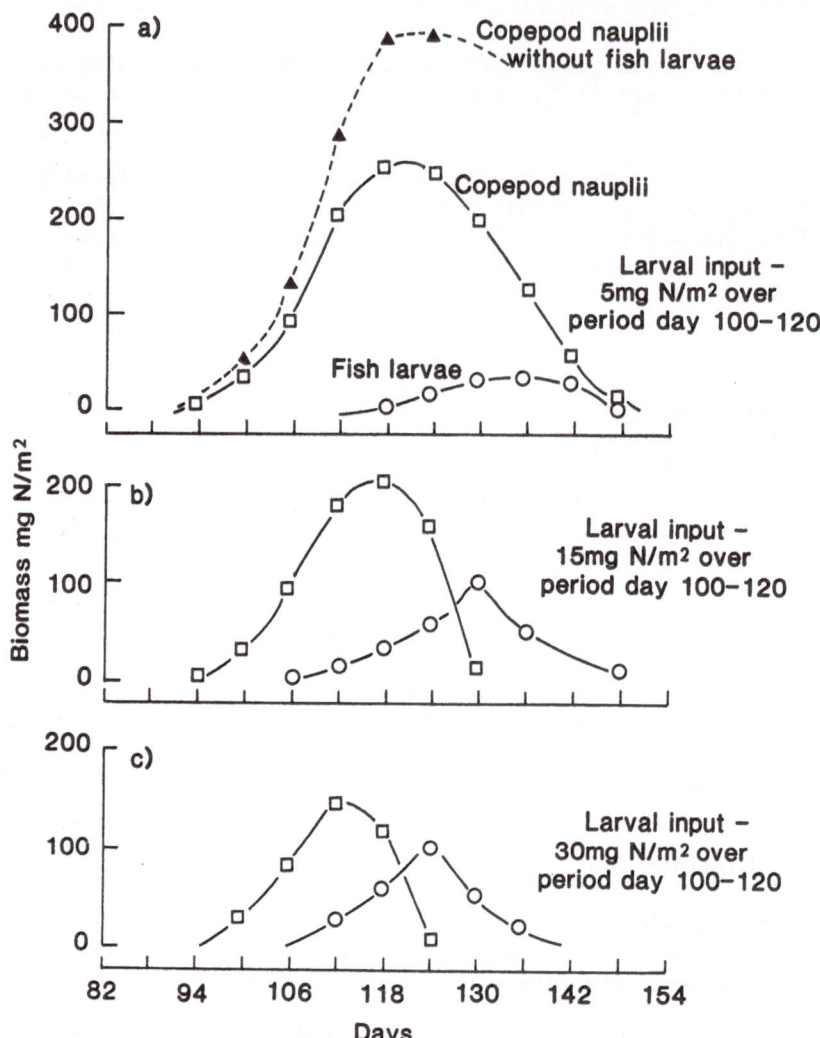

Figure 7. Three examples of the relationship between the biomasses of copepod nauplii and fish larvae from the simulation model when the input biomass of first feeding larvae was varied.

a) larval biomass - 5 mg N/m² over the period from day 100 to day 120.
b) as (a) but larval biomass equals 15 units.
c) as (a) and (b) but larval biomass equals 30 units.

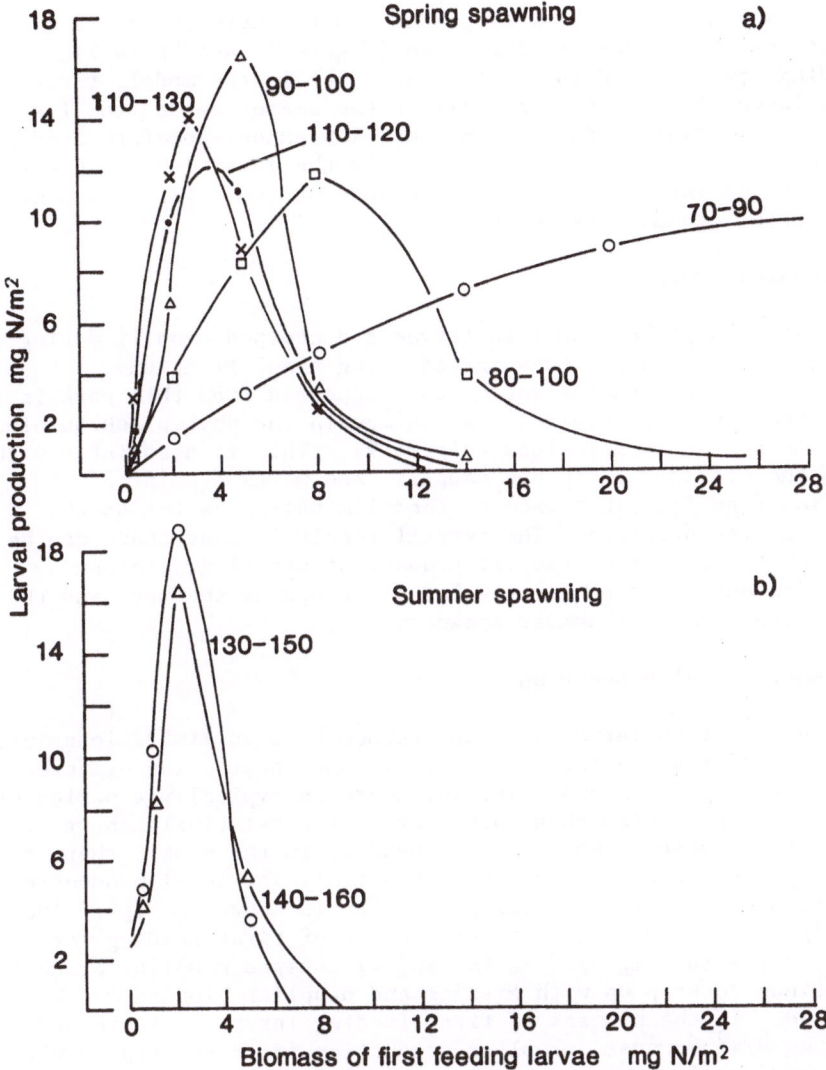

Figure 8. The cumulative production of larval biomass for different biomasses of first feeding larvae and different spawning periods - from the simulation model.

a)   at the time of the spring bloom
b)   during the summer

For very early first feeding periods (over days 60-80 and 70-90) there was a tendency for larval production to increase as the biomass of first feeding larvae was increased (Figure 8a and Table 2). This is misleading however, and due to the fact that in the model, first feeding larvae introduced too early in the season tended to di.e. of starvation. Larval production in these instances therefore tended to depend mainly on the small proportions of the initial biomasses that happened to be introduced towards the ends of these first feeding periods (i.e. around days 80-90).

3.6 Summer Spawners

Interrelationships between fish larvae and copepod nauplii during the summer months were also investigated using simulation runs.

One important generalisation that appeared from this work is that in the model there is typically a minimum in the phytoplankton biomass at the end of the spring bloom (Figure 9). This is associated with a peak in the biomass of copepod nauplii, and marks a point of transition from food abundance to food limitation as far as the herbivores are concerned. The overall result is that there can be a discontinuity in copepod nauplii production creating, a clear distinction between food for the larvae of spring spawners and the food for the larvae of summer spawners.

3.7 Summer Larval Production

Summer spawned fish larvae are constrained by food limitations just as much as spring spawned larvae. There is one interesting difference however. For spring spawned larvae, there is typically a period of high food concentration that lasts for only a relatively short time period. For summer spawned larvae however, in the model, there can be a longer period of continuous, but relatively low level production.

Just how this might larval production is shown in Figure 10a and 10b. Figure 10a shows that if the biomass of first feeding larvae is relatively low (0.5 mg N/m^2 in the model) copepod naupliar production can continue to keep up with grazing and naupliar biomass can be maintained. If the biomass of first feeding larvae is increased to about 2 mg N/m^2 however, naupliar production fails to keep up with grazing and naupliar biomass declines rapidly. This suggests that if egg production is relatively high, only fast growing larvae will eventually survive, as in the case of spring spawned larvae. If egg production is relatively low however, the effective growing period for larvae might not be curtailed so drastically as it is in the spring, and the expectation of survival of an individual larvae might be correspondingly greater.

3.8 Optimum Biomasses of First Feeding Larvae

If the values in Figure 8a and Table 1b for very early first feeders are neglected, as being misleading, the results of this study show

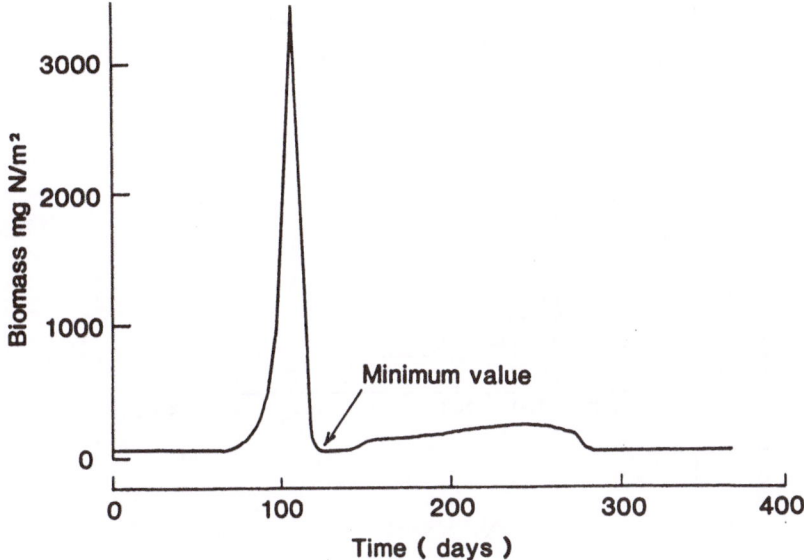

Figure 9. Showing the minimum value in phytoplankton biomass immediately after the spring bloom that was typically in very many of the simulation runs. This was due to grazing by zooplankton nauplii.

that optimum larval production occurs at biomasses of first feeding larvae around 2-5 mg N/m^2.

No direct estimates are available of the biomasses of first feeding larvae, but some idea of possible values can be deduced from estimates of the numbers of eggs produced. Some details are given by Jones 1973 and values for cod and haddock eggs throughout the north Atlantic range from 1-40 x 10^8 per 3 km^2. This is equivalent to 33 1320 eggs/m^2. The average calorific value of a gadoid egg is about 0.00055 Kcal (Hislop and Brown, in press) equivalent to about 0.055 mgC or 0.01 mg N. In nitrogen units, the biomass of eggs produced by an average year class is therefore about 0.3 -13 mg N/m^2.

Without an estimate of the mortality rate of eggs and of non-feeding yolk sac larvae, these estimates cannot be compared directly with the theoretical estimates obtained in this paper. It is interesting to note however that the theoretical values are within the observed range of values. Other theoretical estimates of the optimum number of first feeding larvae have been obtained by Jones (1973) based on a rather different computational approach. The optimum

358

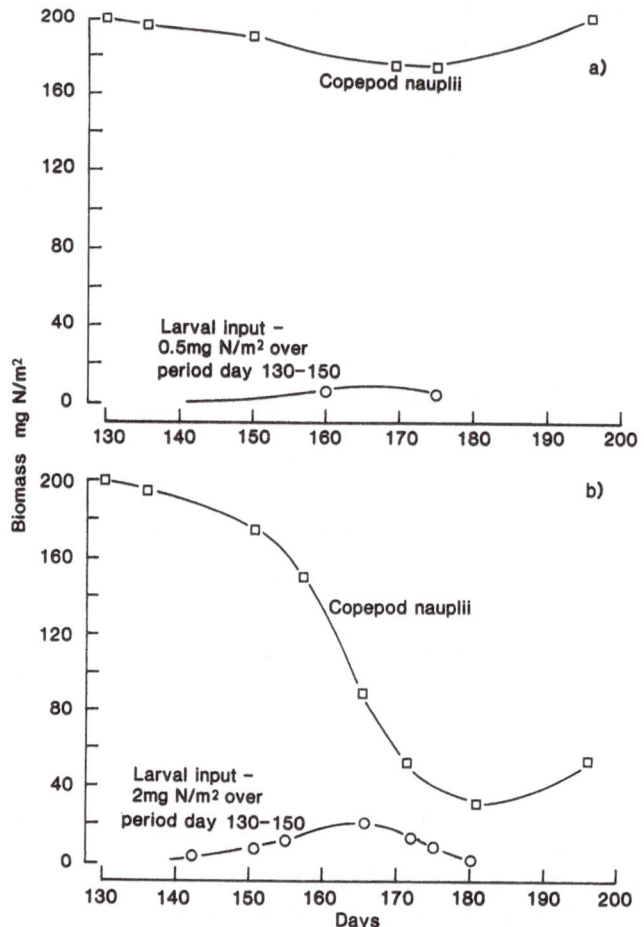

Figure 10. Biomass curves for fish larvae and copepod nauplii during the summer period - results from the simulation model.

a) biomass of first feeding larvae - 0.5 mg N/m^2 over a period from day 130 to day 150.
b) biomass of first feeding larvae - 2 mg N/m^2 over the same period.

values obtained in that paper were about $2\text{-}4 \times 10^8$ per 3 km^2, equivalent to about 0.6-1.3 mg N/m^2, which is somewhat lower than the values of 2-5 mg N/m^2 obtained in this paper.

The overall conclusion is that levels of egg production in North Atlantic gadoids are such that in the pre-exploited state at least, there would probably have been significant resource limitation at the larvae stage.

3.9 The Shape of the Larval Survival Curves

In practice the exact form of the relationship between initial biomass and biomass recruiting to the juvenile stage will depend on various factors. For example if the larvae and their prey were distributed uniformly, and if all larvae grew at the same rate, the relationships shown in Figures 8a and 8b might be expected.

In the real world, larval and prey biomasses are not likely to be uniformly distributed and all larvae do not grow at the same rate. In reality therefore the relationships in Figures 8a and 8b are likely to be "stretched out" along the x axis. Thus, even when the initial biomass of larvae is high, some larvae will be fortunate enough to be located in a region of relatively low larval density and high food concentration. Conversely, even if the initial biomass of larvae is low, some larvae will be unfortunate enough to be located in a region of high larval density and low food concentration. These factors can therefore be expected to affect the relationships in Figure 8 by reducing the peaks, and extending the right hand parts of the curves further to the right.

4. FACTORS LIKELY TO AFFECT LARVAL SURVIVAL

The main conclusions from the simulation studies is that in a seasonally varying world, a fish larva feeding on rapidly growing food organisms, is likely to experience a relatively high concentration of food but only for a relatively short period of time. To survive, it is presumably essential therefore that while food is abundant, a larva is able to grow large enough to be able to transfer eventually to feeding on other and larger food organisms.

From the point of view of an individual larva therefore it appears that there are two important criteria for survival during the early stages. Firstly, it should be in the right place, at the right time, to start feeding just as the biomass of suitable food organisms starts to increase. This conclusion is consistent with views put forward by many people, (particularly Hjort 1914 and Cushing 1986). The possession of a yolk sac therefore, may be important in this context, since this should permit an individual to remain alive for some time without feeding.

This in turn should increase the probability of an individual still being alive, at the right place and time, to start feeding when conditions do happen to be suitable for feeding.

Secondly, it should not be one of a large number of competing individuals, with a combined grazing power large enough to shorten the time during which the consideration of food organisms is above some critical value.

4.1 Variations in Larval Growth Rate

Variations in larval growth rate, are likely to influence the probability of survival, by generating relatively large variations in the length range of the larvae after a certain time. For example, after 80 days, a larva growing at 10% per day will increase its body weight over 2000-fold. A larva growing only a little more slowly however, at 8% per day, would only increase its body weight 472-fold.

Figure 11 indicates that by the end of the very rapid growth stage, which is about 80 days for a haddock (Jones, 1973), there is every likelihood that the range of larval size could be quite large. Many larvae therefore may fail to become large enough to make the transition to the juvenile stage before the end of the productive season. They may then fail to survive, irrespective of whether they are eaten or not.

Given the relatively high growth rate of larvae, and the potential for food limitation described above it is likely that in practice, some larvae will grow well, with a high probability of survival, whilst others will grow more slowly, with a much lower probability of continuing to survive.

Figure 12 shows one way of depicting the situation, with just two categories of larvae. One group (labelled "fast growers") is assumed to be growing normally and is composed of individuals that are destined to attain juvenile status, if they are not eaten. The other group (labelled "slow growers") is composed of individuals that are assumed to be growing too slowly to attain juvenile status, whether they are eaten or not.

A simple mathematical representation for such a system is given by the expression:-

$$N = No \exp-(D + P) t \qquad\qquad (1)$$

Where No = initial number of larvae
  N = number of surviving larvae, in the "fast growing" category, after a period of t.
  P = instantaneous rate of removal of "fast growers" due to predation.
  D = instantaneous rate of "displacement" of "fast growers" into the "slow growing" category.

The important feature of the equation is that the coefficient D is a function of the availability of food for larvae, and therefore not necessarily likely to be something that is constant with time. The equation may not be particularly useful for computational purposes therefore, but it clearly illustrates how food limitation could affect survival, even when the apparent cause of death is predation.

An alternative method of representation, and one that has been adopted in the simulation model, is to subdivide the fish larval "pool" into a number of stages as indicated in Figure 3. The passage of biomass from first feeding stage to juvenile stage, is then dependent on the rate of transfer through each of there intermediate stages. Within the model, this rate has been made a function of food concentration at each step.

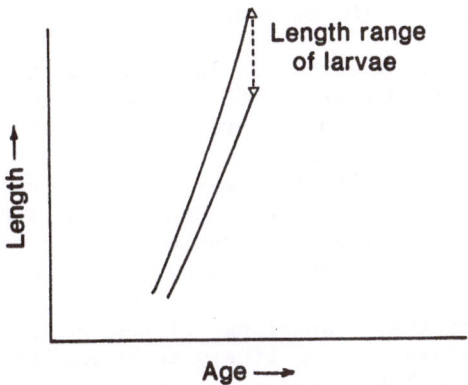

Figure 11. Showing that for rapidly growing organisms like fish larvae, even a small variation in growth rate can quickly generate a relatively large variation in length range.

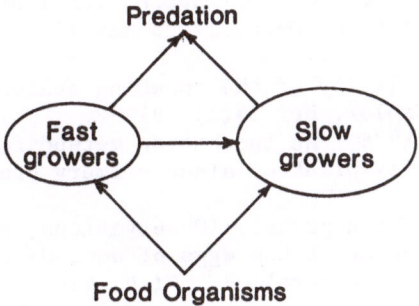

Figure 12. A subdivision of fish larvae into fast growers and slow growers as deduced from the simulation results.

4.2 Predation and Food Limitation as Causes of Mortality

Although larvae that do not survive may be presumed to die of predation or starvation, it is not likely that the effects of these two causes of mortality can be separated in any simple way, in the field. Equation 1) for example, demonstrates the important fact that to calculate the number of surviving larvae, it is necessary to know not only the number of predators (to determine the value of P), but

also the rate of "displacement" (D) of individuals into the slow
growing category.

Once they become resource limited, it is possible that
individuals may become weakened sufficiently to become vulnerable to
organisms that could not capture non-resource limited larvae. In this
way they might become eliminated relatively rapidly, and before there
was time for obvious signs of starvation among the survivors to become
evident. It is quite possible therefore to conceive of a period in
which mortality was principally due to predation, but in which the
actual numbers dying were largely determined by larval abundance and
food availability.

The main conclusion is that a knowledge of the number of larvae,
and of the number of predators, is not likely to be sufficient for
calculating the overall larval survival rate.

5. REFERENCES

Anon: 1980, 'Workshop on the Effects of Environmental Variation on the
    Survival of Larval Pelagic Fishes', Lima, 20 April - 5 May 1980.
    IOC Workshop Report No. **28**, UNESCO.
Blaxter, J.H.S. (ed.): 1973, *Report of the International Symposium on
    the Early Life History of Fish*, Oban, 17-23 May 1973. Springer
    Verlag, Berling, 765p.
Cushing, D.H.: 1969, 'The regularity of the spawning season of some
    fishes. *J. Cons. int. Explor. Mer* **33(1)**, 81-92.
Gargas, E. and C.S. Nielsen: 1976, 'An incubation method for
    estimating the actual daily plankton algae primary production.
    Water Res. **10**, 853-860.
Hislop, J.R.G. and M.A. Brown: (In press), 'Observations on the size,
    dry weight and energy content of the eggs of some demersal
    species from British marine waters', *J. Fish. Biol.*
Hjort, J.: 1914, 'Fluctuations in the great fisheries of Northern
    Europe', *Rapp. P.-v. Cons. int. Explor. Mer* **20**: 1-228.
Hunter, J.R.: 1976, 'Report of a Colloquium on Larval Fish Mortality
    Studies and their Relation to Fishery Reserach, January 1975',
    NOAA Tech. Rep. NMFS CIRC-395, Seattle, WA.
Jones, R.: 1973, 'Density dependent regulation of the numbers of cod
    and haddock', *Rapp. P.-v. Reun. Cons. Perm. Int. Explor. Mer*,
    164, 156-173.
Jones, R.: 1981, 'Simulation studies of the larbal stage and
    conclusions relating to the first year of life with particular
    reference to the haddock', *Rapp. P.-v. Reun. Cons. Perm. Int.
    Explor. Mer*, **178**, 15-16.
Jones, R: 1982, 'Species interactions in the North Sea', in M.C.
    Mercer (ed.), Multispecies approach to fisheries management
    advice, *Can. Spec. Pub. Fish. Aquat. Sci.* **59**, pp. 48-63.
Jones, R. and W.B. Hall: 1973, 'A simulation model for studying the
    population dynamics of some fish species', in M.S. Bartlett and
    R.W. Hiorns, (eds.), *The Mathematical Theory of the Dynamics of
    Biological Populations*, Academic Press, 347p.

Jones, R. and E.W. Henderson: (In press), 'The dynamics of the nutrient cycle and simulation studies of the nutrient cycle', *J. Cons. int Explor. Mer.*

Jones, R. and E.W. Henderson: (In press), 'The dynamics of energy transfer in marine food chains', *S. African J. Mar. Sci.*

Mullin, M.M. and E.R. Brooks: 1970, 'The effect of concentration of food on body weight, cumulative ingestion, and rate of growth of the marine copepod *Calanus helgolandicus*', *Limnol. Oceanogr*, **15**, 748-755.

Robb, A.P. and J.R.G. Hislop: 1980, 'The food of five gadoid species during the pelagic O-group phase in the North Sea', *J. Fish. Biol.* **16**, 199-217.

Parrish, B.B.: 1973, 'Fish stocks and recruitment', *Rapp. P.-v. Reun. Cons. Perm. Int. Explor. Mer*, **164**, 372p.

Steele, J.H.: 1974, *The Structure of Marine Ecosystems*, Oxford: Blackwell Scientific Publications.

APPENDIX 1

Details of simple ecosystem model

Main Equations

This appendix gives details of the equations used for simulating the
simple model shown in Figure 4. For greater generality, the dead
organic pool has been expanded into several pools and these, along
with descriptive labels are shown in Figure 13. Nitrogen pools are
labelled with capitals. Flows and constants are labelled with small
letters. All units are in terms of quantities of nitrogen.

$$NU_2 = NU_1 + (exc + bad.DOM(5) + bar. REF - (1+g).PHgrph) \Delta t . \quad 1)$$
$$PH_2 = PH_1 + (grph.PH - ps - phgr) \Delta t \dots\dots\dots\dots\dots 2)$$
$$REF_2 = REF_1 + (xref - bar. REF) \Delta t \dots\dots\dots\dots 3)$$
$$DOM(1)_2 = DOM(1)_1 + (detr - bad(1).DOM(1)) \Delta t \dots\dots\dots 4)$$
$$DOM(I)_2 = DOM(I)_1 + (bad (I-1). DOM(I-1)-bad(I).DOM(I)) \Delta t \dots\dots 5)$$
$$\text{for } I = 2 \text{ to } 5$$
$$AN_2 = AN_1 + (phgr(1- zaf) - fim - exc) \Delta t \dots\dots\dots\dots 6)$$

where

NU	=	Total quantity of nitrogen in the inorganic nitrogen pool
PH	=	Biomass of phytoplankton
REF	=	Total quantity of nitrogen in the refractory pool
$DOM(1)$	=	Total quantity of nitrogen in the first of the seasonal pools of dead organic nitrogen
$DOM(I)$	=	Total quantity of nitrogen in the remaining four pools of dead organic nitrogen
AN	=	Biomass of animals

The suffices 1 and 2 refer to the values in each of the above pools at
the beginning and end respectively of an interval Δt.

Living micro-organism biomass is not included explicity, since, at all
times of the year this appears to be insignificant compared to the
magnitudes of the other pools.

Divison by 365 is introduced, to make each step equivalent to one day.

Principal Flows

grph	=	net particulate primary production due to phytoplanton growth - based on the Michaelis-Menten relationship ie
grph	=	$a.mg.NU/(kn. dp + NU)$
where mg	=	maximum rate of phytoplankton growth when nutrient is unlimited.
kn	=	nutrient concentration at which phytoplankton growth rate = mg/2
dp	=	depth of euphotic zone and

a   — light factor. Made to vary seasonally, using a formula
by Gargas and Nielsen (1976), scaled so that at its
maximum, a = 1.

g.grph.PH   — Flow of inorganic nitrogen to bacteria

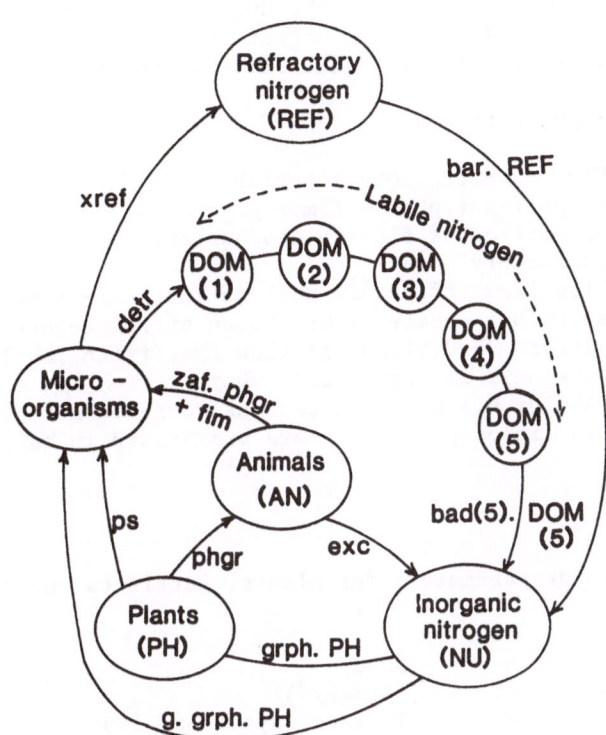

Figure 13. Expansion of model in Figure 4 along with descriptive
labels as used in connection with the equations described in
Appendix 1.

where
g   — bacterial uptake of inorganic nitrogen as a proportion
of particulate primary production.

ps   — Flow of phytoplankton to dead organic matter either in
the water column or on the bottom.
     — sk. PH

sk   — rate of conversion of plants to detritus.

phgr   — Flow of phytoplankton to animals due to grazing in the
water column or on the bottom.

exc   — Excretion of inorganic nitrogen by animals phgr and exc
can be input values, or calculated from further
equations based on a food chain and a
phytoplankton/herbivore relationship.

zaf.phgr. flow of faeces from animals

where zaf	=	proportion of grazing flow that goes to dead organic matter in the form of herbivore faeces.
fim	=	Food induced mortality of animals. Term included to allow for animal deaths due to food limitation.
detr	=	Flow of labile dead organic matter
	=	$(1-rp)(ps + zaf.phgr + fim + g.grph.PH)$

  where rp = input parameter

xref	=	Flow of refractory material
	=	$rp(ps + zaf.phgr + fim + g.grph.PH)$
rp	=	proportion of flux in dead organic pool that becomes refractory.
bar.REF	=	Flow from refractory pool to inorganic nitrogen pool, where bar = rate of breakdown of refractory organic matter. bad(5).DOM(5) flow from fifth labile pool to inorganic nitrogen pool, where
bad(I)	=	rate of breakdown of seasonal organic matter.

Values used for the input parameters are summarised in Table 3.

Table 3.

Values of input parameters for plants, nutrients and organic matter

dp	=	50 m
kn	=	$14(mgN/^3)$
sk	=	12 from days 85 to 145. Otherwise sk = 5 (units gN/gN/yr)
mg	=	maximum value 365-720 (gN/gN/yr)
g	=	1.3
zaf	=	0.3
rp	=	0.3
bad(I)	=	11-29 (gN/gN/yr at Tm°C)
bar	=	0.3 (gN/gN/yr at Tm°C)

constants varied seasonally, using a temperature factor Tf:

Tf	=	$Exp(0.081(T-Tm)$
Tm	=	8°C
T	=	Temperature varied sinusoidaly from 5 in spring to 13°C in autumn

Total nitrogen = 20 gN/m^2

APPENDIX 2

Equations for determining the rate of change of biomass of a single animal life stage, with reference to Figure 14.

Basic Equations for Each Animal Life Stage

$$LF2 = LF1 + (LF1.(fe-ex-fim) + r1 - r2 - fep) \Delta t \dots (1)$$

where
- LF_1 = Biomass of life stage at beginning of interval Δt
- LF_2 = Biomass of life stage at end of interval Δt
- r_1 = Rate of flow to life stage from previous life stage
- r_2 = Rate of flow to next life stage
- fep = Rate of consumption of life stage by predators
- fe = Rate of food consumption by life stage per unit biomass of life stage.
- ex = Rate of loss in weight per unit biomass of life stage, due to excretion/respiration.
- fim = Rate of food induced mortality per unit biomass of life stage.

Calculation of Terms in Equation (1)

Food eaten by predators (fep)

The biomass of a life stage consumed by predators is assumed to depend on the concentration of animals in the life stage (in weight units), and the vulnerability of the individuals that make up the life stage. The basic equation is:

$$fep = BP.x1 \ (gp/\varepsilon + mp) \dots \quad (2)$$

where
- BP = Biomass of predators
- gp = Maximum growth rate per unit biomass of predators (including gonad growth if applicable) when food for predators is unlimited.
- mp = Rate of food consumption per unit biomass of predators needed to satisfy the maintenance requirements of the predators.
- x_1 = Food consumption index that depends on biomass and vulnerability of the individuals in the life stage on which the predators are feeding (see below).
- ε = Efficiency of conversion of food available for growth, into growth.

The term x_1 is calculated from information about the animals in the life stage on which the predators are feeding. A possible expression for x_1 is given by:

$$x_1 = LF_1/(LF_1 + bl \ (1-vl) \dots \quad (3)$$

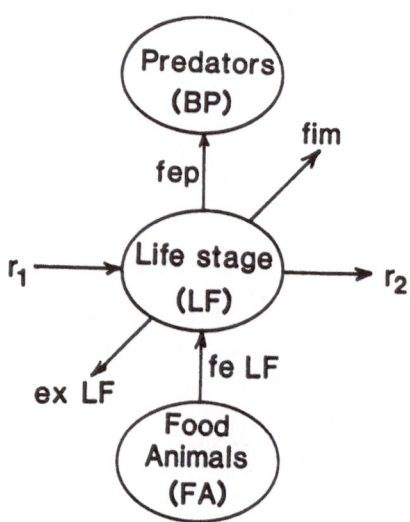

Figure 14. Relationship of animals in a life stage to their predators and food animals plus descriptive labels as used with the equations in Appendix 2.

where b_1 — Biomass of life stage (LF_1) at which a predator succeeds in capturing just half of what it would be capable of eating if its food was unlimiting.

   vl   — Vulnerability factor for individuals in life stage (see below).

The vulnerability of animals in the life stage (LF_1) is assumed to depend on the biomass of the food organisms on which they are feeding.

$$i.e. \qquad vl = BFA/(FA + bfa) \quad \ldots \ldots \qquad (4)$$

where   FA   — Biomass of food animals

   bfe   — Biomass of food animals for which vl is equal to 0.5

Rate of Food Consumption (fe) of Food Animals by per unit biomass of Animals in the Life Stage

Given unlimited food, the rate of food consumption per unit biomass of animal, needed to satisfy maintenance and growth, would be equal to $gp/\varepsilon + mp$ where gp and mp are now equal to the growth and maintenance coefficients for the animals in the life stage. To allow for food limitation, a term x_2, analogous to x_1 above, can be introduced so that:

$$fe = x_2 \ (gp/\varepsilon + mp) \ \ldots \ldots \qquad (5)$$

where x_2 =   Food consumption index that depends on the biomass and vulnerability of the food animals eaten by the life stage.

           ε =   Efficiency of conversion food available for growth, into growth - here applicable to animals in the life stage.

The term x_2 is calculated from information about the food animals on which the life stage is feeding and by analogy with equation 3), a possible expression is given by:

$$x_2 = FA/(FA + bp(1-v_p)) \ \ldots \ldots \qquad (6)$$

where   FA   =   Biomass of food animals

           bp   =   Biomass of food animals for which an individual in the life stage succeeds in capturing just half of what it would be capable of eating if its food was unlimiting.

           v_p =   Vulnerability factor for the food animals.

By analogy with 4), the term vp is assumed to depend on the amount of food available for the food animals.

$$i.e. \qquad v_p = cppr/(FFA + cppr) \ \ldots \ldots \ (7)$$

where   FFA   =   Biomass of the food of the food animals.

          cppr   =   Biomass of the food of the food animals for which vl is equal to 0.5.

As one proceeds to lower trophic levels, the "food animals" referred to in this context may be phytoplankton or even nutrients. If the assumption of variation in vulnerability seems biologically unrealistic, the vulnerability term can be made equal to zero.

Excretion Term (ex)

The rate of excretion per unit biomass of animals is given by the difference between food eaten and actual growth per unit biomass.

Actual growth is given by $\varepsilon(fe-mp)$.
If actual growth is positive, (*i.e.* if fe-mp> = 0)

$$ex = fe-\varepsilon(fe-mp) \ \ldots \ldots \qquad (8a)$$

If actual growth is negative however, it is preferable to write something like:

$$ex = mp \ldots\ldots \qquad (8b)$$

Alternative expressions can be tried if desired.

Food Induced Mortality (fim)

The rate of food induced mortality per unit biomass of animal in the life stage can be represented by writing:

$$fim = mr \ (1 - x_2) \ \ldots\ldots \qquad (9)$$

where   mr   = mortality rate (directly or indirectly due to starvation) when $x_2 = 0$.

The flux generated by food induced mortality can be directed to the detritus pool, or to the predators as seems appropriate.

Lateral Rates of Transfer (r_1 and r_2)

The rate of flow from one life stage to the next will depend on the actual rate of growth of the animals in the life stage concerned. For example, let dur be the duration in days of a life stage when food is unlimiting and growth is at its maximum rate. To a first approximation one can say that the proportion of individuals transferring to the next life stage, on any one day, will be equal to 1/dur. In terms of biomass this becomes LF_1/dur.

More generally, for a time interval equal to Δt years, one can write

$$r_2 \ \Delta t = 365.LF_1.\Delta t/dur$$
so that $\qquad r_2 = 365 \ LF_1/dur \ \ldots\ldots \qquad (10)$

This formulation, assumes a uniform distribution of biomass by age, within a life stage. For young, rapidly growth individuals, this assumption is likely to be unrealistic and the model can be made more realistic by sub-dividing the youngest stages into many very short stages, as shown in Figure 3. In these simulations this was done by allowing one transistion stage per day's growth so long as food was unlimiting *i.e.* if the larval stage was meant to represent 50 days, there would be 50 sub-stages in the larval stage. If food was unlimiting, it would then take 50 days for larval biomass to be transferred through the whole series of stages and to attain juvenile status. Whenever food became limiting however, a reduced growth rate was allowed for by transferring only a proportion of what was in each stage to the next stage at each time step.

r_2 for the oldest life stage (which is the same as r_1 for the youngest life stage) can be given as input data or calculated from some estimate of the amount of growth that might have gone into a reproductive model.

Total Nitrogen and Negative Biomasses

Within an enclosed system, total nitrogen must remain consistent, and it was found useful to keep a running check on the cumulative nitrogen total for all pools. A departure from a constant value, after a program change, was a sure sign of an error.

The programme also included a sub-routine making sure that in no pool was biomass allowed to become negative.

Table 4.

Sample set of values of input parameters - for the equations in Appendix 2.

gp - growth rates per unit biomass when food is unlimiting 1)

Herbivores	120	50	0	) (gN/gN/yr)
Fish	40	20	2	)

mp - maintenance food per unit biomass

Herbivores	12	12	12	) (gN/gN/yr)
Fish	4	4	4	)

mr - food induced mortality rates when food intake = zero

Herbivores	0	0	0	)
Fish	40	20	0	)

dur - durations of stages when food is unlimiting   2)

Herbivores	7	23	) (days)
Fish	60	53	)

bp - grazing coefficients

Herbivores on phytoplankton	1000	600	100	) mgN/m2 (in 50 m
Fish on herbivores	20	200	250	) deep water)

egg production 3)

Herbivores	6	gN/gN/day	
Fish	0.5	mgN/m^2/day	during productive season from day 100 to day 120

v - vulnerability factors. For simplicity, they can be made equal to zero initially

$\Delta t = 1/365$ years

1. values for each of the three life stages per trophic level.
2. values for each of the two younger life stages per trophic level.
3. just one way in which herbivore and fish egg production can each be allowed for - used for calculating values of r_1 for the youngest life stages.

Where appropriate rates were varied seasonally using the temperature factor Tf (see Table 3).

THE "RECRUITMENT PROBLEM" FOR MARINE FISH POPULATIONS WITH EMPHASIS ON GEORGES BANK

E.B. Cohen, M.P. Sissenwine
United States Department of Commerce
National Oceanic and Atmospheric Administration
National Marine Fisheries Service
Northeast Fisheries Center
Woods Hole, Massachusetts 02543
U.S.A.

and G.C. Laurence
United States Department of Commerce
National Oceanic and Atmospheric Administration
National Marine Fisheries Service
Northeast Fisheries Center
RR 7, South Ferry Rd.
Narragansett, Rhode Island 02882
U.S.A.

ABSTRACT. The factors that determine recruitment of marine fish populations have been the focus of research for nearly a century. Much research on Georges Bank and around the world has been guided by hypotheses that relate the size of a year class to (a) the availability of prey during the larval period and (b) the physical transport of eggs and larvae to unfavorable locations. On Georges Bank water column stability enhances feeding conditions for cod and haddock larvae by concentrating their prey at the thermocline. Consequently, larval growth is higher over the stratified portion of the Bank than in the central, well mixed region. Although starvation mortality does not appear to be population limiting over the range of prey densities observed in the thermocline, storm events can disrupt stratification and potentially cause starvation. Eggs and larvae may also be advected off the Bank during storms or as a result of warm core rings. Additionally, abnormally low temperature may threaten survival of haddock eggs and larvae in some years. Pre recruit fish suffer about the same cumulative mortality during the post-larval stage as they do as eggs and larvae. Predation is the most likely cause of post larval mortality and it has the potential to affect recruitment and species composition of fish populations. A more comprehensive investigation of recruitment processes from the gonad development stage of adults to recruitment will be necessary.

B. J. Rothschild (ed.), Toward a Theory on Biological-Physical Interactions in the World Ocean, 373–392.

1. INTRODUCTION

Variability of recruitment is responsible for most of the population
fluctuations affecting major fish stocks around the world. The factors
responsible for recruitment fluctuations are still largely unknown
despite the fact that recruitment has been studied for over a century.
Researchers working on Georges Bank have applied a multifaceted
approach including laboratory experiments, mesocosm techniques, field
investigations, biochemical studies, and simulation modelling. This
paper reviews research concerning the "recruitment problem", for
Georges Bank and vicinity, in order to assess what aspects of the
problem are solvable and what redirection of research is appropriate.
    Georges Bank is a northeastward extension of the middle Atlantic
shelf, approximately 150 miles off of Cape Cod (Figure 1). Its area
within the 200 m depth contour is about 53,000 km^2. The circulation
on Georges Bank is characterized by high tidal energy and a "leaky
gyre" (Butman et al. 1982), which dominate the production at the lower
trophic levels (Davis 1982; Schlitz and Cohen 1984; Cohen and
Grosslein 1987; O'Reilly et al. 1987) and may strongly affect the
growth and survival of young fish in some years. Georges Bank is one
of the most productive marine environments in the world, O'Reilly and
Busch (1984) reported a primary productivity of 470g C m^{-2} yr^{-1}
for the shoal well mixed part of the Bank and 300g C m^{-2} for the
deeper seasonally stratified zone. The biomass of fish and squid on
the bank is about 1 million metric tons (Grosslein et al. 1980) with a
maximum sustainable yield (MSY) of about 400,000 tons (Cohen and
Grosslein 1987).

2. RELATIONSHIP BETWEEN PREY ABUNDANCE AND LARVAL GROWTH AND SURVIVAL

It has been hypothesized that prey abundance and successful feeding
are critical to growth and survival and strongly influence recruitment
(e.g. Hjort 1914). Examining this hypothesis has been the focus of
most recruitment research on Georges Bank.
    Laboratory studies on larval cod and haddock established the
critical ranges of food requirements (Laurence 1974), growth and
metabolic rates (Laurence 1978), and the influence of temperature and
salinity (Laurence and Rogers 1976). Mesocosm techniques were used
as a transitional approach between laboratory studies and field
research. An "in situ" environmental chamber demonstrated that at
field concentrations of prey organisms, larval growth and survival
could be maintained and measured (Laurence et al., 1979). A land
based mesocosm was used to test the effects of a stratified and a
well mixed water column on successful feeding and growth of gadid
larvae. Results strongly indicated that an environment similar to
the region on Georges Bank with a seasonal thermocline, was
beneficial for larval survival (Bergen et al. 1985).
    Simulation modelling evolved from a simple mass balance approach
based on deterministic bioenergetic measurements to stochastic

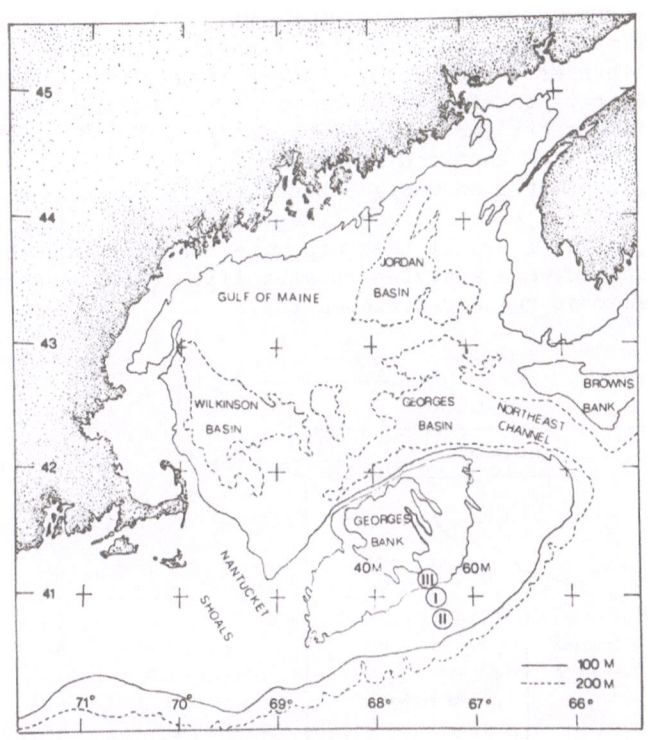

Figure 1. Map of Georges Bank and vicinity. The location of the study sites represented in Figures 2-3 (Buckley and Lough 1987) area indicated by the symbols I, II, III on eastern south of Georges Bank.

population simulations, culminating in a multi-factor, stochastic model of starvation, growth and mortality during the larval stage of Georges Bank cod and haddock (Laurence, 1985a). The conclusions based on the larval population responses were: (1) starvation is undoubtedly one of the largest components of total larval mortality, (2) starvation mortality is most significant in the first 2-3 weeks after hatching and (3) starvation mortality does not appear to be population limiting or the single controlling mortality factor under the normal range of prey densities.

Information from the laboratory, mesocosm, and modelling studies was used to design a series of cruises to Georges Bank concentrating on the feeding dynamics of larval cod and haddock. Special emphasis was given to a description of the spatial-temporal variability of the larval and prey distributions on scales from meters to kilometers and minutes to weeks and to the identification and measurement of both small and large scale physical process affecting the biological distributions. Results indicated that larval food was contagiously

376

distributed and that larvae and prey co-occurred vertically in the
water column (Laurence et al, 1984; Lough 1984). The co-occurrence of
larvae and their prey seems to depend on the existence of a
thermocline, but can be disrupted by a storm (Lough 1984, Figure 2)
with possible deleterious effects on the larvae. Buckley and Lough
(1987) used a biochemical indicator of recent growth and condition
(RNA-DNA ratio), to show that cod and haddock larvae collected at two
thermally stratified sites on Georges Bank were in better condition
than those collected at a well mixed site (Figure 3-4). Less than 5%
of the larvae at sites I and II combined were starving (RNA-DNA ratio
<3.5) while about 50% were starving at site III. Prey abundance was
considerably higher at the stratified sites.

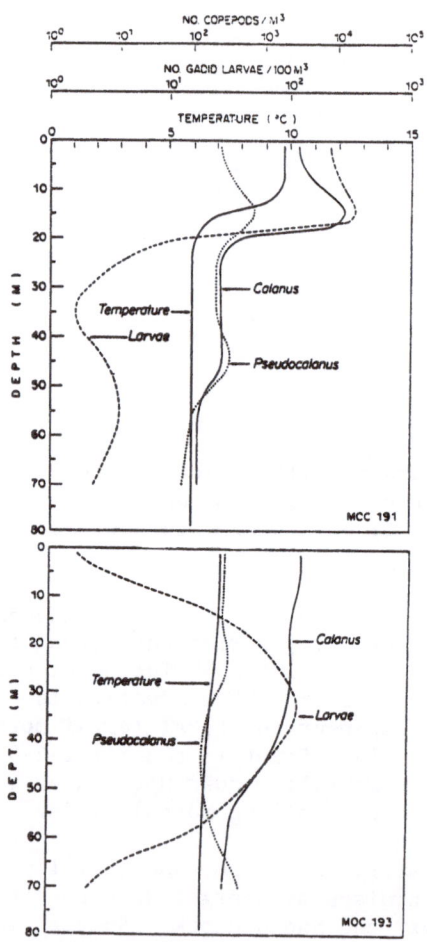

Figure 2. Vertical distribution of temperature, gadid larvae and prey
at a location on the southern flank of Georges Bank before (May 21,
1981) and after (May 24, 1981) a storm.

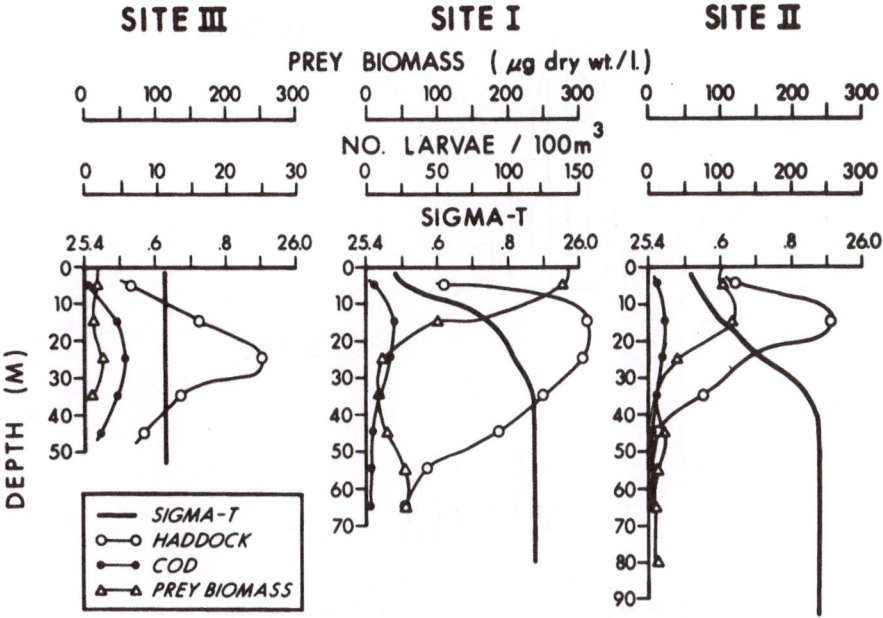

Figure 3. Vertical distribution of temperature, haddock larvae, cod larvae and prey biomass at three sites on Georges Bank after Buckley and Lough 1987. See Figure 1 for the location of the sites.

3. COMPETITION BETWEEN LARVAE

So far, we have focused on the effect of prey density on the growth and starvation of larvae as predators. Another aspect of the question is the effect of the larvae on their prey. Unless larval predators are abundant enough to reduce the concentration of their prey, the growth and starvation related mortality rate of the larvae is independent of larval abundance. Density dependence is assumed in most models of the spawner-recruit relationship so it is important to verify a density dependent biological mechanism.

Cushing (1983) analyzed data from Laurence (1985b) and concluded that late stage larvae, as they approach metamorphosis, may affect the density of their prey. Cohen et al. (1984) also analyzed the data for Georges Bank cod and haddock and determined that a cohort of haddock larvae consumed about 8% to 25% of the available microzooplankton production. The amount of the microzooplankton production consumed was about 60% in a larval patch. A similar calculation for post larval fish indicated that they could consume about 30% of the microzooplankton production if microzooplankton were their exclusive prey. The effect of post larval fish on microzooplankton is actually less because they also eat larger zooplankton and benthic invertebrates as well. The situation is

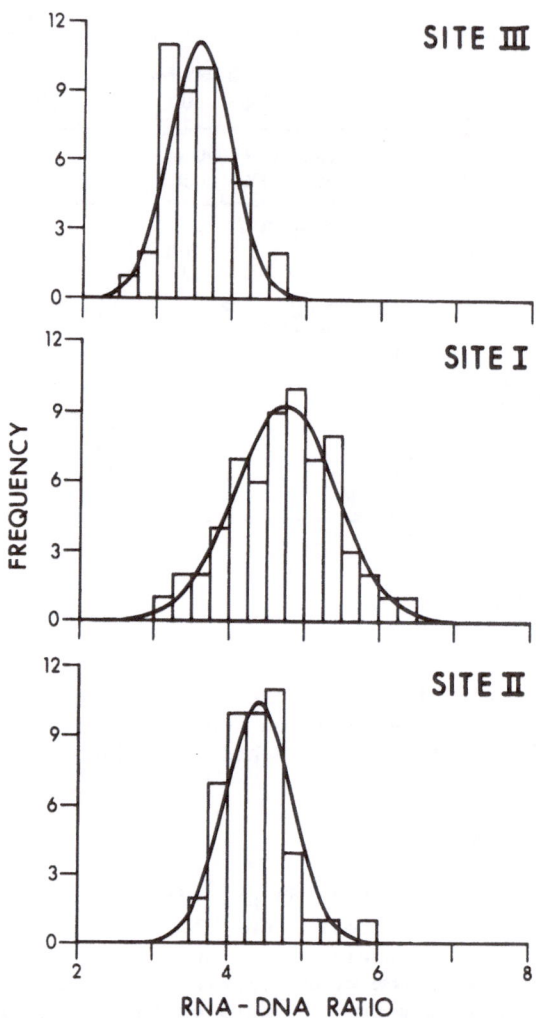

Figure 4. RNA-DNA ratio for larval haddock collected of three sites on Georges Bank after Buckley and Lough 1987. A ratio of <3.5 indicates starvation. See Figure 1 for the location of the sites.

similar for cod. The question is however, whether the consumption of 25% or 30% of the microzooplankton production by a cohort, or the consumption of 60% in a patch of larvae is significant. Sissenwine et al. (1984) concluded that the energy budget for Georges Bank left little excess production, *i.e.*, the system was tightly coupled in terms of available energy, particularly during periods of high fish abundance. This indicates that competition for food may exist at the community level, and the potential certainly exists for density dependence through competition for prey.

4. PHYSICAL TRANSPORT OF EGGS AND LARVAE

Offshore transport has been hypothesized as a process affecting the recruitment of Georges Bank populations. Walford (1938) collected larval haddock outside the 200 m isobath in the slope water. It appears that investigators sometimes found evidence for advective loss of haddock (Bigelow 1926; Chase 1955; Colton 1959). Colton and Temple (1961) questioned how haddock could sustain themselves on the bank with respect to advective loss. Advective losses could be caused by either storms or warm core rings (WCR's).

Interestingly, later investigators (Bumpus 1976; Laurence and Burns 1982; Smith and Morse 1985) did not find evidence for the advective loss of cod or haddock eggs and larvae from the bank. While it is possible that conditions have changed since the late 1950's an alternative explanation for not observing eggs and larvae off of the bank is the small number of stations in recent years outside 200m and the short time that eggs and larvae persist after being advected off the bank.

Flierl and Wroblewski (1985) compared data on WCR's in the vicinity of Georges Bank with subsequent cod and haddock year class strength for 1975-1979. They concluded that the number of WCR's near the bank was inversely correlated with recruitment success. Cohen et al. (1982) examined the same data set but used the "ring weeks" (the number of weeks a ring was in close proximity to the bank) as an indicator of the effects of a ring on cod and haddock recruitment. They found no correlation for the same recruitment data. Unfortunately, there is no measure available of the actual entrainment caused by a WCR.

5. THE IMPORTANCE OF THE POST-LARVAL PERIOD

Most recruitment research has focused on eggs and larvae. However, the age or life stage at which recruitment is determined is not clear. While year class strength may be decided, in some years, during the egg and larval period (Koslow et al., 1985b; Cohen et al., 1986), this need not be the case in general (Sissenwine 1984; Sissenwine et al. 1984).

What does it mean to state when "year class strength is determined"? For the purposes of this discussion, we define the age at which the size of a year class is determined as the age when recruitment is predictable from the number of pre-recruits alive at that age. The concept is illustrated in Figure 5 which gives the correlation [r(t)] between the number of pre-recruits at age t [N(t)] and the number at the age of recruitment (t=r). In general, the correlation is very low (statistically insignificant) at spawning (t=0.0). The implications of a "critical period" is that at some age there is a brief period of highly variable mortality. Following this period of highly variable mortality, recruitment is predictable. If the critical period were during the early larval stage (as hypothesized by Hjort (1914) and as implied by the focus of most research on recruitment processes), the r(t) function would look like curve A in Figure 5. If there was no a critical period the curve would resemble curve B. If there was a critical period late in the pre-recruit stage, the function would resemble curve C. Despite the focus of most of the recruitment research to date, there are reasons to believe that curves B or C are more likely than curve A, particularly for Georges Bank. For example, haddock recruitment is not predictable from larval abundance, but it is from the abundance of demersal juveniles six months later.

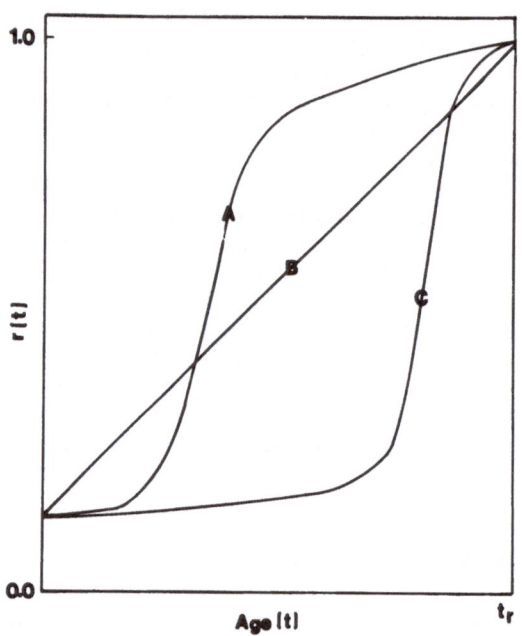

Figure 5. The hypothetical correlation (r(t)) between the number of individuals in a year-class at age t and at recruitment (age t=r) when there is a critical period during the larval stage (A), late in the pre-recruit period (C) and when there is not a critical period (B).

In addition, the cumulative mortality (*i.e.* mortality rate times its duration) during the post-larval period is comparable to the egg and larval cumulative mortality (Cohen and Grosslein 1982; Sissenwine *et al* 1984; Sissenwine 1984) for several species on Georges Bank (*i.e.* cod, haddock, silver hake and herring). Rothschild (1986) also concluded that the cumulative mortalities for the egg and larval stages and for the post-larval period are similar. It is the cumulative mortality during a life stage that determines its potential contribution to recruitment variability. Rothschild argued that the egg and larval stages have greater potential to contribute variability than the post-larval period because there are many more deaths in the earlier stages. While this argument is intuitively appealing, it is inconsistent with the exponential model that is virtually always applied to fish populations.

Predation is almost certainly a major cause of post-larval mortality. Several energy budgets (Sissenwine *et al* 1984; Cohen and Grosslein 1987) have been used to examine the role of predation on Georges Bank. The energy budgets indicate that most fish production is by pre-recruits. Cohen and Grosslein (1987) compared fish production and consumption and found a reasonable balance. They found that fish consumed about 70% of their own production. The rest of the fish production was consumed by man, marine mammals, sea birds and large oceanic fishes.

Predation could cause recruitment variability since both the abundance and diet of the predators varies over time. For example, the percentage of fish by weight in the stomach contents of silver hake during the spring of 1973-1976 was 24%, 58% 88%, and 69% respectively (Bowman 1980). Some of this variability might account for year to year changes in recruitment of the prey species. On the other hand, the differences in predator diet may be due to changes in the relative abundance of fish prey or the relative abundance of alternative (*i.e.* non fish) prey. Analyses to determine the actual prerecruit mortality caused by predation on an annual and species specific basis are lacking. These are necessary in order to determine to what degree predation by fish explains recruitment variability, it certainly influences average levels of recruitment and species composition.

6. CASE STUDY OF THE 1982 HADDOCK YEAR CLASS ON GEORGES BANK, WITH REFERENCE TO ADJACENT YEARS AND BROWNS BANK

The potential role of many of the processes that have been discussed so far in this paper are illustrated by examining the 1982 haddock year class on Georges Bank. It is also interesting to compare events on Georges Bank in 1982 with 1981 and 1983 and with Browns Bank which is about 100 km to the northeast.

In 1982 the only significant concentration of haddock larvae was observed in mid-April on the southeastern side of the bank. Frontal analysis charts issued by the Atlantic Environmental Group (AEG, NMFS, NEFC, Narragansett RI) showed that for the month preceding the collection of the larvae WCR 82-A was nearly stationary south of the bank near 67° W. A large amount of water from Georges Bank appeared to be entrained off of the bank by the ring from March through the time of the larval observations (Figure 6). The larval distribution during mid-April is shown in the figure as there are no data available from March but haddock larvae are generally found in that area of the bank in March. The contours of larval density are open at the edge of the survey area, suggesting that the distribution may have extended off of the southern side of the bank. In fact, the most seaward station had the highest number of larvae caught. The water entrained by the ring (stippled in Figure 6) appeared to be leaving the bank in

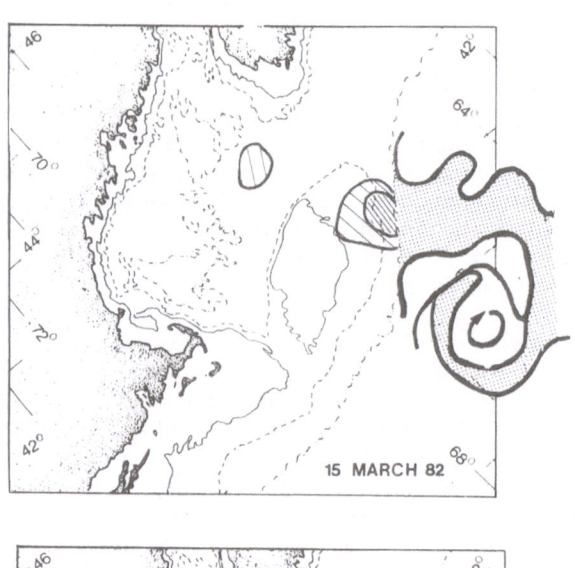

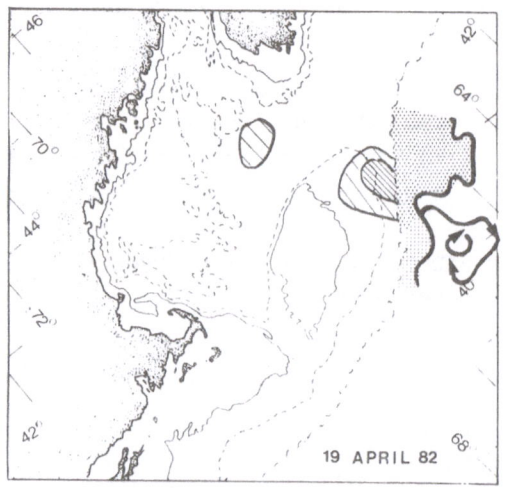

Figure 6. Warm Core Ring WE 82-A with associated entrainment of shelf water (stippled area) and contours of haddock larval densities in mid-April on Georges Bank. The larval densities are 11-100 larvae per 10 m^2 for the more widely spaced lines.

the same area that the larval distribution would extend off the bank.

Unfortunately, no haddock larvae were observed in the entrainment feature in April-May although shelf zooplankton were collected (Laurence and Burns 1982). However, very few haddock larvae were captured on the bank at that time either. Leblanc (1986) reported finding the larvae of other shelf fish species in the entrainment

feature, although these were from species such as Gulfstream flounder
and butterfish which spawn in the slope water as well as on the shelf.

The 1982 year-class of haddock turned out to be very weak and
there is some evidence, although inconclusive, that a WCR may have
contributed to the poor recruitment. But just the occurrence of WCR's
near Georges Bank does not constitute evidence of a relationship
between WCR's and recruitment. For example, in March, 1983 a large
WCR entrainment feature was observed off the southern side of the
bank. It was in the vicinity of significant larval haddock
concentrations observed between 6-19 April. Again as in 1982 there
was no cruise in March so the distribution of larvae in April was used
(Figure 7). By mid-April both the ring and the entrainment were much
smaller than in March, and by the end of April, only a small
deformation of the shelf-slope front remained. The larval
distribution in 1983 did not extend to the edge of the bank and the
larvae did not appear to be removed by this entrainment. Year class
strength in 1983 was among the strongest in recent years.

While the paucity of haddock eggs and larvae on Georges Bank in
1982 may have been due to WCR entrainment there are other
possibilities. There was a very large storm 6-9 April 1982 with
steady winds in excess of 60 knots and gusts of about 90 knots (Ramp
1982). This storm resulted in the surface edge of the shelf-slope
front (determined from satellite images) moving offshore about 25 km
from 3 April to 10 April along the southern side of Georges Bank
(Figure 8). During the storm there was advection to the southeast
according to the computed Ekman transport computation (R. Armstrong,
AEG, NMFS, NEFC, Narragansett RI) and current meter data (Ramp 1982).
While there was a significant movement of water off Georges Bank
caused by the storm, there is no indication of the quantity of water
permanently exchanged or mixed into the slope region.

It is interesting to compare the situation for haddock on Browns
Bank (to the northeast of Georges Bank, Figure 1) in 1983 with the
Georges Bank situation in 1982. The 1983 haddock year class on
Browns Bank was very poor; the cause seems to have been very high
mortality at the egg or early larval stages (Koslow et al. 1985b).
There was an advective event due to a storm on Browns Bank during the
spring of 1983 (Smith 1983) which may have been responsible for the
loss of eggs and larvae.

The occurrence of storms, as is the case with WCR's, does not
always result in a loss of eggs and larvae. In 1981, April was stormy
with Ekman transport of moderate intensity consistently to the south.
There were two significant storms, 1-5 and 21-24 May, with Ekman
transport vectors to the west and northwest, however this Ekman
transport would aid larval retention. Although the latter storm
resulted in mixing of the water column to at least 80m and affected
the vertical distribution of gadid larvae and their zooplankton prey
(Lough 1984) it does not appear to have affected larval abundance, as
the number of larvae sampled from February through June in 1981 was
not unusually low. However, the 1981 year class was even smaller than
the 1982 year class, presumably due to high mortality after the larval
period.

384

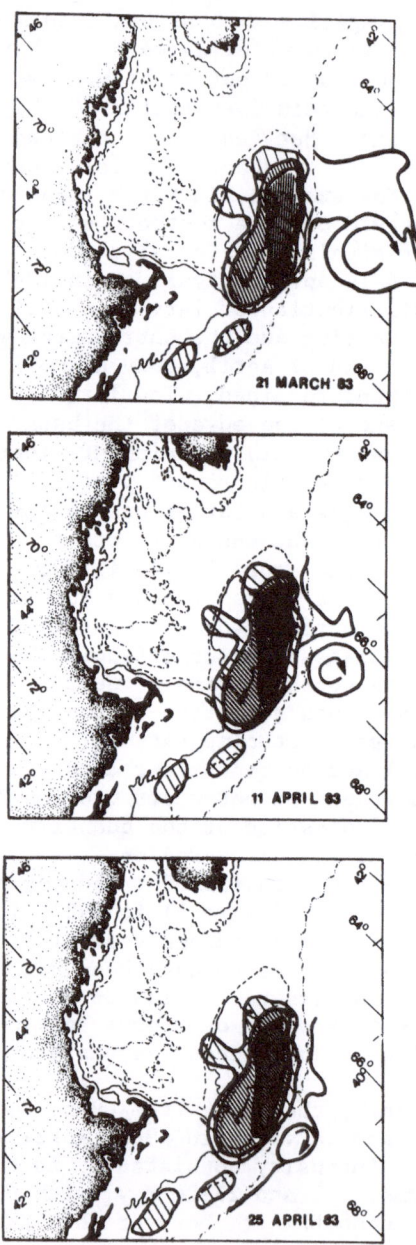

Figure 7. Warm Core Ring 83-B with associated entrainment of shelf water (stippled area) and contours of larval haddock densities in mid-April on Georges Bank. The densest shading represents an abundance of 101-1000 per 10 m^2 to other two shadings represent 11-100 per 10 m^2 and 1-10 per 10 m^2 with decreasing density.

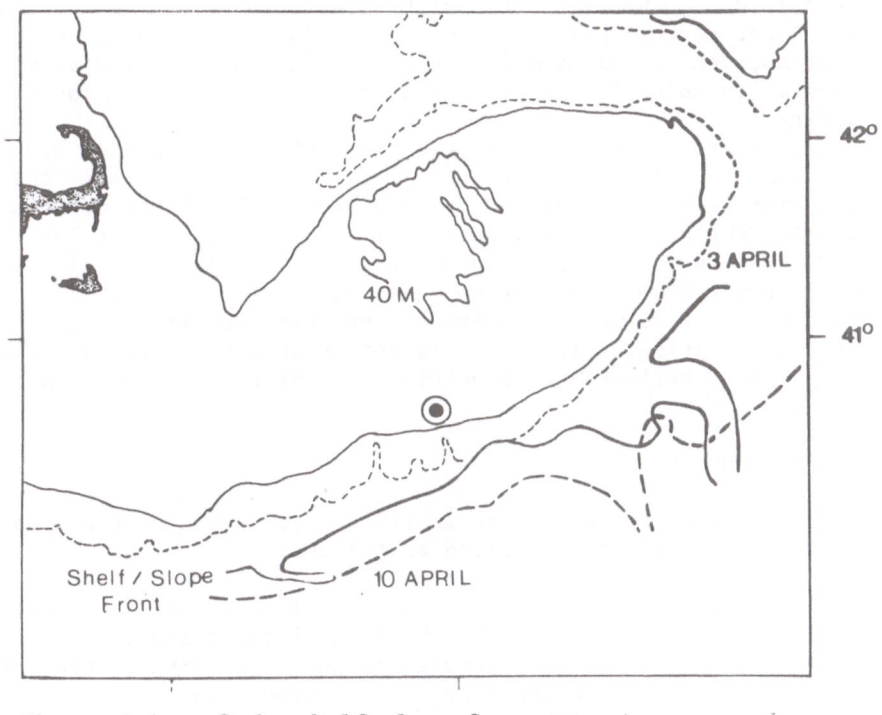

Figure 8. The position of the shelf-slope front near Georges Bank
before (3 April 1982) and after (10 April 1982) a major storm.

While advective loss of larvae caused by storms and/or a WCR may
have been a contributing cause of the poor 1982 year-class of haddock,
there is also evidence that events prior to the larval stage must have
been important. Early stage cod and haddock eggs are
indistinguishable. The total number of these eggs was similar in 1982
to the number during 1979-1981 (P. Berrien, NMFS, NEFC, Sandy Hook
Lab., Highlands NJ personal communication). There was however, a
dearth of haddock eggs at the stage when they could be identified,
indicating that the loss occurred sometime in the egg stage, or that
haddock did not spawn. There is some evidence that gonad development
of Georges Bank haddock appeared normal (based on observations made
during resource surveys in the autumn of 1981 and spring of 1982; W.
Overholtz, NMFS, NEFC, Woods Hole, MA personal communication).
Although there is not direct evidence of it, of course, advective loss
of eggs is a possible cause.

The unusually low temperature during the time of spawning may
have also caused high egg mortality (see discussion below). Laurence
and Rogers (1976) and Laurence (1978) have demonstrated that
temperatures below 4° C. are lethal to haddock eggs and larvae. The

temperature on Georges Bank during the early spring of 1982, a year of very poor recruitment was about 3⁰ C. Interestingly the period of poor haddock recruitment in the mid to late sixties was also abnormally cold with temperatures of 4⁰ C. or less in the early spring.

The effect of temperature on the 1983 year-class of Browns Bank haddock is unclear. Koslow *et al.* (1985 a) reported that the mean temperature in the upper 50m ranged from 3.0⁰ - 6.0⁰C between the end of February and early June. If the temperature was less than 4⁰C. at the time of peak spawning in April (Koslow *et al.* 1985 a) then it may have been a cause of mortality. Frank (1986) indicated that the surface waters on Browns Bank were warmer during the winter and early spring of 1983. It is not known how the sea surface temperature reflects the conditions in the upper 50m on Browns Bank.

7. DISCUSSION

The results of research that is related to recruitment processes on Georges Bank can be summarized as follows:

1. Primary productivity is high all over the Bank, but particularly in the shoal, well-mixed zone;
2. Cod and haddock larvae and their zooplankton prey are most abundant in the deeper stratified zone;
3. Both the fish larvae and their prey are concentrated at the thermocline in the stratified waters of the bank;
4. Based on laboratory experiments and simulation models, the local concentration of prey when the water column is stratified is high enough to support rapid larval growth;
5. Biochemical indices confirm that growth is high for fish larvae in the thermocline, but that up to fifty percent of the fish larvae sampled from a well mixed area of the bank were starving;
6. Storm events have the potential to mix the water column, thus reducing the local concentration of prey in the vicinity of larvae;
7. Both storm events and warm core rings have the potential to cause advective loss of eggs and larvae;
8. Laboratory studies indicate that the temperature on Georges Bank may be low enough in some years to be a direct cause of haddock egg and larval mortality;
9. The cumulative mortality during the post-larval stage is high enough so that year class strength may not be determined until late in the pre-recruit period;
10. Intra- and inter-species predation by fish plays and important role in determining recruitment as well as species composition.

The complexity of the recruitment problem is apparent. There are multiple factors that operate over the entire pre-recruit period

which vary in importance between species, locations and years. For a
system as complex as Georges Bank, which is not dominated by any
particular physical process, it is premature to predict recruitment on
an annual basis. Rothschild (1986) identified a more appropriate
research goal which is to determine what is feasible to learn and what
is not. In this regard, there are several future research
opportunities.

It is time to study the processes that determine the size of an
individual year class in a much more comprehensive manner. Figure 9
is a three-dimensional illustration of the recruitment problem space.
Of course there are many more dimensions, but the illustration only
includes species, life stage and year. The x's indicate a
hypothetical distribution of research. Over many years, there have
been studies of a variety of species, life stages and processes (which
are higher dimensions not included in the illustration). But, there
has not been a comprehensive study (as indicated by a slice through
problem space) of a single year class extending from gonad development
of adults to late enough in the pre-recruit period to be beyond the
point when year class strength has been established.

Research on the 1982 year class of haddock on Georges Bank
illustrates the point. A lot is known about the physical factors
(*i.e.* temperature, advection, due to storms and warm core rings) that
might have caused a poor year class. There is some evidence that
gonad development was normal, but a piece of critical information is
missing. Did haddock spawn in 1982? The 1981 year class of haddock
is another good illustration of the problem. It is hypothesized that

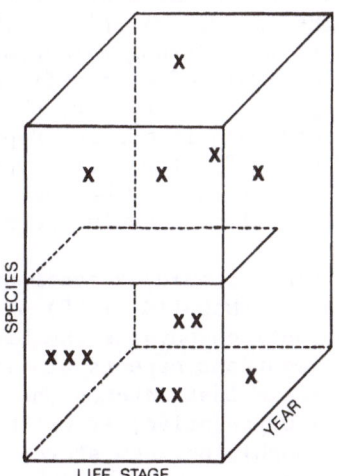

Figure 9. Hypothetical three dimensional problem space for
recruitment research specified by species, life stage and year. The
X's illustrate piecemeal research, while slice through the life-stage
plane indicates a comprehensive study of a year-class or year-classes.

stratification enhances feeding conditions for haddock larvae. In 1981, a storm disrupted stratification. Unfortunately, there were no measurements (*i.e.* biochemical indices) of growth before and after the storm. The 1981 year class turned out to be very poor (even worse than 1982), but the importance of the breakdown of stratification by the storm is unknown.

In order for a comprehensive study of a year class(es) to achieve its potential, new technologies for measurement must be applied. There is a need for a quantitative sampler for post-larvae. Both the 10 m^2 Multiple Opening Closing and Environmental Sampling System (MOCNESS, Wiebe *et al* 1976) and the International Young gadoid Pelagic Trawl (IYGPT) are being tested for Georges Bank cod and haddock. Sampling should be frequently enough to estimate mortality rate and growth rate with much greater temporal resolution than has been possible to date. Acoustic measurements which have the potential of providing much greater three-dimensional spatial resolution (Farmer and Huston 1988) may prove useful. Biochemical methods also have great potential. They are needed to distinguish between early egg stages of cod and haddock. There is also the potential to identify (quantitatively) stomach contents which would be extremely valuable to assessing the role of predation.

Stable isotope analysis can also be used to study trophic aspects of the recruitment problem (*e.g.* Frye and Sherr 1984). By measuring the concentration of stable isotopes, it should be possible to determine the trophic level of each life stage and how it varies. The size and age at which pre-recruits switch trophic levels may be particularly important in determining the success of a year class.

The genetic aspects of the recruitment problem are long overdue for serious consideration. One hypothesis (Powers *et al.*, 1988; Grosslein and Lilly 1987) is that there is a strong selection pressure in favor of a particular genotype as a result of a specific set of environmental conditions during each year. As a result, the gene heterogeneity of a population is reduced when it becomes dependent on a relatively few year classes due to heavy fishing. The hypothesis should be examined by measuring (a) gene frequency between year classes and (b) within a year class as it increases in age during the pre-recruit period.

A comprehensive study of a year class should be accompanied by the appropriate physical measurements. Rothschild (1988) provides a biodynamic theory for the interaction between physics and biology. From the biological perspective, the important aspects are light, temperature and water movements on a scale that matches the size of individual organisms. From a physical perspective, it is important to relate scales (*e.g.* what is the relationship between storms and mixing of the water column; or what is the relationship between water column stability and turbulence in the thermocline?). More sophisticated models and methods of measurement are needed.

What has been described does not constitute an entire research plan. It only indicates several components that have the potential to contribute to our understanding. Two obvious questions arise. Is it worth an undertaking an effort of this magnitude if prediction of annual recruitment is not the immediate goal? And, is an effort of the magnitude feasible?

In fact, there are more important needs than annual predictions of recruitment. It is important to quantify the trophic aspects of the recruitment problem in order to establish the nature of density dependent population regulation. Density dependent mechanisms are responsible for the persistence of populations and their ability to withstand stress (*e.g.* fishing).

Trophic interactions are also a key to understanding the relationship between species. Of course, biological processes, such as predation, are influenced by the environment. When there is a major change in species composition (*e.g.*, an increase in sand lance abundance following the collapse of herring populations off the northeastern U.S.A.), it is important to distinguish between a biological cause or a change in environmental conditions which favors one species.

With regard to the second question, we believe that a much more comprehensive attack on the recruitment problem is feasible. Several papers in this volume (already cited) address the potential for new technology. The Northeast Fisheries Center is already planning a more comprehensive study of Georges Bank haddock. Within the United States, both the National Oceanographic and Atmospheric Administration and the National Science Foundation have shown increased interest in the recruitment problem, as have international organizations such as the International Council for Exploration of the Sea and the International Oceanographic Congress. NATO sponsorship of an advanced workshop "Towards a Theory of Biological and Physical Interactions in the World Ocean" is another indication that the time is right.

8. REFERENCES

Bergan, R.H., G.C. Laurence, C.A. Oviatt, and L.J. Buckley: 1985 'Effect of thermal stratification on the growth and survival of haddock larvae *(Melanogrammus aeglefinus)*', ICES C.M. 1985 Mini-Symposium No. 6.

Bigelow, H.B.: 1926, 'Plankton of the offshore waters of the Gulf of Maine', *Bull. U.S. Bur. Fish.* 40 (Part II), 1-509.

Bowman, R.E.: 1980, 'Silver hake's regulatory influence on the fishes of the northwest Atlantic', NMFS, NEFC, Woods Hole Lab. Ref. Doc. NO. 80-05.

Buckley, L.J. and R.G. Lough: 1987, 'Recent growth, biochemical composition, and prey field of larval haddock (*Melanogrammus aeglefinus*) and Atlantic cod (*Gadus morhua*) on Georges Bank', *Can. J. Fish. Aquat. Sci.* 44, 14-25.

Bumpus, D.F.: 1976, 'Review of the physical oceanography of Georges Bank', *ICNAF Res. Bull.* 12, 119-134.

Butman, B., R.C. Beardsley, B. Magnell, D. Frye, J.A. Vermersch, R.
    Schlitz, R. Limeburner, W.R. Wright and M.A. Noble: 1982,
    'Recent observations of the mean circulation on Georges Bank', *J.
    Phys. Oceanogr.* **12**, 569-591.
Chase J.: 1955, 'Winds and temperatures in relation to the brood-
    strength of Georges Bank haddock', *J. Cons. int. Explor. Mer* **21**,
    17-24.
Cohen, E.B. and M.D. Grosslein: 1982, 'Food consumption by silver hake
    (*Merluccius bilinearis*) on Georges Bank with implications for
    recruitment', in G.M. Calliet and C.A. Simenstadt, eds., *Gutshop
    '81, Fish Food Habits Studies*, Proceedings of the Third Pacific
    Workshop, Wash. Sea Grant., Univ. of Wash., Seattle, pp. 286-294.
Cohen, E.B. and M.D. Grosslein: 1987, 'Production on Georges Bank
    compared with other shelf ecosystems', in R.H. Backus, R.L. Price
    and D.W. Bourne, eds., *Georges Bank*, MIT Press, Cambridge, Mass.,
    pp. 383-391.
Cohen, E.B., G.C. Laurence and W.G. Smith: 1984, 'The role of
    starvation and predation in regulating year-class strength in
    several fish stocks on Georges Bank', *Int. Cons. Explor. Sea*
    1984/G:32.
Cohen, E.B., D.G. Mountain and R.G. Lough: 1986a, 'Possible factors
    responsible for the variable recruitment of the 1981, 1982, and
    1983 year-classes of haddock (*Melanogrammus aeglefinus* L.) on
    Georges Bank', NAFO Special Session on Recruitment, September
    1986, Dartmouth N.S. Canada, NAFO SCR Doc 86/110.
Cohen, E.B., D.G. Mountain, and W. Smith: 1982, 'Physical processes
    and year class strength of commercial fish stocks on Georges
    Bank', *EOS, Trans. Am. Geophys. Union* **63**, 956, Abstract only.
Colton, J.B., Jr.: 1959, 'A field observation of mortality of marine
    fish larvae due to warming', *Limnol. Oceanogr.* **4**, 219-222.
Colton, J.B., Jr. and R.F. Temple: 1961, 'The enigma of Georges Bank
    spawning', *Limnol. Oceanogr.* **6**, 280-291.
Cushing, D.W.: 1983, 'Are fish larvae too dilute to affect the density
    of their food organisms?', *J. Plank. Res.* **5**, 847-854.
Davis, C.S. II: 1982, 'Processes controlling zooplankton abundance on
    Georges Bank', PhD. Thesis Boston Univ. Marine Program, Boston
    Mass.
Farmer, D.M. and R. D. Huston: 1988, 'Novel applications of acoustic
    backscatter to biological measurements', in B.J. Rothschild, ed.,
    *Toward a Theory on Biological-Physical Interactions in the World
    Ocean*, Kluwer Academic Publishers, Dordrecht, pp. 597-614.
Flierl, G.R. and J.S. Wroblewski: 1985, 'The possible influence of
    warm core Gulf Stream rings upon shelf water larval fish
    distribution', *Fish. Bull. U.S.* **83**, 313-330.
Frank, K.T.: 1986, 'Ecological significance of the ctenophore
    Pleurobrachia pileus off southwestern Nova Scotia', *Can. J. Fish.
    Aquat. Sci.* **43**, 211-222.
Fry, B. and E. Sherr: 1984, 'aka[11]<3C measurements as indicators of
    carbon plow in marine and freshwater ecosystems', *Contrib. Mar.
    Sci.* **27**, 13-47.

Grosslein, M.D. and G.R. Lilly: 1987, 'Summary report of the Special
    Session on Recruitment Studies', NAFO Sci. Counc. Studies No. 11.
Grosslein M.D., R.W. Langton and M.P. Sissenwine: 1980, 'Recent
    fluctuations in pelagic fish stocks of the northwest Atlantic-
    Georges Bank Region in relation to species interactions', *Rapp.
    P.-v. Reun. Explor. Mer.* **177**, 374-404.
Hjort, J.: 1914, 'Fluctuations in the great fisheries of northern
    Europe viewed in the light of biological research', *Rapp. P.-v.
    Reun. Cons. int. Explor. Mer* **20**, 1-228.
Koslow, J.A., S. Brault, J. Dugas, R.O. Fournier, and P. Hughes:
    1985a, 'Condition of larval cod (*Gadus morhua*) off southwest Nova
    Scotia in 1983 in relation to plankton abundance and
    temperature', *Mar. Biol.* **86**, 113-121.
Koslow, J.A., S. Brault, J. Dugas, and F. Page: 1985b, 'Anatomy of an
    apparent year-class failure: the early life history of the 1983
    Browns Bank haddock *Melanogrammus aeglefinus*', *Trans. Am Fish.
    Soc.* **114**, 478-489.
Laurence, G.C.: 1985a, 'A report on the development of stochastic
    models of food limited growth and survival of cod and haddock
    larvae on Georges Bank', NOAA Tech. Memo. NMFS-F/NEC-36.
Laurence, G.C.: 1985b. Nutrition and trophodynamics of larval fish--
    review, concepts, strategic recommendations and opinions. NOAA
    Tech. Memo. NMFS-F/NEC-36.
Laurence, G.C.: 1974, 'Growth and survival of haddock (*Melanogrammus
    aeglefinus*) larvae in relation to planktonic prey concentration',
    J. fish. Res. Bd Canada. **31(8)**, 1415-1419.
Laurence, G.C.: 1978, 'Comparative growth, respiration and delayed
    feeding abilities of larval cod (*Gadus morhua*) and haddock
    (*Melanogrammus aeglefinus*) as influenced by temperature during
    laboratory studies. *Mar. Biol.* **50**, 1-7.
Laurence, G.C. and B.R. Burns: 1982, 'Ichthyoplankton in shelf water
    entrained by warm-core rings', *EOS, Trans. Am. Geophys. Union*
    63, 998, Abstract only.
Laurence, G.C., J.R. Green, P.W. Fofonoff, and B.R. Burns: 1984,
    'Small-Scale spatial variability of plankton on Georges Bank with
    particular reference to prey organisms of larval cod and
    haddock', ICES C.M. 1984/L:9.
Laurence, G.C., T. Halavik, B. Burns and A. Smigielski: 1979, 'An
    environmental chamber for monitoring "*in situ*" growth and
    survival of larval fishes', *Trans. Amer. Fish. Soc.* **108**, 197-
    203.
Laurence, G.C., and C.A. Rogers: 1976, 'Effects of temperature and
    salinity on comparative embryo development and mortality of
    Atlantic cod (*Gadus morhua* L.) and haddock (*Melanogrammus
    aeglefinus* L.)', *J. Cons. int. Explor. Mer* **36**, 220-228.
LeBlanc, P.R.: 1986, 'The distribution and abundance of larval and
    juvenile fishes in shelf water entrained by warm-core ring 82-1B
    and contiguous water masses', Masters thesis, Southeastern Mass.
    Univ., 143 pp.

Lough, R.G.: 1984, 'Larval fish trophodynamic studies on Georges Bank:
    Sampling strategy and initial results', in E. Dahl, D.S.
    Danielssen, E. Moksness and P. Solendal, eds., *The Propagation of
    Cod (Gadus morhua* L.), Flodevigen rapportser. 1, 395-434.
O'Reilly, J.E. and D.A. Busch: 1984, 'Phytoplankton primary production
    on the northwestern Atlantic shelf', *Rapp. P.-v. Reun. Cons.
    int. Explor. Mer* **183**, 255-268.
O'Reilly, J.E., C. Evans-Zetlin and D.A. Busch: 1987, 'Primary
    Production', in R.H. Backus, R.L. Price and D.W. Bourne, eds.,
    Georges Bank, MIT Press, Cambridge Mass., pp. 220-233.
Powers, D.A., R. Clapman, T.T. Chen and L. DiMichele: 1988: 'A
    molecular approach to recruitment problems: Genetics and
    physiology', in B.J. Rothschild, ed., *Toward a Theory on
    Biological-Physical Interactions in the World Ocean*,
    Kluwer Academic Publishers, Dordrecht, pp. 411-440.
Ramp, S.: 1982, 'Georges Bank current meter monitoring survey
    scientific report', Appendix, Prepared for Mobil Research and
    Development Corp. Tech. Rept. #3. ENDECO Inc., Marion, MA.
Rothschild, B.J.: 1986, *Dynamics of Marine Fish Populations*, Harvard
    University Press, Cambridge, MA, 227 pp.
Rothschild, B.J.: 1988, 'Biodynamics of the sea: The ecology of high
    dimensionality systems', in B.J. Rothschild, ed., *Toward a Theory
    on Biological-Physical Interactions in the World Ocean*,
    Kluwer Academic Publishers, Dordrecht, pp. 527-548.
Schlitz R.J. and E.B. Cohen: 1984, 'A nitrogen budget for the Gulf of
    Maine and Georges Bank', *Biol. Oceanogr.* 3, 203-222.
Sissenwine, M.P.: 1984, 'Why do fish populations vary?', in R. May,
    ed., *Workshop on Exploitation of Marine Communities*, Springer
    Verlag, Berlin, pp. 59-94.
Sissenwine, M.P., E.B. Cohen and M.D. Grosslein: 1984, 'Structure of
    the Georges Bank ecosystem', *Rapp. P.-v. Reun. Cons. int. Explor.
    Mer.* **183**, 243-254.
Smith, P.C.: 1983, 'Circulation and dispersion on Browns Bank', in
    Southwest Nova Scotia Fishery Ecology Program Steering Committee
    Report No. 3. Bedford Institute of Oceanography, Dartmouth, N.S.,
    Canada, pp. 11-12.
Smith W.G. and W.W. Morse: 1985, 'Retention of larval haddock
    Melanogrammus aeglefinus in the Georges Bank region, a gyre-
    influenced spawning area', *Mar. Ecol. Prog. Ser.* 24, 1-13.
Walford, L.A.: 1938, 'Effect of currents on distribution and survival
    of the eggs and larvae of haddock (*Melanogrammus aeglefinus*) on
    Georges Bank', *Bull. U.S. Bur. Fish.* 49, 1-73.
Wiebe, P.H., K.H. Burt, S.H. Boyd, and A.W. Morton: 1976, 'A multiple
    opening/closing net and environmental sensing system for sampling
    zooplankton', *J. Mar. Res.* 34, 313-326.

ENVIRONMENTAL FACTORS, GENETIC DIFFERENTIATION, AND ADAPTIVE STRATEGIES IN MARINE ANIMALS

Bruno Battaglia and Paolo M. Bisol
Dipartimento di Biologia
Università di Padova
Via Loredan, 10
35131 Padova, Italy

ABSTRACT. Recent advances in oceanographic research have revealed an impressive spatial and temporal heterogeneity in a number of biotic and abiotic features at different scales. However, the possibility of bringing physical and biological observations in the marine realm into harmony is often precluded by the inadequacy of methods for a correct determination of marine species, particularly at the level of secondary producers. Laboratory and field investigations show that many taxonomic entities, previously considered as "good" species, are instead clusters of reproductively isolated sibling species, or species *in statu nascendi* among which gene-flow may be reduced. Certain physical conditions play an active role in creating barriers to gene-flow. Populations of the same species from different geographic and/or ecological origin may exhibit genetic differentiation affecting a broad spectrum of characters. Some of these inherited differences are adaptive and maintained by natural selection. Similar genetic changes may be induced also by the impact of man on the environment. The relevance of both long term and short term effects on recruitment is briefly discussed.

1. INTRODUCTION

In the marine environment, growing attention is being focused on mechanisms controlling recruitment. These are of various kinds and may be largely influenced by a multiplicity of environmental factors, physical as well as biotic. Both sets of agents may not act independently of each other, usually operating through complex interactions: it often happens that the former act on the latter, which thus mediate their effects.

Factors such as temperature, salinity, light, hydrostatic pressure, etc. may directly induce changes in the effective size of a population, thus acting as stimulating factors classified as density-dependent.

393

B. J. Rothschild (ed.), Toward a Theory on Biological-Physical Interactions in the World Ocean, 393–410.
© *1988 by Kluwer Academic Publishers.*

394

In the recruitment process it is important to take into account, in addition to the number, also the quality of the recruits, and the 'gains' become the more significant the greater the role played by genetic factors, or, in other words, 'quality' coincides with darwinian fitness. In this context, it becomes fundamental to identify and estimate as accurately as possible the role played by heredity in the attribution of adaptive properties.

The problem of genetic diversification and related mechanisms of adaptation in marine organisms began to be tackled in a conceptually and methodologically satisfactory way only recently. This was made possible, on the one hand, by the progress in identifying inherited factors responsible for intraspecific differentiation and on the other hand, by recognizing that the marine environment was not as traditionally thought, as a relatively uniform and continuous habitat.

Oceanographic investigations have demonstrated the spatial and temporal heterogeneity in a number of physical and biological features. The high complexity of physical variables has led physical oceanographers to adopt various techniques, including time-series analysis, in order to determine the relative importance of variability at different scales. Biological oceanographers are facing similar problems with patterns concerning the distribution and abundance of organisms. All the evidence indicates that, for instance, plankton communities are patchy on a broad spectrum of scales (Haury et al., 1978). Habitat patchiness, together with certain persistent hydrographic features, such as fronts, rings, gyres, pycnoclines, and others, act as potential barriers to dispersal, similar to those known to be effective in terrestrial environments.

However, bringing physical and biological observations in the marine realm into harmony is as yet a difficult task. We must admit that the considerable progress made by physical oceanographers in understanding the abiotic variables of the ocean has not yet been paralleled by similar advances in biological oceanography. This is in part due to the lack of biological sampling on space and time compared to the physical sampling. Another critical problem is, in fact, the correct determination of species, particularly at the level of secondary producers.

Let us consider, for instance, two or more geographically-remote populations ascribable to the same species according to conventional taxonomy. To what extent is it legitimate to assume that the population belong to the same species? A similar question can be raised for populations sampled from areas exhibiting different environmental conditions. Are we actually dealing with Mendelian populations of a single species? In the following pages we shall try to provide answers to these questions.

The aim of this paper is to illustrate and discuss cases of long term (evolutionary) changes in marine organisms, the underlying causitive mechanisms, whenever possible. A tentative analysis will also be provided of the genotype-environment relationships, and of the kinds of adaptive devices utilized by a variety of animals in the marine realm. Special emphasis will be given to the role of life-

histories in the adoption of specific genetic strategies of
adaptation to the environment. These will include the genetic
changes induced by the impact of man on the environment. The effects
of genetic modifications on the recruitment process will be briefly
discussed.

2. LONG TERM GENETIC MODIFICATIONS

A preliminary approach to the problem of genetic diversification in
marine animals consists in the detection of possible geographic
variations in morphological and physiological traits. A comparison
between geographic populations appears more promising when they
occupy areas with different ecological features. If this is the
case, the opportunity will be provided for identifying and measuring
the selective forces eventually involved in the differentiation
process.

Another good opportunity is offered by the availability of
species suitable for cross-breeding experiments in the laboratory,
which allow possible reproductive barriers to be revealed. Certain
harpacticoid copepods, namely *Tisbe* and *Tigriopus*, offer an
excellent material for this kind of study (Battaglia, 1970).
Observations in the field and laboratory experiments have shown, in
the genus *Tisbe*, an impressive range of inter- and intraspecific
differentiations affecting a multiplicity of characters such as
degree of polymorphism, physiological adaptations, population
dynamics parameters, concealed genetic variability, sex-
determination, and reproductive strategies. In most cases
morphological differentiations are almost absent, even between good
species, or barely conspicuous.

The situation is somewhat different in the genus *Tigriopus*,
which, though widely distributed, is a "specialist" being
exclusively confined to the rock- to tide-pool environment.
Moreover, *Tigriopus* is represented by a much smaller number of
species compared with *Tisbe*.

The genus *Tisbe* exhibits a wide range of reproductive isolation
(Battaglia and Volkmann-Rocco, 1973). From the results of cross-
breeding experiments, three main patterns can be distinguished: 1)
Species, like *T. holothuriae*, characterized by an almost total
absence of barriers to gene-flow, regardless of their geographic
origin. Perfect interbreeding has been observed also in
transatlantic crosses. Moreover, certain Atlantic and Mediterranean
populations are not only interfertile, but produce F_1 hybrids which
are heterotic. 2) Species, such as *T. reticulata*, providing
evidence of an incipient reproductive isolation. Crosses between
Atlantic and Mediterranean populations result in lower fecundity and
viability of F_1 hybrids, and involve strong deviations of sex-ratios
in favor of males (Battaglia, 1957). 3) Species, such as *T.
clodiensis*, showing almost total reproductive isolation even between
populations not too remote from each other. Crosses between several
Mediterranean populations show varying degrees of success, ranging

from full interfertility to complete intersterility. In certain cases, the degree of reproductive success depends on the geographic origin of the female or male ('relative intraspecific incompatibility') (Volkmann *et al.*, 1978; Fara and Battaglia, 1985).

Several cross-breeding experiments have been conducted also in *Tigriopus* which, as already mentioned, occupies the most demanding and severe habitat to be found in the marine environment: that of the rock-pools. This organism is characterized by an extraordinary physiological adaptability, being capable of living in conditions ranging from nearly freshwater up to salinity over-saturation. Crosses were made utilizing geographic strains from the Atlantic and the Mediterranean coasts of Europe, the Pacific coast of North America, and the Southern Indian Ocean (Kerguelen and Crozet Islands).

In *Tigriopus* sp., like in *Tisbe clodiensis*, reproductive success between the different populations may vary to a large extent. Also in this case the effect of crosses, which show a remarkable gradation, appears to depend on the direction of the crosses (Battaglia, 1982).

Both in *Tisbe* and in *Tigriopus*, the detection of reproductive barriers has led to the recognition of increasing numbers of sympatric or allopatric sibling species groups and closely related species within each genus. The occurrence of 'relative intraspecific incompatibility' is interpreted as a sign of speciation *in fieri*.

Differentiations between allopatric or sympatric sibling species can be disclosed by adequate karyological comparisons. A first example is offered by three sibling species of the *Tisbe reticulata* group. One of them, collected from Banyuls-sur-Mer, possesses an aploid set of 11 chromosomes, whereas the populations from Anzio and from the Lagoon of Venice show an aploid set of 12 chromosomes (Lazzaretto-Colombera, 1981). Similar differences in chromosome numbers were detected among several other species of *Tisbe*, where the values of n range from 8 to 12. Other parameters, such as chromosome shape and size, seem to be often species-specific (Lazzaretto, 1983).

An interesting case of chromosomal differentiation concerns two North Adriatic (Venice and Trieste) populations of the super-species *Tisbe clodiensis*, which, though very close geographically, are reproductively isolated from each other. Their karyotypes differ both in morphology and in location of satellites (Lazzaretto and Libertini, in press) (Figure 1).

Another investigation carried out on three Mediterranean populations of the genus *Tigriopus* has revealed the presence in Sicily of a strain characterized by a chromosome morphology very different from that exhibited by the other two populations. These findings support the hypothesis that another species, in addition to *Tigriopus fulvus*, may be present in the Mediterranean (Lazzaretto and Libertini, 1986).

Geographic differentiation concerns also a number of physiological traits. Populations of *Tisbe* sp. from temperate

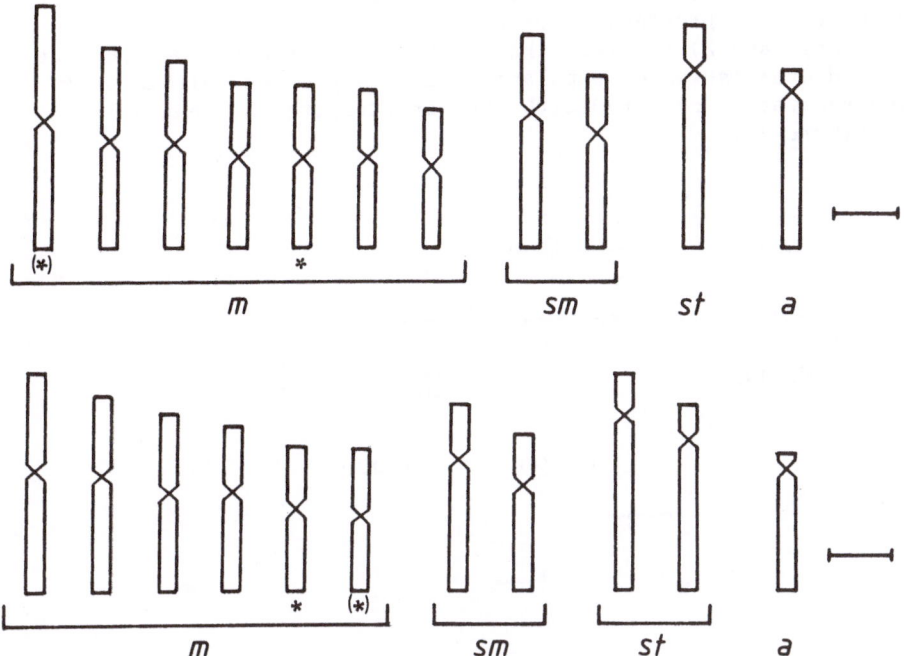

Figure 1. Comparison between karyograms from two North Adriatic populations of *Tisbe clodiensis* (upper row, Venice; lower row, Trieste), based on metaphase plates of dividing eggs (m – metacentric; sm – sub-metacentric; st – sub-telocentric; a – acrocentric; the asterisks indicate the presence of satellites; bar – 1μm). (From Lazzaretto and Libertini, in press).

regions are currently cultivated at the standard temperature of 18°C. This condition proved lethal for populations of the same genus collected from sub-Antarctic areas (Kerguelen Islands), which have to be raised at temperatures not exceeding 10°C. These differences in tolerance persist over years practically unchanged, which indicates their hereditary nature (Battaglia and Varotto, unpublished). Indications of genetic adaptation of the copepod *Acartia tonsa* to temperature increases, were provided by Reeve and Cosper (1970). Gonzalez (1974) showed that critical thermal maximum and upper lethal temperature are related in copepods, at least between locations.

Even more informative, in connection with geographic variability of physiological traits, are the comparisons between geographic populations of *Tisbe holothuriae* from habitats characterized by different salinity regimes. The populations sampled from habitats with higher constant salinities, compared with those from estuaries or brackish-water lagoons, show a much smaller tolerance of diluted sea water. Again in this case, the different responses must be

398

under genetic control, as indicated by 1) their persistence in time (experiments were protracted for a period corresponding to ca. 60 generations), and 2) the fact that hybrids from crosses between the marine and brackish-water populations show a greater tolerance of diluted sea water, compared with the parental strains (Battaglia, 1967) (Figure 2).

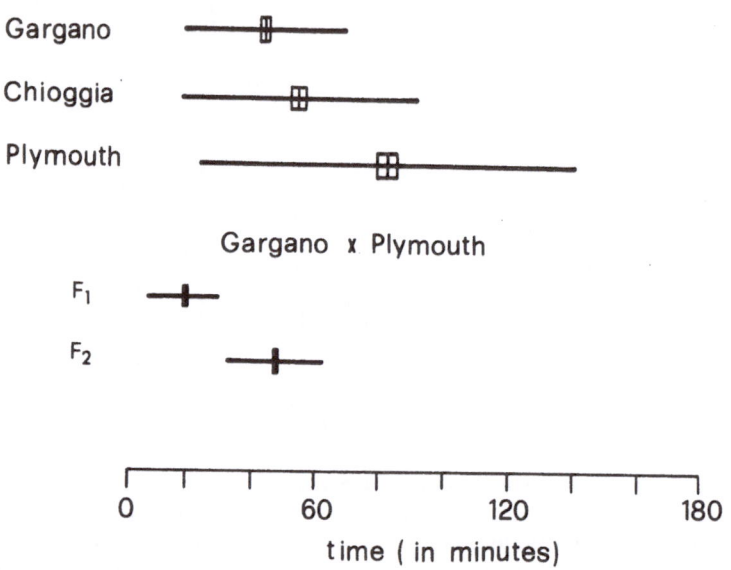

Figure 2. Times of recovery from the shock suffered by individuals belonging to three geographic populations of *Tisbe holothuriae* (Gargano, fully brackish; Chioggia, semi-brackish; Plymouth, marine), and by hybrids of the cross Gargano x Plymouth, transferred into diluted sea water. The vertical bar indicates the value of the mean; the straight line, a standard deviation on each side of the mean; the empty rectangle, twice the standard error on each side of the mean. (From Battaglia, 1967, modified).

Other experiments with copepods, aiming to understand the role of physiological and genetic adaptation to variable environmental regimes in the sea, are reported by Bradley (1982, 1986).

From the above examples it may be inferred that several factors are involved in causing geographic differentiation. Evidence is available to suggest that in some cases (*e.g. Tisbe clodiensis*) ecological features play a determining role in the formation of the isolating patterns, whereas in other cases (e.g. *Tigriopus* sp.) geographic distance *per se* acts as the main diversifying agent.

The real point at issue is the rate of gene-flow between populations, which would retard, prevent, or enhance genetic adaptation to purely local conditions. The results of studies on geographic differentiation provide a first clue to the understanding of geotype-environment relationships in the sea. Local components of natural selection, life-histories, and gene-flow interact in various ways in the process of differentiation of marine species; their joint or separate action may affect considerably recruitment patterns.

3. GENOTYPE-ENVIRONMENT INTERACTION IN THE MARINE ENVIRONMENT

One of the first examples of genotype-environment relationships in marine animals is offered by the copepod *Tisbe reticulata*, characterized by a colour polymorphism under the control of three alleles at the same locus (V-locus). By means of observations in the field and laboratory experiments, it was possible to demonstrate that the different morphs possess different norms of reaction when confronted with varying salinities and temperatures. Some components of their fitness are affected, to varying degrees, by differences in these ecological parameters. Amongst the mechanisms which maintain this adaptive polymorphism in *Tisbe reticulata*, is the higher viability of the heterozygotes compared to the homozygotes (Battaglia, 1958, 1965; Battaglia and Lazzaretto, 1967).

Similar experiments conducted in another species, *Tisbe clodiensis*, suggest that the adaptive value of a genotype may be a function of biotic factors also. A diallelic colour polymorphism, with the alleles P and p, is ascertainable in females (Battaglia and Fava, 1968). Equilibrium experiments have shown that the fitness values at this locus changes as a consequence of the introduction of another species in the culture (Battaglia and Finco, 1969).

Further evidence of interplay of physical and biotic factors in modifying important population parameters, is provided by studies on sex-ratio. Inference from experimental and field studies have led a number of workers to suspect that environmental factors may induce sex change in Calanoid copepods. Some derive from the Ghiselin (1974) and Charnov and Bull's (1977) models for sex change and sex determination, suggesting that larger genetic males are those which most likely turn into females, thus contributing to an above-average production of eggs. Similar events have recently been described by Fleminger (1983) in 14 of the 25 Calanidae he examined.

In copepods of the genus *Tisbe*, population density alone may not affect sex determination and sex-ratio. In this organism, the genetic mechanism for determination of sex makes an environmental choice of the fittest sex-ratio highly probable (Battaglia, 1965 l.c.). Without entering into details, population density, which is often controlled by physical factors, may in turn cause shifts of sex-ratio such as to ensure at any moment the maximization of the reproductive potential (r). An interesting work on quantitative traits (body size, age of maturity, survivorship, and sex-ratio) important in population dynamics and production, is the one conducted by McLaren (1976) in the copepod *Eurytemora herdmani*.

Here, unbalanced sex-ratios appear to be controlled by a major gene, with polygenic overlay of variance.

The study of genes controlling the main population parameters, and of those responsible for sex determination, can be integrated by the analysis of gene-enzyme systems (see Powers et al., this volume, for details). This is achieved by means of electrophoretic techniques, which have proved so informative in providing descriptions of several properties of gene-pools, even in marine organisms. Electrophoretic techniques permit detection of allelic variants in individual genes, thus revealing a number of very useful polymorphisms in otherwise monomorphic organisms. Most of the alleles involved being codominant, the estimation of gene frequencies and the characteristics of the genetic structure of populations become relatively easy.

In the framework of studies aiming to elucidate the mechanisms of genetic differentiation and adaptation in the sea, we have carried out extensive investigations on marine invertebrates, belonging to different taxa and collected from various localities, searching for correlations between environmental regimes and levels of genetic variability. For previous work on this subject, see Gooch (1975) and Nevo (1978).

The main results of our investigations are graphically summarized in Figure 3. At first sight, the distribution of genetic variability does not seem to be correlated with the environment. Let us consider, for instance, the wide dispersion of the heterozygosity values measured in an amphipod (*Gammarus insensibilis*, $\bar{H}$ = .06), a gastropod (*Gibbula adriatica*, $\bar{H}$ = .11), and an ophiuroid (*Ophiothrix fragilis*, $\bar{H}$ = .27), sampled simultaneously in the Lagoon of Venice from the same area of a few square meters. Similar differences in $\bar{H}$ values can be observed whenever comparisons concern phylogenetically distant organisms, such as decapods, gastropods and ophiuroids collected from the deep-sea (Costa and Bisol, 1978), or intertidal isopods and bivalves sampled from a sub-Antarctic station (Battaglia et al., 1985).

In all the above cases, the extremely high dispersion of the heterozygosity values could be in agreement with the hypothesis of neutrality for allozyme polymorphisms (Kimura and Ohta, 1974; Nei, 1975). However, if the comparisons are made between populations belonging to related taxa, a trend can be detected consisting in a drop of genetic variability in ecologically demanding environments, such as those characterized by wide fluctuations of physical parameters. Samples of the harpacticoid copepod *Tigriopus*, which occupies the severe habitat of rock-pools, show $\bar{H}$ values (.043 ± .008) much lower than those measured in another harpacticoid, *Tisbe* (.197 ± .016), inhabiting a less demanding marine region (Battaglia et al., 1978). Similarly, in samples of *Ophiothrix fragilis* from brackish-waters, the level of $\bar{H}$ (.27) is lower than in the closely related *O. quinquemaculata* (.32) collected from a more stable marine habitat (Bisol and Marigo, 1980; Bisol et al., 1984a).

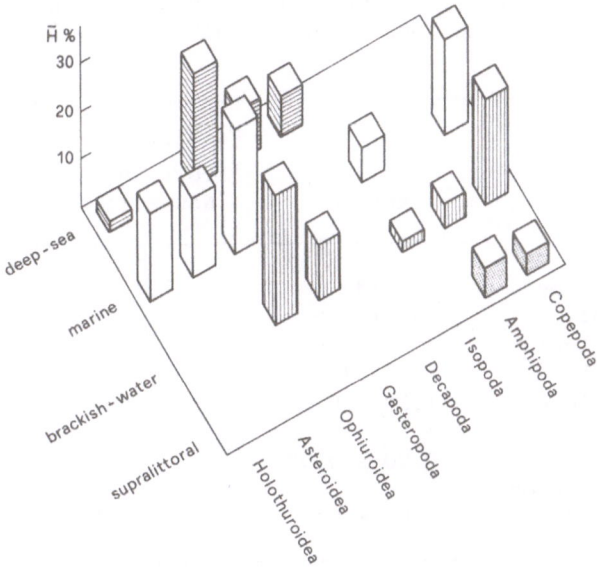

Figure 3. Distribution of genetic variability in marine invertebrates from different Habitats (Ḣ = expected mean heterozygosity).

The case of *Tisbe bulbisetosa* is even more representative of this trend (Bisol *et al.*, 1979). Three populations were sampled from localities exhibiting varying degrees of environmental instability: marine, quasi-brackish, fully brackish (Table 1). Since the sizes of the three populations were quite high, and there was no sign of reduced gene flow between them, the effects of drift appeared negligible. Therefore, the observed decrease of heterozygosity levels along a transect of increasing salinity fluctuations, and the finding of inter-population differences also in life-length and rate of development, suggest the selective nature of this polymorphism.

The 'low' levels of genetic variability which are often detected in fluctuating environments should not be considered in absolute terms but, rather, in comparison with the average values peculiar to each species, or group of closely related taxonomic entities. Life-histories and ecological determinants must play an important role in this connection. Relatively high levels of polymorphism are often found in species with short life-cycles and broad ecological versatility, as indicated by their ability to build up large populations in a variety of habitats (Battaglia *et al.*, 1978; Nevo, 1978; Nevo *et al.*, 1983). On the contrary, narrow ecological specialists tend to possess lower genetic variabilities. This is,

Table 1. Summary of the genetic variability in three populations
of *Tisbe bulbisetosa*. Samples were taken along a gradient
(Banyuls → Leucate) of increasing salinity instabilities.

Parameter	Banyuls	Population Port-La Nouvelle	Leucate
Number of enzymes	13	13	13
Number of loci	18	18	18
Polymorphic loci (5%) p ≤ .95	33.33	33.33	27.28
Alleles per locus	1.89 ± 0.29	1.83 ± 0.29	1.72 ± 0.23
Average hetero- zygosity expected	0.20 ± 0.07	0.17 ± 0.06	0.15 ± 0.06
Minimum generation time	12.03 ± 0.19	13.47 ± 0.34	14.13 ± 0.40

for instance, the case with some strictly abyssal holothurians and
nevertheless represented by large Mendelian populations (Bisol *et
al.*, 1984b). Here, the low levels of genetic polymorphism might
reflect the severity of the environment, subject to high hydrostatic
pressure, rather than the stability of its physical features
(Grassle and Grassle, 1978).

The low genetic diversity in widely fluctuating environments, or
in environments that are stable but constantly rigorous, may be
suggestive of mechanisms which, under demanding conditions, promote
individual homeostasis through the fixation of alleles conferring
greater adaptedness. This strategy does not exclude the possibility
that certain degrees of polymorphism be, nevertheless, preserved.
In some cases polymorphism is restricted to a few loci for which
the maintenance of several genetic variants might prove
advantageous. For instance, in the amphipod *Gammarus insensibilis*,
whose average heterozygosity is quite low, the Pgi-1 locus is
represented by six alleles with frequencies persisting practically
unchanged over years (Bisol *et al.*, in press). Laboratory
experiments conducted at 10°C and 27°C have shown that the higher
temperature induces differential mortality of Pgi genotypes. The
significantly greater tolerance of the heterozygotes compared to the
homozygotes, is in agreement with a mechanism of balanced selection
(Patarnello *et al.*, in press). Other mechanisms responsible for the
maintenance of polymorphism could be based on migration events among
geographic, or microgeographic, populations which are genetically
differentiated.

4. GENETIC CHANGES CAUSED BY THE IMPACT OF MAN

According to Nevo (1978), the normal condition of most outbreeding species is one of relatively high genetic variability. As a consequence, there is considerable room for adaptive shifts in the structure of gene pools in response to possible selective pressures resulting from pollution agents.

A first example is provided by the work of Mitton and Kohen (1975), on the effect of thermal pollution on the fish *Fundulus heteroclitus*. Also the observations by Nevo *et al*. (1977) on the intertidal barnacle *Balanus amphitrite* are indicative of a strong temperature selection on allozyme as well as on size variations. In this case the sampling areas were the open Mediterranean sea-water canals of the Haifa Electrical Plant cooling systems.

In the same species of barnacle (Nevo *et al*., 1978), the allelic frequencies at 10 out of the 15 scored loci showed consistent variations in the localities of the Bay of Haifa differing in chemical pollution level. Moreover, especially at the most polluted sites, most loci exhibited significant heterozygote deficiency.

In the preceding section, attention was focused on genetic changes partly due to selective directional forces operating, in particularly demanding environments, in favour of a reduced genetic variability. Declines of a similar kind can be observed also in areas where environmental stress is induced by man.

Examples are given by work on natural populations of the blue mussel, *Mytilus galloprovincialis*, and on experimental populations of the copepod *Tisbe holothuriae* (Battaglia *et al*., 1980; Beardmore *et al*., 1980). The analysis of mussel samples collected in the Lagoon of Venice from a number of stations situated along a gradient of pollution levels, showed a common trend of allozyme variation along the gradient. In some of the seven loci scored there was a significantly higher frequency of the more common allele in the more polluted areas. Moreover, lower average heterozygosities and increased deficiencies of heterozygotes correlated with the higher levels of pollution. These results indicate that a mussel population in the most polluted site is under a strong selective effect, although it is not known whether the loci assayed are themselves under selection or whether they simply mark the genome sections on which selection is operating.

The observations were made by collecting samples from the same areas in two subsequent years (1978, 1979). The trends appear quite similar in both years (Figure 4) (for statistical treatment, see Battaglia *et al*., 1980). However, by comparing the $\bar{H}$ values obtained at each site in the two years of the experiment, an increase of the average heterozygosities can be observed in the second year of sampling in the less polluted area, clearly due to non-selective recruitment. The $\bar{H}$ values become practically the same in the most polluted site, where natural selection presumably favours some adaptively superior genotypes produced by the mixing of populations.

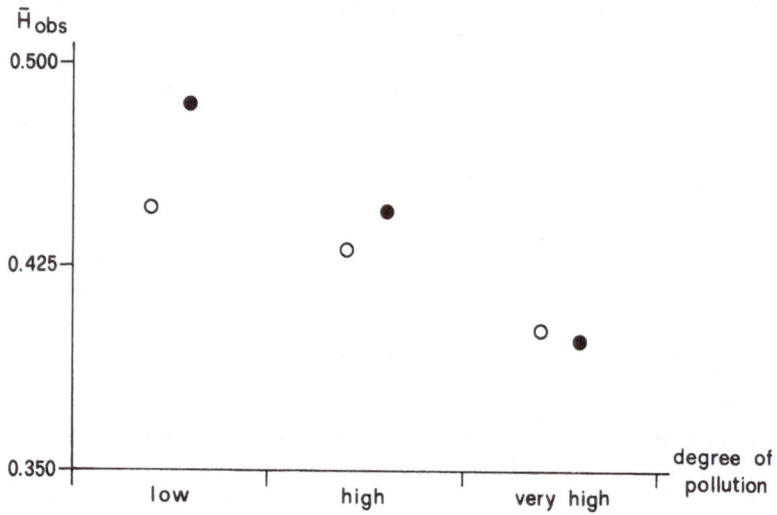

Figure 4. Observed frequencies of heterozygotes in *Mytilis galloprovincialis* from areas of the Lagoon of Venice with different pollution levels. Samples were taken in two subsequent years (1978, empty circles; 1979, solid circles). (From Battaglia *et al.*, 1980, modified.)

Changes in genetic structure occurred also in laboratory populations of the copepod *Tisbe holothuriae* reared under various experimental conditions. In particular, certain alleles of the loci Pgi-1 and Ap-1 exhibited an increase in frequency, especially in populations cultured at different concentrations of hydrocarbon pollution. This trend appeared more significant for the locus Pgi-1. The fact that equilibria are reached and maintained for several generations, and that the less favoured alleles are nevertheless kept in the populations, even at very low frequencies, provides further evidence of the balanced nature of this polymorphism.

The presence of oil in the water seems to affect the adaptive value of the locus Pgi (or if other genes closely linked with it). However, this effect is not necessarily a specific one since, as shown in previous experiments, other environmental conditions lead to similar results. In other words, other environmental conditions lead to similar results. In other words, the presence of the pollutant may create, as do low temperatures and salinities, a demanding environment where the more 'generalist' alleles are at a premium.

In a series of more recent contributions by Nevo and coworkers, the problem of the genetic effects of pollution has been approached in some marine gastropods also. Differential tolerance of electrophoretical Pgi allozyme genotypes to zinc and copper pollution was observed in two species of *Monodonta* (Lavie and Nevo, 1982). Other results obtained in *Cerithium scabridum* indicate differential survivorship of allozyme genotypes specific for cadmium and mercury, and for their interaction (Baker *et al.*, 1985; Lavie and Nevo, 1986). These results appear to reflect the adaptive nature of at least some allozyme polymorphisms observed in marine organisms.

5. DISCUSSION AND CONCLUSIONS

A considerable amount of evidence is now available on the role played by abiotic environmental factors in shaping the gene pools of marine populations. Due to these factors, barriers can be established leading to the formation of isolation and speciation patterns similar to those occurring in terrestrial environments. Nowadays, seas and oceans are viewed as a complex of highly diversified habitats, even at micro-geographic scales. Areas such as estuaries, brackish-water lagoons, rings, fronts, convergencies, may provide powerful barriers to gene flow. To date several studies on pelagic biogeography have led to the formulation of general concepts concerning the problem of transition zones and faunal boundaries in relation to physical features (see references in Pierrot-Bults *et al.*, 1986). In the contribution by C.-t. Shih (see Pierrot-Bults *et al.*, l.c.), the hypothesis is proposed that a unique type of speciation of planktonic animals may sporadically take place in areas where stability of oceanic conditions is altered by the admixture of waters from different origins.

Similar areas of transition, such as the one from Antarctic to subantarctic close to Kerguelen, appear "likely to be rewarding... for cooperation in physical and biological studies and... in which satellite radiation monitoring would give rewarding information on variability" (Deacon, 1983).

The detection of genetic diversification in marine organisms poses the problem of its possible adaptive significance. Although stochastic factors cannot be neglected, there is growing evidence of genetic adaptation. However, the diverse biological complexities of organisms might lead to incorrect interpretations of the genotype-environment relationships. Similar environmental stimuli may induce dissimilar genetic reactions in organisms characterized by different life-histories. A reasonable assumption is that they may perceive the environmental texture in different, often contrasting ways. Hence the illegitimacy of drawing general conclusions about mechanisms of adaptive strategies, by pooling together data drawn from phylogenetically remote species, without taking into account the features of the niche and the biological peculiarities of the species.

The detection of the trend we have indicated, of a reduced genetic variability in demanding environments, was made possible by addressing adequate attention to life-histories components.

Until recently, much of the research on responses to stress did not distinguish between adaptation viewed as the end-product of a physiological process considered to be of survival value for the individual, and adaptation based on selection operating on genotypes having different fitness values (Bayne, 1985).

In the marine realm, this distinction has been permitted by the utilization of species suitable to analysis of the role played, in the process of adaptation, by genetic factors. However, in most cases it appears difficult to discriminate between non-genetic (compensation) and genetic adaptation, since selective agents may be involved also in the first kind of adaptation. This is, for instance, the case when selection operates in favour of alleles, or combinations of alleles, conferring higher levels of individual physiological homeostasis. A mechanism as such is clearly based on a compromise between genetic and non-genetic adaptation, and might somehow compensate the loss of genetic variability. In this regard, a molecular approach appears very promising. Interesting results have already been achieved by research on single or multiple loci which encode for enzymes characterized by more than one aggregational or conformational state, and whose function in controlling adaptive processes has been clearly evidenced (for references, see Powers, this volume).

The reported differences in ecological versatility between *Tigriopus* and *Tisbe*, as well as between *Tisbe holothuriae* and *Tisbe clodiensis*, suggest that a variety of adaptive strategies may be at work even within closely related taxonomic units.

In this paper, we did not intend to review the numerous and often controversial hypotheses and theories concerning the strategies of adaptation in the marine environment (see, for references, Battaglia and Beardmore, 1978; Nelson and Hedgecock, 1980; Hedgecock *et al.*, 1982; Burton, 1983). Our purpose was, rather, to show how the various strategies adopted, not mutually exclusive and often based on a multiplicity of interacting mechanisms, may provide additional evidence of biological diversification. We must admit that we are a long way from knowing what species are really present in the marine environment, and to what extent the nature, structure and functioning of the living material in the oceans reflect the physical complexities of their habitat. Much field and laboratory work is needed, and adequate research programs should be designed to fill major gaps in knowledge. A broad range of organisms should be sampled, and sampling should be intensified, particularly in transition zones, to provide sufficient resolution of biotic gradients.

The understanding of the biological/physical interplay in the sea, could perhaps be enhanced by concentrating efforts on problems such as those inherent in recruitment patterns and mechanisms, especially in areas showing sharp discontinuities in environmental factors. The recruitment process reflects the relations between the ecology of larval, or juvenile stages and the sizes and structure of the adult populations. As already reported in this paper, when dealing with causes and mechanisms of genetic differentiation, correspondence between the two sets of population stages may be upset even by polluting agents.

An important objective is the understanding of the inter-annual changes in size and structure of recruited populations, due to environmental changes. Correct estimates of migration rates among populations, paralleled by monitoring of genetic modifications, are strongly required in order to evaluate the extent to which gene flow may be altered by distance or other kinds of barriers.

Studies on recruitment should take into account the mechanisms of the kinds described in this paper, in the attempt to predict which of them should be expected in action under the various local conditions, and to identify the adjustments of gene-pools which might have taken place. There is a challenge here for the marine scientist to reconcile the physical factors of the environment with the population structure and dynamics.

6. ACKNOWLEDGEMENTS

This work was supported by grants from the Consiglio Nazionale delle Ricerche and the Ministero della Pubblica Istruzione of Italy. We are grateful to Vittorio Varotto, Tomaso Patarnello and Renzo Mazzaro for their help in the preparation of the manuscript and drawings, and to our colleague Giancarlo Fava for discussion of these topics.

7. REFERENCES

Battaglia, B.: 1957, 'Ecological differentiation and incipient intraspecific isolation in marine copepods', Année Biol., 33, 259- 268.

Battaglia, B.: 1958, 'Balanced polymorphism in Tisbe reticulata, a marine copepod', Evolution 12, 358-364.

Battaglia, B.: 1965, 'Advances and problems of ecological genetics in marine animals', Proceedings XI Int. Conf. Genetics, Pergamon Press 2, 451-463.

Battaglia, B.: 1967, 'Genetic aspects of benthic ecology in brackish water', in G.H. Lauff (ed.), Estuaries, A.A.A.S., Washington, D.C., pp. 574-577.

Battaglia, B.: 1970, 'Cultivation of marine copepods for genetic and evolutionary research', Helgoländer wiss. Meeresunters 20, 385-392.

Battaglia, B.: 1982, 'Genetic variation and speciation events in marine copepods', in C. Barigozzi (ed.), Mechanisms of Speciation, A.R. Liss, Inc., New York, pp. 377-392.

Battaglia, B. and I. Lazzaretto: 1967, 'Effect of temperature on the selective value of genotypes of the copepod Tisbe reticulata', Nature 215, 999-1001.

Battaglia, B. and G. Fava: 1968, 'Prime osservazioni sulla genetica di popolazioni naturali e sperimentali di Tisbe clodiensis n. sp. (Copepoda, Harpacticoida)', Riv. Biol. 61, 3-19.

Battaglia, B. and G. Finco: 1969, 'Fattori biotici e selezione naturale in copepodi del genere Tisbe', Atti Ist. veneto Sci. (Sci. mat. nat.) 127, 363-370.

Battaglia, B. and B. Volkmann-Rocco: 1973, 'Geographic and reproductive isolation in the marine harpacticoid copepod *Tisbe*', *Mar. Biol.* **19**, 156-160.

Battaglia, B. and J.A. Beardmore (eds.): 1978, *Marine Organisms: Genetics, Ecology, and Evolution*, Plenum Press, New York, 757pp.

Battaglia, B., P.M. Bisol and G. Fava: 1978, 'Genetic variability in relation to the environment in some marine invertebrates', in B. Battaglia and J.A. Beardmore (eds.), *Marine Organisms: Genetics, Ecology, and Evolution*, Plenum Pres, New York, pp. 53-70.

Battaglia, B., P.M. Bisol, V.U. Fossato and E. Rodinò: 1980, 'Studies on the genetic effects of pollution in the sea', *Rapp. P.-v. Reun. Cons. int. Explor. Mer* **179**, 267-274.

Battaglia, B., P.M. Bisol, G. Fava, E. Rodinò and V. Varotto: 1985, 'Genetic variability and geographic differentiation in some benthic invertebrates from the Kerguelen region', in J.S. Gray and M.E. Christiansen (eds.), *Marine Biology of Polar Regions and Effects of Stress on Marine Organisms*, J. Wiley and Sons, London, pp. 299-311.

Baker, R., B. Lavie and E. Nevo: 1985, 'Natural selection for resistance to mercury pollution', *Experiments* **41**, 697-699.

Bayne, B.L.: 1985, 'Responses to environmental stress: Tolerance, resistance and adaptation', in J.S. Gray and M.E. Christiansen (eds.), *Marine Biology of Polar Regions and Effects of Stress on Marine Organisms*, J. Wiley and Sons, London, pp. 331-349.

Beardmore, J.A. (Chairman), C.J. Parker, B. Battaglia, R.J. Berry, A. Crosby Longwell, J.F. Payne, and A. Rosenfield: 1980, 'The use of genetical approaches to monitoring biological effects of pollution (Genetics panel report), *Rapp. P.-v. Reun. Cons. int. Explor. Mer* **179**, 299-305.

Bisol, P.M., V. Varotto, and B. Battaglia: 1979, 'Variabilità genetica di tre popolazioni del copepode arpacticoide *Tisbe bulbisetosa*, *Atti Soc. Tosc. Sci. Nat.*, Mem., ser. B, **86**, suppl., 357- 359.

Bisol, P.M. and N. Marigo: 1980, 'Variabilità genetica in *Ophiotrix fragilis* (Ophiuroidea, Echinodermata), *Atti Assoc. Genet. Ital.*, **25**, 46-48.

Bisol, P.M., R. Costa, O. Franceschini and M. Vignaduzzi: 1984a, 'Variabilità genetica in Echinodermi dell'Alto Adriatico', *Nova Thalassia*, 6, suppl., 699.

Bisol, P.M., R. Costa, and M. Sibuet: 1984b, 'Ecological and genetical survey on two deep-sea holothurians: *Benthogone rosea* and *Benthodytes typica*', *Mar. Ecol. Prog. Ser.* **15**, 275-281.

Bisol, P.M., T. Patarnello and B. Battaglia: (in press), 'Variabilità genetica in anfipoda del genere *Gammarus* di ambienti salmastri', *Rend. Acc. Naz. Lincei*.

Bradley, B.P.: 1982, 'Models for physiological and genetic adaptation to variable environments', in H. Dingle and J.P. Hegmann (eds.), *Evolution and Genetics of Life-Histores*, Proceed. in Life Sciences, Springer-Verlag, New York, pp. 33-50.

Bradley, B.P.: 1986, 'Traits, problems and methods in copepod life history studies', in G. Schriever, H.K. Schminke and C.-t. Shih (eds.), *Proceedings II Int. Conf. on Copepoda*, Syllog. 58, Nation. Museum of Nat. History, Ottawa, pp. 247-253.

Burton, R.: 1983, 'Protein polymorphisms and genetic differentiation of marine invertebrate populations', *Mar. Biol. Lett.* 4, 193-206.

Charnov, E.L. and J.J. Bull: 1977, 'When is sex environmentally determined?', *Nature* 266, 828-830.

Costa, R. and P.M. Bisol: 1978, 'Genetic variability in deep-sea organisms', *Biol. Bull. mar. biol. Lab.*, Woods Hole, 155, 125-133.

Deacon, G.E.R.: 1983, 'Kerguelen, Antarctic and subantarctic', *Deep-Sea Res.* 30(1), 77-81.

Fava, G. and B. Battaglia: 1985, 'Processes of differentiation between Mediterranean populations of the super-species *Tisbe clodiensis* Battaglia and Fava (1968) (Copepoda)', in M. Moraitou-Apostolopoulou and V. Kiortsis (eds.), *Mediterranean Marine Ecosystems*, Plenum Press, New York, pp. 333-346.

Fleminger, A.: 1985, 'Dimorphism and possible sex change in copepods of the family Calanidae', *Mar. Biol.* 88, 273-294.

Ghiselin, M.T.: 1974, *The Economy of Nature and the Evolution of Sex*, University of California Press, Berkeley, 346 pp.

Gonzalez, J.: 1974, 'Critical thermal maxime and upper lethal temperatures for the calanoid copepods *Acartia tonsa* and *A. clausi*', *Mar. Biol.* 27, 219-223.

Gooch, J.L.: 1975, 'Mechanisms of evolution and population genetics', in O. Kinne (ed.), *Marine Ecology: A Comprehensive Integrated Treatise on Life in Oceans and Coastal Waters*, J. Wiley and Sons, London, 2(1), 349-409.

Grassle, J.F. and J.P. Grassle: 1978, 'Life histories and genetic variation in marine invertebrates', in B. Battaglia and J.A. Beardmore (eds.), *Marine Organisms: Genetics, Ecology, and Evolution*, Plenum Press, New York, pp. 347-364.

Haury, L.R., J.A. McGowan and P.H. Wiebe: 1978, 'Patterns and processes in the tiime-space scales of plankton distributions', in J.H. Steele (ed.), *Spatial Pattern in Plankton Communities*, Plenum Press, New York, pp. 277-327.

Hedgecock, D., M.L. Tracey, and K. Nelson: 1982, 'Genetics', in D.E. Bliss (ed. in chief), *The Biology of Crustacea*, Academic Press, New York, L.G. Abele (ed.), Vol. 2, pp. 284-403.

Kimura, M. and T. Ohta: 1974, 'On some principles governing molecular evolution', *Proc. Natl. Acad. Sci.* 71, 2848-2852.

Lavie, B. and E. Nevo: 1982, 'Heavy metal selection of Phosphoglucose Isomerase allozymes in marine gastropods', *Mar. Biol.* 71, 17-22.

Lavie, B. and E. Nevo: 1986, 'The interactive effects of cadmium and mercury pollution on allozyme polymorphisms in the marine gastropod *Cerithium scabridum*', *Mar. Poll. Bull.* 17, 21-23.

Lazzaretto-Colombera, I.: 1981, 'Karyological comparison between three sibling species of the *Tisbe reticulata* group (Copepoda, Harpactacoida), *Zool. Scripta* 10, 33-36.

Lazzaretto, I.: 1983, 'Karyology and chromosome evolution in the genus *Tisbe* (Copepoda)', *Crustaceana* **45(1)**, 85-95.

Lazzaretto, I. and A. Libertini: 1986, 'Karyological comparison among different Mediterranean populations of the genus *Tigriopus* (Copepoda, Harpacticoida), *Boll. Zool.* 53, 197-202.

Lazzaretto, I. and A. Libertini: (in press), 'Chromosomal differentiation in two reproductively isolated populations of the superspecies *Tisbe clodiensis* (Copepoda, Harpacticoida), *Biol. Zentralblatt*.

McLaren, I.A.: 1976, 'Inheritance of demographic and production parameters in the marine copepod *Eurytemora herdmani*', *Biol. Bull. mar. biol. Lab.*, Woods Hole, 151, 200-213.

Mitton, J.B. and R.K. Kohen: 1975, 'Genetic organization and adaptive response of allozymes to ecological variables in *Fundulus heteroclitus*', *Genetics* 79, 97-111.

Nelson, K. and D. Hedgecock: 1980, Enzyme polymorphisms and adaptive strategy in the decapod crustacea', *Am. Nat.* 116. 238-280.

Nei, M.: 1975, *Molecular Population Genetics and Evolution*, North-Holland Publication Co., Amsterdam.

Nevo, E.: 1978, 'Genetic variation in natural populations: Patterns and theory', *Theor. Popul. Biol.* 13, 121-177.

Nevo, E., T. Shimony, and M. Libni: 1977, 'Thermal selection of allozyme polymorphisms in barnacles', *Nature* 267, 699-701.

Nevo, E., T. Shimony, and M. Libni: 1978, 'Pollution selection of allozyme polymorphisms in barnacles', *Experientia* 34, 1562-1564.

Nevo, E., B. Lavie, and R. Ben-Shlomo: 1983, 'Selection of allelic isozyme polymorphisms in marine organisms: Patterns, theory, and application', *Isozymes* 10, 69-92.

Patarnello, T., P.M. Bisol and B. Battaglia: (in press), 'Sopravvivenza genotipica differenziale in realizione alla temperatura in *Gammarus insensibilis* (Crustacea, Amphipoda), *Rend. Accad. Naz. Lincei*.

Pierrot-Bults, A.C., S. van der Spoel, B. Zahuranec, and R.K. Johnson (eds.): 1986, *Pelagic Biogeography*, Proc. of an Int. Conference, UNESCO Tech. Papers in Mar. Sci. N. 49, 295pp.

Powers, D.A., R. Chapman, T.T. Chen, and L. Di Michele: 1988, 'A molecular approach to recruitment problems: Genetics and physiology', in B.J. Rothschild (ed.), *Toward a Theory on Biological-Physical Interaction in the World Ocean*, Kluwer Academic Publishers, Dordrecht, pp. 411–440.

Reeve, M.R. and E. Cosper: 1970, 'The acute effects of heated effluents on the copepod *Acartia tonsa* from a subtropical bay and some problems of assessment', *FAO Technical Conference on Marine Pollution and its Effects on Living Resources and Fishing*, Rome, Italy, 9-18 Dec. 1970 (Ref. FIR: MP/70/E-59), 4pp.

Volkmann, B., B. Battaglia, and V. Varotto: 1978, 'A study of reproductive isolation within the super-species *Tisbe clodiensis* (Copepoda, Harpacticoida)', in B. Battaglia and J.A. Beardmore (eds.), *Marine Organisms: Genetics, Ecology, and Evolution*, Plenum Press, New York, pp. 617-636.

A MOLECULAR APPROACH TO RECRUITMENT PROBLEMS: GENETICS AND PHYSIOLOGY

Dennis A. Powers,[1,2,3] Robert Chapman,[2] Thomas T. Chen[3]
Leonard DiMichele[2], and L. Irene González-Villaseñor[1]
[1]Department of Biology, The Johns Hopkins University,
Baltimore, MD 21218 U.S.A.
[2]Chesapeake Bay Institute, The Johns Hopkins University,
Baltimore, MD 21218 U.S.A.
[3]Center of Marine Biotechnology, University of Maryland,
Baltimore, MD 21202 U.S.A.

ABSTRACT. Recruitment applies to the transition between life history stages. These transitions entail a degree or mortality. Recruitment related mortality is a function of a number of biological and physical parameters. While a large fraction of mortality arises from random stochastic processes, a number of recent studies indicate that some genotypes have a greater probability of survival than others. Two major obstacles to recruitment studies have been the detection of genetic variation within and between species and the assessment of the physiological status of different life stages. Molecular techniques can be particularly useful in resolving these problems. For example, molecular methodologies are useful in the identification of morphologically indistinguishable larvae of different species. Analysis of restriction patterns of mtDNA can help establish the existence of previous genetic isolation and even permit the differentiation between primary and secondary integradation. Immunological techniques can be used in food chain studies, the identification of minute larvae and the analysis of population. The applications of these and other molecular methodologies are exciting research tools that can open doors to the study of recruitment that were previously unapproachable.

1. INTRODUCTION

Since a recruit is a member of any new group, recruitment is the process, action or state of bringing new individuals into that group. Biological oceanographers are often concerned with the number of juveniles of a species that become adults. However, in a broader sense, recruitment applies to the transition between any life history stage and the next, e.g., transitions from eggs to larvae, larvae to juveniles, and juveniles to adults. There are a number of different approaches to the study of recruitment. Some biologists are concerned

411

B. J. Rothschild (ed.), Toward a Theory on Biological-Physical Interactions in the World Ocean, 411–440.
© 1988 by Kluwer Academic Publishers.

with the density dependent spatial and temporal variation of marine organisms. There are others that focus on the role of environmental variation, still others have turned their attention toward the analysis of field data by novel mathematical treatments like spectral and fractal analysis, and some biological and physical oceanographers have combined their talents to address the role of ocean turbulence and other factors in the recruitment process. Recently a few scientists have become fascinated by the genetic and physiological mechanisms that may participate in the recruitment process.

Transitions between life history stages entail a degree of mortality. Often the magnitude of this mortality can be staggering. In fact, the probability of survival between egg and adult is often less than a fraction of one percent. From probability theory we know that recruitment is a sampling without replacement design and the hypergeometric distribution describes the array of genotypes passing from one life history stage to another. It is usually assumed that death is random relative to genetic background and that survival is primarily a stochastic process. From this assumption, it follows that the frequency of a particular genotype should be maintained between life history stages. While a large fraction of recruitment mortality may be due to random stochastic processes, a number of recent studies have clearly indicated that some of this death is not random and some genotypes have a greater probability of survival than others.

Some of the major difficulties that must be confronted when studying recruitment are: (i) the identification of species, (ii) the detection of genetic variation within a species, (iii) the assessment of the physiological, nutritional and reproductive status of organisms, and (iv) the role of physical and chemical variables on the above parameters. The application of molecular techniques can help resolve some of these problems.

2. GENETIC VARIATION WITHIN AND BETWEEN SPECIES

The analysis of RNA, DNA and proteins have been useful in distinguishing between species and detecting genetic variation within species.

2.1. The Use of 5S and 16S RNA to Distinguish Species

Analysis of genetic variation of RNA has become a very useful tool for the identification of different species. Macromolecular comparisons can infer phylogenetic relationships among microorganisms. In addition, quantitative analysis can provide information on species composition of microbial communities (Pace *et al.*, 1985; Olsen *et al.*, 1986). Comparisons may be based either on measurements of molecular similarity by immunochemical reactions, DNA-RNA hybridization or DNA-DNA hybridization, or on mathematical analyses of nucleotide sequence data.

2.1.1. *The Use of 5S RNA.*   In recent years, due to rapid advances in nucleic acid sequencing and recombinant DNA methodologies, ribosomal RNAs and their genes have been chosen for intensive phylogenetic explorations. With some exceptions, the 5S rRNA from most microbial species contains about 115-120 nucleotides which can be directly sequenced by cleavage with base-specific enzymes (Donis-Keller *et al.*, 1977) or chemicals (Peattie, 1979). Briefly, determination of 5S rRNA sequence involves isolation by gel fractionation, ^{32}P-end labelling, cleavage with base-specific enzymes or reagents, and finally separation of RNA fragments on high resolution polyacrylamide gels. About 250 5S rRNA sequences have been collected and the data indicate that they are highly conserved molecules. When variation does occur, it is not randomly distributed, *i.e.*, some positions drift more freely than others. Although this method has been used successfully to investigate the phylogenetic relationships of several types of microbial communities (Stahl *et al.*, 1984, 1987; Lane *et al.*, 1985; McDonell *et al.*, 1986), the paucity of independently varying nucleotide positions of 5S rRNA limits its phylogenetic usefulness.

2.1.2. *The Use of 16S RNA.*   The 16S rRNA (about 1600 nucleotides) is an appropriate size for broader phylogenic analysis. Due to its size, the complete sequence of 16S rRNA cannot be determined easily by the methods described above. Before the development of DNA cloning and modern nucleotide sequencing methodologies, the sequence of 16S rRNA was partially determined by the so called "oligonucleotide cataloging" method using RNase T1 (Fox *et al.*, 1977). Employing this method, Woese and colleagues have characterized the 16S-like rRNAs from over 300 organisms and organelles.

Recently, by the use of advanced nucleotide sequencing technologies, the full nucleotide sequences of the 16S rRNAs have been determined from about 25 diverse organisms. This information, coupled with the partial 16S sequences accumulated by Woese and collaborators, provides a reasonably detailed picture of the molecule in terms of its primary and secondary structures (Figure 1). Examination of the 16S rRNA sequences, shows that regions of constant or nearly constant nucleotide sequences exist across taxa. Three oligonucleotides of these conserved regions have been synthesized. These oligonucleotides are used as primers to provide a fast and direct approach for the determination of 16S rRNA partial sequences. Through comparison of partial 16S rRNA sequences of *Camplobacter pylori*, generated by the above sequencing method, Romaniuk *et al.* (1987) showed that *C. pylori* is not a true *Camplobactin* species.

To analyze the population contents of microbial communities by 16S rRNA sequence comparisons, the strategy of shotgun cloning of 16S rRNA genes has been attempted (Figure 2). In this method, DNA is purified from collected biomass, shotgun cloned into the bacteriophage lambda, and individual rRNA genes are isolated as recombinant bacteriophages by hybridization with the "mixed kingdom" 16S rRNA probe. Using one of the three synthetic primers in the dideoxynucleotide-chain-termination reactions, the nucleotide sequences of individual rRNA genes are determined. By referring to existing collections of complete and

414

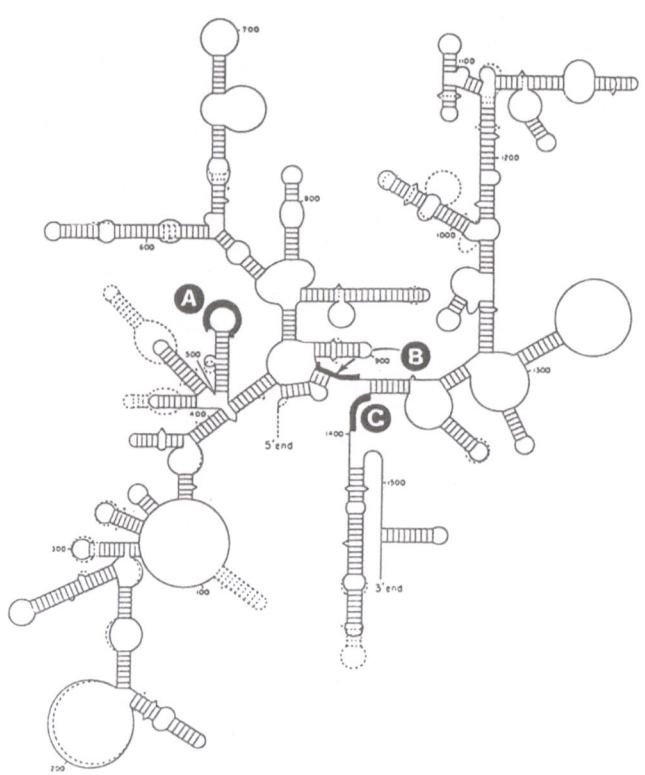

Figure 1. Bacterial 16S rRNA. A line diagram of the 16S rRNA
sequence of *Halobacterium volcanii* (solid line) is superimposed on
that of *E. coli* (dotted line). Crossbars indicate hydrogen bonds
ordering secondary structure. The heavy bars, denoted A, B, and C,
indicate constant sequences useful for sequencing primers (From Olsen
et al, 1986).

partial sequences, it is possible to infer the phylogenetic affinities
of the organisms in the original community (Vaughn *et al*., 1984). In
addition, the individually cloned rRNA genes can also be used as in
hybridization probes to quantitate the species composition of the
mixed community.

       A novel alternative approach for counting and identifying species
in a community is the application of an *in situ* hybridization method
to detect the presence of organism-specific or group-specific
sequences. This method involves hybridization of radio-labelled
nucleic acid probes with the complementary sequences inside fixed
cells or in fixed thin sections of tissues, and visualization of the
radioactive hybrid molecule by autoradiography. The bound probes can

also be quantitated by counting the number of silver grains in each sample.

In fishery recruitment studies, a common problem is species or stock identification of larval stages. Nucleic acid hybridization may be the method of choice provided suitable "signature gene probes" are available. These "signature gene probes" could be portions of the

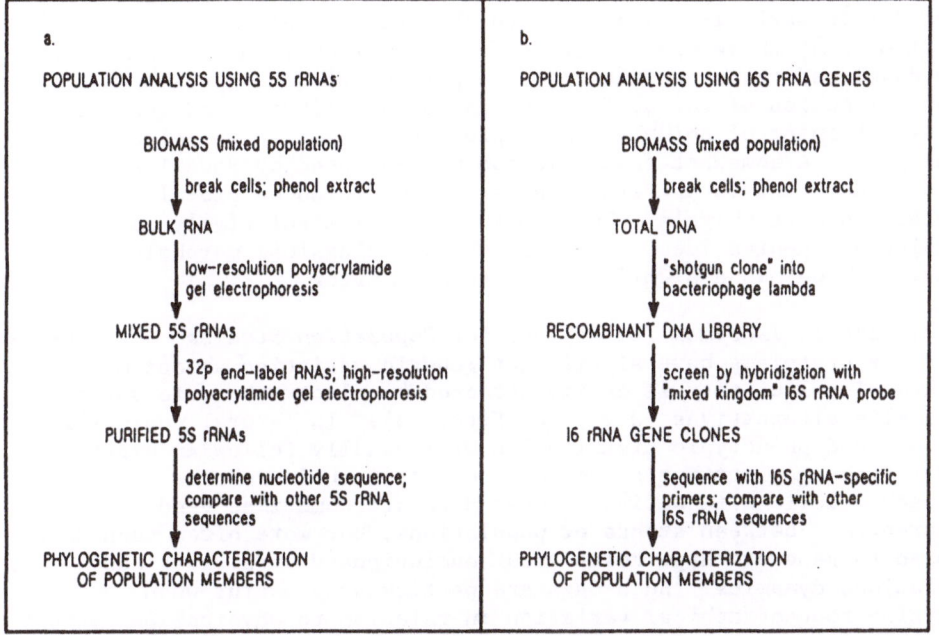

Figure 2. Flow chart for analysis of (A) 5S rRNAs and (B) 16S rRNA genes from natural populations (from Olsen *et al.*, 1986).

2.2. Multilocus Isozymes

During the last two decades, the introduction of electrophoresis coupled with histochemical staining for the identification of protein electromorphs has uncovered a wealth of genetic variation. This genetic variation is reflected at two fundamental levels: multilocus isozymes and allelic alternatives of specific loci. Multilocus isozymes reflect one or more evolutionarily ancient gene duplications followed by the genetic divergence of protein coding loci. For example, lactate dehydrogenase is a multilocus system with at least three different loci and three gene products that are differentially expressed; e.g., in marine fishes the Ldh-A locus is expressed in white muscle, Ldh-B locus is expressed in red muscle, and Ldh-C is expressed in eye and/or nervous tissue (see Figure 3). organism-specific or species-specific 16S rRNA, mitochondrial genes, or specific nuclear genes.

2.2.1. *Isozymes for Distinguishing Between Species.* When there is no genetic variation within a species but substantial variation between species, these isozymes can be used to detect species differences between larvae, or other life stages, that may be ambiguous from a morphological perspective. For example, the white perch, *Morone americana*, is extremely difficult to distinguish from *Morone saxatilis* in early larval stages and thus studies of larval recruitment of these species are difficult. Biochemical methods have been developed to assess the relative portion of each species in a given collection of larval fish. Morgan *et al.* (1975) used general protein staining of soluble muscle proteins separated on polyacryl-amide gels. A somewhat faster method was advanced by Sidell *et al.* (1978), where enzyme specific stains were employed on proteins separated on starch gels. These methods provide relatively fast and unambiguous species identifications whereas classical morphological identification has a significant associated error.

2.2.2. *Allelic Isozymes (Allozymes) For Population Studies.* A number of marine organisms have significant genetic variation at enzyme synthesizing loci that is easily detected by electrophoretic variation of allelic alternatives (e.g., see Figure 3). In natural poplulations the observed phenotypic distribution will usually follow an expected Hardy-Weinberg distribution so that gene frequencies can be calculated for each allelic alternative. These data are sometimes used to differentiate between stocks or populations, but more often such data is used to generate hypotheses about evolutionary strategies and populations dynamics. Such data are particularly useful when analyzing zoogeographical variation in relation to physical parameters like salininty, temperature, oceanic circulation, etc. The extensive work on *Mytilus edulis* illustrates this point.

Mytilus edulis has been the subject of genetic, physiological, and biochemical studies for almost two decades. The first work on this species was done by Milkman (1970), but it was the discovery of a geographical change in Leucine amino peptidase (Lap-1) gene frequency (*i.e.*, cline) in Long Island Sound, N.Y. (Koehn *et al.* 1976) that led to intensive investigations of the population genetics and possible adaptive significance of the Lap-1 locus. The gene frequency cline discovered by Koehn and colleagues (1976) was virtually a complete substitution of one Lap-1 allele for an alternative allele in Long Island Sound.

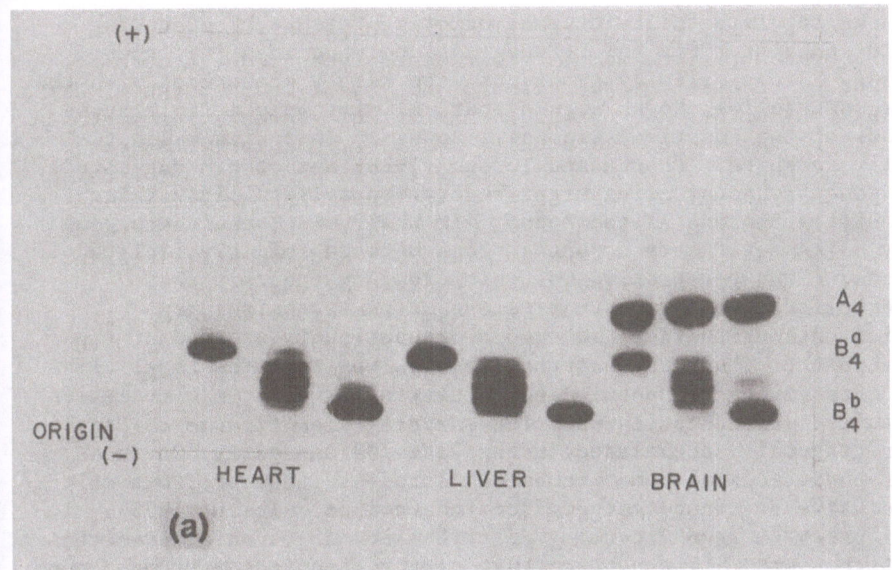

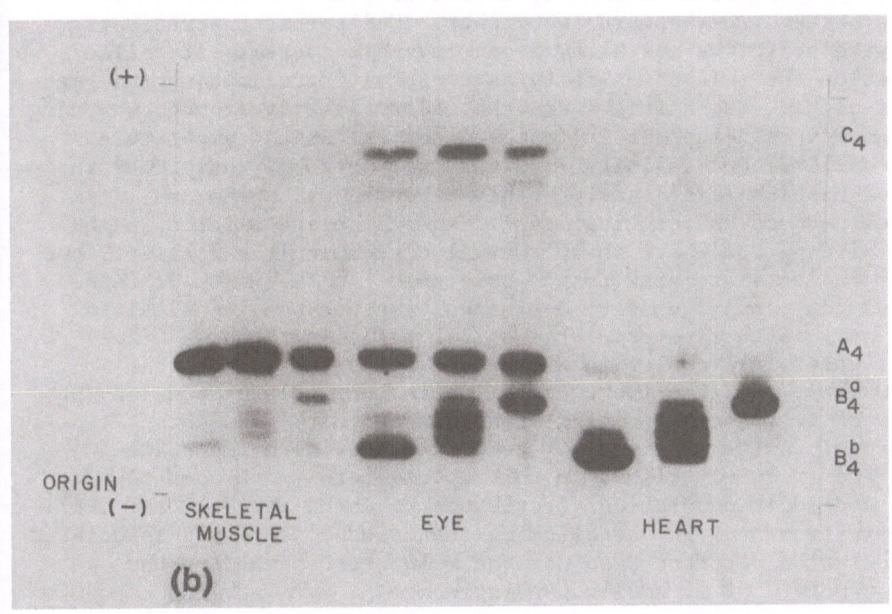

Figure 3. Tissue specificity for the <u>Ldh</u> loci in *F. heteroclitus*. As described for many other teleost species, little hybridization occurs between the A- and B-type LDF. The more anodally migrating LDH-C_4 isozyme does, however, hybridize with LDH-B_4 isozyme as evident in eye tissue. The phenotypes for the polymorphic B locus, as seen in heart or liver tissue are designated B^aB^a, B^aB^b, and B^bB^b, respectively. The trace of LDH-A_4 activity seen in the heart extract is due to contamination. (From Place and Powers, 1978).

Mollusks regulate their internal osmotic pressure by adjusting amino acid concentrations and LAP degrades polypeptides into these amino acids. Since salinity gradients were nearly concordant with the Lap-1 frequency cline, Koehn argued that salinity adaptation must be at the core of the adaptive response. However, this view was not universally accepted. Lassen and Turano (1978) and others suggested that the gene frequency cline might be attributable to circulation discontinuities in Long Island Sound. In their view, the sharp gene frequency cline was due to a contact zone between formerly isolated populations. This perspective was reinforced by the fact that environmental gradients in estuaries are often established by circulation discontinuities and such a discontinuity existed at the clinal midpoint. Thus, the alternative view was that the Lap-1 cline resulted from mixing of oceanic and estuarine larvae at a circulation discontinuity. Evidence favoring the adaptive significance of Lap-1 allozymes gradually accumulated in the late 70s and early 80s in a series of physiological and biochemical studies. However, some of the most conclusive evidence emerged from the work of Hilbish (1985). In a seminal paper on gene frequency distribution of juveniles over the course of the year, it was shown that oceanic genotypes intruded upon "estuarine" populations during larval settlement. However, these genotypes abruptly disappeard in October. While this paper clearly demonstrated selective mortality among juvenile mussels, it called into question the role of salinity as a causal force molding the gene frequency cline. If salinity were the major selective force, why did oceanic genotypes disappear in October when estuarine temperatures begin to decline, but salinity gradients were strongly enforced? The results of Hilbish (1985) was probably expected by Koehn and colleagues because an identical gene frequency cline existed in the vicinity of Cape Cod where sharp thermal discontinuities persist, but salinity gradients are weak to non-existent. To the north of Cape Cod, M. edulis populations are dominated by genotypes identical to those found in estuarine populations in Long Island Sound. Since estuarine waters are cooler in winter and warmer in summer than oceanic waters, the gene frequency cline in Long Island Sound may be more related to seasonal temperatures than salinity gradients. Regardless of whether salinity or temperature is more important in shaping gene frequency clines in this species, the work comparing juvenile and adult populations (reviewed by: Koehn and Hilbish, 1987) has been instrumental in establishing that random processes associated with circulation patterns were not the major forces molding the genetic structure of M. edulis populations.

2.3. Mitochondrial DNA (mtDNA) For Intra- and Inter-Species Studies

The study of allelic isozymes have been useful for the analysis of some species, but the lack of such genetic variation in a host of important commercial species has dampened its wide application to problems of stock assessment and recruitment. This was partly due to the fact that such methods underestimate the extent of genetic variability due to their inability to detect isopolar amino acid changes in protein structure and to detect nucleotide changes in the DNA that are not reflected in the protein sequence.

Recent advances in biotechnology and the commercial availability of restriction enzymes and other molecular tools have made it possible for population biologists to examine changes in DNA, revealing a host of previously hidden genetic variability. Not only do these methods permit the study of a broader array of genetic diversity, but their sensitivity makes it possible to study egg and juvenile stages as well as tissue biopsies of adults.

Perhaps the most useful of these DNA methods at the present time, is the use of endonuclease restriction digests of mitochondrial DNA (mtDNA) followed by electrophoresis and the construction of mtDNA restriction maps (Figure 4). Since mtDNA is usually between 16 and 19 kilobases, it is small enough to map with an array of 4 and 6 base sequence-specific endonuclease restriction enzymes. The most parsimonious assumption for a change in a restriction site is a minimum of a single nucleotide change (*i.e.*, a point mutation). Because mtDNA is maternally inherited, individuals within and between populations can be studied and various matriarchal lineages can be followed geographically, including contributions to new lineages, genetic exchange between stocks, and magnitude of contributions by a particular population to a fishery. Let us consider an example where mtDNA helped resolve an issue when allelic isozyme analysis was inadequate.

The striped bass, *Morone saxatilis*, is an important commercial and recreational species found along the East Coast of the U.S. The Atlantic race of *M. saxatilis* historically ranged from the St. Lawrence River, Canada to the St. Johns River, Florida. Females of the Atlantic population undertake oceanic migrations beginning at age 3-4 and males begin ocean migrations at age 5-6. A Gulf of Mexico race of *M. saxatilis* ranges from the Swannee River, Florida to Lake Pontchartrain, Louisiana. This race was nearly extinct by the 1960's, but stocking efforts have restored and actually broadened its former range. Most Gulf of Mexico populations rarely, if ever, migrate to the ocean (Rulison *et al*. 1982).

Beginning in the 1950's, fisheries scientists attempted to define various stocks within and among the major Atlantic spawning grounds which are: 1) The Hudson River, N.Y.; 2) The Chesapeake Bay; 3) Albemarle Sound, N.C.; and 4) The Santee-Cooper River System, S.C. Early workers found evidence of morphological differences between populations in the major areas and to a lesser extent among spawning grounds in the Chesapeake Bay. The first study of *M. saxatilis* populations in the Chesapeake Bay using genetic markers (Morgan *et al*. 1970), suggested that the populations defined by morphological characters and mark-recapture data might have a genetic basis, but the differences reported were very small compared to those found in other species. In a more extensive survey of *M. saxatilis* populations in the Chesapeake Bay, Hudson River and Albemarle Sound, Grove *et al*. (1976) found some allelic isozyme variation, but little or no evidence for genetic differentiation within Chesapeake Bay populations, and only

420

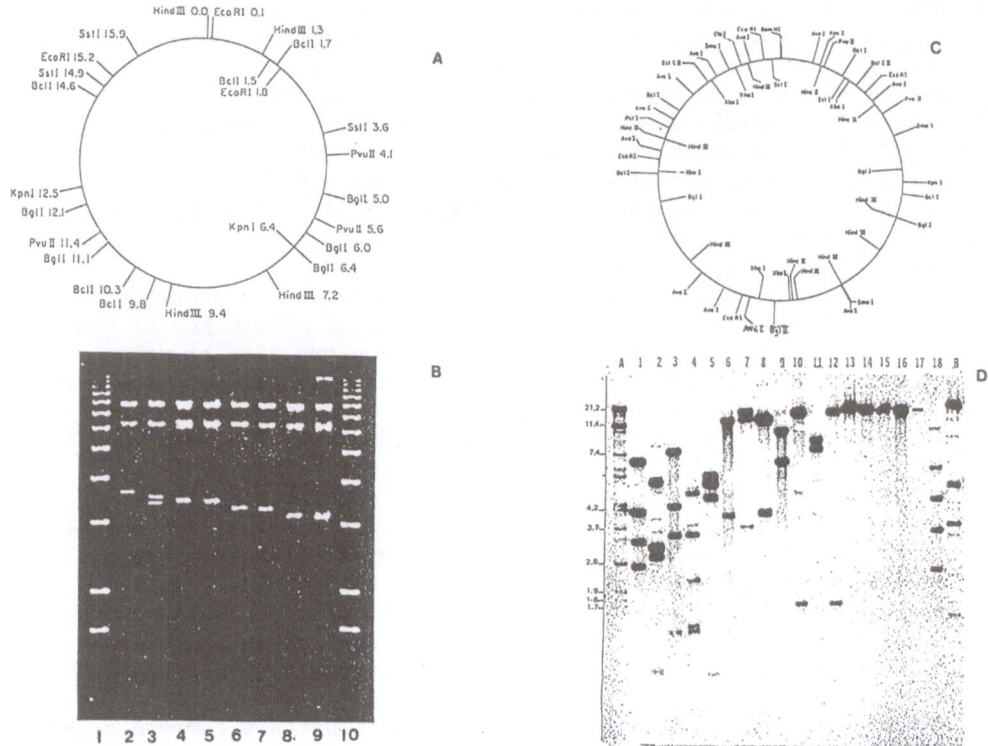

Figure 4. Restriction maps and electrophoretic patterns of digested
mtDNA. A. Map for striped bass mtDNA, B. Bcl-digests of striped bass
(*Morone saxatilis*) mtDNA showing six sizes (lanes 2 thru 9) with
standards in lanes 1 and 10. C. Restriction map for cloned *Fundulus
heteroclitus* mtDNA. (From: Gonzáles-Villaseñor *et al.*, 1986). D.
Autoradiograph of a single fish's mtDNA electromorph patterns
generated by digestion with each of 18 restriction endonucleases.
The visualization of the mtDNA fragments was attained by Southern
analysis followed by hybridization with the complementary DNA of the
radiolabeled probe. Bacteriophage lambda fragments produced by
digestion with HindIII (IA) and EcoRI + ClaI (B) were used as
markers. Restriction endonucleases: 1, EcoRI; 2, HindIII; 3, BclI;
4, AvaI; 5, HincII; 6, BstEII; 7, KpnI; 8, Sma; 9, XhoI; 10, SstI;
11, BglI; 12, PvuII; 13, PstI; 14, BamHI; 15, BglII; 16, ClaI; 17,
SalI; 18, XbaI.

limited differences among major spawning areas. A major problem confronting students of *M. saxatilis* population biology was that the species was among the most homozygous vertebrates known and, thus, limits were placed upon one's ability to descriminate between stocks.

Investigations of *M. saxatilis* populations (Chapman and Powers, in press; Chapman, 1987) using restriction endonuclease digests of mtDNA have provided a partial resolution to this problem. In these studies, evidence supporting the existence of discrete populations of *M. saxatilis* within the Chesapeake Bay has been uncovered and in general agrees with most of the delineations suggested by morphological studies (see Figure 5).

The work of Chapman and Powers (in press) also uncovered information that provides new insight into *M. saxatilis* biology. For example, in the Maryland portion of the Chesapeake Bay, it was noted that 15% of the 1982 year class males taken on the spawning grounds in 1984 possessed an mtDNA genotype, "C" (Figure 4B) that was not found in the adult females. Since many of these females were old enough to have spawned in 1982, the source of the "C" genotype was unknown. This uncertainity was exacerbated by a survey of 1982 year class females taken in February, 1985 from Hart-Miller Island (near Baltimore, MD), where it was found that 53% of the individuals were "C" genotypes. Moreover, the 1985 collection of 1982 year class females contained genotypes that were considered unique to the Chesapeake Bay. Overall, the collection of 1982 year class females at Hart Miller Island seemed to be an aggregate of individuals from many areas within the Chesapeake Bay, but dominated by genotypes considered rare in the spawning populations. In a second collection of 1982 year class males in 1986, Chapman (1987) found mtDNA genotype frequencies among 1982 year class females that were identical to those in the 1982 year class females. The mtDNA frequency differences among spawning areas noted in 1984 could not be duplicated using the 1986 data from 1982 year class males. These observations suggested that the 1982 year class was either produced by a small fraction of the adult females and/or that substantial immigration into the Maryland portion of the Chesapeake Bay occurred between May of 1984 and February of 1985.

Recent work by Carol Furman (unpublished) of the Virginia Institute of Marine Sciences has suggested that the James and Rappahannock Rivers are likely sources for the genotypes so common in 1982 year class individuals but absent in adult females spawning in the Maryland portion of the Chesapeake Bay. These genotypes were present in spawning females (1980 year class and older) taken from the Rappahannock River in 1985, but even here were not as common (ca. 15%) as in the 1982 year class (>50%). Thus, migration of two year old fish may account for some of the differences between various collections of the 1982 year class. However, even if this migration was substantial, frequency differences of the "C" genotype between the 1982 year class and older females strongly indicate that a substantial portion of the 1982 year class was produced by a relatively small fraction of the adult female stock.

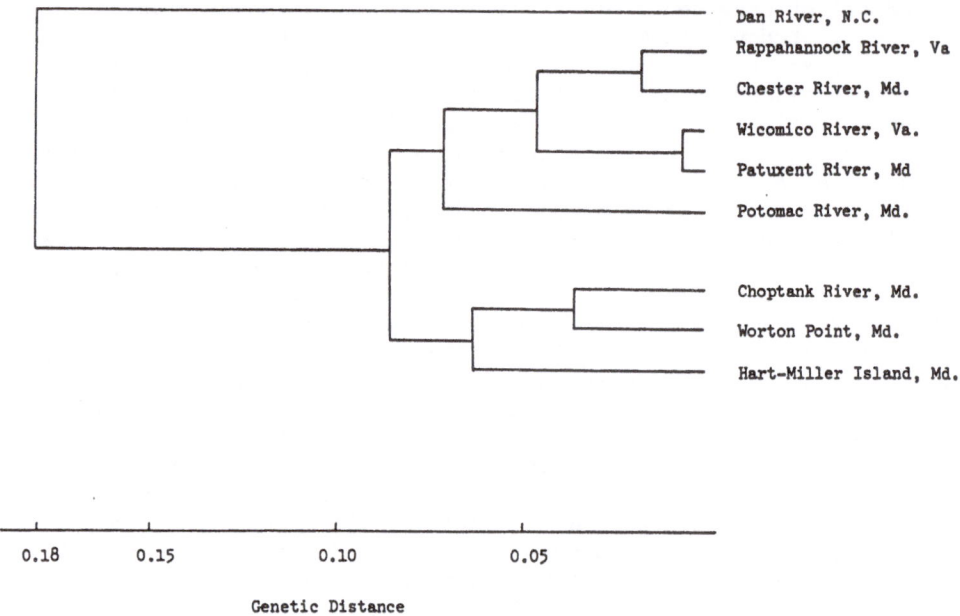

Figure 5. A UDGMA clustering of genetic distances (from Chapman and Powers, in press).

These hypotheses present some problems. If migration is extensive within the Chesapeake Bay, how is it that Chapman and Powers (in press) found differences among spawning stocks (see Figure 5)? Perhaps the answer is that males move extensively among spawning grounds, while females are more faithful to their natal areas even though females may form mixed schools during the winter. If this is true, surveys of the 1982 year class females during their first spawning runs (in 1987) should reveal differences among spawning aggregations. The samples necessary to examine this potential have been collected, but the data are not yet available.

The hypothesis that many individuals in a year class are progeny of a small portion of the available females may, in conjunction with population declines over the past 15 years, explain the disparity between the data of Morgan et al. (1973) and that of later workers; Grove et al. (1976) and Sidell et al. (1980). Both of these factors would tend to reduce the effective population size and genetic variabiity. Since the data of Morgan et al. (1970) was taken prior to the decline and the other studies were coincident with falling populations (Grove et al., 1976; Sidell et al., 1980), it is quite possible that the conclusions of all of these studies were correct, but the genetics of these populations changed between one study and the next.

To summarize, individuals recruited into spawning populations often do not reflect the composition of the parental stock. This difference could arise from several sources including: 1) stochastic variation in larval and juvenile survival; 2) immigration from genetic distinct populations; and 3) sexual asymmetry in migration patterns and reproductive cycles. A study on mtDNA frequency distributions in adult spawning stocks and resulting juveniles is currently in progress and should answer some of these questions.

2.4. Combining Studies of Allelic Isozymes and mtDNA

There are a number of examples where the use of allelic isozymes in combination with mtDNA analysis provide special insight into natural populations and closely related species. *Anguilla* are known to spawn in the Sargasso Sea and the leptocephalus larvae migrate thousands of kilometers to metamorphose in estuarine waters. Given this life history, it would seem a foregone conclusion that genetic uniformity would be expected over vast regions of the Americas and probably in Europe as well. Historically these fish have been considered distinct species, based largely upon morphological criteria. There has always been some doubt about the systematic status of these "morphs" because both migrate to the Sargasso Sea to spawn and the resulting progeny undergo metamorphosis after migrating thousands of kilometers to North American and Europe. It seemed almost impossible that there was no mixing of North American and European populations either through exchange of larvae or mating of the adults. The morphological differences could be due to differences in thermal regimes during metamorphosis in Europe vs. North America or to the duration of the larval stage.

In a study of allozyme variation along the East Coast of the U.S. and Canada, Williams et al. (1973) found little interlocality variation among elvers but some interlocality differences among adults. The investigators claimed that the general pattern was for interlocality differences among adults only, but this observation was highly locus specific. In more extensive studies (Koehn and Williams, 1978; Williams and Koehn, 1984), it was claimed that this general pattern was repeated, and argued that the small gene frequency differences among populations were due to natural selection operating on these allozymes. The data failed to replicate some of the patterns evident in the early work and the investigators argued that this was due to complex patterns of selection that varied over time. While the evidence favoring selection may be somewhat overstated, the approach of these elegant studies are important in understanding recruitment in natural populations.

The first attempt to resolve the systematic status of *A. anguilla* and *A. rostrata* using genetic markers (Sola *et al.*, 1980) revealed no significant differences between the karyotypes of European and American Atlantic populations. However, in the mtDNA study of Avise *et al.* (1986), it was found that the European and American species were highly divergent at a number of mtDNA restriction sites and thus, species recognition was well justified. Moreover, Avise *et al.* (1986) uncovered an extraordinary degree of sequence identity among American populations. A number of mtDNA clones were identified, but the clones were widely dispersed over North America. This pattern of variation was anticipated based upon existing knowledge of *A. rostrata* reproduction, dispersal and the allozyme data aluded to above. These results raise many questions concerning the recruitment of larvae into adult populations, especially concerning the developmental program of closely related species.

3. AN INTEGRATED EXAMPLE

Studies on the fish, *Fundulus heteroclitus*, are an excellent example of integrating the molecular techniques above with the physiological and environmental factors that affect recruitment.

3.1. Geographical Distribution of Allelic Isozymes

Examination of the geographical distribution of 16 polymorphic allelic isozyme coding loci have uncovered significant directional changes with latitude in gene frequencies (*i.e.*, clines) and in degree of genetic diversity. The distribution of these gene frequencies have been divided arbitrarily into four classes. Class I loci are clinal, having two predominant alleles, one fixed in northern populations, the other in southern populations. Figure 6 illustrates Ldh-B as a typical Class I locus. Class II loci are fixed for a single allele at the northern extreme of species range but have substantial genetic variation at other latitudes. Class III loci are fixed for an allele having the same electrophoretic mobility at both the northern and southern extremes, but show variability at middle latitudes. Class IV loci are not clines, as defined by Huxley (1938). Rather, they show no significant change in gene frequency with latitude.

Directional changes in genetic characters with geography (*i.e.*, clines) have classically been described by two general models: primary and secondary intergradation (reviewed by Endler, 1977). In the primary intergradation model (Figure 7), adaptation to local conditions along an environmental gradient or genetic drift may lead to genetic differences along the gradient. Gene flow may not eliminate these differences either because it is too small or because of nonrandom dispersal along the gradient. In the secondary intergradation model (Figure 7), populations are first separated by some barrier that prevents gene flow. Next, either adaptation to local conditions or genetic drift produces genetic differences between these disjunct populations. Finally, when the barrier is removed, the formerly

disjunct populations interbreed, producing a cline in gene frequencies
between them. Therefore, the main difference between these two models
is the need for the previous existence of isolating barriers to gene
flow in the latter. Figure 7 diagrammatically ilustrates these two
models and some of the driving evolutionary forces.

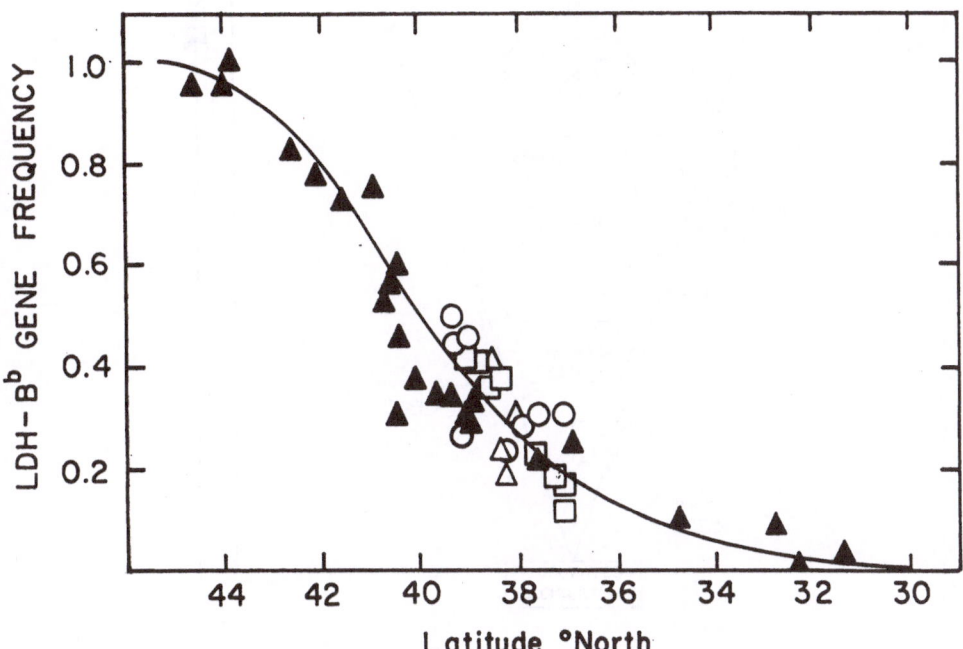

Figure 6. Frequency of Ldh-B^b allele versus latitude ^{o}N. Solid
symbols are for the coastal areas while open symbols represent the
Chesapeake Bay.

The present day spatial patterns of *Fundulus heteroclitus* allelic
isozyme gene frequencies could have arisen by either type of
intergradation. One cannot distinguish between these on the basis of
classical zoogeographical data unless it is available within a few
hundred generations of an alleged secondary contact (Endler, 1977).
However, direct analysis of mitochondrial DNA (mtDNA) can allow the
distinction between the primary and secondary models at much greater
intervals after a secondary contact.

3.2. Mitochondrial DNA to Distinguish Between Primary and Secondary
Integration

Fundulus heteroclitus populations were analyzed by studying mtDNA
fragments obtained by digestion with each of 17 restriction
endonucleases (Powers *et al.*, 1986; González-Villaseñor and Powers,

426

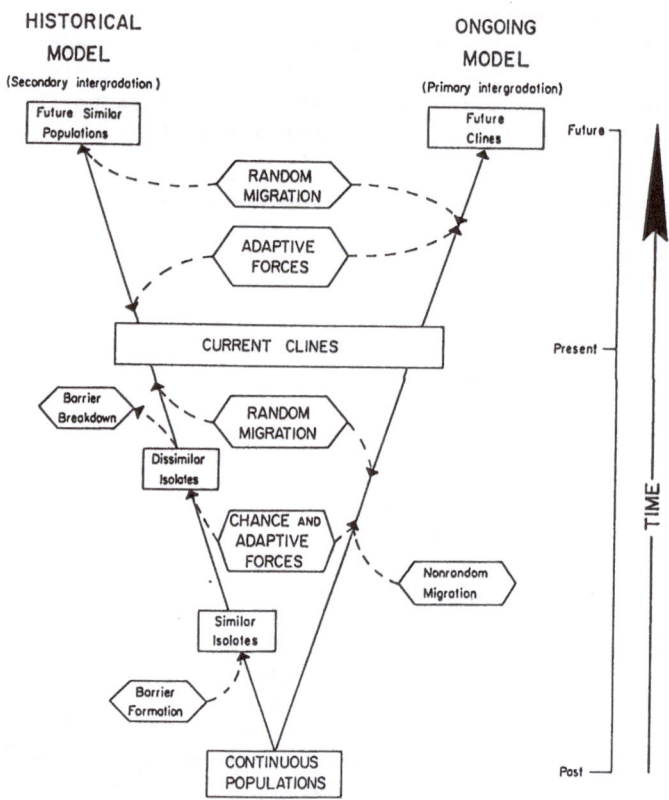

Figure 7. Model of the forces and routes of cline formation for primary and secondary integradation.

submitted). The mtDNA from individuals representing four localities were analyzed by the general procedure of González-Villaseñor *et al.*, (1986) using a radiolabeled clone of Fundulus mtDNA as a probe (Figure 4C and 4D). Analysis of the mtDNA restriction fragment data indicated that a previous barrier to gene flow existed some time ago. This conclusion was based on the fact the mtDNA restriction patterns of fish at specific localities could be interrelated by a network of single nucleotide base changes. However, populations on each side of the contact zone required many nucleotide changes (Figure 8). The extent of those differences indicated that the populations had to have been separated for several thousand years. The presence of a migrant derived genotype in one of the populations suggested that gene flow was probably reestablished sometime during the last 10,000 years. Those data clearly and unequivocally support the secondary intergradiation model of cline formation for *Fundulus heteroclitus*. However, it does not provide insight concerning the various driving evolutionary forces.

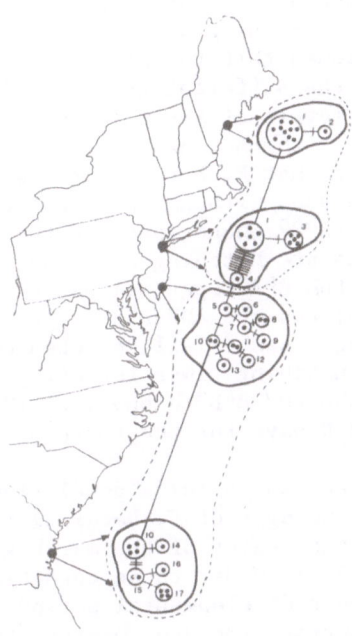

Figure 8. Phylogenetic network of the 17 composite mtDNA restriction phenotypes of *Fundulus heteroclitus*. Each circle indicates a different mtDNA clone. Dots inside circles indicate the number of fishes sharing an identical mtDNA phenotype. Clones are interconnected by branches. Solid lines crossing branches of the network indicate the number of base substitutions necessary to account for the differences in the mtDNA clones.

3.3 Genetic Basis for Differential Recruitment of Larvae

Reproductive capacity of individuals with different phenotypes within and between populations can have a tremendous impact on differential recruitment of populations into fisheries. Since only a small fraction of the eggs spawned by oviparous animals survive to the adult stage, reduction of reproductive capacity can have a tremendous evolutionary impact over time.

It has also been shown that rate of development and hatching of several fish species is directly related to the genetic background of the species. Early hatching often increases an individual's ability to compete for food. Genetic background can be studied relative to developmental rate and hatching time by fertilizing eggs *in vitro* and scoring the genotypes of the fish as they hatch. Such studies can provide significant insight into the role of genetic background in the recruitment process. Let us use the model fish *Fundulus heteroclitus* once again to illustrate this point.

DiMichele and Taylor (1980,1981) have shown that respiratory stress triggers the hatching mechanism in F. *heteroclitus*. In view of this, DiMichele and Powers (1982a) reasoned that the hatching rate of LDH-B phenotypes should differ because the different LDH-B allelic isozymes caused different steady-state levels of ATP which modified hemoglobin-oxygen affinity. Consistent with that expectation, DiMichele and Powers (1982a; 1984) found that F. *heteroclitus* embryos hatched at rates that were highly correlated with LDH-B phenotype. LDH-B^{a}B^a individuals hatched before LDH-B^{b}B^b phenotypes and the heterozygotes (LDH-B^{a}B^b) had an intermediate hatching distribution.

Figure 9 shows that hatching among LDH-B^{a}B^a eggs dominated the first three days of the hatching period, and the LDH-B^{b}B^b eggs dominated the last half. The LDH-B^{a}B^b eggs essentially hatched over the entire time span. The overall mean hatching times for offspring of 20 random crosses were: 11.9 days for the LDH-B^{a}B^a phenotype, 12.4 days for the LDH-B^{a}B^b phenotype, and 12.8 days for the LDH-B^{b}B^b phenotype.

The time of hatching is very important to *Fundulus* populations because of its reproductive strategy. The eggs of F. *heteroclitus* are laid in empty mussel shells or between the leaves of the marsh grass, *Spartina alterniflora* (Taylor *et al.*, 1977). Under these conditions, the eggs incubate in air for most of their developmental period. Hatching occurs when eggs laid on one spring tide are immersed in water by the following spring tide. As the water covers the eggs, there is a drop in environmental oxygen at the egg surface, which is the hatching cue for the embryo (DiMichele and Taylor, 1980). Hatching at the correct time would seem to be important for survival of the fry. Therefore, overall plasticity of hatching times may be important in protecting F. *heteroclitus* populations that live under variable environmental conditions. For example, premature hatching cues (*e.g.*, rainstorms) select mostly against LDH-B^{a}B^a individuals, while late hatching (*i.e.*, after the tide has retreated) primarily selects against the LDH-B^{b}B^b phenotypes (DiMichele *et al.*, 1986). This argument is particularly compelling in light of the recent finding of Meredith and Lotrich (1975) that the mortality of F. *heteroclitus* in age class zero (eggs to fry of 59 mm) is greater than 99.5%.

Since extremes in hatching time are selected against, there should be a net heterozygote advantage in a variable or uncertain environment. Such an advantage should result in the maintenance of genetic variability at the Ldh-B locus as well as a stability in gene frequency at those localities where such selection operates. This is consistent with the temporal stability in Ldh-B gene frequencies at several localities (Powers and Place, 1978).

When other loci were examined, it was evident that several genes affect developmental rate (DiMichele *et al.*, 1986). Table 1 shows the hatching times of the genotypes of 3 loci. The alleles of each locus significantly affected developmental rate and thus the additive effects of the multilocus genotypes (Table 2) leads to the smooth continuum of hatching times that is essential to the successful recruitment of larvae into the salt marsh community.

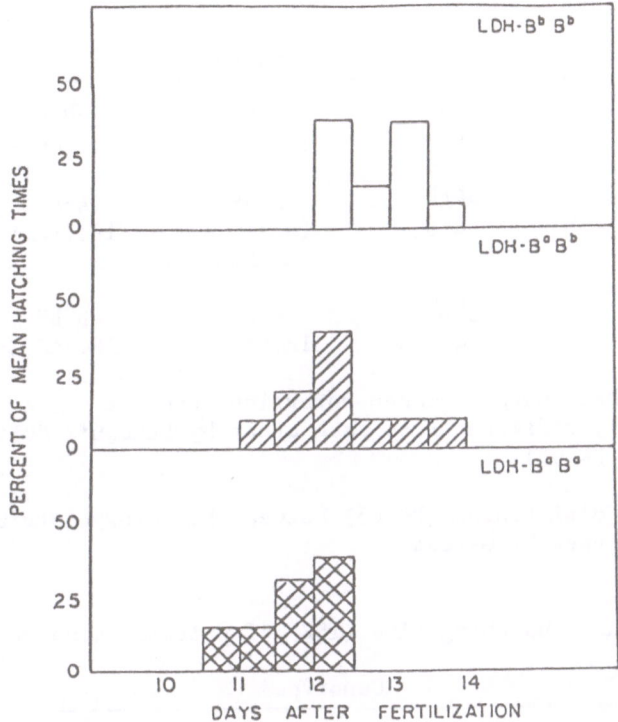

Figure 9. Distribution of mean hatching times among three LDH-B
genotypes from 20 random crosses. Analysis of variance indicated that
there were differences between the crosses (P<.05). Duncan's Multiple
Range Test showed that some of this variation was due to LDH-B
genotype (e.g., all of the LDH-B^{a}B^a crosses hatched before all of the
LDH-B^{b}B^b X LDH-B^{b}B^b crosses, P<.05). (From DiMichele and Powers
(1982a).

3.4. The Physiological Basis for Swimming Endurance Differences Between Ldh-B Genotypes

Powers *et al.* (1979) showed that intraerythrocyte ATP concentrations differed between Ldh-B genotypes of *Fundulus*. Since ATP had
been shown to be an allosteric modifier of fish hemoglobin, Powers *et
al.* (1979) correctly predicted differences in hemoglobin-oxygen
affinity between the Ldh-B genotypes, *i.e.*, the genotype with the
lowest ATP should have the highest oxygen affinity.

An analysis of the purified LDH-B allelic isozymes from *Fundulus
heteroclitus* (Place and Powers, 1979; 1984) indicated that the
greatest catalytic differences existed at high (40°C) and low (10°C)
temperatures while no significant differences were detectable at

Table 1. Mean hatching Times (± SE) of LDH-B, MDH-A, and GPI-B Genotypes.

Locus	Genotype*		
LDH-B	B^aB^a 17.8±0.1	B^aB^b 18.4±0.2	B^bB^b 19.0±0.3
MDH-A	A^bA^b 18.4±0.1	A^aA^b 18.7±0.1	A^aA^a 18.7±0.1
GPI-B[#]	B^bB^b 18.4±0.3	B^bB^c 18.5±0.1	B^cB^c 18.7±0.2

*Genotypes that share a common underline were not significantly different from each other by Duncan's Multiple Range test (P<.05).

[#]Significant differences (P<.05) between hatching distributions of the genotypes by G-test.

Table 2. Mean hatching time (±SE) of pairwise genotypes.

Loci	Genotype*								
LDH-B	aa	ab	aa	ab	aa	ab	bb	bb	bb
MDH-A	bb	ab	aa	bb	ab	aa	bb	ab	aa
	18.3 ±0.2	18.4 ±0.2	18.4 ±0.3	18.5 ±0.2	18.7 ±0.2	18.7 ±0.5	18.7 ±0.2	19.2 ±0.2	19.2 ±0.2
LDH-B	aa	ab	ab	bb	aa	aa	ab	bb	bb
GPI-B	bb	bb	cc	bb	cc	bc	bc	bc	cc
	18.1 ±0.4	18.3 ±0.5	18.5 ±0.3	18.5 ±0.8	18.6 ±0.2	18.7 ±0.2	18.8 ±0.2	18.8 ±0.5	19.0 ±0.6
MDH-A	aa	ab	bb	aa	ab	bb	ab	aa	bb
GPI-B	bb	bb	bc	cc	bc	cc	cc	bc	bb
	18.1 ±0.6	18.1 ±0.4	18.3 ±0.2	18.5 ±0.5	18.6 ±0.2	18.6 ±0.2	18.7 ±0.2	18.9 ±0.4	18.9 ±0.5

*Genotypes that share a common underline were not significantly different from each other by Duncan's Multiple Range test (P<.05).

intermediate temperatures (25°C). On the basis of this combined
physiological and biochemical information, it was predicted that
swimming endurance differences between Ldh-B genotypes should exist at
extreme temperatures but no significant differences at intermediate
temperatures. DiMichele and Powers (1982b) tested this expectation.
They reported that swimming performance is higly correlated with
genetic variation at the Ldh-B locus for *Fundulus* acclimated to 10°C,
while no such differences exist for 25°C acclimated fish.

After an acclimation period, fish of each of the two homozygous
LDH-B phenotypes were swum to exhaustion in a closed water tunnel. The
exhausted fish were sacrificed immediately and the appropriate
biochemical and physiological parameters determined.

Among resting fish acclimated to 10°C, hemocrit, blood pH, blood
oxygen affinity, serum lactate, liver lactate, and muscle lactate were
not significantly different between the two LDH-B homozygous
phenotypes. Exercising fish, acclimated to 10°C, to the point of
fatigue caused a significant change in all of these parameters. The
LDH-B^bB^b phenotype was able to sustain a swimming speed 20% higher
than that of LDH-B^aB^a phenotype. Blood oxygen affinity, serum
lactate, and muscle lactate also differed between the two phenotypes.
Since the rate of lactate accumulation was the same for the LDH-B
phenotypes, LDH-B^bB^b fish accumulated more lactate in the blood and
muscle simply because they swam longer.

In an extensive analysis of the binding of ATP to carp deoxyhemo-
globin, Greaney *et al.* (1980) have shown that the organophosphate-
hemoglobin affinity constants change by two orders of magnitude
between pH 8 and pH 7. The same general phenomenon appears to be true
for *F. heteroclitus* hemoglobins (Powers, 1980; Greaney and Powers,
unpublished). In resting Fundulus at 10°C, the blood pH was about 7.9
At this pH, the difference in erythrocyte ATP between LDH-B phenotypes
(ATP/Hb were 1.65 ± 0.12 and 2.11 ± 0.22 for LDH-B^aB^a and LDH-B^bB^b,
respectively) is not reflected as a significant difference in blood
oxygen affinity. However, as blood pH falls with increasing exercise,
the organophosphate-hemoglobin affinity increases, and differences in
oxygen affinity between homozygous LDH-B genotypes become apparent.
As blood pH is lowered, ATP amplifies the dissociation of oxygen from
F. heteroclitus hemoglobin; the more ATP, the greater the effect.
This difference is translated into a differential ability to deliver
oxygen to muscle tissue, which in turn affects swimming performance
(DiMichele and Powers, 1982b).

In fish acclimated to 25°C, there were no significant differences
in erythrocyte ATP concentrations. The ATP/Hb ratios were 1.45 ± 0.24
and 1.65 ± 0.31 for LDH-B^aB^a and LDH-B^bB^b, respectively. In addition,
there were no significant differences between LDH-B phenotypes in any
of the other parameters. Since there were differences at 10°C, but
none at 25°C, the data are consistent with the hypothesis that the
LDH-B isozyme influences red cell ATP levels and thereby affects
swimming endurance.

4. MOLECULAR METHODS FOR ASSESSING PHYSIOLOGICAL STATUS

Recruitment related mortality, as illustrated by the examples above, is a result of the physiological status of marine organisms and is highly dependent on both physical and biological parameters. In order to study the status of individuals within and between populations, one needs a series of techniques that will reflect an organism's nutritional, homeostatic, and reproductive status. Some of these methods are currently available, others are being developed that may be even more useful, and in some cases there is a need for sensitive, accurate and practical techniques to address specific problems associated with recruitment.

It is of particular interest to consider that most physiological conditions are related to the environment by the regulation of gene products in response to environmental stimuli. Since it is possible to test for the expression of these genes and their products (i.e., proteins and mRNA), the development of molecular techniques to quantitate such changes as indices of physiological status would greatly facilitate recruitment studies. Techniques that would assess growth and reproductive status would be particularly useful.

4.1. A Potential Technique for Assessing Growth Status

Growth rate studies are classically done in the laboratory by regulating food intake and measuring size and/or weight on a periodic basis. However, these are time consuming studies that are not practical for field analysis. In recent years, a number of attempts have been made to develop a simple biochemical assessment of growth that could be applied to field samples. RNA-DNA ratios, RNA content per individual and tissue specific RNA concentration have all been used as indicators of growth, metabolism, and physiological condition. While these techniques have limitations, they are extremely useful if proper controls and standards are employed. For example, the most popular technique, RNA-DNA ratios, is useful when it is restricted by species, size, life history stage, environmental temperature and physical activity of the individuals being tested (reviewed by: Bulow, 1987). Thus, while these techniques have limited potential for broad interspecies application they can be relatively useful if limited to intraspecies studies of similar life history stage.

When laboratory growth rate studies are done on fish with different amounts of growth hormone, growth rates are easily differentiated. It has been shown (Danzman, unpublished) that increased growth rate is directly related in an increased RNA/DNA ratio. However, these studies were done under a defined feeding regime and at defined environmental conditions (e.g., temperature). The obvious question is, can one estimate the relative growth rate or growth potential of an organism sampled from the field by quantitating the growth hormone level and/or RNA/DNA? If so, can such data be related to food assimilated? Although a complete answer is not yet available, a first step in that direction looks promising.

433

It has been shown by Agellon *et al.* (1987) that application of recombinant growth hormone to juvenile rainbow trout resulted in accelerated growth (Figure 10). These results suggest that the level of serum growth hormone may reflect the status of growth of individual fish in a population. Levels of growth hormone in serum samples can be determined by radioimmunoassay using a polyclonal or monoclonal antibody to growth hormone as a probe. Alternatively, the status of growth hormone production in a given fish can be assessed by determining the amount of growth hormone mRNA in the pituitary gland, since cDNA to growth hormone has been isolated from several fish species (Agellon and Chen, 1986). Such an assay can be achieved by the use of dot blot hybridization. However, the most obvious general indicator is RNA/DNA.

4.2. A Potential Technique for Detecting Female Reproductive Status

Classically, gametogenic capacity is estimated by counting changes in gamete numbers or gonad weight. While these are useful for estimating reproductive capacity, gamete counts are excessively labor intensive and gonad weight indices are often misleading. A possible alternate method for detecting female reproductive status might be the assessment of vitellogenin.

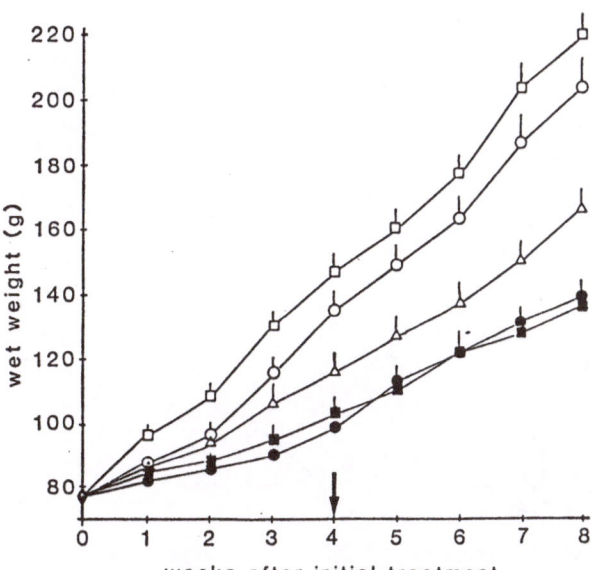

Figure 10. Effect of multiple recombinant GH treatment on growth of yearling rainbow trout. Open symbols, GH-treated fish: (□), 0.2 µg/g; (○), 1.0 µg/g; (Δ), 2.0 µg/g body weight. Closed symbols, control fish: (□), mock-treated fish; (○) untreated fish. The arrow indicates the time of the last hormone treatment (From Agellon *et al.*, 1987).

During oogenesis in fish, the egg-yolk precursor protein (vitellogenin) is synthesized in the liver, secreted into the vascular system, and then deposited in the developing oocytes as lipovitellin and phosvitine (Chen, 1983). Therefore, the reproductive status of female fish in a population can be determined by measuring the levels of vitellogenin in the serum, the rate of vitellogenin synthesis in the liver or the accumulation of vitellogenin mRNA in the liver. Levels of vitellogenin in serum samples can be determined quantitatively by the rocket immunoelectrophoresis (Figure 11). This method can detect vitellogenin levels as low as 0.05 mg/ml in serum, and give reliable quantitation of vitellogenin in both hormone-induced and reproductively active female fish. The rates of vitellogenin synthesis in livers of reproductively active females can be determined by either a radioimmunoprecipitation method (Chen, 1983) or by RNA-DNA hybridization (Figure 12).

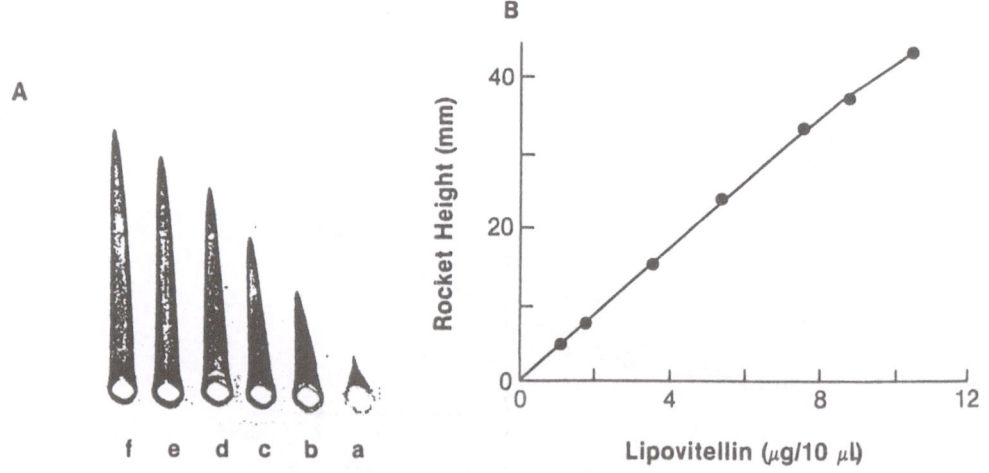

Figure 11. (A) Quantitative rocket immunoelectrophoresis to determine the amount of rainbow trout vitellogenin; (B) a standard curve (from Chen *et al.*, 1986).

435

4.3. The Potential of Immunochemical Techniques

In addition to using immunochemical methods to quantitate vitellogenin, described above, such an approach can also be applied to a host of other specific proteins. We briefly mentioned its potential for quantitating growth hormone levels and other serum protein but there is a much wider potential for immunochemical methods in recruitment studies. Polyclonal, or even better, monoclonal antibodies can be used in food chain studies, the identification of minute larvae, and the analysis of population structure.

Polyclonal antibodies are prepared by immunizing an animal with a protein that is specific for a particular species. The serum of the immunized animal can then be used to detect small amounts of the species in question. This method is useful in studying predator-prey interactions by analyzing the gut contents of predators. Moreover,

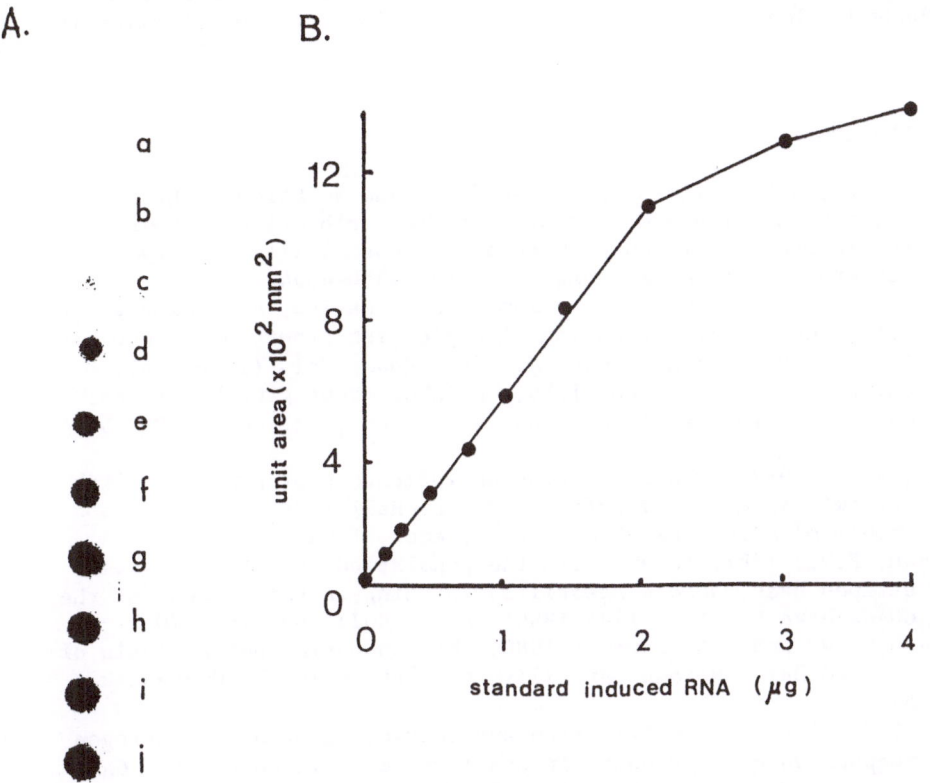

Figure 12. A representative standard curve of the quantitive RNA dot blot hybridization. (A) RNA blot; (B) a linear relationship of the amount of RNA on each spot (from Chen *et al.*, 1987).

proteins isolated from large adults can be used as antigens to
generate antibodies that are useful in detecting and quantitating
species specific larvae when the antigenetic determinants are shared
between adults and larvae. This approach can allow the identification
of minute larvae that are undetectable by less sensitive methods like
mtDNA endonuclease restriction analysis.

Since polyclonal antibodies are often of inadequate specificity
and less sensitive than monoclonal antibodies, the latter are
preferred, when possible, in studies like those alluded to above.
Preparation of a series of specific monoclonal antibodies generally
involves the following steps: (1) Several mice are immunized with
desired antigens (purified or partially purified), and spleen cells
are prepared from these mice several weeks after immunization; (2) The
spleen cells are fused to mouse myeloma cells and the hybrid cells
are selected and propagated on HAT selective medium; (3) Monoclonal
antibody-producing hybridoma clones are screened by an immuno-
binding assay, using purified antigen as a probe. Each monoclonal
antibody is then characterized extensively in order to determine its
specificity to the respective antigen.

5. REFERENCES

Avise, J.C., G.S. Helfman, N.C. Saunders, and S. Hales: 1986,
'Mitochondrial DNA differentiation in North Atlantic Eels:
Population genetic conquences of an unusual life history
pattern', *Proc. Natl. Acad. Sci.* **83**, 4350-4354.

Agellon, L.B., C.J. Emery, J.M. Jones, S.L. Davies, A.D. Dingle, and
T.T. Chen: 1987, 'Promotion of rapid fish growth by a recombinant
fish growth hormone', *Can. J. Fish. Aqua. Sci.* (in press).

Agellon, L.B. and T.T. Chen: 1986, 'Rainbow trout growth hormone:
molecular cloning of cDNA and expression in *E. coli*', *DNA* **5**, 463-
471.

Bulow, F.J.: 1987, 'RNA-DNA ratios as indicators of growth in fish: A
review', in R.C. Summerfelt and G.E. Hall (eds.), *The Age and
Growth of Fish*, Iowa State Press, Ames, Iowa, pp 45-64.

Chapman, R.W.: 1987, 'Changes in the population structure of male
striped bass, *Morone saxatilis*, spawning in three areas of the
Chesapeake Bay from 1984-1986', *Fish. Bull.* **85**, 167-170.

Chapman, R.W. and D.A. Powers: 1988, 'Mitochondrial DNA analysis of
striped bass, *Morone saxatilis*, population in the Chesapeake
Bay', (in press).

Chen, T.T.: 1983, 'Identification and characterization of estrogen-
responsive gene products in the liver of rainbow trout', *Can. J.
Biochem. Cell Biol.* **61**, 802-810.

Chen, T.T., P.C. Reid, R.J. van Beneden, and R.A. Sonstegard: 1986,
'Effect of Aroclor 1254 and Mirex on estrogen-induced
vitellogenin production in juvenile rainbow trout (*Salmo
gairdneri*)', *Can. J. Fish. Aqua. Sci.* **43**, 169-173.

DiMichele, L. and D.A. Powers: 1982a, 'LDH-B genotype-specific hatching times of *Fundulus heteroclitus* embryos', *Nature* **296**, 563-564.

DiMichele, L. and D.A. Powers: 1982b, 'Physiological basis for swimming endurance difference between LDH-B genotypes of *Fundulus heteroclitus*', *Science* **216**, 1014-1016.

DiMichele, L. and D.A. Powers: 1984, 'Developmental and oxygen consumption rate differences between Ldh-B genotypes of *Fundulus heteroclitus* and their effect on hatching time', *Physiol. Zool.* **57**, 52-56.

DiMichele, L. and M.H. Taylor: 1980, 'The environmental control of hatching in *Fundulus heteroclitus*', *J. Exp. Zool.* **214**, 181-187.

DiMichele L., and M.H. Taylor: 1981, 'The mechanism of hatching in *Fundulus heteroclitus*: Development and physiology', *J. Exp. Zool.* **217**, 73-79.

DiMichele, L., D.A. Powers, and J. DiMichele: 1986, 'Developmental and physiological consequences of genetic variation at enzyme synthesizing loci in *Fundulus heteroclitus*', *Amer. Zool.* **26**, 201-208.

Donis-Keller, H., A. Maxam, and W. Gilbert: 1977, 'Mapping adenines guanines and pyrimidines in R or A', *Nucleic Acid Res.* **4**, 2527-2538.

Endler, J.A.: 1977, *Geographical variation, speciation and clines*, Princeton University Press.

Fox, G.E., K.R. Pechman, and C.R. Woese: 1977, 'Comparative cataloging of 16 S ribosomal ribonucleic acid: molecular approach to prokaryotic systematics', *Lit. J. Syst. Bacteriol.* **27**, 44-57.

Gill, J.A., J.P. Sumpter, E.M. Donaldson, H.M. Dye, L. Souza, T. Berg, J. Wypych, and K. Langley: 1985, *Biotechnology* **3**, 643-646.

González-Villaseñor, L.I., A.M. Burkhoff, V. Corces, and D.A. Powers: 1986, 'Characterization of cloned mtDNA from the teleost *Fundulus heteroclitus* and its usefulness as an interspecies hybridization probe', *Can. J. Fish. and Aquat. Sci.* **43**, 1866-1872.

González-Villaseñor, L.I. and D.A. Powers: 1987, 'mtDNA restriction site polymorphisms in the teleost *Fundulus heteroclitus* supports secondary intergradation', Submitted.

Greaney, G.S., M.K. Hobish, and D.A. Powers: 1980, 'The effect of temperature and pH on the binding of ATP to carp hemoglobin (Hb-I)', *J. Biol. Chem.* **255**, 445-453.

Grove, T.L., T.J. Berggren, and D.A. Powers: 1976, 'The use of innate tags to segregate spawning stocks of striped bass (*Morone saxatilis*)', in M. Wiley, ed., *Estuarine Processes*, Vol. 1. Academic Press, New York, pp. 166-176.

Hilbish, T.J.: 1985, 'Demographic and temporal structure of an allele frequency cline in the mussel *Mytilus edulis*', *Mar. Biol.* **86**, 163-171.

Huxley, J.S.: 1938, 'Clines: An auxiliary taxonomic principle', *Nature* **142**, 219-220.

Koehn, R.K.: 1972, 'Genetic variation in the eel: A critique', *Mar. Biol.* **14**, 179-181.

438

Koehn, R.K. and T.J. Hilbish: 1987, 'The adaptive importance of genetic variation', *Amer. Sci.* 75, 134-141.

Koehn, R.K. and G.C. Williams: 1978, 'Genetic differentiation without isolation in the American eel, *Anguilla rostrata*. II. Temporal stability of geographic patterns', *Evolution* 32, 624-637.

Koehn, R.K., R. Milkman, and J.B. Mitton: 1976, 'Population genetics of marine pelecypods. IV. Selection, migration and genetic differentiation in the blue mussel *Mytilus edulis*', *Evolution* 30, 2-32.

Lane, D.J., D.A. Stahl, G.J. Olsen, D.J. Heller, and N.R. Pace: 1985, 'Phylogenetic analysis of the genera *Thiobacillus* and *Thiomicyospira* by 5S rRNA sequences', *J. Bacteriol.* 163, 75-81.

Lassen, H.H. and F.J. Turano: 1978, 'Clinical variation and heterozgyote deficit at the LAP locus in *Mytilus edulis*', *Marine Biol.* 49, 245-254.

Lotrich, V.A.: 1975, 'Summer home range movements of *Fundulus heteroclitus* in a tidal creek', *Ecology* 56, 191-198.

MacDonell, M.T., D.G. Swartz, B.A. Ortiz-Conde, G.A. Last, and R.R. Colwell: 1986, 'Ribosomal RNA phylogenies for the vibrio-enteric group of enbacteria', *Microbiol. Sci.* 3, 172-178.

Manseuti, R.J.: 1961, 'Movement, reproduction and mortality of the white perch, *Roccus americanus*, in the Patuxent estuary, Maryland', *Chesapeake Sci.* 2, 142-205.

Milkman, T. and L.D. Beaty: 1970, 'Large scale electrophoretic studies of allelic variation in *Mytilus edulis*', *Biol. Bull.* 139, 430.

Morgan, R.P. II, T.S.Y. Koo, and G.E. Krantz: 1973, 'Electrophoretic determination of populations of striped bass, *Morone saxatilis*, in the Chesapeake Bay', *Trans. Amer. Fish. Soc.* 102, 21-32.

Morgan, R.P. II.: 1975, 'Distinguishing larval white perch and striped bass by electrophoresis', *Chesapeake Sci.* 16, 68-70.

Mulligan, T.J. and R.W. Chapman: 1986, 'Stock identification of white perch, *Morone americana*, based on mitochondrial DNA analyses', ICES 1986/M:20.

Olsen, G.J., D.J. Lane, S.J. Giovannoni, and N.R. Pace: 1986, 'Microbial ecology and evolution: a ribosomal RNA approach', *Am. Rev. Microbiol.* 40, 337-65.

Pace, N.R., D.A. Stahl, D.J. Lane, and G.T. Olsen: 1986, 'The analysis of natural microbial populations by ribosomal RNA sequences', *Adv. Microbiol. Ecol.* 9, 1-55.

Peattie, D.A.: 1979, 'Direct chemical method for sequencing RNA', *Proc. Natl. Acad. Sci USA.* 76, 1760-17645.

Place, A.R. and D.A. Powers: 1978, 'Genetic bases for protein polymorphism in *Fundulus heteroclitus*', *Biochem Genet.* 16, 577-591.

Place, A.R. and D.A. Powers: 1979, 'Genetic variation and relative catalytic efficiencies: The LDH-B allozymes of *Fundulus heteroclitus*', *Proc. Natl. Acad. Sci. USA* 76, 2354-2358.

Place, A.R. and D.A. Powers: 1984, 'Kinetic characterization of the lactate dehydrogenase (LDH-B) allozymes of *Fundulus heteroclitus*', *J. Biol. Chem.* 259, 1309-1318.

Powers, D.A., I. Ropson, W.C. Brown, R. Van Beneden, R. Cashon, L.I. González-Villaseñor, and J. DiMichele: 1986, 'Genetic variation in *Fundulus heteroclitus*: Geographic Distribution', *Amer. Zool.* **26**, 131-144.

Powers, D.A., G.S. Greaney, and A.R. Place: 1979, 'Physiological correlation between Ldh-B genotypes and hemoglobin function in killifish', *Nature* **277**, 240-241.

Romaniuk, P.J., B. Zoltowska, T.J. Trust, D.J. Lane, G.J. Olsen, N.R. Pace, and D.A. Stahl: 1987, '*Campylobacter pylori*, the spiral bacterium associated with human gastritis, is not a true *Campyylobacter* sp.', *J. Bacteriol.* **169**, 2137-2141.

Rothschild, B.J. and A.J. Mullen: 1985, 'The information content of stock-and-recruitment data and its non-parametric classification', *J. Cons. int. Explor. Mer* **42**, 116-124.

Rulison, R.A., M.T. Huish, and R.W. Thoesen: 1982, 'Status of anadromous fishes in southeastern U.S. estuaries', in *Estuarine Comparisons*, Academic Press, New York. pp. 413-425.

Sidell, B.D., R.G. Otto, D.A. Powers, M. Karweit, and J. Smith: 1980, 'A reevaluation of the occurrence of subpopulations of striped bass (*Morone saxatilis*, Walbaum) in the upper Chesapeake Bay', *Trans. Amer. Fish. Soc.* **109**, 99-107.

Sidell, B.D., R.G. Otto, and D.A. Powers: 1978, 'A biochemical method for distinction of striped bass and white perch larvae', *Copeia* 2, 340-343.

Sola, 1., G. Gentili, and S. Cataudella: 1980, 'Eel chromosomes: Cytotaxonomical interrelationships and sex chromosomes', *Copeia* 4, 911-912.

Stahl, D.A., D.J. Lane, G.T. Olsen, and N.R. Pace: 1984, 'Analyses of hydrothermal vent-associated symbiontsby ribosomal RNA sequences', *Science* **224**, 409-411.

Stahl, D.A., D.J. Lane, G.J. Olsen, D.J. Heller, T.M. Schmidt, and N.R. Pace: 1987, Phylogenetic analysis of certain sulfide-oxidizing and related morphologically conspicuous bacteria by SS ribosomal ribonucleic acid sequences, *Intl. J. Syst. Bacteriol.* 37, 116-133.

Taylor, M.H., L. DiMichele, and G.J. Leach: 1977, 'Egg stranding in the life cycle of the mummichog, *Fundulus heteroclitus*', *Copeia* 1977, 397-399.

Vaughn, J.C., S.J. Sperbeck, and J.M. Hughes: 1984, 'Molecular cloning and characterization of ribosomal RNA genes from the brine shrimp', *Biochem. Biophys. Acta* **783**, 144-151.

Williams, G.C. and R.K. Koeh: 1984, 'Population genetics of North Atlantic catadromous eels (*Anguilla*)', in B.J. Turner, ed., *Evolutionary Genetics of Fishes*, Plenum Press, New York, pp. 529-560.

Williams, G.C., R.K. Koehn, and J.B. Mitton: 1973, 'Genetic
    differentiation without isolation in the American eel, *Anguilla
    rostrata*', *Evolution* 27, 192-204.
Woolcott, W.S.: 1962, 'Intraspecific variation in the white perch,
    Roccus americanus (Gmelin)', *Chesapeake Sci.* 3, 94-113.
Zola, H. and D. Brooks: 1983, 'Techniques for the production and
    characterization of monoclonal hybridoma antibodies', in J.G.R.
    Hurrell, ed., *Monoclonal Hybridoma Antibodies: Techniques and
    Applications*, CRC Press, Inc., Boca Raton, Florida.

THE REPLACEMENT CONCEPT IN STOCK RECRUITMENT RELATIONSHIP

Niels Daan
Department of Aquatic Ecology
University of Amsterdam
Kruislaan 320
1098 SM Amsterdam
The Netherlands

ABSTRACT. Results are presented of population experiments with guppies to study the stock recruitment relationship. Inferences are made on the likely nature of such relationship in exploited fish populations. Over a wide range of population sizes and exploitation levels there is probably no density-dependent control of subsequent recruitment. Also density-dependent control appears not to be a prerequisite in order to maintain a steady state: when recruitment is determined by the carrying capacity of the egg and larval environment, then a stable population size can be controlled entirely by adjusting adult mortality rate. The only critical landmark of the traditional stock recruitment relationship refers to the threshold level of biomass below which recruitment must be negatively affected.

1. INTRODUCTION

Since the concept of a stock-recruitment relationship was first introduced in the dynamics of exploited fish populations by Ricker (1954, 1958) and Beverton and Holt (1957), density dependent regulation has become a major issue in fisheries research (Parrish, 1973). The theoretical basis is relatively simple and draws from the observation that unexploited populations do not increase beyond limits, nor do they collapse to the point of extinction. It would appear that in order to maintain stability, a density-dependent population response is required, either in terms of growth or mortality of the adults or in terms of juvenile survival. Any relationship between stock and recruitment should also cut the stock axis close to zero, because, if there are no parents, there will be no juveniles.

The originators of the concept showed that in order to ensure resilience of a stock against random fluctuations in recruitment due to density independent environmental variations, recruitment and stock must be linked by a curvilinear function with a gradually decreasing slope. The equilibrium population size in an unexploited population is

441

B. J. Rothschild (ed.), Toward a Theory on Biological-Physical Interactions in the World Ocean, 441–440.

given by the cut-off point of the replacement line, which is defined as the set of points that characterize the number of recruits required to keep the population at a steady density, when there is only mortality due to natural causes. The models proposed by Ricker (1954) and Beverton and Holt (1957) differed in respect of the magnitude of the density-dependent response, but as shown by Paulik (1973) and Shepherd (1982) their equations belonged to the same family of possible formulations for an universally applicable stock-recruitment curve (see also Rothschild, 1986:108). Shepherd (1982) draws attention to the fact that there cannot exist one unique replacement line, but that each level of exploitation is characterized by a specific replacement line and an associated equilibrium. When exploitation changes over time, the replacement value of batches of eggs of similar size laid by multiple spawners in consecutive years will vary, which will obviously complicate the interpretation of actual data in this respect.

From the fisheries-management point of view quantitative knowledge of the effect of biomass reductions due to fishing on subsequent recruitment is of primary importance, because this information is prerequisite for making long term catch predictions as a function of optional exploitation rates. Over the years numerous exercises have attempted to fit empirical data sets on stock size and corresponding recruitment values according to these models and to estimate parameter values (e.g. Cushing and Harris, 1973). As pointed out by Rothschild (1986), however, the success of the great majority of these trials has been very poor indeed and he concludes that there is actually a seeming paradox: 'how can there be so little evidence for a relation of recruitment to parent stock when the very existence of such a relationship is so critical to population stability?'.

In addressing this question, he investigates three possible conclusions that (a) the relationship is not very precise but the density dependent factors still work sufficiently well to regulate the population; (b) the relationship is relatively precise but its precision is masked by density independent effects; or (c) any relationship is occluded by measurement error. Ultimately, Rothschild concludes that 'low variability is not a particular requirement of the recruitment-stock relationship, while curvilinearity is'. There might, however, be another conclusion that (d) the model is ill defined. In this paper I will explore this latter possibility with particular reference to the replacement concept.

Stabilization of stock size around some equilibrium would mean that under unexploited conditions biomass and recruitment should vary stochastically, and maybe also cyclically, within well defined levels. Most fish species exploited by human consumption fisheries in temperate waters are long-lived and therefore the variance of recruitment should be considerably larger than the variance of biomass, the latter being strongly buffered against annual variations in year class strength. Exploitation will negatively affect biomass with associated increases in variance. At the same time the equilibrium must change as a consequence of higher mortality and lower biomass per recruit (Shepherd, 1982). Actually, we are then

considering a population characterized by different dynamics and it
might be too simplistic to suppose that there must be a continuous
deterministic relationship relating recruitment to stock size over the
full range of hypothetical biomasses, independent of exploitation
level (cf. Rothschild and Mullen, 1985). Since density dependence
appears to be particularly relevant in relation to stabilization
around equilibria, it would seem more appropriate to try to address
the stock recruitment problem in relation to replacement levels as
affected by exploitation rate.

From a scientific point of view fisheries can be considered as
large scale population experiments, but as such they have the
disadvantage that they are generally carried out beyond the control of
the scientist, or even of the manager for that matter. Exploitation
patterns change continuously and interpretation of stock recruitment
data in terms of equilibria is made extremely difficult. Therefore, I
will first present some preliminary results from population
experiments with guppies, which were specifically designed to study
stock and recruitment. Next I will discuss possible ways to describe
variability of replacement levels in some exploited fish stocks and
the potential impact on fisheries management advice and ultimately I
will discuss stock and recruitment relationship in more general
terms.

2.   STOCK AND RECRUITMENT IN EXPERIMENTAL GUPPY POPULATIONS

2.1 Objectives

To study the stock-recruitment relationship, I set up experimental
populations of the guppy, *Poecilia reticulata* (Peters). Guppies were
chosen primarily because of their short generation time, but this
species appeared to be particularly well suited for the purpose,
because its cannibalistic behaviour had resulted in stable equilibria
in various previous experiments (*e.g.* Breder and Coates, 1932).
Therefore strong density dependence could be expected in the
regulation of their numbers. Although extensive population
experiments with guppies have been described in the literature (Rose,
1952; Silliman and Gutsell, 1958; Silliman, 1968; Warren, 1973;
Yamagishi, 1976), the stock-recruitment problem appears to have never
been addressed specifically. Fecundity is generally believed to add
an additional density dependent control factor (Bagenal, 1973) and
for guppies fecundity has been shown to depend on food supply by
Hester (1964). Therefore it seemed worthwhile to study whether
fecundity could regulate population size independently of
cannibalistic behaviour. These considerations led to the
establishment of three controlled populations: two populations (B and
C) were allowed to cannibalize their young and served as duplicates
to describe the 'normal' situation, whereas the third one (A) was
manipulated in such a way that cannibalism was virtually excluded as
a population control mechanism.

Rose (1952) and Warren (1973) have shown that crowding may result in specific dissolved substances, which affect population development. In order to make sure that differences among test populations could be attributed solely to either cannibalism or fecundity, a common recirculating water system was developed to supply all experimental containers with identical water. To reduce potential sources of differences between populations no vegetation was provided and *ad libitum* food conditions were presented in all instances in order to ensure that density dependent feeding conditions did not hamper interpretation of the results.

Apart from Yamagishi (1976), who later used 1000 L tanks, most experimenters had so far used rather small aquaria to raise their populations (*e.g.* 5.5 L by Breder and Coates, 1932; 17 L by Silliman and Gutsell, 1958). The equilibrium populations maintained were correspondingly small and death of a single adult or survival of a single juvenile represented a significant change in population size. Since space limitations in general may result in atypical border effects on population development, considerably larger basins of 300 L were chosen to host our populations.

Data on population structure were obtained for subsequent analysis of various population dynamic parameters by carrying out regular censuses.

Unfortunately, just after the populations with cannibalism appeared to stabilize at their ultimate level (after 65 weeks) an epidemic disease (*Microbacterium*) struck the system, affecting all populations equally. It was then decided to stop the special treatment for the population without cannibalism after week 71, but to continue the censuses at less regular intervals in order to follow the expected decline in population sizes. The populations survived for another 2 years, after which the experiments were stopped. Although some interesting results were obtained, the main question of whether it is possible to construct a relatively precise stock-recruitment relationship could not be answered, because of the interference of the disease.

In 1983 similar experiments were attempted to study regulation of numbers in more detail. Since it was obvious from the former exercises that fecundity did not provide a mechanism in this species to control population size and that cannibalism was essential to keep populations from exploding, it was decided to study predefined environmental effects on equilibrium population size : (1) the influence of the amount of food and (2) the effect of nursery size.

In respect to the feeding experiments, the rationale was that the rate of cannibalism depends *ceteris paribus* on the available amount of external food. However, among populations of equal size less external food should lead to increased cannibalism, less recruitment, and therefore to smaller populations. This mechanism should tend to level off the differences in amount of external food available per individual fish and, because smaller populations produce less young to be preyed upon, the ultimate effect of the external food resource was expected to be limited to population number and biomass. It was hypothesized that the intrinsic rates of cannibalism in the different populations should be rather similar.

The nursery experiments addressed the issue of possibilities for the juveniles to escape predation (Yamagishi, 1976). A reduced nursery was expected to lead to higher rates of cannibalism and therefore reduced recruitment. This process should continue until again some equilibrium was reached between rate of cannibalism, recruitment and parent stock.

Obviously, I cannot go into much detail here and I will restrict myself to the significant methodical aspects and then discuss the overall features of the first series of experiments and some major results of the second series in respect of the effect of food only.

2.2. Methods

2.2.1 *First series*. The experimental design involved 6 barren square basins of 300 L without any vegetation. A dense vegetation was only allowed to develop in the reservoir to satisfy oxygen demands. The vegetation was regularly cropped as a means to reduce the nutrient load in the system. The fish populations were maintained within bowl shaped nets fitting the basins, which were held open by a construction of bricks on the bottom. These bricks served as an abiotic nursery area, where juveniles could escape predation pressure to some extent. During winter water temperature could be maintained at 23-25 °C, but no cooling facilities were available and summer temperatures might temporarily increase to 28 °C. A 12 h light/12 h darkness regime was maintained throughout the year. Excess food was provided daily in the form of some kind of artificial dry food, frozen daphnids and artemia.

During the censuses every 4-6 weeks the populations were lifted in their nets from the basins and sieved through a 2 mm square mesh size to separate postrecruits from prerecruits. The postrecruits were then split upon individual inspection into subadults, males and females and of each category a photograph was taken of for subsequent analysis of the size distributions. Nets and bricks were cleaned from algal growth and detritus before the fish were returned to their home basin.

In population A in which cannibalism was to be excluded a weekly search was carried out for all pregnant females expected to give birth shortly. These were put into individual jars and a daily check was made for the presence of juveniles. The young mothers were measured and returned to the population and the juveniles were counted and put into a separate nursery. In this nursery there were 6 divisions and all young born in the same week were combined into one division. After having been reared for 6 weeks, when they had grown beyond the size where they could be effectively preyed upon, the survivors were added to the parent population. In one of the duplicate populations with cannibalism (B) the same procedure of separating pregnant females was followed except that the juveniles born in the jars were immediately returned to the parent population after counting. From week 71 onwards, after *Microbacterium* had infected the entire system, no pregnant females were separated any longer. Consequently, from this date onwards the conditions for cannibalism are equal in all three populations. The censuses were maintained at less regular intervals until week 160, when the experiment was stopped.

2.2.2. *Second series*. The same basic set up was used, but smaller 200 L basins were employed to obtain rather smaller equilibrium populations and correspondingly reduce the workload. For our discussion here only the three populations receiving different amounts of food are relevant, the other data series not yet being analysed. On the basis of the required amount of calories for some 300 adult guppies according to data provided by Silliman and Gutsell (1958), a standard amount of artificial dry food was calculated and this amount was doubled, respectively halved in order to establish 3 feeding regimes. Censuses were held every 6 to 8 weeks according to the methods described for the first series, but in addition males and females were weighed in order to obtain a direct measurement of biomass. No effort was made to get a direct count of the numbers of young guppies born per week, because according to the results in the first series of experiments the average production of juveniles could be reliably estimated from the female biomass.

2.3. Results

2.3.1. *First series*. Figure 1 shows the development in the number of post recruits and the average weights per adult female in the three populations. During the first 47 weeks of the experimental period there were regularly system failures of some sort, which resulted in losses of animals through overflows or increased mortality as a consequence of breakdown in thermostatic control. Only from week 50 onwards the populations were able to develop without major interferences until the outbreak of the infectious disease in week 65. This is obviously too short for an extensive population dynamic analysis of the stock recruitment problem, but still some interesting features are apparent.

All populations started to increase from week 50 onwards, but the populations with cannibalism (B and C) appeared to stabilize within 10 weeks at a level of approximately 900 individuals, whereas the population without cannibalism (A) exploded to over 4000 individuals in week 67 with some 1500 prerecruits still waiting in the nursery. At this stage the search for pregnant females had become an almost impossible task and most likely some have been overlooked. Still, among the pregnant females isolated no reduction in the fecundity length relationship was observed. However, as can be deduced from the fact that average weight per female in this population remained far below those in populations B and C, individual growth was greatly reduced. Undoubtedly, the reduced growth rate affected fecundity secondarily, but the effect has not nearly been large enough to stabilize the population. Actually, densities of 13 guppies per L must be considered to be unnatural and artificial: population A gave the impression of a merry-go-round, the fast swimming fish causing a small gyre in the middle of the basin. This suggests that maintenance requirements had enormously increased as a consequence of crowding and

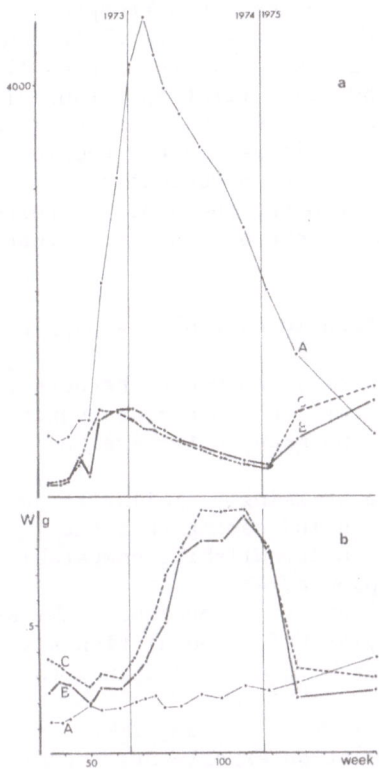

Figure 1. Population development in first series of experiments: a. Number of subadults and adults present (N); b. Mean weights per female (W in g). In population A cannibalism was excluded up till week 71; populations B and C are duplicates with cannibalistic control.

that these higher costs of swimming were limiting the scope for growth.

Following the outbreak of the disease all populations started to decrease, although the effect was not nearly as dramatic as we had expected at the onset. During this phase nearly all fish were mature and subadults were virtually absent indicating that production of juveniles was almost completely absorbed by cannibalism. This can also be deduced from the rapidly increasing weights per female in populations B and C. In population A mean weights increased at a much

reduced rate, but also in this case recruitment had stopped
completely. In the populations allowed to cannibalize their young the
first recruits appeared again in significant numbers in week 120, when
population number had fallen below 350. This first appearance is
followed by a rapid increase in stock size, which might herald a new
cycle. The third population contained still nearly 600 guppies at the
end of the experiment and as yet there were no subadults.

Although of limited utility for studying the stock-recruitment
relationship, some important conclusions can be drawn from these
experiments:

1) cannibalism is an essential feature in controlling population
   size in guppies;
2) fecundity is only secondarily density dependent, because it is
   related to growth and growth is retarded as a consequence of
   crowding; still, fecundity is by no means effective in
   controlling population size;
3) the duplicate populations show a strikingly similar response
   throughout the experimental period indicating that the
   experimental setup was adequate in establishing comparable
   environments for the different populations;
4) the populations with cannibalism appear to overshoot the steady
   state so that it takes about a year before population size is
   reduced to a size where juveniles have again a chance to survive
   cannibalistic pressure.

2.3.2 *Second series*. From these population experiments rather more
detailed data are at hand to investigate the dynamics over three years
of population development. Figure 2 provides plots of adult population
size in numbers, mean weights per adult, estimated rates of mortality
due to cannibalism during the prerecruit stage, rate of mortality
among adults and estimated recruitment. Apparently, the amount of food
has a direct effect on adult population size in terms of numbers. The
biomass curves are not shown, but they exhibit virtually the same
pattern as can be deduced from the fact that there are no consistent
differences between the mean weights per adult.

In these experiments also a time lag of approximately 50 weeks is
observed before the populations started to increase to their ultimate
levels. Again this is associated with system failures but probably it
has also something to do with ageing of the system. In particular, the
pH of the system was unstable during the first half year and
presumably the conditions were suboptimal.

After this initial phase, there is clearly an annual cycle
superimposed upon the numbers of adults present, which is in phase for
all populations. This cycle is absent from the estimated prerecruit
mortalities due to cannibalism, which seem to fluctuate around a
constant level independent of feeding conditions. However, the adult
natural mortality follows also an annual cycle, increases in
population size coinciding with low mortalities and vice versa. The
trends in and level of mortality within the three populations appear
not to be significantly different. On average the estimated natural

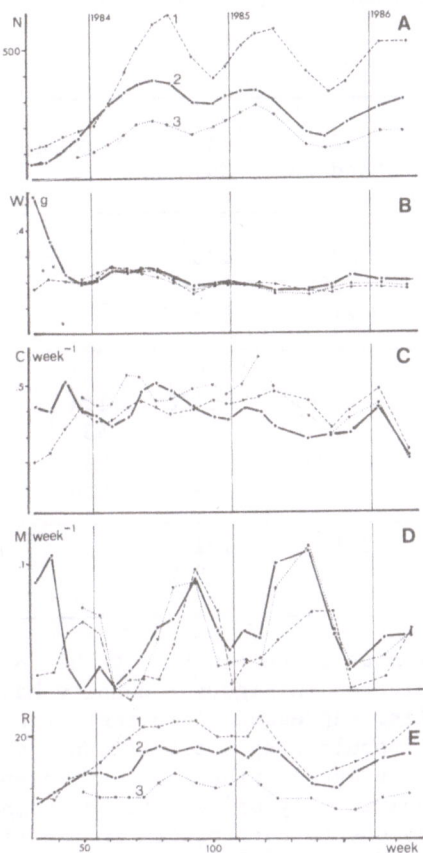

Figure 2. Population development in second series of experiments: A. Number of adult fish present (N); B. Mean weights per adult (W in g); C. Rate of prerecruit mortality due to cannibalism (C in 1/week); D. Rate of adult mortality (M in 1/week); E. Recruitment per week (R). Populations 1, 2 and 3 were maintained on double, standard and half rations respectively.

mortalities in this experiment were approximately 0.04 1/week, which is twice as high as in the the first series of experiments (0.02 1/week for both the period prior to the infection and thereafter).

Estimated recruitment to the population older than 6 weeks shows no particular annual trend, but recruitment in the population at double rations is consistently higher and in the population at half rations consistently lower than in the population at standard rations.

In Table I various population parameters are averaged over the period beginning from week 50 to week 176.

Table I. Average population parameters in second series of
          experiments.

Population Food ration	1 Double	2 Standard	3 Half	Average
Prerecruits	651	491	393	
Subadults	38	30	18	
Adults	432	272	175	
Adult biomass	82.6	55.5	32.5	
Mean weight adults	.192	.202	.188	.194
Fraction males	.54	.54	.50	.53
Natural mortality	.032	.047	.042	.040
Cannibalistic mort.	.484	.427	.457	.456
Nr of births/week	459	291	191	
Nr of recruits/week	19.3	15.4	9.2	

Mortality due to cannibalism would be expected to be correlated
to adult stock size, but as indicated in Figure 3 density has only a
very limited effect on cannibalism expressed as an instantaneous rate
of mortality. Within individual populations a significant correlation
($P<0.025$) is observed, but the average cannibalism coefficients
between populations are not significantly different. For comparison
available estimates of mortality due to cannibalism in the first
series of experiments are included in this Figure. Even though the
experimental setup of the two series is not directly comparable, the
average rate of mortality due to cannibalism is in the same order of
magnitude and also in this case there is a weak positive correlation
with density.

Figure 4 presents the actual stock recruitment plot for each of
the three populations. The averages within each population, which can
be considered to represent equilibria, are indicated. According to the
results obtained there is obviously no dome shaped stock recruitment
relationship and within each population a significant positive
correlation between recruitment and stock size is found. Though
plotted in the same graph, the points for the different populations
should obviously not be interpreted as parts of the same curve,
because they refer to different environmental conditions in terms of
amount of external food.

Since the experiments were carried out in a dark room under
constant conditions, it is not directly obvious why an annual cycle
would be maintained in these populations. The only clue we have is
that temperature during summer could not always be maintained within

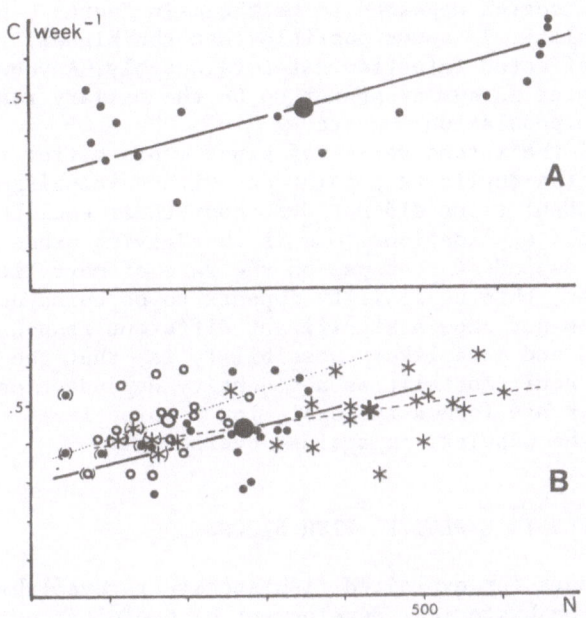

Figure 3. Relationship between instantaneous rate of mortality due to cannibalism (C) and stock size (N): A. First series - populations with cannibalism combined; B. Second series - populations on double (*), standard (o) and half (o) rations.

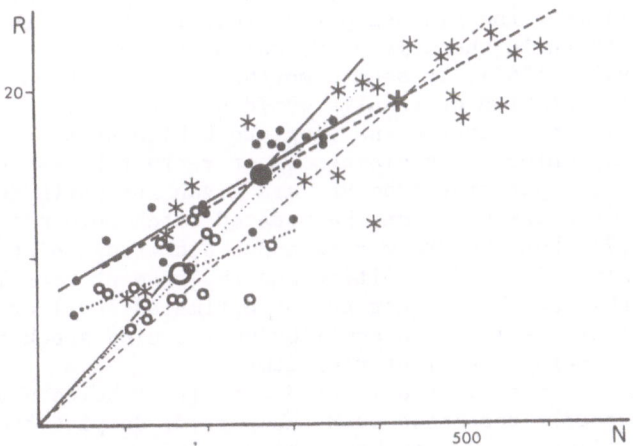

Figure 4. Stock recruitment plots for second series populations (for symbols see fig 3B). Equilibria, replacement lines and regression lines are also shown.

the range aimed for during prolonged periods of time. Since also in
this case various diseases appeared to be the main factor causing
adult natural mortality, it seems possible that the higher
temperatures have affected infection rates favourably. Anyhow, the
higher mortality rates in summer appear to be the primary reason for
the annual cycle in population numbers.

The results of the second series of experiments differ in one
major respect from the duplicate populations with cannibalism in the
first series: the populations did not overshoot their equilibria.
There are two obvious explanations. One is that giving excess food
spoils the density dependent response on the rate of mortality due to
cannibalism. However, this possibility appears to be borne out by
Figure 3, which does not show a significant different response in this
respect. The second and more likely possibility is, that the interplay
of cyclically high adult mortalities and density dependent cannibalism
in the second series has formed the basis for a more timely control
and has protected the populations against overshoots.

3. REPLACEMENT LEVELS IN EXPLOITED FISH STOCKS

Stock recruitment data for exploited fish stocks are available as time
series describing the historical development of exploitation. Since
replacement lines change in relation to changes in exploitation and
since it is not directly obvious that the exploited population has
some memory of the pristine equilibrium, stock recruitment
relationship should probably not be interpreted in relation to the
steady state associated with the unexploited stock, but in relation to
the replacement line applicable to the conditions of the stock at the
time the recruitment is created. The recruitment required to replace a
particular biomass in a particular year can be directly estimated
(Shepherd, 1982) from the biomass per recruit related to the
exploitation pattern in the same year, which can be easily calculated
(Beverton and Holt, 1957). It seemed worthwhile to work out one
example and see what kind of results would be obtained.

In Table II the estimated spawning stock biomasses, recruitments
and calculated spawning stock biomasses per recruit based on the
annual exploitation patterns from VPA are given for North Sea cod in
each year. The data are based on the updated catch information in 1987
(Anonymous, 1987), but the VPA was extended backwards to cover the
entire period since 1966. In addition the replacement recruitment
required to stabilize the biomass at the estimated level at each point
of time is given as well as the equilibrium spawning stock biomass
related to the observed level of recruitment.

These data are are plotted in Figure 5. It is quite obvious that
biomass has gradually decreased over the period, despite the fact that
recruitment has fluctuated without a major trend over the entire
period. The situation is even worse, however, than the estimated
biomasses would indicate, because the biomass per recruit has
proportionally dropped considerably faster. The data indicate that in
order to maintain the population at a decreasing size consistently

Table II.   Spawning stock biomass (SSB) and recruitment (R)
            parameters of North Sea cod.

Year	SSB	R	SSB/R	Replacement R	Equivalent SSB
1966	222	479	.497	447	238
1967	244	461	.394	620	182
1968	252	185	.258	978	48
1969	251	197	.400	627	79
1970	271	729	.308	879	225
1971	269	847	.193	1393	163
1972	225	160	.138	1627	22
1973	197	293	.158	1244	46
1974	210	234	.211	998	49
1975	190	426	.208	915	89
1976	163	208	.161	1016	33
1977	142	710	.198	718	141
1978	142	427	.114	1253	49
1979	145	454	.163	890	74
1980	160	800	.128	1248	103
1981	173	271	.127	1367	34
1982	167	556	.067	2483	37
1983	134	276	.063	2134	17
1984	114	552	.078	1461	43
1985	104	93	.097	1064	9
1986	95	729	.093	1031	67

increasing recruitment would have been required. Another way of
expressing the same thing is that equilibrium biomasses related to
estimated recruitment levels are well below the biomasses from which
the recruiting year classes originated.

It then appears that stock recruitment plot (Figure 6) does not
allow any qualified statements about the density dependent population
control in the population, recruitment being just variable and biomass
being largely depending on exploitation rate. The absence of any
density dependent control is better illustrated in a plot of
equilibrium biomass related to observed recruitment against observed
biomass present in each year (Figure 7), indicating that under the
present conditions of exploitation there is no expectancy of a steady
state in the cod population. It would appear that , if no major
management actions are taken, sooner or later the spawning stock will
drop close to extinction.

454

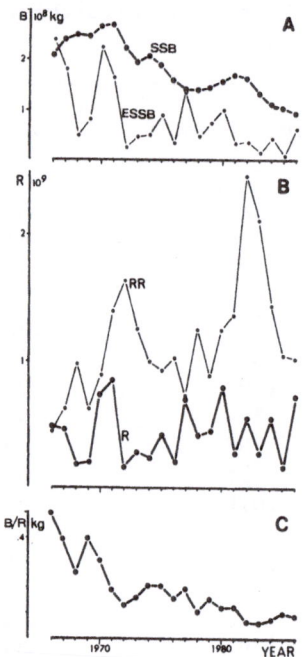

Figure 5. Stock recruitment parameters of North Sea cod plotted as time series: A. Estimated spawning stock biomass (SSB) and equivalent spawning stock biomass (ESSB); B. Estimated recruitment (R) and replacement recruitment (RR); C. Spawning stock biomass per recruit (B/R).

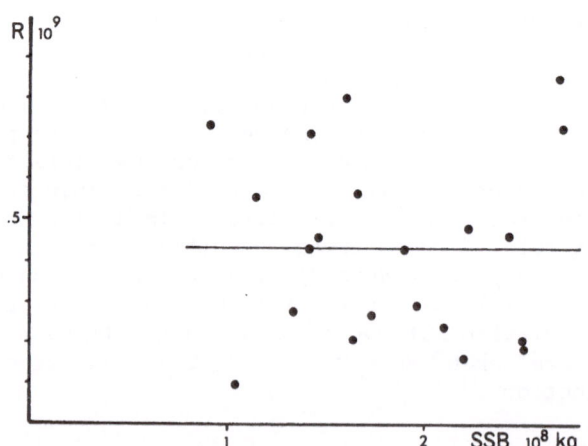

Figure 6. Recruitment (R) plotted vs spawning stock biomass (SSB) in North Sea cod.

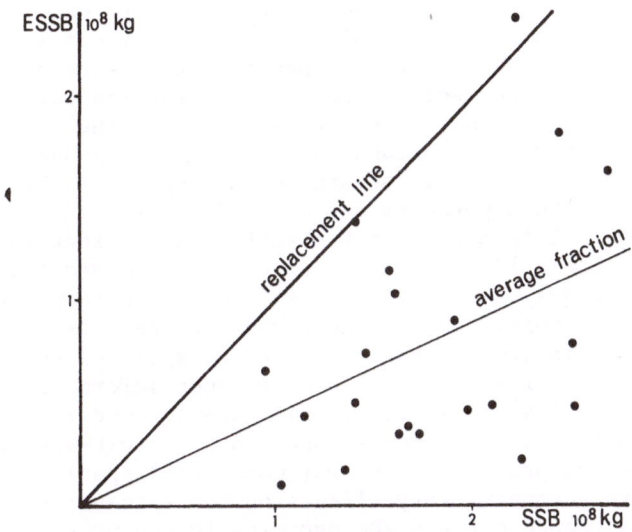

Figure 7. Equivalent spawning stock biomass (ESSB) plotted vs observed spawning stock biomass (SSB) in North Sea cod.

4. DISCUSSION

The dome shaped Ricker-type stock recruitment relationship is generally believed to stem from strong compensatory population control mechanisms and among the latter cannibalism is at least potentially a very powerful one. Indeed, when cannibalistic control was excluded, our guppy populations exploded. Paradoxically, in food limited experiments the correlation between stock and recruitment is positive in the region of the equilibrium, suggesting that in practice cannibalism does not necessarily result in a strong compensatory effect, even when it represents a crucial element in population control. A relatively minor effect of stock size on rate of mortality due to cannibalism is apparently enough to maintain the population around some equilibrium value. In the experiments there is no true stability, because the cyclical change in adult mortality leads to cyclical changes in the biomass per recruit and consequently in the replacement line. These shifts appear to be the primary reason, why a global equilibrium is maintained in all populations of the second series. We must conclude that, although cannibalism is an essential feature of population regulation, population equilibrium is reached by the intricate interplay of population size, external food, size of the nursery, cannibalism and adult mortality. Of these, the latter is probably the most significant steering factor.

The equilibrium between stock and recruitment depends directly on the available amount of external food. There is, however not a one to one relationship. As can be deduced from table I, doubling the amount of food yields approximately only an 1.5x increase in stock size and recruitment, so that overall a 4-fold increase in external food results in a 2-fold increase in population size. It seems likely that this a consequence of other limiting aspects of the carrying capacity of the environment created. For instance, the nurseries were structurally identical in the three populations, but the functional significance of the nursery is probably affected by the number of births, because that number will determine the degree of shelter being used. Nevertheless, the hypothesis set out originally, that varying the amount of external food would not significantly affect the rate of mortality due to cannibalism and population structure and that the effects would be largely restricted to level of recruitment and population size, is supported by the experimental results.

Guppy experiments can of course not be extrapolated to commercially exploited fish stocks. However, some inferences can be made. First of all, if a cannibalistic fish species under controlled conditions does not exhibit a dome-shaped stock recruitment curve, there would appear no primary reason why commercial species should be controlled by strong compensatory effects on recruitment: there is obviously less coherence between the dynamics in the prerecruit and postrecruit phase of a species producing vast numbers of very small eggs than there is between the juveniles and adults of a viviparous fish inhabiting largely the same environment. Consequently any density effects in the former must be less direct, even for a species like cod, which to some extent at least feeds on its own offspring (Daan, 1973, 1983).

Fecundity did not emerge as a strong regulating factor from our experiments. It should be borne in mind, however, that guppies are viviparous, producing small numbers of offspring. This reproduction strategy is quite different from the one commonly observed among most of the important marine and fresh water fish species, which produce enormous numbers of very small eggs, which are subsequently subject to very high mortalities. If there would be only a small density-dependent effect on the quality of the eggs produced, the ultimate regulatory force might well be significant in these species, because mortality might act as an amplifier.

In ecological theory r- and K-strategies are distinguished to differentiate between short-lived animals producing large numbers of offspring and long- lived animals producing few juveniles (for a review see Stearns, 1976). The former represent the opportunists increasing rapidly under favourable conditions and the latter the conservationists building up stable populations. Many fish species do not seem to fit very nicely in this respect, because they are relatively long-lived and unexploited stocks should have been fairly stable, but at the same time they produce a tremendous number of offspring in the form of very small eggs. The opportunists and conservationists generally exploit different environments, the former being associated with the unpredictable and the latter with the predictable environment. This corresponds largely with their positioning in the food pyramid. The r-selected species can be found at the bottom and the K-selected species at the top. Again fish seem to play a double game, the adults largely exploiting the tertiary

consumers at the top, but their offspring forage on the secondary
consumers at the bottom. In a population dynamic sense, one might
perhaps better think of fish as representing two species: the larvae
represent the r-part, which yield a highly variable number of
survivors at the age of recruitment depending on the conditions
encountered; the adult population has to cope with variable
recruitment but succeeds in following a K-strategy by longevity. In
this scenario there is not necessarily a feed back of biomass to
recruitment, as long as some minimum conditions of egg production are
met. The carrying capacity of the environment defines the number of
recruits and the average number of recruits defines along with the
adult mortality and the associated biomass per recruit the equilibrium
size of the stock. There is no *a priori* need to assume some density
dependent control mechanism beyond some threshold value of stock size
in order to maintain stability, because longevity itself tends to
level off variations due to variable recruitment. The ultimate stock
recruitment relationship falls apart in two parts : a horizontal limb
describing the density independent part and an ascending limb
describing a depensatory part. The recruitment problem would be
reduced to a study of the factors determining carrying capacity of the
environment for larvae and early juveniles from year to year and the
search for clues for the biomass threshold level.

Above the threshold value, the main effect of fishing is largely
restricted to destabilization of the adult stock, because increased
fishing mortality results in smaller biomasses per recruit and less
year classes in the stock. Considering numerous plots of recruitment
versus stock size of many exploited fish species in the past, my
conclusion would be that this scenario fits the available data much
better than any theoretical supposition of a density dependent control
of the parent stock on subsequent recruitment. In accordance with
Shepherd's (1982) conclusion, it would follow that for assessment
purposes the biomass per recruit remains a parameter of great concern,
because it is indicative of the average stock size that can be
expected in the future on the basis of the observed recruitments
rather than of the expected recruitment on the basis of present stock
size.

Turning to the paradox outlined by Rothschild (1983), the
solution appears to lie in the proposition that there must be some
density-dependent control of the adult population over subsequent
recruitment in order to ensure resilience. Resilience may have been
acquired by selection for longevity, allowing populations to survive
even prolonged periods of poor recruitment. The factors determining
survival of larvae and recruitment represent an entirely different
matter. Therefore, the traditional stock recruitment model might well
be ill defined.

5. REFERENCES

Anonymous: 1987, 'Report of the North Sea Roundfish Working Group',
    ICES C.M.. 1987/Assess:15.

Bagenal, T.B.: 1973, 'Fish fecundity and its relation with stock and recruitment', *Rapp. P.-v. Reun. Cons. int. Explor. Mer* **164**, 186-198

Beverton, R.J.H. and S.J. Holt: 1957, 'On the dynamics of exploited fish populations', *Fish. Invest. Lond. Ser.* **2**, 19.

Breder, C.M. and C.W. Coates: 1932, 'A preliminary study of population stability and sex ratio of Lebistes', *Copeia* **3**, 147-155.

Cushing, D.H. and J. Harris: 1973, 'Stock and recruitment and the problem of density-dependence', *Rapp. P.-v. Reun. Cons. int. Explor. Mer* **164**, 142-155.

Daan, N.: 1973, 'A quantitative analysis of the food intake of North Sea cod, *Gadus morhua*', *Neth. J. Sea Res.* **6**, 479-517.

Daan, N.: 1983, 'Analysis of the cod samples collected during the 1981 stomach sampling project', ICES C.M. 1983/G:61.

Hester, F.J.: 1964, 'Effects of food supply on fecundity in the female guppy, *Lebistes reticulatus* (Peters)', *J. Fish. Res. Bd. Can.* **21**, 757-764.

Parrish, B.B. (ed.): 1973, 'Fish stocks and recruitment', *Rapp. P.-v. Reun. Cons. int. Explor. Mer* **164**, 1-372.

Paulik, G.J.: 1973, 'Studies on the possible form of the stock-recruitment curve', *Rapp. P.v. Reun. Cons. int. Explor. Mer* **164**, 303-315.

Ricker, W.E.: 1954, 'Stock and recruitment', *J. Fish. Res. Bd. Can.* **11**, 559-623.

Ricker, W.E.: 1958, 'Handbook of computations for biological statistics of fish populations', *Bull. Fish. Res. Bd. Can.* **119**, 1-300.

Rose, S.M.: 1959, 'Population control in guppies', *The American Midland Naturalist* **62**, 474-481.

Rothschild, B.J.: 1986, *Dynamics of marine fish populations*, Harvard University Press, Cambridge, Massachusetts, 277pp.

Rothschild, B.J. and A.J. Mullen: 1985, 'The information content of stock-and-recruitment data and its non-parametric classification', *J. Cons. int. Explor. Mer* **42**, 116-124.

Shepherd, J.: 1982, 'A versatile new stock-recruitment relationship for fisheries, and the construction of sustainable yield curves', *J. Cons. int. Explor. Mer* **40**, 67-75.

Silliman, R.P.: 1968, 'Interaction of food level and exploitation in experimental fish populations', *Fishery Bull. Fish Wildl. Serv. U.S.* **66**, 425-439.

Silliman, R.P. and J.S. Gutsell: 1958, 'Experimental exploitation of fish populations', *Fishery Bull. Fish Wildl. Serv. U.S.* **58**, 215-252.

Stearns, S.C.: 1976, 'Life-history tactics: a review of the ideas', *Q. Rev. Biol.* **51**, 3-47.

Warren, E.W.: 1973, 'The establishment of a 'normal' population and its behavioural maintenance in the guppy - *Poecilia reticulata* (Peters)', *J. Fish Biol.* **5**, 285-304.

Yamagishi, H.: 1976, 'Experimental study on population dynamics in the guppy, *Poecilia reticulata* (Peters), Effect of shelters on the increase of population density', *J. Fish Biol.* **9**, 51-65.

LARGE MARINE ECOSYSTEMS AS GLOBAL UNITS FOR RECRUITMENT EXPERIMENTS

Kenneth Sherman
National Marine Fisheries Service
Northeast Fisheries Center
Narragansett Laboratory
Narragansett, RI   02882-1199
U.S.A.

ABSTRACT. Over the last million years, fish species presently of significant economic importance (including species of cod, haddock, hake, herring, mackerel, and others) have evolved feeding, spawning and recruitment patterns that are difficult to understand unless observed throughout the ranges of the populations under investigation. Over the past decade a research strategy has been developed based on the recognition that a large number of marine fish species have adapted to extensive marine systems of unique bathymetry, circulation, productivity, and trophic interrelationships.

Within the Exclusive Economic Zone (EEZ) of the United States, seven such systems have been identified--the East Bering Sea, Gulf of Alaska, California Current, Insular Pacific, Gulf of Mexico, Southeast Atlantic Shelf, and Northeast Atlantic Shelf. Each of these large marine ecosystems (LMEs) extends over a geographic range exceeding 200,000 km^2 within which systematic studies of the recruitment processes are presently underway. Within these LMEs and others around the globe, including the Scotian Shelf Ecosystem, the Tasman Sea Ecosystem, the Gulf of Thailand Ecosystem, the Iberian Peninsula Ecosystem, the Kuroshio Current Ecosystem, the Oyashio Current Ecosystem, the Yellow Sea (Huanghai Sea) Ecosystem, the Baltic Sea Ecosystem, and the North Sea Ecosystem large scale biomass flips in the fish component of the systems have occurred in which a dominant species within a relatively short span of less than 10 years has been reduced to a subordinate level within the system. Strategies are described for measuring biological and physical variability relating to recruitment success and factors controlling biomass flips within LMEs at several scales from the very small (<10 km) to large (100's to 1000's km) in a manner that is both tractable and cost-effective.

B. J. Rothschild (ed.), Toward a Theory on Biological-Physical Interactions in the World Ocean, 459-476.
© 1988 by Kluwer Academic Publishers.

1. INTRODUCTION

For most stocks of fish, large numbers of eggs and larvae are produced, but only a small fraction survive to become reproducing adults. It is not clear what factors are principal causes of this large-scale mortality, nor is there presently a satisfactory explanation of the relationship between the size of a spawning stock and the size of subsequent new year-classes. A better understanding of the recruitment process is needed to improve forecasts of expected population levels of fishery resources. For purposes of this discussion, the term recruitment is defined as the process by which a population passes from the unexploited early life stages to the stages most usually targeted by the fisheries. Changes in recruitment levels (e.g., year-class strength) of marine fish populations have not, as yet, been successfully forecast on a routine basis. Although some study of the process is underway, much of the effort is short-term, poorly-funded, and of limited scope. In recognition of the fragmentary nature of the present effort, the International Oceanographic Commission (IOC), International Council for the Exploration of the Sea (ICES), and other international bodies have encouraged international programs that will provide scientific, technological, and logistical support for greater cooperation and initiative among marine scientists in the public and private sectors for implementing research to overcome present deficiencies in understanding the recruitment process. The results of related research have been discussed at special symposia and presented in volumes dealing with international fisheries. Notable among the volumes are those of the International Council for the Exploration of the Sea (e.g., 1978a; 1978b; 1980; 1984), the Food and Agriculture Organization of the United Nations (e.g., 1983; 1985), and other investigators (Cushing, 1975; Longhurst, 1981; Laevastu and Larkins, 1981; Rothschild, 1983, 1987; May, 1984). Examination of these volumes reveals an emerging pattern of research from studies of single-species demographics in relation to the fisheries and ocean variability, to multispecies interactions of fish and fisheries from an ecosystem perspective (Zijlstra, 1984). The latter topic has been the subject of more recent symposia sponsored by the American Association for the Advancement of Science (Sherman and Alexander, 1986, 1987). This report is intended as an initial characterization of the predominant variables controlling large-scale fisheries fluctuations within the spatial boundaries of marine ecosystems. It focuses on large marine ecosystems (LME's) as macroscale units for investigating the principal spatial and temporal events controlling the recruitment of fish stocks on a global basis. The LME's are extensive areas, generally greater than 200,000 km^2, in which biological communities have evolved together in response to unique bathymetry, hydrography, and circulation (Sherman and Alexander, 1986). The designation of LME's is based, not only on biological and physical criteria, but also on the basis of geopolitical (Alexander,

1987; Morgan, 1987; Prescott, 1987), legal (Belsky, 1986), and economic (Christy, 1986) considerations.

The research approach to fishery recruitment studies in LME's is based on the argument that marine fish species have evolved and adapted reproductive strategies to geographic areas of unique bathymetry, circulation, biological productivity, and trophodynamic relationships among the populations. Most populations of finfish are highly mobile, migrating hundreds to thousands of kilometers within relatively large ocean areas that they inhabit and within which they grow, reproduce, and die. The LME's extend over broad geographic areas, within which unique predator-prey and environmental relationships have developed since the last ice age. The samples of fish collected by fisheries scientists represent a slice through evolutionary time in which economically important species such as herrings, mackerels, cods, hakes, and others have evolved spawning, migration, and feeding patterns that are difficult to understand unless observed throughout population ranges of the stocks under investigation.

2. RECRUITMENT STUDIES IN LARGE MARINE ECOSYSTEMS

Although the designation of global LME's is, at present, an evolving scientific and socioeconomic process, sufficient progress has been made to allow for useful comparisons to be made of the different processes influencing large-scale changes in the biomass of fish species among LME's. Populations within LME's have been altered significantly by natural and anthropogenic changes, resulting in negative economic impacts. By matching sampling efforts to the time and space scales of the processes that are of most direct influence to recruitment, abundance forecasts can be improved in LME's. Studies of changes in abundance and population renewal of fish species on a large marine ecosystem scale is consistent with the proposition by Ricklefs (1987) that ecologists should begin to address critical community processes on a regional basis. Ricklefs argues that "The regional-historical viewpoint provides a fundamental challenge to ecologists. Broadened concepts of the regulation of local community structure, incorporation of historical, systematic, and biogeographic information into the phenomenology of community ecology, and expanded investigations that address global variation in local species richness will help unite local and regional perspectives" (Ricklefs, 1987).

Greater emphasis has been focused over the past decade within the National Marine Fisheries Service (NMFS) on approaching fisheries recruitment research from an ecosystems perspective in seven LME's within the exclusive economic zone of the United States--the Northeast Shelf, the Southeast Shelf, the Gulf of Mexico, the California Current, the Gulf of Alaska, the East Bering Sea, and the Insular Pacific, including the Hawaiian Islands marine ecosystem. These ecosystems, in 1986, supported 3 million metric tons (mmt) of catches valued at $2.8 billion as commercial landings and more than $10 billion to the national economy.

Two assessment strategies are used by NMFS to provide information on the recruitment process in LME's: (1) Fisheries-independent surveys of fish eggs and larvae and bottom and pelagic fish on mesoscale grids of 20-100 km at frequencies of two to twelve times per year to obtain estimates of the size of fish stocks, and (2) Bioenvironmental studies within the mesoscale matrix aimed at discovering the process controlling recruitment of new year classes (Sherman *et al.*, 1983b). Processes under investigation include growth and mortality of eggs and larvae under variable density-dependent, predator-prey interactions and density-independent influences of changes in circulation, water-column structure, biological production, and pollution. In general, the strategies employed include efforts to obtain measurements directly by conducting multidisciplinary surveys of fish and changes in their physical, chemical, and biological environments (*e.g.*, primary productivity, chlorophyll, nutrients, predators, prey, growth, mortality, distribution, spawning temperatures, salinities, density fields, climate/weather, currents, water masses). The information obtained is augmented from academic, private sector, and international cooperative research. Special effort is made to examine the effects of changes in plankton production with the growth and survival of the pelagic early life stages of fish in relation to recruitment as depicted in the heuristic temporal, spatial, and functional scale of biological and physical events first depicted by Steele (1980). Among these studies are the fisheries-independent ichthyoplankton surveys, acoustic surveys, and bottom trawl surveys in the East Bering Sea Ecosystem and Gulf of Alaska Ecosystem, where hypotheses are being tested of the bioenvironmental variables controlling the recruitment success of walleye pollock, cod, and king crab (Wilson *et al.*, 1986). In the California Current Ecosystem, plans have been developed to conduct recruitment studies on sardines and anchovies as part of the International Recruitment Program (IREP) of the Intergovernmental Oceanographic Commission (IOC) (Lasker, 1986). The fisheries-independent measures of fish stock abundance underway in the California Current LME are the classic CalCOFI ichthyoplankton surveys. Studies in the Insular Pacific are on reef fish communities where ichthyoplankton surveys are used to measure changes in community structure and dominance (Boehlert and Bakun, 1986). Fisheries-independent surveys in the Gulf of Mexico Ecosystem include ichthyoplankton surveys of the entire Gulf, and bottom trawling surveys of the Gulf shelf (Richards and McGowan, 1986). Within the Southeast Continental Shelf Ecosystem, from the Florida Keys to Cape Hatteras, the fisheries-independent measure of interannual variability in fish stocks is obtained by systematic surveys of reef fish species and ichthyoplankton collections. The Northeast Continental Shelf Ecosystem is monitored for interannual and decadal changes in fish stocks by conducting fisheries-independent bottom trawl surveys and ichthyoplankton surveys from the Gulf of Maine to Cape Hatteras. This large survey effort provides multiple data sets of juvenile and adult bottom and pelagic fish, eggs and larvae, and other ecosystem measurements, including information on marine mammals, marine birds, currents, water masses, nutrients, and productivity, when augmented with information from academic institutions conducting studies in the same regions (Sherman *et al.*, in press). Recruitment studies are focused on testing hypotheses relating to the growth and survival of cod and haddock during the first-year-of-life (Laurence *et al.*, 1987).

3. MAJOR FACTORS PERTURBING LME'S

Increasing attention has been focused over the past few years on synthesizing the available biological and environmental information for fish stocks within LME's as a way of understanding recruitment mechanisms and other important biological processes aimed at improving the management of living marine resources from an ecosystem perspective (Sherman and Alexander, 1986, 1987). Of the 20 LME's for which recent syntheses have been reported, initial determinations have been made on the variables controlling large-scale changes in fish species abundance (Figure 1). In four--the Yellow Sea, Gulf of Thailand, Great Barrier Reef and Northeast Continental Shelf--the controlling variable appears to be predation. Major changes in the Great Barrier Reef ecosystem have been attributed to the predation

LARGE MARINE ECOSYSTEMS

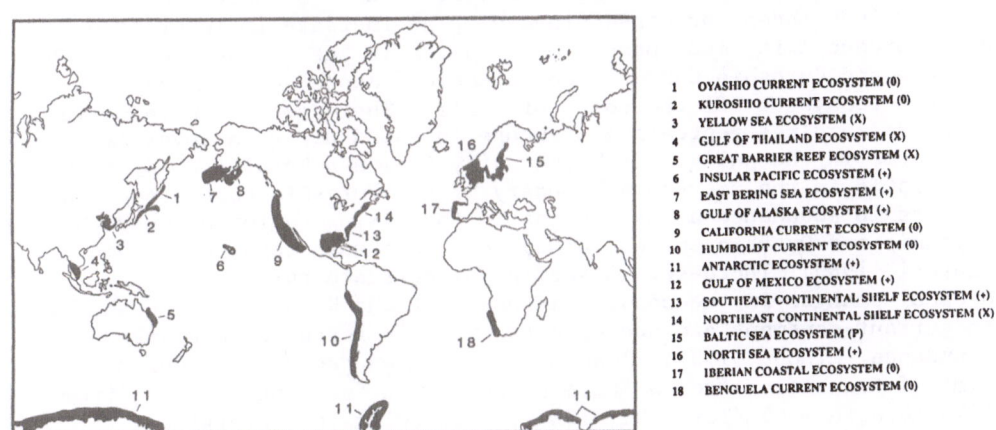

Figure 1. Predominant variables influencing changes in fish species biomass in large marine ecosystems. Predominant variable--Predation (X); Environment (O); Pollution (P); Insufficient Information (+).

effect of the crown of thorns starfish (Bradbury and Mundy, 1987). The principal variable in the other three is recruitment over-fishing, considered for purposes of this discussion as human predation (Tang,

1987; Piyakarnchana, 1987; Sissenwine, 1986). For six other LME's the predominant variable is environmental change--Oyashio, Kuroshio, (Minoda, 1987; Terazaki, 1987), California Current (McCall, 1986), Humboldt Current (Canon, 1986), Iberian Coastal (Wyatt and Gandaras, 1987), and Benguela Current (Crawford et al., 1987). It appears that the dominant influence on the Baltic is coastal pollution (Kullenberg, 1986). For the remaining LME's, the information is insufficient to make an initial determination.

4. ENVIRONMENTALLY PERTURBED LME'S

Management of stocks responding to strong environmental signals will be enhanced by improving the understanding of physical factors forcing biological change. This should lead to improved forecasting of perturbations affecting large-scale recruitment responses, and allow more lead time to make necessary economic adjustments. Large-scale changes in species biomass resulting from environmental perturbations have been reported for LME's around the globe. The most dramatic recently is the spectacular increase in the abundance of sardines off the coasts of Japan and Chile. The global impact of this biomass increase is significant. In 1974 the annual total yield of the global fisheries was approximately 65 million metric tons (mmt). By 1984 it had risen to 75 mmt. The additional biomass yield resulted from increases in landings of about 4 mmt each of Japanese and Chilean sardines (FAO, 1985). The increased yields have been attributed to density-independent processes involving an increase in lower food-chain productivity made possible by shifts in the mixing areas of the Oyashio and Kuroshio Current ecosystems off Japan and the Humboldt Current ecosystem off the coast of Chile. The dramatic increase in Japanese sardines represents a change in abundance to a level of dominance in the pelagic fish community from a subordinate level in the 1950's in the nearshore boundary of the Kuroshio ecosystem. The changes in abundance among species in the mixing region between the Kuroshio and Oyashio Current ecosystems have been reported previously (Hayasi, 1983). Increases in sardine biomass are correlated with decadal changes in the coastal meanders of the Kuroshio. The shifts in current movement are accompanied by an increase in zooplankton abundance (Minoda, 1987), leading to enhanced recruitment during a time of transition from a "warm period" to a "cooling period" from 1975 through 1985 (Terazaki, 1987). Following the initial environmental impact, Terazaki (1987) noted a major secondary effect attributed to predation on anchovy by jack mackerel. An increase in jack mackerel abundance was observed following the oceanographic perturbation in the Oyashio/Kuroshio frontal zone and a subsequent decline in anchovy attributed, in part, to mackerel predation (6.7×10^4 tons) which exceeds the mean catch of anchovy (0.3×10^4 tons) in the area (Terazaki, 1987).

Changes within the California Current Ecosystem (a reported mmt decline of Pacific sardines (*Sardinops sagax caerulea*) and the subsequent increase in anchovy (*Engraulis mordax*) abundance) are considered to be the result of natural environmental change rather than from any significant competition between the two species (MacCall, 1986). In the Iberian Coastal Ecosystem, characterized as an upwelling regime, the fisheries yield ranges between 400,000 to 500,000 metric tons. Nearly 50% of the total catch is sardine *Sardina pilchardus*, followed by horse-mackerel, *Trachurus* sp., (11%); anchovy, *Engraulis encrasicholous*, (10%); blue whiting, *Micromesistius poutassou*, (6%); and hake, *Merluccius merlucius*, (5%). The greatest variability among species is in the alternation of abundance between sardine and horse-mackerel. The changes observed are attributed to natural environmental perturbation rather than to any competition between species (Wyatt and Perez-Gandaras, 1987). In the Benguela Current Ecosystem an upwelling-type cool water regime is positioned between warm waters to the north and south. Although overall yield of the system remains between 2 and 3 mmt, considerable variability in the form of species biomass flips have been observed during the 1960's and 1970's. Natural environmental perturbations in the form of "cool" and "warm" periods have resulted in species replacement of pilchard in the late 1960's by anchovy in the southern area of the system and horse-mackerel in the north. In general, it appears that "cool" conditions within the Benguela Current Ecosystem favor the recruitment of groundfish species, while "warm" conditions favor epipelagic stocks (Crawford *et al.*, 1987). The long-term shifts in the abundance of fish species in the Benguela Current Ecosystem are attributed to environmental signals as the major factor responsible for changes in the ecosystem (Crawford *et al.*, 1987).

In the Humboldt Current Ecosystem off the coasts of Chile and Peru, the increase in yield of sardine (*Sardinops sagax musica*) from a few thousand metric tons in 1970 to approximately 3 mmt in 1983 is attributed to a persistent warming trend first observed in 1973 (Canon, 1986). The shift in predominant species from anchovy to sardine off the coasts of Chile and Peru is shown in Figure 2.

5. PREDATOR PERTURBED LME'S

Those LME's controlled by natural and fishing predation offer more options for increasing yields through directed fishing effort, or species enhancement, than exist for stocks in environmentally controlled LME's.

The biomass of commercially important fish stocks of the Northeast Continental Shelf Ecosystem declined by approximately 50% between 1968 and 1975. The principal cause of the loss of biomass is thought to be excessive fishing mortality on juvenile and adult stocks (Clark and Brown, 1977). The predominance of the Atlantic herring, *Clupea harengus*, of Georges Bank "flipped" with sand eel, *Ammodytes* spp. (Figure 3). A biomass flip occurs when the population of a dominant species rapidly drops to a very low level and is replaced by a second species. Sand eel, herring, and mackerel inhabit, at least for part of the year, the same areas on Georges Bank and the Southern

466

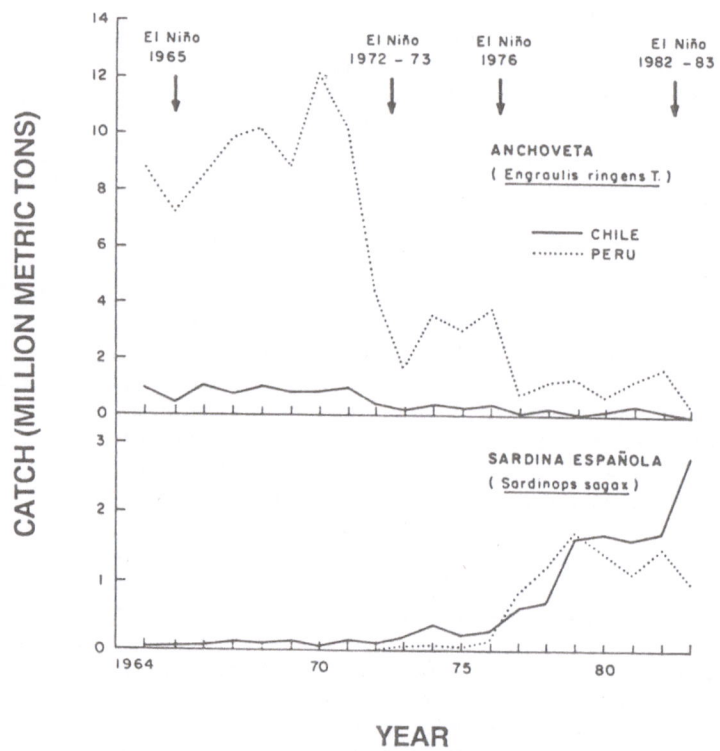

Figure 2. Fluctuations in the catch of anchovies and sardines from the waters of the Humboldt Current ecosystem off the coasts of Chile and Peru. Adapted from Canon (1986).

New England continental shelf. Evidence of herring predation on sand eel, sand eel predation on herring larvae, and mackerel predation on the early developmental stages of both species has been observed on the Northeast Continental Shelf (M. D. Grosslein, personal communication), and for the North Sea, where the distribution of the three species overlap. It is possible, in the absence of any prolonged environmental signal, that the decline in both herring and mackerel stocks during the mid-1970's released predation pressure on sand eel and allowed the population to explode (Sherman et al., 1981). Fishing mortality has been reduced on herring and mackerel stocks since the mid-1970's. No fishery exists for sand eel. It appears that the reduction of fishing mortality on mackerel and herring has allowed the stocks to begin a recovery trend. The present biomass of mackerel is estimated at 1.4 mmt and is increasing. Also, evidence of herring returning to Georges Bank has been reported from the recent discovery of juveniles in stomachs of spiny dogfish, *Squalus acanthius*, and other predators. Unlike the North Sea, for which

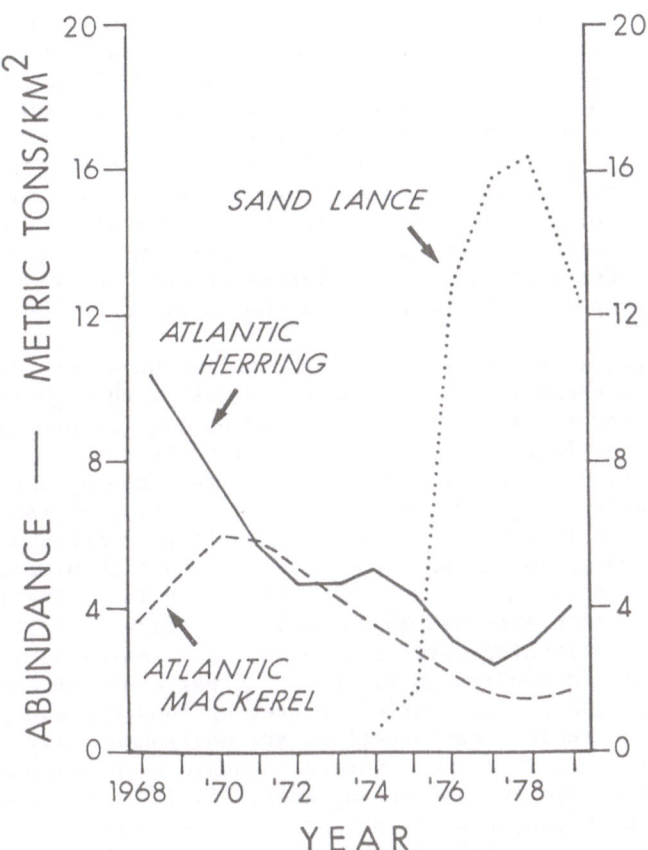

Figure 3. Decline of Atlantic herring and Atlantic mackerel and apparent replacement by the small, fast-growing sand eel within the Northeast Shelf ecosystem, 1968-1979.

recent climatic changes have been reported (Garrod and Colebrook, 1978), no long-term climatic change has been observed for the Northeast Continental Shelf Ecosystem. No declining trend in primary production, or zooplankton biomass, has been detected (Sherman *et al.*, 1983a). The energetics of the shelf ecosystem appear tightly bound in relation to fish production. Recent estimates on the Georges Bank subarea place fish predation on fish at 70% of annual production, followed by approximately 10% as fisheries yield, 10% consumed by marine mammals, 5% by marine birds, and 5% by apex predators, including sharks, tunas, and billfish (Sissenwine, 1986).

Another example of biomass flips occurred in the Northeast Atlantic, where the dominance of species among the finfish stocks of

the North Sea Ecosystem changed during the 1960's. The pelagic
herring and mackerel yields decreased from approximately 5 mmt to 1.7
mmt; whereas, sand eel, Norway pout, and sprat increased by about 1.5
mmt along with an approximate 36% increase in gadoid yields. Both
density-dependent predation and density-independent environmental
changes have been proposed as processes controlling recruitment and
causing the biomass flips (Hempel, 1978). However, none of the
arguments can be considered more than speculative at this time without
a better understanding of the recruitment process within the North Sea
Ecosystem. The need for more systematic measurements of variability
within the North Sea Ecosystem from a fisheries perspective has been
acknow ledged (Hempel, 1978; Daan, 1986) and is presently underway
(ICES, 1987).

Major changes in the biomass of fish species have also occurred
in the Yellow Sea Ecosystem. From the early 1960's through the early
1980's the fish invertebrate stocks declined by 40% (Figure 4).
During this period, large-size and commercially important species
including croaker, *Pseudosciaena crocea*, were replaced by smaller-
size, less-valuable forage species, including anchovy, *Engraulis
japonica*. The principal cause of the biomass flip is attributed
principally to overexploitation. About 70% of the fish biomass in
1985 was attributed to small fish, below 20 cm, while in the 1960's
the mean size of catch exceeded 20 cm in length (Tang, 1987). Marine
ecosystems in which fishing mortality (predation), or natural
predation, has been demonstrated as the principal cause of a change in
the dominance of fish species, offer greater options for management
than LME's where principal perturbations are environmentally induced.
Recognizing that the larger, carnivorous, benthic fish component of
the Yellow Sea Ecosystem was depleted, an artificial enhancement
program for fleshy prawn, *Penaeus orientalis*, was successfully
implemented. From an initial attempt at the grow-out of 4 million
juveniles introduced into the southern coastal waters of the Shandong
Peninsula in 1984, a total of 1,000 metric tons was caught 3 months
later. From a release of 9 million juveniles in 1985, a total of
2,000 metric tons was caught (Tang, 1987). In the Gulf of Thailand
Ecosystem, the yield of finfish in the inner Gulf declined from 150
kg/hr in 1963 to less than 50 kg/hr in 1984 (Figure 5). As in the
Yellow Sea model, the average catch of fish declined to less than 15
cm length in 1984 from larger sizes that predominated in 1963, when
more than 90% of the catch exceeded 15 cm. The major decline is in
the demersal component of the system. Two of the smaller pelagics now
appear to be increasing in importance in fisheries yield of the
ecosystem--*Rastrelliger brachysoma* and *Stolephorus heterolobus*. The
depleted status of demersal stocks has been attributed to excessive
fishing predation in the Gulf of Thailand that has resulted in a
redirection of fishery effort by large trawlers to fishing grounds
outside of the Gulf (Piyakarnchana, 1987).

For the Great Barrier Reef, the major changes in biomass are
induced by periodic outbreaks of predation by the crown-of-thorns
starfish on coral polyps. The effect on the fish community is
presently under investigation. Preliminary findings based on studies

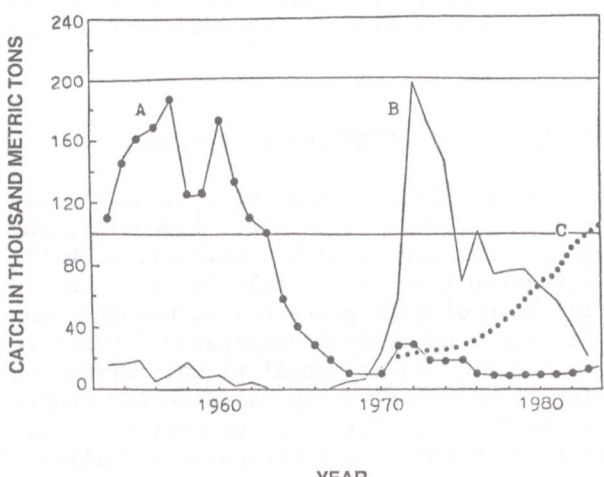

Figure 4. Trends in the annual catch of dominant species of fish from the Yellow Sea ecosystem. (A) small yellow croaker and hairtail; (B) Pacific herring and Japanese mackerel; (C) *Setipinna taty*, anchovy, and scaled sardine. From Tang (1987).

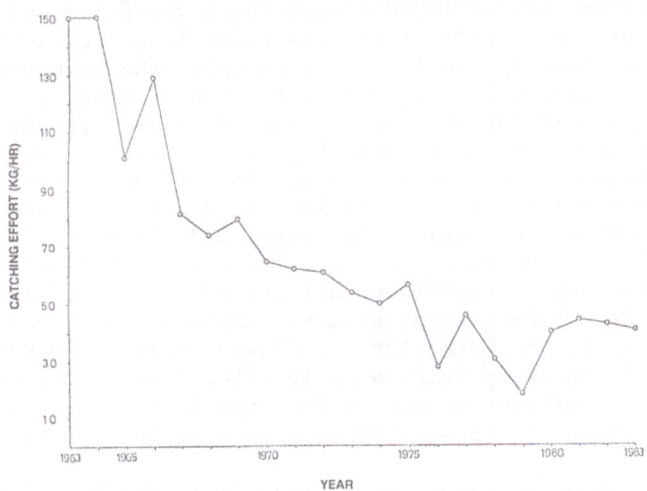

Figure 5. Annual average catch-per-unit-effort (kg/h) in the inner Gulf of Thailand from 1963 to 1983. From Piyakarnchana (1987).

of individual reefs indicate that scarids and acanthurids, both algal grazers, showed no significant change in abundance. In contrast,

species of chaetodontids and pomocentrids that are dependent on corals
for food showed significant declines in abundance (Bradbury and Mundy,
1987).

6. LME'S AS GLOBAL UNITS FOR RECRUITMENT STUDIES

Resources that were previously shared among nations are now under EEZ
jurisdiction subject to national regulation, licensing, and other
restrictions on users. The new "wealth", however, is being subjected
to erosion and dissipation from heavy exploitation. On a global
scale, the potential loss of fish protein from mismanagement and
overexploitation has not been fully investigated. Considerable
controversy surrounds estimates of annual global harvesting levels of
the marine fisheries. The annual yield based on FAO statistics in
1985 was approximately 74 mmt. Fisheries projections given in the
Global 2000 Report (U.S. Council on Environmental Quality, 1980)
indicated that the world harvest of fish was expected to rise little,
if at all, by the year 2000 from the level of 60 mmt reached in the
1970's. In contrast, the predictions in *The Resourceful Earth* (Wise,
1984) argue for an annual yield of 100-120 mmt per year of
conventional species by the year 2000. This kind of controversy is
not unexpected when one considers the meager efforts presently
underway on a global scale to improve the information base for
estimating fishable biomass and probable levels of annual sustained
yields, from an ecosystem perspective.

The strategies now employed around the globe for a better
understanding of the process of biomass renewal, or recruitment, in
the sea are surprisingly similar. For example, the recruitment
research strategy recently proposed by the Soviet Union to the
International Council for the Exploration of the Sea (ICES) calls for:
a long time-series of no less than 15 years of observations on
physical and chemical oceano-graphic conditions during the spawning
and drift of early life stages; and long-term measurements of
variability in growth of early life stages and their predators and
prey (ICES, 1986). The Japanese argue for similar studies with a
focus on environmental controls on biomass production (Watanabe,
1983). Scientists from Canada, Denmark, England, the Federal Republic
of Germany, France, the Netherlands, Norway, Poland, Portugal,
Scotland, and Spain are participants in efforts to foster coordinated
international recruitment studies within the ICES community, where
considerable effort is already underway on recruitment studies (ICES,
1985, 1986).

The approach to biomass recruitment within the United States EEZ
is conducted with a similar focus on the early life stages of fish
species. Given the similarities in research strategy among virtually
all of the major fishing nations, it would appear that a comparative
global study of the factors within LME's that control recruitment
variability resulting in multimillion-metric-ton changes in fish
biomass would be most useful. Through international efforts in this
study it should become possible to improve forecasting of population

471

trends in the large marine ecosystems depicted in Figure 1. The topic
is of particular interest in relation to a growing awareness of global
fragility, based on recent concerns of probable ozone depletion
(Cicerone, 1987) and the greenhouse effect (National Academy of
Sciences, 1979).

In an effort to monitor long-term global change, NOAA, along with
the National Science Foundation and the National Aeronautics and Space
Administration, agreed in 1986 to develop and implement program of
global research aimed at studying the globe as an ecosystem. The
effort, to be conducted as part of the International Geosphere-
Biosphere Program (IGBP) (Abelson, 1987), will require the use of a
permanent network of satellites, new instruments for measuring changes
on land and in the sea; and new high-speed data from satellites,
buoys, acoustic systems, electronic pattern recognition systems, and
other technological advances that will allow for systematic
measurements of important environmental variables in the world oceans
of particular relevance to recruitment studies.

On a global basis, the LME's represent geographically distinct
units for the conservation and management of fisheries resources
(Belsky, 1987; Alexander, 1987; Morgan, 1987; Prescott, 1987) for
which the IGBP observations in the ocean will be most useful. Within
several of the LME's, hypotheses thought to be controlling recruitment
variability are under investigation. They include the Oyashio/Kuroshio
Current, California Current, the Eastern Bering Sea, the Gulf of
Alaska, the Humboldt Current, and Peru Current ecosystems in the
Pacific rim area, and the Northeast U.S. Continental Shelf, Southeast
U.S. Continental Shelf, the Gulf of Mexico, the Scotian Shelf,
Norwegian Sea/Barents Sea, North Sea, the Iberian Coastal and the
Benguela Current ecosystems in the Atlantic rim area.   Within the
LME's recruitment studies are focused on upwelling, frontal zones,
continental shelf systems, and pelagic systems. By focusing on
comparisons among these systems, and among reproductive strategies of
the fish species under investigation, it should be possible to
accelerate progress toward understanding recruitment mechanisms and
improving forecasts of probable fisheries yields.   New hypotheses are
being developed and tested to determine the relationship between
environmental and biological factors on variability in the abundance
of fish pcpulations (Rothschild, 1985, 1986; ICES 1987). The
comparisons among LME's should allow for narrowing the problem and
avoiding duplication of effort. By comparing different systems and
life histories of fish it should be possible to develop
generalizations useful for understanding recruitment in previously
unstudied systems. The intellectual and cost-saving advantages of the
comparative approach in studying the recruitment process from an
ecosystem perspective is recognized as an important research strategy
by the International Oceanographic Commission (Bakun, 1985), and
should be encouraged.

7. REFERENCES

Abelson, P. H.: 1986, 'The International Geosphere-Biosphere Program', *Science* **234**, 657pp.

Alexander, L. M.: 1987, 'LME's as global management units', in K. Sherman and L. M. Alexander (eds.), *Biomass and Geography of large marine ecosystems*, Symposium sponsored by AAAS, February 16-17, 1987, at Chicago, IL. Westview Press, Boulder, CO. (in prep.)

Bakun, A.: 1985, 'Comparative studies and the recruitment problem: Searching for generalizations', *CalCOFI* Rep. **26**, 30-40.

Belsky, M.: 1986, 'Legal constraints and options for total ecosystem management of large marine ecosystems', in K. Sherman and L. M. Alexander (eds.), *Variability and management of large marine ecosystems*, AAAS Selected Symposium 99, Westview Press, Boulder, CO., pp. 145-174.

Belsky, M.: 1987, 'Developing a management regime for large marine ecosystems', in K. Sherman and L. M. Alexander (eds.), *Biomass and geography of large marine ecosystems*, Symposium sponsored by AAAS, February 16-17, 1987, at Chicago, IL., Westview Press, Boulder, CO. (in prep.)

Boehlert, G. W., and A. Bakun: 1986, 'Local recruitment in isolated bank and island ecosystems', LO42-01 EOS, *Trans. Am. Geophys. Union* 1986 **67(44)**, 991pp.

Bradbury, R., and C. N. Mundy: 1987, 'Large scale shifts in biomass of the Great Barrier Reef ecosystem', in K. Sherman and L. M. Alexander (eds.), *Biomass and geography of large marine ecosystems*, Symposium sponsored by AAAS, February 16-17, 1987, at Chicago, IL, Westview Press, Boulder, CO. (in prep.)

Canon, J. R.: 1986, 'Variabilidad ambiental en relacion con la pesqueria neritica pelagica de la zona Norte de Chile', in P. Arana (ed.), *La pesca en Chile*, Escuela de Ciencias del Mar, Facultad de Recursos Natjurales, Universidad Catolica de Valparaiso, pp. 195-205.

Christy, F. T., Jr.: 1986, 'Can large marine ecosystems be managed for optimum yield?', in K. Sherman and L. M. Alexander (Eds.), *Variability and management of large marine ecosystems*, AAAS Selected Symposium 99, Westview Press, Boulder, CO., pp. 263-267.

Cicerone, R. J.: 1987, 'Changes in stratospheric ozone', *Science* 237, 35-42.

Clark, S. H., and B. E. Brown: 1977, 'Changes of biomass of finfishes and squids from the Gulf of Maine to Cape Hatteras, 1963-74, as determined from research vessel survey data', *Fish. Bull. U.S.* 75, 1-21.

Crawford, R. J. M., L. V. Shannon, and P. A. Shelton: 1987, 'Characteristics and management of the Benguela as a large marine ecosystem', in K. Sherman and L. M. Alexander (eds.), *Biomass and geography of large marine ecosystems*, Symposium sponsored by AAAS, February 16-17, 1987, at Chicago, IL., Westview Press, Boulder, CO. (in prep.)

Cushing, D. H.: 1975, *Marine Ecology and Fisheries*, Cambridge Univ. Press, London. 278 pp.
Daan, N.: 1986, 'Results of recent time-series observations for monitoring trends in large marine ecosystems with a focus on the North Sea', in K. Sherman and L. M. Alexander (eds.), *Variability and management of large marine ecosystems*, AAAS Selected Symposium 99, Westview Press, Boulder, CO., pp. 145-174.
FAO [Food and Agriculture Organization of the United Nations]: 1983, *Proceedings of the expert consultation to examine changes in abundance and species composition of neritic fish resources*, edited by G. D. Sharp and J. Csirke, San Jose, Costa Rica, 18-29 April 1983, *FAO Fisheries Report* No. 291, Vol. 2 and 3.
FAO: 1985, 'Selected world fisheries landings, 1938-84', Yearbook of Fishery Statistics, Vol. 58 and preceding volumes. Excludes marine plants after 1970 and marine mammals.
Garrod, D. J., and J. M. Colebrook: 1978, 'Biological effects of variability in the North Atlantic Ocean', *Rapp. P.-v. Reun. Cons. Int. Explor. Mer* 173, 128-144.
Hayasi, S.: 1983, 'Some explanation for changes in abundances of major neritic-pelagic stocks in the northwestern Pacific Ocean', *FAO Fisheries Report* 291(2), 37-55.
Hempel, G.: 1978, 'Synopsis of symposium. North Sea fish stocks: Recent changes and their causes', *Rapp. P.-v. Reun. Cons. Int. Explor. Mer* 172, 445-449.
ICES [International Council for the Exploration of the Sea]: 1978a, 'North Sea fish stocks--recent changes and their causes', A Symposium held in Aarhus, 9-12 July 1975, edited by G. Hempel. *Rapp. P.-v. Reun. Cons. int. Explor. Mer* 172, 449 pp.
ICES: 1978b, 'Marine ecosystems and fisheries oceanography', Four Symposia held during the Joint Oceanographic Assembly in Edinburgh 13-24 September 1976, edited by T. R. Partons, B. O. Jansson, 1A. R. Longhurst, and G. Saetersdal. *Rapp. P.-v. Reun. Cons. int. Explor. Mer* 173, 240 pp.
ICES: 1980, 'The assessment and management of pelagic fish stocks. A Symposium held in Aberdeen, 3-7 July 1978, edited by A. Saville, *Rapp. P.-v. Reun. Cons. int. Explor. Mer* 177, 517 pp.
ICES: 1984, 'The biological productivity of North Atlantic shelf areas. A Symposium held in Kiel, 2-5 March 1982, edited by J. J. Zijlstra, *Rapp. P.-v. Reun. Cons. int. Explor. Mer* 183, 284 pp.
ICES: 1985, 'Report of IREP Study Group', Copenhagen, 10-13 1985, ICES C.M.1985/Gen:4, Session Q.
ICES: 1986, 'Report of IREP Steering Group, ICES Headquarters, 25-27 June 1986, ICES C.M.1986/Gen:7.
ICES: 1987, 'Report of the Working Group on Larval Fish Ecology to the Biological Oceanography Committee of ICES', Hirtshals, Denmark 17-19 June 1987. ICES C.M.1987/L:28.
Kullenberg, G.: 1986, 'Long-term changes in the Baltic ecosystem', in K. Sherman and L. M. Alexander (eds.), *Variability and management of large marine ecosystems*, AAAS Selected Symposium 99, Westview Press, Boulder, CO, pp. 19-31.

474

Laevastu, T., and H. A. Larkins: 1981, *Marine fisheries ecosystem, Its quantitative evaluation and management*, Fishing News Books Ltd., Farnham, Surrey, England, 162 pp.

Lasker, R.: 1986, 'Proposal for a sardine/anchovy recruitment program (SARP) in California waters, FY-86 and 87', Southwest Fisheries Center, LA Jolla, CA. Jan. 1985. Annex III to the Report of IREP Steering Group, ICES C.M. 1986/Gen:7.

Laurence, G. C., E. Cohen, M. Grosslein, R. G. Lough: 1987, 'Proposed strategies for recruitment research on haddock and cod within the Northeast Continental Shelf ecosystem', (MARMAP Contribution No. FED/NEFC 87-03), Nat. Mar. Fish. Serv., Northeast Fish. Ctr. Narragansett Lab. Ref. No. 87-02.

Longhurst, A. R.: 1981, *Analysis of marine ecosystems*, Academic Press, London, 741 pp.

MacCall, A. D.: 1986, 'Changes in the biomass of the California Current ecosystem', in K. Sherman and L. M. Alexander (Eds.) *Variability and management of large marine ecosystems*, AAAS Selected Symposium 99, Westview Press, Boulder, CO, pp. 33-54.

May, R. M.: 1984, *Exploitation of marine communities*, Dahlem Konferenzen, Springer-Verlag, Berlin, 366 pp.

Minoda, T.: 1987, 'Oceanographic and biomass changes in the Oyashio Current ecosystem', in K. Sherman and L. M. Alexander (eds.), *Biomass and geography of large marine ecosystems*, Symposium sponsored by AAAS, February 16-17, 1987, at Chicago, IL. Westview Press, Boulder, CO. (in prep.)

Morgan, J.: 1987, 'Large marine ecosystems in the Pacific', in K. Sherman and L. M. Alexander (eds.), *Biomass and geography of large marine ecosystems*, Symposium sponsored by AAAS, February 16-17, 1987, at Chicago, IL, Westview Press, Boulder, CO. (in prep.)

Nakai, Z.: 1962, 'Studies relevant to mechanisms underlying the fluctuation in the catch of the Japanese sardine Sardinops melanosticta (Temminck and Schlegel)', *Jap. J. Ichthy.* 9(1-6):1-115.

National Academy of Science: 1979, *Carbon dioxide and climate: A scientific assessment*, Climate Research Board, National Academy of Sciences.

Piyakarnchana, T.: 1987, 'Yield dynamics as an index of biomass shifts in the Gulf of Thailand Ecosystems', in K. Sherman and L. M. Alexander (eds.), *Biomass and geography of large marine ecosystems*, Symposium sponsored by AAAS, February 16-17, 1987, at Chicago, IL, Westview Press, Boulder, CO. (in prep.)

Prescott, V.: 1987, 'The political division of large marine ecosystems in the Atlantic Ocean and some associated seas', in K. Sherman and L. M. Alexander (eds.), *Biomass and geography of large marine ecosystems*, Symposium sponsored by AAAS, February 16-17, 1987, at Chicago, IL, Westview Press, Boulder, CO. (in prep.)

Richards, W. J., and M. F. McGowan: 1987, 'Biological productivity in the Gulf of Mexico: Identifying the causes of variability, in K. Sherman and L. M. Alexander (eds.), *Biomass and geography of large marine ecosystems*, Symposium sponsored by AAAS, February 16-17, 1987, at Chicago, IL, Westview Press, Boulder, CO. (in prep.)

Ricklefs, R. E.: 1987, 'Community diversity: Relative roles of local and regional processes', *Science* 235, 167-171.

Rothschild, B. J. (ed.): 1983, *Global fisheries. Perspectives for the 1980s*, Springer-Verlag, New York. 289 pp.

Rothschild, B. J.: 1985, 'Feasibility of relating recruitment to environmental variation', ICES C.M. 1985/L:38.

Rothschild, B. J. (ed.): 1987, 'Recruitment processes and ecosystem structure--biodynamics of the sea', Prepared for the Ocean Studies Board, Commission on Physical Sciences, Mathematics, and Resources. Nat. Acad. Press, Washington, DC.

Rothschild, B. J., and T. R. Osborn: 1986, 'Biodynamics of the sea: Preliminary observations on high dimensionality and the effect of physics on predator-prey interrelationships', ICES C.M. 1986/L:25.

Sherman, K., and L. M. Alexander (eds.): 1986, *Variability and management of large marine ecosystems*, AAAS Selected Symposium 99, Westview Press, Boulder, CO (MARMAP Contribution No. FED/NEFC 85-23) 319 pp.

Sherman, K., and L. M. Alexander (eds.): 1987, *Biomass and geography of large marine ecosystems*, Symposium sponsored by AAAS, February 16-17, 1987, at Chicago, IL, Westview Press, Boulder, CO. (in prep.).

Sherman, K., J. R. Green, J. R. Goulet, and L. Ejsymont: 1983a, 'Coherence in zooplankton of a large Northwest Atlantic ecosystem', *Fish. Bull., U.S.* 81(4), 855-862.

Sherman, K., M. Grosslein, D. Mountain, D. Busch, J. O'Reilly, and R. Theroux: (in press), 'The continental shelf ecosystem off the northeast coast of the United States', in J. Zijlstra (ed.), *Ecosystems of the World*, Vol. 27, Elsevier Press, The Netherlands.

Sherman, K., C. Jones, L. Sullivan, W. Smith, P. Berrien, and L. Ejsymont: 1981, 'Congruent shifts in sand eel abundance in western and eastern North Atlantic ecosystems', *Nature* 291, 486-489.

Sherman, K., R. Lasker, W. Richards, and A. W. Kendall, Jr.: 1983b, 'Ichthyoplankton and fish recruitment studies in large marine ecosystems', *Mar. Fish. Rev.* 45(10-11-12), 1-25.

Sissenwine, M.: 1986, 'Perturbation of a predator-controlled continental shelf ecosystem', in K. Sherman and L.M. Alexander (eds.), *Variability and management of large marine ecosystems*, AAAS Selected Symposium 99, Westview Press, Boulder, CO, pp. 55-85

Steele, J. H. (Chairman): 1980, *Fisheries ecology: Some constraints that impede advances in our understanding*, An Ad Hoc Group of the Ocean Sciences Board, National Academy of Sciences, Washington, D.C. 1980.

Tang, Q.: 1987, 'Changes in the biomass of the Huanghai Sea ecosystem, in K. Sherman and L. M. Alexander (eds.), *Biomass and geography of large marine ecosystems*, Symposium sponsored by AAAS, February 16-17, 1987, at Chicago, IL, Westview Press, Boulder, CO. (in prep.)

Terazaki, M.: 1987, 'Recent large-scale changes in the biomass of the Kuroshio Current ecosystem', in K. Sherman and L. M. Alexander (eds.), *Biomass and geography of large marine ecosystems*, Symposium sponsored by AAAS, February 16-17, 1987, at Chicago, IL, Westview Press, Boulder, CO. (in prep.)

U.S. Council on Environmental Quality and the Department of State, Gerald O. Barney (Director): 1980, 'The global 2000 report to the president: Entering the twenty-first century', Vols. I-III, U.S. Government Printing Office, Washington, D.C.

Watanabe, T.: 1983, 'Stock assessment of common mackerel and Japanese sardine along the Pacific coast of Japan by spawning survey', *FAO Fisheries Report* **291(2)**, 57-81.

Wilson, J. G., L. S. Incze, S. A. Macklin, and J. D. Schumacher: 1986, 'FOX 1985--The Northwest Gulf of Alaska Fishery Oceanography Experiment', NOAA Data Report ERL PMEL-15, 133pp.

Wise, J. P.: 1984, 'The future of food from the sea', in J. L. Simon and H. Kahn (eds.), *The resourceful earth*, Basil Blackwell Inc., New York, pp. 113-127.

Wyatt, T., and G. Perez-Gandaras: 1987, 'Biomass changes in the Iberian ecosystem', in K. Sherman and L. M. Alexander (eds.), *Biomass and geography of large marine ecosystems*, Symposium sponsored by AAAS, February 16-17, at Chicago, IL, Westview Press, Boulder, CO. (in prep.)

Zijlstra, J.: 1984, 'Symposium on biological productivity of continental shelves in the temperate zone of the North Atlantic', *Rapp. P.-v. Reun. Cons. int. Explor. Mer* **183**, 5-7.

WHY STUDY FISH POPULATION RECRUITMENT?[1]

J.-P. Troadec
*French Research Institute for the Exploitation
of the Sea (IFREMER)
66, avenue d'Iéna
75116 Paris
FRANCE*

ABSTRACT. Recent changes in world fisheries and in the Ocean Regime
have substantially modified the scope of fishery science.
Variability in fish stock abundance, a new concern for small-scale
fisheries and conservation of littoral environments, and the weight
of extensive mariculture have enhanced the priority of investigations
on the ecology and dynamics of early stages of aquatic populations.
Quantitative understanding of the regulatory mechanisms through which
fish stocks regulate their numbers in response to environment
fluctuations would contribute to the progress of marine ecosystems
exploitation and coherent use.

1. INTRODUCTION

At the turn of this century, variability in fish stock abundance
became a matter of concern for fishery science in the North Sea:
Heincke (1898) demonstrated the existence of self-sustaining
populations and Hjort (1914) hypothesized that fish population
variability originates in processes taking place in the early,
pelagic, phase of stock cohorts. Their work had a great influence
on the development of fishery science, systematics and evolutionary
biology (Sinclair and Solemdal, 1987).

However, attention to this topic dropped after World War I.
Investigations on population recruitment variability were then
restricted to assessing the effects of reduction in spawning biomass
induced by fishing (Ricker, 1954; Beverton and Holt, 1957).
Attention to fish population variability and its environmental causes
has been recently revived at national and international levels (IOC,
1983; ICES, 1985), primarily among marine scientists and biological
oceanographers. Such concern is justified by the role, played by

---

[1]A preliminary version of this paper has been published in mimeo
form (Troadec, 1985).

B. J. Rothschild (ed.), Toward a Theory on Biological-Physical Interactions in the World Ocean, 477–500.
© 1988 by Kluwer Academic Publishers.

the processes determining the recruitment success, in the regulation
and evolution of aquatic population (Rothschild, 1986; Sinclair, in
press).

This historical sequence raises the question of the timeliness
and relevance of that topic for fishery science since academic
interest is not in itself a sufficient motivation for targeted
research: as potential contributions to the economic sector need to
be taken into account when ranking research priorities .

The purpose of the present paper is to analyze to what extent a
better understanding of the recruitment processes and sources of
variability would improve fishery science potential contribution to
the further development and better management of exploited fish
populations--including the development of "fish ranching"
opportunities and the need to conserve the carrying capacity of
coastal environments. This assessment will be conducted by comparing
developments in fishery science concerns to the specific issues
associated with the successive steps observed in fishery development.
This comparison is expected to indicate to what extent current
scientific paradigms and research strategies are suited for
investigating questions which have emerged with the full exploitation
of world fishery resources and the adoption of a new Ocean Regime
(see *e.g.* Rothschild 1972, 1983).

2. RESOURCE EVALUATION AND CONSERVATION

2.1 Historical Overview

Heincke's and Hjort's initial concern for fish population structure
and fluctuations was clearly motivated by the intensive exploitation
of North Sea stocks of major economic significance. The Western
Norway herring fishery collapsed, for example, from 1870 to 1895;
the Northern Norway segment did too in 1875; the Bohüslan stock was
extremely low from 1810 to 1880 and, again, in 1980 (Devold, 1963).
Herring population structure and interannual fluctuations in
landings were directly connected issues and perceived as such, as
shown by the two interpretations of divergent economic and political
implications which were opposed one another at that time: that of a
single stock migrating across the Northeast Atlantic against that of
several discrete populations of fluctuating size spawning year after
year in the same locations (Sinclair and Solemdal, *op. cit.*). There
was also concern for the consequences of overfishing on demersal
stocks subsequent to the introduction of steam propulsion and otter
trawl (Went, 1972): as was later determined, signs of biological
overfishing appeared first in 1890 on North Sea plaice and in 1905
on North Sea haddock.

A similar interest for the determinants of resource fluctuations
was also noted in the French pilchard fishery which occurred as far
back as the end of last century: "(il est) urgent de savior si
l'irrégularité des rendements de la pêche est la résultat, comme
certains l'affirment sans preuve, de l'action de l'homme ou échappe à

son influence" (Oudin, 1896, quoted by Durand, in press).

Investigations on recruitment variability stopped with World War I and were not resumed immediately afterwards for reasons that are not clear (Sinclair *et al.*, 1987). The likely explanation is that the rapid development of long-distance fisheries and the corresponding expansion through geographical expansion and stock diversification diverted attention away from the scientific issues associated with the politically less palatable task of resource conservation and fishery management. Confirmation of such hypothesis would require more detailed historical analysis for long-distance fishing started some centuries earlier and concerned small-scale as well as large-scale operations, in both developed and developing countries. Consequently, stagnant fisheries and expanding ones coexisted, *e.g.*, in inshore and offshore segments or in different regions. However, it is during the inter-War period and, even more so, after World War II that long-range fishing expanded most rapidly, initially in the North Atlantic and Pacific and, then, throughout the World Ocean, as shown by the rise in world landings from about 7 million in 1920 to 72 million metric tons in 1972.

This expansion was paralleled by the development of the "theory of fishing" in three steps: (i) early observations and concepts relative to the effects of fishing upon stock abundance (*e.g.*, Peterson, 1894; Garstant, 1900-3 and Heincke, 1913); (ii) formulation of a theory for rational exploitation of stock cohorts after recruitment (Russell, 1931 and Graham, 1935); and (iii) the development of computational models of operational applicability for stock assessment (Schafer, 1954 and 1957; Beverton and Holt, *op. cit.*; Ricker, 1958 and Gulland, 1965(2).

These works provided the paradigm and the tools for assessing potential yields of stocks exploited by selective--in terms of target species landings, if not of catches--and capitalistic fisheries-- inasmuch as the amount of fishing effort and balance between capital, manpower and resource were assumed to be easily modifiable. (It should be noted that the paradigm was oriented toward addressing only single species rather than the complex of species generally and taken any particular fishery).

Under the paradigm, opportunities for investments could be assessed and the conditions for coherent exploitation, in terms of fishing effort and size at first capture, determined. This phase of scientific development and application to fishery development planning culminated with the assessment of the world fishery potential yield (Gulland, ed., 1971).

The paradigm, however, disregarded the dynamics of cohorts during the early stages of their life history and the effects of environment upon their recruitment success. This exclusion was then considered as operationally appropriate for only spawning biomass can be

---

[2]Baranov's precursory contributions (1918) remained unknown and largely unused, even in his own country, before being unearthed in 1938 by the Western scientific community (Cushing, ed., 1983).

manipulated through the amount of fishing. Moreover, assuming only average environmental conditions and recruitment greatly facilitated the modeling of cohort yield as a function of fishing mortality and its distribution throughout the exploitated phase. The recruitment/stock theory formulated by Ricker (*op. cit.*) and Beverton and Holt (*op. cit.*) can be interpreted as a straightforward extension of the dominant paradigm of the time as was, later, the distinction between two causes of biological overfishing, namely: (i) growth overfishing, *i.e.*, the decline in yield per recruit consecutive to the excessive fishing of recruited cohorts at young ages, and (ii) recruitment overfishing, *i.e.*, the fall in average recruitment and the increased risk of stock "collapse" as a result of excessive reduction by fishing of the parental biomass (Cushing, 1977; see however, Rothschild, 1986:244).

Variability in stock abundance did not initially raise excessive difficulties to long-distance fleets as fishing for accumulated biomass and the practice of pulse fishing provided alternatives for upsetting temporary and localized stock declines. Eastern European countries had a systematical recourse to that strategy by assessing cohort abundance before their entry into the exploited phase and optimizing year after year the geographical distribution of state-owned fleets over major world fish stocks.

As opportunities for expansion through geographical deployment faded away, long-distance fleets diversified the species basis of their harvests. Progressively, the single-species approach for stock assessment showed limitations. Already, analysts of limnetic, inshore and tropical fisheries were aware of the need for multispecies resource assessment, but no fully satisfactory methodology was available. The situation did not simply reflect the complexity of the ecological issue. It resulted also from the limited original research devoted to assessing small-scale fisheries at all latitudes. That this sector produced 27 million out of total 59 million metric tons used for human consumption and supports 10 million fishermen as compared to 0.5 million fishermen in the large scale sector (UNIDO, 1986) was largely overlooked. Eventually, the issue acquired recognition among fishery biologists (FAO, 1978), as did the trophic relationships among species components of exploited ecosystems (*e.g.*, Anderson and Ursin, 1977; Hempel, 1978a, 1978b; Laevatsu and Larkin, 1981).

In the early 1970's, the state and political context of world fisheries changed radically. From 1972--the very year the Peruvian anchovy fishery collapsed--until 1983, the annual increase rate of world landings dropped from an average of 6.5% which was maintained since World War II to a mere 1%. After 1983, this rate increased again, but in a chaotic fashion (FAO, 1987a): in addition to localized developments in aquaculture, the recent improvement is due essentially to the episodic recovery of certain coastal pelagic stocks, such as the Japanese sardine, which yielded 5.9 millions tons in 1986 whereas its annual production remained below 10 thousands tons throughout the 1960's.

Despite this recent increase, the state of world fisheries has

not basically changed: most resources, notably the highly-valued
ones, remain fully or excessively exploited; among those, some
undergo fluctuations of large and unpredicted amplitude. Exhaustion
of accumulated biomass and the restrictions now imposed by the New
Ocean Regime against geographical deployments expose fishing fleets
to the full amplitude of stock fluctuations, which tend to be
magnified by the reduction in the number of year-classes in the
catch. The adoption of annually-set catch quotas as the main measure
for limiting the amount of fishing, i.e., without the simultaneous
adequate adjustment of the overall fishing capacity to medium-term
productivity of accessible resources, does nothing to mitigate the
detrimental effects of unexpected medium-term fluctuations.

Fishery collapses with catastrophic economic and social hardship
were experienced apparently more frequently, notably in fisheries
based upon coastal pelagic and bivalve molluscs. In some regions,
fish stocks exploded, such as trigger fish or cephalopods off West
Africa (Gulland and Garcia, 1984). The list and history of such
dramatic events are now well documented (i.e., Saville, ed., 1980;
Sharp and Csirke, eds., 1984). What emerges from that history is
the limited ability of fishery science to provide meaningful
explanations of such phenomena, notably regarding the relative roles
of natural and man-made causes of fluctuations and the implications
on the relevance of scientific advice provided as a basis for
designing fishing strategies likely to maximize the probability of
stock recovery or to minimize future collapse risk.

For example, it was after the North Sea herring stocks had
shrunk, and then recovered, and after a moratorium on their fishing
had been adopted, and then relaxed, that the simultaneous occurrence
of a hydrological anomaly was noted: the latter may have affected
the drift and survival of herring larvae across the North Sea
(Corten, 1984). Even if this explanation is not fully established
as yet, natural factors seem to have played a critical role in the
collapse of North Sea herring stocks. When the fishery was open
again, deprived of supply, certain domestic markets had collapsed.

When in the mid 1970's, the Namibian pilchard stock began to
collapse, the purse seine fleet was unable to harvest annual quotas
(Troadec et al., 1980): the stock declined faster than forecast by
projections based on standard yield-per-recruit computations and
prior recruitment trends.

As a consequence, for example, confronted by the simultaneous
increase of the North-west African pilchard stock and its
equatorwards extension, the Moroccan fishery administration could not
be advised objectively on the respective merits of building new
harbours and processing factories further south or seiners of longer
operating range, for biologists were unable to assess the likely
duration of the current stock high and southern extension.

Several stocks (e.g., the Japanese sardine, the California
sardine, the Peruvian anchoveta, the Greenland cod, the Bay of Brest
(France) scallop, etc.) remained at extremely low levels for long and
lasting periods after fishing has been banned or kept at minimum
levels. Conversely stocks, such as the Japanese sardine, reached

482

abundances and yields never observed in the documented history of corresponding their fisheries.

As summarized by Hempel (*Op. cit.*): "The specific causes of fluctuations in fish population abundance are supported by a long list of speculations and a short list of facts."

2.2 State-of-the-art

From this brief historical overview, progress in the exploitation and conservation of fishery resources can be related to three basic sets of issues: (i) the yield of cohorts after recruitment and the condition for their rational exploitation; (ii) the variability in initial cohort number and partition of the cohort between natural and man-made causes of variability; and (iii) the trophic relationships within multispecies resources. This section reviews the present state-of-the-art respective to these issues and the suitability of available methods to investigating them.

2.2.1 *Yield-per-recruit*. Yield-per-recruit models are conceptually identical. They differ by the mathematical treatment, and its sophistication, of input data. Extensive routine applications shows that, provided adequate data are available, they give for many stocks operationally good answers to question (i) above.

Applications to short-term forecasts could be somewhat simplified, the more so that appreciable changes in the rate of fishing are not considered. Difficulty in accurately estimating natural mortality is, their major weakness, but its effects are less important for short term, *status quo* assessments (Shepherd, 1984).

In routine assessments, density-dependent and density-independent effects on growth and natural mortality of exploited cohorts are usually not taken in to account. The good fits generally observed between predicted and observed stock states confirm that, for a majority of finfish stocks, population regulatory processes are not concentrated on the adult phase. (This may not be the case for all taxonomic groups, notably bivalve molluscs). This does not exclude the possibility that appreciable changes cannot occur occasionally in the adult fish condition (fat content, mean weight at age, etc.) with parallel modifications in growth, fecundity and natural mortality rates. Such changes are usually observed in relation to sudden modifications in large marine ecosystems (*e.g.*, El Nino; SELA-BID, 1984) or with changes in their boundaries (*e.g.*, Icelandic summer spawning herring, Jakobsson and Halldorsson, 1984).

Thus Y/R models are efficient tools for short-term, operational forecasts but the undetermination of natural mortality and, more important of future recruitments limit their usefulness for medium-term, strategical projections as required for fishery policy making: assessments will be biased in case the error structure of future recruitments distribution does not replicate the one derived from past observations (Rothschild, *op. cit.*).

2.2.2 *Stock and recruitment*. As indicated earlier, S/R theory
considers explicitly the effects of changes in parental stock size
and disregards those in environmental conditions. Based on
theoretical considerations, two mathematical functions have been
proposed to represent the relation between recruitment and parent
stock, namely:

- an asymptotic function (Beverton and Holt, *op. cit.*) and
- a dome-shaped function (Ricker, *op.* cit.).

A mathematical generalization of these two functions was later
developed (Shepherd, 1982).
However, plots of annual recruitments versus parental spawning
biomasses reveal for most stocks considerable dispersion (*e.g.*,
Garrod, 1982). Implications are twofold: (i) year-specific
environment conditions, prevailing from spawning to recruitment,
affect markedly the recruitment success, *i.e.*, the number of recruits
per unit spawning biomass; and (ii) as such, the S/R plots have in
most cases no operational utility.
Furthermore, recruitment strength distribution is not
necessarily normal. Examination of time series of annual
recruitments (*e.g.*, Cushing, 1982; Rothschild, *op. cit.*) reveals the
frequent superposition of:

- trends of varying durations and slopes,
- noise of moderate amplitude, and
- isolated explosions or drops of higher amplitude (Figure 1).

As for many climatological phenomena, the distribution of annual
recruitment reflects the superposition of processes of different time
and space scales: *e.g.*, recruitment (Garcia and Le Reste, 1981);
transoceanic events, such as ENSO or the North Atlantic low salinity
anomaly (Corten, *op. cit.*); shelf and oceanic signals of smaller
scale, such as storms affecting the stratification and density of
plankton forage locally available for fish larvae (Lasker, 1975), etc.
Attempts have been made to relate recruitment success to certain
environmental proxies (*e.g.*, Bakun and Parrish, 1980). Occurrence of
successive periods of relative environment stability enabled, for
example, Watanabe (1977) to adjust three S/R curves of different
heights to describe variations in Japanese mackerel recruitment as a
function of parental biomass and environmental conditions.
Along with Kuhn's theory (1962), the introduction of environment
indicators in S/R models can be interpreted as an attempt to
internalize processes external to the dominant paradigm. As an
investigation strategy, such an approach has little heuristic power.
Aside from describing past states of fisheries, its efficiency is
limited as it is not specifically tailored to the relevant processes,
variables, and scales. Even for identifying environmental "signals",
its discriminatory power is markedly reduced by the use of proxies
measured on a time scale appreciably longer than that of the
processes involved.

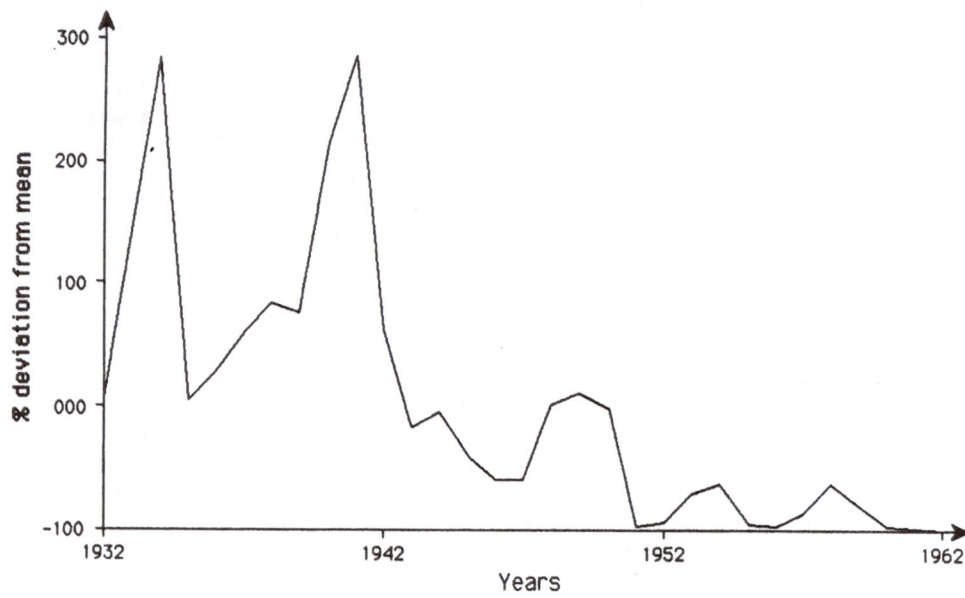

Figure 1. California sardine: fluctuations of annual recruitment.
(after Garrod, 1982).

It is, therefore, no surprise if attempts at correlating
recruitment and stock, annual estimates, derived as by-products of
recruitment-stock assessments, have met little success. However
sophisticated, subsequent mathematical processing cannot compensate
for the lack of appropriate information in input data. Not backed by
adequate field observations, mathematical functions proposed for
representating certain past events (e.g., depensatory mechanisms to
explain stock collapses) appear largely speculative for explaining
stock fluctuations of large amplitude. For example, to validate
theory on "replacement" of one stock by another, processes involved
between trophically related stocks as well as with environment
signals need to be taken into account; similarly, recruitment
overfishing cannot be ascertained as long as effects of medium-term
environment fluctuations on recruitment are not accounted for.
    These observations do not imply that investigating the S/R
relationship is not important since even limited density-dependent
effects would affect recruitment year after year in the same fashion.
Knowing the shape of the S/R relationship will indicate whether it
would be beneficial to maintain the biomass within certain limits to
maximize average recruitment or above a minimum threshold to limit
the risk of future declines in average recruitment.

2.2.3 *Surplus production and regenerative models.* Surplus
production models (Schaefer, *op. cit.*) are an empirical procedure for
representing the combined effects of fishing and past recruitments on
stock size and yield. *De facto*, the model reflects past fluctuations
in recruitment, incorporating density-dependent or density-
independent factors. It may provide an operational tool for
projecting the effects of changes in the amount of fishing upon
multispecies communities whose components are trophically
interrelated (May, ed., 1984).

Regenerative models combine a Y/R function and a S/R
relationship with annual recruitments randomly generated according
to the statistical distribution adopted for representing the
recruitment probability (Beverton and Holt, *op. cit.*; Walters, 1969;
Laurec *et al.*, 1980).

Both approaches depend on annual estimates of exploited
population parameters; consequently, they are ill-suited for
investigating recruitment processes. Attempts have been made to
incorporate climatic proxies into production models (*e.g.*, Binet,
1982; Fréon, 1984, for West African Sardinella). Similar comments as
those made in the previous section regarding Watanabe's treatment of
Japanese sardine recruitment data also apply.

Although, in regenerative models, the analytical treatment of
input data is precise, the choice of the mathematical function to
represent the S/R relationship remains arbitrary as long as it has
not been objectively determined through specifically designed
investigations. On the other hand, regenerative models enable to
incorporate selected recruitment distribution patterns for the years
to come, which the production model does not.

As long as climate cannot be predicted with a sufficient lead
time, the question of climatic effects on recruitment are difficult
to deal with. Thus, Sissenwine (1974) and Doubleday (1976) stressed
the risk of overoptimistic assessments of maximum potential yields
for pulse fisheries, since those were generally initiated after a
succession of above average recruitments.

Depending on whether future error structure replicates that which
applies to past observed recruitments, requires consideration of two
situations.

In the first, past recruitment frequency distribution could be
used in regenerative models, with conventional Y/R and knowingly
adjusted S/R models (which can be expected from investigations on
recruitment). Historical information on past recruitment is
therefore essential for determining the future recruitment
probability distribution. Three sources of data, of varying accuracy
and time coverage, can be used for that purpose: (i) scientific
assessments of fish stocks provide yearly estimates of standing stock
and recruitment; they are available for a growing number of fish
stocks and a few cultivated ones; they seldom go more than a few
decades backwards; (ii) historical fishery records give yearly
landings; in a few cases, they extend backwards over a few centuries;
they are limited to a small number of fisheries in Europe and Japan;
and (iii) paleo-indices of stock abundance, such as those derived

from scale depositions in anoxic sediments in upwelling areas off California and Peru (Soutar and Isaacs, 1974; DeVries and Pearcy, 1982).

In the second case, applications of improved knowledge on processes linking environmental regime to recruitment success will depend on future developments in forecasting hydrodynamic events affecting recruitment at time scales matched to recruitment biology and major economic investments in the production sector, problems affecting the basic uncertainty affecting potential application of findings on recruitment.

2.2.4 *Multispecies resource assessment and ecosystem modelling.*
Appreciable research effort has been devoted to multispecies resource assessment (*e.g.*, FAO, *op. cit.*; Anderson and Ursin, *op. cit.*; May, ed., *op. cit.*). Attempts at modelling whole marine ecosystems have also been made (*e.g.*, Laevatsu and Larkin, *op. cit.*). However, developments of operational applicability remain so far limited. Reasons for slow progress are several: (i) trophic relationships vary markedly according to the successive life history stages of preys and predators making up ecosystems; (ii) strong divergences of opinions exist regarding the major processes determining recruitment success with strong supporters of effects of hydrodynamic processes on egg and larvae distribution and dispersion (*e.g.*, Sinclair, *op. cit.*), predation (for a review, see Bailey and Houde, 1987), starvation, etc.; this may result from differences in the trophodynamics of particular ecosystems; in any case, it reflects a general lack of quantitative information, at the population level, needed for ecosystem modelling; and (iii) there are indications of various kinds suggesting that trophic relationships are less tight, at least for youngs and adults than for larvae and juveniles:

- as indicated earlier, estimates derived form Y/R projections, which do not take into account interspecies trophic relationships, fit usually well with observed stock states;

- this is supported by experiences from ranching (see next section);

- large-scale adult stock fluctuations are not accompanied by changes of comparable amplitude in their food abundance;

- according to Sinclair's "member/vagrant" theory, adult stock sizes would be determined predominantly by spatial processing taking place at early stages; in addition, adult populations are not necessarily food limited.

Such observations and considerations suggest that: (i) trophic relationships have probably not, for adult stocks, the importance it is frequently assumed; and (ii) progress in ecosystem modelling would depend on prior advance in the understanding and quantification of

regulatory mechanisms occurring during early stages at the population
level.

3. RANCHING

Notwithstanding attempts towards intensification, the bulk of world
aquaculture production still comes from extensive forms of
cultivation in which depend primarily on the seeding of wild-caught
or hatchery-raised juveniles (Table 1).

Table 1. World aquaculture production (in millions metric
tons) in 1983 (FAO, 1987b).

Finfish	Molluscs	Crustaceans	Seaweeds	Total
4.45	3.25	0.12	2.39	10.21

Since the end of last century, various attempts have been made,
mainly in the Northern hemisphere, to extend ranching to new species
and areas. Efforts concentrated mostly on the production of young or
juveniles called "seed." Comparatively less attention was devoted to
the ecological and population aspects of such endeavours, i.e., to
the survival of seed stocked in natural systems. In some cases, the
strategy was directly inspired from the S/R paradigm: the objective
was to artificially create a minimum spawning stock with the
expectation that it will later further develop on its own. Most
projects were approached in an empirical fashion; many failed. To
quote Larkin (1977) in an assessment of the British Columbian salmon
programmes: "Attempts to enhance salmon production by various "fish-
cultural" activities were initiated almost a century ago in a wave of
enthusiasm for hatcheries that was generated by the discovery that it
was easy to collect salmon eggs, fertilize them and rear the
progeny"... "However, it is almost impossible to document in detail
the success with which... these efforts were attended, for most of
them were on a small-scale and without a large research machinery for
evaluation. Their effects were lost in the much larger-scale
fluctuations of the natural system."
Considering the complexity of population regulatory mechanisms
and their concentration on early stages on which we know the least,
a better understanding of the population dynamics of recruitment
appears the condition for apprising the ecological feasibility and
determining the technical modalities for optimizing seed survival
and return rates. In Japan, where such endeavours have been the
most extensive in terms of areas and species (salmon, sea bream,
scallop, abalone, penaeid shrimp, blue crab, etc.), remarkable
success have been obtained for some species (salmon, scallop,
abalone) while remaining to be demonstrated for others (Hénocque,
1984).

In this respect, the example of salmon ranching development in Hokkaîdo (Kobayashi, 1980) is interesting. Catches of chum salmon oscillated between 6 and 8 million fish, with an average of 7, between 1878 and 1892; then, they declined to an average of 3 million (ranging from 1 to 5) between 1905 and 1970; following an increase in the number of fry released and improvements in rearing techniques and releasing modalities, catches raised to an average of 9.6 millions (ranging from 7 to 16 millions) from 1971 to 1977. Simultaneously the S/R relationship changed radically with a threefold increase in return rate and, still, no apparition of saturation effects relative to the carrying capacity of the marine ecosystem for the adult stock (Figure 2). This corroborates experimentally an observation already made by Larkin (*op. cit.*) regarding Northern Pacific salmon stocks: "Apparently the various stocks and species collectively do not exploit the marine environment as to show striking relationships."

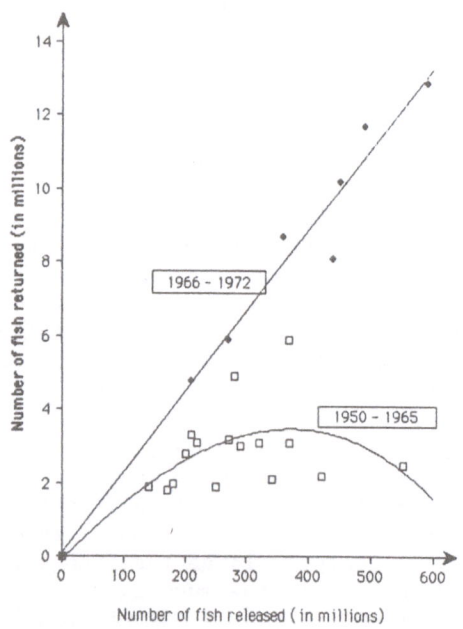

Figure 2. Development of salmon ranching in Hokkaîdo. (modified from Kobayashi, 1980).

In Mutsu Bay (Japan), the recruitment pattern of the scallop stock has been radically modified following the initiation, in the early 1970's, of systematic actions of various kinds: the setting of spat collectors of different types, the regular seeding on the bottom with spat harvested on collectors as well as artificially reared and the beginning of suspended culture. Simultaneously, the annual yield which previously exceeded only occasionally 10 thousand tons, being very low the rest of the time, reached and remained between 20 and 40 thousand tons since (Figure 3). Unfortunately, as adequate observations were not carried out simultaneously, it is not possible to determine the respective roles of increased spawning stock, of the setting of collectors and of the regular addition of artificially-raised spat. Consequently, it cannot be established whether the enhanced stock can be self- sustainable only by the modification of its habitat (*i.e.*, the setting of collectors), or whether regular additional stocking of artificially-produced seed is required. This understanding is critical both on scientific grounds as well as for further stock enhancement purposes.

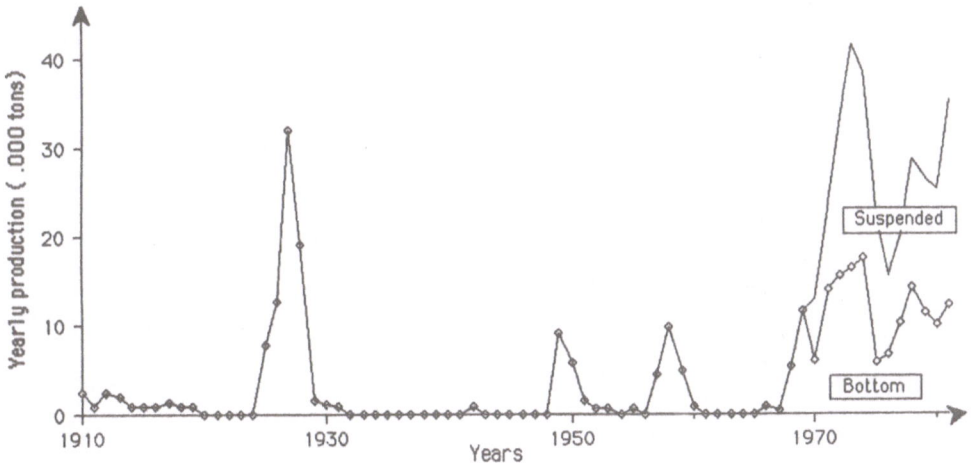

Figure 3. Scallop production from capture and culture in the Bay of Mutsu (Japan). (from data available at the Laboratory of Moura (Japan).

The above example shows that, even in a semi-closed environment as a bay, natural ecosystems can sustain standing stocks of much higher biomass than usually observed under wild conditions.

In the Marennes-Oléron Bassin (France) where oyster production is ancient and intensive, the trophic saturation of a coastal marine ecosystem has been modeled (Figure 4). With Portuguese oyster (*Crassostrea angulata*), the standing stock regularly increased to reach almost 200,000 tons. The stock was close to that level when, from 1968 to 1971, it was decimated by a disease. Production resumed with Pacific osyter (*C. gigas*), following the same production/biomass relationship as observed for the development of Portuguese oyster. Biomass did exceed largely the level of about 80,000 tons corresponding to maximum annual yield (about 40,000 tons) and considerably more than existed under natural conditions (IFREMER, 1986).

490

These ranching examples are interesting because: (i) under
natural conditions, natural mortality before recruitment can be so
high that the recruited standing stock does not saturate the carrying
capacity of natural systems; (ii) thus, high fluctuations of the
recruited biomass do not depend on comparable fluctuations in

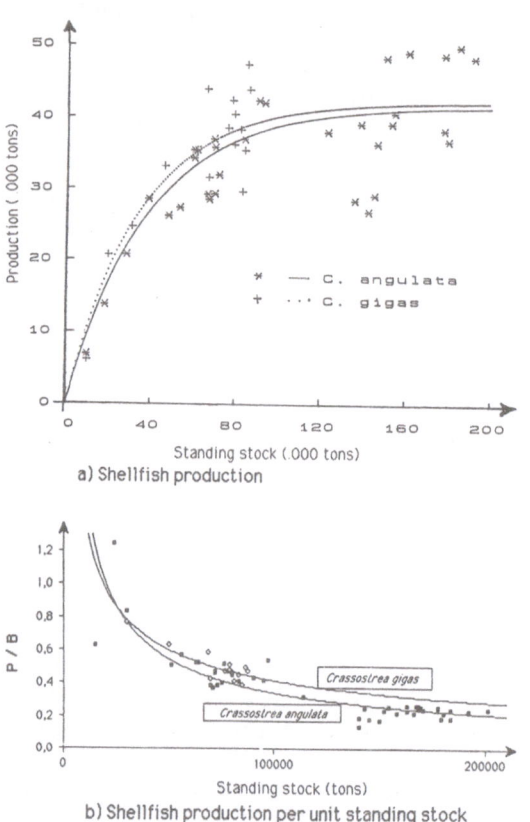

Figure 4. Yearly production and production per unit standing stock
of oyster in the Marennes-Oléron Bassin (France). (after IFREMER,
1986).

plankton forage production; (iii) existence of different processes at
the egg/larvae and juvenile/adult stages could explain why trophic
relationships and energy transfer efficiency show little consistency
at the upper trophic levels of the food web; if confirmed as a
general principle, this would reduce the importance to be attached to
trophic relationships within adult components of multispecies
resources; and (iv) in the Marennes-Oléron oyster example, P/B

declines from 1.0 to 0.2 while standing stock (B) increased from 0 to 200,000 tons and yield (P) from 0 to 40,000 tons: the relationship between P/B and B changes markedly as B increases, P being more stable over a certain biomass treshold than P/B which declines monotonously; such ranges of variations are greater than those currently assumed in production analyses of ecosystem food webs.

Thus, investigations in ranching development offer facilities for experimental manipulation of natural ecosystems. Fishery science and oceanography are generally restricted to observing natural changes in systems or, in the former case, to observations on the effects of changes in the harvesting pattern but, with usually no possibility left to the scientist to direct changes. Well-selected experimental ranching projects would enable manipulation of habitat features, predator densities, etc. in addition to recruitment levels and, thus, help clarify the respective roles of spawning stock size, habitat modifications, predator removal, carrying capacities of the environment for particular stages in the life history, etc.

It is also worth noting that the "member/vagrant" theory on marine population structure and variability (Sinclair, *op. cit.*), makes homing a general feature in aquatic population behavior: otherwise, it is difficult to understand how those stocks can visit the same hatchery grounds year after year and succeed in maintaining their genetic pool within highly diffusive environments. If confirmed, this feature would widen perspectives regarding opportunities in ocean ranching.

Investigations on ranching offer the possibility of combining *in vitro* experiments with *in vivo* observations and experiments and of benefitting for that purpose from the sets of information, methods and expertise already accumulated in aquaculture research which will be long and costly to develop separately. Selecting coastal stocks of limited size can facilitate observations by making sampling efforts and experimental manipulations cheaper and easier. Also the existence of several discrete stocks of restricted geographical extension can provide replicates for comparative observations.

4. HABITAT MANAGEMENT AND ECOSYSTEM CONSERVATION[3]

Detection of anthropogenic impacts on marine populations are commonly attempted through observations and monitoring of environment changes supposedly induced by the development of certain human activities (*e.g.*, nuclear plants). The implicit strategy is that impacts can be detected through correlations at the organism level (lethal effects, disease symptoms, contaminant contents, etc.).

Such an approach has a low discriminatory power for impacts on the abiotic environment usually generate weak signals as compared to the noise associated with the various time and space scales of natural ocean variability. In addition, such investigations do not say much about effets at fish population level, whereas it is those effects--through changes in growth, "natural" mortality and recruitment--which matter directly for fishing and culture

activities. Estimations of anthropogenic effects at fish stock level
requires prior evaluations of population productivity undertain
certain conditions of reference.

Quantitative measurements are few, except for impacts of large
amplitude which can be estimated directly through simple correlations
(e.g., the collapse of the Nile *Sardinella* stock in relation to the
building of the Assouan dam and the subsequent reduction in
freshwater outflow; Shaheen, 1975).

For similar reasons, positive effects of habitat modifications on
fish stock productivity are hard to establish unless habitat
management projects are "large-scale." With the methodology
available, effects of artificial reefs on stock production and their
differentiation from mere fish attraction are difficult to
demonstrate unless they extend over large geographical areas. Owing
to the current methodological shortcomings, this is seldom possible
and biological effects and economic profitability cannot be assessed
prior to implementations of costly projects.

Because aquatic populations concentrate their regulatory
mechanisms on the early stages of their life history and it is at
such stages that populations are most sensitive to environment
modifications, improvement in the discriminatory power of analytical
methods should result from progress in understanding and modelling
of recruitment processes. Accurate determination of critical phases
at which cohort survival is most sensitive to environment
alterations would enable to refine standards for environment uses or
for maximizing expected impacts of habitat management (e.g.,
localization of reefs).

The decline of oyster production in the Arcachon lagoon (France)
was directly caused by a collapse in spat collection. This lasted
for five years in succession (1977-1981) whereas, over almost a
century, spat settlement never failed more than once every decade, on
average. This was associated with mass mortalities of oyster larvae,
directly and through a sharp reduction in their plankton forage,
originally caused by organotin compounds in antifouling paints, (His
and Robert, 1985 and 1986). There is, thus, a need to develop models
interrlating recruitment success and environmental factors likely to
be altered by human activities (e.g., temperature; Mclean et al., 1981).

---

[3]Potential applications of early-stage dynamics of exploited
populations to habitat management and ecosystem conservation would
justify a development as extensive as the one devoted to ranching or
fishery management. For keeping the present paper within an
acceptable length, this section is restricted to an identification of
major potential developments.

5. DISCUSSIONS

Interesting comparisons have been made between historical developments in fisheries and progress in fishery science.

Under the "theory of fishing", a powerful set of tools has been developed to assess the yield of cohorts after recruitment and the conditions for their exploitation. The theory provided answers to issues associated with the growth of world fisheries, notably the ocean wide spread of large-scale fishing operations. The underlying paradigm was based on the single-stock Y/R assessment approach, the hypothesis of average environment and recruitment conditions, the concept of stable maximum yield and the prevalence of fishing effort among the variables determining, recruitment, the state of fish stocks and their yields. This conceptual framework represented "a considerable advance that did much to organize and systematize thinking about mortality rates, growth, production, stock and recruitment, yield per recruit,..." (Rothschild, *op. cit.*).

However, since the 1960's, progress in fishery science has been "of computational rather than conceptual importance" (Rothschild, *op. cit.*). This situation reflects the very heavy involvement of fishery biology, initially in the assessment of world fishery resources, today in repetitive applications to fish stock monitoring and year-to-year yield projections for operational fishery regulation purposes. This workload is made heavier by current shortcomings in fishery management practices which do not tackle properly the economic and political causes of overfishing (Troadec, in press).

This situation has impeded fishery biology to give adequate attention to the issues which have emerged in relation with the full exploitation of world fishery resources and the restrictions imposed to the geographical deployment of fishing operations, the recognition of the importance and specific features of small-scale fisheries, the development opportunities offered by extensive mariculture or the need to assess the effects, on fish stock productivity, of anthropogenic modifications of coastal environments.

Under the new Ocean Regime, the importance for fishery management of medium-term projections has increased: medium-term trends in fish stock abundance are needed for determining adequate balances between average fish stock productivity and fishing capacities operating in major fisheries. Simultaneously, a progressive reduction in the demand for short-term stock monitoring and support for operational fishery regulation can be expected from progress in fishery management, for resource conservation will be less demanding when excessive fishing capacities will be removed. On this aspect, the example of fishery management in Japan is relevant. Better performances are achieved on several aspects, notably the internalization of the management issue among fishermen, the reduction of conflicts, the limitation and distribution of fishing activities. These are not independent from the arrangements adopted in response to the political dimensions of the fishery management issue, *i.e.*, the regulation of access (Ruddle, 1987).

494

However, as compared to Western practices, less attention is given to short-term stock monitoring.

With more attention being given to recruitment variability, a change in the prevailing paradigm has already occurred. The importance of environment factors in stock variability, as opposed to spawning stock size, has acquired a more general recognition. Interpretations of certain collapses by recruitment overfishing have been reevaluated. Quantification of the respective roles of environment and spawning biomass, which can be expected from recruitment investigations, should improve long-term strategies for minimizing the risk of stock depletion as well as for stock rebuilding. For example, the principle of maintaining spawning stock size above a certain minimum level (which cannot be objectively determined with the methodology available) may appear unjustified for certain fish stocks. Such improvements can be achieved without the ability to forecast the hydroclimatic factors influencing most the recruitment success.

The question of the prospects for forecasting climatic changes on a scale matched to the span of major investments in the fishing industry is a major one when assessing efforts to be devoted to recruitment investigations. Until such forecasts are feasible, trade-offs will have to be made between the amplitude and scales in stock variability, the forecasting capability, the time lags required by the industry to adjust its capacities and to open new markets, the costs involved for acquiring scientific information and optimizing the industry flexibility. These elements need to be assessed. Meanwhile, the means available to dampen the detrimental effects of stock variability have to be optimized: increasing the number of year-classes making up standing stocks; regulating the rate of fishing through the licensing of adequate fishing capacities; greater flexibility of fishing fleets to shift between stocks within fishing areas accessible to them, of the industry to process new species, to develop new products and markets; monetary compensation schemes, etc.

Ranching can be a major beneficiary from progress in recruitment studies. Challenging parallels have been made between certain theoretical developments concerning population structure and variability, and evidence drawn from ranching.

Performance of hatchery techniques are now such that the requirement of raising spat or fingerlings at an economically acceptable cost given the expected rates of return can be met for several species of molluscs, shrimp and finfish. Likely, more complex are the ecological aspects of stock enhancement and the institutional difficulties related to coastal ecosystem uses management.

In many instances, an empirical approach is likely to be insufficient: specifically tailored, innovative, scientific programs will be needed. Moreover, their size will have to be matched with that of the population or ecosystem to be manipulated. This is another argument for selecting, initially, stocks of limited geographical extension.

These considerations have a direct bearing on institutional and legal aspects. Effective development implies that action can also be taken on the modalities of exploitation at a scale corresponding to that of the resource. This may not be the least constraint. However, the issue is not specific to ranching. It is shared with fisheries, where some progress is currently observed. In a context of overall stagnation in production, advance towards production intensification (per unit sea area) could promote changes in legislation. Already, administrative and legal arrangements regarding, for example, site allocation and transfer are comparatively more advanced and making more progress in production systems, such as shellfish culture, which are well-developed and less complicated to administrate (sedentary organisms, self- contained eco- and socio-systems facilitating the internalization of the free competition).

Coherent use of exploited aquatic ecosystems depends also urgently on the ability to assess and allocate changes in fish stock recruitment among their respective potential causes: harvesting, fluctuations of climatic origin, restocking, pollution. So far, recruitment variability has received little direct attention except, at the turn of the century, in North Sea fisheries already confronted by overfishing. The methodology developed under the "theory of fishing" is not appropriate for investigating the ecology and dynamics of early stages. It is, therefore, no surprise if attempts made at approaching those issues within the traditional methodology have met little success. Probability of success will be substantially improved by adopting research strategies and protocols well-matched to the nature of the issue including: formulation of theories from which efficient research protocols can be designed; identification of appropriate time-and-space scales for observing processes; conduct of parallel field and laboratory activities; and modelling complex, multivariate phenomena.

Recruitment investigations fall predominantly in the field of ecology: the respective roles of spatial versus energetic processes in determining recruitment success, the coupling of hydrodynamic processes and larval behaviors, the role of discontinuities and heterogeneties in abiotic (fronts, upwellings, bottom, estuaries, etc.) and biotic (populations) structures of aquatic ecosystems, etc. are some of the topics to be investigated. This requires inputs from climatology, physical oceanography, planktology, population dynamics - incuding genetics, marine biology and aquaculture, in jointly designed multidisciplinary programs.

The historical review presented in the first part of this paper has shown that lags frequently appear between the emergence of fishery issues, the apparition of scientific concerns and the development of appropriate methods. One and a half decades ago, Kesteven (1972) noted already: "In its inability to predict, fisheries biology sits squarely and sadly beside ecology... In the terminology of Kuhn (op. cit.), fisheries biology is now operating with a paradigm whose potential of creativity is almost exhausted; employing its practices and models can serve only to confirm and refine the results already established; its practitioners are mostly engaged on convergent research. And this is where I am finding justification in revolution: fisheries biology now needs divergent research."

Similar lags occur also between scientific understanding, the development of appropriate methodology and their application to fishery management. With respect to fishery management, the issue is not new: "During the first half on this century, a curious dichotomy existed between the emerging "theory of fishing" and the management measures then in vogue, which were mostly concerned with trying to ensure adequate reproduction and recruitment... For most species this preoccupation with recruitment seemed directly opposed to the accumulating scientific evidence of the time" (Ricker, 1977).

Qualitative progress in fishery science concepts and methods have been made generally when fishery biology addressed issues of strategic significance for fishery development and management and relied for their solution upon scientific findings in ecology and oceanography. Investigating the history of fisheries can help in comprehending long-term forces affecting the production sector and, thus, contribute to a better and more rapid adjustment of scientific concerns to emerging fishery issues. Now that physical growth is achieved in most capture fisheries, it is hard to see how fishery science will sustain further progress without specifically addressing the ecology and dynamics of early stages of populations in exploited ecosystems.

6. REFERENCES

Anderson, K.P. and E. Ursin: 1977, 'A multispecies extension to the Beverton and Holt theory of fishing, with accounts on phosphorus circulation and primary production', *Meddr. Danm. Fish. -og Havunders*. N.S., 7, 319-435.

Bailey, K.M. and E.D. Houde: 1987, 'Predators and predation as a regulatory force during the early life of fishes', *Int. Cons. Explor. Sea*. CM/mini, Z:30.

Bakun, A. and R. Parrish: 1980, 'Environmental inputs to fishery population models for Eastern Boundary Current', in G.D. Sharp (ed.), *Workshop on the Effects of Environmental Variations on the Survival of Larval Pelagic Fishes*, Intergovernmental Oceanographic Commission, UNESCO, pp. 67-104.

Baranov, T.I.: 1918, 'On the question of the biological basis of fisheries', *N. -i Ikhtiologicheskii Int.*, I, 1, 81-128.

Beverton, R.J.H. and S.J. Holt: 1957, *On the Dynamics of Exploited Fish Populations*, Fish. Invest. London, Ser. 2, 19, 533.

Binet, D.: 1982, 'Influence des variations climatiques sur la pêcherie des *Sardinella aurita ivoiro-ghanéennes*: relations sécheresse-surpêche', *Oceanol. Acta* 5(4), 443-452.

Corten, A.: 1984, 'The recruitment failure of herring in the Central and Northern North Sea in the years 1974/78 and the mid 1970's hydrography anomaly', *Int. Cons. Explor. Sea*, CM/GEN = 12:18 (mimeo).

Cushing, D.H.: 1977, 'The problem of stock and recruitment', in J.A. Gulland (ed.), *Fish Population Dynamics*, New York, Wiley.

Cushing, D.H.: 1982, *Climate and Fisheries*, London, Academic Press.

Cushing, D.H. (ed.): 1983, *Key Papers on Fish Populations*, IRL Press, Oxford, Washington, D.C. 405.

Devold, F.: 1963, 'The life history of the Atlanto-Scandian herring', *Rapp. P.-v. Réun. Cons. int. Explor. Mer*, 154, 98-108.

DeVries, T.J. and W.G. Pearcy: 1982, 'Fish debris in sediments of the upwelling zone off central Peru: a late Quaternary record', *Deep-Sea Res.* 28, 87-109.

Doubleday, W.C.: 1976, 'Environmental fluctuations and fishery management', *ICNAF Sel. Pap.* 1, 141-150.

Durand, M.H.: La crise sardinière francaise: les premières recherches autour d'une crise économique et sociale, in press.

FAO: 1978, 'Some scientific problems on multispecies fisheries', Report of the Expert Consultation of Multispecies Fisheries, *FAO Fish. Tech. Pap.* 181.

FAO: 1987a, 'The state of world fishery resources', *FAO Fish. Circ.* 710, rev. 5.

FAO: 1987b, 'Thematic evaluation of aquaculture', Joint study by UNDP, the Norwegian Ministry of Development Cooperation and FAO, 85 + ann.

Fréon, P.: 1984, 'Des modèles de production appliqués à des fractions de stocks dépendants des vents d'upwelling: la pêche sardinière au Sénégal', *Océanogr. trop.* 19(1), 67-94.

Garcia, S.: 1983, 'The stock-recruitment relationship in shrimps: reality or artefacts and misinterpretations?', *Océanogr. trop.* 18, 25-48.

Garcia, S. and L. Le Reste: 1981, 'Life cycles, dynamics, exploitation and management of coastal penaeid shrimp stocks', *FAO Fish. Tech. Rep.* 203.

Garrod, D.J.: 1982, 'Stock and recruitment - again', *Fish Res. Tech. Rep.*, MAFF Direct. Fish. Res., Lowestoft 68.

Garstang W.: 1900-03, 'The impoverishment of the sea', *J. Mar. Biol. Assoc. UK, NS*, 6, 1-70.

Graham, M.: 1935, 'Modern theory of exploiting a fishery, and application of North Sea trawling', *J. Cons. int. Expl. Mer* 10, 263- 274.

Gulland, J.A.: 1965, 'Estimation of mortality rates', Annex to Arctic Fisheries Working Group Report, *Int. Counc. Explor. Sea.*, Ann. Meeting (mimeo).

Gulland, J.A. (ed.): 1971, *The Fish Resources of the Oceans*, Fishing News (Books), West Byfleet.

Gulland, J.A. and S. Garcia: 1984, 'Observed patterns in multispecies fisheries', in R.M. May (ed.), *Exploitation of Marine Communities*, op. cit.

498

Heincke, F.: 1898, 'Naturgeschichte des Herings. I. Die
Lokalformen und die Wanderungen des Herings in den europäischen
Meeren', *Abh. D; Seef. Ver.* 2 S CXXX VI u. 128 S
Heincke, F.: 1913, 'Investigations on the plaice. General
report. I. The plaice fishery and protective regulations',
First part. *Rapp. Cons. Expl. Mer.* **17A**, 1-153.
His, E. et R. Robert: 1985, 'Développement des véligères de
Crassostrea gigas dans le bassin d'Arcachon', *Rev. Inst. Pêches
Marit.* **47(1 et 2)**, 63-88.
His, E. et R. Robert: 1986, 'Observations complémentaires sur les
causes possibles des anomalies de la reproduction de *Crassostrea
gigas* (Thunberg) dans le bassin d'Arcachon', *Rev. Trav. Inst.
Pêches Marit.* **48(1 et 2)**, 45-54.
Hempel, G.: 1978 a, 'Symposium on North Sea fish stocks - recent
changes and their causes', *Rapp. P.-v. Cons. int. Explor. Mer*
172, 5-9.
Hempel, G.: 1978 b, 'North Sea fisheries and fish stocks - a
review of recent changes', *Rapp. P.-v. Cons. int. Explor. Mer*
173, 145-167.
Hénocque, Y.: 1984, 'Aménagement de la ressource côtière au
Japon: effets des repeuplements marins', *Rapport technique
ISTPM*, 11, 137 (mimeo).
Hjort, J.: 1914, 'Fluctuations in the great fisheries of northern
Europe viewed in the light of biological research', *Rapp. P.-v.
Réun. Cons. int. Explor. Mer* **20**, 1-228.
ICES: 1985, 'Report of the IREP Study Group', *Int. Cons. Explor.
Sea* CM/Gen = 4, 43.
IFREMER: 1986, 'Evolution et état du cheptel ostrécole dans le
bassin de Marennes-Oléron, intérêt d'une régulation', *Document
Technique IFREMER*. DRV. 86-6, AQ/TREM, 35 (mimeo).
IOC: 1983, 'Workshop on the IREP component of the IOC Programme
on Ocean Science in relation to Living Resources (OSLR),
Halifax, N.S. (Canada), 26 Sept. 1983 IOC *Workshop Rep.*, **33**,
17.
Jakobsson, J. and O. Halldorsson: 'Changes in biological
parameters in the Icelandic summer spawning herring', *Int. Cons.
Explor. Mer* CM/H, 43.
Kesteven, G.L.: 1972, 'Management of the exploitation of fishery
resources', in B.J. Rothschild (ed.), *World Fisheries Policy.
Multidisciplinary Views*, The University of Washington Press,
Seattle.
Kobayashi, T.: 1980, 'Salmon propagation in Japan', in J.E.
Thorpe (ed.), *Salmon Ranching*, Academic Press, pp. 91-107.
Kuhn, T.S.: 1962, *'The Structure of Scientific Revolutions*l, The
University of Chicago Press, Chicago.
Laevatsu, T. and H.A. Larkin: 1981, *Marine Fisheries Ecosystem:
its Quantitative Evaluation and Management*, Fishing News
(Books), Farnham, England, 162.
Larkin, P.A.: 1977, 'Pacific Salmon', in J.A. Gulland (ed.), *Fish
Population Dynamics*, A Wiley-Interscience Publication, pp. 156-
186.

Lasker, R.: 1975, 'Field criteria for survival of anchovy larvae: the relation between inshore chlorophyll maximum layers and successful first feeding', *Fish. Bull.* 73, 453-462.

Laurec, A., A. Fonteneau and C. Champagnat: 1980, 'A study of the stability of some stocks described by self-generating stochastic models', in A. Saville (ed.), *The Assessment and Management of Pelagic Fish Stocks'*, op. cit.

Maclean, J.A., B.J. Shuter, H.A. Regier, and J.C. MacLeod: 1981, 'Temperature and year-class strength of "Smallmouth Bass"', *Rapp. P.- v. Cons. int. Explor. Mer* 178, 30-40.

May, R.M. (ed.): 1984, *Exploitation of Marine Communities, Report of the Dalhem Workshop*, Berlin, 1984, April 1-6, Life Sciences Research Report, 32, Springer-Verlag, 366.

Petersen, C.G.J.: 1894, 'On the biology of our flatfishes and the decrease of our flatfisheries', *Rep. Dansk. biol. Stat.* 4, 146.

Ricker, W.E.: 1954, 'Stock and Recruitment', *J. Fish. Res. Board Can.* 11, 559-623.

Ricker, W.E.: 1958, 'Handbook of computations for biological statistics of fish populations', *Bull. Fish. Res. Board Can.* 119, 300.

Ricker, W.E.: 1977, 'The historical development', in J.A. Gulland (ed.), *Fish Population Dynamics*, A Wiley-Interscience Publication, 372.

Rothschild, B.J., ed.: 1972, *World Fisheries Policy: A Multidisciplinary View*, Univ. Washington Press, Seattle and London, 272 pp.

Rothschild, B.J., ed.: 1983, *Global Fisheries: Perspectives for the 1980's*, Springer-Verlag, New York.

Rothschild, B.J.: 1986, *Dynamics of Marine Fish Populations*, Harvard University Press, 277.

Ruddle, K.: 1987, 'Administration and conflict management in Japanese coastal fisheries', *FAO Fish. Tech. Pap.* 273, 93.

Russell, E.S.: 1931, 'Some theoretical considerations on the "over- fishing" problem', *J. Cons. int. Explor. Mer* 6, 3-20.

Saville, A. (ed.): 1980, 'The assessment and management of Pelagic fish stocks', *Rapp. P.-v. Réun. Cons. int. Explor. Mer* 177.

Schaefer, M.B.: 1954, 'Some aspects of the dynamics of populations important to the management of the commercial marine fisheries', *Bull. Inter-Am. Trop. Tuna Comm.* 1, 27-56.

Schaefer, M.B.: 1957, 'A study of the dynamics of fishery for yellowfin tuna in the eastern tropical Pacific Ocean', *Bull. Inter-Am. Trop. Tuna Comm.* 2, 247-285.

SELA-BID: 1984, 'Evaluacion de los recursos sardina, jurel y caballa en el Pacifico Suroriental', Informe Regional Final, 7a; *Reunion del Grupo Tecnico Cientifico*, Lima, Peru.

Shaheen, A.H.: 1976, 'La pêcherie de sardinelles sur le littoral méditerranéen de l'Egypte', in Groupe de travail sur l'évaluation des ressources et les statistiques de pêche du Conseil Général des Pêches pour la Méditerranée (CGPM), Rome, 10-14 novembre 1975, *Rapp. FAO Pêches* 182, 35-36.

Sharp, G.D. and J. Csirke (eds.): 1983, 'Proceedings of the
    Expert Consultation to Examine Changes in Abundance and Species
    of Neritic Fish Resources', *FAO Fish. Rep.* 291, 1, 2, and 3.
Shepherd, J.G.: 1982, 'A versatile new stock-recruitment
    relationship for fisheries and the construction of sustainable
    yield curves', *J. Cons. int. Explor. Mer* 40(1), 67-75.
Shepherd, J.G.: 1984, 'The availability and information contents
    of fisheries data', in R.M. May (ed.), *Exploitation of Marine
    Communities'*, op. cit.
Sinclair, M.M. 'Marine Populations - An essay on population
    regulation and speciation in the ocean', in press.
Sinclair, M.M. and P. Solemdal: 1987, 'The development of
    "population thinking" in fisheries biology between 1878 and
    1930', *Int. Cons. Explor. Sea* CM/L = 11, 54.
Sinclair, M.M., J.W. Loder, D. Gascon, E.P. Horne, I. Perry and
    E.J. Sandeman: 1987, 'Fisheries needs for physical
    oceanographic information within the Atlantic Zone', *Can. Tech.
    Rep. Fish. Aquat. Sci.* 1568: VIII + 166.
Sissenwine, M.P.: 1974, 'Variability in recruitment and
    equilibrium catch of the Southern New Engl

ZOOPLANKTON - THE CONNECTING LINK: A HISTORICAL PERSPECTIVE

Michael R. Reeve
Ocean Sciences Division
National Science Foundation
Washington D.C. 20550
U.S.A.

ABSTRACT. Marine zooplankton, like most other marine animals including fish, produce many more offspring than eventually grow to maturity. Zooplankton encompass a wide range of taxonomic diversity. We know little about their early life history during which most of their mortality occurs, because they are very small (less than 1 mm), transparent and very difficult to observe. By definition, they cannot navigate or traverse significant distances at will. They have widely ranging fecundities from tens, hundreds, thousands to millions. The early life history stages of zooplankton are almost universally microscopic, indistinguishable from related species, are dependent on adequate food supplies within hours to a day or two at most for survival, and have innumerable predators. Very little is known of the ecology and behavior of most of these early life history stages, yet they provide the food, and in some cases are the predators, for virtually all larval and post-larval fish. I review, from a historical perspective, the slow development of our understanding of this connecting link, attributable to the enormous difficulties in working both in the laboratory and the environment with a huge diversity of very small and very delicate organisms for which an adequate sampling technology has yet to be developed.

1.   INTRODUCTION

This paper is one of at least three (two others in this volume are by Mullin and Lasker) in which there is a strong sense of frustration expressed in having spent over half of a working career on experimental zooplankton ecology, without significantly increasing our ability to predict population variability beyond seasonal generalities. This is the major midlife crisis probably experienced by all marine ecologists studying organisms with planktonic life cycles. We know, or at least we still profess our belief, that the key to understanding more than the mere generalities of the relationship between primary production and adult fish stocks is understanding complex connections through which these are modulated

B. J. Rothschild (ed.), Toward a Theory on Biological-Physical Interactions in the World Ocean, 501–512.
© 1988 by Kluwer Academic Publishers.

by the zooplankton. These connections, which result in the year-to-year success or failure of recruitment, have yet to be well defined.

In this paper I review the major approaches and technology applied to understanding variability of marine zooplankton populations, and in discussing their shortcomings, suggest some directions of future research, based on hindsight and our obvious lack of success until now.

2. THE PAST

I define the past as the first 100 years, as summarized in the first edition of *"Plankton and Productivity in the Oceans"* a milestone compendium produced as a labor of love by my guide and mentor, John Raymont (1963). It documents just about all of the significant zooplankton experimental and environmental studies up to about 1960. The major tools of all these heroic efforts of data collection and analysis were ships, plankton nets and microscopes. Single samples were collected at single points of space and time. Even if samples were collected sequentially in the same place, it was certain that the water body was not the same, hence the zooplankton population was not the same. How different it was depended a multitude of physical factors, most of which were completely unknown. Monthly sampling intervals were standard, and a year's worth of collections was the norm. Because zooplankton populations vary by orders of magnitude over a year, monthly sampling often produced some relatively clear "seasonal" patterns. The second season was rarely visited. As sampling became more frequent, violent fluctuations between successive sampling periods became much more apparent, because patchiness in time and space obscured smaller scale real population changes.

Since nets are the heart of the traditional sampling system, the reality of the sample is only the reality of the design of the net, and particularly the size of mesh employed. At least up to 25 years ago, most nets were towed in a manner which gave ample warning to all but the old and infirm that they should move out of the way. Only the tiniest individual, whose absolute ability to move through water is severely limited, could be captured with any degree of certainty. Unfortunately, most of those could be swept both into, and then out of the net, because standard net meshes were at least 300 um. A quick review of published data shows that inevitably, in any collection of copepods for instance, there were few early life history stages and adults of very small copepods, such as *Oithona*. In the few studies which attempted to categorize a copepod species by stage, there were few or no nauplii, and not many early copepodites. Although most reports implicitly or explicitly assumed loss of smaller stages, they nevertheless had no practical measure of them, and so no feel for the mortality of early life history stages of copepods. If this were true for copepods, virtually nothing was known of early life history stages of soft-bodied and gelatinous organisms, because the few that might be retained in the net were distorted beyond recognition in this process, and by application of formalin for preservation.

Up to this time, therefore, only seasonal cycles were identified, and sometimes successive generations within a season. The general impression of the generation time of a zooplankton population varied from more than a month in the warmer temperate waters, to up to two years at high latitudes.

The other major phenomenon of water-column biological oceanography which had been established by this time was that of vertical migration. Hypotheses were advanced to provide general and plausible explanations, and even significant experimental work was done (Hardy and Bainbridge, 1954), which once and for all disposed of the notion that zooplankton were "passive drifters", and demonstrated that larger copepods could swim tens, if not hundreds of meters in a few hours, extra-ordinary feats for their size. Hardy (1956) also introduced the concept that zooplankton had considerable control of their geographical location, horizontally as well as vertically.

Up until 1937, laboratory studies on zooplankton had been descriptive. Visual observations on behavior and feeding were recorded from animals collected in nets, returned to the laboratory and observed until they died. Fuller (1937) published the first quantitative study on feeding rates for laboratory confined copepods. Such studies continue today, and it is clear that we still have little quantitative idea about how copepods feed in the environment. More on this below. Parenthetically, a tribute should be paid to those plankton ecology pioneers, Sheina Marshall and A. P. Orr, who produced such prodigious amounts of data on *Calanus* that they were able to write a small book on *The Biology of a Marine Copepod* (1955). This feat has never been duplicated for another marine zooplankton species. Amongst other efforts, they sampled a relatively enclosed Scottish marine Loch on a frequent basis to document successive generations of *Calanus*. They identified all nauplii and copepodite stages and counted their abundance (a feat rarely repeated), and they were the first to introduce radio-tracers (32P) to label phytoplankton cells, which were subsequently ingested and assimilated by *Calanus*.

Marshall and Orr also demonstrated the complexity and non-uniformity of vertical migration throughout the season, and as a function of life history stage. They, and other significant experimentalists, however, worked largely on *Calanus* feeding on diatoms, which resulted in a dogma being established by the sixties that copepods consumed all the diatoms in the ocean but little else. Subsequent workers were slow to rediscover classic earlier observations, such as those of Marie Lebour (1923), demonstrating carnivory.

It has long been recognized that the tiny scale of the laboratory bench experiment bears little resemblance to nature. On the other hand nature is impossible to control and very difficult to monitor. Raymont (1947) tried to resolve this dilemma when he experimentally fertilized a Scottish marine Loch in order to monitor the effects progressively up the food chain.

It had already become clear to Gordon Riley (1947) that there was a great need for more than field observations and laboratory experiments, and he produced the first great synthesis of the two for Long Island Sound, developing simple models relating primary and secondary production in an attempt to demonstrate that seasonal cycles were predictable.

3. THE PRESENT

I take the present to extend back over about 25 years. The vast increases of United States federal funds in the early sixties recruited a new generation of graduate students in that country, who evolved into the researchers of the seventies. These in turn fueled the graduate schools for ocean sciences, which had proliferated as a result of the great sixties expansion.

Recognition of the many inadequacies of plankton nets led to an array of designs, to overcome one or more of these. Most earlier workers had no alternative to making relative estimates of densities of organisms, based on distances or duration of tows. As small, relatively inexpensive current meters became available in the sixties, their installation became *de rigeur* in the net mouth, particularly for the younger generation of scientists. The "Discovery"-type net, with its three progressively smaller net meshes, and ability to provide the best general impression of the widest range of the total zooplankton community, was abandoned. In its place came "numbers per cubic meter" often extended out to one or more decimal places, as the electronic calculator replaced the slide rule.

A single net tow usually extended over hundreds of meters. A single numerical value for organisms per cubic meter resulted, useful for some gross scale comparisons but little else. Studies using paired nets, repetitive sampling in the same location, small scale pump sampling, and the Longhurst-Hardy plankton recorder, gradually built up the overwhelming case for meter to kilometer-scale patchiness. The idea that no number is different from another unless it is less than half or greater than double, was simply accepted as somehow unavoidable, for whatever reason. A disturbing phenomenon most people cared not to think about was finally made respectable by the Steele (1978) edited NATO workshop proceedings on patchiness. This was the realization that zooplankton population distributions are patchy over all scales.

The most significant advance in net systems was the Bongo-type frame of McGowan and Brown (1964). It substantially reduced the pressure wave in front of the net mouth. During the seventies, the ultimate expression of the plankton net (or at least I hope there are no further expressions) was seen in the development of multiple, opening/closing, instrumented, heavy and bulky metal-framed monsters, capable of obtaining up to 9 consecutive samples, and being monitored and controlled from the ship. At least three lineages of these currently exist. Although they take advantage of the micro-computer

era, they are at heart, curiously old-fashioned. They return larger
quantities of net-collected plankton samples to accumulate ever more
rapidly on shelves in laboratories, with no hope of more than a
cursory examination of most of them.

Attempts to take advantage of modern technology to advance our
understanding of small-scale distribution and movement of zooplankton
came around the beginning of the seventies, when the Food Chain Group
at Scripps Institution of Oceanography experimented with laser
holography in an imaginative effort to photograph a cubic meter of
sea water and reconstruct it three-dimensionally. Probably orders of
magnitude more funding would have been required to make the project a
success. Ortner *et al.* (1981) substituted film and a parallel light
source for plankton netting to develop a meter-scale plankton
recorder capable of obtaining hundreds of "samples". Other workers
experimented with a TV camera which produces a digital signal ready
to be electronically processed.

The most promising technology to date makes use of acoustics,
which as a means of detecting plankton distributions, goes back at
least to the forties, and the acoustic discovery of the "deep
scattering layer". The major step forward represented by the
prototype design of Pieper and Holliday (1984), is its multi-
frequency capability, micro-scale profiling and real-time data
analysis ability, providing the power to resolve organisms in several
size categories down to about 0.1 mm. It has the potential,
therefore, to co-exist with physical and chemical profiling
instrumentation. This, combined with real-time analysis capability,
is absolutely essential for the future.

The greatest need in zooplankton ecology has always been the
ability to map distributions *i.e.* take very rapid synoptic or at
least sequential samples, and perform analyses in real-time or at
least at a sampling rate that matches the sampling rate for other
environmental variables. Net systems that collect plankton samples
cannot be analyzed in real-time, and have little possibility of being
carefully analyzed even within a few hours. The ability to take very
high speed flash photographs of unpreserved samples on shipboard
might help eventually if pattern recognition techniques are
successfully developed. The processing of acoustic samples is a
realistic goal, particularly with the continued rapid advancement of
microcomputer technology.

Advances have been made in the last few years on TV-image based
image analysis and pattern recognition of preserved zooplankton
samples. Such an instrument is currently in use at the National
Marine Fisheries Laboratory in Rhode Island (Jeffries *et al.*, 1984).
It is a prototype, and is slow; the sample is mechanically scanned,
and only major categories (copepod, fish larva, chaetognath) can as
yet be resolved. The further development of such technology could
well be laser-based, ultimately for application *in situ*, the laser
beam "recognizing" the pattern of the particle passing through its
range, and identifying it.

The vast body of our information base comes from studies of
crustacean zooplankton. Their hard external skeleton preserves their

characteristic appearance through the rigors of the net collection
and formaldehyde fixation process. In contrast, much less is known
about the very delicate "gelatinous" zooplankton, because collection
and fixation can render many of them unrecognizable, and totally
destroy some, leaving no trace of their existence.

Coastal swarms of jellyfish and ctenophores have long been known,
but largely ignored. Only recently has the vast range of larvae and
exotic gelatinous zooplankton been documented in the open ocean
surface waters (e.g. Hamner et al., 1975), by direct observation using
SCUBA. The most recent observations, using research submersibles,
have shown such swarms to be characteristic of the deeper ocean layers
as well.

We have a better idea of adult population distributions of these
non-crustacean zooplankton in coastal waters, where gentle towing can
preserve of less delicate organisms to allow at least a biomass
estimate based on volume. We know virtually nothing, however, about
larval stages. Reeve and Baker (1975) documented the population
structure of chaetognaths and were able to collect organisms at near
their newly-hatched size. Collections of ctenophores were made which
recorded young animals as small as about 2 mm (although in much
reduced numbers). There are no field observations of ctenophores
less than 2 mm, despite the fact that newly-hatched animals (at 0.3
mm) go through about 6 biomass doublings to reach that size. In
terms of numbers of doublings and time to reach maturity, they have
already lived half of their lives by the time they reach 2 mm. In
fact, virtually nothing is recorded in terms of field observations
about the early life history of most zooplankton, except of course,
crustacean zooplankton and fish. This period of time in the very
early life history is the time when most prey/predator interactions
are taking place and the abundance of adult populations is largely
determined.

It would be hard to claim that zooplankton-distribution data
sets have increased in sophistication in any qualitative sense over
the last 25 years. Nevertheless, because of the magnitude of the
effort expended, some long-term data sets are adding to our
understanding of how much inter-annual as well as seasonal and
shorter scale variability exists. Foremost in this regard are the
continuation of the Hardy Plankton Recorder data sets, the CALCOFI
program off California, and the initiation of the National Marine
Fisheries Service MARMAP program to accumulate a monthly grid of
stations over the north-east shelf of the United States, extending
now for over 10 years.

It is in the laboratory, perhaps, where the most effort to
understand the ecology of zooplankton has been taking place over the
last 25 years. The long-term culture of marine zooplankton, thought
to be the keystone to significant advances, seemed to become a real
possibility when the small neritic copepod Acartia was first
maintained in continuous culture 20 years ago. Eventually, Calanus
was maintained throughout whole generations, as well as a small
neritic chaetognath, and then a coastal ctenophore. The effort of
maintaining such populations is enormous, and few researchers

routinely attempt to do it, even today. Nevertheless, expertise at
handling living organisms has been increasing dramatically, paving
the way for better and more sophisticated experimental techniques.

Central questions of zooplankton feeding ecology handed down
from before the sixties were those of food selection, efficiency of
conversion into biomass, and feeding rate, particularly as a function
of food concentration. These were the essential ingredients of early
deterministic models of how the ecosystem worked, particularly of
calculations of how much biomass was available to successive levels
of the food chain, and calculations of system efficiency. Another
vital question concerned how much the particular experimental
conditions of confinement and other non-natural influences, were
controlling the outcomes of experiments. Many chapters of review
could be written on this subject. There are no definitive
conclusions.

Besides the obvious limitations of confinement in small
containers, there are two major concerns. In the first place, it is
impossible to duplicate the natural environment in terms of densities
of the copepod grazer and its phytoplankton food, or the predator and
its prey. The time varying nature of each population is unknown, and
even if it were, it would be technically impractical to duplicate.
In the second place, a choice must be made between either (1) using
the broad range of food organisms available from the environment at
any particular time, and sacrificing the ability to exactly repeat
any experiment, and have any consistency of food quality, or (2)
alternatively, pre-selecting the food organism in some way for
consistency and repeatability.

These problems of experimental artifacts are overwhelming and
have never been satisfactorily solved in the laboratory. I have
documented growth efficiencies in ctenophores in the laboratory in
experiments three years apart varying by 30% or more. I do not know
if this is real variability between populations separated in time, or
an artifact of the experimental system.

We seek generalizations of behavior, for instance in the feeding
behavior of all copepod species. The solution to the question of
whether a copepod demonstrates threshold feeding at low food
concentrations, or satiation at high food concentrations, can have as
many different outcomes as there are life history stages, species of
copepods, kinds of food, previous nutritional status, and other
factors. A single numerically accurate relationship between feeding
rate and food concentration is not likely to be available from a few
laboratory experiments. Because of this it is problematical how such
data might be used in a mathematical model. This challenges the
predictive utility of models based upon parameter estimates with an
unreasonable amount of real variability in nature. The same can be
said for estimates of such ratios as growth efficiency, the subject
of much experimentation over the last 25 years. Determining growth
efficiency requires longer-term experiments, which multiply
uncertainties, and experimental and statistical error.

Laboratory experiments do provide useful background information for recruitment processes. Particularly document basic behavior and life-history traits, and those which establish ranges between which parameters can vary. Culture studies in the seventies, for instance, with food provided "in excess", demonstrated for the first time that zooplankton generation times could be a lot faster than previously thought. Smaller copepods could pass through generations in about a week, coastal ctenophores in two weeks and chaetognaths in less than three weeks. The potential growth rate of larval forms, including fish, could approach daily doubling of biomass.

Another important laboratory-derived parameter connected estimate with recruitment is potential egg production. The old idea of single egg broods for zooplankton was called into question by Marshall and Orr (1955). It has been subsequently shown for the few species (some copepods, a chaetognath, several ctenophores) that when animals become mature they can produce successive batches of eggs, often on a daily basis, which is largely controlled by food availability. We now know that while copepod and chaetognath egg production can be measured in the tens to a hundred or two, ctenophore egg production over a few days can be in the thousands to tens of thousands. Some benthic in vertebrates and fish release offspring into the water-column numbered in the millions. Animals amenable to laboratory maintenance provide valuable information on the biology of recruitment through careful observation. Copepods need only be fertilized once, for production of successive batches of eggs. Chaetognaths and ctenophores, which are hermaphrodites, are capable of self fertilization, and will produce viable eggs and healthy, fast-growing larvae.

The food selection question for copepods - at least in the context of "do they or don't they select food?" - has been solved by painstaking efforts of several researchers using high speed micro-cinematography. Copepods do indeed coordinate their feeding appendages at millisecond frequencies, and can select or reject many particles per second. Copepods do not normally ingest non-nutritive particles.

Important aspects of predator/prey behavior based on laboratory studies over the past few years provide insight into the complexity of early life-history interaction of marine zooplankton. Adult copepods can consume eggs and nauplii, though usually of other species. Copepods prey upon fish larvae. Copepods can influence a population outburst of ctenophores. Large copepods "destroy" newly-hatched ctenophores probably by damaging their tentacles, and causing them to starve (Stanlaw and Reeve, 1981). In this way copepods have been shown to effectively protect themselves from one of their major predators, ctenophores, in coastal waters.

To summarize the fore-going discussion, we know so little of the basic biology of most zooplankton, particularly their early life history stages, that much useful direct laboratory observational work remains to be done in this area. However, in quantitative experimental work, on the other hand, the law of diminishing returns has long been operating. This is because, as noted above, rates and

ratios (*e.g.* feeding rates and growth efficiency ratios) are entirely
dependent on the particular experimental regime, which never
resembles nature, particularly in the dynamics of change
characteristic of the environment.

Since nature itself is so difficult to observe on the
microscale, a third approach has been tried sporadically, over the
past two decades. The approach involves what have come to be known
as "mesocosms" following the volume of contributed papers on large
scale marine enclosures (Grice and Reeve, 1982).

Mesocosms are represented by a range of enclosures, defined only
as larger than can be accommodated on a laboratory bench, and smaller
than a natural sub-unit of the environment. Their most well-known
representatives are the land-based 15 m3 MERL towers, still in
operation, and the 1300 m3 plastic tubes floating at the sea surface
during the CEPEX experiments of the seventies. As Banse (in the
above-cited volume: p. 15) explained, in discussing the history of
such efforts, captured water columns were "an attempt to bridge the
gap between the rigorous, reductionist laboratory experiments and
traditional field investigation of real ecosystems, where an ever
increased and refined number of mere observations of successive
states may not reveal the processes at work".

Such mesocosms, while permitting some population manipulation,
such as nutrient fertilization, or removal of predators, cannot be
rigorously controlled. On the other hand, mesocosms can sustain
reproducing and interacting populations at several trophic levels
over periods of weeks, or even months. Though mesocosms can be
criticized for compromised control and reproducibility from the
laboratory experimentalist's viewpoint, or compromised naturalness
compared to the real environment, their strength lies in combining
some elements of both (see p. 389 et seq. in Grice and Reeve). In
one such "experiment" described by Reeve *et al*. (1982) which studies
two water columns for more than several weeks, we noted that the
suite of physical, chemical and biological data obtained in daily
sampling of the confined water column far surpassed any data set
accumulated previously over a 100-day time scale either in the
natural environment or laboratory. The most striking observation was
that although the two water columns were manipulated very differently
at the outset, biological events were qualitatively similar.
Populations would rise and fall at about the same time and the same
species become dominant in each column at the same time. The
implications of this to me are that large-scale external forces were
controlling population interactions in both columns, rather than
differences between the columns in terms of predator/prey
concentrations, growth efficiencies etc.

Modelling efforts over the past 25 years have been largely
disappointing. Deterministic models require the kind of biological
information which is hard to obtain experimentally for the reasons
belabored above. The kind of physical input needed to develop
realism has yet to be achieved. It is certain that if the level of

computing power required to physical models requires
"supercomputers", then models of biological systems will require at
least as much computing power.

4. THE FUTURE

How far have we advanced in the last 25 years in understanding
zooplankton population dynamics, and in particular the special case
of recruitment processes? Certainly both the intellectual effort and
financial resources have been several times greater in the last 25
years than in all the time before that. Just about every significant
study up to 1960 is included in Raymont's first edition on *Plankton
and Productivity in the Oceans*. We have certainly raised many
important questions which were hardly even issues in those days, and
settled some of those existing then, as unimportant. Some, which
seemed to be single questions capable of single answers, such as how
do copepods feed, or what is the reason for vertical migration, or
what is the magnitude of secondary production are now known to be
very complex, and do not have single answers. We still cannot
provide details of the zooplankton linkages which control adult stock
biomass variability.

Just about all zooplankton, including fish larvae, begin life
somewhere between 0.05 and 2 mm in size. Most predator/prey
interactions in the water column must take place in this size range
arena. Death of early life history stages is almost inevitable. The
only circumstances when it might fail to happen is when growth can
occur close to its theoretical maximum, and predation is minimal.
The controversy as to whether starvation or predation are the major
causes of death are not real questions. Predation and starvation
are, as many have pointed out, inseparable. What matters is to try
to define approximately the conditions at spawning which create those
rare windows of time for any species when food is maximum and
predation is minimum, to identify the physical forces which
predispose such conditions, and to develop techniques to predict the
occurrence of such windows.

I believe some of the factors involved will be the development of
new theory which interrelates biology and physics, the development of
new technology to provide much better observational tools, and the
integration of biological with physical models. I doubt that it will
involve primarily laboratory scale experimental techniques.

5. REFERENCES

Fuller, J.L.: 1937, 'Feeding rate of *Calanus finmarchicus* in
   relation to environmental conditions', *Biol. Bull.* 72, 233-246.
Grice, G.D. and M.R. Reeve: 1982, *Marine Mesocosms*, Springer-Verlag.
Hamner, W., L.P. Madin, A.L. Alldredge, R.W. Gilmer and P.P. Hamner:
   1975, 'Underwater observations of gelatinous zooplankton:
   Sampling problems, feeding biology, and behavior', *Limnol.
   Oceanogr.* 20, 907-917.
Hardy, A.C.: 1956, *The Open Sea, Its Natural History: The World of
   Plankton*, Collins.
Hardy, A.C. and R. Bainbridge: 1954, 'Experimental observations on
   vertical migrations of plankton animals', *J. Mar. Biol. Assoc.
   U.K.* 33, 409-448.
Jeffries, H.P., M.S. Berman, A.D. Poularikas, C. Katsinis, I. Melas,
   K. Sherman, and L. Bivins: 1984, 'Automated sizing, counting and
   identification of zooplankton by pattern recognition', *Mar. Biol.*
   78, 329-334.
Lebour, M.V.: 1923, 'Food of plankton organisms II', *J. Mar. Biol.
   Assoc. U.K.* 13, 70-92.
Marshall, S.M. and A.P. Orr: 1955, *The Biology of a Marine Copepod
   Calanus finmarchicus (Gunnerus)*, Oliver and Boyd.
McGowan, J.A. and D.W. Brown: 1964, 'A new opening-closing paired
   zooplankton net', *Scripps Inst. Oceanogr. Ref.* 66-23, 1-56.
Ortner, P.B., L.C. Hill, and H.E. Edgerton: 1981, '*In-situ* silhouette
   photography of Gulf Stream zooplankton', *Deep Sea Res.* 28, 1569.
Pieper, R.E. and D.V. Holliday: 1984, 'Acoustic measurements of
   zooplankton distributions in the sea', *J. Cons.* 41, 226-238.
Raymont, J.E.G.: 1947, 'An experiment in marine fish cultivation IV',
   Proc. Roy. Soc. Edinb. B63, 34-55.
Raymont, J.E.G.: 1963, *Plankton and Productivity in the Oceans*,
   Pergamon.
Reeve, M.R. and L.D. Baker: 1985, 'Production of two planktonic
   carnivores (chaetognath and ctenophore) in south Florida inshore
   waters', *Fish. Bull.* 73, 238-248.
Reeve, M.R., G.D. Grice, and R.P. Harris: 1982, 'The CEPEX approach
   and its implications for future studies in plankton ecology', in
   Grice, G.D. and M.R. Reeve, eds., *Marine Mesocosms*, Springer-
   Verlag.
Riley, G.A.: 1947, 'A theoretical analysis of the zooplankton
   population of Georges Bank', *J. Mar. Res.* 6, 104-113.

512

Stanlaw, K.A., M.R. Reeve, and M.A. Walter: 1981, 'Growth and
    vulnerability to damage of the ctenophore Mnemiopsis mccradyi in
    its early life history stages', *Limnol. Oceanogr.* **26**, 224-234.
Steele, J.H. (ed.): 1978, *Spatial Patterns in Plankton Communities*,
    Plenum.

SCALE SELECTION FOR BIODYNAMIC THEORIES

John H. Steele
Woods Hole Oceanographic Institution
Woods Hole, MA 02543
U.S.A.

ABSTRACT. Although the primitive (Navier-Stokes) equations for ocean physics are known, any useable model simplifies them, normally by eliminating terms and selecting a specific range of space and time scales. Processes outside these scale ranges are parameterized (or ignored). For a joint physical-biological theory, an essential assumption is that the same scale divisions can be made and the problem of parameterizing the neglected biological scales can be solved. Similarly the kinds of data must correspond. Possible subdivisions of the space and time scales will be discussed.

1. INTRODUCTION

The aim of this meeting is to consider the relation between biological and physical dynamics in the ocean. Are there any general principles? Or do the processes of evolution lead to the adaptation of individual species to the particular features of a local or regional environment--an environment that includes the specializations of other organisms?

Thus the null hypothesis is that the observed patterns of behavior, in a broad sense, are unique at the individual, population or community levels, providing natural histories without any global generalizations. I shall suggest that there are necessary patterns but these are associated with the scale of events rather than with the location, and that these correspond to similarities in physical patterns.

There is good evidence at the molecular and cellular level for general processes which have survived evolutionary diversification. This is the basis for a biochemical view of marine and terrestrial organisms. It can allow us to pursue a global view of chemical fluxes in the ocean mediated by a combination of physical and biological processes. For logistic and conceptual simplicity such programs must focus on the larger space and time scales and make largely, empirical assumptions about biological factors such as fecal production in mid-water. Implicitly these require generalizations about energy transformations in food webs. At the other extreme there is a long

B. J. Rothschild (ed.), Toward a Theory on Biological-Physical Interactions in the World Ocean, 513–526.
© 1988 by Kluwer Academic Publishers.

tradition using simple deterministic and apparently very general models to explore the internal dynamics of prey-predator and competitive relations--the Lotka-Volterra view of the world. This tradition has had limited success as a method for broad generalization. The best uses of this approach are in specific applications such as epidemiology (Anderson and May, 1982).

For the larger and more complex interactions that we call ecosystems, we have seen the failure to construct effective large deterministic numerical models simulating each component as explicitly as possible. The problem is that the response of the model may depend on numerical intricacies rather than on the biological assumptions. This approach has given way to more elementary structures where the elements in a system have simple links and where descriptions and tests of the system are in terms of numbers of species at each trophic level and the numbers of links between levels (Newman and Cohen, 1986).

Most of these theories and their applications derive from terrestrial ecology and the emphasis is on the internal dynamics of the ecological system. I have argued elsewhere (Steele, 1985) that this is a basic assumption, often implicit, in terrestrial studies. At time scales of years to decades we assume that the variability we see in populations or communities is a result of interactions between components of the system rather than response to external forces. The underlying assumption is that terrestrial organisms, through their evolution, have adapted as far as possible to eliminate the effects of random and unpredictable variations in their physical environment, although sporadic disturbance may play some part in maintaining species diversity (Connell, 1978).

For marine communities in the open sea, the argument is usually in terms of variability being dependent on changes in the physical environment. There is, necessarily, internal community or population structure which prevents the system from taking random walks to extinction, but our search for causes of change looks to patterns in ocean dynamics. The questions then concern the appropriate physical events and the nature of the interactions. Again these questions relate to the ability to generalize. The assumption here will be that such generalizations relate to the inherent space and time scales of ocean dynamics. The problem is to select critical and relevant scales for particular ecological phenomena.

2. PHYSICAL AND BIOLOGICAL SCALES

In relation to general ecology, the best known marine work is fish population dynamics represented by the early studies of Beverton and Holt (1957) and Schaefer (1954). Their approach was similar in principle to terrestrial studies of age structured or logistic populations. Spatial scales arise only indirectly through the concept of a discrete reproducing stock. In Beverton and Holt, particularly, the use of "yield per recruit" as the basic measure removed the immediate effect of temporal variability on reproductive success. The extension of community structure was made by introducing the ideas of energy flow (*e.g.*, Steele, 1965). Again these derived from terrestrial work (Lindemann, 1942), focussed on internal properties of the system and were unsuitable for considering temporal or spatial variability. More recently the concept of "large marine ecosystems" (Sherman and Alexander, 1986) defines the unit of study as being of the order of 1000km in spatial scale. This scale is appropriate to many fish populations and so the implicit assumption is that of a "top-down" approach to factors determining variability within the ecosystem. By implication large-scale changes in circulation would be critical for major variations in ecosystem structure (*e.g.*, Corten, 1986).

At the other extreme studies of phytoplankton dynamics have attempted, with some degree of success, to include vertical structure on the scale of meters and varying with time on scales of hours or days (Marra, 1980). Such studies have demonstrated that even in apparently uniform environments such as oligotrophic subtropical gyres the cell distribution is not uniform or smooth and the cells are not near biochemical equilibrium even on a diurnal cycle (Goldman, 1984). Goldman (this volume) has proposed that the cells may be aggregated in micro-patches (flakes) with significantly lower viscosity that allows predation and nutrient cycling within the flakes. At a larger vertical scale Klein and Coste (1984) propose that nutrient input from below the euphotic zone is episodic and of the order of days or weeks. Also mixing events may be spatially discrete depending on the interaction of small scale turbulence patches with internal waves. Jenkins and Goldman (1985) have suggested that the possible resolution of the different estimates of primary production may depend on the different scales at which phenomena are observed.

Between these scales for fish and phytoplankton dynamics, these are even more difficult questions about appropriate scales for intermediate links in the food web such as herbivorous copepods. The "Stommel diagram" of Haury *et al* (1978) illustrated the wide range of relevant scales for zooplankton--from meters and hours for plankton patches to ocean basin and evolutionary time scales. In a specific study Mackas (1984) has shown that coherence scales for zooplankton biomass are of the order of a few kilometers whereas community structure has coherence at the order of 50km. He associated the former with daily feeding aggregation and the latter with the longer (30-50 day) reproductive cycles.

Thus for any species there are at least four relevant space/time scales: (I) for the individual organism acquiring particular food items or avoiding a predator; (II) for a group (a patch or shoal) searching for food concentrations and moving vertically or horizontally as an aggregation; (III) for a breeding stock or population; and (IV) for the overall species distribution. There is obviously a vast range of scales involved in all these activities so that it is impossible to integrate all four patterns of behavior into a single conceptual framework. Can the implicit separation into these four aspects represent a hierarchy in the sense that the dynamics within each level are more closely linked than the connections between them?

Much of the past and present research could possibly be defined as existing within one or other of these four categories. Undoubtedly we need more such studies. Yet there is now great emphasis on work which potentially crosses these boundaries. We wish to understand, and even to forecast, events in one category based on information from others. As a particular example there is a special interest in population variability between generations (III). It is possible to look for causes in each of the other categories. For pelagic fish stocks, explanations can be given in terms of: (a) larval survival as a function of individual food search (Lasker, 1975); category I; (b) interactions of fishing methods with behavior of shoal size and location (Ulltang, 1980); category II; and (c) large-scale transport of larvae/juveniles as a function of climatically changing physical current systems (Corten, 1986); category IV.

In each example there is some mix of biological interactions and physical mechanisms. Because of the present importance attached to an understanding of population variability, the emphasis in this paper will be on this factor, especially interactions between scales.

For the study of physical factors in the ocean, it is usual to separate processes by selecting a range of scales such as (1) microstructure, (2) internal waves, (3) eddies and (4) basin (gyre) circulation. The typical problems involve the methods of parameterizing finer structures than those under study, and how to embed the model in a larger system with empirically defined boundary conditions. Much recent work concerns interactions between these scales, e.g., turbulence and internal waves (Toole and Schmitt, 1987), or eddy and gyre kinetic energy (Schmitz and Holland, 1982).

Do these physical scale divisions correspond in any way to the separation of ecological processes described earlier? Can such comparisons be used to define components for study? Even more important, can the interactions between physical scales be used to determine links between the ecological categories. For example it appears that the episodic turbulence--internal wave relation is relevant to the primary production--nutrient uptake questions discussed earlier. Also, the relation of eddy scales to larger motions is certainly significant in redistribution of species (Wiebe *et al*, 1985) and may affect recruitment. Thus the issue is how generally can scale properties be linked to ecological questions.

3. SPATIAL STRUCTURE

Satellite images have revolutionized our perception of near-surface conditions in the ocean. At all realizable scales we see complex patterns in surface temperature, color, and topography. Thus there is no preferred scale at which the ocean can be divided into provinces, nor is there only one way to represent the patterns. Numerical models of an idealized ocean basin can include the random turbulent structure of the larger eddies (Holland, 1978). Or, the system can be represented as a finite number of relatively uniform regions (sub-Arctic, sub-tropical, etc.) separated by relatively sharp fronts (Roden, 1975). In an ecological context, the usefulness of alternative pictures of the ocean depends on the questions being asked. Thus the former turbulent approach is valuable in considering the fine or mesoscale relation of physics to phytoplankton productivity; the latter is relevant to the depiction of zoogeographical regions or to large scale fish migration routes.

A detailed example concerns the shelf fronts generated by the relation of depth to tidal energy (h/u^3) as defined by Simpson (1981). Recent studies of a front off Scotland by Richardson *et al* (1986) show that phytoplankton and herring larvae are concentrated at such a front, Figure 1. The picture for zooplankton as food for the larvae is less clear (Kiørboe and Jøhansen, 1986). The authors state that "it may be hypothesized that herring larvae, or rather the spawning adults, react to long-term average conditions (at fronts) when selecting spawning grounds." This type of hypothesis has been used by Iles and Sinclair (1982) as a basis for depicting the discrete herring stocks and their average abundance in regions around the North Atlantic. Thus the division of the shelf regions into stratified and unstratified domains, ignoring other factors, may provide a valuable simplification of the complexities of coastal dynamics, for the purpose of explaining average stock sizes of certain fish species. This does not explain, as the authors recognize, the variability in these stocks. Does the variance depend on smaller scale or episodic events determining, say, the location and density of larval food items; or does the variance depend on larger scale and longer-term trends in circulation beyond the spawning grounds (Corten, 1986)? These suggestions would imply that one selected set of scales of physical process may define the average state but quite other scales are relevant for the variability.

4. TEMPORAL VARIANCE

There is a comparable situation when we examine changes with time. It is obvious that organisms adapt their feeding, breeding or migration cycles to physical factors with regular periodicities. Apart from the diurnal, lunar, and seasonal cycles, there are also quasi-cycles associated with geostrophic dynamics and a rough periodicity of about seven days associated with weather events.

These features can be seen in the analysis of variance as a function of frequency. Such spectral analysis, however, also shows that there is a general increase in variance per unit frequency as one goes to longer-term periods (Wunsch, 1981). Thus the longer-time scales accumulate variance comparable to that associated with the regular higher-frequency cycles. Further, the longer period variance is normally associated with larger spatial scales. An example is the "southern oscillation" at 5-10 year scales with effects at least over ocean basins. But annual breeding patterns often occur over short periods and appear to be triggered by some specific feature of the

518

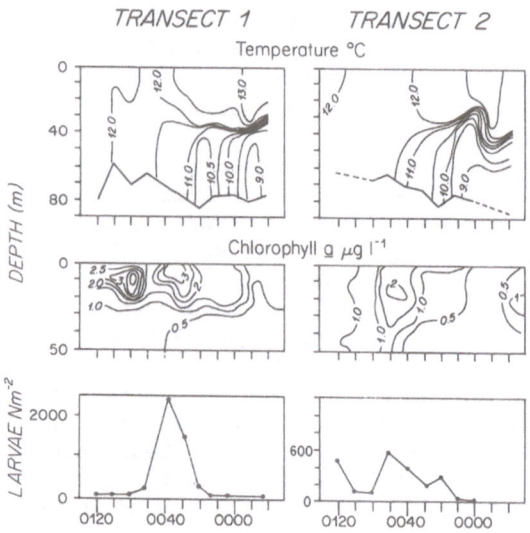

TRANSECT 1   TRANSECT 2

Figure 1. Data collected on two transects off the north-east of
Scotland on 17-19 September (left) and 27-29 September (right), 1984:
(a) temperature profiles, (b) chlorophyll concentration, (c) larval
herring concentrations in plankton net hauls (see Richardson *et al*,
1986).

annual cycle. Then year-to-year differences in recruitment might be
associated with higher frequency variance, such as weather. On the
other hand major population changes would appear to require both the
larger and longer period variability as well as the larger spatial
scales associated with these events. The implication of the general
red spectrum in the ocean is that populations can absorb the higher
frequency noise but will respond to the larger variance by changes in
population or community structure (Steele, 1985).
      Again, the general assumption is that the ecological systems are
closely linked to the scales of the physical processes but variations
are determined by events at greater or smaller scales. Further,
because of the peaks on this red spectrum, we can expect these factors
to occur at discrete frequencies associated with other components of
the physical dynamics. As an example, Figure 2, there is a two-
dimensional peak in the frequency/wave number presentation related to
the eddy field at scales of about 50 days/50km. It is an interesting
but speculative question whether certain zooplankton species that
breed several times per year may be associated with this peak. Such
spectral representations are a useful way of describing the broad
range of processes which need to be considered--processes covering
many temporal scales. However, they are descriptions of the phenomena
and do not explain the underlying causal factors. If we wish to

consider the basic physical mechanisms and their interactions with the
biology, then the scale relations are much more restrictive.

5. PHYSICAL-ECOLOGICAL MODELS

It has been pointed out that studies of ocean physics must select
certain scale ranges both for conceptual reasons and also for
numerical modelling. As a rough guide it is usual for such studies to
be limited to about two orders of magnitude in both horizontal spatial
and in temporal scales such as uniform areas separated by fronts and
with fast and slow variables. Also, there is an association between
the space and time scales since these tend to increase together. Thus
the *range* of space/time scales can be considered as one parameter or
"dimension" in selecting or designing a theoretical model.
    When ecological components are added there are two extra
dimensions to be considered. There is the grouping of organisms
within a species in terms of the individual, the patch, the stock and
finally the species distribution. Nearly all considerations of
population dynamics now recognize the need to consider variability at
one level of this hierarchy as important to the response at other
levels. Yet, as Haury *et al* (1978) point out, the range of

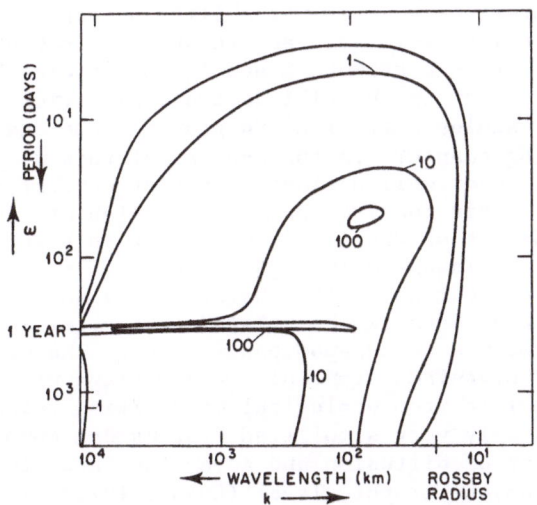

Figure 2.   Sketch of the frequency-wave number spectrum of the general
circulation at mid-latitudes with arbitrary contour units (from
Satellite Measurements of the Ocean).

spatial/temporal scales can cover four or more orders of magnitude.
The definition of a population process is explicitly a matter of
selection of scales.

Similarly, there is a trophic hierarchy going through nutrient requirements, prey and predators to community dynamics. Attempts to include all trophic levels become too unmanageable conceptually and numerically. Yet, explanations of ecological dynamics usually require at least some definition of three levels--prey and predators as well as the population itself.

These different components can be represented as three dimensions, Figure 3. For any model, increasing complexity in structure along one dimension will result in a decreased capability along the others. In this sense we can regard the limitations on our ability to model a system, conceptually or numerically, as corresponding to a limitation on the volume we can consider within these three dimensions. Increasing the ecological complexity will result in an inability to handle space/time variations. Oppositely, a desire to examine such scale variance will require a greater simplified statement of the ecological relations.

As idealized examples, Figure 4, consider (a) the age-structured models of fish populations which define the equilibrium state subject to parameters defining recruitment and fishing stress but largely ignoring the spatial and temporal scales of variation. Then (b) there are programs such as the Global Ocean Flux Study (GOFS) which must eliminate most of the ecological detail in order to represent the space and time changes in fluxes of organic matter, from short-term production cycles to decadal trends.

The focus of interest here is in variability of populations and communities. But to include relevant factors appears to require some representation both of trophic structure and of the species hierarchy. Can this be done in a manner which still covers a significant portion of the spatio-temporal scales sufficient to provide some description of the processes causing change? In the context of this diagram, can we select volumes that cover well-defined, distinct ecological relations. Essentially, this requires parameterization of the processes at the boundaries of the intermediate box in Figure 4.

Can we construct models which incorporate some of each dimension? In a few cases this does seem possible. Davis (1984) has shown how the circulation on Georges Bank can be related to spatial and temporal changes in the age-structure of copepod populations. The modelling of temporal dispersion of organisms from Gulf Stream rings was valuable in separating physical loss from ecological mortalities (Wiebe *et al*, 1985). Often, however, only very simplified systems can compare in broad terms the relation of diffusion and advection rates to populations at more than one trophic level (Evans, 1978; Klein and Steele, 1985).

6. SELECTION OF SCALES

We assume that the relations between physical and ecological processes can be described in terms of scale effects. The general ecological relations can be described as lying along a diagonal in the space-time domain. Further this diagonal corresponds to the locus of the

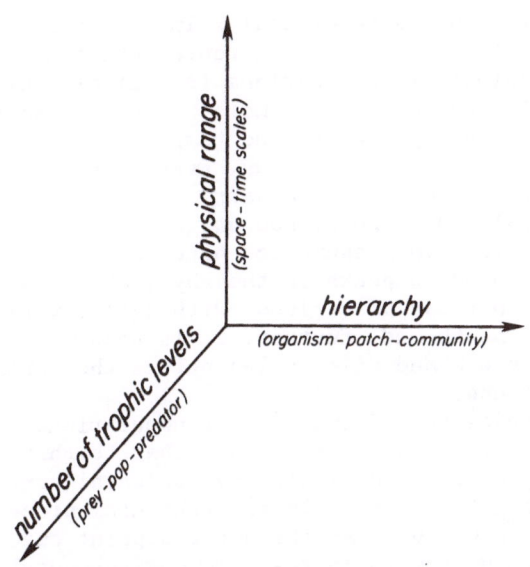

Figure 3. Three "dimensions" which are required to express the main features of an ecological system (see text).

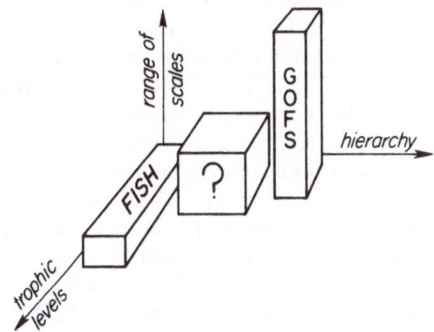

Figure 4. Idealized representation of a steady state fishery model and a model of biochemical fluxes covering a range of scales (see text).

physical processes. For trophic interactions I produced (Steele, 1980) a simple example, Figure 5, to emphasize the scale relations between populations of phytoplankton (P), zooplankton (Z) and pelagic fish (F). This, however, does not include those special features responsible for the spatial and temporal variability outside the regular cycles. Examples are frontal processes with small cross-axis dimension and large persistence (X) which affect spatial variability, and weather events which can have short episodic effects at large

spatial scales (Y). From the range of scales in this diagram it is
apparent that a full analytic or numerical representation of the
average physical and ecological interactions is unlikely, even for
this simple three-level food chain. Even more so, any attempt to
include major processes causing spatial and temporal variability takes
us beyond the limits of tractability. Thus, even in this relatively
simple conceptual model, it would be necessary to parameterize
conditions at the spatial and temporal boundaries. For each component
population or trophic level, the assumption here is that the
space/time scales correspond to peaks in the physical variance. Also,
because of the general increase of variance with scale, variability at
smaller scales appears smooth and, further, it is necessary to go to
much larger and longer space and time scales before the variance will
affect population abundance.

In a review of changes in cod and haddock populations in the
north west Atlantic, Koslow *et al* (1987) state that "either biological
processes are smoothing some of the local environmental variability or
recruitment is responding to large-scale environmental processes." I
would suggest that both are true. As the authors point out, the
comparison of population data with large-scale environmental
parameters is "unlikely to resolve the basis of recruitment
variability in the northwest Atlantic." We may consider that the main
results of trophic interactions occur along the diagonal but this
assumption ignores the effects of spatial and temporal processes
occurring off the axis. As we now observe from more dense data sets,
the oceans are variable at all space and time scales. There is
increasing evidence that populations have adapted or evolved to take
advantage of these processes. The simplified idea of a diagonal
representation between trophic levels can explain some of the steady
state features but not their resilience to most variability, nor their
occasional large and often abrupt changes. Mechanisms for adaptation
and response should be sought off the diagonal. Thus persistent
features such as fronts or eddies, combined with behavioral patterns
can alter the way in which we represent, say, feeding intensity as a
function of overall population density. In the sea such spatial
patchiness rather than purely behavioral mechanisms (Holling, 1965)
can determine what type of response curve is appropriate--linear,
hyperbolic, s-shaped or threshold (Steele and Henderson, 1984). Thus
to parameterize the prey-predator response on the diagonal and its
effect at large scales, requires detailed knowledge of the adaptations
in the spatial domain.

Similarly the short period, unpredictable or episodic events
generated by internal ocean dynamics or by weather events, will affect
the population. It may be appropriate to parameterize this by some
stochastic input to a model but the exact character of this input is
critical. It can be shown in simple models (Steele, 1985) that such
inputs with different power spectra produce very different patterns,
in the ways in which populations absorb the high-frequency variability
and respond by changes at longer periods.

The general implications from field observation and theoretical
models is that we should not expect to see simple statistical

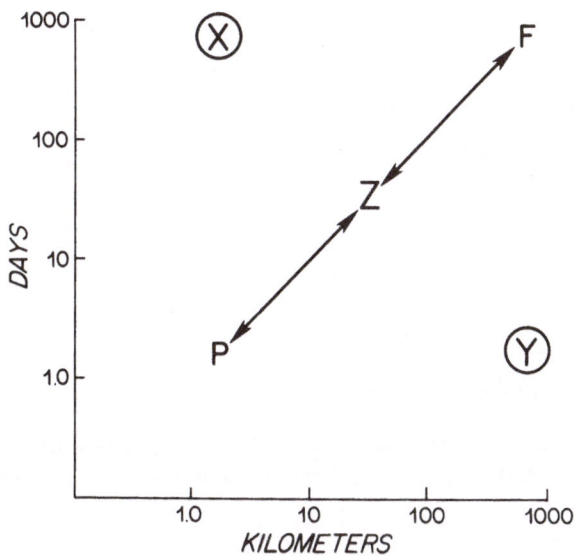

Figure 5. A simple set of scale relations for the food web P
(phytoplankton), Z (zooplankton) and F (pelagic fish). Two physical
processes are indicated by X, predictable fronts with small cross-
front dimensions and (Y) unpredictable weather events occurring on
relatively large scales.

relations between the biology and physics at particular selected
scales. Rather the "output" at longer and larger scales is a complex
(and highly non-linear) response to shorter period and smaller scale
features. I am suggesting that, although these interactions are
complex, it may be possible to consider general rules for simplifying
and parameterizing these relations.

     As a summary of these speculations and a general hypothesis,
Figure 6 indicates how, for any stock, population, or trophic group
(X), the spatial and temporal "environment" may be systematized. This
environment appears *smooth* at smaller scales in both dimensions. The
physical fine structure at longer time scales provides features to
which populations *adapt* through patchiness, migration or choice of
breeding areas. The episodic features at smaller time scales are
absorbed. However, because of the non-linear nature of these
interactions they can contribute significantly to, and possibly
determine, the longer term and larger scale trends in population
abundance so that the populations may track environmental change but
not in any simple linear fashion.

     In particular, research programs to understand links between
physical and biological events in the upper right quadrant of Figure 6
will not be successful if they address solely events or data at these
scales.

524

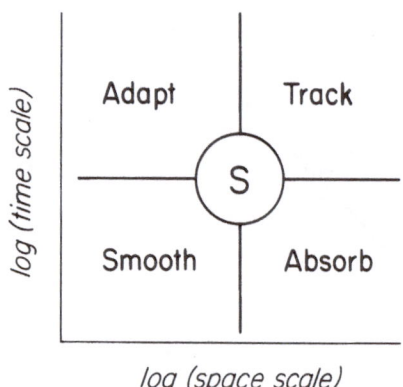

log (space scale)

Figure 6. A summary of the discussion of the responses of a
population, S, to variability at different space and time scales (see
text).

7. CONCLUSIONS

What generalizations, or speculations, are possible from this review
of variability as a function of space and time scales?

(1) It is inadequate for these purposes to have models or concepts
     which consider only the dominant cycle in time and take population
     density as an average over one particular spatial scale.
(2) It is too simple to consider variability as a random noise that
     can be added linearly to such models.
(3) On the other hand, the full range of scale effects cannot be
     incorporated because of the physical complexities as well as the
     ecological interactions.
(4) We may assume that there are close correspondences between
     regular or quasi-cyclical features of ocean dynamics and the
     space/time scales of specific populations.
(5) The study of population variability requires the selection of an
     appropriate region of the space/time domain.
(6) Within this domain it will still be necessary to parameterize the
     responses to physical processes and some of the ecological
     interactions, particularly those at smaller scales.
(7) It appears possible that these simplifying processes may be
     similar for different populations or communities.
(8) Thus generalizations between different studies may be in terms of
     the type of scale relations illustrated conceptually in Figure 6.

8. REFERENCES

Anderson, R. M. and R. M. May: 1982, 'Directly transmitted infectious
    diseases: control by vaccination,' *Science* **215**, 1053-60.

Beverton, R. J. H. and S. J. Holt: 1957, 'On the dynamics of exploited fish populations,' *Fish. invest. Lond. Ser.* 2, **19**, 533.

Connell, J. H.: 1978, 'Diversity in tropical rain forests and coral reefs,' *Science* **199**, 1302-1310.

Corten, A.: 1986, 'On the causes of the recruitment failure of herring in the central and northern North Sea in the years 1972-1978,' *J. Cons. int. Explor. Mer* **42**, 281-294.

Davis, C. S.: 1984, 'Interaction of a copepod population with the mean circulation on Georges Bank,' *J. Mar. Res.* **42(3)**, 573-590.

Evans, G. T.: 1978, 'Biological Effects of Vertical-Horizontal Interactions,' in J.H. Steele (ed.), *Spatial Pattern in Plankton Communities*, pp. 157-179.

Goldman, J.: 1984, 'Oceanic Nutrient Cycles,' in M.J.R. Fasham (ed.), *Flows of Energy and Materials in Marine Ecosystems: Theory and Practice*, Plenum Press, NY, pp. 137-170.

Haury, L. R., J. A. McGowan and P. H. Wiebe: 1978, 'Patterns and processes in the time-space scales of plankton distributions,' in J.H. Steele (ed.), *Spatial Pattern in Plankton Communities*, pp. 277-327.

Holland, W. R.: 1978, 'The role of mesoscale eddies in the general circulation of the ocean--Numerical experiments using a wind-driven quasi-geostrophic model,' *J. Phys. Oceanogr.* **8**, 363-392.

Holling, C. S.: 1965, 'The functional response of predators to prey density and its role in mimicry and population regulation,' *Mem. Entom. Soc. Can.* **45**, 1-60.

Iles, T. D. and M. Sinclair: 1982, 'Atlantic Herring: Stock Discreteness and Abundance,' *Science* **215**, 627-633.

Jenkins, W. J. and J. C. Goldman: 1985, 'Seasonal oxygen cycling and primary production in the Sargasso Sea,' *J. Mar. Res.* **43**, 465-491.

Kiørboe, T. and K. Jøhansen: 1986, 'Studies of a larval herring (*Clupea harengus* L.) patch in the Buchan area. IV. Zooplankton distribution and productivity in relation to hydrographic features,' *Dana* **6**, 37-51.

Klein, P. and B. Coste: 1984, 'Effects of wind-stress variability on nutrient transport into the mixed layer,' *Deep-Sea Res.* **31**, 21-37.

Klein, P. and J. H. Steele: 1985, 'Some physical factors affecting ecosystems,' *J. Mar. Res.* **43**, 337-350.

Koslow, J. A. and K. R. Thompson: 1987, 'Recruitment to Northwest Atlantic Cod (*Gadus morhua*) and Haddock (*Melanogrammus aeglefinus*) Stocks: Influence of Stock Size and Climate,' *Can. J. Fish. Aquat. Sci.* **44**, 26-39.

Lasker, R.: 1975, 'Field criteria for survival of anchovy larvae: The relation between inshore chlorophyll maximum layers and successful first feeding,' *Fish. Bull.* **73(3)**, 543-462.

Lindemann, R. L.: 1942, 'The trophic dynamic aspect of Ecology,' *Ecology* **23**, 399-418.

Mackas, D. L.: 1984, 'Spatial autocorrelation of plankton community composition in a continental shelf ecosystem,' *Limnol. Oceanogr.* **29(3)**, 451-471.

Marra, J.: 1980, 'Time course of light intensity adaptation in a marine diatom,' *Marine Biology Letters* **1**, 175-183.

526

Newman, C. M. and J. E. Cohen: 1986, 'A stochastic theory of community food webs IV. Theory of food chain lengths in large webs,' *Proc. R. Soc. Lond. B*, **228**, 355-377.

Richardson, K., M. R. Heath, and N. J. Pihl: 1986, 'Studies of a larval herring (*Clupea harengus* L.) patch in the Buchan area. I. The distribution of larvae in relation to hydrographic features,' *Dana*, A Journal of Fisheries and Marine Research, **6**, 1-10.

Roden, G. I.: 1975, 'On North Pacific temperature salinity, sound velocity, and density fronts and their relation to the wind and energy flux fields,' *J. Phys. Oceanogr.* 5, 557-571.

Satellite Altimetric Measurements of the Ocean, 'Report of the TOPEX Science Working Group,' March 1, 1981, Jet Propulsion Laboratory, California Institute of Technology Pasadena, CA, 1-78.

Schaefer, M. B.: 1954, 'Some aspects of the dynamics of populations important to the management of the commercial fish populations,' *Inter. Amer. Trop. Tuna. Commn Bull.* 1(2), 27-56.

Schmitz, W. J. and W. R. Holland: 1982, 'Numerical eddy-resolving general circulation experiments: Preliminary comparison with observation," *J. Mar. Res.*, (to be published).

Sherman, K. and L. M. Alexander (eds.): 1986, *Variability and Management of Large Marine Ecosystems*, AAAS Selected Symposium, May 24-29, 1984, New York City, NY, Westview Press, Inc., Boulder, CO., 319pp.

Simpson, J. H.: 1981, 'Shelf sea fronts, implications of their existence and behavior,' *Philos. Trans. R. Soc. London, Ser. A.* **302**, 531-546.

Steele, J. H.: 1965, 'Some problems in the study of marine resources,' *Spec. Publ. Int. Comm. Nthw. Atlant. Fish.* 6, 463-476.

Steele, J. H.: 1978, 'Some Comments on Plankton Patches,' in J.H. Steele (ed.), *Spatial Pattern In Plankton Communities*, pp. 1-20.

Steele, J. H.: 1980, 'Some varieties of biological oceanography,' in B.A. Warren and C. Wunsch (eds.), *Evolution of Physical Oceanography*, MIT Press, Cambridge, MA, pp.376-383.

Steele, J. H.: 1985, 'A comparison of terrestrial and marine ecological systems,' *Nature* **313(6001)**, 355-358.

Steele, J. H. and E. W. Henderson: 1984, 'Modeling Long-Term Fluctuations in Fish Stocks,' *Science* **224**, 985-987.

Toole, J. and R. Schmitt: 1987, 'Small scale structures in the north-west Atlantic subtropical front,' *Nature* 327, 47-49.

Ulltang, O.: 1980, 'Factors affecting rhe reaction of pelagic fish stocks to exploitation and requiring a new approach to assessment and management,' *Rapp. P.-v. Reun. Cons. Int. Explor. Mer* 177, 489-504.

Wiebe, P. H., G. R. Flierl, C. S. Davis, V. Barber, and S. H. Boyd: 1985, 'Macrozooplankton Biomass in Gulf Stream Warm-Core Rings: Spatial Distribution and Temporal Changes,' *Jour. Geophys. Res.* **90(C5)**, 8885-8901.

Wunsch, C.: 1981, 'Low-frequency variability of the sea.' in B.A. Warren and C. Wunsch (eds.), *Evolution of Physical Oceanography*, MIT Press, Cambridge, MA, pp. 342-374.

BIODYNAMICS OF THE SEA: THE ECOLOGY OF HIGH DIMENSIONALITY SYSTEMS

B. J. Rothschild
University of Maryland
Center for Environmental and Estuarine Studies
Solomons, Maryland 20688
U.S.A.

ABSTRACT. Understanding variability in the biological production of the sea is complicated by the very-high dimensionality of the biodynamic system. Addressing highly-dimensional systems requires attention to theory, identification of the "right" dimensions, and concentration on the shortest possible causal chains. Recentering investigations to meet these aims might be accomplished by considering the statistics of individual trophic transactions in a signal-theoretic context.

1. INTRODUCTION

Explanation of the biodynamics of the sea requires transcending disciplinary boundaries; focusing on variability rather than on average relationships; and understanding components of variability which are are generally suppressed by highly aggregated ecosystem models yet representative of important sources of variation.

Two interlocked problems make the task formidable. The first is that the biodynamic system's high dimensionality makes it difficult to formulate a explanatory theory regarding its operation. The second is that the absence of a theory consonant with the complex nature of the system makes it difficult to address the complexity of the high-dimensionality system.

The nature of the first problem is that the number of important dimensions at any instant of time is far greater than the number of dimensions that can be considered in a single analysis. Analogously any theory can only deal with a small subset of the total possible number of dimensions. Thus, the criteria for choosing the particular subset of dimensions is of considerable concern because if the criteria are somehow inadequate, the theory will not be useful for explaining observed phenomena. The nature of the second problem, the converse of the first, relates to the problem of defining criteria in the absence of an organizing theme or theory. Clearly, an adequate theory will facilitate further theoretical development and empirical verification while an inadequate theory will not serve either

B. J. Rothschild (ed.), Toward a Theory on Biological-Physical Interactions in the World Ocean, 527–548.

purpose. A lack of an adequate theory makes the analysis of a
complex system virtually impossible.

The interlocked problems suggest that a) a full set of desirable
measurements will always be difficult or impractical to obtain; b)
criteria for selection of a subset of dimensions for analysis,
particularly taking account the fact that the important dimensions
change with time requires special concentration; and c) correlational
and empirical schemes, not consonant with a high-dimensional setting
are likely to be inefficient, at best.

In order to address these difficulties it is important to
control dimensionality. Dimensionality control has two aspects.
The first relates to identifying the proper dimensions while the
second relates to minimizing the number of dimensions that need to be
considered. For example, an analysis of variance table listing all
important sources of variation contributing to population fluctuation
must include population-dynamic, density-dependent, population
responses interacting with temperature, motion, and light. Yet these
dimensions are hardly ever included in most analyses of biodynamic
variability, showing that dimensionality can be reduced, or at least
controlled, by focusing on those dimensions known to be associated
with major sources of biodynamic variation.

With regard to causal-chain length, many analyses of biodynamic
variability consider causal chains that are relatively long and skip
over events intervening between cause and effect. For instance, the
cube of wind velocity is frequently correlated with biological
phenomena in the evident hope that a simple relation between wind-
velocity cubed and a complex biological event such as year-class
strength in a fish stock exists. The causal chain in such situations
is quite long, the intervening events are difficult to describe and
as a result large unaccounted sources of variation "leak" into the
data set making evaluation of specific causal phenomena difficult.

This paper considers the problem of controlling dimensionality
by focusing attention on the most likely sources of variation and
causes that are most proximal to effects by sketching linkages among
traditionally isolated problems: variability in primary and
secondary production and the physical environment. Defining the
linkages require a metric that can be related to all forms of life,
as without such a metric a theory that relates to all living
organisms and their physical environment cannot be contemplated. To
develop the metric, unified theories of biological productivity are
used as a point of departure to show that averaging or integration
across populations tends to be driven by the nature of mathematical
techniques (e.g., differential equations), obscuring density-
dependent population interactions, and the effects on these
interactions of temperature, motion, and light. We proceed by
considering what seem to be the most fundamental interactions between
population dynamics and physics in a multipopulation setting.

2. STRUCTURE OF UNIFIED THEORIES

A small segment of the biological-oceanographic literature has been devoted to what might be called unified theories of biological productivity. Unified theories attempt to account for all components of biological production.

Unified theories have been available for several decades (see, for example, Riley, 1963). As these theories deal with complex systems and are highly aggregative, alternative interpretations of analyses based on unified theories are usually possible. Nevertheless a rejuvenated interest in the unified-theory approach is warranted in the context of the study of global biological production.

As a point of departure, it seems that all unified theories are built on a differential-equation template (DET),

$$\dot{N}_1 = G_1(N_1, N_2, \ldots, N_M)$$

$$\dot{N}_2 = G_2(N_1, N_2, \ldots, N_M) \tag{1}$$

$$\vdots \qquad \qquad \vdots$$

$$\dot{N}_M = G_M(N_1, N_2, \ldots, N_M)$$

where $\dot{N}_i$ is the derivative of population number or weight with respect to time; G_i is the functional relationship between $\dot{N}_i$, N_i, and all other N_j; and M is the number of populations in the ecosystem.

Equation 1 contains the important properties of any population-dynamic system: a) $\dot{N}_i$ depends on N_i; b) all M populations are linked, that is, $\dot{N}_i$ also depends on all N_j, $i \neq j$; and c) each population is affected by its own set of environmental conditions. Unfortunately, the DET is not analytically tractable. For $M = 3$ the equations are quite complex. For $M > 5$ analytic results seem unattainable (Nisbet and Gurney, 1982).

Three methodologies have been used to ameliorate the intractability of the M-population DET. The first methodology involves partitioning the M equations according to broad taxonomic categories and then constructing a much smaller number of equations, one for each category. For example, phytoplankton, zooplankton, and fish populations might be represented by three interrelated equations, one for phytoplankton, one for zooplankton, and one for fish. A related approach is to consider only a particular subset of the M populations to the exclusion of others. In taking this tack, perhaps only one population might be studied, or perhaps a group of populations might be considered: typical examples involve the study of "copepods" or "primary production" or "groundfish". The second methodology involves using either alternatives to, or proxies for, the M equations. For example, the use of carbon flux might be used to represent partitions of the DET. Another proxy for the DET is to

modify the *M* equations to represent size categories of organisms
rather than populations (see *e.g.*, Ulanowicz and Platt, 1985), a form
of aggregation included as a proxy methodology because it replaces
taxonomic classification with ataxonomic classification. The third
methodology involves replacing the DET with a simulation model
containing many of the DET or DET-like equations. But the
simulations are often, in effect, as complicated as Nature defeating
their intended purpose.

The various simplifying schemes have their pros and cons. In
terms of pros, the extant simplifications make dealing with a
multitude of populations manageable. But in terms of cons, the
schemes obliterate the critical sources of population variability:
the density-dependent, population-dynamics responses of individual
populations and the relation of these responses to the physical
environment. If these phenomena are not included in a theory, then
how can the theory explain population-dynamic variability?

3. PARTICLES AND TROPHIC TRANSACTIONS

As an alternative, let us try to identify the most fundamental
ecological transformations in the biodynamic system in the hope that
the study of these transformations will minimize the number of
dimensions that need to be studied and also account for critical
sources of variation.

To begin, we define each living organism and each speck of
particulate matter as a "particle." The definition enables defining
the state of each particle in terms of its time-space position; age;
genesis, or reproduction rate; and dissolution, or death rate (see
Rothschild *et al.* 1982). Particles can be associated with particular
populations in the sense that those within any population are more
alike than those in different populations. Particle populations can
be defined in different ways. For example, some populations can be
defined in terms of genetic affinity, while others can be defined in
terms of organism size.

The definition of a "particle" provides a common framework for
describing and interrelating the dynamics of all organisms in the
sea. The dynamics of the particles may be considered in terms of the
state of each particle and transitions among states. Ecological
transitions are defined as the predator-prey interaction in which a
predator consumes or destroys a prey. The specific destruction of a
prey by a predator can be called a *trophic transaction*. The notion
of a trophic transaction applies to all trophic levels. For example,
the situations where an animal eats another animal (carnivory), an
animal eats a plant (grazing), a plant eats a nutrient molecule, or a
bacterium eats an organic molecule are all examples of trophic
transactions.

The establishment of the notion of a trophic transaction enables
considering the *statistics* of prey-predator interactions, not only in
terms of the conventional transfer of energy or material, but in terms
of genetic and population stabilization information critical to the

maintenance of stability and variability in the population. In terms
of genetics, when a predator destroys a prey, it destroys prey genes
and acquires a component of prey biomass, possibly increasing its own
genetic fitness. If the predator is nonselective with respect to the
prey population, then the genetic structure of the prey-population
remains essentially unchanged. If, however, the destruction of
individual prey is selective with respect to the varying genetic
structure of each prey, then the genetic structure of the prey
population changes, evidencing a change in information. In terms of
population stabilization, each population is regulated by a set of a
population-regulatory modules, one for each life-history stage (these
are discussed in detail in Rothschild, 1986:Chapter 8). The modules
comprise a) density-enhancing mechanisms (growth and reproduction),
and b) density-dampening mechanisms (mortality). The intensity of
density-enhancing and density-dampening mechanisms and hence the
stability of the population depends upon the trophic interactions
among the predators of the population of interest, the individuals in
the population of interest, and the prey of the population of
interest. Put another way, a population at a particular level of
abundance is exposed to a particular quantity of food and predation.
As population abundance increases, food becomes more sparse and
predation more intense. As the population abundance decreases, food
becomes more abundant and predation less intense. These immediate
effects are, of course, modified by a complex of nonlinear
interactions, but nevertheless the fact remains that population
stabilizing information is transferred among populations.

Thus, a trophic transaction, an ecological quantum jump, is the
fundamental unit that transfers energy from donor to acceptor and
modifies both genetic and population dynamic information. Trophic
transactions in any one population might be thought of as carrying
information on the abundance of the population of interest. In fact
the collection of trophic transactions might be thought of as signals.
Population abundance changes "normally" with normal signal
transmission where the predator perceives prey abundance to be in
direct proportion to the relative abundance of predator and prey.

However, populations exhibit unusual fluctuations (that is, they
tend to increase when abundant or decrease when depleted) when food
signals, become enveloped in "noise" or otherwise become aberrant
owing to variations in the physical environment. This sets the stage
for the study of trophic-transaction statistics, a prerequisite for
the study of normal and aberrant food signals.

4. SIGNAL GENERATION

This section develops the idea of trophic-transaction statistics in a
signal-theoretic context. To do this it is necessary to define
contact among particles, as contact is a necessary prerequisite to
ingestion. "Contact" is, of course, a special case of time-and-space
distribution of particles. In other words if we know the time-space

distribution of organisms, then we could define "contacts" as a special case of this distribution.

To develop the framework, we can think of a predator particle centered in its own coordinate system so that all prey particles move relative to the predator particle even though the predator particle may have a velocity "with respect to the ground". In the simplest case, the predator can be observed, and the instants of time when prey pass within striking range of the predator can be noted. An observation of this system constitutes a single realization of a stochastic process, motivating the study of the statistics of many realizations. As the statistics on the arrival times or contact rates of prey involve the temporal distribution of prey relative to predators, the distribution of prey relative to predators can be thought of as a signal.

Thus a food signal carries energy (in prey calories) and information on prey abundance. The predator, the receiver, is "tuned" to those components of the signal it can actually or potentially consume. The physical environment modifies transmission or reception of the signal and the transmission medium as well.

At "normal" levels of predator-and-prey abundance, the predator perceives prey abundance to be in direct proportion to the numbers of prey per predator, maximizing the efficiency of the population-regulatory process. However, as prey abundance becomes abnormally high or low, the actual abundance or *frequency* (a scarcity of food might be represented by a relatively low-frequency signal while an abundance of food might be represented by a relatively high-frequency signal) of prey is modified. This is because in general, a predator can only "receive" a fixed number of prey per unit of time before it becomes satiated. Put another way, as the density of prey increases above a particular threshold, the predator becomes inefficient at capturing the prey, the well known "functional response."

4.1 Food Signal Generation

Food signals are a function of the generalized distance between prey particles and the cross-sectional area swept by the predator (Figure 1). The generalized distance between prey particles is defined as having two components: a) the metric distance between particles, and b) the relative velocity of the particles.

The metric distances among prey can be deduced from the density of prey and converted into frequency. In the simplest deterministic case, the distance between particles is $N^{-1/3}$, where N is the density per-cubic-unit (*e.g.*, cubic centimeter). However, as we know, it is unlikely that particles are deterministically distributed. As a first simple assumption, we can assume that the particles are poisson distributed. Under this assumption, the mean distance to the nearest-neighbor particle in one, two or three-dimensional space can be deduced. In three-dimensional space we have (Chandrasekhar, 1943),

$$D_3 = .55 \ N^{-1/3} \qquad\qquad (2)$$

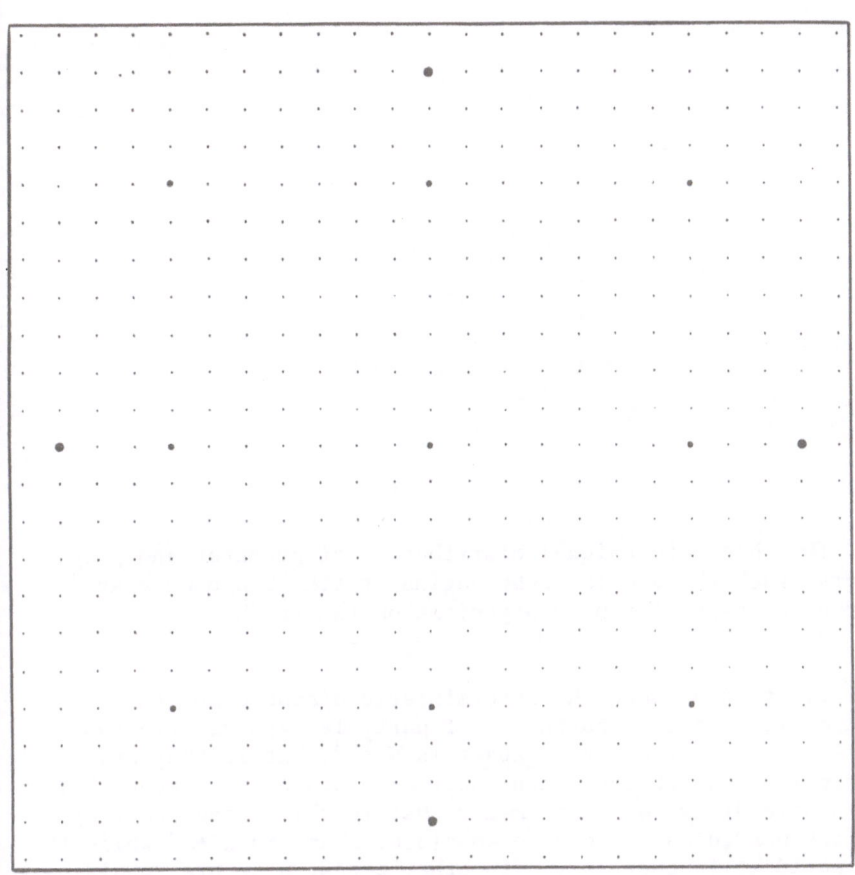

· = Bacteria   • = Nannoplankton   ● = Phytoplankton

Figure 1a.   Diagrammatic representation in two-dimensional space of
plankton.   Densities, diameters, and deterministic distances are
based on Table 1.   The two-dimensional representation is a slight
distortion of three-dimensional space but provides a point of
departure for appreciating the distances among particles given that
actual distributions are probabilistic and the particle possesses
velocities.   The distance taking account of velocity is the
generalized distance.   The generalized distance is the critical
ecological distance since it is an essential component of contact
among organisms.

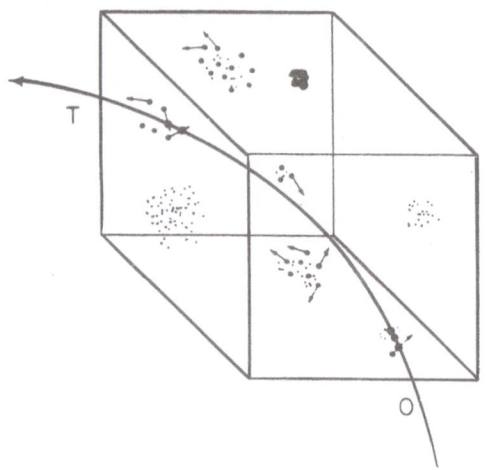

Figure 1b. The three dimensional distribution of predator and prey.
The predators track through the cube begins at time 0 and ends at
time T. It encounters five prey organisms on the track.

     The distinction between the deterministic distance and the
probabilistic distance is important. If particles are distributed
deterministically, their distance apart is $N^{-1/3}$, but if they are
probabilistically distributed, then their mean distance is about half
the deterministic distance. This means that in the poisson setting
about half the particles would be even closer than $.55\ N^{-1/3}$ while the
other half would be farther away. In other words, some prey would be
much closer to the predator than would be otherwise thought.
     However, the likelihood that a poisson distribution, or more
accurately, a distribution that *seems* like a poisson distribution,
actually exists depends upon scale. It is conceivable that a poisson-
like distribution always exists at very small scales but as
scales get larger and larger, the likelihood that the poisson
properties hold diminishes.

Figure 1c. The transformation of the predator's track into a food
signal. This is an individual realization. If the predator is a
high-frequency predator it can ingest every particle. If the
predator is a low-frequency predator it can only ingest one prey per
cluster. In the former case, the predator "thinks" there are five
prey per T. In the latter case the predator thinks there are only two
prey per T. Physical forcing can change these second order-
properties giving the predator a different impression of the
abundance of prey and eliciting different population-dynamics
responses, all other things being equal.

    A number of distributions could represent the situation where the
poisson distribution does not apply and since there does not seem to
be any theoretical basis for any particular distribution, the
poisson-poisson distribution is more attractive than the more
traditionally used negative binomial as the poisson-poisson seems more
analytically tractable (see *e.g.*, discussion of compound and doubly
stochastic distributions by Fasham 1978).
    The relative velocity of the particles is the other important
component generating the food signal. For example, fish larvae swim
at velocities greater than 1 cm sec^{-1} with velocities that appears to

increase considerably with size (*e.g.*, Hunter, 1972). Flagellate
bacteria swim at 200-300 um sec^{-1} (reviewed by Goldman 1984). In
addition immotile phytoplankton cells possess sinking velocities which
can be generate a velocity which is large relative to other plankters.

The relative velocity of a prey particle to a predator particle
is given by Gerritsen and Strickler (1977) as,

$$\frac{u^2 + 3v^2}{3v}, \text{ for } v > u$$

$$\frac{v^2 + 3u^2}{3u}, \text{ for } u > v \tag{3}$$

Contact rate is derived by multiplying the relative velocity by the
cylindrical volume swept by the predator and the number of prey
contained in that volume, $N \pi r^2$, where N is the density of prey and r
is the radius of the cross-sectional area swept by the predator. This
provides "first-cut" statistics for a food signal given certain
assumptions on prey distribution, predator and prey velocity, and the
cross-sectional area swept by the predator. As Mangel and Clark
(1986) have shown other formulations of the search problem warrant
study as other search configurations are conceivable (see also,
Koopmans 1980). As we point out later, second order distributional
properties can often be more important than first order properties
testifying to the need for additional work.

5. SIGNAL RECEPTION--THE EFFECT OF THE OCEANIC ENVIRONMENT

Up to this point we have assigned a common metric to all living
organisms and have sketched the idea that this metric can be used to
define conditions for the occurrence of trophic transactions. In this
section we sketch the idea that the physical structure of the sea is
an important source of variation contributing to the variability of
trophic transactions.

The ether and the oceanic environment are analogous in the sense
that they both contain energy. The ether contains electromagnetic
radiation at various frequencies, while the ocean contains biological
energy allocated among living and some non-living particles existing
at various frequencies relative to a reference time-space point. The
energy in the ether and in the ocean represent information. That is
to say, both the electromagnetic and biological energy are not
distributed at random, they possess *inter alia* spatial-temporal, time-
space and spectral properties.

However, a untuned broad-band receiver, or otherwise crude device
which would sample either electromagnetic radiation or biological
energy would lead to the conclusion that the radiation or the
biological energy was random or at least very noisy; static conveys no
information.

This is because the maximum information content of signals can
only be extracted with a receiver of complexity consonant with that of

the signal. Furthermore, the "meaning" or information is a function of not only the transmitter but of the state of the transmission medium and the receiver. For example, suppose station WJPT transmits Bach's Toccata and Fugue. The music can only be heard with a properly tuned receiver. However, even with a properly tuned receiver, reception requires that conditions in the transmission medium must not envelope the signal in noise that could be generated by an environmental event such as an electrical storm, for example.

Analogously, the biological signal is a function of the temporal and spatial distribution of biological energy (the prey); the quality of the transmission medium (e.g., kinetic structure and temperature); and the tuning and selectivity properties of the receiver (the predatory characteristics of the predator).

Defining the predator as a receiver enables defining reception as the temporal distribution of ingested prey. So from the biological "noise" we extract a predator-specific food signal in terms of the temporal contact between predator and prey (where a prey is an organism or a particle that has a probability of being destroyed by the predator).

Clearly, the analysis needs to be extended to considering the functional response of predator to prey. This would reflect that if prey occur at frequencies that are too high, then consumption will not be proportional to abundance, pointing toward the importance of the second order distributions of prey relative to predators (note that in Figure 1c, the first-order distributions of 5 prey per unit T are identical, but the second-order distributions are quite different).

5.1 The Transmission Medium

We can now consider the effects of physical properties, motion, temperature, and light on the food signal. Put another way, we can consider how these properties effect the transmitter(s), the receiver, and the transmission medium.

In actuality, these effects are quite well known for temperature, less well known for light (except in the case of primary production), and least known for kinetic structure operating on feeding scales (with the possible exception of low-Reynolds number flow in very standardized situations (Figure 2).

There are basically two sorts of physical effects. The first involves those effects that modify the properties of the transmitter (the prey) and receiver (the predator) *per se*. The second involves the kinetic modification of the transmission medium (clearly kinetic and thermal structure can be correlated, but this is ignored for simplicity sake). With regard to the first, temperature affects the swimming velocity of predator and prey perhaps in the same direction, perhaps in opposite directions, for example. Many organisms are visual feeders, and hence light is an important component of the trophic transaction, either directly in the case of phytoplankton or indirectly in the case of light necessary to see the prey. In some instances only one organism in the prey-predator pair is a visual feeder generating additional sources of variability.

With regard to kinetics, the distribution of first and second order properties of prey distribution can be strongly influenced by turbulence. Rothschild and Osborn, 1988, consider the RMS uncorrelated relative velocity of two particles as a function of ambient turbulent kinetic energy and distance between the particles. They modify (3) to obtain the RMS velocity difference of a predator and prey given the given the turbulent energy dissipation rate and the distance between the particles,

$$\frac{u^2 + 3v^2 + 4w^2}{3(v^2 + w^2)^{1/2}} \tag{4}$$

where the predator is faster than the prey and w is the RMS velocity difference of two points in a homogenous isotropic turbulent medium.

Rothschild and Osborn suggest that taking account of small-scale turbulent motions requires reinterpretation of published statistics on plankton food requirements derived from laboratory studies. In addition, they point out that introducing turbulent motion effects into the trophodynamic calculus suggests certain conceptual revisions involving so-called "optimum foraging theory", patch formation and dissipation mechanisms, nutrient donor-acceptor relations, and methodology for linking global- or regional-scales with microscale events. They point out that taking account of small-scale turbulent motion opens new questions regarding a) most conventional notions of optimum foraging because the energetics of foraging can depend as much on water motion as on biological properties of the prey and predator, b) patch formation and dissipation because taking account of small-scale motion can create patches in uniform densities of prey-and-predator on one hand, and on the other hand, obliterate density-defined patches, and c) linkages between micro and large scale physical events .

As another example, affecting both large and small organisms, consider that all larger organisms begin their life as very small organisms, hence all organisms are at some point subject to low-Reynolds-number conditions. Of particular interest are the effects of micro-changes in temperature which would have a micro-effect on viscosity. But this change in viscosity could have a substantial effect on fertilization success, particularly taking into account that realizations of fertilization events are extremely large (e.g., 10^{12} in a cod population).

In sum, then, the physical structure can have a profound effect on the food signal. Temperature change, for example, tends to affect either the predator or the prey, or both, directly; it can increase or decrease frequency and affect the second-order properties of the

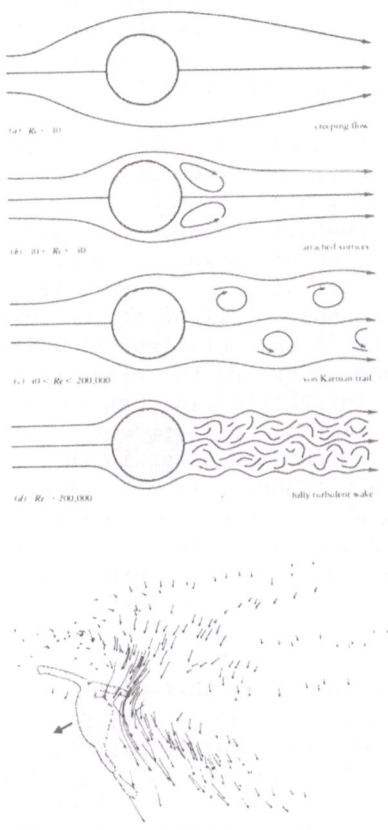

Figure 2. Contrast between "nice" low Reynolds number flow about a cylinder (Re < 10) and the flow field of *Eucalanus crassus* during feeding. Each arrow shows the pathline of an alga during a 0.4 sec time interval. (The upper panel is from Vogel 1981:Figure 5.5; the lower is from Strickler 1985:Figure 2).

signal without changing the relative abundance of either predator or prey, distorting a population's population-dynamic regulatory response to its own abundance. On the other hand, changes in the kinetic environment tend to effect the transmission medium and thereby alter the second-order properties of the signal which can appear to the predator as a first-order change affecting again the "normal population dynamics" response.

6. EXAMPLES

Up to this point, the food-signal notion has been used as a picturesque way of depicting the interaction between predator and prey

540

in a population-dynamic and physical setting. The role of the food-signal notion can be sketched with a few examples that show how physical variability can be taken into account and how the length of the causal chain can be reduced, at least relative to conventional approaches.

As a first example, motivated by Woods and Onken (1982), consider the exposure to light of a single particle which oscillates between the surface to the bottom of a mixed layer. Consider also that the intensity of irradiance declines exponentially from the surface to the deepest particle excursion. Now, if up-and-down excursions of the particle can be considered cosinusoidal, and the exponential decline of light is modulated by the daily change of light at the surface, then the particle will be exposed to flashes of light at a frequency determined by the depth of the mixed layer and the kinetics that drive the particle up-and-down. What is interesting is that in many instances, the production of phytoplankton cells is more responsive to "flashes" of light than to constant light and hence in this case, the magnitude of primary production could, under a robust set of circumstances, relate as much to high frequency (one day or less) physical events as to biological properties of the phytoplankton cell. Further, as many organisms require some minimum threshold of light to feed, physically forced, up-and-down motions would have a powerful effect on the food signal.

As a second example, consider the trophic relations among organisms on Georges Bank (zooplankton, large phytoplankton, heterotrophs, autotrophs, and bacteria) in the context of trophodynamic contact rates. In order to study this problem, consider the temporal distribution of organisms on Georges Bank. Davis (1987) reviews the copepods, Cura (1987) the phytoplankton, and Hobbie (1987) discusses the autotrophs and heterotrophs. Impression of organism temporal densities gleaned from these papers is such that the abundance of the major organisms is highly periodic within a year; most organisms being abundant for only the same two or three months each year. Because of these periodicities, an annual average of these interactions would be difficult to interpret. Because of the difficulty of computing annual averages, we use the statistics for May at the end of the spring bloom. These statistics are given in Table 1.

In order to use equation (4), we also need to develop some notion of the turbulent velocities of particles relative to one another at feeding scales. These can be derived from Rothschild and Osborn (1988); a reasonable range of values are given in Table 2. It is important to notice that these values can vary considerably as a function of the distance between particles and the turbulent dissipation rate.

Table 1. Georges Bank scenario. Densities of organisms approximate those reported for May at the terminus of the spring bloom.

	# per cc	Diameter um	Velocity in cm hour^{-1}
Bacteria	10^6	1	.1
Phytoplankton	100	20	.1
Nannoplankton*	2000	10	36**
Pseudocalanus	.002	1000	720***

*mostly at surface and heterotrophs
**100 μm/sec
***.2 cm sec

Table 2. A range of turbulent RMS velocities, *w*, given wind velocities and particle separation distances. The wind velocities are only given as examples as tides or other sources might also generate turbulent motions (based on Rothschild and Osborn, 1988).

Particle Separation Distance Meters	Wind Velocity	
	10 Knot Wind	20 Knot Wind
10^{-4} - 10^{-3}	.005 cm/sec	.01 cm/sec
10^{-2} - 10^{-1}	.1 cm/sec	.2 cm/sec

Employing the particle separation velocities in equation 4 and the swimming velocities and densities in Table 1 gives the relative velocity and contact rates among the various classes of organisms in the Georges Bank scenario. The results of the computations are given in Table 3.

Features of the scenario bear comment. These relate to the nature of the turbulence effect and the magnitude of contact rates. As we can see, turbulence has an important effect on contact rate. Again the opportunities for variability are substantial as these are a function of the variability in the statistical spacing between organisms, organism motility, and the statistical distribution of turbulence-induced motion.   Clearly a shift in the persistence of one set of turbulence-related conditions to another would have a remarkable effect on shifting predation intensities within the ecosystem explaining how subtle shifts in physical conditions can set in motion a whole chain of population-dynamic responses.

With regard to the magnitude of contact rates we find that for some pairs of organism's contact rates are surprisingly high even for the highly productive spring-bloom situation. Of particular note, the high-contact rates between bacteria and copepods are remarkable. To use signal theory jargon, the contact rate might be as high as about 5,000 herz.

Table 3. Relative velocity and number of contacts among different kinds of organisms based on Table 1 and equation 4. A zero, low (.005 cm sec) and high turbulence (.2 cm/sec) rms velocity, w is incorporated in the computation.

		Bacteria		Phytoplankton		Nannoplankton	
	Turbulence	Relative Velocity	Contacts Per Hour	Relative Velocity	Contacts Per Hour	Relative Velocity	Contacts Per Hour
Pseudocalanus	zero	720	22×10^6	720	2262	721	4.5×10^4
	low	720	22×10^6	720	2262	721	4.5×10^4
	high	1187	37×10^6	1187	3729	1188	7.5×10^4
Nannoplankton	zero	36	113	36	–		
	low	43	135	43	–		
	high	961	3019	961	0.03		
Phytoplankton	zero	0.1	1				
	low	24	301				
	high	960	12064				

Contact rates may reflect on the mechanisms by which autotrophs obtain "old" nitrogen in oligotrophic situations. Some arguments minimize the amount of nitrogen obtained from heterotroph excretion because excreted nitrogen diffuses so rapidly that it is too dilute to be utilized by autotrophs. On the other hand, a high frequency of "contact" between bacteria and autotrophs might suggest an alternative mechanism for transferring nitrogen from heterotrophic donors to autotrophic acceptors. Whether this actually works depends on the metabolism of the involved organisms. The acquisition of nitrogen by autotrophs by "contact" suggests that measurements of nitrogen flux need to stress bacterial metabolism, the effects of turbulence; and the contact notion (where contact of course means "in very close proximity"). These observations suggest reexamining the work of Gavis (1976) for example, which is developed in a setting where nutrient molecules reach their acceptors over relatively long, diffusional distances, rather than over the very short distances implied by contact rather than diffusion-related transfer. In considering Goldman's 1984 discussion of autotrophic nitrogen acquisition, it appears that motion-enhanced contact strengthens the argument for relatively large nitrogen interchange between autotrophs and heterotrophs, however, this would depend upon the density, motility, and kinetic structure of the environment.

7. DISCUSSION

The dynamics of populations in marine ecosystems are characterized by large variability superimposed upon an over-all stability. Accordingly, an understanding of these biodynamics requires an understanding of the mechanisms associated with stability and departures from stability. In fact, understanding the interaction between population stability, population change, and dynamics of resulting biological productivity is the key scientific question in biological oceanography.

Addressing these global-context issues is difficult: The high-dimensionality setting requires a) a theory that matches the problem, b) identification of the "right" dimensions, and c) focus on minimal-length causal chains. The right dimensions involve density-dependent population dynamics processes and physical structure and its variability. Minimal-length causal chains involve examining trophic transactions in a Lagrangian setting.

To carry this idea to a useful stage, it is necessary to consider the best way to average, or otherwise integrate, the individual trophic-transaction realizations (recall that the DET's are one form of averaging or integration). The best approach for integration is not yet clear. However, the consequences of accreting data along traditional lines is evident. A lack of theory will result in treating every problem as a special case; not taking account of physics and population-dynamic density dependence misses the opportunity to account for substantial known sources on variation; and working with long causal chains such as the relations between wind-speed and recruitment, or primary production and fish abundance, or productivity and various structural features such as domes or fronts generates relations concerning variability which can never be verified owing to variance leakage typical of high-dimensionality systems.

A totally empirical approach assumes that all components of the system can be "seen" and leads to a philosophy of correlational empiricism. Strict empiricism leads away from the most critical problems in understanding biodynamic variability which are rooted in the high dimensionality of the ecosystem. Without a theoretical approach those components of the system which cannot be observed, cannot be addressed. For example, we have shown how the interrelationships among predator-and-prey organisms can depend as much on the physical setting as on the relative density of predator and prey. As another example, consider the important intertwining

between first and second-order statistical properties. On one hand,
food abundance can be unusually low, but if the variance of abundance
is unusually high, then some predators will be well fed. Consider
also that the relationship between first- and second-order properties
devolves from the statistical distribution of the properties. An
examination of the distribution of a property such as food abundance
suggests certain events are extremely rare, that is, they have a low
probability of occurrence. However, what is critical is not the
probability of an event, but its expected or average value. In the
sea, the expected value associated with very small probabilities can
be of considerable magnitude (e.g., the probability of survival of a
cod egg might be .001, but there are 10^{12} eggs produced annually by a
typical population). In other words, the very large numbers of
individuals that occur in most marine populations generate seemingly
deterministic properties in what would ordinarily be probabilistic
settings.

This says something about what can be observed and what cannot be
observed because ordinarily we would think it more efficient to sample
a group of organisms for survival rate to determine whether they were
in salubrious environmental conditions. But the detection of tiny
statistical differences in survival rate could require an impractical,
or even impossible sample size, even though the miniscule differences
would have a substantial effect on population size. Because of this
difficulty, the measurement of the abundance of the survivors might be
attempted. But the survivors might have a temporal-spatial
distribution which is too extensive, or with second order properties
that challenge the efficiency of ordinary sampling techniques.

As another critical problem, the high-dimensionality of the
ecosystem is interrelated with problems of scale, particularly the way
that events on different time-space scale are interrelated (Rothschild
and Rooth 1982:10-11):

> Events that affect the survival of larvae operate in
> the micro- and fine-scale. Yet, for any single
> recruitment year, there must be trillions of micro-
> and fine-scale events entrained in relatively large
> time-space scales (weeks and tens-of-miles). Thus,
> mechanisms must exist which represent an integration
> or averaging of the micro- and fine-scale events.
> The nature of these mechanisms may be one of the key
> questions which, when answered, will lend
> considerable understanding of the dynamics of the
> micro- and fine-scale ecosystem.

But it appears that this problem too has its anthropogenic
components as the definition of appropriate scale can depend on
viewpoint. Our viewpoint is that the trophic-transactions
incorporated in food signals comprise the most fundamental
organization of the biodynamic system. The food signal, or the
temporal distribution of food, possesses certain statistical
properties defined in terms of the relative velocity and density of

predator and prey. These properties define a particular scale. This scale defined on the basis of food signals is in general quite different than a scale defined in terms of a sampling procedure, such as sampling using a plankton net. In other words, the questions of scale that deal solely with anthropocentric metric systems (*e.g.*, minutes, weeks, months, and mm, cm, and M) are only descriptive of the interaction of human perceptions and sampling systems juxtaposed with the distribution of organisms.

In fact, the food signal itself defines an important integral scale, the exact form of which depends on the probabilistic food signal's autocorrelation function. The important point is that the notion of scale in the context of our discussion must be defined relative to the food signal. In other words feeding scales are the critical unit for averaging ecosystem events. Further, as we have shown in Table 3, scale in any particular situation is not a constant but a dynamically changing variable.

In order to account for these criticisms a different setting is considered where the trophic transaction is the fundamental unit of ecological interaction. As trophic transactions are chance events, their statistical properties are important. Structuring analysis around the averaging or integration of trophic-transactions enables the direct study of the interaction of density-dependent population-regulating phenomena and variability in the temporal-spatial distribution of kinetics, temperature, and light.

The quantum change/Lagrangian view of biodynamics is not only consonant with interpreting the stabilizing-destabilizing phenomena as we understand them but amenable to analysis, using a large body of extant statistical and signal-processing theory. In fact the analysis of predator-prey relationships in the context of signals could open new vistas of understanding.

This notion of particle dynamics, even at this sketchy stage of analysis sheds some light on requisite theory. First, it must be obvious that there can be no single canonical representation of the productive processes of the sea. The representation must depend upon the questions asked. We can divide questions into those that relate to single-population effects and to those that relate to multiple-population effects. With regard to a single population the population receives its integrated food signal and at the same time it constitutes a food signal for its predators. If the population abundance is slightly perturbed, plus-or-minus 10 percent for example, effects on prey may not be noticed because of compensatory mechanisms in the prey populations. Also the effect on its predator might not be noticed because in general, predators feed on many different populations of prey. In other words, small changes in the population of interest tend to be absorbed rather than propagated.

Suppose, however, we increase or decrease our target population by a moderate amount plus-or-minus 30 or 40 percent, for example. We move into a range where the abundance of both the prey and the predator would be arguably affected, the food signals modified, and changes in population abundance become evident. However, even in this multiple population setting, the perturbation may not propagate among many populations because of cascading compensatory effects. In fact, aside from special cases, there would need to be an extraordinarily large disturbance in any one population to have a large effect on the entire ecosystem. Such a scenario implies that analysis based upon conventional aggregation of populations would be inefficient for interpretation of dynamic change.

If the dominant source of change is the physical environment, then the physical environment can be used to partition the populations into one group that would be dynamically benefited by a physical change, and another group that would be depreciated by the same physical change. The approach would be to identify the environment as a vector of functions depending on motion, temperature, and light. We identify a particular vector under which the mix of populations have known dynamic properties in terms of mean values and higher moments as inferred by the affects of the physical environment on the collection of food signals. Now we change the value of the physical-environment vector and if we change it enough, one collection of populations will be dynamically advantaged and another will be dynamically disadvantaged and the quantity and quality of biological production will change.

What does the vector of functions look like? How do various populations respond dynamically to change in the physical-environment vector; that is, which populations increase and which decrease? How do the changes in populations affect production, not only the elaboration of biomass-per-unit time, but the production of catabolites which can have an important effect on other populations?

On one hand, such an approach seems quite complex. On the other hand, it is quite natural. If we were to measure wind velocity over a large area of the sea, we could ultimately make inferences on how wind-velocity changes would affect small-scale water motion enabling inferences as to which groups of organisms had enhanced or diminished contact rates thereby becoming susceptable to enhanced or diminished predation.

Such an approach would need to utilize the quantum/Lagrangian format. Any particular vector would imply a particular food signal structure given a constant density of food. Now we change the vector and we change the higher order properties of food structure and in some instances the mean food abundance. For example, kinetic changes change signal-transmission properties, change temperature and change contact rates.

The test and application of such a model really depends upon available technology. We anticipate a pilot study conducted in a relatively small region that is physically well understood and variable. In other words, the same population of organisms would be subject to widely different magnitudes of the physical variables given equivalent average densities of organisms. By hypothesis, the food signals should vary as a function of physical structure. In order to measure both the signals and response to the signals, it will be necessary to study a) the density of the organisms, b) the relative velocity of the organisms as a function of their own motility and environmentally induced velocity, and c) the feeding and genetic history as inferred by the well-known biochemical changes induced by feeding which occurs on various time scales.

9. ACKNOWLEDGEMENTS

John Woods set me thinking about the Lagrangian setting. Joel Goldman provided many stimulating conversations on life at the microscale. John Steele has also been a stimulating influence. Much of this paper was written while I was on sabbatical leave at Woods Hole Oceanographic Institution.

11. REFERENCES

Chandrasekhar, S.: 1943, 'Stochastic problems in physics and astronomy', *Reviews of Modern Physics* 15(1), 1-89.

Cura, J.J., Jr.: 1987, 'Phytoplankton', in R.H. Backus (ed.), *Georges Bank*, MIT Press, pp. 213-218,

Davis, C.S.: 1987, 'Zooplankton life cycles', in R.H. Backus (ed.), *Georges Bank*, MIT Press, pp. 256-267.

Fasham, M.J.R.: 1978, 'The application of some stochastic processes to the study of plankton patchiness', in J.H. Steele (ed.), *Spatial Pattern in Plankton Communities*, Plenum Press, New York, pp. 131-156.

Gavis, J.: 1976, 'Munk and Riley revisited: nutrient diffusion transport and rates of phytoplankton growth', *J. Mar. Res.* 34, 161-179.

Gerritsen, J. and J.R. Strickler: 1977, 'Encounter probabilities and community structure in zooplankton: a mathematical model', *J. Fish. Res. Board Can.* 34, 73-82.

Goldman, J.C.: 1984, 'Conceptual role for microaggregates in pelagic waters', *Bull. Mar. Sci.* 35(3), 462-476.

Hobbie, J.E., T.J. Novitsky, P.A. Rublee, R.L. Ferguson, and A.V. Palumbo: 1987, 'Microbiology', in R.H. Backus, (ed.), *Georges Bank*, MIT Press, pp. 247-251.

Hunter, J.: 1972, 'Swimming and feeding behavior of larval anchovy, *Engraulis mordax*', *Fish. Bull.*, *U.S.* 70, 821-838.

Koopman, B.O.: 1980, *Search and Screening, General Principles with Historical Applications*, Pergamon Press, 369pp.

Mangel, M. and C.W. Clark: 1986, 'Towards a unified foraging theory', *Ecology* 67(5), 1127-1138.

Nisbet, R.M. and W.S.C. Gurney: 1982, *Modelling Fluctuating Populations*, New York, Wiley.

Riley, G.A.: 1963, 'Theory of food chain relations in the ocean', in M.N. Hill (ed.), *The Sea*, Vol. 2, New York, Wiley.

Rothschild, B.J.: 1986, *Dynamics of Marine Fish Populations*, Harvard University Press, 277pp.

Rothschild, B.J. and T.R. Osborn: 1988, 'Small-scale turbulence and plankton contact rates', *J. Plankton Res.*, In press.

548

Rothschild, B.J. and C. Rooth (eds.): 1982, *Fish Ecology III: A Foundation for REX, A Recruitment Experiment*, University of Miami Technical Report No. 82008.

Rothschild, B.J., E.D. Houde and R. Lasker: 1982, 'Causes of fish stock fluctuation: problem setting and perspectives', in B.J. Rothschild and C. Rooth (eds.), *Fish Ecology III: A Foundation for REX, A Recruitment Experiment*. University of Miami Technical Report No. 82008.

Ulanowicz, R.E. and T. Platt (eds.): 1985, 'Ecosystem theory for biological oceanography, *Can. Bull. Fish. Aquat. Sci.* **213**, 1-260.

Woods, J.D. and R. Onken: 1982, 'Diurnal variation and primary production in the ocean - preliminary results of a Lagrangian ensemble model', *J. Plankton Res.* **4(3)**, 735-756.

BRINGING PHYSICAL AND BIOLOGICAL OBSERVATIONS INTO HARMONY

Walter H. Munk
University of California, San Diego
Scripps Institution of Oceanography
La Jolla, California   92093
U.S.A.

and David M. Farmer
Institute of Ocean Sciences
P.O. Box 6000
9860 West Saanich Road
Sidney, B.C.   V8L 4B2
CANADA

ABSTRACT.   Global Satellite measurements of winds, sea level and surface temperature and the computer application to modern linear estimation theory have brought a century of undersampling to an end. We can now provide significant maps of the physical (and some of the chemical) parameters, unlike the dream-like displays of the past.   This provides a unique opportunity for the study of biological-physical interactions, provided some of the fields of biological significance are sampled and mapped in a commensurate fashion.

Satellite measurements of physical parameters (wind, SST,..) are ambiguous and inaccurate as compared to classical *in situ* point measurements, but this inadequacy is more than compensated for by the enhanced space-time sampling density.   We propose that similar trade-offs prevail for the biological parameters (color,..), so that these measurements are extremely useful even if the relation to productivity (say) is not unique.

A second consideration involves the representation of the measured physical parameters by just a few representative fields, using the methods of *principal value decomposition* (also called *empirical orthogonical functions*).   Progress in studies of primary production and recruitment will depend strongly on whether the high dimensionality of the biological distributions can be similarly reduced to just a few phyto- and zoo-plankton assemblages.

A third consideration is the extent to which the sampling strategy is to be governed by the existing *models* of physical-chemical-biological interactions.   Model testing and model improvement is the preferred procedure if it converges.   In physical oceanography, most pre-satellite models did not survive under the stringent tests of the dense global sampling.   We suggest that

B. J. Rothschild (ed.), Toward a Theory on Biological-Physical Interactions in the World Ocean, 549–554.

future strategy must allow for some biological *agnostic* sampling, in addition to model testing.

A final consideration is desirability of *selective* sampling of such areas (fronts, eddies,..) in space-time that are considered of particular consequence in the biological processes. We have in mind *in situ* measurements from research vessels that are positioned in response to real-time satellite images. We believe these shipboard measurements could be enhanced by further developments in acoustic remote sensing.

1. INTRODUCTION

The title of this paper expresses the long term goal of every true oceanographer, we all agree. But there are differences in opinion of how to attain this goal.

Some twenty years ago, Henry Stommel commented on the "dream-like" character of the existing models of ocean circulation. Why? It is our view that the fields of density and motion were so *undersampled* that they failed to provide guidance to reality. A century of undersampling (since the *Challenger* expedition) finally came to an end with the advent of global satellite sampling. We now have realistic pictures of the near-surface ocean circulation down to mesoscale (100 km and 100 days) elements. It is a lucky accident that orbital dynamics provided just the scales in space-time required for resolution of mesoscale dynamics. The emerging circulation pattern bears little resemblance to the classical concepts; it is far more convoluted and variable than any of us had imagined.

2. SATELLITE SAMPLING

The essential contribution of satellite observation is the available sampling density, 25 km and 10 days! It is *not* the precision of the data themselves. Surface temperatures are obtained to 1° accuracy, compared to $0.01^{\circ}C$ available for hydrocasts for a century. Scatterometry winds are good to 2 m/s and 25° in bearing, vastly inferior to anemometer measurements. Furthermore, surface roughness and wind velocity do not bear a one-to-one correspondence (*e.g.*, increased roughness may be the result also of flow convergence). Yet the satellite images have provided us with patterns of realism that were previously unattainable. Evidently the lack of precision and the existing ambiguity of the satellite measurements is more than made up for by the dense sampling. In this same spirit we can accept satellite measurements which do not bear a precise quantitative relation, nor a one-to-one correspondence, to biological processes. Color and productivity are a case in point.

3. COMPUTER ANALYSIS

The development of computers has brought many vital benefits to the
study of oceanography. We emphasize here just one of the benefits:
the opportunity for applying modern linear estimation theory to a
myriad of disparate datasets. A case in point is the blending of
the traditional, precise but undersampled observations with the new,
well-sampled data of lesser precision and of greater ambiguity.

4. DATA REDUCTION

For illustration, consider the tide level $\eta_i(t)$ for tide guages at
x_i, y_i, $i = 1, 2 .. I$. We have in mind I of order 10^3. By the method
of empirical orthogonical functions (eof's) the basin-wide
oscillations in sea level can be represented by

$$\eta(x,y,t) = C_1(t)F_1(x,y) + C_2(t)F_2(x,y)+.. \qquad (1)$$

where C_j are the time-variable amplitudes of the eof's F_j). The
remarkable finding is that in some of the applications just two or
three eof's could account for something like 80% of the η variance.
For illustration, one eof could represent the north-south variation
of the seasonal type, a second eof the east-west oscillation
associated with El Niño type phenomena, a third eof the migration of
water mass from the gyre center to the gyre periphery. The
discussion here is deliberately vague (note the omission of any
references).
    The orthogonality condition

$$\int \int F_1(x,y)F_k(x,y) = 0 \; \text{for} j \neq k$$

and 1 otherwise permits us to determine the eof's and their
amplitudes from the data $\eta_i(t)$. The time-dependent amplitudes are
weighted sums of the data:

$$C_j(t) = \sum_i a_{ij}\eta_i(t).$$

One expects that nearby stations give nearly the same information,
and that just a few stations determine C_1, another few stations
determine C_2, a third small set of stations determine C_3. If this
is so, then something like ten tide stations suffice to give a rough
estimate of C_1, C_2, C_3, and this in turn gives a rough (but useful)
estimate of $\eta(x,y,t)$.

Suppose $\eta_n(x,y,t)$ now represents the distribution of phytoplankton species n, with $\eta = \sum_n \eta_n$ the total phytoplankton. It is too much to hope for that $\eta(x,y,t)$ can be written as a sum of just a few terms as in (1), with the C_j's each being governed by a few dominant species? This would lead to a tremendous simplification in the problems of sampling and analysis.

5. ORTHODOX VERSUS AGNOSTIC SAMPLING

Space platforms have introduced two new elements: a *global* coverage and a new level of (space-time) sampling *density*. Among the biologically oriented fields, the measurements of ocean color covers both elements, and already has had a profound influence on our understanding of some biological processes.

In situ measurements will never be global, but they can cover an ocean eddy, or might be of gyre dimension. They can, however, provide measurements with a sampling density adequate for the processes under consideration. We recommend that the development of high-density automated biological sampling techniques be given priority. Only in this way can we take advantage of the plans now underway (WOCE, TOGA,..) for the 1990's. The 3-dimensionality of the physical-chemical-biological interactions makes it imperative to augment the space-based surface observations by shipboard measurements.

6. MODEL TESTING

We now realize that leading classical oceanographers, Nansen, Helland-Hanson, Sverdrup, men whose uncanny intuition had developed by a lifetime of observing at sea and thinking of ocean processes, had often failed to come up with realistic pictures of the ocean circulation. It is very difficult to short-cut the demanding requirements of the Nyquist sampling theorem (two samples per shortest significant wavelength). We are not convinced, therefore, that the myriad of existing hypotheses concerning productivity, population dynamics, and the physical-chemical-biological interaction will outlast the advent of adequate biological sampling.

There is then a valid argument as to the extent to which *in situ* measurements should be fashioned to test existing models of the interaction processes. This is the traditional approach of scientific inquiry, and the preferred approach if the models are good. We call it *orthodox* sampling: you believe in the model. (Disbelief and demolishment of a model is a form of orthodoxy.) On the basis of historical precedent one might expect that some of the interaction models may not survive the revolution of observing techniques, and we recommend that *agnostic* sampling of biologically related fields should have a place in future planning.

7. ACOUSTICS

The fact that the ocean is transparent to sound and opaque to
electromagnetic radiation argues for the increased application of
acoustic methods. The suggestion is not new, and it can be said
that acoustic efforts over the last few decades directed at
biological problems have been useful, but not extraordinarily
useful. Yet it is our view that the potential is very large, and
has not been exploited. Most of the work has been by people
interested primarily in acoustics *per se*, rather than by those using
acoustics as a tool for studying ocean processes. In the spirit of
satellite sampling discussed above, we suggest an extensive use of
ships-of-opportunity to obtain continuous acoustic transects across
the oceans with high frequency echo-sounders. The goal here would
be to provide the extra dimension, vertical as well as horizontal,
in the description of biological activity. Striking acoustic images
of biological acoustic backscatter have recently been obtained in
the vicinity of fronts, at the edge of the Gulf Stream, and so
forth. Such measurements can provide insight to the biological
response to these features, and provide motivation and guidance for
appropriate *in situ* measurements. The need here is for straight
forward echo-sounder systems operating in a suitable frequency range
(50-100 kHz), with the emphasis on systematic coverage of large
sections of the world's oceans. Such records would not in
themselves provide taxonomic discrimination; their value would lie
in identification of patterns which could in turn be related to the
two-dimensional patterns of satellite imagery. Other suggestions
for acoustic sampling on finer scales appear elsewhere in this
volume.

8. CONDITIONAL SAMPLING

There is some evidence that most of the significant physical-
biological interaction occurs in a very limited fraction of the
total space-time observation space. The critical areas may be fixed
(above sea mounts) or moving (warm core rings, meandering fronts).
to those who believe in the "catastrophe theory" of productivity and
recruitment (*catastrophe* in the sense that the significant
information is at the *singularities*), there is increased hope that a
timely analysis of satellite imagery will provide the opportunity
for positioning research vessels in the singular areas. The
movements are slow and generally predictable for a few weeks in
advance.

9. DISCUSSION

We look forward to the 1990's with great expectation! Satellite observations and computer analysis now offer the opportunity of realistic mapping of the physical ocean environment; we have not previously had this luxury. For the first time since the IDOE days there is a concerted and coordinated effort to take advantage of a technological quantum jump, as evidenced by the *Global Change* initiative in the United States. To bring physical and biological observations into harmony is, in our view, an urgent, challenging and achievable requirement.

RECENT ADVANCES AND FUTURE DIRECTIONS IN MULTI-DISCIPLINARY *IN SITU* OCEANOGRAPHIC MEASUREMENT SYSTEMS

T D. Dickey
Ocean Physics Group
Department of Geological Sciences
University of Southern California
Los Angeles, California 90089-0741
U.S.A.

ABSTRACT. A remarkable number of advances in multi-disciplinary *in situ* measurement systems have been made during the past decade. The number of types of multi-disciplinary measurements and their temporal and spatial scale ranges have been expanded significantly. Particularly important developments have occurred in optical and acoustical instrumentation. These have enabled many biological measurements on scales comparable to those of CTD's and current meters. In addition, new systems which collect essentially co-located and concurrent multi-disciplinary data have been developed. Further advances in multi-disciplinary measurement systems will be dependent upon new sensor and system technology and the availability of economical bio-optical and acoustical sensors, fiber optics and telemetry. The study of global scale oceanic problems will require both satellite remote sensing and selective *in situ* observations. Moored, drifting and expendable bio-optical and physical measurement systems are quite attractive for the latter purpose because of their potential for expanding the ranges of temporal and spatial measurements and greatly increasing the number of observations.

1. INTRODUCTION

There is currently considerable interest in multi-disciplinary oceanographic measurements. This interest is driven in large part by the recognition that many interesting and important oceanographic problems require multi-disciplinary data sets and modeling approaches. Included among these problems are biomass determination, phytoplankton primary productivity, secondary productivity, biogeochemical fluxes and cycles, gas exchange across the air-sea interface, larval fish recruitment, optical variability and bioluminescence. The effects of physical and optical variability have been considered to be important for the understanding of biological systems for some time. However, the relevance of

B. J. Rothschild (ed.), Toward a Theory on Biological-Physical Interactions in the World Ocean, 555–598.
© 1988 by Kluwer Academic Publishers.

biological processes to ocean physics has been recognized only recently.

The two primary methods which will be available to ocean scientists for studying the oceanic ecosystem within the foreseeable future are *in situ* sampling and remote sensing. The advantages and disadvantages of these two approaches have been discussed extensively (*e.g.*, Esaias, 1981). Global observations must include remote sensing with satellites (*e.g.*, Esaias, 1981; MAREX, 1982; McCarthy *et al.*, 1986; Brewer *et al.*, 1986) but the restriction of these measurements to near surface observations of a limited number of variables with relatively coarse spatial and temporal resolution will probably persist. Thus, it is likely that *in situ* observations will continue to be needed for some time. Complementary remotely sensed and *in situ* data have been used most effectively and it is likely that this hybrid approach will continue to be necessary for most large scale studies.

Multi-disciplinary measurement systems have been motivated by the desire to understand complicated oceanic ecosystems. A conceptual model (Figure 1) may be used to illustrate some of the aspects of these ecosystems (Dickey and Siegel, manuscript). External forcing

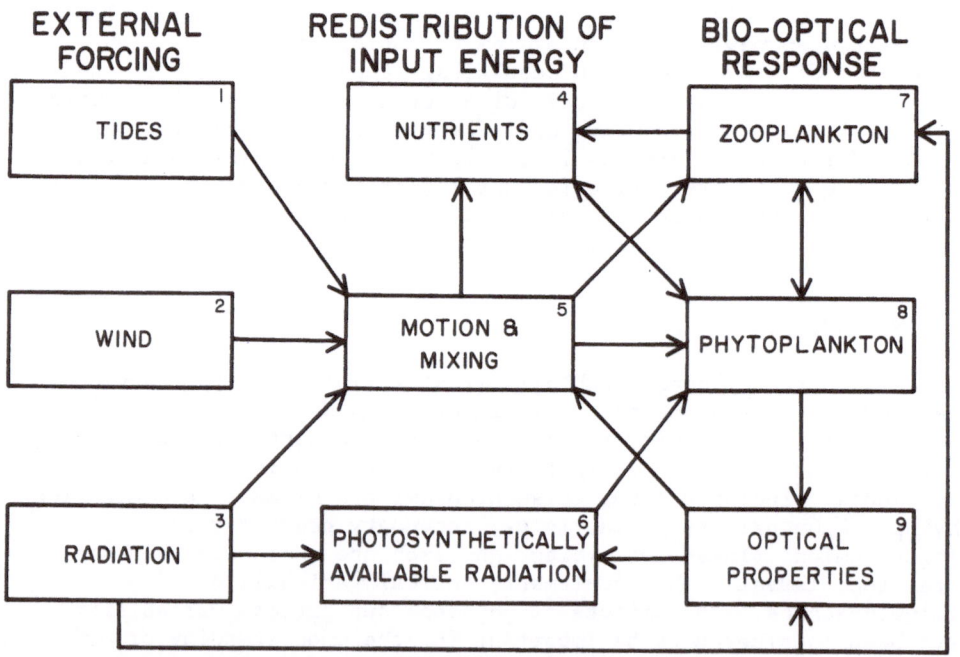

Figure 1. Conceptual model of the oceanic ecosystem. The arrows indicate functional relationships among the various major components.

of the ecosystem occurs through inputs of energy from tides, wind and solar radiation. Each of these affects the motion and mixing of the upper ocean. The motion and mixing in turn act to redistribute nutrients, phytoplankton and zooplankton. The solar radiation affects the phytoplankton and zooplankton abundances and their distributions in both space and time. The light available at depth is controlled in large part by the phytoplankton and their associated products as well as the incident solar radiation. The concentrations and distributions of the nutrients, phytoplankton and zooplankton are clearly intimately related through growth, grazing, photoadaptation, nutrient cycling, etc.

One of the more intriguing feedback loops in the conceptual model is between the phytoplankton, the light field and the effects of radiant heating upon mixing. This aspect suggests that bio-optical processes can indeed affect upper ocean physics. The importance of optical water type for mixed layer heating was noted by Kraus (1972) and Denman (1973). Later, the sensitivity of upper ocean thermal structure and heat content to various water types and parameterizations was examined (Simpson and Dickey 1981a, 1981b; Zaneveld et al., 1981). The relationship of heating to phytoplankton distributions was studied by Lewis et al. (1983). The dependence of the upper ocean thermal structure on optical water type has been modeled by Dickey and Simpson (1983) and Price et al. (1986) for the diurnal cycle, by Martin (1985) for the seasonal cycle and by Woods et al. (1984) who considered both diurnal and seasonal cycles. Clearly, there are many other processes and interactions which are relevant to the system as well. It is apparent that many studies of the oceanic system require multi-disciplinary measurements.

The intent here is to provide a summary of some recent advances (emphasizing the last decade) and to suggest some future directions in the study of the upper ocean with multi-disciplinary in situ oceanographic measurement systems. The following review will include discussions of time and space scales, sensor technology, profiling devices, underway (subsurface) sampling systems, moored systems and finally directions for future developments.

2. TIME AND SPACE SCALES

Abundance, variability and diversity of marine populations are affected by biological, chemical, optical and physical processes and their interactions. The determination of the temporal and spatial scales of variability of these processes is vital to our understanding of marine populations. Haury et al. (1978) have provided an in-depth review of time and space scales and emphasized the importance of detecting the temporal and spatial changes in abundances and distributions of plankton. The relative merits of various platforms (e.g., ships, buoys, satellites, airplanes, etc.) for studying particular temporal and spatial scales have been reviewed by Esaias (1981). For a common point of reference, representations of some of the more important open ocean physical, optical and biological time and space scales are shown in Figures 2 and 3. The physical time scales of relevance correspond to the physical forcing and responses indicated in Figure 1 and include: tides, storm mixing events and inertial motions, internal waves, diurnal and seasonal incident irradiance (I_0) and heating cycles, etc. The phytoplankton distributions are related in time to the physical forcing (i.e., through light and nutrient availability and motion), particularly on storm event, tidal, diurnal and seasonal scales. The time scales for the zooplankton are in turn dependent upon phytoplankton time scales. Additionally, more subtle physiological effects such as photoadaptation and photoinhibition (e.g., Kiefer, 1973; Marra, 1978; Denman and Gargett, 1983; Lewis et al., 1984a), behavioral effects such as food perception and feeding selectivity (e.g., Strickler, 1985) and diurnal migration of zooplankton populations (e.g., Wiebe et al., 1985) cannot be neglected.

Important interdisciplinary studies have used a broad range of multi- disciplinary instrumentation to collect time series data in bays (e.g., Haury et al., 1979; Brandt et al., 1986) and coastal environments (e.g., Armstrong and La Fond, 1968; Whitledge and Wirick, 1982; Cullen et al., 1983; Booth, personal communication). However, only a few multi-disciplinary studies have been devoted to open ocean observations spanning time scales ranging from a few hours to weeks or seasons. Among these are the Optical Dynamics Experiment (ODEX) (Dickey and Siegel, manuscript) and the Biowatt (Bioluminescence and Optical Variability in the Sea) experiment (Marra and Hartwig, 1984). Two of the objectives of ODEX were to determine the temporal and vertical variability of physical and bio-optical parameters in the open ocean and to develop a model to simulate the observations. Goals of the Biowatt experiment include: 1) the identification of the dominant absorbers, scatterers and emitters of light in the upper ocean and 2) the determination and modeling of the temporal evolution of their distributions. Previously, seasonal cycle observations in the open ocean have been limited to relatively coarse (e.g., bi-weekly) observations (e.g., Menzel and Ryther, 1960; 1961). Among the disadvantages of coarsely spaced observations is the aliasing of time series. However, the currently ongoing Biowatt II experiment (to be described later) will resolve time scales of phenomena encompassing high frequency internal waves and the seasonal heating cycle (Figure 2 and Dickey et al., 1986a).

The dominant spatial scales of variability and their associated physical processes are shown in Figure 3. It is well-known that small time scale processes correlate well with small spatial scale processes and that vertical scales are smaller than horizontal scales. The horizontal patch scales for plankton have been studied extensively (e.g., Denman and Mackas, 1977; Herman, 1985). Additionally, interdisciplinary studies have been devoted to

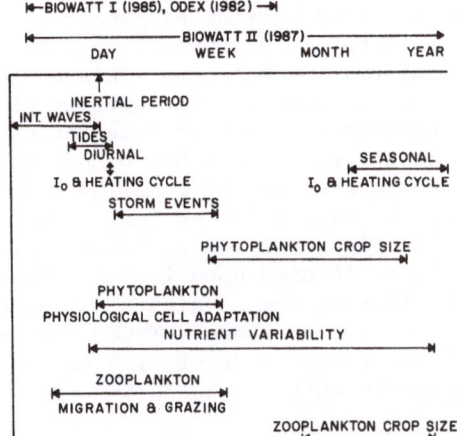

Figure 2. A summary of some of the more important physical, chemical and bio-optical time scales of variability. The time scales resolved by the multi-disciplinary field studies ODEX, Biowatt I and Biwatt II are also indicated.

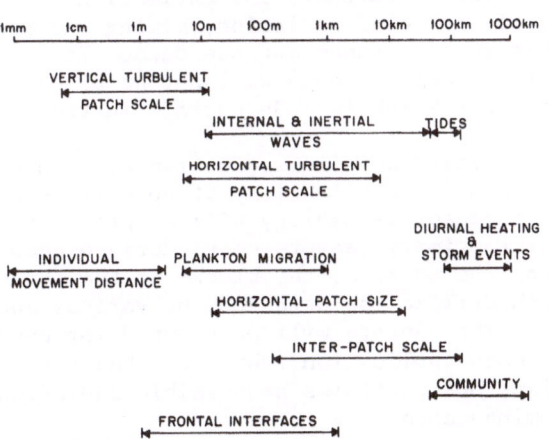

Figure 3. A summary of processes and their associated spatial scales. This figure is analogous to the time scale representation shown in Figure 1.

mesoscale rings and eddies (*e.g.*, Joyce, 1985; Simpson *et al.*, 1984) and to frontal interfaces (*e.g.*, Aiken, 1985; Arnone *et al.*, 1986). Nonetheless, there have been few multi-disciplinary investigations of the very small scales relevant to individual zooplankton ambits or to scales greater than the mesoscale (restriction by synoptic sampling).

To date, the most progress in obtaining synoptic and broad spectral (in space and time) information has been achieved in physical oceanography. For example, physical variability with spatial scales as small as a centimeter and with time scales as short as a few seconds and as long as several months has been sampled. The collection of concurrent and co-located interdisciplinary data with comparable resolution and range is imperative as indicated by the time and space diagrams shown in Figures 2 and 3. Recent progress toward achieving this goal will be focused upon in the following discussions. The optimization of sampling remains a dominant consideration for observational oceanographers. Thus, the need to minimize aliasing and to maximize the periods of time series observations as well as to maximize spatial resolution and synopticity is implicit.

3. SENSOR TECHNOLOGY

A broad range of measurements is required to analyze, interpret and model the oceanic ecosystem system as illustrated in Figure 1. Surface measurements of wind stress and solar radiation are required for external forcing (see blocks 2 and 3 in Figure 1), or more specifically for boundary conditions for fluxes of momentum and heat across the air-sea interface. Further, the estimation and modeling (or parameterization) of the solar radiation, mixing and currents within the water column require these surface data. The sensors required for these surface measurements will not be discussed here, however, particular systems which include surface instrumentation will be noted.

The development of multi-disciplinary instrumentation systems has been highly dependent upon the availability of specific sensors. Thus, a brief review of sensor technology will be presented. Table 1 lists sensors with their primary measurements, derived quantities which require data from the sensors, the blocks in the conceptual model of Figure 1 which utilize the data and the various modes of deploying the sensors. The sensors will be grouped for convenience according to their primary application: physics, chemistry, bio-optics (*e.g.*, phytoplankton, particles and visible radiation), zooplankton and bioluminescence.

3.1. Physics

Fundamental information concerning tides (block 1 in Figure 1) and motion and mixing (block 5 in Figure 1) requires current measurements (U, zonal current and V, meridional current). Some of

the devices developed recently for use in upper ocean studies are
vector measuring current meters (*e.g.*, Weller and Davis, 1980),
acoustical current measuring systems (ACM) (*e.g.*, Pinkel, 1980) and
drifters and drogues (to be discussed later). The data acquisition
systems of some vector measuring current meters may be utilized for
multi-disciplinary bio-optical and physical systems (Dickey *et al.*,
1986b, 1986c). The derived quantities which need current data
include: horizontal advection (Adv.), shear or the vertical gradient
of the horizontal currents and the gradient Richardson number (Ri)

Table 1. A summary of sensors including their primary measurements,
quantities which may be derived from the measurements, relevant
blocks in Figure 1 and modes of deployment where P=profile,
M=moored, S=submarine, T=towed and X=expendable.

Sensor	Measurement	Derived Quant.	Block	Deployment
Physical:				
Current Meter	U,V	Shear, Ri, Adv.	1,5	PMSX
ACM	U,V	Shear, Ri, Adv.	1,5	PMS
Pressure	P	Depth	4-9	PMTS
Thermistor	T	ρ, Strat., Ri	5,7,8	PMTSX
Conductivity	C	S, ρ, Strat., Ri	5,7,8	PMTX
Chemical:				
Dissolved O_2	Dissolved O_2	Water mass, Prod.	7,8	PM
Autoanalyzer	Nutrients	Water mass, Prod.	4	PMT
Bio-optical:				
PAR	PAR	$z_{1\%PAR}$, K_{PAR}	6,7,8	PMT
Spect. Irr.	Up/Dn.Irr.(λ)	$K(\lambda)$, $a(\lambda)$	6,7,8,9	PMT
Spect. Rad.	Up Rad.(λ)	Chl-a	8,9	PMT
Beam Trans.	c(660nm)	Particle Conc.	7,8,9	PMT
Fluorometer	Fluorescence	Chl-a	8	PMT
Zooplankton:				
Multiple Nets	Zoos: Size/Type/Amt.	Biomass/Grazing	7	T
Plankton Recorder	Zoos: Size/Type/Amt.	Biomass/Grazing	7	TS
Light Imaging	Zoos: Size/Type/Amt.	Biomass/Grazing	7	PT
Holography	Zoos: Size/Type/Amt.	Biomass/Grazing	7	P
Particle Counter	Particle Size/Amt.	Particle Size Distribution	7	PT
Acoustics	Zoos: Size/Amt.	Biomass/Grazing	7	PT
Bioluminescence:				
Bathyphotomter	Stimulated Biolum.	Biolum. Potential	7,9	PMTS
Photometer	Natural Biolum.	Natural Biolum.	7,9	M
CCD	Natural Biolum.	Natural Biolum.	7,9	M

which is the ratio of stratification to the square of shear. Often,
vector measuring current meters are equipped with temperature sensors
(thermistors). Some of these instruments may be used in either fixed
depth moored or vertical profiling modes and can resolve time scales
from minutes to several months or a few meters to a few hundred
meters in the vertical depending on deployment strategy.

Velocity fields have also been measured using several acoustical methods (Pinkel, 1980). Doppler acoustic systems have been used for interdisciplinary studies by Joyce and Kennelly (1985) and Barth and Brink (1987) to study Gulf Stream rings and coastal upwelling, respectively. Real-time acoustic sampling has provided especially interesting, although sometimes ambiguous, records of various phenomena in that sound energy can be reflected by fish (e.g., Holliday and Pieper, 1980), sediment particles (e.g., Orr and Hess, 1978), bubbles (e.g., Thorpe, 1982) and temperature microstructure (e.g., Thorpe and Brubaker, 1983). Some of the processes studied with acoustics include: biomass variability (Pieper et al., manuscript),internal waves and mixing events (e.g., Proni and Apel, 1975; Haury et al., 1979; Farmer and Smith, 1980; Joyce and Stalcup, 1984), turbulence (e.g., Thorpe, 1982 and 1984), microstructure (e.g., Proni and Apel, 1975; Thorpe and Brubaker, 1983), near surface waves (e.g., Farmer and Lemon, 1984) and circulations (e.g., Thorpe, 1985) and gas transfer (e.g., Thorpe, 1982).

Special velocity (shear) microstructure devices capable of resolving vertical scales on order of 2cm have also been developed. These can be used to determine the dissipation rate of turbulent kinetic energy, ε (see review by Yamazaki and Osborn, this volume). This quantity is useful because it can, in principle, be related to vertical eddy viscosity and diffusivity along with vertical turbulent fluxes (e.g., Lueck and Osborn, 1986). Recently, Lewis et al. (1986) have utilized measurements of ε to estimate the vertical turbulent flux of nitrate and Lewis et al. (1984b) have related e to algal photosynthetic rates and photoadaptation processes. Velocity microstructure probes can be used in either profiling (e.g., Caldwell et al., 1985; Gregg et al., 1986; Lueck and Osborn, 1986), horizontally towed (Osborn and Lueck, 1985a) or submarine mounted (Osborn and Lueck, 1985b) modes.

The CTD (conductivity, temperature, depth measurement system) technology has advanced at a rapid pace. The primary measurements of the CTD profiler enable the determination of conductivity (for salinity), density and stratification to scales of order 1m or less. While the CTD is typically used for standard vertical profiling, it may also be used for towed studies as described later. When these data are combined with current data, gradient Richardson numbers may be determined. Special microstructure probes are often used to resolve vertical scales in temperature down to ~1cm for temperature and conductivity and can be used to estimate vertical diffusivity and turbulent fluxes. Washburn (1986, 1987) has tow-yoed a CTD package equipped with a microconductivity probe to do horizontal-depth mapping of regions of mixing activity. The more conventional CTD sensors can also be used for long-term (months) mooring studies (examples will be given later). Using this latter mode, data are typically block averaged over several minutes which limits the short time scale resolution. In addition, the sensor suite can be deployed beneath a drifter or on a submarine. Expendable temperature and conductivity profiling probes are useful for rapid surveys of large horizontal areas.

3.2. Chemistry

For some time, CTD's have been used with rosette water sampling bottles of various volumes (usually 1.7l to 5.0l) for discrete determinations of nutrients, pigment concentrations, dissolved oxygen, etc. However, several continuous or pulsed sampling sensors of chemical, bio-optical and bioluminescent interest have been developed (Table 1). Many of these can be interfaced to CTD's with auxiliary data channels (hereafter CTD+'s) and particular current meters for various deployment modes. Some of these will be described below.

Dissolved oxygen determinations have been been done for water mass analysis and for the estimation of the net photosynthetic production (Prod.) of the upper ocean (block 8, Figure 1). The chemical titration method has been commonly used for discrete and pumped sampling determinations of dissolved oxygen. This method is not suitable for *in situ* measurements; therefore, special probes have been developed. Unfortunately, some of these probes have relatively poor reproducibility and drift characteristics at present. However, a new dissolved oxygen sensor based upon pulsed voltommetry which utilizes a membrane covered polaragraphic sensor has been developed and tested recently by Langdon (1984). This sensor has improved reproducibility and stability characteristics. The device uses discrete pulses, and thus its resolution is dependent upon the pulsing rate. Its spatial and temporal resolution are suitable for many profiling, towed and moored applications. Another recently developed device is a thick-film three electrode oxygen sensor (Karagounis *et al.*, 1986). This instrument, which can sample on vertical scales of 2 to 5cm, was used recently to study oxygen patches in an estuary (Atkinson *et al.*, 1987).

Measurements of plant nutrients (block 4 in Figure 1) have been motivated by water mass movement studies and by their role in phytoplankton primary productivity (block 8 in Figure 1). Automated chemical analysis (with autoanalyzers) has made possible rapid, accurate and precise determinations of the major nutrients including nitrate, nitrite, silicate, phosphate and ammonia. The principle techniques are segmented continuous flow analysis (see Whitledge *et al.*, 1981) and flow injection analysis (see Johnson *et al.*, 1985). The components of the former system include a sampler which acts as a timer and a sample changer, a peristaltic pump, a manifold where the samples are mixed with reagents and segmented with air bubbles, a colorimeter which measures the spectral light transmittance at specific wavelengths and a data acquisition and recording system. Between the samples, distilled water is introduced (in the sampler) so that a baseline appears between absorption peaks. The flow injection system is similar, but periodic mixing through injection of the sample (or a reagent) into a laminar flow is used. Automated nutrient analysis has been used for both profiling and towing observations.

3.3. Bio-optics

The importance of optical measurements in the ocean has been recognized for some time by biological oceanographers. The bulk of ocean optical data has been obtained from Secchi disk observations which have considerable information content (*e.g.*, Preisendorfer, 1986; Hojerslev, 1986) but also possess several well-known limitations and deficiencies. In fact, two recent compilations of world and coastal ocean optical data have been based upon these measurements (Simonot and Le Treut, 1986; Arnone *et al.*, 1984, respectively). Fortunately, within the past decade bio-optical instrumentation with sampling capabilities generally comparable to those of the CTD has been developed to determine variability of bio-optical properties. The theoretical bases and applications of many of the recent bio-optical measurements are presented in reviews by Jerlov (1976), Kirk (1983), Gordon *et al.* (1984), Yentsch and Yentsch (1984) and Blizard (1986).

In situ bio-optical measurements have several important functions. Two of the principal functions are 1) to enable the determination of the intensity and quality (wavelength) of light available for photosynthesis at depth and 2) to facilitate the identification and the quantification of phytoplankton populations (and their growth rates) and their products. The sensors described below may be used for one or both of these purposes and provide virtually continuous sampling capability (*e.g.*, Yentsch and Yentsch, 1984). Thus, vertical resolution comparable to CTD's (few meters or less) and temporal resolution comparable to moored current meters (few minutes) may now be attained for the sampling of several bio-optical water properties.

Photosynthetically available radiation (PAR) sensors measure the flux of quanta or the wavelength weighted integral of spectral scalar irradiance in the visible waveband (~350-700nm) using a spherical light collector. One of the more commonly used PAR sensors is described by Booth (1974). Also, a logarithmic PAR sensor has been developed by Aiken and Bellan (1986a). The importance of the measurement of PAR is that it quantifies the amount of radiation (number of quanta per unit area per unit time) received at a given depth in the water column. This is important for both describing and modeling photosynthetic processes (Siegel and Dickey, manuscript). A common, although crude, figure of merit which may be derived from PAR profiles is the depth where the *in situ* PAR is reduced to 1% of its surface value or the euphotic zone depth ($z_{1\%PAR}$). Another important derived parameter is the diffuse attenuation coefficient for PAR, K_{PAR}, which quantifies the rate of loss of quantum flux with depth. This has been considered in more detail by Siegel and Dickey (1987 and manuscript).

A more sophisticated optical instrument for quantifying the oceanic photoenvironment is the multiwavelength spectroradiometer. One of the commonly used spectroradiometers is described by Smith *et al.* (1984). This particular spectroradiometer measures either downwelling or upwelling vector irradiance in ~12 wavebands of ~10nm bandwidth ranging from ~380-770nm. Data taken with this type of instrument have been presented recently (e.g., Smith and Baker, 1984; Siegel and Dickey, 1987). The diffuse attenuation coefficient for each of the wavebands, $K(\lambda)$, can be used to estimate specific light absorption, a, as a function of wavelength and depth (e.g., Morel and Prieur, 1977). For particular wavebands, profiles of these coefficients have been shown to be well-correlated with important bio-optical variables including fluorescence, chlorophyll-a and phaeopigments (e.g., Siegel and Dickey, 1986). However, due to the limitations of irradiance filters, the depth to which $K(\lambda)$ can be estimated is restricted (Tyler, 1959; Siegel et al., 1986). An instrument which simultaneously measures both scalar and vector (upwelled and downwelled) irradiance spectrally (11 wavebands between 380 and 700nm) has been developed by Spitzer and Wernand (1981) to determine absorption spectra based upon the conservation of energy. This general principle has been utilized by Booth (personal communication) for moored (see Section 6.2) and profiling instrumentation.

Data obtained from upwelling spectral (~440-550nm) radiance sensors are used for groundtruthing remotely sensed color (e.g., airborne oceanographic lidar, AOL, and satelliteborne Coastal Zone Color Scanner, CZCS). Satelliteborne imaging sensors have been used to estimate pigment (chlorophyll-a and phaeopigment) concentrations (Gordon and Morel, 1983). The specific wavebands chosen for the sensors are based upon absorption spectra of chlorophyll-a, phaeopigments, dissolved organic matter (DOM, e.g., yellow substance), etc. (e.g., Yentsch and Yentsch, 1984).

It should be noted that optical devices which depend upon natural sunlight (measure apparent optical properties and downwelled light) are vulnerable to natural variability caused by clouds, surface wave refraction, etc., which can occur during the course of profiling observations. Thus, observations analogous to those done in the water column should be done at the surface if possible (e.g., Smith and Baker, 1984). Other complications can arise because of platform or ship shadowing (e.g., Gordon, 1985; Voss et al., 1986) and non-vertical orientation of instruments such as the multiwavelength spectroradiometer which uses a cosine collector. In principle, these sensors can be deployed in any of the various modes discussed previously (except for expendable) with resolution comparable to the standard CTD variables. However, the aforementioned difficulties, along with biofouling for moored operations, must be circumvented. This group of recently developed sensors is particularly important in that data derived from these sensors are directly relevant to five of the blocks (3, 6, 7, 8 and 9) of the conceptual model in Figure 1.

Fluorometers are used to obtain nearly continuous records of fluorescence in order to estimate chlorophyll-a concentration and to infer phytoplankton pigment biomass. This measurement utilizes the fact that chlorophyll-a, which is the major light sensitive pigment used in photosynthesis, is a fluorescent molecule. The fluorometer's blue light source illuminates a test volume of seawater containing the phytoplankton with their chlorophyll-a cells. The cells fluoresce red light which is detected and the signal is processed. Fluorometers are used both with pumping systems and *in situ*. *In situ* fluorometers are used in profiling and towed modes (*e.g.*, Platt, 1972) and most recently in moored mode (Whitledge and Wirick, 1983; Dickey *et al.*, 1986d). Provision for biofouling must be made for the latter mode however. Their sampling is not strictly continuous because of the required excitation pulses and thus the sampling rate leads to spatial and temporal resolutions which are slightly poorer than those of standard CTD variables. Interpretation of the fluorometric measurement as phytoplankton concentration (through a standard calibration) is complicated by several factors. For example, the carbon to chlorophyll-a ratio is not constant because of environmental stresses such as light and nitrogen deficiency (*e.g.*, Yentsch and Yentsch, 1984). Other factors include: the presence of accessory pigments, species specific variation, light history and growth state. While these effects complicate the relationship between fluorescence and phytoplankton pigment biomass, it has been suggested that spectral signatures may be used to derive important information such as taxonomic grouping based upon pigments (Yentsch and Phinney, 1985). A multichannel *in situ* fluorometer capable of measuring variables related to bioluminescence, several pigments and chlorophyll-a is described by Yentsch and Yentsch (1984). Also, dissolved organic material (DOM, *e.g.*, yellow substance) specific fluorometers have been in use for some time (*e.g.*, Jerlov, 1976).

When used in the laboratory for discrete samples or with pumping systems, fluorometers can be used to determine concentrations of phaeopigments by acidifying samples and repeating the measurements since phaeopigments are degradation products of chlorophyll-a. The degradation occurs within the guts of zooplankton which have grazed upon the phytoplankton. Phaeopigments absorb light in the blue as do the chlorophyll-a pigments, thus relatively great spectral resolution is required to distinguish the two. In fact, the CZCS with a limited number of spectral bands could not be used to distinguish the two; hence, the pigment concentrations measured were actually comprised of both chlorophyll-a and phaeopigment concentrations (Smith and Baker, 1978; MAREX, 1982). In addition, dissolved organic material (DOM, *e.g.*, yellow substance) can further confound the determination of chlorophyll-a concentrations. The relative contributions will vary but the chlorophyll-a and phaeopigment concentrations may be of the same order of magnitude in the open ocean and yellow substance may be quite important in estuaries and coastal regions. The estimation of phytoplankton biomass distributions (block 8, Figure 1) is important for the light field (block 9, Figure 1) since the phytoplankton and their detrital products (in the form of phaeopigments) are often the primary attenuators of light. Phytoplankton biomass is also important for the consumption of plant nutrients (block 4, Figure 1) and for a source of food for higher trophic levels (*e.g.*, grazing zooplankton; block 7, Figure 1).

Beam transmissometers measure an inherent optical property of seawater, the beam attenuation coefficient, c. One of the more

commonly used transmissometers (Bartz *et al.*, 1978) will be
described. The light source for the device is a light emitting diode
and the beam of collimated light is received by a silicon
photodetector. The typical wavelength used (~660nm) is chosen to
minimize the absorption of light due to humic acids (yellow
substance). The beam attenuation coefficient relates to the volume
of suspended matter or particle concentration in the water column
(blocks 7, 8 and 9 in Figure 1) and is useful for water mass analysis
(Spinrad, 1986a). The instrument can be used in most deployment
modes and its vertical resolution and temporal response are
comparable to those of the optical sensors described above.
Biofouling may be a problem for long term deployments. Performance
characteristics and recent data obtained with beam transmissometers
may be found in Bishop (1986) and Spinrad (1986b), respectively.

3.4. Zooplankton

To determine speciation and abundances of zooplankton and
micronekton (block 7, Figure 1), multiple opening and closing net
systems are frequently used (*e.g.*, see Wiebe *et al.*, 1976 and Ortner
et al., 1982). Two of these systems are the MOCNESS (Wiebe *et al.*,
1976) and the BIONESS (Sameoto et al., 1980), which follow a system
developed by Frost and McCrone (1974). These utilize direct methods
of sampling zooplankton opposed to the indirect methods which will be
discussed later. Briefly, nets are opened at up to eight depths as
the system is towed. Mesh sizes range from ~0.06mm to over 1mm and
vertical and horizontal scales as small as a few meters and ~100m can
be sampled, respectively. Important sampling parameters such as net
speed, filtering efficiency, net depth, net orientation, etc. are
measured. In addition, other system sensors may include thermistors,
conductivity sensors, fluorometers, dissolved oxygen sensors and
light sensors. The advantages of these systems include: capability
of species determination, real-time data acquisition and minimal
intersample contamination. Towed deployments are required. Thus,
vertical profiles are precluded and net avoidance is of concern.
    Another direct zooplankton observation method is based upon the
Hardy Continuous Plankton Recorder which utilizes a sequentially
stepped gauze roll to enable the collection of up to 1000 samples
(see review by Haury *et al.*, 1976). The method can resolve vertical
scales of a few meters and horizontal scales of tens of meters.
Disadvantages of the method include sample integrity and difficult
analysis. This technique has been utilized by Haury *et al.* (1979).
    Pump-based systems have also been used to directly sample
zooplankton (*e.g.*, Ortner *et al.*, 1982). With this method, samples
are pumped from known depths (determined with a pressure transducer)
to the surface for analysis. Relatively high vertical and horizontal
resolution (a few meters or less in the vertical and tens of meters
in the horizontal) can be achieved in profile or tow modes. The pump
system can also be utilized for various other measurements
(nutrients, fluorescence, particle sizes and concentrations, etc.).

Disadvantages include difficulty in sampling large, fragile and motile organisms.

Light imaging systems for zooplankton observations have been developed relatively recently. An *in situ* photographic system consisting of a 35mm camera and a strobe light has been described by Ortner *et al.* (1981). Spatial resolution of ~1m may be achieved and organisms as small as ~0.1mm may be identified. Another system developed by Shulenberger and Lange (see description in Ortner *et al.*, 1982) employs a towed body with a strobe light, a video camera, CTD, a fluorometer and an acoustic current meter. This system is capable of recording an extremely large amount of data in real- time. One of the system's limitations is its technical complexity which impedes its general usage.

Electronic particle counters have been used to determine zooplankton abundance and physical dimensions indirectly (Herman and Dauphinee, 1980). A sampling net is used to funnel zooplankton into a cell where the change in electrical conductance relating the displacement of the seawater electrolyte by the zooplankton is detected. Spatial patterns of ~1m can be resolved for copepods ranging in size from about 0.5 to 3.0mm. One of the problems with this technique is net clogging by the zooplankton. To circumvent this problem, Herman (1986) has developed an optical plankton counter which can also be used to measure light attenuation and chlorophyll-a fluorescence. With this modification, information related to blocks 7, 8 and 9 can be obtained.

Sound scattering has been used to locate fish populations for some time. However, only recently has this principle been exploited with sophisticated acoustical techniques. Pieper *et al.* (manuscript) have described the Multi-frequency Acoustic Profiling System (MAPS), which is capable of resolving small spatial scales (a few meters in the vertical) of zooplankton by size class distributions. The MAPS system will be described in more detail later. In addition to sound scattering methods, Farmer and Huston (this volume) have suggested that coherence and phase acoustical information may be useful for the study of physical and biological fine scale processes.

3.5. Bioluminescence

Most oceanic species of fish (approximately 65-75%) bioluminesce (Nealson and Arneson; 1985); yet our knowledge of bioluminescent organisms (blocks 7 and 9, Figure 1), their behaviors, their distributions and their mechanisms for bioluminescent stimulation is quite limited. One indirect measurement of bioluminescence is the determination of the luciferin molecule using an *in situ* multichannel fluorometer (described earlier, Yentsch and Yentsch, 1984). Another bioluminescence measurement technique entails the artificial stimulation of bioluminescence. The measurement of artificially stimulated bioluminescence or bioluminescence potential in units of photons emitted per cell has been described recently by Losee *et al.* (1985) and Swift *et al.* (1985). Both use bathyphotometer devices which utilize submersible pumps for generating flow through a constriction or an impeller which results in turbulence and stimulated bioluminescent light which is measured with photomultiplier tubes. These systems, which have been used with CTD sensors along with fluorometers and PAR sensors, may be deployed in profile or tow modes with resolution comparable to CTD's. A moored system designed for natural as well as stimulated bioluminescence measurements will be described in Section 6.3.

Stimulated bioluminescence typically produces several orders of magnitude of light more than natural bioluminescence (Swift *et al.*, 1985), thus measurements of natural bioluminescence are quite demanding. The use of photometers (6 wavebands) and charge coupled devices (CCD's with up to 64 wavebands) for the sampling of bioluminescent light from a nearshore mooring (with current, pressure and temperature sensors) in Scripps Canyon near La Jolla, California has been described by Nealson *et al.* (1984) and Nealson and Arneson (1986). They ascertained that characteristic light signatures can be distinguished for individual species.

4. PROFILING SYSTEMS AND PLATFORMS

Many of the sensors described above may be considered to be modules, which can be interfaced with submersible packages including data acquisition systems and microprocessors. The primary goals of the multi-disciplinary *in situ* measurement devices are to: 1) sample with the various complementary, interdisciplinary sensors as closely in space and time as possible and 2) resolve temporal and spatial scales of the ecosystem so as to avoid aliasing (Figures 2 and 3). A variety of systems will be described below. The difficult choice of which systems to describe was based upon representativeness, but is inherently arbitrary and biased toward applications to bio-optical and physical processes. Pertinent references to data taken with each of these systems are provided in the text below.

4.1. R/P FLIP

As an introduction to several profiling devices, the R/P FLIP component of ODEX (Dickey and Siegel, manuscript) will be described. The R/P FLIP was used to make concurrent observations of bio-optical, physical and meteorological variables (see Figure 4 and Table 2). The data obtained from the R/P FLIP observations are relevant to questions concerning air-sea exchanges of heat and momentum, upper ocean dynamics and thermodynamics, the underwater photoenvironment and the upper ocean ecosystem in general (all blocks in Figure 1 except for block 7, zooplankton). During ODEX (October 20 through November 12, 1982), the R/P FLIP was used to deploy three separate profiling packages to a nominal depth of 250m in the central North Pacific Ocean (~32N 142W). An extensive set of meteorological data was also taken concurrently. The CTD+ package included a rosette water sampler, a beam transmissometer, a fluorometer and a dissolved oxygen sensor. Profiles were made at intervals ranging from ~15min

to ~4h. Using water samples from the rosette, discrete
determinations of salinity along with dissolved oxygen, nutrient and
pigment concentrations were made. An autonomous profiling package
(modified cyclesonde; Van Leer, 1980) was used to determine the
vertical shear of horizontal currents, temperature and depth on an
hourly cycle. An optics package consisting of a spectroradiometer
(for 12 wavebands with bandwidths of 10nm ranging from 410 to 767nm
in the visible spectrum) for downwelling spectral irradiance, a
thermistor, a pressure sensor and a beam transmissometer was used to
obtain vertical profile data every few hours during daylight. The
time and vertical scales of sampling resolution were somewhat
variable but nominally a few hours or less and a few meters (after
bin averaging) in the vertical, respectively. Some of the
interesting results of this experiment include a high correlation
between the depth of the particle maximum and the depth of the

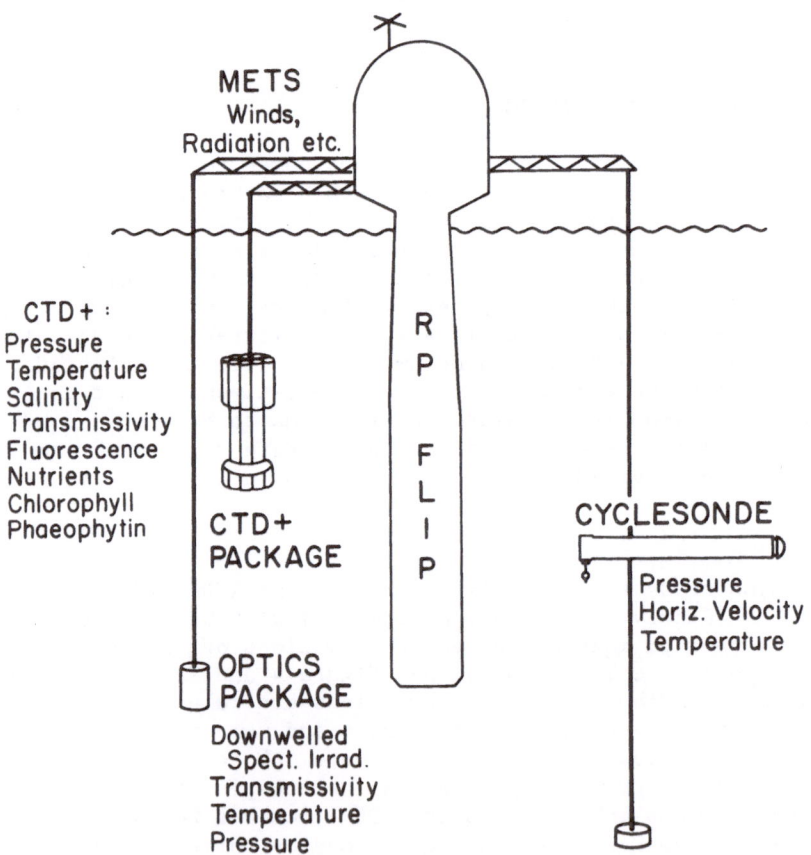

Figure 4. A depiction of the instrumentation deployed from the R/P
FLIP which was used for the multi-disciplinary experiment ODEX.

mixed layer and observations of coincident fine scale structure of bio-optical and physical variables (Dickey et al., 1986d). In addition, correlations between vertical profiles of the spectral diffuse attenuation coefficient and other bio-optical variables such as beam transmission, fluorescence and chlorophyll-a and phaeopigment concentrations were evaluated (Siegel and Dickey, 1987). Other results from this study will be presented in the near future. The R/P FLIP was also used by Weller et al. (1985) to do physical and biological measurements which will be described later.

4.2. Bio-optical Profiling System (BOPS)

The bio-optical profiling system (BOPS; Smith et al., 1984) was designed to collect concurrent meteorological, optical, biological and physical data in the upper 200m of the ocean in order to study radiant transfer processes, coupling of physical and bio-optical mechanisms, primary productivity and optical variability (blocks 2,3,4,6,8 and 9 in Figure 1). The system has been used for real-time synoptic sampling of the upper ocean photoenvironment and for providing surface data for groundtruthing remotely sensed optical observations (CZCS and AOL) related to pigment concentrations (Smith et al., 1984; Smith and Baker, 1985). The BOPS (see Figure 5 and Table 2) consists of a CTD with a 12-bottle rosette, a PAR sensor, a

Table 2. Summary of several multi-disciplinary platforms and systems including measurement capabilities, quantities which may be derived from measurements, relevant blocks in Figure 1 and deploymentr modes. Abbreviations and symbols are given in the text and in caption for Table 1.

System	Measurement	Derived Quant.	Block	Deploy
FLIP	Mets., C, T, P, c, Fl, Dn Spect.Irr.(12λ's), Ros.Samples, Horiz.Curr.	S, ρ, Strat., Part.Conc, K(λ), K_{PAR}, Shear, Adv., Ri	1,2,3,4,5,6,8,9	P
BOPS	Mets., C, T, P, c, Fl, PAR, Up/Dn Spect. Irr.(12λ's), Up Spect.Rad.(4λ's), Ros. Samples	S, ρ, Strat., Part.Conc., 1% Light, K(λ), K_{PAR}, a(λ), Chl-a.	2,3,4,6,8,9	P
MVP	C, T, P, Fl, PAR, Horiz.Currents	S, ρ, Strat., Shear, Ri, Adv., 1% Light, K_{PAR}, Chl-a	1,5,6,8,9	P,M
Batfish	C, T, P, Fl, Zoo.Samples	S, ρ, Strat., Chl-a, Biomass	7,8,9	P,T
UOR	C, T, P, Fl, Up/Dn Spect.Irr.(4λ's), PAR, Zoo.Sampling	S, ρ, Strat., Chl-a, Zoo Size & Amt., K_{PAR}, K(λ)	6,7,8,9	P,T
MAPS	Zoo's Using 21 Acoust. Freq., PAR,C, T, P, Fl	S, ρ, Strat., chl-a, Zoo Size & Amt., Biomass, K_{PAR}	6,7,8	P,T
SCMS	T, P, Horiz. Curr., Fl, c, Up/Dn Scal.Irr. (3λ's), Up Vect.Irr.(8λ's), Dn Vect.Irr. (13λ's), Up Rad.(8λ's), PAR	Chl-a, Reflectance Spectra, Particle Conc., Various Optical Quantities, Adv., Natural Fluor.	1,5,6,8,9	M
Biowatt Mooring Program	Mets.& Sfc.Rad., C, T, Fl, Horiz.Curr., c, DO_2, Up/Dn Irr. (5λ's), Up Rad.(6λ's), PAR, Biolum.	S, ρ, Strat., Shear, Ri, Adv., K(λ), K_{PAR}, Part.Conc.,Chl-a, Resp., Natural and Stimulated Biolum.	1,2,3,5,6,7,8,9	M

spectroradiometer for downwelling irradiance (12 wavebands with bandwidths of 10nm ranging from 410 to 767nm), a spectroradiometer for upwelling irradiance (same characteristics as those of the downwelling irradiance spectroradiometer), a spectroradiometer for upwelling radiance (wavebands of 441, 488, 520 and 550nm with bandwidths of 10nm; except for the 488nm waveband, these wavebands match those of the CZCS onboard the Nimbus-7 satellite), a beam transmissometer and a fluorometer. The spectroradiometers are microprocessor controlled. The system also includes a deck mounted spectroradiometer for downwelling irradiance (12 wavebands ranging

4.3. Multi-Variable Profiler (MVP)

Another system, the multi-variable profiler (MVP), was developed by our group in collaboration with John Marra, John Van Leer and Chuck Abbott for Biowatt I to do measurements similar to those made from the R/P FLIP during ODEX. In order to achieve some of the physical objectives for Biowatt I, the cyclesonde used for ODEX was reconfigured as shown in Figure 6 (also see Table 2). An electromagnetic vector measuring current meter, an additional

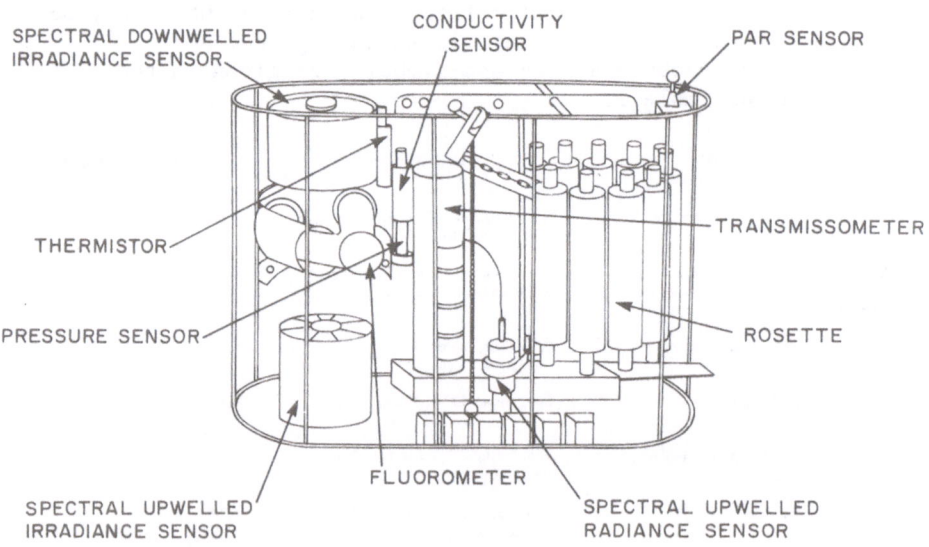

Figure 5. The bio-optical profiling system (BOPS) developed by Smith *et al*. (1984).

temperature sensor, a conductivity sensor, a PAR sensor and a fluorometer were added to the sensor suite. The MVP may be tethered to a surface buoy and can operate independently of an attending ship in order to enable other shipboard profiling to be done concurrently. Data are recorded internally on magnetic tape and simultaneously transmitted via radio transmissions to an attending vessel for real-time data acquisition. The MVP makes hourly profiles through the upper ~200m of the water column using helium gas as a buoyancy medium. The vertical resolution is a few meters and depends on the vertical profiling speed, the sampling frequency and vertical averaging. The MVP was tested in the Tongue of the Ocean (Dickey *et al.*, 1986b) and was used subsequently during Biowatt I in the Sargasso Sea (Dickey *et al.*, 1986c).

The MVP can also be used in a moored mode in a coastal environment (*e.g.*, Dickey and Van Leer, 1984; Stacey *et al.*, 1986) where data tapes and batteries can be exchanged periodically (~monthly). In another study, Van Leer and Villanueva (1986) used a conventional cyclesonde (rotor current sensors, pressure sensor, thermistor, etc.) equipped with a beam transmissometer and a PAR sensor in the Fram Strait marginal ice zone to obtain vertical profiles of physical and bio-optical properties to a depth of ~200m.

5. UNDERWAY DEVICES

5.1. Batfish

The Batfish system is one of the better known towed systems. It was designed to sample vertical and horizontal scales of variability of physical and biological parameters (see review by Herman, 1985 and Figure 1 blocks 7, 8 and 9). The Batfish vehicle (see Figure 7 and Table 2) is towed behind a ship, resulting in a sawtooth undulating pattern to depths up to ~400m. Continuously sampled variables include: pressure, temperature, conductivity and fluorescence. Zooplankton sampling has been accomplished with a net sampler, an *in situ* electronic particle counter (Herman and Dauphinee, 1980) and most recently with an optical plankton counter (Herman, 1986). The latter two devices were described earlier. The horizontal patch scales of interest for Batfish studies are ~5-10km and thus require successive vertical profiles on horizontal scales less than ~1km. The spatial resolution is a few meters or less. Sampling considerations and criteria have been reviewed by Denman and Mackas (1978) and many of the important results obtained with the Batfish have been summarized by Denman and Mackas (1978) and more recently by Herman (1985).

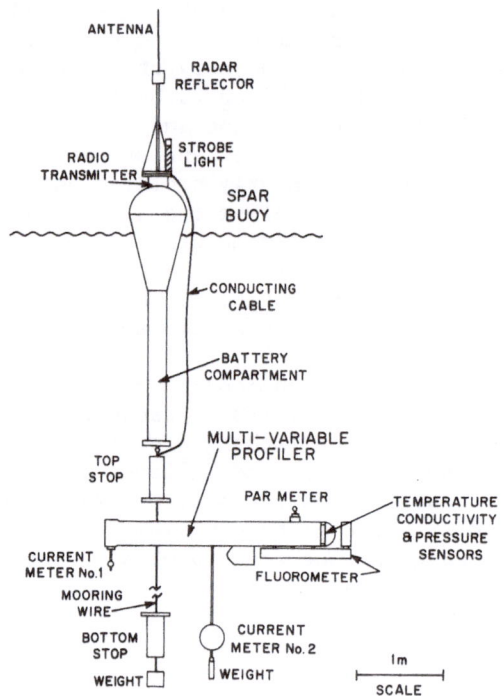

Figure 6. The multi-variable profiler (MVP) as deployed from a free drifting spar buoy.

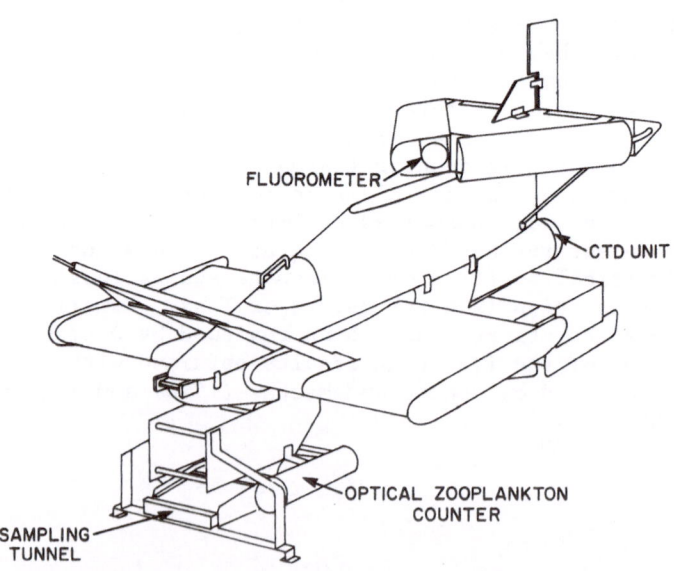

Figure 7. The Batfish system as modified recently by Herman (1986).

Another system, the Kiel Sea Rover, includes the Tow-Fish which is similar to the Batfish. It has been developed for seasonal and regional studies (*e.g.*, Fischer, *et al.*, 1985; Woods *et al.*, 1986; Strass and Woods, this volume). Variables measured with the Tow-Fish include: pressure, temperature, conductivity, solar radiation and fluorescence. Data taken with this system are complemented with shipboard derived acoustic Doppler current profiles.

5.2. Undulating Oceanographic Recorder (UOR)

The original undulating oceanographic recorder (UOR) was developed at approximately the same time as the Batfish and also has continued to evolve. The UOR (see Figure 8 and Table 2) differs from the Batfish in that an extensive set of optical sensors is included. The UOR is designed to make observations relevant to studies of primary productivity, ichthyoplankton, physical and optical properties and remote sensing of ocean color (Figure 1, blocks 6-9). The most recent version of the UOR (Mark 2) vehicle (Aiken, 1981; Aiken, 1985; Aiken and Bellan, 1986b) undulates from near the surface to depths of ~200m with successive undulating horizontal scales of ~800- 4000m. The system design was motivated by the intent to operate similar systems using ships of opportunity. The UOR sensor suite includes: a pressure sensor, a thermistor, fluorometers for chlorophyll-a fluorescence (ambient light and dark adapted), downwelling and upwelling hemispherical irradiance sensors (450, 520, 550 and 670nm; near the wavebands of the CZCS) and a PAR sensor. Other sensors which can be accommodated by the system are dissolved oxygen sensors, pH electrodes and particle counters. Recent data taken with the UOR in the English Channel have been presented by Aiken and Bellan (1986b).

5.3. Multi-Frequency Acoustic Profiling System (MAPS)

The Multi-frequency Acoustic Profiling System (MAPS) is a novel system which was developed to examine scales of variability and distributions of several biological and physical parameters (blocks 6, 7 and 8 in Figure 1) simultaneously (*e.g.*, Holliday and Pieper, 1987; Pieper *et al.*, manuscript). The potential application of the system to the fish recruitment problem has also been discussed by Holliday and Pieper (1986). The system (see Figure 9 and Table 2) may be deployed in either profiling or towing (sawtooth) mode. The system consists of a submersible vehicle with an array of 21 side-mounted acoustic transducers (frequencies range from 0.1 to 10MHz), a pressure sensor, a thermistor, a conductivity sensor, a light sensor, a submersible pump and hose (2.5cm diameter) for chlorophyll-a fluorescence sampling onboard ship and another hose (7.5cm diameter) which is connected to a shipboard pumping system for sampling zooplankton. The pumping systems are used to measure the chlorophyll-a concentrations and to determine the taxonomic composition and abundances of the zooplankton sampled with the acoustic sensor array. The acoustic transducers sample organisms ranging in equivalent

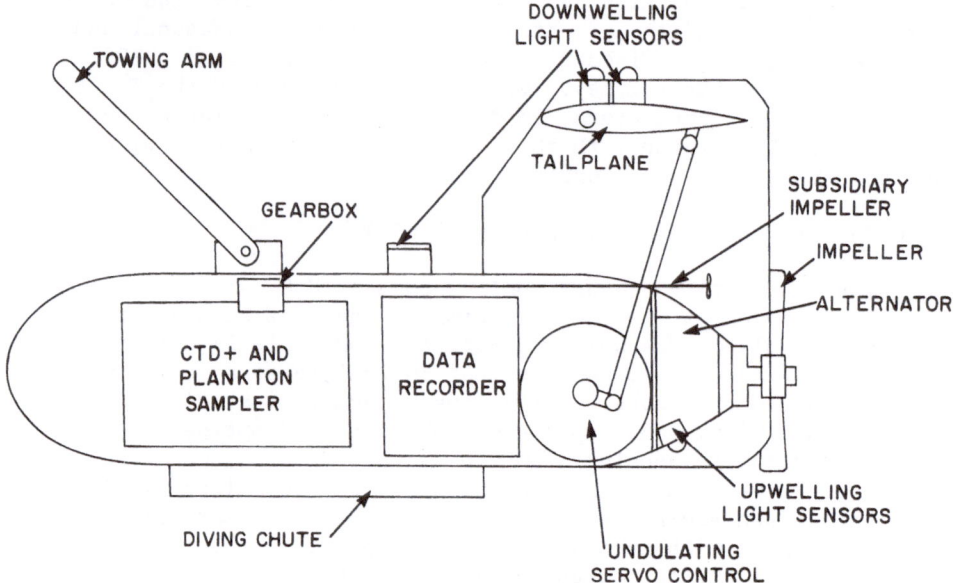

Figure 8. The Mark 2 undulating oceanographic recorder (UOR)
developed by Aiken (1985).

spherical radii from roughly 0.1 to 1.0mm (or animal lengths of ~0.4
to ~4mm) in a test volume of ~0.01m3 at a range of 1 to 2m from the
sensors every 0.67sec. Data are collected in real-time using two
computers, one for data acquisition and another for real-time data
processing (binning, contouring, etc.). The data are averaged in 2m
vertical intervals and the acoustic volume scattering data are
transformed into size-class estimates of zooplankton abundance and
biomass.

A study conducted with MAPS in Santa Monica Bay, California in
1982 revealed that the variability in plankton was often coherent
with physical features; however, departures were also observed
(Pieper *et al.*, manuscript). It was suggested that these may be
caused by complexity within the physical environment or by
behavioral responses of the plankton. The MAPS system along with an
acoustic Doppler current measuring system, a MOCNESS, an *in situ*
photographic system and a CTD was used for recent studies in the
Gulf Stream off the coast of Palm Beach, Florida and off the coast
of southern California. Preliminary results of the first 1985 study
have been presented by Ortner *et al.* (1986) and Nieman (1986).

6. MOORED SYSTEMS

The systems described in the previous section were primarily
designed to resolve spatial scales of variability. Relatively high
frequency time series interdisciplinary sampling has been done by
repeated profiling from a shallow water platform (*e.g.*, Armstrong
and La Fond, 1968), using anchored ships (*e.g.*, Brandt *et al.*,
1986), in combination with a ship and a current meter mooring (*e.g.*,
Haury *et al.*, 1979; Cullen *et al.*, 1983), from R/P FLIP (described
in Section 4.1 and Weller *et al.*, 1985) and by using the MVP (also
described above). The duration of the sampling for these studies,
which is limited by the availability of a ship or platform, is
typically a few days to a few weeks. Thus, phenomena with longer
time scales (*e.g.*, synoptic, seasonal and interannual variability)
cannot be studied in this manner. To increase the time domain of
bio-optical sampling, moored instrumentation may be used. However,
the selection of appropriate sensors and inherent constraints such
as power consumption, data storage, biofouling, etc. are of special
concern for this method of deployment.

6.1 Nearshore Moored Fluorometers

The potential of obtaining moored bio-optical data has only begun to
be realized. Moored observations of chlorophyll-a fluorescence at
one depth were made during 10 and 11 day observational periods in a
nearshore continental shelf region by Whitledge and Wirick (1983). A
separate mooring with current meters along with temperature and
conductivity sensors was located ~30m away from the fluorometer
mooring. Considering both moorings, data relevant to blocks 1, 5 and
8 in Figure 1 were taken. The fluorescence time series was well-
correlated with temperature at lower frequencies and with currents at
higher frequencies. It was suggested that diel chlorophyll-a
concentration variability may have resulted from a combination of
nighttime grazing and daytime phytoplankton growth. Other
complicating factors, as mentioned earlier, include phytoplankton
photoadaptations, which can change the amount of chlorophyll-a per
cell and the fluorescence yield per cell (Falkowski and Kiefer,
1985). Recently, the Shelf Edge Exchange Processes (SEEP) study near
the shelf break off the coast of Nantucket Island, Massachusetts has
utilized moorings instrumented with current meters, thermistors, beam
transmissometers, fluorometers and sediment traps (Aikman, personal
communication).

6.2. Nearshore Bio-optical and Physical Mooring

A subsurface spectroradiometer-based monitoring system has been
developed by Booth (personal communication) who is continuing to test
the system at Scripps Canyon, California. The Scripps Canyon moored

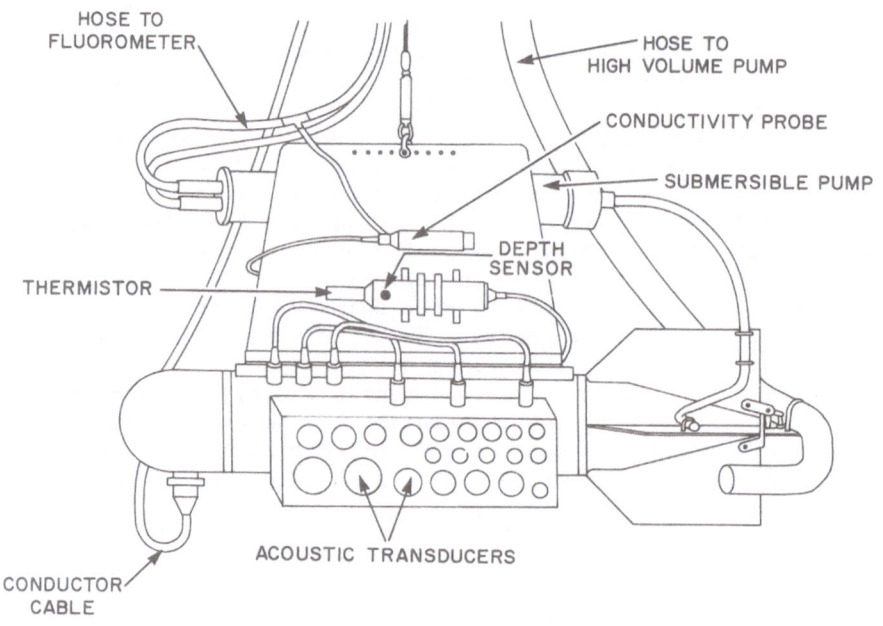

HOSE TO
FLUOROMETER

HOSE TO
HIGH VOLUME PUMP

CONDUCTIVITY PROBE

SUBMERSIBLE PUMP

DEPTH
SENSOR

THERMISTOR

CONDUCTOR
CABLE

ACOUSTIC TRANSDUCERS

Figure 9. The multi-frequency acoustical profiling system (MAPS)
developed by Holliday and Pieper (1987).

system (SCMS in Table 2) is deployed at a depth of ~8m and includes
sensors to measure: spectral upwelling and downwelling scalar
irradiance (3 wavebands each), spectral upwelling and downwelling
irradiance (8 and 13 wavebands, respectively), spectral upwelling
radiance (8 wavebands), PAR, pressure (for depth), platform
orientation, conductivity, temperature at 6 depths, beam
transmission, active and passive (natural) fluorescence and currents
(blocks 1, 3, 5, 6, 8 and 9 in Figure 1). A sea cable link to shore
is used to obtain real-time data. Periodically, vertical profiling
with similar sensors and discrete sampling are done at the site. The
latter samples are used for the following analyses: pigments,
nutrients, particles, primary productivity rates, total seston, total
particulate organic carbon and nitrogen, cyanobacteria, particulate
absorption and fluorescence spectra. These on-going observations
provide information concerning several bio-optical properties on
time scales heretofore unattainable and have the potential for
groundtruthing remotely sensed ocean color observations.

6.3. Biowatt Open Ocean Bio-optical and Physical Mooring

The mooring programs described above have been conducted in nearshore environments and with a limited number of instrument packages for determination of vertical structure. However, there is clearly a need for open ocean long-term time series data as well. As mentioned earlier, the importance of longer (seasonal and interannual) time series data for biological productivity studies was recognized by Menzel and Ryther (1960, 1961) who conducted bi-weekly experiments at Ocean Station S near Bermuda for approximately 3 years. One of the primary objectives of Biowatt is to develop predictive models of temporal and spatial variability of optical properties and bioluminescence in the open ocean. A time series of bio-optical and physical variables spanning the seasons is thus needed to provide a broad dynamic range in variables and to observe major changes which often occur in episodic fashion. To this end, a strategy to obtain bio-optical and physical data on time scales ranging from a few minutes to about 10 months (spanning the seasons; see Figure 2) was developed for the Biowatt II field study (Dickey et al., 1986a). The only economically feasible approach was to develop a set of moored instruments to measure the relevant bio-optical and physical parameters at as many depths as possible in order to maximize both temporal and vertical resolution.

Three separate types of in situ instrument packages have been developed for the mooring, which is located in open ocean waters of the eastern North Atlantic Ocean (34N 70W). The Multi-variable Moored System (MVMS) (Figure 10 and Table 2) is the first type of instrument package (8 total on mooring). It utilizes a vector measuring current meter (VMCM; Weller and Davis, 1980) for data acquisition and system control as well as horizontal current and temperature sensors. The availability of auxiliary channels enabled the interfacing of other sensors to the system. In addition to the current and temperature sensors, the MVMS system includes a fluorometer, a PAR sensor, a beam transmissometer, a dissolved oxygen sensor and a conductivity sensor. During the past year, the system was developed by our group in collaboration with a group from Lamont-Doherty Geological Observatory led by John Marra. The system has been tested at the Scripps Canyon mooring site, on the continental shelf near Palos Verdes, California, and near Santa Catalina Island, California. The second instrument package type (developed by C. R. Booth and Ray Smith) is a bio-optical measurement system (BOMS), which consists of a microprocessor and a data acquisition system with spectral downwelling irradiance sensors (410, 441, 488, 520 and 560nm), spectral upwelling radiance sensors for the same wavelengths along with 683nm for passive (natural) chlorophyll-a fluorescence, a thermistor, a pressure sensor and sensors for two axes of orientation. The final in situ package (developed by C. R. Booth and Elijah Swift) is the bioluminescence moored system (BLMS). This package also utilizes a microprocessor and data acquisition system. Two photomultiplier tubes are used to sample unstimulated bioluminescent light during nighttime. Stimulated bioluminescence

MULTI-VARIABLE MOORED SYSTEM
MVMS

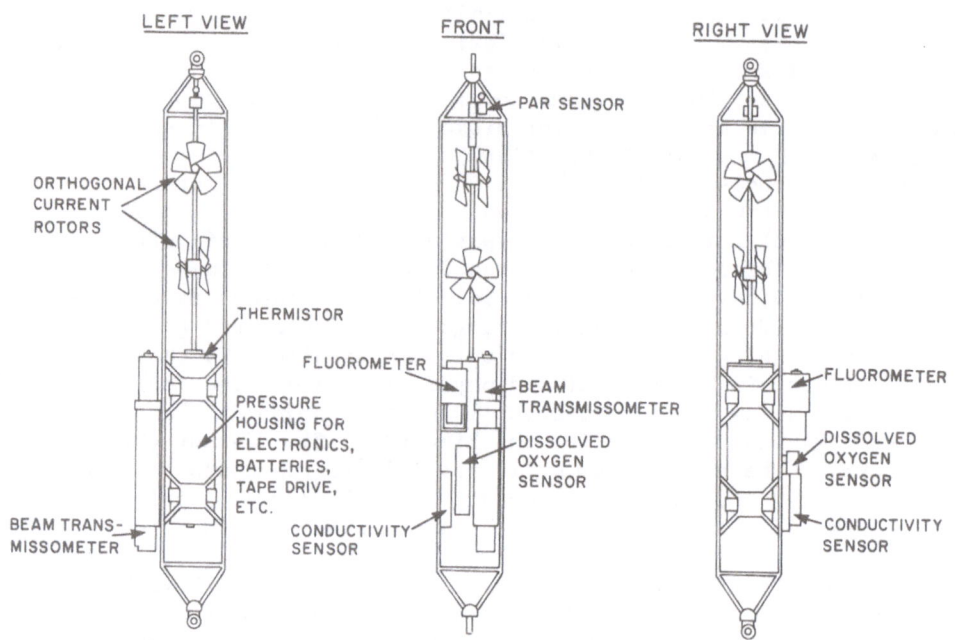

Figure 10. The multi-variable moored system (MVMS) currently being used for the Biowatt II experiment.

for the determination of bioluminescent potential is measured with a silicon photodiode after the organisms are pumped through an impeller into a viewing chamber. These measurements are also done during the nighttime at two depths.

The instrumentation is located in the upper ~160m and the vertical spacing of the packages is ~10m in the upper ~70m and ~20-40m below (see Table 1 of Dickey *et al.*, 1986a). Data are recorded every 4min (in order to reduce aliasing) for each of the packages except for the BLMS. The Biowatt mooring also has a surface meteorological package with sensors to determine wind speed and direction, barometric pressure, sea and air temperature, relative humidity and solar insolation. In addition, a surface spectroradiometer is mounted on the buoy to facilitate interpretation of the subsurface light measurements. Data relevant to all blocks in Figure 1 except for nutrients are collected from the mooring.

The first of three deployments was done in February 1987. Shipboard measurements were done at the time of the first deployment and will also be done at the time of each servicing of the mooring (May, August, and November) in order to obtain detailed bio-optical, physical and bioluminescence profile data at the mooring site. Complementary sea surface temperature (with satelliteborne Advanced Very High Resolution Radiometer, AVHRR) and chlorophyll-a fluorescence (aircraft AOL and ocean color scanner) data are also being collected in order to characterize the horizontal variability in the region of the mooring. In addition, horizontal mapping of surface temperature and chlorophyll-a concentrations and subsurface temperature with XBT's and AXBT's is being done during the cruises. Measurements will continue through November 1987.

7. FUTURE DIRECTIONS

7.1. Sensors

Technological advances during the past decade have enabled the rapid expansion of the sampling domain of many of the important bio-optical and physical variables. In particular, new optical and acoustical sensors, along with microprocessors, have enabled biological and optical oceanographers to begin to study phenomena on scales comparable to those studied by physical oceanographers. The integration of these sensors into CTD and current meter types of systems has made it possible to investigate bio-optical and physical interactions through co-located and concurrent observations. Many of the future advances will most likely depend on continued progress in the development of fundamental instrumentation and *in situ* sampling systems which utilize this instrumentation. In the following discussion, some of the more promising directions in sensor technology and systems will be discussed. It should also be re-iterated that additional satellite derived oceanographic data will become available for regional and ocean basin scale studies and that *in situ* measurements should be designed to complement these data in terms of types of observations, particularly in depth and time.

The use of satellite altimetry to determine basin scale surface general circulation is planned for the near future (Wunsch, 1981). Similar data obtained from the oceanographic satellite Seasat have been used to observe features such as the Gulf Stream and its rings (Cheney and Marsh, 1981). These data can also be used to provide surface boundary conditions so that geostrophic currents at depth (based upon CTD data) may be computed without the ambiguity of the level of no motion. The need for more direct mapping of subsurface current structure has motivated the application of another methodology, acoustic tomography (*e.g.*, Munk and Wunsch, 1979; Behringer *et al.*, 1982). This technique involves the measurement of the field of sound speed fluctuations within a control volume by transmitting acoustic signals along several diverse paths. Inverse theory calculations are used to transform the acoustical data into the desired current field. A field experiment utilizing this methodology was executed in a 300km square of the Sargasso Sea in 1981 (*e.g.*, Behringer *et al.*, 1982; Cornuelle *et al.*, 1985). One of the attractive features of acoustic tomography is that a relatively large volume of the ocean can be sampled synoptically. It is difficult to estimate how soon the tomographic technique will be generally available to the oceanographic community.

Vertical velocities have been most difficult to measure but have been accomplished. For example, shipboard measurements of vertical velocities have been done by using a four beam 300kHz acoustic Doppler system (*e.g.*, Joyce and Stalcup, 1984). Schott (1987) has recently reported measurements taken with a buoy-mounted, upward looking acoustic Doppler current profiler, which is capable of both horizontal and vertical current measurements with vertical resolution of ~9m. Interestingly, the latter data suggest that the measured vertical current may be dominated by biological scatterers during active vertical migration periods. Weller *et al.* (1985) have reported vertical velocity measurements taken with a real-time profiler (RTP), based on a VMCM design. The R/P FLIP was used to deploy this device which was also interfaced with a fluorometer. The routine measurement of vertical velocities remains as an important goal.

Techniques for the determination of oceanic fluxes are highly desirable; however at present there are no direct methods for their determination. Fluxes of material may be estimated by using modeled eddy diffusivity (*e.g.*, Munk and Anderson, 1948; Henderson-Sellers, 1982; Mellor and Yamada, 1982; Rahm and Svensson, 1986) or by using microstructure data with an eddy diffusivity parameterization (Osborn and Cox, 1972; Gregg *et al*, 1986; Lewis *et al.*, 1984b, 1986). Both methods also require vertical gradients of the mean material concentrations. It appears that the measurement of oceanic microstructure will become relatively common within the next decade. The use of these data to infer vertical fluxes of materials (such as nutrients and plankton) will be important for multi- disciplinary oceanography. Atmospheric scientists have developed instrumentation for the estimation of vertical fluxes using the eddy correlation technique (*e.g.*, Friehe and Schmidt, 1976; Smith and Jones, 1985) and other techniques involving lasers have been used in the laboratory for simultaneous measurements of fluctuations of velocity and fluorescing dye concentrations (Hannoun *et al.*, 1987). Unfortunatley, such techniques will be more difficult to utilize for ocean applications.

The development of chemical sensors for deployment from CTD and current meter types of packages has begun with dissolved oxygen sensors, however comparable sensors for pH, total carbon dioxide and other specific ions (*e.g.*, nutrients) need to be developed. In principle, continuous *in situ* nutrient autoanalyzers can be developed as well. In the meantime, pumping systems will continue to be quite useful, particularly with improved vertical resolution capabilities. Quite recently, a moored chemical sampling system has been developed (Codispoti, personal communication). The system can collect twenty pairs of chemical or biological samples sequentially (intervals ranging from ~1 hour to ~10 days) using a timing system and appropriate preservatives. In addition, a low cost chemical time-series sampler has been developed for use in estimating benthic fluxes (Berelson and Hammond, 1986) These approaches enable the acquisition of coarse time series chemical data.

The utility of satellite measurements of surface bio-optical properties with the CZCS has been discussed previously (*e.g.*, Esaias, 1981). Although the CZCS is no longer operational, it is likely that other satelliteborne color scanners will provide data in the future. The need for complementary *in situ* data has driven and will continue to drive much of the work in ocean optics. Nonetheless, many important and fundamental problems related to productivity, underwater visibility and light propagation also necessitate improved bio-optical measurements.

The further advancement of bio-optical instrumentation will require a variety of sensors which measure the complete optical data set so that inherent and apparent optical properties may be related (*e.g.*, Blizard, 1986). Devices which are needed to better characterize the inherent optical properties are absorption meters (*e.g.*, Zaneveld and Bartz, 1984) and scattering meters (*e.g.*, Wells, 1984). The theoretical bases for these devices appear to be reasonably well-developed, however the technological difficulties are non-trivial. Several investigators are attacking these problems with a variety of methods (*e.g.*, Zaneveld and Bartz, 1984; Caimi et al., 1984; Bennett *et al.*, 1986; Wyatt, 1986; Caimi, 1986). In addition to measurement systems for inherent optical properties, inverse methods have been proposed by Preisendorfer and Mobley (1984) to calculate inherent optical properties from irradiance data.

Information concerning the optical characteristics of individual particles are also needed to relate inherent with apparent optical properties. One technique is flow cytometry (*e.g.*, Yentsch and Yentsch, 1984; Spinrad and Brown, 1986). Several optical properties of particles ranging in size from ~1 to 150μ may be measured quite rapidly (several thousand particles per sec, Spinrad, 1984). Some of the properties include: particle index of refraction, forward scattering for particle sizing and spectral fluorescence for determination of physiological states. Recently, flow cytometers have been used at sea. Another instrument which is designed to measure scattering from individual particles (but *in situ*) is the Laser Troller (Wyatt, 1986). This device uses an array of detectors which intercept laser light which is scattered from a particle. In addition, the system has the potential for doing bioluminescence measurements by using natural bioluminescent light emitted by organisms instead of the laser light. Recently, the optical properties of single cells have been studied using microphotometry techniques with a computer controlled microscope (Iturriaga *et al.*, 1987).

Spectroradiometers have been particularly valuable research instruments. The development of a spectroradiometer which measures wavelengths from 400 to 700nm using acousto-optical tunable filters has been reported by Caimi (1986). Using a different technology, Carder *et al.* (1984) have developed a solid-state spectral transmissometer and radiometer using a 256 channel charge coupled array and a diffraction grating. Also, a fish-eye lens radiance video system has been developed (Voss, 1986). This system uses two electro-optic CID cameras to sample the entire radiance distribution so that both inherent (*e.g.*, $a(\lambda)$) and apparent (*e.g.*, $K(\lambda)$) optical

584

properties may be determined. Beyond the development of
spectroradiometers, the interpretation of the data collected with
these instruments is important and will require careful pigment
analyses with instruments such as high precision liquid
chromatographs (HPLC's) and consideration of dissolved materials.

Recently, it has been suggested that a relatively simple
relationship may exist between the rate of photosynthesis and natural
chlorophyll-a fluorescence (Falkowski and Kiefer, 1985; Topliss and
Platt, 1986). A new sensor to determine natural fluorescence
(described earlier) by measuring upwelling radiance at 685nm has been
developed. This measurement may be useful in determining primary
productivity rates (Booth and Kiefer, 1985). Upwelling radiance
sensors are being used with the BOMS sensor suite deployed as part of
the Biowatt mooring (Section 6.3) and the SCMS sensor suite (Section
6.2). Also, Demers et al. (1985) have proposed a technique involving
two flow-through fluorometers and an intense light source which may
be used for continuous in situ sampling of phytoplankton responses to
the photo-environment.

The use of fiber optics to bring light signals from depth to the
surface for shipboard (or surface buoy) signal processing and data
analysis appears to be a viable option for several physical and bio-
optical applications. Seaver (1986) has developed a critical
wavelength refractometer using fiber optics to measure the seawater's
index of refraction. Kakui et al. (1985) have developed a system
which connects a submersible sensory unit through an optical fiber
cable to a shipboard argon ion laser source and data acquisition
system. The underwater unit is towed and the sensors are used to
measure particle size using fluorescence pulses and excitation
spectra for chlorophyll-a emission. Another fiber optic based system
(FOSSUM - Fiber Optic Spectrometer System for Underwater
Measurements; Moore et al., 1984) has been used to measure the
upwelling and downwelling spectral attenuation of light by seawater.

Promising optical methods, which have potential application for
in situ small scale predator-prey studies, include Schlieren video
systems (e.g., Strickler, 1985) and holography (e.g., O'Hern et al.,
1986). The former system has been used to investigate feeding
behaviors of copepods in the laboratory. From this work, unique
information concerning the currents set up by zooplankton and their
interaction with algae has been obtained. It seems plausible that an
in situ version of this system can be developed (Strickler, personal
communication). Swimming velocities of motile phytoplankton have
been measured in the laboratory using laser Doppler spectroscopy
(Bauerfeind et al., 1986). A holographic system (Figure 11), which
was developed for studies of bubble nucleation (O'Hern et al., 1986),
has been used to record remarkably clear images of several
zooplankton (size scales of ~100μm to ~800μm) during a study near
Santa Catalina Island, California. One of the advantages of this
method is that three-dimensional analyses of the individual organisms
and their spatial relations can be conducted. The development of the
next generation of bioluminescence sensors is closely related to
several of these devices.

7.2. Platforms and Systems

Stationary platforms and floating platforms such as R/P FLIP
are particularly advantageous for many types of measurements which
require stability and minimal motion and light shading. Platforms
such as these are particularly well-suited for time series and
process oriented studies which require extensive suites of bio-
optical, chemical and physical data that would exceed the capacity of
moored systems.

The utilization of shipboard profiling and towing will continue
to be important. In addition, towed arrays (comparable to towed
thermistor chains) with bio-optical sensors have been used by a few
investigators (*e.g.*, Moore *et al.*, 1984) and can provide yet another

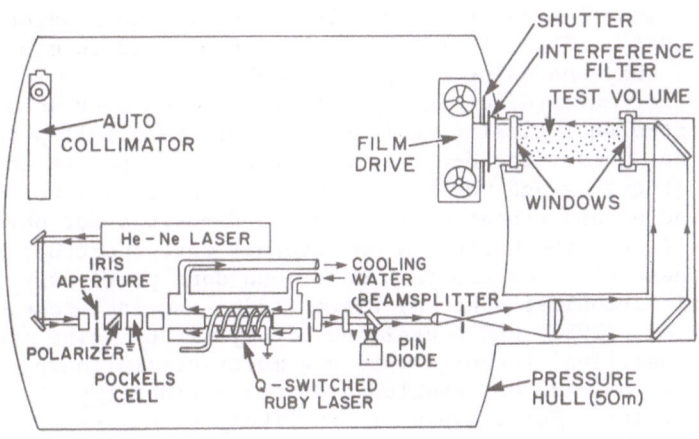

Figure 11. An underwater holographic system developed by O'Hern *et
al.* (1986).

data type which would be useful for several types of studies (*e.g.*,
internal gravity waves, frontal zones, etc.). The development of
expendable bio-optical probes similar to XBT's would greatly improve
our global data bases and enable rapid mapping for process oriented
studies. The use of fiber optics for such measurement devices seems
to be a feasible approach.

Long time series data with relatively high temporal resolution
have only begun to be taken with moored bio-optical and physical
packages. Thus, the continued development of these systems is
desirable in order to broaden geographical coverage for global
studies (*e.g.*, GOFS) and to enable specialized process oriented
studies. With the availability of instrumentation which can be used
to telemeter mooring data in real-time, the mooring data sets will
become even more valuable (*e.g.*, for data assimilation and predictive
models).

For several years, drifters and drogues have been utilized by physical oceanographers for current measurements (*e.g.*, Davis, 1985), however their integration into biological studies is relatively recent (*e.g.*, Brink *et al.*, 1981; MacIsaac *et al.*, 1985; Barth and Brink, 1987; Wilkerson and Dugdale, 1987; Jones *et al.*, manuscript). The MVP, described earlier, has been deployed from a surface drifting buoy and an array of multi- disciplinary sensors is to be deployed beneath a drifter as part of the forthcoming Coastal Transition Zone Experiment (Abbott, personal communication). One of the principal attractions of drifters and drogues, which are equipped with bio-optical instrumentation, is that a broad geographical extent can be sampled. In order for this approach to be viable for general usage, satellite telemetry of data and production of sensors of moderate cost will be required. Other ways to deploy bio-optical and physical instrumentation packages include submarines (Osborn and Haury, personal communication) and remotely controlled underwater vehicles (Billet, 1984). The latter vehicles could be used to minimize the problem of ship shadowing of optical packages.

These new multi-disciplinary instrumentation systems will require sophisticated analytical techniques so that their full potential may be realized. Also, the dependence of the power spectra and cospectra (in both frequency and wavenumber) of various bio-optical variables may be studied and compared with similar functions for physical variables (*e.g.*, the Garrett-Munk internal wave spectrum, Munk, 1981). These kinds of analyses have been done previously in horizontal wavenumber space by Platt (1972) and in frequency space by Cullen *et al.* (1983) and Siegel and Dickey (1986). The application of these analytical techniques to new multi-disciplinary data sets can greatly increase our knowledge of the variability of the upper ocean ecosystem. For example, it is likely that considerable progress may be made in quantifying the effects of biological, optical and physical processes related to plankton distributions (*e.g.*, Marra, 1978; Denman and Gargett, 1983).

8. CONCLUSIONS

A significant number of advances in multi-disciplinary *in situ* measurement systems have been made during the past decade. The number of types of multi-disciplinary measurements and their temporal and spatial extents have increased. Important developments have occurred in optically and acoustically based instrumentation. Thus, many biological and optical measurements may be made on scales comparable to those of CTD's and current meters. Further advances will be dependent upon new sensor and system technology and the availability of economical sensors, fiber optics and telemetry. Global scale oceanic problems will require both satellite remote sensing and selective *in situ* observations. Moored, drifting and expendable bio-optical and physical measurement systems, which could provide subsubsurface as well as surface data, are attractive for the latter purpose because the ranges of temporal and spatial observations could be expanded and the numbers of observations could be greatly increased.

9. ACKNOWLEDGEMENTS

I would like to thank Dr. Burt Jones, Dr. Dale Kiefer, Dr. John Marra, Mr. David Siegel and Dr. Libe Washburn for their comments and suggestions. My appreciation is also extended to Professor Claes Rooth who originally stimulated my interest in multi- disciplinary oceanographic observations. Much of the work described in this review has been sponsored by the Office of Naval Research (ONR) Ocean Optics and Oceanic Biology Programs. Travel support for the NATO workshop was provided by NATO.

10. REFERENCES

Aiken, J., 1981: 'The Undulating Oceanographic Recorder Mark 2', *Plankton Res.*, **3**, 551-560.

Aiken, J.: 1985, 'The Undulating Oceanographic Recorder Mark 2: A multirole oceanographic sampler for mapping and modeling the biophysical marine environment', in A. Zirino (ed.), *Mapping Strategies in Chemical Oceanography*, American Chemical Society, Washington, D.C., pp. 315-332.

Aiken, J. and I. Bellan: 1986a, 'A simple hemispherical, logarithmic light sensor', in M.A. Blizard (ed.), Ocean Optics VIII, Proc. SPIE, 637, International Society for Optical Engineering, Bellingham, Washington, pp. 211-216.

Aiken, J. and I. Bellan: 1986b, 'Synoptic optical oceanography with the undulating oceanographic recorder', in M.A. Blizard (ed.), *Ocean Optics VIII*, Proc. SPIE, 637, International Society for Optical Engineering, Bellingham, Washington, pp. 221-230.

Armstrong, F.A.J. and E.C. LaFond: 1968, 'Chemical nutrient concentrations and their relationship to internal waves and turbidity off Southern California', *Limnol. Oceanogr.* **11**, 538-547.

Arnone, R.A., R.R. Bidigare, C.C. Trees and J.M. Brooks: 1986, 'Comparison of the attenuation of spectral irradiance phytoplankton pigments within frontal zones', in M.A. Blizard (ed.), *Ocean Optics VIII*, Proc. SPIE, 637, International Society for Optical Engineering, Bellingham, Washington, pp. 126-130.

Arnone, R.A., S.P. Tucker and F.A. Hilder: 1984, 'Secchi depth atlas of the world coastlines', in M.A. Blizard (ed.), *Ocean Optics VII*, Proc. SPIE, 489, pp. 126-130, International Society for Optical Engineering, Bellingham, Washington.

Atkinson, M.J., T. Berman, B.R. Allanson and J. Imberger: 1987, 'Fine-scale oxygen variability in a stratified estuary: Patchiness in aquatic environments', *Mar. Ecol. Prog. Ser.* **36**, 1-10.

588

Barth, J.A. and K.H. Brink: 1987, 'Shipboard acoustic Doppler profiler velocity observations near Point Conception: Spring 1983', *J. Geophys. Res.* **92**, 3925-3943.

Bartz, R., J.R.V. Zaneveld and H. Pak: 1978, 'A transmissometer for profiling and moored observations in water', in *Ocean Optics V*, Proc. SPIE, 160, International Society for Optical Engineering, Bellingham, Washington, pp. 102-108.

Bauerfeind, E., M. Elbrachter, R. Steiner and J. Throndsen: 1986, 'Application of laser Doppler spectroscopy (LDS) in determining swimming velocities of motile phytoplankton', *Mar. Biol.* **93**, 323- 327.

Behringer, D., T. Birdsall, M. Brown, B. Cornuelle, R. Heinmiller, R. Knox, K. Metzger, W. Munk, J. Spiesberger, R. Spindell, D. Webb, P. Worcester and C. Wunsch: 1982, 'A demonstration of ocean acoustic tomography', *Nature* **299**, 121-125.

Bennett, G.T., E.S. Fry and F.M. Sogandares: 1986, 'Photothermal measurement of the absorption coefficient of water at 590nm', in M.A. Blizard (ed.), *Ocean Optics VIII*, Proc. SPIE, 637, pp. 172-180, International Society for Optical Engineering, Bellingham, Washington.

Berelson, W.M. and D.E. Hammond: 1986, 'The calibration of a new free vehicle benthic flux chamber for use in the deep sea', *Deep-Sea Res.* **33**, 1439-1454.

Billet, A.B.: 1984, 'Optimizing optics for remotely controlled underwater vehicles', in M.A. Blizard (ed.), *Ocean Optics VII*, Proc. SPIE, 489, International Society for Optical Engineering, Bellingham, Washington, pp. 399-406.

Bishop, J.: 1986, 'The correction and suspended particulate matter calibration of Sea Tech transmissometer', *Deep-Sea Res.* **91**, 7761- 7764.

Blizard, M.A.: 1986, 'Ocean optics: Introduction and overview', in M.A. Blizard (ed.), *Ocean Optics VIII*, Proc. SPIE, 637, International Society for Optical Engineering, Bellingham, Washington, pp. 2-17.

Booth, C.R.: 1976, 'The design and evaluation of a measurement system for photosynthetically active quantum scalar irradiance', *Limnol. Oceanogr.* **19**, 326-335.

Booth, C.R. and D.A. Kiefer: 1985, 'Natural fluorescence: A field study of its relationship to chlorophyll and rates of photosynthesis', *Eos*, **66**, 1333.

Brandt, A., C.C. Sarabun, H.H. Seliger and M.A. Tyler: 1986, 'The effects of the broad spectrum of physical activity on the biological processes in the Chesapeake Bay', in J.C.J. Nihoul (ed.), *Marine Interfaces Ecohydrodynamics*, Elsevier, Amsterdam, pp. 361-384.

Brewer, P.G., K.W. Bruland, R.W. Eppley and J.J. McCarthy: 1986, 'The Global Ocean Flux Study (GOFS): Status of the U.S. GOFS Program', *Eos*, **67**, 827-832, 835-837.

Bricaud, A., A. Morel and L. Prieur: 1983, 'Optical efficiency factors of some phytoplankton', *Limnol. Oceanogr.* **28**, 816-832.

Brink, K. H., B.H. Jones, J.C. Van Leer, C.N.K. Mooers, D.W. Stuart, M.R. Stevenson, R.C. Dugdale and G.W. Heburn: 1981, 'Physical and biological structure and variability in an upwelling center off Peru near 15°S during March, 1977', in *Coastal Upwelling, Coastal and Estuarine Sciences I*, American Geophysical Union, 473-495.

Caimi, F.M.: 1986, 'Ocean optical measurements using acousto-optic filtering', in M.A. Blizard (ed.), *Ocean Optics VIII*, Proc. SPIE, 637, International Society for Optical Engineering, Bellingham, Washington, pp. 181-185.

Caimi, F.M., R.F. Tusting and G. Kennedy: 1984, 'In situ forward scatter and transmittance measurement using a low power laser diode', in M.A. Blizard (ed.), *Ocean Optics VII*, Proc. SPIE, 489, International Society for Optical Engineering, Bellingham, Washington, pp. 364-374.

Caldwell, D.R., T.M. Dillon and J.H. Moum: 1985, 'The rapid-sampling vertical profiler: An evaluation', *J. Atmos. Ocean Tech.* 2, 615- 625.

Carder, K.L., R.G. Steward and P.R. Payne: 1984, 'A solid-state spectral transmissometer and radiometer', in M.A. Blizard (ed.), *Ocean Optics VII*, Proc. SPIE, 489, International Society for Optical Engineering, Bellingham, Washington, pp. 325-334.

Cheney, R.E. and J.G. Marsh: 1981, 'Seasat altimeter observations of dynamic ocean currents in the Gulf Stream region', *J. Geophys. Res.* 86, 473-483.

Cornuelle, B., C. Wunsch, D. Behringer, T. Birdsall, M. Brown, R. Heinmiller, R. Knox, K. Metzger, W. Munk, J. Spiesberger, R. Spindel, D. Webb and P. Worcester: 1985, 'Tomographic maps of the ocean mesoscale. Part 1: Pure acoustics', *J. Phys. Ocean.* 15, 133-152.

Cullen, J.J., E. Stewart, E. Renger, R.W. Eppley and C.D. Winant: 1983, 'Vertical motion of the thermocline, nutricline and chlorophyll maximum layers in relation to currents on the Southern California Shelf', *J. Mar. Res.* 41, 239-262.

Davis, R.E.: 1985, 'Drifter observations of coastal surface currents during CODE: The method and descriptive view', *J. Geophys. Res.* 90, 4741- 4755.

Demers, S., J.-C. Therriault, L. Legendre and J. Neveux: 1985, 'An in vivo fluorescence method for the continuous *in situ* estimation of phytoplankton photosynthetic characteristics', *Mar. Ecol.* 27, 21-27.

Denman, K.L.: 1973, 'A time-dependent model of the upper ocean', *J. Phys. Oceanogr.* 3, 173-184.

Denman, K.L. and A.E. Gargett: 1983, 'Time and space scales of vertical mixing and advection of phytoplankton in the upper ocean', *Limnol. Oceanogr.* 28, 801-815.

Denman, K.L. and D.L. Mackas: 1978, 'Collection and analysis of underway data and related physical measurements', in J.H. Steele (ed.), *Spatial Patterns in Plankton Communities*, Plenum, New York, pp. 85-109.

Dickey, T., E. Hartwig and J. Marra: 1986a, 'The Biowatt bio-optical and physical moored measurement program', *Eos* **67**, 650.

Dickey, T., J. Marra, D. Siegel, C. Abbott and J. Van Leer: 1986b, 'Measurements in the Tongue of the Ocean with the multi-variable profiler', *Biowatt News* **6**, 1-6.

Dickey, T.D. and D.A. Siegel: 'Physical, optical, biochemical and meteorological variability in the eastern North Pacific Ocean', Manuscript.

Dickey, T.D., D.A. Siegel, A. Bratkovich and L. Washburn: 1986d, 'Optical features associated with thermohaline structures', in M.A. Blizard (ed.), *Ocean Optics VIII*, Proc. SPIE, 637, International Society for Optical Engineering, Bellingham, Washington, pp. 308-313.

Dickey, T.D., D.A. Siegel, J. Marra and M.K. Hamilton: 1986c, 'Variability of physical and optical properties in the Sargasso Sea during the Biowatt I experiment', *Eos* **67**, 968.

Dickey, T.D. and J.J. Simpson: 1983, 'The influence of optical water type on the diurnal response of the upper ocean', *Tellus* **35B**, 142-154.

Dickey, T.D and J.C. Van Leer: 1984, 'Observations and simulations of a bottom Ekman layer on a continental shelf', *J. Geophys. Res.* **89**, 1983- 1988.

Esaias, W.E.: 1981, 'Remote sensing in biological oceanography', *Oceanus* **24**, 32-38.

Falkowski, P. and D.A. Kiefer: 1985, 'Chlorophyll a fluorescence in phytoplankton: Relationship to photosynthesis and biomass', *J. Plankton Res.*, 7, 715-731.

Farmer, D.M. and R.D. Huston: 1988, 'Novel applications of acoustic backscatter to biological measurements', in B.J. Rothschild (ed.), *Toward a Theory on Biological Physical Interactions in the World Ocean*, Kluwer Academic Publishers, Dordrecht, pp. 597–614.

Farmer, D.M. and D.D. Lemon: 1984, 'The influence of bubbles on ambient noise in the ocean at high wind speeds', *J. Phys. Oceanogr.* **14**, 1762-1778

Farmer, D.M. and J.D. Smith: 1980, 'Tidal interaction of stratified flow with a sill in Knight Inlet', *Deep-Sea Res.* A27, 239-254.

Fischer, J., C. Meinke, P.J. Minnett, V. Rehberg and V. Strass: 1985, 'A description of the Ifm-Schleppfisch-System', Tech. Rep. Inst. f. Meereskunde, Abt. Reg. Oz., 2nd ed., Kiel, West Germany.

Friehe, C.A. and K.F. Schmitt: 1976, 'Parameterization of air-sea interface fluxes of sensible heat and moisture by the bulk aerodynamic formulas', *J. Phys. Oceanogr.* 6, 801-809.

Frost, B.W. and L.E. McCrone: 1974, 'Vertical distribution of zooplankton and myctophid fish at Canadian weather station P, with description of a new multiple net trawl', *Proc. of Int. Conf. on Eng. in the Ocean Environ.*, 159-165.

Gordon, H.R.: 1985, 'Ship perturbation of irradiance measurements at sea. 1: Monte Carlo simulations', *Appl. Optics* 24, 4172-4182.

Gordon, H.R. and A.Y. Morel: 1983, *Remote Assessment of Ocean Color for Interpretation of Satellite Visible Imagery: A Review*, Springer- Verlag, New York, NY, 114p.

Gordon, H.R., R.C. Smith and J.R.V. Zaneveld: 1984, 'Introduction
    to ocean optics', in M.A. Blizard (ed.), *Ocean Optics VII*, Proc.
    SPIE, 489, pp. 2-41, International Society for Optical
    Engineering, Bellingham, Washington.
Gregg, M.C., E. D'Asaro, T.J. Shay and N. Larson: 1986,
    'Observations of persistent mixing and near-inertial internal
    waves', *J. Phys. Oceanogr.* 16, 856-885.
Gytre, T.: 1980, 'Acoustic travel time current meters', in F.
    Dobson, L. Hasse and R. Davis (eds.), *Air-Sea Interaction:
    Instruments and Methods*, Plenum, New York, pp. 155-170.
Hannoun, I.A., H.J.S. Fernando and E.J. List: 1987, 'Turbulence
    structure near a sharp interface', *J. Fluid. Mech.*, in press.
Haury, L.R., M.G. Briscoe and M.H. Orr: 1979, 'Tidally generated
    internal wave packets in Massachusetts Bay', *Nature* 278, 312-317.
Haury, L.R., J.A. McGowan and P.H. Wiebe: 1978, 'Patterns and
    processes in the time-space scales of plankton distributions', in
    J. H. Steele (ed.), *Spatial Patterns in Plankton Communities*,
    Plenum, New York, pp. 277-327,
Haury, L.R., P.H. Wiebe and S.H. Boyd: 1976, 'Longhurst-Hardy
    Plankton Recorders: Their design and use to minimize bias',
    Deep-Sea Res. 23, 1217-1229.
Henderson-Sellers, B.: 1982, 'A simple formula for vertical eddy
    diffusion coefficients under conditions of non-neutral stability',
    J. Geophys. Res. 87, 5860-5864.
Herman, A.W.: 1985, 'Biological profiling in the upper oceanic
    layers with a Batfish vehicle: A review of applications', in A.
    Zirino (ed.), *Mapping Strategies in Chemical Oceanography*,
    American Chemical Society, Washington, D.C., pp. 293-314.
Herman, A.W.: 1986, 'A new optical zooplankton counter measuring
    simultaneous profiles of zooplankton and light attenuance', *Eos*
    67, 970.
Herman, A.W. and T.M. Dauphinee: 1980, 'Continuous and rapid
    profiling of zooplankton with an electronic counter mounted on a
    "Batfish" vehicle', *Deep-Sea Res.* 27, 79-96.
Hojerslev, N.K.: 1986, 'Visibility of the sea with special
    references to the Secchi disc', in M.A. Blizard (ed.), *Ocean
    Optics VIII*, Proc. SPIE, 637, International Society for Optical
    Engineering, Bellingham, Washington, pp. 294-306.
Holliday, D.V. and R.E. Pieper: 1980, 'Volume scattering strengths
    and zooplankton distributions at acoustic frequencies between 0.5
    and 3MHz', *J. Acoust. Soc. Am.* 67, 135-146.
Holliday, D.V. and R.E. Pieper: 1986, 'Potential uses of underwater
    acoustics in recruitment studies', *Eos* 67, 987.
Holliday, D.V. and R. E. Pieper: 1987, 'Applications of underwater
    acoustics to the study of micronekton and zooplankton', *Proc. Int.
    Conf. on Mar. Sci. of the Arabian Sea*, Karachi, Pakistan, Amer.
    Inst. Biol. Sci., in press.
Iturriaga, R., B.G. Mitchell and D.A. Kiefer: 1987, 'Microphotometric
    analysis of spectral absorption for individual phytoplankton
    cells', Submitted to *Limnol. Oceanogr.*
Jerlov, N.G.: 1976, *Marine Optics*, Elsevier, Amsterdam, 231p.

592

Johnson, K.S., R.L. Petty and J. Thomsen: 1985, 'Flow-injection
  analysis for seawater micronutrients', in A. Zirino (ed.), *Mapping
  Strategies in Chemical Oceanography*, American Chemical Society,
  Washington, D.C., pp. 7-30.
Jones, B.H., L.P. Atkinson, D. Blasco, K.H. Brink and S.L. Smith:
  'The asymmetric distribution of chlorophyll associated with a
  coastal upwelling center', Submitted to *Cont. Shelf Res*.
Joyce, T.M.: 1985, 'Gulf Stream warm-core ring collection: An
  introduction', *J. Geophys. Res*. 90, 8801-8802.
Joyce, T.M. and M.A. Kennelly: 1985, 'Upper-ocean velocity
  structure of Gulf Stream Warm-Core Ring 82B', *J. Geophys. Res*. 90,
  8839-8844.
Joyce, T.M. and M.C. Stalcup; 1984, 'An upper ocean current jet and
  internal waves in a Gulf Stream warm core ring', *J. Geophys. Res*.
  89, 1997- 2004.
Kakui, Y., A. Nishimoto, J. Hirono and M. Nanjo: 1985, 'Underway
  analysis of suspended biological particles with an optical fiber
  cable', in A. Zirino (ed.), *Mapping Strategies in Chemical
  Oceanography*, American Chemical Society, Washington, D.C., pp.
  211-234.
Karagounis, V., L. Lun and C.C. Liu: 1986, 'A thick film multiple
  component cathode three-electrode oxygen sensor', *IEEE Trans. on
  Biomed. Eng*. 33, 108-113.
Kiefer, D.A.: 1973, 'Fluorescence properties of natural
  phytoplankton populations', *Mar. Biol*. 22, 263-269.
Kirk, J.T.O.: 1983, *Light and Photosynthesis in Aquatic Ecosystems*,
  Cambridge Univ. Press, Cambridge, U.K., 401p.
Kraus, E.B.: 1972, *Atmosphere-Ocean Interaction*, Clarendon Press,
  Oxford, U.K., 275p.
Langdon, C.: 1984, 'Dissolved oxygen monitoring system using a
  pulsed electrode: Design, performance and evaluation', *Deep-Sea
  Res*. 31, 1357-1367.
Lewis, M.R., J.J. Cullen and T. Platt: 1983, 'Phytoplankton and
  thermal structure in the upper ocean: Consequences of
  nonuniformity in chlorophyll profiles', *J. Geophys. Res*. **88**, 2565-
  2570.
Lewis, M.R., J.J. Cullen and T. Platt: 1984a, 'Relationship between
  vertical mixing and photoadaptation of phytoplankton: Similarity
  criteria', *Mar. Ecol. Prog. Ser*. 15, 141-149.
Lewis, M.R., W.G. Harrison, N.S. Oakey, D. Hebert and T. Platt:
  1986, 'Vertical nitrate fluxes in the oligotrophic ocean', *Science*
  234, 870-873.
Lewis, M.R., E.P.W. Horne, J.J. Cullen, N.S. Oakey and T. Platt:
  1984b, 'Turbulent motions may control phytoplankton photosynthesis
  in the upper ocean', *Nature* 311, 49-50.
Losee, J., D. Lapota and S.J. Lieberman: 1985, 'Bioluminescence: A
  new tool for oceanography', in A. Zirino (ed.), *Mapping Strategies
  in Chemical Oceanography*, American Chemical Society, Washington,
  D.C., pp. 211-234.

Lueck, R. and T. Osborn: 1986, 'The dissipation of kinetic energy in a warm-core ring', *J. Geophys. Res.* **91**, 803-818.

McCarthy, J.J., P.G. Brewer and G. Feldman: 1986, 'Global ocean flux', *Oceanus* **29**, 16-26.

MacIsaac, J.J., R.C. Dugdale, R.T. Barber, D. Blasco and T.T. Packard: 1985, 'Primary production cycle in an upwelling center', *Deep-Sea Res.* **32**, 503-529.

MAREX: 1982, 'The Marine Resources Experiment Program (MAREX)', in J. Walsh (ed.), *Report of the Ocean Color Science Working Group*, Goddard Space Flight Center, Greenbelt, Maryland.

Marra, J.: 1978, 'Phytoplankton photosynthetic response to vertical movement in a mixed layer', *J. Mar. Res.* **46**, 203-208.

Marra, J. and E. Hartwig: 1984, 'Biowatt: A study of bioluminescence and optical variability in the sea', *Eos* **65**, 732-733.

Martin, P.: 1985, 'Simulation of the mixed layer at OWS November and Papa with several models', *J. Geophys. Res.* **90**, 903-916.

Mellor, G. and T. Yamada: 1982, 'Development of a turbulence closure model for geophysical problems', *Rev. Geophys. Space Phys.* **20**, 851-875.

Menzel, D.W. and J.H. Ryther: 1960, 'The annual cycle of primary production in the Sargasso Sea off Bermuda', *Deep-Sea Res.* **6**, 351-367.

Menzel, D.W. and J.H. Ryther: 1961, 'Annual variations in primary production of the Sargasso Sea off Bermuda', *Deep-Sea Res.* **7**, 282-288.

Moore, C.A., R.C. Honey, D.M. Hancock, S. Damron, R. Hilbers and S.P. Tucker: 1984, 'Instrumentation for measuring in-situ sea truth for laser radar applications', in M.A. Blizard (ed.), *Ocean Optics VII*, Proc. SPIE, 489, International Society for Optical Engineering, Bellingham, Washington, pp. 343-354.

Morel, A. and L. Prieur: 1977, 'Analysis of variations in ocean color', *Limnol. Oceanogr.* **22**, 708-722.

Munk, W.: 1981, 'Internal waves and small-scale processes', in B.A. Warren and C. Wunsch (eds.), *Evolution of Physical Oceanography*, MIT Press, Cambridge, MA, pp. 264-290.

Munk, W. and E.R. Anderson: 1948, 'Notes on a theory of the thermocline', *J. Mar. Res.* **1**, 276-295.

Munk, W. and C. Wunsch: 1979, 'Ocean acoustic tomography: A scheme for large scale monitoring', *Deep-Sea Res.* **26**, 123-156.

Nealson, K.H. and A.C. Arneson: 1985, 'Marine bioluminescence: About to see the light', *Oceanus* **28**, 13-19.

Nealson, K.H., A.C. Arneson and A. Bratkovich: 1984, 'Preliminary results from studies of nocturnal bioluminescence with subsurface moored photometers', *Mar. Biol.* **83**, 185-191.

Nieman, D.R.: 1986, 'Spatial distributions of Gulf Stream zooplankton relative to acoustic Doppler current profiles', *Eos* **67**, 1029.

O'Hern, T.J., J. Katz and A.J. Acosta: 1985, 'Holographic measurements of cavitation nuclei in the sea', ASME Cavitation and Multiphase Flow Forum, Albuquerque, NM.

Orr, N.H. and F.R. Hess; 1978, 'Remote acoustic monitoring of natural suspensate resuspensions and slope/shelf water intrusions', *J. Geophys. Res.* **83**, 4062-4068.

Ortner, P.B., L.C. Hill and H.E. Edgerton: 1981, 'In situ silhouette photography of Gulf Stream zooplankton', *Deep-Sea Res.* **28A**, 1569- 1576.

Ortner, P.B., R.E. Pieper and D.L. Mackas: 1982, 'Advances in zooplankton sampling', in B.J. Rothschild and C.G.H. Rooth (eds.), *Fish Ecology III*, University of Miami Tech. Rep. No. 82008, Miami, Florida, pp. 355-379.

Ortner, P.B., M.R. Reeve and D.V. Holliday: 1986, 'Diel variability in Gulf Stream epizooplankton', *Eos* **67**, 1029.

Osborn, T.R. and R.G. Lueck: 1985a, 'Turbulence measurements from a towed body', *J. Atm. Ocean. Tech.* **2**, 517-527.

Osborn, T.R. and R.G. Lueck: 1985b, 'Turbulence measurements from a submarine', *J. Phys. Oceanogr.* **15**, 1502-1520.

Pieper, R.E., D.V. Holliday and G.S. Kleppel, 'Small-scale spatial distributions of zooplankton resolved by multi-frequency acoustics', Manuscript.

Pinkel, R.; 1980, 'Acoustic Doppler techniques', in F. Dobson, L. Hasse and R. Davis (eds.), *Air-Sea Interaction: Instruments and Methods*, Plenum, New York, pp. 171-199.

Platt, T.: 1972, 'Local phytoplankton abundance and turbulence', *Deep-Sea Res.* **19**, 183-188.

Price, J.F., R.A. Weller and R. Pinkel: 1986, 'Diurnal cycling: Observations and models of the upper ocean response to diurnal heating, cooling, and wind mixing', *J. Geophys. Res.* **91**, 8411-8427.

Preisendorfer, R.W.: 1986, 'Secchi disk science: Visual optics of natural waters', *Limnol. Oceanogr.* **31**, 909-926.

Preisendorfer, R.W. and C.D. Mobley: 1984, 'Direct and inverse irradiance models in hydrologic optics', *Limnol. Oceanogr.* **29**, 903-929.

Proni, J.R. and J.R. Apel: 1975, 'On the use of high-frequency acoustics for the study of internal waves and microstructure', *J. Geophys. Res.* **80**, 1147-1151.

Rahm, L.-A., and U. Svensson: 1986, 'Dispersion of marked fluid elements in a turbulent Ekman Layer', *J. Phys. Oceanogr.* **16**, 2084-2096.

Sameoto, D.D., L.O. Jaroszynski and W.B. Fraser: 1980, 'The BIONESS, a new design in multiple net zooplankton samplers', *Can. J. Fish. Aquat. Sci.* **37**, 722-724.

Schott, F.: 1987, 'Half-year long measurements with a buoy-mounted acoustic Doppler current profiler in the Somali Current', *J. Geophys. Res.*, in press.

Seaver, G.: 1986, 'A new refractometer for uses in oceanography', in
    M.A. Blizard (ed.), *Ocean Optics VIII*, Proc. SPIE, 637,
    International Society for Optical Engineering, Bellingham,
    Washington, pp. 217-220.
Siegel, D.A., C.R. Booth and T.D. Dickey: 1986, 'Effects of sensor
    characteristics on the inferred vertical structure of the diffuse
    attenuation coefficient spectrum', in M.A. Blizard (ed.), *Ocean
    Optics VIII*, Proc. SPIE, 637, International Society for Optical
    Engineering, Bellingham, Washington, pp. 115-123.
Siegel, D.A. and T.D. Dickey: 1986, 'Statistical analysis of
    fluorescence variability in the upper ocean during ODEX', *Eos* 67,
    969.
Siegel, D.A. and T.D. Dickey: 1987, 'Observations of the vertical
    structure of the diffuse attenuation coefficient spectrum', *Deep-
    Sea Res.* 34, 547-563.
Siegel, D.A. and T.D. Dickey: 'On the parameterization of irradiance
    for open ocean photoprocesses', *J. Geophys. Res.*, in press.
Simonot, J.-Y and H. Le Treut: 1986, 'A climatological field of mean
    optical properties of the world ocean', *J. Geophys. Res.* 91,
    6642-6646.
Simpson, J.J. and T.D. Dickey: 1981a, 'The relationship between
    downward irradiance and upper ocean structure', *J. Phys.
    Oceanogr.* 11, 309-323.
Simpson, J.J. and T.D. Dickey: 1981b, 'Alternative parameterizations
    of downward irradiance and their dynamical significance', *J.
    Phys. Oceanogr.* 11, 876-882.
Simpson, J.J., C.J. Koblinsky, L.R. Haury and T.D. Dickey: 1984, 'An
    offshore eddy in the California Current System', *Prog. in Ocean.*
    13, 1-111.
Smith, R.C. and K.S. Baker: 1978, 'The bio-optical state of ocean
    waters and remote sensing', *Limnol. Oceanogr.* 23, 247-259.
Smith, R.C. and K.S. Baker: 1984, 'Analysis of ocean optical data', in
    M.A. Blizard (ed.), *Ocean Optics VII*, Proc. SPIE, 489,
    International Society for Optical Engineering, Bellingham,
    Washington, pp. 119-126.
Smith, R.C. and K.S. Baker: 1985, 'Spatial and temporal patterns in
    pigment biomass in Gulf Stream Warm-Core Ring 82B and its
    environs', *J. Geophys. Res.* 90, 8859-8870.
Smith, R.C. and K.S. Baker: 1986, 'Analysis of ocean optical data,
    II', in M.A. Blizard (ed.), *Ocean Optics VIII*, Proc. SPIE, 637,
    International Society for Optical Engineering, Bellingham,
    Washington, pp. 95-107.
Smith, R.C., C.R. Booth and J.L. Star: 1984, 'Oceanographic biooptical
    profiling system', *Appl. Optics* 23, 2791-2797.
Smith, S.D. and E.P. Jones: 1985, 'Evidence for wind-pumping of air-
    sea gas exchange based on direct measurements of CO_2 fluxes', *J.
    Geophys. Res.* 90, 869-875.
Spinrad, R.W.: 1984, 'Flow cytometric analysis of the optical
    characteristics of marine particulates', in M.A. Blizard (ed.),
    Ocean Optics VII, Proc. SPIE, 489, International Society for
    Optical Engineering, Bellingham, Washington, pp. 335-342.

596

Spinrad, R.W.: 1986a, 'A calibration diagram of specific beam attenuation', *J. Geophys. Res.* **91**, 7761-7764.

Spinrad, R.W.: 1986b, 'An optical study of the water masses of the Gulf of Maine', *J. Geophys. Res.* **91**, 1007-1018.

Spinrad, R.W. and J.F. Brown: 1986, 'Relative real refractive index of marine microorganisms: A technique for flow cytometric estimation', *Appl. Optics* **25**, 1930-1934.

Spitzer, D. and M.R. Wernand: 1981, 'In situ measurements of absorption spectra in the sea', *Deep-Sea Res.* **28A**, 165-174.

Strass, V. and J.D. Woods: 1988, 'Horizontal and seasonal variation of density and chlorophyll profiles between the Azores and Greenland, in B.J. Rothschild (ed.), *Toward a Theory on Biological-Physical Interactions in the World Ocean*, Kluwer Academic Publishers, Dordrecht, pp. 113-136.

Strickler, J.R.: 1985, 'Feeding currents in calanoid copepods: Two new hypotheses', in M.S. Laverack (ed.), *Physiological Adaptations of Marine Animals*, Symposia of the Society for Experimental Biology, 23, Pinder Group of Companies, Scarborough, North Yorkshire, pp. 459-485.

Swift, E., W.H. Biggley and E. Lessard: 1985, 'Distributions of epipelagic bioluminescence in the Sargasso and Caribbean Seas', in A. Zirino (ed.), *Mapping Strategies in Chemical Oceanography*, American Chemical Society, Washington, D.C., pp. 235-258

Thorpe, S.A.: 1982, 'On the clouds of bubbles formed by breaking wind-waves and their role in air-sea gas transfer', *Phil. Trans. Roy. Soc. Lond.* **A304**, 155-210.

Thorpe, S.A.: 1984, 'On the determination of K_v in the near-surface ocean from measurements of bubbles', *J. Phys. Oceanogr.* **14**, 855-863.

Thorpe, S.A. and J.M. Brubaker: 1983, 'Observations of sound reflection by temperature microstructure', *Limnol. Oceanogr.* **28**, 601-613.

Thorpe, S.A., A.J. Hall, A.R. Packwood and A.R. Stubbs: 1985, 'The use of a side-scan sonar to investigate processes near the sea surface', *Cont. Shelf Res.* **4**, 597-607.

Topliss, B.J. and T. Platt: 1986, 'Passive fluorescence and photosynthesis in the ocean: Implications for remote sensing', *Deep-Sea Res.* **33**, 849-864.

Tyler, J.E.: 1959, 'Natural water as a monochromator', *Limnol. Oceanogr.* **6**, 451-456.

Van Leer, J.C.: 1980, 'Profiling Devices', in F. Dobson, L. Hasse and R. Davis (eds.), *Air-Sea Interaction: Instruments and Methods*, Plenum, New York, pp. 701-724.

Van Leer, J.C. and J.Z. Villanueva: 1986, 'Observations of potentially double-diffusive convective layers in Fram Strait', *Eos* **67**, 1014.

Voss, K.J.: 1986, 'Electro-optic radiance distribution camera system', *Eos* **67**, 969.

Voss, K.J., J.W. Nolten and G.D. Edwards: 1986, 'Ship shadow effects on apparent optical properties', in M.A. Blizard (ed.), *Ocean Optics VIII*, Proc. SPIE, 637, International Society for Optical Engineering, Bellingham, Washington, pp. 186-190.

Washburn, L.: 1987, 'Two-dimensional observations of temperature microstructure in a coastal region', *J. Geophys. Res.*, in press.

Washburn, L. and T.K. Deaton: 1985, 'A simple system for mapping conductivity mixing and dissipation in a shallow water coastal region', *J. Atmos. Ocean Tech.* 3, 345-355.

Weller, R. and R.E. Davis: 1980, 'A vector measuring current meter', *Deep-Sea Res.* 27, 565-582.

Weller, R.A., J.P. Dean, J. Marra, J.F. Price, E.A. Francis and D. Boardman: 1985, 'Three-dimensional flow in the upper ocean', *Science* 227, 1552-1556.

Wells, W.H.: 1984, 'Scattering meters for light in the sea', in M.A. Blizard (ed.), *Ocean Optics VII*, Proc. SPIE, 489, International Society for Optical Engineering, Bellingham, Washington, pp. 308-317.

Whitledge, T.E., S.C. Malloy, S.C. Patton and C.D. Wirick: 1981, 'Automated nutrient analyses in seawater', Brookhaven National Laboratory Report BNL 51398, 216p.

Whitledge, T.E. and C.D. Wirick: 1983, 'Observations of chlorophyll concentrations off Long Island from a moored *in situ* fluorometer', *Deep-Sea Res.* 30, 297-309.

Wiebe, P.H., K.H. Burt, S.H. Boyd and A.W. Morton: 1976, 'A multiple opening/closing net and environmental sensing system for sampling zooplankton', *J. Mar. Res.* 34, 313-326.

Wiebe, P.H., G.R. Flierl, C.S. Davis, V. Barber and S.H. Boyd: 1985, 'Macrozooplankton biomass in Gulf Stream warm-core rings: Spatial and temporal changes', *J. Geophys. Res.* 90, 8885-8901.

Wilkerson, F.P. and R.C. Dugdale: 1987, 'The use of large shipboard barrels and drifters to study the effects of coastal upwelling on phytoplankton dynamics', *Limnol. Ocean.* 32, 368-382.

Woods, J.D., W. Barkmann and A. Horch: 1984, 'Solar heating of the oceans - diurnal, seasonal and meridional variation', *Quart. J. Roy. Met. Soc.* 112, 29-42.

Woods, J.D., R.Onken and J. Fischer: 1986, 'Thermohaline intrusions created isopycnically at oceanic fronts are inclined to isopycnals', *Nature* 322, 446-449.

Wunsch, C.: 1981, 'The promise of satellite altimetry', *Oceanus* 24, 17-24.

Wyatt, P.J.: 1986, 'A new instrument for the *in situ* measurement of individual marine particulate', in M.A. Blizard (ed.), *Ocean Optics VIII*, Proc. SPIE, 637, International Society for Optical Engineering, Bellingham, Washington, pp. 95-107.

Yamazaki, H. and T. Osborn: 1988, 'Review of oceanic turbulence: Implications for biodynamics', in B.J. Rothschild (ed.), *Toward a Theory on Biological-Physical Interactions in the World Ocean*, Kluwer Academic Publishers, Dordrecht, pp. 215-234.

598

Yentsch, C.S. and D.A. Phinney: 1985, 'Fluorescence spectral signatures for studies of marine phytoplankton', in A. Zirino (ed.), *Mapping Strategies in Chemical Oceanography*, American Chemical Society, Washington, D.C., pp. 259-274

Yentsch, C.M. and C.S. Yentsch; 1984, 'Emergence of optical instrumentation for measuring biological properties', *Oceanogr. Mar. Biol. Ann. Rev.* 22, 55-98.

Zaneveld, J..V. and R. Bartz: 1984, 'Beam attenuation and absorption meters', in M.A. Blizard (ed.), *Ocean Optics VII*, Proc. SPIE, 489, International Society for Optical Engineering, Bellingham, Washington, pp. 318-324.

Zaneveld, J.R.V., J.C. Kitchen and H. Pak: 1981, 'The influence of optical water type on the heating rate of a constant depth mixed layer', *J. Geophys. Res.* 86, 6426-6428.

NOVEL APPLICATIONS OF ACOUSTIC BACKSCATTER TO BIOLOGICAL MEASUREMENTS

David M. Farmer and R. Del Huston
Institute of Ocean Sciences
P.O. Box 6000
9860 West Saanich Road
Sidney, B.C. V8L 4B2
Canada

ABSTRACT. Acoustic remote sensing of plankton has normally been restricted to observations of the amplitude or intensity of backscattered sound. Here we suggest some novel possibilities associated with two different applications of phase measurement. First, random components of the plankton motion can be derived from the Doppler spectrum using coherent techniques that have been developed for other applications. A second approach exploits the statistics of backscattered phase. A coherent summation of the signals from successive transmissions is used to develop a stable absolute phase measurement as a function of range and time. It is shown that this summation is equivalent to raising the ratio of coherent to incoherent signal energy as successive targets are included in the observed volume. The concept can be applied to the measurement of population densities as well as the acoustic scattering properties of the individual targets.

1. INTRODUCTION

As the study of physical-biological interactions extends to finer scales and more fundamental processes, the demands increase for improved observational approaches. Remote sensing techniques must play a major role in such studies, both because of the desirability for non-invasive measurement and because of the required range of observational scales in both space and time. At one end of the spectrum satellite images of ocean colour patterns provide an opportunity for studying the interaction of biological productivity with mesoscale features. At the other end, remote sensing of fine and micro-scale processes, as well as almost all processes occurring more than a few tens of metres beneath the ocean surface, are most effectively approached with acoustic techniques. While there is a large literature on the topic of acoustic backscatter from plankton (*i.e.* Andersen and Zahuranec, 1977), we suggest here alternative approaches offering quite different possibilities for studying their distributions and behaviour. These concepts have as yet no basis in

599

B. J. Rothschild (ed.), Toward a Theory on Biological-Physical Interactions in the World Ocean, 599–614.

systematic biological studies, having arisen from efforts to develop
novel schemes for measuring physical properties (and which are
described in detail in a Ph.D. thesis, Huston, 1987). Nevertheless
the physical and statistical concepts have equal application to
biological measurement and are described here in the hope of
stimulating interest in their potential for remote sensing of
plankton.

Most acoustic remote sensing measurements of plankton have made
use of 'target strength', or backscatter intensity, in assessments of
population density. This approach has been developed through the use
of multifrequency systems that exploit the frequency dependence of
the signal so as to place bounds on the size distribution of the
targets (Holliday and Pieper, 1980). More recently the statistics of
volume reverberation have been used to make inferences about the
density of fish distributions (Clay and Heist, 1984; Stanton and Clay,
1986; Wilhelmij and Denbigh, 1984). While these are undoubtedly
powerful techniques, they nevertheless fail to make use of information
contained in the phase of the acoustic signal. In this paper we
explore some of the possibilities of backscatter phase measurement for
the study of plankton.

For simplicity we consider a monostatic echo-sounder
configuration, as shown in Figure 1, although our own work has been
carried out with a more complex bistatic device. The only difference
between this system and a standard echo-sounder is that the received
signal is complex demodulated, so as to recover both in-phase and
quadrature components for recording and analysis. The full complex
signal is used, rather than the 'hard limited' signal, which has been
considered by Spindel and McElroy (1973), but which is subject to
additional assumptions in its interpretation. For correlation
measurements it is also useful to be able to control precisely the
time of transmitted pulses.

Echo-sounders with complex demodulation are equivalent to the
type of acoustic systems used for Doppler measurement of current
profiles. The Doppler frequency, or first moment of the Doppler
spectrum, is linearly related to the mean flow speed resolved along
the axis of the acoustic beam.

While the mean velocity component is of great interest, the first
moment of the Doppler spectrum provides no information about the
random component of motion of the plankton, or of their spatial
density; neither does it provide, apart from the overall measure of
backscatter strength, an effective means for assessing their acoustic
properties which might be used in species identification. The random
component of motion, or Doppler spreading, has been studied in
relation to turbulence measurements (Lhermitte and Serafin, 1984) and
will not be discussed in detail here except to emphasize essential
differences between scatter from turbulent fluctuations and from
swimming zooplankton. Measurement of the absolute orientation of the
acoustic signal vector however, which can contain useful information
about the scatterers, requires a new approach and is described in more
detail.

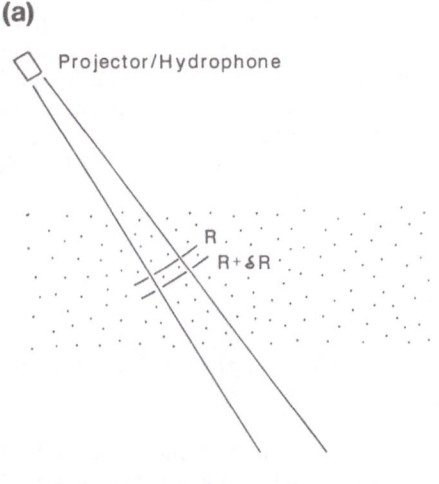

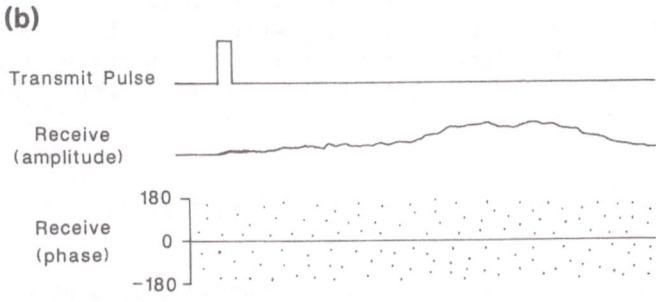

Figure 1. (a) Schematic of archetype echosounder configuration.
(b) Representative transmit pulse, received amplitude and phase
signals with the (x) axis representing time or range.

2. RANDOM COMPONENTS OF MOTION

Standard incoherent Doppler sonar systems are normally unable to
resolve subtle variations associated with random motions of the
scatterers. This difficulty arises from the contributions to Doppler
bandwidth of other factors, especially the pulse modulation. The
limitation is avoided through coherent measurement, whereby coherent
processing of many closely spaced short pulses defines the Doppler
spectrum, with high resolution (Lhermitte, 1983). Lhermitte's
observations were obtained in a tidal current in a shallow estuary;
the physical interpretation of statistical properties in this case is
based on the assumption that the targets are passive tracers of water
motion.

Even in quiescent water however, we can expect the plankton to move as they feed. The resulting motion will be random and will have the effect of broadening the Doppler spectrum or equivalently, reducing the time for decorrelation of the acoustic signal. These random components of motion can be detected in the same way as Doppler measurement of fluid velocity fluctuations.

Consider, for example, the complex autocorrelation function $R_r(\tau)$, at range r,

$$R_r(\tau) = \langle S_r(t)S_r^*(t + \tau)\rangle \tag{1}$$

where,

$S_r(\tau)$ = time series of the complex echo at range r.
τ = time lag.

The motions of the ith scatterer within a range gate of the transmit pulse can be represented by a mean velocity vector $\vec{v}$ which gives the mean advection of the scatterers through the range gate and an individual velocity vector $\vec{u}_i$ with random orientation. For a monostatic echo-sounder the direction of transception (x component) determines those components v_x, $(u_i)_x$ which contribute to decorrelation of the signal. For two successive pulses τ represents the time lag. Farmer et al. (1981) give the resulting normalised correlation:

$$\frac{R_r(\tau)}{R_r(0)} = \exp[-i4\pi v_x\tau/\lambda]\exp[-8\pi^2(u_x^2)\tau^2/\lambda^2], \tag{2}$$

where (u_x^2) is the second moment of the random component of motion resolved along the acoustic (x) axis. τ must be sufficiently small for significant correlation to occur. Equivalently the bandwidth of the Doppler spectrum is increased by a term $[-8\pi^2(u_x^2)/\lambda^2]^{1/2}$ as a result of the random motion. Thus both the autocorrelation and Doppler functions allow separation of mean and random components. In practice there are advantages to operating in the time domain and allowance must also be made for the presence of noise in the signal (see detailed discussion in Lhermitte and Serafin, 1984, who also discuss appropriate pulse repetition strategies, and Lhermitte and Poor, 1983 who comment on other factors contributing to decorrelation that may require data correction).

Some care is required in the interpretation of the second moment (u_x^2). If we represent an individual plankton motion u as having a Gaussian probability distribution with mean scalar component u_o and standard deviation σ, then the mean value of u^2 will be

$$(u_{\bar{x}}^2) = \frac{1}{2\pi\sigma} \int_0^\infty u^2 \exp[-(u - u_o)^2/2\sigma^2]\,du, \qquad (3)$$

$$= \frac{3}{\sqrt{2\pi}}\, u_o\sigma\exp[-u_o^2/2\sigma^2]+\frac{3}{2}(u_o^2+\sigma^2)(1+\mathrm{erf}(\frac{u_o}{\sqrt{2\sigma}})). \qquad (4)$$

For small fluctuations in velocity about u_o,

$$(u_{\bar{x}}^2) \sim \frac{u_o^2}{3} \approx \frac{(u^2)}{3} \qquad (5)$$

so that, combining (4) and (5)

$$(u_{\bar{x}}^2) = \frac{1}{\sqrt{2\pi}}\, u_o\sigma\exp[-u_o^2/2\sigma^2]+\frac{1}{2}(u_o^2+\sigma^2)(1+\mathrm{erf}(\frac{u_o}{\sqrt{2\sigma}})). \qquad (6)$$

There would seem to be a case for initially exploiting this technique in water that is known to be quiescent, in order to acquire basic information about the acoustic signature of zooplankton behaviour in the absence of complications due to turbulence. Multicomponent acoustic systems could be used to test for isotropy in the plankton movements. Factors that are known to perturb the random grazing motions of plankton, such as the influence of artificial illumination suddenly switched on at night (Sameoto *et al.* 1985), would provide opportunities for exploring the acoustic signature of different behaviour patterns.

These results show that an analysis of the Doppler signal and complex autocorrelation can provide three biologically useful things: (i) measure u_x; (ii) measure $(u_{\bar{x}}^2)$ which yields u_o^2 from (5); (iii) compute σ/u_o from (6). In addition, the ratio of rms u_x/v_x from orthogonal beams would provide an indication of anisotropy.

3. STATISTICS OF STOCHASTIC SCATTERING FROM PLANKTON AND THE USE OF COHERENT PROCESSING

Under normal circumstances zooplankton are randomly located and tend to be far apart from each other relative to the wavelength of sound most useful for detecting their presence (*i.e.* $\lambda = 0.5 - 3$ cm). A coherent pulse of acoustic energy is therefore scattered at random ranges (Figure 1) yielding a random phase on reception. If the scatterer positions remain sufficiently coherent over some short period, typically a few ms, then components of their motion will contribute to the rate of phase change, whose first and second moments provide useful information on their motion, as discussed above. If the separation between pulses is greater than the 'decorrelation time', the scatterer positions are again random, and a new random phase will be detected.

It might seem unrealistic to expect to acquire useful information about the plankton distribution from the acoustic phase under these circumstances; nevertheless it turns out that the statistical properties of the signal allow us to learn about their concentration, or equivalently, the 'mean target spacing', and provide a means of characterising their size and structure.

The statistics of sonar volume backscatter have been discussed by Clay and Heist (1984) and Stanton and Clay (1986), especially in connection with scatter from various types of fish. A concept that finds application in these studies is the simultaneous presence in the signal of both a coherent and an incoherent component. Previous work has focussed on the use of amplitude statistics; however we shall show that the phase statistics can provide a sensitive probe of the distribution and properties of the acoustic targets.

3.1. Ricean Phase Statistics

Lord Rayleigh (Rayleigh, 1877) was the first to show that if n sources with uniform amplitude have random phases then in the limit $n \to \infty$ the amplitude probability distribution function (henceforth, pdf), $P(A)$ is given by,

$$P(A) = \frac{A}{\sigma^2} \exp\left[-\frac{A^2}{2\sigma^2}\right]. \tag{7}$$

$P(A)$ is known as the Rayleigh distribution. It has the unique characteristic of being completely defined by the single parameter σ^2 which is equivalent to the variance of the received amplitude.

The Rayleigh distribution also applies to a uniform spatial distribution of discrete scatterers with a mean spacing greater than the insonifying wavelength which is typical for acoustic scattering from zooplankton. When these targets are randomly distributed in space the phase from each target will have a uniform pdf. This will be referred to as the incoherent field. The signal scattered by a large motionless target, or a tightly packed group of small targets, will tend to maintain a constant phase. This is the coherent field. The statistics of this type of problem were first developed by Rice (1945). When a coherent signal is present in an incoherent field the statistical properties will be altered. The problem of amplitude statistics for this case has been widely discussed in the literature; however in the following discussion we develop the interesting case of phase statistics, which have special application to zooplankton studies.

The signal scattered by a group of targets about $\vec{r} = (x,y,z)$, for a narrow band signal with center frequency can be represented by,

$$u[\vec{r}(t)] = A_{in}[\vec{r}(t)]e^{iwt} + A_{co}[\vec{r}(t)]e^{iwt}. \tag{8}$$

The $A[\vec{r}(t)]$ terms are complex phasor amplitudes with modulus $A[\vec{r}(t)]|$ and argument $\phi(t)$. The subscripts *in* and *co* refer to the incoherent and coherent signals respectively. In general,

$$A[\vec{r}(t)] = \sum_{j=1}^{N} |A_j[\vec{r}(t)]| \exp[i\phi_j(t)] \tag{9}$$

where N represents the total number of targets from the insonified volume at r. This summation can best be understood as a random walk in the complex plane.

For the incoherent signal each term is statistically independent. The resulting amplitude (envelope of the received signal) will follow Rayleigh statistics for large N and the phase will be uniformly distributed. A pure coherent signal however will have a constant phase value $\phi_j - \phi_o$ so that the resulting amplitude becomes the sum of the individual values. When the coherent properties are stable (reflective properties remain constant) both the resulting amplitude and phase of the received signal will have Dirac delta functions for their pdfs (assuming the system noise is negligible).

The in-phase and quadrature components of the received signal, X and Y respectively, can be represented by,

$$X = a_o\cos\phi_o + \sum_{j=1}^{N} a_j\cos\phi_j = x_{co} + x_{in}$$

(10)

$$Y = a_o\sin\phi_o + \sum_{j=1}^{N} a_j\sin\phi_j = y_{co} + y_{in}$$

where the amplitude and phase of the coherent signal are represented by a_o and ϕ_o. When the terms in the summation are independent random variables, Gaussian statistics can be used to describe x_{in} and y_{in}. As $N \to \infty$ the central limit theorem (c.f. Davenport and Root, 1958) states that the probability distribution of a sum of N independent random variables approaches a normal distribution. Since quadrature components are uncorrelated, the joint pdf of x_{in} and y_{in} will be given by,

$$P(x_{in},y_{in}) = P(X,Y) = \frac{1}{2\pi\psi} \exp[\frac{-1}{2\psi}((X - x_{co})^2 + (Y - y_{co})^2)] \quad (11)$$

where, $\psi = ((X - x_{co})^2) = ((Y - y_{co})^2)$. It is more desirable to transform these statistics into amplitude A and phase Φ space:

$$\begin{align} X &= A \cos\Phi \\ Y &= A \sin\Phi \end{align} \quad (12)$$

Then,

$$P(A,\Phi)dAd\Phi = \frac{A}{2\pi\psi} \exp[- \frac{1}{2\psi}(A^2 + a_o^2 - 2Aa_o\cos(\Phi - \phi))]dAd\Phi. \quad (13)$$

A diagram of $P(X,Y)$ and $P(A,\Phi)$ is shown in Figure 2.

For the case of an incoherent signal the corresponding phase pdf is uniform (Ishimaru, 1978). However a pure coherent signal will have a delta function phase pdf centered about the reference phase ϕ_o.

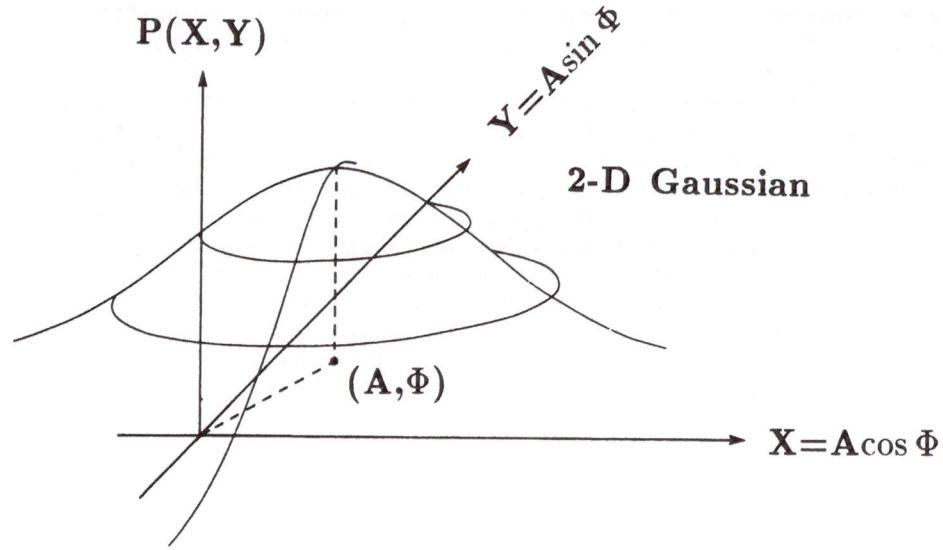

Figure 2. Probability distribution for In-phase (X) and Quadrature (Y) components of the backscattered signal.

$P(\Phi)$ is determined by finding the probability of an event occurring within an angular sector defined by Φ and $\Phi + d\Phi$. This corresponds to integrating the joint pdf (13) over all possible values of A

$$P(\Phi)d\Phi = \int_0^\infty P(A,\Phi)\,dA\,d\Phi. \tag{14}$$

The resulting integration yields,

$$P(\Phi) = \frac{2}{2\pi}e^{-\gamma} + \frac{\sqrt{\gamma}}{2\sqrt{\pi}}\cos\Phi\,\exp[-\gamma\,\sin^2\Phi]\left[1 + \frac{\cos\Phi}{|\cos\Phi|}\mathrm{erf}(\sqrt{\gamma}\cos\Phi)\right], \tag{15}$$

where γ is the ratio of coherent to incoherent energy in the signal. A plot of $P(\Phi)$ for $-\pi \le \Phi \le$ and various values of γ is given in Figure 3. A continuous transition from a uniform distribution when $\gamma = 0$ to a Gaussian centered about $\Phi = 0$ for larger γ is observed. Equation (15) is evaluated for a zero phase offset $\phi_o = 0$. For $\phi_o = 0$ a simple transformation is applied. These results are also consistent with the interpretation of a pure incoherent signal when $\gamma = 0$ and a pure coherent signal when $\gamma^{-1} = 0$.

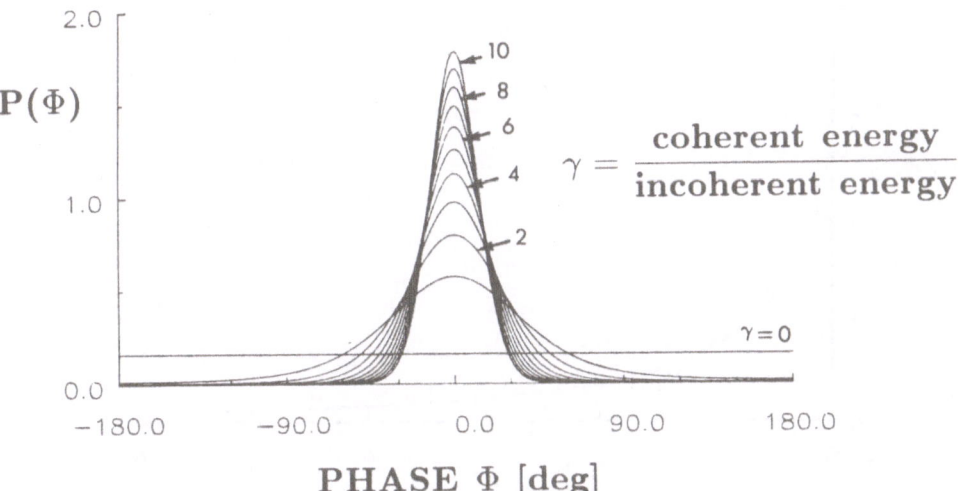

PHASE Φ [deg]

Figure 3. Plot of the phase probability function $P(\Phi)$ for different values of the ratio γ of coherent to incoherent energy.

The νth phase moment is generated by evaluating,

$$\langle \Phi^\nu \rangle = \int_{-\pi}^{\pi} \Phi^\nu P(\Phi)\,d\Phi \tag{16}$$

For simplicity we will assume $\langle \Phi^1 \rangle = 0$ in this analysis although, in general, the appropriate phase offset ϕ_0 must be included.

An analytic evaluation of (16) leads to multiple infinite sums of transcendental functions, so that it is more useful to insert (15) into (16) and evaluate numerically. Figure 4 shows the result. The second moment of phase decreases sharply as the ratio of coherent to incoherent signal γ increases in the range $0 < \gamma < 1$.

An empirical expression for this relationship is,

$$\langle \Phi^2 \rangle = \frac{\pi^2}{3} e^{-f_2(\gamma)} \tag{17}$$

where $f(\gamma)$ is a 3rd order polynomial,

$$f(\gamma) = 0.329 + 1.34\gamma - 0.239\gamma^2 + 0.0186\gamma^3$$

and has a variance of ± 0.0143 for $\gamma < 5$.

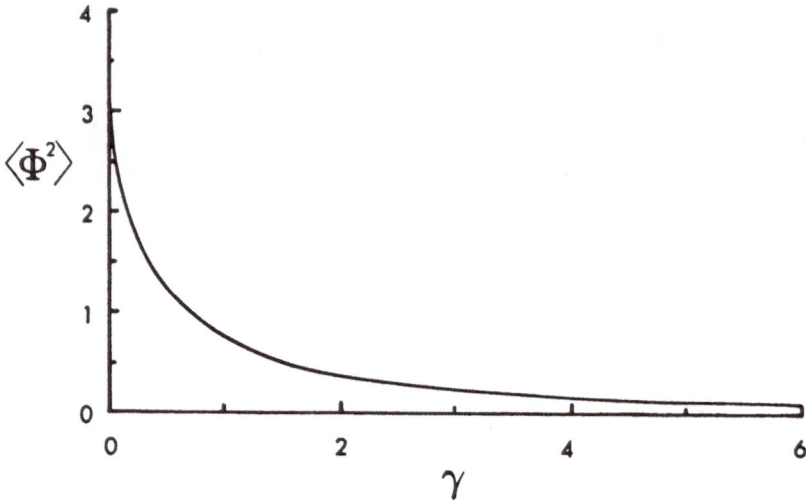

Figure 4. Numerically evaluated relationship between the second moment of the phase (Φ^2) and γ.

3.2. Coherent Processing

Coherent processing involves the coherent summation of the complex signal over successive transmissions. We show that this procedure is equivalent to increasing the ratio γ of coherent to incoherent energy in the acoustic signal. The targets are assumed to be in different positions after each transmission by a period long compared to the decorrelation time.

The echo from each transmission is a one-dimensional distribution. This may be represented, for example, by a binary distribution in which the absence or presence of a target at a particular range R, of width δR, can be represented by a 0 or 1 respectively. Since the number of targets in a particular sequence is random, the binary representation is a Poisson process (Miller and Freund, 1985).

For a large number of digits, the Poisson process is described by the Poisson distribution and the probability of k events (*i.e.* targets) occurring, is

$$P(k;\mu) = e^{-\mu}\frac{\mu^k}{k!} \tag{18}$$

where μ is the mean number of events. Thus, for a large number of ranges, with μ representing the mean number of targets between R and $R + \delta R$, the probability of k targets occurring is $P(k;\mu)$.

Coherent processing of successive echoes simulates higher scatterer densities, because it incorporates the superposition of the scattered signal from each target. This can also be represented by the addition of binary strings. The sums $S_1, \ldots, S_n$ for each range gate will then correspond to the number of targets in those range gates when summed over the total (l) number of processed echoes. When each echo is independent of previous echoes (*i.e.* the delay between successive transmissions exceeds the decorrelation time), the statistics at each range correspond to a Poisson process. As the number of processed signals (l) increases, the Poisson distribution becomes a good approximation. For $\mu > 30$ the Poisson distribution approximates a Gaussian distribution (Barford, 1967) with,

$$P(k,\mu) \simeq \frac{1}{\sqrt{2\pi\mu}} \exp\left[\frac{(k-\mu)^2}{2\mu}\right] \text{ for } k \geq 0. \tag{19}$$

This transition from a Poisson to a Gaussian distribution has the same physical interpretation as the Ricean distribution discussed earlier. The mean corresponds to the specular component and the standard deviation corresponds to the incoherent component, with

$$\gamma = \frac{\mu}{2}. \tag{20}$$

Thus a superposition of discrete systems of random events evolves from a random sequence (Poisson distribution) to a Gaussian distribution. As the number of superpositions increases, so also does the ratio between the mean and second moment. This corresponds to an increase in the ratio of coherent to incoherent signals γ.

The relevance of these results to acoustic studies of zooplankton is that (i) the rate at which γ changes with successive superpositions is related to the number of targets in the scattering volume, and (ii) for appropriately large numbers of superpositions, stable phase measurements can be achieved, which can be used in estimates of the phase spectrum with which to characterize their acoustic properties.

4. APPLICATIONS

4.1. Mean Target Spacing

Convergence of phase measurement with successive superposition of random target distributions can be demonstrated numerically. Figure 5 shows successive phases for modelled echoes from a cube of side 28.6λ. With each superposition the effective scatterer density increases; the phase is therefore plotted as a function of mean target spacing. For large separations the phase is essentially random. As the mean target separation decreases through $\sim 0.5\lambda$, the phase rapidly converges to a stable value.

Calculations of this type can be repeated several times for random target distributions, so as to determine the way in which the second moment of phase, or equivalently the ratio γ of coherent to incoherent signal, varies with mean target separation. Figure 6 shows

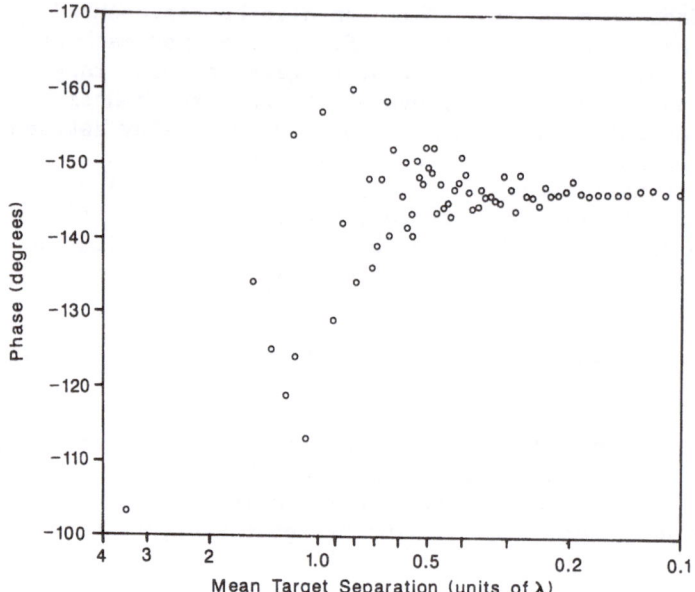

Figure 5. Composite phase determined by numerical evaluation of backscatter from a model cube with sides of 28.6λ, for different mean target spacings. The mean spacing *l* of each realization is constant but the equivalent *l* decreases with coherent superposition.

the result of such calculations, based on 10 simulations for each value of target separation *d*. A linear fit through these points yields

$$\gamma = 0.0813(d/\lambda)^{-3} \tag{21}$$

or equivalently,

$$\gamma = 0.0575\rho \tag{22}$$

where ρ is the number of targets per cubic wavelength.

As with any acoustic representation of biological targets, a proper interpretation requires use of an appropriate model of the scatterer. If the wavelength is much greater than a typical length scale for the target, the use of a point representation, as in Figures 5,6 and in the empirical fit (21) and (22), will be appropriate. If, on the other hand, the target length is a significant fraction of the acoustic wavelength, then an alternative representation, such as a short string of points or a fluid sphere, may be more appropriate. Transmission at more than one acoustic

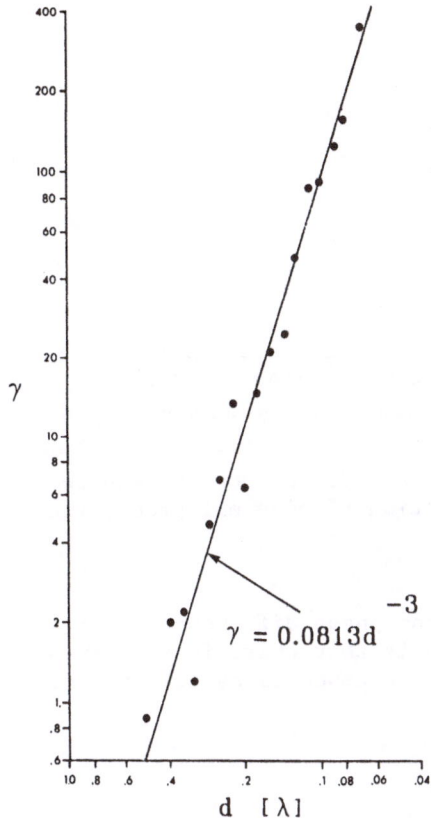

Figure 6. Numerical calculations of γ as a function of mean target separation expressed in units of λ, based on 10 estimates for each point.

frequency will yield different phase convergence rates (more rapid convergence for lower frequencies). This frequency dependence however, will depart from that implied by (21) at higher frequencies, for which the point scatterer model breaks down. The resulting discrepancy provides a basis for testing alternative models, such as finite linear arrays, which can be used to characterise the plankton.

Figure 7 shows an example of phase convergence for backscatter data obtained in Saanich Inlet, B.C., at a frequency of 215 kHz. The bistatic system that was used in this experiment is sensitive to refractive effects (in fact the goal of the measurements was to detect these); but these effects are removed by forming two interleaved data sets from the original time series and calculating the rms phase difference. Initially the phase is essentially random, but after about 300 superpositions the rms phase difference decreases, reaching a minimum of 85°, corresponding to an accuracy bound for the $2N$ data set of ±42.5°. For larger numbers of superpositions, heterogeneity in

612

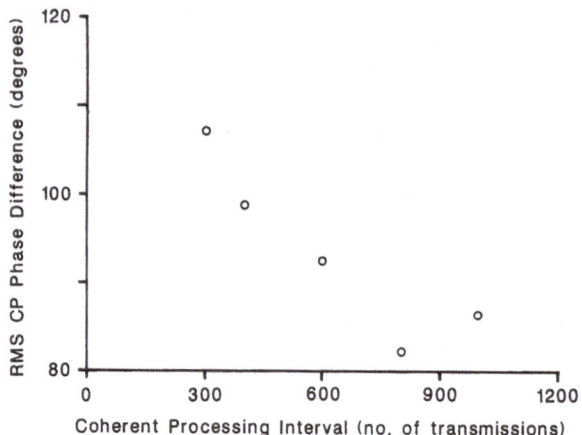

Figure 7. Observations of rms phase difference between two
interleaved data sets, as a function of coherent processing
interval.

the target distribution causes the phase difference to diverge again.
The important point to note here is that there is a coherent
processing interval for which convergence is systematic and can be
used to estimate γ and hence d.

4.2. Complex Scattering Spectrum

As mentioned earlier, there have been substantial efforts to use
acoustic target strength as a function of frequency to characterise
and classify zooplankton. The coherent processing scheme discussed
above allows this concept to be extended to a phase measurement, thus
permitting calculation of the complex spectrum. The complex
scattering spectrum of a plankton depends in a subtle way on its
acoustic properties and in principle provides a much more sensitive
probe than target strength alone.

We have shown that coherent superposition can lead to stable
phase measurements. The phase can also be calculated as a function of
frequency. The resulting phase spectrum, together with the
corresponding amplitude spectrum, can then be used to test specific
models of plankton, such as the fluid cylinder model recently proposed
by Stanton (1986).

It is necessary that coherent superposition of many echoes be
used for this purpose, since even if a common projector is used for
each frequency, small differences in beam properties as a function of
frequency, and hence in the precise number of targets included in each
transmission, are unavoidable. The superposition leads to an average
phase orientation of the acoustic signal vector that is independent of
precise beam properties and thus allows comparison as a function of
frequency.

5. CONCLUSIONS

The methods discussed here all relate to the use of acoustic phase.
The statistical basis for volume phase measurements has been developed
so as to show how absolute orientations of the acoustic signal vector
can be recovered, even though the targets are typically randomly
positioned and sparse relative to the acoustic wavelength.
   The process of volume scatter is analogous to scattering from a
roughened mirror. Coherent superposition is thus equivalent to
polishing the mirror: as the target density rises with successive
coherent summations the fraction of coherent energy in the received
signal also rises. Acoustic measurements confirm the validity of this
concept. Our purpose in outlining these ideas is to show their
potential for new acoustic remote sensing applications in biological
oceanography. Their future success depends upon careful °ground
truth' comparisons with independent biological observations. Further
experiments will be carried out and a more detailed analysis published
subsequently.

6. REFERENCES

Andersen, R. and B.J. Zahuranec, (eds.): 1977, *Oceanic Sound
   Scattering Predictions*, New York: Plenum.
Barford, N.C.: 1967, *Experimental Measurements: Precision, Error, and
   Truth*, Addison-Wesley, London.
Clay, C.S. and B.G.Heist: 1984, 'Acoustic scattering by fish--Acoustic
   models and a two-parameter fit', *J. Acous. Soc. Am.* 75, 1077-
   1083.
Davenport, W.B., Jr. and W.L. Root: 1958, *An Introduction to the
   Theory of Random Signals and Noise*, McGraw-Hill, New York.
Farmer, D.M., A.D. Booth, and G. Kamitakahara: 1981, 'Preliminary
   considerations in the design of a correlation sonar for remote
   velocity profile measurements in the ocean', *Proceedings of
   International Symposium on Acoustic Remote Sensing of the
   Atmosphere and Oceans*, University of Calgary Printing Services.
Holliday, D.V. and R.E. Pieper: 1980, 'Volume scattering strengths and
   zooplankton distributions at acoustic frequencies between 0.5 and
   3MHz', *J. Acous. Soc. Am.* 67, 135-146.
Huston, R.D.: 1987, *Acoustic Phase Measurements From Volume Scatter in
   the Ocean*, Ph.D. Thesis, University of Victoria.
Ishimaru, A.: 1978, *Wave Propagation and Scattering in Random Media*,
   Vol. 1, Academic Press, New York.
Lhermitte, R.: 1983, 'Doppler sonar observations of tidal flow', *J.
   Geophys. Res.* 88, 725-742.
Lhermitte, R. and H. Poor: 1983, 'Multi-beam Doppler sonar
   observations', *Geophys. Res. Lett.* 10, 717-720.

Lhermitte, R. and R. Serafin: 1984, 'Pulse-to-pulse coherent Doppler sonar signal processing techniques', *J. Atmos. Ocean. Tech.* 1, 293-308.

Miller, I. and J.E. Freund: 1985, *Probability and Statistics for Engineers*, Prentice-Hall, NJ.

Rayleigh, Lord [J.W. Strut] (2nd Eds., 1894 and 1896): 1945, *The Theory of Sound Vol. 1*, Dover, New York.

Rice, S.O.: 1945, 'Mathematical analysis of random noise II', *Bell System Tech. J.* 24, 46.

Sameoto, D., N.A. Cochrane, and A.W. Herman: 1985, 'Response of biological acoutic backscattering to ship's lights', *Can. J. Fish. Aquat. Sci.* 42, 1535-1541.

Spindel, R.C. and P.T. McElroy: 1973, 'Level and zero crossings in volume reverberation signals', *J. Acous. Am.* 53, 1417-1462.

Stanton, T.K.: 1986, 'Sound scattering by fluid cylinders of finite length', Submitted to JASA Jan. 1986.

Stanton, T.K. and C.S. Clay: 1986, 'Sonar echo statistics as a remote-sensing tool: volume and seafloor', *IEEE Journal of Oceanic Engineering* **Vol. OE-11**, 79-96.

Wilhelmij, P. and P. Denbigh: 1973, 'A statistical approack to determining the number density of random scatterers from backscattered pulses', *J. Acous. Am.* 76, 1810-1818.

LIST OF PARTICIPANTS

A. Bakun
National Marine Fisheries Service
Pacific Fisheries Environment Group
Monterey, California 93942
U.S.A.

B. Battaglia
Universita Degli Studi di Padua
Dipartimento di Biologia
35131 PD, Padua
Italy

W.C. Boicourt
University of Maryland
Horn Point Environmental Laboratories
Cambridge, Maryland 21613
U.S.A.

J. Boucher
Centre de Brest
P.O. 337
29273 Brest Cedex
France

D.H. Cushing
198 Yarmouth Road
Lowestoft, Suffolk
England NR32 4AB

N. Daan
Netherlands Institute for Fishery Investigation
Postbus 68
1970 A B IJmuiden
The Netherlands

Y. Dandonneau
Centre ORSTOM
B.P. 45
Noumea
New Caledonia

616

T.D. Dickey
Ocean Physics Group
Department of Geological Sciences
University of Southern California
Los Angeles, California 90089-0741
U.S.A.

G. Evans
Fisheries and Oceans, Science Branch
P.O. Box 5667
St. Johns, Newfoundland
Canada A1C 5X1

D. Farmer
Institute of Ocean Sciences
P.O. Box 6000
9860 West Saanich Road
Sidney, B.C. V8L 4B2
Canada

S. Frontier
Laboratoire d'Écologie Numérique
Université des Sciences et Techniques de Lille (SN3)
59655 Villeneuve d'Ascq Cedex
France

J. Goldman
Woods Hole Oceanographic Institution
Woods Hole, Massachusetts 02543
U.S.A.

A. Herbland
IFREMER
BP 1049, Nantex
France

W. Jenkins
Woods Hole Oceanographic Institution
Woods Hole, Massachusetts 02543
U.S.A.

R. Jones
Department of Agriculture and Fisheries for Scotland
Marine Laboratory
P.O. Box 101, Victoria Road
Aberdeen, Scotland

T.M. Joyce
Woods Hole Oceanographic Institution
Woods Hole, Massachusetts 02543
U.S.A.

J. Le Fèvre
Laboratoire d'Océanographie Biologique
Université de Bretagne Occidentale
29287 Brest Cedex
France

L. Legendre
GIROQ, Département de biologie
Université Laval
Québec, Québec
Canada G1K 7P4

K. Mann
Bedford Institute of Oceanography
Box 1006
Dartmouth, Nova Scotia
Canada B27 4A2

J.J. McCarthy
Museum of Comparative Zoology
Harvard University
Cambridge, Massachusetts 02543
U.S.A.

M.M. Mullin
Institute of Marine Resources, A-018
Scripps Institution of Oceanography
University of California, San Diego
La Jolla, California 92093
U.S.A.

W. Munk
University of California, San Diego
Scripps Institution of Oceanography
La Jolla, California 92093
U.S.A.

P. Nival
Station Zoologique
Villefranche sur mer
France

T.R. Osborn
Chesapeake Bay Institute
The Johns Hopkins University
Suite 315/The Rotunda
711 W. 40th Street
Baltimore, Maryland 21211
U.S.A.

D.A. Powers
Department of Biology
The Johns Hopkins University
Baltimore, MD 21218
U.S.A.

M.R. Reeve
Ocean Sciences Division
National Science Foundation
Washington, D.C. 20550
U.S.A.

B.J. Rothschild
University of Maryland
Center for Environmental and Estuarine Studies
Solomons, Maryland 20688
U.S.A.

M.P. Sissenwine
United States Department of Commerce
National Oceanic and Atmospheric Administration
National Marine Fisheries Service
Northeast Fisheries Center
Woods Hole, Massachusetts 02543
U.S.A.

J.H. Steele
Woods Hole Oceanographic Institution
Woods Hole, Massachusetts 02543
U.S.A.

V. Strass
Institut fuer Meereskunde an der Universitaet Kiel
Duesternbrooker Weg 20
D-2300 Kiel 1
F.R. Germany

U. Wolf
Institut fuer Meereskunde an der Universitaet Kiel
Duesternbrooker Weg 20
D-2300 Kiel 1
F.R. Germany

J.D. Woods
Robert Hooke Institute
Oxford University
England

LIST OF CONTRIBUTORS

Bruno Battaglia
Dipartimento di Biologia
Università di Padova
Via Loredan, 10
35131 Padova, Italy

Paolo M. Bisol
Dipartimento di Biologia
Università di Padova
Via Loredan, 10
35131 Padova, Italy

William C. Boicourt
Horn Point Environmental Laboratories
University of Maryland Center for
Environmental and Estuarine Studies
Cambridge, MD 21613
U.S.A.

J. Boucher
IFREMER
Laboratoire Pêche
BP 337
29273 Brest Cedex
France

F. Carlotti
Station Zoologique
Villefranche sur mer
France

Robert Chapman
Chesapeake Bay Institute
The Johns Hopkins University
Baltimore, MD 21218
U.S.A.

Thomas T. Chen
Center of Marine Biotechnology
University of Maryland
Baltimore, MD 21202
U.S.A.

E.B. Cohen
United States Department of Commerce
National Oceanic and Atmospheric Administration
National Marine Fisheries Service
Northeast Fisheries Center
Woods Hole, MA 02543
U.S.A.

David H. Cushing
198 Yarmouth Road
Lowestoft
Suffolk NR32 4AB
England

Niels Daan
Department of Aquatic Ecology
University of Amsterdam
Kruislaan 320
1098 SM Amsterdam
The Netherlands

Y. Dandonneau
Centre ORSTOM
B.P. A5, Noumea
New Caledonia

T.D. Dickey
Ocean Physics Group
Department of Geological Sciences
University of Southern California
Los Angeles, CA 90089-0741
U.S.A.

Leonard DiMichele
Chesapeake Bay Institute
The Johns Hopkins University
Baltimore, MD 21218
U.S.A.

David M. Farmer
Institute of Ocean Sciences
P.O. Box 6000
9860 West Saanich Road
Sidney, B.C. V8L 4B2
Canada

Serge Frontier
Laboratoire d'Ecologie Numérique
Université des Sciences et Techniques de Lille (SN3)
59655 Villeneuve d'Ascq Cedex
France

Joel C. Goldman
Woods Hole Oceanographic Institution
Woods Hole, MA   02543
U.S.A.

L. Irene González-Villaseñor
Department of Biology
The Johns Hopkins University
Baltimore, MD   21218
U.S.A.

E.W. Henderson
Department of Agriculture and Fisheries for Scotland
Marine Laboratory
P.O. Box 101, Victoria Road
Aberdeen, Scotland

A. Herbland
IFREMER
BP 1049
Nantes
France

R. Del Huston
Institute of Ocean Sciences
P.O. Box 6000
9860 West Saanich Road
Sidney, B.C.   V8L 4B2
Canada

R. Jones
Department of Agriculture and Fisheries for Scotland
Marine Laboratory
P.O. Box 101, Victoria Road
Aberdeen, Scotland

Terrence M. Joyce
Woods Hole Oceanographic Institution
Woods Hole, MA   02543 U.S.A.

Reuben Lasker
Southwest Fisheries Center
NOAA/NMFS
La Jolla, CA   92038
U.S.A.

G.C. Laurence
United States Department of Commerce
National Oceanic and Atmospheric Administration
National Marine Fisheries Service
Northeast Fisheries Center
RR 7, South Ferry Road
Narragansett, RI   02882
U.S.A.

Jacques Le Fèvre
Laboratoire d'Océanographie Biologique
Université de Bretagne Occidentale
29287 Brest Cedex
France

Louis Legendre
GIROQ, Département de biologie
Université Laval
Québec, Québec
Canada   G1K 7P4

M.M. Mullin
Institute of Marine Resources, A-018
Scripps Institution of Oceanography
University of California, San Diego
La Jolla, CA   92093
U.S.A.

Walter H. Munk
University of California, San Diego
Scripps Institution of Oceanography
La Jolla, CA   92093
U.S.A.

P. Nival
Station Zoologique
Villefranche sur mer
France

Thomas R. Osborn
Chesapeake Bay Institute
The Johns Hopkins University
711 W 40th Street
Baltimore, MD   21211
U.S.A.

Dennis A. Powers
Department of Biology
The Johns Hopkins University
Baltimore, MD   21218
U.S.A.

Louis Prieur
Laboratoire de physique et chimie marines
CEROV, B.P. 8
06230 Villefranche-sur-Mer
France

Michael R. Reeve
Ocean Sciences Division
National Science Foundation
Washington, D.C.   20550
U.S.A.

B.J. Rothschild
University of Maryland
Center for Environmental and Estuarine Studies
Solomons, MD   20688
U.S.A.

A. Sciandra
Station Zoologique
Villefranche sur mer
France

Kenneth Sherman
National Marine Fisheries Service
Northeast Fisheries Center
Narragansett Laboratory
Narragansett, RI   02882-1199
U.S.A.

M.P. Sissenwine
United States Department of Commerce
National Oceanic and Atmospheric Administration
National Marine Fisheries Service
Northeast Fisheries Center
Woods Hole, MA   02543
U.S.A.

John H. Steele
Woods Hole Oceanographic Institution
Woods Hole, MA   02543
U.S.A.

V. Strass
Institut fuer Meereskunde an der Universitaet Kiel
Duesternbrooker Weg 20
D-2300 Kiel 1
F.R. Germany

J.-P. Troadec
French Research Institute for the Exploitation
of the Sea (IFREMER)
66, avenue d'Iéna
75116 Paris
France

K.U. Wolf
Institute fuer Meereskunde an der Universitaet Kiel
Duesternbrooker Weg 20
D-2300 Kiel 1
F.R. Germany

John Woods
Robert Hooke Institute
Oxford University
England

Hidekatsu Yamazaki
Chesapeake Bay Institute
The Johns Hopkins University
711 W 40th Street
Baltimore, MD 21211
U.S.A.

isozymes, 416
Ivory Coast, 248
jack mackerel, 464
James River, 185, 205, 421
Jan Mayen, 236
Japan, 152, 464, 485, 487
Japanese, 464; mackerel, 469, 483; sardine, 464, 480, 481, 485
jellyfish, 506
jets, 36
juvenile, 190, 306, 347, 411, 419, 441, 445, 447, 448, 456, 457, 466,
    486; fish stage, 352; stage, 349, 350, 359, 406, 419, 490;
    stocks, 465; survival, 423, 441
k-selected, 456, 457
Kelvin waves, 221
Kiel "Sea Rover" system, 22, 24, 30, 116, 132, 133, 575
kinematic: effect, 227; viscosity, 219, 290
kinetic, 538, 540; changes, 547; energy, 9, 11, 216, 218-220, 562;
    energy dissipation, 229; energy field, 219; environment, 539;
    factors, 540; modification of the transmission medium, 538;
    structure, 211, 528, 537, 538
Kolmogorov wavenumber, 219
Lagoon of Venice, 396, 400, 403, 404
Lagrangian, 16-19, 39, 51, 57, 184, 187, 529; integration, barotropic,
    19; conservation law, 33; ensemble method, 11, 30, 52, 53, 58,
    67; integration, 18; setting, 544; time series, 196; view, 18
Lamont-Doherty Geological Observatory, 579
land runoff, 254
Langmuir circulation, 224
large-volume pumping, 196
larva, 344, 345, 350, 359, 505
larvae, 188, 238, 311, 324, 325, 344, 345, 347, 360, 362, 374, 376,
    383, 388, 416, 486, 508, 529; benthic organisms, 322;
    distribution, 486; environment, 334; fish, 346, 348-350, 353,
    354, 356, 358; fish, 361, 386, 508, 510; hatching, 339;
    trochophore, 325
larval: abundance, 193, 362, 380; biomass, 355; cod, 375; cohorts,
    211; cumulative mortality, 380; density, 359, 381; development,
    241, 322, 339; dispersal region, 195; distributions, 185, 192,
    194, 375; drift, 193, 238; dynamics, 210, 211; fish, 193, 298,
    303, 347, 348, 352,377, 416; fish recruitment, 555; fish
    survival, 174; food items, 517; gadids, 374; growth, 334, 386;
    growth and survival, 374; growth rate, 344, 359; haddock, 375,
    377; haddock densities, 384; herring concentrations, 518; mean-
    depth and current speed, 186; mean-depth time series, 186;
    mortality, 241, 375; nutrition, 198; production, 352, 353, 356;
    retention, 185; sampling techniques, 195; stage, 322, 325, 344,
    350, 370, 380, 406, 423; survival, 359, 423; survival curves,
    359; transport, 187, 192, 193, 195, 197; transport paths, 192;
    transport process, 193
LDH, 416, 424, 428, 429, 431; loci, 417; -A locus, 416
leptocephalus, 423